RFID SYSTEMS

RESEARCH TRENDS AND CHALLENGES

Edited by

Miodrag Bolić
University of Ottawa, Canada

David Simplot-Ryl
INRIA, France and University of Lille, France

Ivan Stojmenović
University of Ottawa, Canada

WILEY
A John Wiley and Sons, Ltd., Publication

This edition first published 2010

Registered office
John Wiley & Sons Ltd, The Atrium, Southern Gate, Chichester, West Sussex, PO19 8SQ, United Kingdom

For details of our global editorial offices, for customer services and for information about how to apply for permission to reuse the copyright material in this book please see our website at www.wiley.com.

Library of Congress Cataloging-in-Publication Data

RFID systems : research trends and challenges / edited by Miodrag Bolic, David Simplot-Ryl, and Ivan Stojmenovic.
p. cm.
Includes index.
ISBN 978-0-470-74602-8 (cloth)
1. Radio frequency identification systems. I. Bolic, Miodrag. II. Simplot-Ryl, David. III. Stojmenovic, Ivan.
TK6570.I34R4868 2010
658.7′87–dc22

2010003318

A catalogue record for this book is available from the British Library.

ISBN 978-0-470-74602-8 (H/B)

Set in 10/12 Times by Laserwords Private Limited, Chennai, India
Printed and Bound in Singapore by Markono Print Media Pte Ltd

RFID SYSTEMS

To my wife Andjelka and children Marija,
Natasa and Katarina.
Miodrag Bolić

To Isabelle, my wife.
David Simplot-Ryl

To my wife Natasa and children
Milos and Milica.
Ivan Stojmenović

Contents

About the Editors

Miodrag Bolić, mbolic@site.uottawa.ca, www.site.uottawa.ca/~mbolic

Miodrag Bolić received his B.S. and M.S. degrees in electrical engineering from the University of Belgrade, Serbia in 1996 and 2001, respectively, and his Ph.D. degree in electrical engineering from Stony Brook University, NY, in 2004. Since 2004, he has been with the University of Ottawa, Canada, where he is an associate Professor at the School of Information Technology and Engineering. His current research interests include computer architectures, biomedical signal processing and RFID. He has eight years of industrial experience from the US and Serbia related to digital signal processing and embedded system design. He is a co-founder of a start-up Astraion Inc., NY, that develops novel RFID systems. He is a founder and director of Computer architecture research group and RFID research group at the University of Ottawa. He has been a principal investigator on a number of projects funded by NSERC, Canada, Ontario Centres of Excellences and industry. Dr. Bolić has been involved in a number of research service activities including: chair of the joint chapter of signal processing, oceanic engineering, geosciences and remote sensing for the IEEE Ottawa section, and associate editor of Telecommunication Systems journal, Springer.

David Simplot-Ryl, David.Simplot-Ryl@inria.fr, http://www.lifl.fr/~simplot

David Simplot-Ryl received the Graduate Engineer degree in computer science, automation, electronic and electrical engineering, and M.Sc. and Ph.D. degrees in computer science from the University of Lille, France, in 1993 and 1997, respectively. In 1998, he joined the Fundamental Computer Science Laboratory of Lille (LIFL), France, where he is currently professor. He received the Habilitation degree from the University of Lille, France, in 2003. His research interests include sensor and mobile ad hoc networks, mobile and distributed computing, embedded operating systems, smart objects and RFID technologies. Recently, his main occupation is contributing to international standardization on RFID tag identification protocols in partnership with Gemplus and TagSys companies. He has written scientific papers, book chapters and patents and received the Best Paper award at the 9th International Conference on Personal Wireless Communications (PWC 2004) and at the 2nd International Conference on Mobile Ad-hoc and Sensor Networks (MSN 2006). He is an associate editor of *Ad Hoc and Sensor Wireless Networks: An International Journal* (Old City Publishing) and a member of the editorial board of *International Journal of Computers and Applications* (Acta Press), the *International Journal of Wireless and Mobile Computing* (Inderscience), and *International Journal of Parallel, Emergent and Distributed Systems* (Taylor & Francis).

Ivan Stojmenović, ivan@site.uottawa.ca, www.site.uottawa.ca/~ivan

Ivan Stojmenović received the Ph.D. degree in mathematics. He has held regular and visiting positions in Serbia, Japan, the USA, Canada, France, Mexico, Spain, the UK (as Chair in Applied Computing at the University of Birmingham), Hong Kong, and Brazil, and is a Professor at the University of Ottawa, Canada. He has published over 250 different papers, and has edited four books on wireless, ad hoc and sensor networks and applied algorithms with Wiley/IEEE. He is the editor of over a dozen journals, is editor-in-chief of *IEEE Transactions on Parallel and Distributed Systems* (from January 2010), and founder and editor-in-chief of three journals (*Multiple-Valued Logic and Soft Computing; Parallel, Emergent and Distributed Systems*; and *Ad Hoc and Sensor Wireless Networks*). Dr. Stojmenović has h-index 35 and >6000 citations. One of his articles was recognized as the Fast Breaking Paper, for October 2003 (the only one for the whole of computer science), by Thomson ISI Essential Science Indicators. He is the recipient of the Royal Society Research Merit Award, UK. He was elected to IEEE Fellow status (Communications Society, class of 2008), and is a recipient of Excellence in Research Award of the University of Ottawa, 2008–2009. He has chaired and/or organized >50 workshops and conferences, and served on over 100 program committees. Among others, he was/is program co/vice-chair at IEEE PIMRC 2008, IEEE AINA-07, IEEE MASS-04&07, EUC-05&08, WONS-05, MSN-05&06, ISPA-05&07, has founded workshop series at IEEE MASS, IEEE ICDCS and IEEE DCOSS, and been Workshop Chair at IEEE MASS-09, ACM Mobicom/Mobihoc-07 and Mobihoc-08.

Preface

RFID networks are currently recognized as one a research area of priority. Research activities related to RFID technology have been booming recently. A number of ongoing projects are being funded in Europe, Asia, and North America. According to leading market analysts, the development of the RFID market is projected to increase from approximately $3 billion in 2005 to $25 billion in 2015. Several countries have dedicated innovation programs to support and develop RFID systems and related technologies: the RFID initiative in Taiwan, Ubiquitous Japan and the NSF SBIR program in the USA. The EU has recently advertised its Strategic Research Roadmap concerning the Internet of Things, which first of all refers to the RFID technology before being extended to communicating devices as in M2M (Machine to Machine). In this roadmap, several application domains have been identified:

- Aerospace and aviation
- Automotive
- Telecommunications
- Intelligent buildings
- Medical technology, healthcare
- Independent living
- Pharmaceutical
- Retail, logistics, supply chain management
- Manufacturing, product lifecycle management
- Oil and gas
- Safety, security and privacy
- Environment monitoring
- People and goods transportation
- Food traceability
- Agriculture and breeding
- Media, entertainment and ticketing
- Insurance
- Recycling

The potential of RFID technology is huge. Contrary to popular belief, RFID technology is not recent and the delay in its deployment in commercial applications is not only due to its excessive cost. Ten years ago, standardization activities were insufficiently developed to allow the emergence of one standard which guarantees interoperability. In the meantime, ISO and worldwide organizations such as GS1 have proposed solutions, but new problems have arisen such as privacy issues and reading accuracy in proximity of certain materials such as water. The integration of RFID data in information systems is also a non-trivial problem. In the vision of the Internet of Things, future applications bring scalability and programmability issues.

The book is intended to cover a wide range of recognized problems in RFID protocols and low-level research challenges, striking a balance between theoretical and practical coverage. The theoretical contributions are limited to the scenarios and solutions that are believed to have some practical relevance. This book is unique in addressing RFID protocols and communication issues in comprehensive manner.

The book is divided into four parts. Part I provides an introduction and describes architectures of both passive UHF readers and tags. In addition, it defines performance metrics and introduces different classifications of RFID systems. Part II is related to networking protocols that involve one reader and multiple tags with the goal of resolving tag-to-tag interference. Tag identification protocols are covered in a systematic way. They include Aloha-based and tree-based protocols, which are the most popular. In addition tag-talks-first and tag-talks-only protocols are discussed and compared with reader-talks-first protocols. Part III provides coverage of networking protocols that involve a host and multiple readers. First, the interface between the host and the readers is considered. Next, MAC layer solutions for reducing reader-to-tag interference are discussed. In addition, the redundant reader elimination problem and delay-tolerant networks are covered. In Part IV, several major research challenges in the RFID field are presented, such unsatisfactory read accuracy even in the most favorable RF environments, low read ranges, security problems, localization of tags, energy harvesting and simulators and emulators for RFID systems. Some of these challenges are so serious that they are preventing the widespread use of RFID technology (e.g. low read accuracy and security). Therefore, a number of these challenges and potential solutions are analyzed in this part of the book.

At the end of most chapters, problems are presented and the solutions to some of the problems are provided on the book's website http://www.wiley.com/go/bolic_rfid.

We believe that this book is an appropriate and timely forum, where industry, and academics from several different areas can learn more about the current trends in RFID networking and become aware of the protocols and current issues in RFID networks. It is well recognized that RFID technology will become a part of everyday life soon. Additionally, we believe that, given the huge interest in this topic shown by the industrial and academic worlds, this book can become a standard guide to modern RFID systems.

Miodrag Bolić
University of Ottawa, Canada

David Simplot-Ryl
INRIA, France and University of Lille, France

Ivan Stojmenović
University of Ottawa, Canada

Acknowledgements

We would like to express our gratitude to the authors of book chapters who not only contributed a book chapter but also reviewed one additional chapter. In addition, we would like to thank a number of people who helped us review this book as shown below (the reviewers are not listed in any specific order).

Gustaw Mazurek (Warsaw University of Technology), Daniel M. Dobkin (Enigmatics), Timo Kasper (Ruhr-Universität Bochum), Venkatesh Sarangan (Oklahoma State University), Carlisle Adams (University of Ottawa), Zhang Xiong (Beihang University), Jeffrey S. Fu (Chang Gung University), Masahiro Miyakawa (Tsukuba University of Technology), Justin Wenck (University of California, Davis), Pradeep Shah (Texas MicroPower Inc.), Christoph Angerer (Vienna University of Technology), Seok Joong Hwang (Korea University), Ilker Onat (University of Ottawa), Md. Suruz Miah (University of Ottawa), Lin Wang (University of Pittsburgh), Fusheng Wang (Emory University), Junho Yeo (Daegu University), Stevan Preradovic (Monash University), Nicolas Pauvre (GS1 France), Mustapha Yagoub (University of Ottawa), Rony Amaya (Carleton University), Francesca Lonetti (ISTI-CNR), Francesca Martelli (Università di Pisa), Gaetano Marrocco (Università di Roma), Michael E. Knox (Mode1corp), Pankaj Mishra (University of Ottawa), Nemai Chandra Karmakar (Monash University), Zhou Yuan (Nanyang Technological University), Guan Yong Liang (Nanyang Technological University), Petar M. Djuric (Stony Brook University), Ali Miri (University of Ottawa), Mohamad Forouzanfar (University of Ottawa), Daniel Shapiro (University of Ottawa), Qinghan Xiao (Defence Research and Development Canada), Qiang Guan (Chinese Academy of Sciences), Bela Stantic (Griffith University), Xianjin Zhu (Stony Brook University) and many others.

We greatly appreciate the support, guidance and encouragement given by Wiley's team including Sarah Tilley, Anna Smart and Tiina Ruonamaa.

Part One

Components of RFID Systems and Performance Metrics

1

Performance of Passive UHF RFID Systems in Practice

Miodrag Bolić[1], Akshay Athalye[2], and Tzu Hao Li[1]

[1]*School of Information Technology and Engineering, University of Ottawa, Canada*

[2]*Astraion LLC, NY, US*

1.1 Introduction

1.1.1 Overview

Radio Frequency Identification (RFID) is a technology that has risen to prominence over the past decade. The clear advantages of this technology over traditional identification methods, along with mandates from supply chain giants like Wal-Mart and the Department of Defense, led to a large number of research and commercialization efforts in the early 2000s. However, almost a decade on, the early promise of widespread, ubiquitous adoption of RFID is yet to materialize. This is due to a combination of several technical and commercial factors. The technical imperfections and shortcomings existing in present-day RFID systems pose a very significant obstacle to the widespread adoption of RFID. Overcoming some of these challenges would amount to a very significant step forward towards realizing the tremendous potential of RFID technology. This book describes the ongoing efforts of some of the leading researchers in the field towards tackling the most challenging issues in today's RFID systems. With this in mind, the aim of this chapter is to clearly demonstrate, through experimentation, some of these technical challenges faced by RFID systems in practice. This chapter will enable the reader to better recognize the shortcomings of today's RFID systems and will allow for a better understanding and appreciation of the research efforts described in the rest of the book.

In this chapter, we focus on passive RFID systems operating in the Ultra High Frequency (UHF) band and adhering to the popular EPC Global Class 1 Generation 2 (Gen 2)

RFID Systems: Research Trends and Challenges Edited by Miodrag Bolić, David Simplot-Ryl, and Ivan Stojmenović

standard [1]. We begin with the characterization of a hypothetical "ideal" RFID system. We then proceed to examine the performance of practical RFID systems through simple experiments and point out the non-idealities and problems that arise in practical systems. We begin this examination by considering a simple system involving a single stationary reader and a single stationary tag in free space. We then examine systems with increasing degrees of complexity with multiple (possibly mobile) readers and tags in more challenging deployment environments. As complexity of RFID systems increase, more problems (non-idealities) are observed in the performance while problems identified with simpler systems remain. We believe that the approach of analyzing RFID systems with an increasing degree of complexity and identifying challenges as they appear will give the reader a sound understanding of the challenges facing real-world RFID systems.

Please note that this book chapter represents our viewpoint on imperfections of RFID systems. We have tried to point out some of the major issues in existing UHF RFID systems. This is not meant to be an exhaustive listing of all the possible challenges in practical UHF RFID systems, and there may be some problems and issues that have not been addressed here.

1.1.2 Background

RFID is a wireless technology that allows for automated remote identification of objects [2]. The major components of an RFID system are tags or transponders that are affixed to objects of interest and readers or interrogators that communicate remotely with the tags to enable identification. RFID systems exist in various flavors that can be classified based on the frequency of operation, power source of the tag and the method of communication between the reader and the tags. A detailed classification of the commercial RFID systems based on the above criteria is presented in Chapter 2. In addition, the overview of RFID technology is presented in a number of publications including [3, 4]. In this introductory chapter, we focus on passive RFID systems operating in the 860–960 MHz band. Passive RFID tags draw the power required for operation from the radio wave transmitted to them by the reader and communicate with the reader by controlled reflection of a portion of this incident wave. This technique of communication by controlled reflection is referred to as *backscatter modulation*. Although this technique was used as early as World War II, RFID transponders were expensive, large devices that remained confined to military applications. However, the tremendous progress in VLSI technology along with the establishment of standards in the early 2000s, enabled RFID tags to be manufactured in high volumes resulting in a price point that initiated numerous commercial applications. The main goal of commercial RFID systems is to automate and enhance asset management by providing global asset visibility. This ability of RFID systems finds various applications in diverse fields such as supply chain management, indoor asset and personnel tracking, access control, robotics and many more.

The immense commercial potential of RFID is mainly due to the numerous advantages that the technology possesses over traditional identification mechanisms such as barcodes. Some of these advantages are: (i) passive RFID tags can be read at much greater distances than barcodes; (ii) there is no need for a line of sight between the reader and tag; (iii) multiple tags can be read at much higher rates than barcodes; (iv) RFID tags have much larger memory than barcodes which allows storage of a lot more information than just the ID; and (v) the information contained in the RFID tag can be modified dynamically using the interrogator.

As mentioned earlier, in order to harness the advantages of RFID technology to build viable commercial solutions, a number of technical challenges needs to be overcome. Some of these challenges are common to other wireless technologies while others are unique to the RFID system to hand. Each RFID technology, including passive, semi-passive and active RFID systems operating at different frequencies, poses a unique set of challenges to obtain the desired performance. In addition, design requirements, performance specifications and protocols for active, passive and semi-passive systems are also very different. Therefore, in this chapter, we will limit our discussion to long-range passive backscatter-based UHF RFID systems operating in the 860–960 MHz band. In our opinion, this type of RFID has the most potential for significant commercial impact. As a result, it has seen the most research and standardization activity in recent years, more than other types of RFID systems. Today commercial systems of this type adhere to the EPC Global Class 1 Generation 2 (Gen 2) standard that has been in effect since 2005 [1]. Gen 2 compliant readers and tags are readily available in today's market from several vendors all over the world including Alien Technology, Impinj Inc., Motorola and others.

There have been several approaches in characterizing RFID systems. They are mainly based on (1) experimental characterization; and (2) mathematical modeling and simulation-based analysis. Experimentation is either performed in a controlled environment such as anechoic chamber, in the laboratory environment [5], [6], [7], [8] or in the application-specific setups such as conveyer applications [8]. An example of modeling and simulation-based analysis is performed in [9] where tag characteristics, propagation environment, and RFID reader parameters have been modeled and simulated. In this chapter, RFID systems are characterized through experimentation in an anechoic chamber and laboratory environment.

1.2 Ideal RFID System

We begin our analysis of practical RFID systems by presenting the characteristics of a hypothetical ideal RFID system. Of course, like most other ideal systems, this RFID system would be unrealizable in practice. However, formulating such a system will give us a better understanding of the problems faced by real-world RFID systems. Once again, we point out that this ideal system is formulated in the context of UHF passive RFID systems.

Since passive tags do not have a battery, they need to receive enough energy to turn on the tags' integrated circuit. Therefore, in order for a passive RFID system to operate, the tag needs to receive enough power to wake up, and its backscattered response needs to be correctly received and decoded by the reader. In addition to this basic functionality, an RFID system has several other requirements for efficient operation that will be described later as desired features. The characteristics of an ideal RFID system, that mainly correspond to the basic functionality, can be summarized as follows:

1. There exists a well-defined, controllable read zone for each reader. For every tag within its read zone, each reader has a 100% read rate or read accuracy and for tags outside its read zone, each reader has a 0% read rate.[1]

[1] For this chapter, we define read rate as the fraction of the number of times the reader is able to read a tag over the number of queries it sends to a tag. Please note that in RFID literature, the term read rate often refers to the speed at which the reader and tag communicate.

2. Performance is insensitive to the physical orientation of tags.
3. Performance is insensitive to the nature of the object on which the tag is placed.
4. Performance is insensitive to the environment in which the system is deployed.
5. Multiple tags communicate with the reader in a collision-free manner and the time for reading a fixed number of tags is a deterministic function of the number of tags while utilizing the maximum allowable bandwidth.
6. Performance is unaffected by the presence of multiple readers with overlapping read zones or of multiple tags within a read zone.
7. Performance is unaffected by relative motion between the readers and tags as long as the tags remain within the read zone of the reader.

In the context of above characterization, read rate or read accuracy for each tag is defined as the percentage of times a reader is able to correctly read the tag's ID. Hence for the ideal system if a reader sends out N queries to a population of tags that are all within its well-defined read zone, it receives N responses from *each* of the tags containing their respective IDs. Similarly if a reader sends out N queries to a population of tags that are all outside its read zone, it will receive no response. In the ideal system, this holds true for multiple readers with overlapping read zones, that is, if multiple readers send out N queries to a population of tags that are simultaneously in the read zone of all the querying readers, then each of the readers will receive N responses from each of the tags containing the respective tag IDs. As mentioned in the above characterization, the behavior of the ideal system is unaffected by factors such as orientation, environment and relative motion.

Besides these ideal characteristics, there are several other desired features that researchers are trying to bring to practical UHF RFID systems. Some of these include:

1. High level of security of an RFID system.
2. Localization of each tag within the read zone with high level of accuracy.
3. Low cost of RFID components and high return of investment.
4. Easy integration of RFID software into existing application software.
5. Simple deployment and networking of multiple readers.
6. Simple synchronization of multiple readers.

Items 4-6 are related mainly to complex RFID systems with a large number of readers.

Although the ideal RFID system is unrealizable in practice, the above characterization provides a useful reference against which to measure various performance metrics of practical systems. Moreover, one can view most of the research efforts described in the rest of the book, whether at the physical, protocol or software level, as attempts to bring practical systems as close as possible to the aforementioned ideal or desirable RFID system.

We now proceed with the analysis of practical systems by considering deployments with increasing levels of complexity, and pointing out the divergences from the ideal system that occur in practice.

1.3 Practical RFID Systems

1.3.1 Complexity of RFID Systems

Figure 1.1 shows simple block diagrams of several RFID systems. The complexity of the systems increases from left to right. Systems consisting of stationary readers and tags are shown in Figure 1.1(a), (b), (c), while the mobile systems are shown in Figure 1.1(d), (e). Stationary readers can be attached to walls, portal constructions or ceilings resulting in fixed deployments, while mobile readers are attached to moving objects such as forklifts, hand carts or are carried by people (handheld readers) and robots. Stationary tags are attached to the objects that usually do not move during the query round, for example, tags attached to shelved items. Mobile tags are attached to mobile objects or people and they move relative to the reader antenna during the query round. Large RFID systems like the one outlined in Figure 1.1(e) require the consideration of several other issues such as networking, synchronization, data processing software and middleware in order to enable an end user to reap the benefits of a deployed RFID system.

1.3.2 Single Reader, Single Tag

The simplest practical RFID system from an analysis viewpoint is the one that consists of a single stationary reader and a single stationary tag (Figure 1.1(a)). Our setup for examining the performance of such a system consists of a Gen 2 compliant tag and a

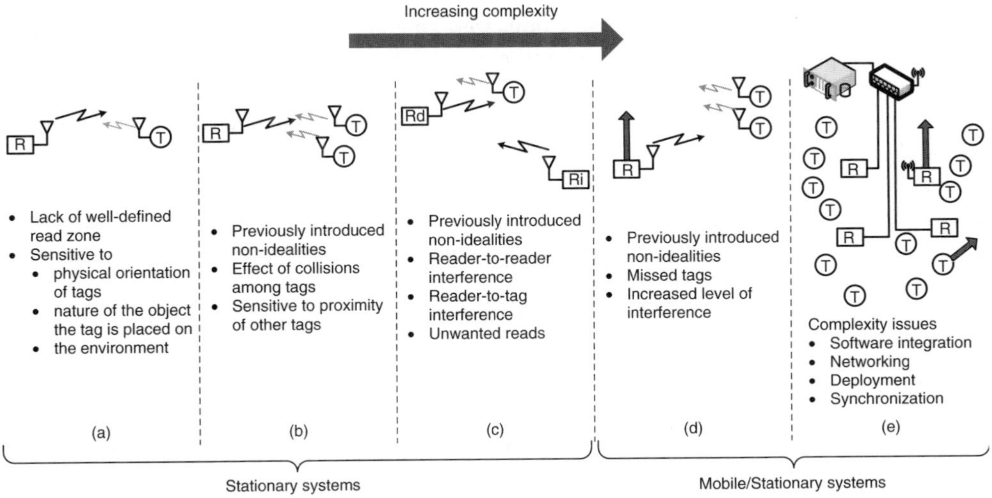

Figure 1.1 RFID systems of increasing complexity together with non-idealities encountered in each system with (a) single reader R single tag T system, (b) single reader multiple tag system, (c) multiple readers single tag system which includes interfering reader R_i and reader that communicates with the tag R_d, (d) mobile reader and multiple tags, (e) complex system that includes many readers and tags where readers are connected to the router and the computer.

Gen 2 compliant reader. The Gen 2 protocol uses a dynamic frame slotted Aloha-based anticollision protocol that enables multiple readers to communicate with a single reader. In this protocol, the reader requests tags to reply to the reader commands in defined time slots. The reader specifies a fixed number of *slots* in a so-called *Inventory Round* or *Query Round*. An inventory round is defined as a single cycle of an algorithm by which a reader attempts to singulate the tags within its environment. Singulation is defined as a process of identifying a single tag and reading its ID number. A Query Round begins with a *Query command* which species a so-called *Q* parameter which indicates the number of slots in a query round. Each tag in a population then selects one random slot out of these slots to communicate with the reader. The reader then sends out successive *Query Rep* commands which designate the start of each slot. A reader may also send *Query Adjust* commands that dynamically increase or decrease the number of slots in the round. In its chosen slot, the tag replies with a 16-bit random number (RN16) using backscatter modulation. Upon successful reception of this RN16, the reader sends out an *Acknowledge* command with the same RN16 back to the tag. If this number matches the number that the tag originally sent out, the tag backscatters a Protocol Control (PC) header, followed by its EPC ID and a 16 bit CRC. A more detailed explanation of the Gen 2 standard is given in Chapter 5. The description of the Aloha-based anticollision algorithm used is presented in Chapter 7 of this book.

In our experimental setup, the Gen 2 reader is attached to a host computer that is capable of monitoring the read rate of a detected tag, that is, the ratio of the number of times a tag responded successfully to the number of Query Rounds sent out by the reader. We are using commercial UHF Gen 2 passive dipole tag. The reader transmits at a power of 30 dBm over a 6 dBil gain circularly polarized antenna. In addition we deploy a sniffer device in the proximity of the tag so as to examine the actual communication happening over the wireless channel. This sniffer device is connected to an oscilloscope that stores the snapshots of the baseband signals going between the tag and the reader. Figure 1.2 shows the captured waveform for single reader, single tag scenario. Figure 1.2 shows the Query Rep commands that a reader sends out designating the tag communication slots. In its selected slot, the tag backscatters an RN16 as shown in the Figure 1.2. This is then followed by transmission of the Acknowledge command by the reader and the subsequent backscattering of the EPC ID by the tag.

Upon examination of the performance, we see that even this simple system differs from the ideal system described in Section 1.2 in several ways. The deviations are presented in Figure 1.1(a) and analyzed below.

Lack of Well-defined Read Zone: The practical single reader-single tag system does not have a specific read zone with respect to the reader antenna wherein the tag exhibits a 100% read rate and outside which the tag exhibits a 0% read rate. It has been shown in several independent works that, in the case of passive RFID systems, such as the one currently under consideration, the read range depends mostly on the power in the forward link which is needed in order to power tags IC [4, 5, 8, 9]. In free space, as a tag moves away from a reader, the mean value of the power it receives drops off as per the Friis equation describing the wireless link. As this power drops off, so does the tag's read rate until the tag reaches a place where it is unable to receive sufficient power and the read rate drops to zero. Figure 1.3 shows the read rate of a single tag as a function of the

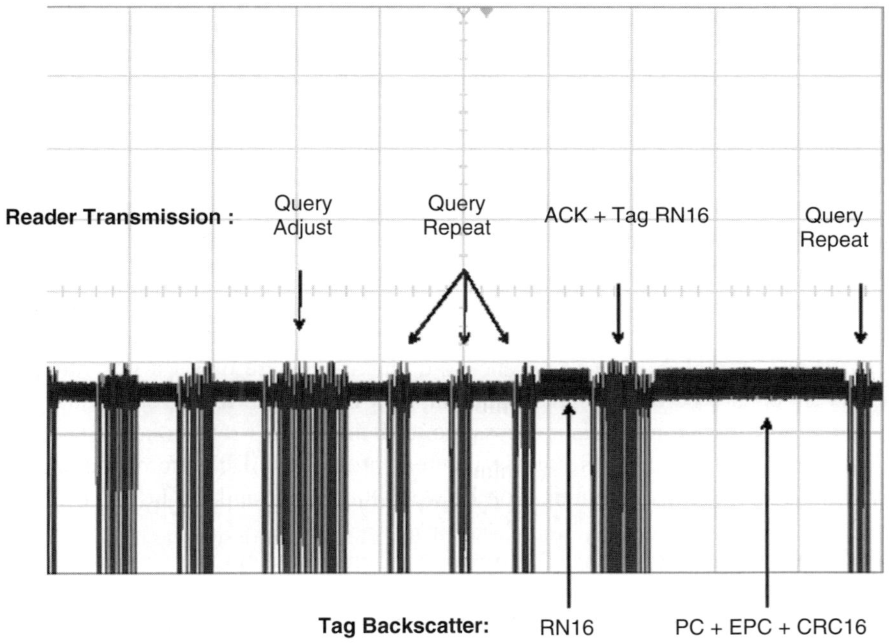

Figure 1.2 Communication between a single Gen 2 reader and single Gen 2 tag.

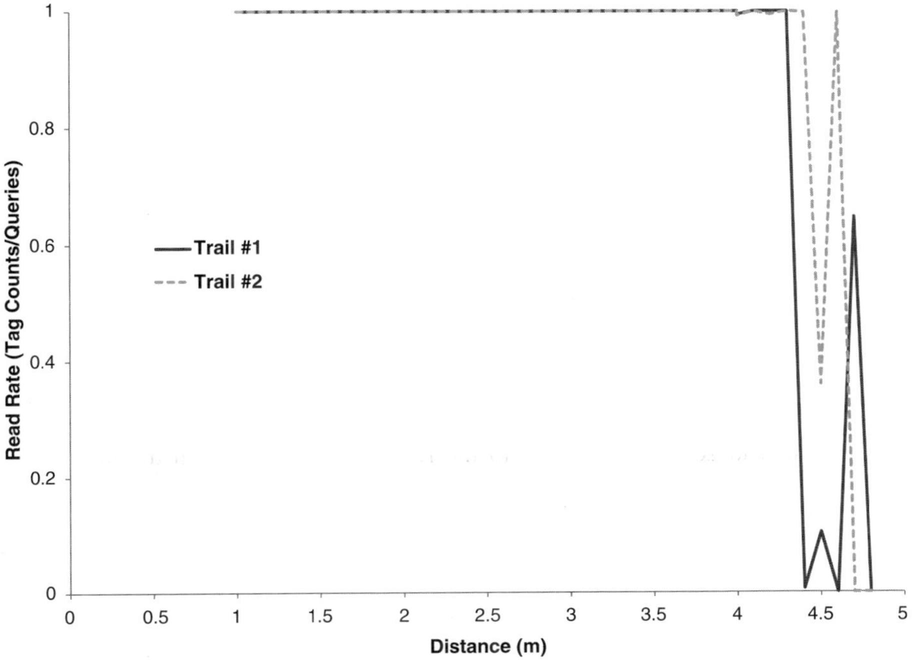

Figure 1.3 Read rate for a single tag in an anechoic chamber.

distance from the reader. Two curves on the graph are two different trials with the same setup. The experiment is carried out in an anechoic chamber (dimensions of chamber are: length 5 m, width 5 m and height 2 m). For this experiment, the tag is placed in the best possible orientation with respect to the reader antenna. By performing the experiment in the anechoic chamber and by fixing the orientation of the tag and reader antennas, we wanted to avoid influence of other parameters to read rate besides the distance. The experiment was started by placing the tag at a distance of 1 meter from the reader antenna. A total of 500 Query rounds were sent by the reader, and the number of responses from the tag was noted. The steps were repeated while increasing the distance between the tag and the reader antenna in 0.5 meter steps in the range between 2 m and 4 m where the read rate was maximum, and in 0.1 m steps in the zone when read rates start to drop. As seen from Figure 1.3, even in an anechoic chamber, there exists a gray area around the reader antenna wherein a tag may or may not be read in a particular query round. Thus, in the practical system it is not possible to define a clear read zone as described for the ideal system. This is because of the inherent properties of electromagnetic radiation which is the basis of the communication between the reader and the tag.

Sensitivity to tag orientation: The relative orientations of the tag and the reader significantly affect the performance of a practical system. Figure 1.4 shows the effect that the tag orientation has on the read performance even when the tag is within the read range of the reader antenna. In order to collect the data we performed the experiments in an anechoic chamber with the similar setup as presented above. In this experiment,

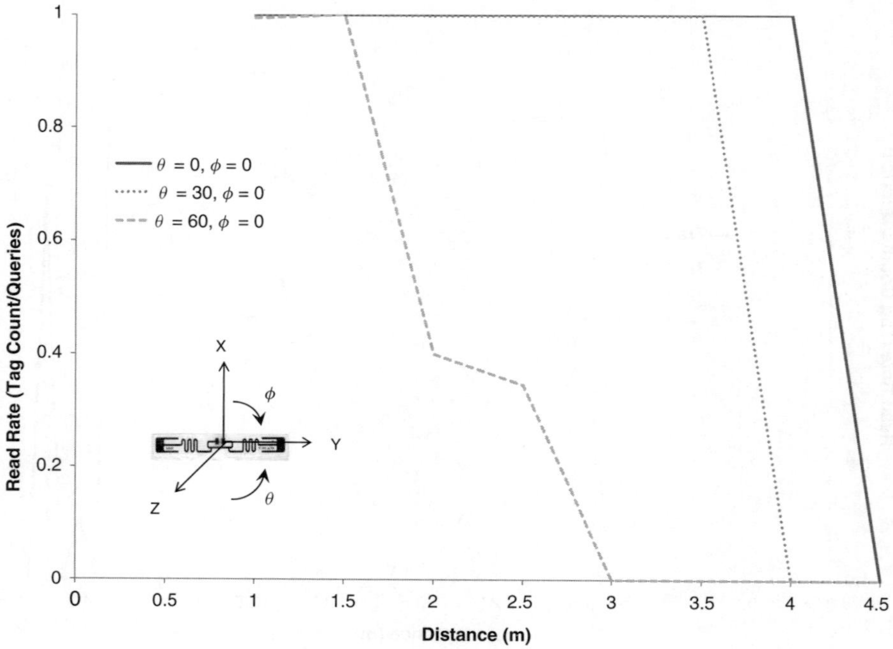

Figure 1.4 Read rate for a single tag with different orientations in an anechoic chamber.

the reader is fixed at the position $(x, y, z) = (0, 0, 0)$ and tag is moved in 0.5 m steps in y-direction. Measurements are repeated three times: each time a tag was oriented differently relative to the orientation of the reader antenna. Read rates are recorded in the same way as in the previous experiment and presented in Figure 1.4. From Figure 1.4, it is obvious that the read range drops significantly when the orientation becomes less favorable for the tag because the tag is not able to collect enough energy to turn on the IC. In addition, Figure 1.4 demonstrates that the read rates depend on the relative orientation of the tag and reader antennas and on the distance between tag and reader.

The problem of orientation sensitivity can be handled by innovative tag antenna designs involving multiple dipoles or monopoles. In fact, ensuring orientation insensitivity is one of the most important goals in designing antennas for tags. The latest research efforts in designing efficient tag antennas are presented in Chapter 3. Experimentation with different tag orientation are published in several different works including [10]. Please note that experimentation of dependency of orientation to read rate in [10] is performed with EPCGlobal Class 1 tags (generation before Gen 2).

Sensitivity to deployment environment: The performance of a practical system is highly dependent upon the environment in which the system is deployed. Like any other wireless system, the nature of the environment affects the multipath and fading properties of the channel. This effect is even more pronounced in RFID systems due to the passive nature of the tag operation and the inherently low signal to noise ratio (SNR) of the weak backscatter signal. In order to examine the effect of environment on performance, we repeated the experiments mentioned previously in the significantly cluttered environment of a normal computer lab room (dimensions of lab room are: length 10 m, width 6 m and height 4 m). Figure 1.5 shows the read rate performance in the cluttered environment of a computer lab. As we can see, the deployment environment hampers the read performance of the system and also introduces some blind spots/null spots due to multipath interference and channel fading. In comparison to the read rates in the anechoic chamber, it is obvious that even for a fixed relative orientation of the reader and tags antenna it is difficult to specify the range in which the read rate is maximum.

It is clear from the above-mentioned experiments, that the read rate and read range of a tag are not solely dependent on the distance between the reader and tag, but are also affected by factors such as orientation and environment. A similar inference has been drawn in [9] and [11] wherein the authors suggested defining read range as the range in which a pre-defined read rate or accuracy of reading a tag can be achieved.

The reading accuracy problem is presented in Chapter 14 of this book together with the directions for improving reading accuracy.

Sensitivity to the nature of the object on which the tag is placed: It is known that UHF RFID systems do not perform well when attached to objects that contain metals and fluids. These materials not only attenuate the signal when placed between the tag and the reader, but also result in detuning of the tag antenna. Detuning occurs if the materials are in close proximity to the tag. The effect of metal and water on read range and read accuracy was analyzed in several publications including [5]. Read ranges were reduced up to three times in proximity of water and metal in comparison with the read range in free space.

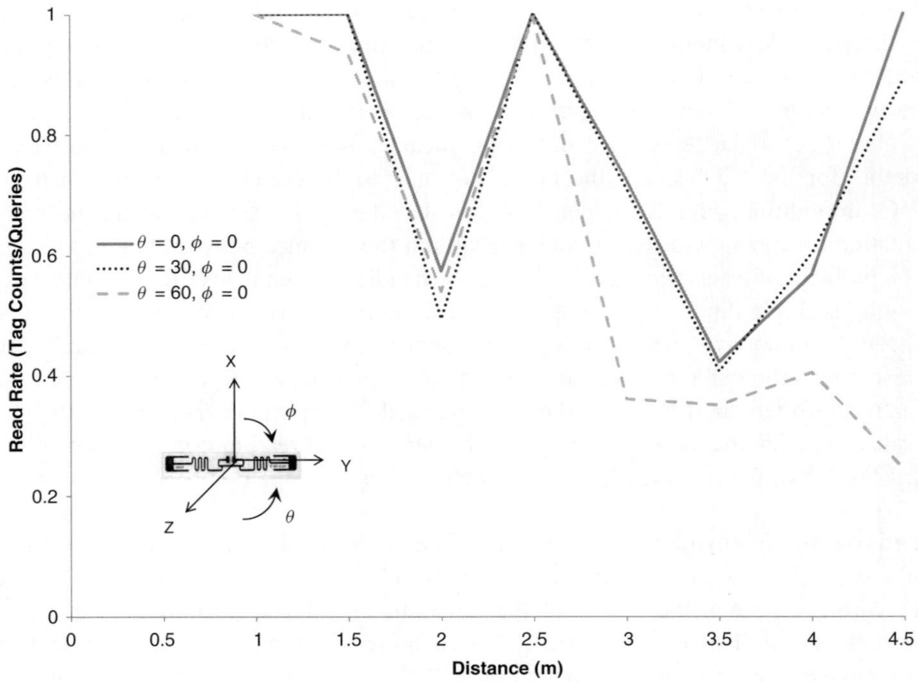

Figure 1.5 Read rate for a single tag with different orientations in a computer lab room.

Approaches to handling this problem are based on modification of tag antennas to work specifically in the proximity of metals and liquids [12]. Design of such tags is described in Chapter 3 as well.

We will now move on to more complicated practical scenarios involving multiple readers and tags. Note again, that all the non-idealities and problems identified with the simpler systems are carried forward to the more complicated systems.

1.3.3 Single Reader, Multiple Tags

As mentioned earlier, Gen 2 tag anticollision protocol is based on a slotted Aloha wherein multiple tags can communicate with the reader in separate time slots. The following non-idealities can be identified in the practical situations based on an examining performance.

Collision between tags: In the practical system, unlike the ideal system, collisions can occur between tags that are trying to communicate with the reader within the same query round. In the context of Gen 2, a collision occurs when two tags select the same slot number to backscatter. In this case, the reader is often unable to decipher tag transmissions and the communication attempt is unsuccessful. Figure 1.6 shows a Query round involving 2 tags where frequent collisions occur.

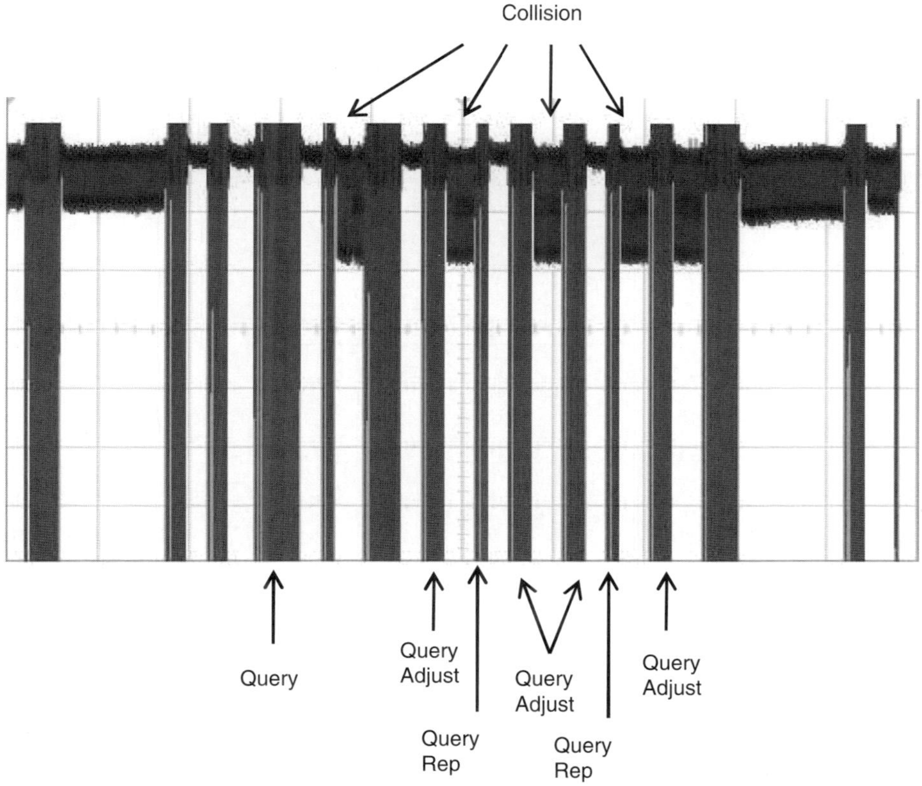

Figure 1.6 Collisions in Gen 2 query round.

Tag collision adds a significant overhead to the time needed to read a population of tags. This non-ideality is further exacerbated in scenarios involving a large number of tags simultaneously in the field of view of the reader as is the case in many commercial applications of RFID. However, these collisions are result of the air protocol selected in Gen 2 standard which is based on the Aloha paradigm. The standard does not specify implementation of the algorithm for anticollision and only gives recommendations. Chapter 7 describes different slotted Aloha anticollision algorithms with the emphasis on the methods for dynamic estimation of the number of tags to be singulated. The time to query population of tags can be minimized if the number of tags in the population is known or if it is correctly estimated.

Effect of tags in proximity to each other: In an ideal RFID system, neighboring tags will have no effect on each other's performance. In reality, however, this is not the case. In most practical applications, RFID tags are placed on objects that are densely co-located. Hence it is very important to understand the effect that the tags have on each other. The effect of proximity of the other tags to the read range and read accuracy is experimentally analyzed in [11] and [4].

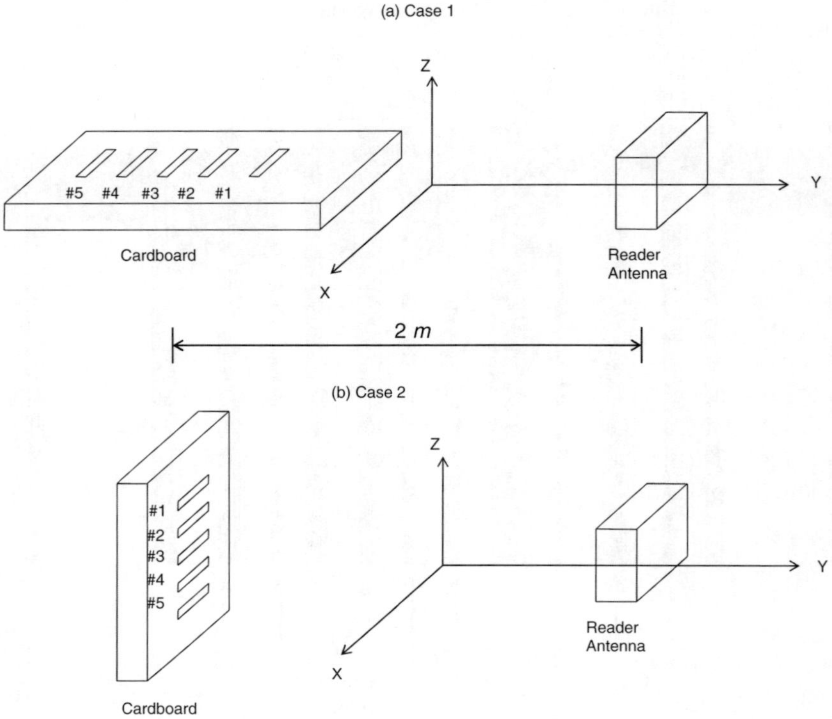

Figure 1.7 Experimental setup for examining effect of tag proximity on performance. Tags are spaced 1 cm apart and parallel to one another and (a) orthogonal to the reader antenna, (b) parallel with the reader antenna.

The presence of multiple tag antennas in close range alters the current distribution, the radiation pattern and introduces mutual impedance. As a result, tags in close proximity tend to cast a shadowing effects on the neighboring tags [4]. This shadowing effect increases as the number of tags increases and as the inter-tag distance decreases. The effect causes tags to receive less power than when no other tags are present in the proximity. This leads to a drop in the read rate or read accuracy of the tag.

In order to examine the effects of tag proximity, we use an experimental setup consisting of a Gen 2 reader with a circularly polarized patch antenna with a 6 dBil gain and five Gen 2 tags with dipole antennas. The tags are placed 2 m away from reader antenna and the reader output power is set to 23 dBm. All tags are placed in the same plane on a single cardboard platform with the best possible orientation angle to the reader. This is done so as to eliminate the influence of orientation sensitivity on the measurements. The experimental setup is shown in Figure 1.7. We consider three cases; in Case 1, tags are spaced 1 cm apart and parallel to one another. The cardboard platform is placed along the $X-Y$ plane and the reader antenna is placed along the $X-Z$ plane. For Case 2, we place the cardboard platform along the $X-Z$ plane while keeping the position of the reader antenna unchanged. Case 3 is similar to Case 1, but we increase the inter-tag spacing to 30 cm. The read rate results are shown in Table 1.1. Case 3 shows that when the tags

Table 1.1 Effect of tags in proximity on tag read rate.

Read Rate (#reads / #queries)	Case 1	Case 2	Case 3
Tag 1	0.98	0.45	0.98
Tag 2	0.98	0.096	0.98
Tag 3	0.56	0.02	0.97
Tag 4	0.38	0.178	0.98
Tag 5	0.11	0.87	0.98

are spaced far enough apart, all of them have a pretty good read rate. Case 1 and Case 2 results clearly demonstrate the significant impact of tag proximity on the read rates of a tag population. This problem becomes very important in practical applications involving a large number of densely packed objects, each having a separate RFID tag.

1.3.4 Multiple Readers, Single or Multiple Tags

In this subsection we will consider a situation in which there are multiple stationary readers and one or more stationary tags. Having multiple readers introduces diversity and redundancy which help in solving some of the previously recognized issues. If multiple readers are used to read the same tag, the chance that the tag will be in the blind spot for all the readers is reduced. Redundant readers or one reader with multiple antennas are commonly used in industry to improve the accuracy of reading. Localization of tags might also be improved by using multiple readers. Coarse-grained localization information, based on association of tags with the reader that detected it, will be improved if tags are read by two or more readers placed at known location. The tag would then be located in the space that represents the intersection of the read ranges of these readers.

However, the presence of multiple readers in an environment gives rise to the very serious problem of interference caused by their simultaneous operation. We examine the effect of this problem on system performance below:

Interference: A system with two readers and one tag is shown in Figure 1.1(c). In this figure, communication between desired reader R_d and tag is affected by the signal sent by the interfering reader R_i. If multiple readers attempt to singulate tags simultaneously, two types of interferences might occur: reader-to-reader and reader-to-tag interference. Reader-to-reader interference or reader jamming occurs when the interfering reader affects the reception of the tag signal by the desired reader. Reader-to-tag interference or tag jamming can occur when the interfering reader affect tags reception of the signal from the desired reader.

We examine the performance of interfering readers through the following experiment. Two Gen 2 readers from different vendors were programmed to work in a dense reader mode. Dense reader mode is proposed in Gen 2 standard [1] to support large-scale enterprise applications with many readers. Spectral allocation is defined so that reader-to-reader interference is reduced. The reader transmits at a power of 30 dBm over a 6 dBil gain

Table 1.2 Effect of multiple reader interference on tag read rate.

Reader 1 Mode	Reader 2 Mode	Reader 1 read rate	Reader 2 read rate
M = 4, BLF = 256 kHz	OFF	1	–
OFF	M = 4, BLF = 200 kHz	–	1
M = 4, BLF = 256	M = 4, BLF = 200	0.355	0.39
M = 8, BLF = 256	M = 4, BLF = 200	0.372	0.38
M = 0, BLF = 640	M = 4, BLF = 200	0.261	0.35

circularly polarized antenna. We are using a Gen 2 dual-monopole tag that is orientation insensitive. The tag is placed in between the readers at a distance of 1.5 m from each reader. Read rates of each reader are recorded when the other reader is off as well as when both readers operate simultaneously. As we see from Table 1.2, the read rate of both readers dropped significantly when both of them are querying simultaneously. When both readers send their commands at same time, the tag cannot decode the collided reader signals. The more readers broadcast at same time, the harder it is to read passive tag.

There are multiple challenges in resolving interference problems that are unique to RFID systems [13]. Passive UHF tags have limited tuning capacity and their reception will be affected by signals from multiple readers even if the readers operate at different frequency channels. In addition, readers transfer high amounts of power that is needed to power up the tag. This high power worsens the interference problems.

Reader-to-reader interference problem is addressed in the Gen 2 standard by introducing the dense reader mode. In this mode, lower data rates are specified, and Miller Subcarrier encoding is used so that the channel for reader and tag transmission can be well defined and separated. Dense reader mode is also analyzed in Chapter 5 of this book. For regions with only narrow frequency band available in UHF band such as Europe, the listen before talk approach together with applying spectral mask constraints are viable solutions [1, 14, 15].

Even though readers transmit in separate frequency channels, they can still cause reader-to-tag interference. In [16], it has been shown that the tag might function even if there are interfering readers. If desired and interfering readers transmit simultaneously, tags will be able to detect the signal from the desired reader if that signal is much stronger than the signal from the interfering reader. Experimentation was performed at the frequency of 866 MHz. Tags can detect the signal from the desired reader if the following conditions are met: (1) the signal power from the desired reader needs to be 6 dB higher than the signal power from the interfering reader when the difference in reader's carrier frequencies is more than 800 kHz; and (2) the signal power from the desired reader needs to be 13–20 dB higher than the signal power from the interfering reader when the difference in readers carrier frequencies is less than 800 kHz [16].

1.3.5 Mobile Readers and/or Mobile Tags

Both reader and tags can be mobile. Figure 1.1(d) shows an example in which the reader is mobile and the tags are stationary. This corresponds, for example, to the situation when the reader is attached to a forklift and the tags are placed on items that are on the shelves. In mobile systems, it will not happen that the tag remains in the blind spot. Due to the mobility of tags and/or readers, the reader will eventually be able to read the tag. While mobility can aid in handling some of the identified issues, it also introduces new problems into the system. Centralized algorithms for reducing interference might fail if readers are mobile and some tags might not be read because they appear in front of the reader for the limited amount of time.

Missed tags: Let us consider the case in which tags are mobile and the reader is stationary. As pointed out in [17], RFID tag antennas designed for stationary applications operate reliably when tags move at low speed. For example, a speed of about 16 km/h is considered low enough for reliable operation. In [18] the effect of accuracy is examined when ten single-dipole tags are moved at the speed of 1 m/s and were kept at 1 m from the reader. Different relative orientations between tags are explored. It was determined that the minimum distance between tags should be 4 cm in order to be able to detect all the tags. Other experimentation is described in [19], where an RFID reader is placed on the side of the conveyer belt whose speed is 2 m/s and tags are attached on boxes placed on the conveyer belt. The system is first calibrated to avoid unwanted reads. Excellent read rate is achieved even in cases when 50 tags are placed on the same box and the speed of the conveyer belt is 2 m/s and when the reader is 44 cm away from the tags. However, the high reading rate in both papers might be due to the fact that the reader was placed very close to the tags.

Simulations with detection of high speed objects are described in Chapter 9 of this book. It has been shown that it is possible to read four tags at a distance of 4 m at speeds of 250 km/h with appropriate setting of parameters of Gen 2 protocol. It also shows that much better performance for high speed tags can be achieved using tag talk first protocols. The experimentation with mobile tags attached to a vehicle is performed in [20]. In the experiment, six tags are attached to the windshield and two antenna are placed at the height of 4.2m with an angle of 45 degrees. High read rate is achieved at the speed of 10 km/h, while very low read rates are achieved for speeds above 70 km/h. The method for predicting the read rate is presented in [20] and it relies on the support vector machine model.

Mobile readers are used in combination with stationary tags in several different applications for localization, mapping and navigation of robots and people. In almost all applications stationary tags are used as landmark tags to assist in positioning of the mobile readers and their carriers. In [21] it was also confirmed that the major problem is how one can cope with low reading accuracy. If the landmark tag is not read, the position cannot be determined. One possible solution, self-localization of robots that takes into account unreliable RFID readings, is proposed in [21]. There, the robot first collects information about the read landmark tags and the number of their detection during the training phase. This information is stored and used to compare with the real measurement results during operation of the robot.

1.3.6 Large Deployments Including Many Readers and Tags

Figure 1.1(e) shows an RFID system consisting of a large number of readers and tags. The figure shows RFID system with four readers of which three are stationary and connected using Ethernet to the switch. One reader is mobile and it communicates with the switch using a wireless link. The switch is connected to the host that runs middleware software. The role of the middleware is to aggregate the data from the deployed readers, filter and process it as per case-specific requirements and present it to the higher level software. In addition to the physical non-idealities that we have described, this large deployment of multiple readers and tags gives rise to a whole new set of challenges related to the middleware, application software and network management.

Synchronization of many readers: In large deployments, implementing the synchronization scheme is a significant challenge. Many algorithms have been developed to reduce interference in large-scale deployments (Chapter 11 of this book). Centralized synchronization is impractical for large-scale solutions and should be avoided [22]. In addition, in multivendor environments, such as shopping centers, receiving dock doors might belong to different organizations and synchronization is again impractical [15]. In these situations, readers will need to be able to operate asynchronously of each other and be truly event driven [15].

Easy integration of RFID software into existing application software: Full benefit of an RFID system can only be utilized if data collected from the RFID system is used for decision-making [23]. Therefore, the RFID reader system needs to be integrated into the existing application software. RFID middleware software is used to provide, besides other functions, easy integration with legacy applications. If the applications have proprietary interface, then the integration becomes complicated [23]. RFID Middleware as a part of EPCGlobal architecture is described in Chapter 6. The architecture of RFID system is presented in Chapter 10.

Installation and tuning: In deployments with a large number of readers, installation of the RFID components can be complicated. It requires installation of power connections and network cabling. In addition, readers and antennas have to deployed properly to reduce interference among the readers and other wireless devices or machineries to the RFID networks. Improper deployment can result in some areas being not properly covered by RFID readers. Therefore, the RFID system needs to be tuned by the experienced hardware engineer [23] to assure that required performance is achieved.

1.3.7 Other Desired Features of Practical RFID Systems

As mentioned earlier, in addition to the basic operational characteristics embodied by the ideal RFID system, practical systems have some other features that are necessary in enabling commercial applications. We describe a couple of these below.

Privacy and security in Gen 2: Privacy and security aspects are extremely important for successful deployment and application of RFID systems. A large number of publications have recently appeared that deal with improving or developing security and authentication frameworks for RFID applications [24].

RFID tags based on Gen 2 standard are considered to belong to low-end systems based on their capacity to implement schemes for security. Current security and privacy mechanisms in the Gen 2 standard, although better than previous standards, are still considered weak (Chapter 16 of this book). In the Gen 2 standard, the reader does not transmit the EPC code, parts of the tag memory can be locked, a *kill* command can permanently disable the tag. Functions such as access to special memory, ability to modify tags and the ability to kill tags require a 32-bit access password. However, the 32-bit access password can be eavesdropped and then easily computed. More advanced security and privacy methods are described in Chapters 16 and 17 of this book. As is pointed out in Chapter 2 of this book, adding security features to passive tags would increase the complexity of the tag's hardware and then likely reduce the tag's range and throughput performance. This tradeoff makes the task of designing security mechanisms for passive RFID systems all the more challenging.

Reducing ambiguity in tag location: Unlike an ideal system, the practical reader-tag RFID system is unable to determine the precise location of the tag within the read zone of the reader. As mentioned earlier, one of the main goals of an RFID system is to provide global asset visibility. In most practical applications, precise asset location is an important attribute of asset visibility in order to automate the asset management process. The application of traditional ranging techniques such as those based on received signal strength (RSS), time of arrivals (TOA), time difference of arrivals (TDOA) and phase of arrivals (POA) to passive RFID is very challenging due to the weak backscatter signal used to communicate back to the reader. Localization techniques such as trilateration rely on precision of ranging techniques. This is the reason why many RFID localization solutions are based on using landmark tags or some kind of map matching. Precise localization using passive RFID is an important problem that is the focus of a number of research efforts today. The overview of localization algorithms is presented in Chapter 2 and detailed analysis of algorithms for ranging and localization is presented in Chapter 15.

Reducing the unwanted reads: The fact that the reading zone is not properly defined causes tags to be read by multiple readers. In some applications, this is not desired. For example, in dock door application, each reader covers a different door. It is important to detect which door the pallet went through. If the tags on the pallets are read by multiple readers, this will introduce ambiguity in associating pallets with corresponding dock doors. Reading of the tags by the readers that cover neighboring doors are called unwanted reads. Other names used in practice are unintended reads and cross reads [25]. Solutions against unwanted reads are based on both hardware (shielding) and software (filtering of data).

1.4 Overview of the Book

This chapter presented problems and non-idealities in current passive UHF RFID systems. The rest of the book describes approaches to deal with the defined problems. In addition, the book tackles a number of other issues that have not been described in this chapter. Therefore, we summarize major problems, methods and approaches covered in the book.

Chapters 3–5 deal with design of tag antennas, tag and reader circuit for passive RFID systems operating in UHF band. Chapter 3 introduces the design of antennas for passive UHF tags. The chapter also describes the effects of environmental factors including proximity of dielectric materials and metals on performance of the antennas. Then, several approaches for designing antennas that operate in proximity of metals have been proposed. In Chapter 4, tag's architecture is presented. RF, baseband and memory design are detailed. Low power design issues are pointed out. Chapter 5 focuses on the reader design issues that are unique to RFID systems including handling transmitter leakage during tag reception, frequency generation, transmit linearity and transmit AM noise.

Chapters 6 and 10 cover EPCGlobal architecture framework and protocols. Chapter 10 analyzes services that the RFID system supports and defines the system architectures. The chapter then analyzes a number of EPCGlobal specifications that standardize interfaces to configuration, monitoring and data processing services. Chapter 6 describes RFID middleware in details. It focuses on the following aspects: reader management, data management and storing and sharing data in RFID networks.

Non-idealities of the RFID systems are tackled throughout the book. The problem of read range is considered in Chapters 2 and 14. The parameters that affect read range in a RFID system are pointed out and ways to achieve longer reading ranges are introduced. Parameters that affect the throughput of RFID systems are considered in Chapter 2. Reading accuracy is explained in more detail in Chapter 14.

Localization of passive and active tags is the application that is attracting increasing interest. The localization problem is introduced in Chapter 2 and elaborated in Chapter 15.

Privacy and security aspects of RFID networks are considered in Chapters 16 and 17. Chapter 16 focuses on low level approaches and covers attacks against common RFID components and lightweight cryptography. Chapter 17 describes higher level aspects of security and privacy including description of the cryptographic protocols for privacy, identification with privacy, and authentication without privacy.

Even though the book mainly deals with passive RFID system, active RFID systems are also described in Chapter 2. Some localization solutions based on active RFID systems are mentioned in Chapter 15. Dual frequency active RFID systems are presented in Chapter 18. Chapter 18 is dedicated to novel technologies and it is especially focused on energy harvesting. Energy harvesting is also briefly discussed in Chapter 19.

In Chapter 19, a methodology that allows translation of the models of RFID system automatically onto a hardware platform to enable real-time verification in the target operating environment is presented. A detailed description of the simulator and the emulator is provided.

Tag identification protocols are presented in Chapters 7, 8 and 9. Reader talk first protocols including Aloha and tree-based protocols are described in Chapters 7 and 8 respectively, while tag-talk-first approach is described in Chapter 9. Combinations of tree-based and Aloha-based algorithms are described in Chapter 7. In general, the number of tags to be singulated is not known. For both tree- and Aloha-based protocol, knowledge of the number of tags would reduce the average time it takes for the protocol to singulate all the tags. Therefore, in Chapters 7 and 8 ways to estimate number of tags while the algorithm is running are described. Tag talk first protocols are analyzed in Chapter 9. Detailed comparisons with Aloha-based protocols have been performed regarding throughput and speed of reading.

In the near future, RFID deployments with large number of readers will be common. In these applications, reducing interference among the readers and performing synchronization of the readers will become a major problem. Detailed comparison of reader anticollision algorithms is presented in Chapter 11. Extension of the tree based protocols for multiple collaborative readers is described in Chapter 8. An example of the reader anticollision algorithm is also presented in Chapter 12.

With improving technology and dropping price of RFID readers, dense reader deployments will soon become reality. In addition to improving accuracy, densely deployed readers can be used, for example, to estimate more accurately the location of the tagged items. On the other hand, in order to reduce overall power consumption in the network and reduce amount of interference, it is important to turn off redundant readers when they are not needed. A redundant reader can be safely turned off or removed from the network without affecting the number of tags covered. Detecting redundant readers is a complex problem that is described in Chapter 12.

Several other networking problems are considered in the book. They include a problem when readers that are connected wirelessly and communicate using multi-hop communication need to optimally report detected tags. The solution to eliminate redundant tag reports generated by multiple readers is provided in Chapter 12. Delay-tolerant mobile networks of RFID readers are introduced in Chapter 13. In these networks information among stationary readers is transferred using mobile tags. The many challenges that this type of network introduces are presented in Chapter 13.

1.5 Conclusion

Although radio frequency identification is a rapidly emerging technology, several technical challenges need be to overcome to enable its ubiquitous adoption. In this chapter, we have examined some of these important technical challenges in present-day UHF passive RFID systems. We did this by formulating a hypothetical *ideal* RFID system which exhibits optimal performance in various aspects. We then examined the performance characteristics of practical RFID systems and how they diverge from the ideal system. Overcoming these non-idealities will be the key in enabling RFID technology to achieve its immense potential. We hope that this chapter has given the reader a better understanding of what ails the RFID systems of today. With this understanding, the reader can better appreciate the research efforts being described in the rest of the book and how each of these efforts fits into the bigger picture of enabling ubiquitous adoption of UHF RFID systems.

Acknowledgements

We would like to thank Alexey Borisenko for performing some of the experiments. We would also like to thank Dr. Michael Knox and Dr. Mustapha Yagoub for giving constructive comments that helped improve this manuscript.

References

[1] EPCglobal Inc. (2008) EPC™ Radio Frequency Identification protocols class-1 generation-2 UHF RFID protocol for communications at 860 mhz–960 mhz, Standard Specification version 1.2.0.

[2] Wyld, D. C. RFID (2005): The right frequency for the government. A research monograph from the IBM Center for the Business of Government.

[3] Want, R. (2006) An introduction to RFID technology, *IEEE Pervasive Computing*, **5**(9): 25–33.

[4] Dobkin, D. M. (2007) *The RF in RFID: Passive UHF RFID in Practice*. Oxford: Elsevier-Newnes.

[5] Derbek, V., Steger, C. Weiss, R. Preishuber-Pflugl, J. and Pistauer, M. (2007) A UHF RFID measurement and evaluation test system, *Electrotechnic and Informationstechnik*, **124**(11), 384–390.
[6] Buettner, M. and Wetherall, D. (2008) "An emperical study of UHF RFID systems," in *MobiCom*, San Francisco, California, USA.
[7] Muhlmann, U. and Witschnig, H. (2007) "Hard to read tags": an application-specific experimental study in passive UHF RFID systems, *Electrotechnic and Informationstechnik*, **124**(11). 391–396.
[8] Ramakrishnan, K. N. (2005) Performance benchmarks for passive UHF RFID tags, M.S. thesis, University of Kansas.
[9] Nikitin, P. and Rao, V. (2006), Performance limitations of UHF RFID systems, in *IEEE Antennas and Propagation Symposium*, pp. 1011–1014.
[10] D'Mello, S., Mathews, E., McCauley, L. and Markham, J. (2008) Impact of position and orientation of RFID tags on real time asset tracking in a supply chain, *Journal of Theoretical and Applied Electronic Commerce Research*, **3**(1), 1–12.
[11] Currie, I. A. and Marina, M. K. (2008) Experimental evaluation of read performance for RFID-based mobile sensor data gathering applications, in *Proceedings of the 7th International Conference on Mobile and Ubiquitous Multimedia*, Umea, Sweden, pp. 92–95.
[12] Mitsugi, J. and Hada, H. (2006) Experimental study on UHF passive RFID readibility degradation, in *Proceedings of the International Symposium on Applications and the Internet*.
[13] Impinj. Inc, (2007) RFID communication and interference, White Paper, Grand Prix Application Series.
[14] Leong, K. S., Ng, M. L. and Cole, P. H. (2006) Positioning analysis of multiple antennas in dense reader environment, in *Applications and the Internet Workshops*.
[15] Tuner, C. The dense reader problem in Europe, White Paper, November 2005.
[16] Martinez, R. Interference in RFID systems, Presentation for SG3.
[17] Rao, K. V. S., Nikitin, P. V. and Lam, S. F. M. Antenna design for UHF RFID tags: A review and a practical application, *IEEE Transactions on Antennas and Propagation*, **53**(12).
[18] Rahmati, A., Zhong, L., Hiltunen, M. and Jana, R. (2005) Reliability techniques for RFID-based object tracking applications, in *37th Annual IEEE/IFIP International Conference on Dependable Systems and Networks*, pp. 113–118.
[19] Ren, Z., Tan, C. C., Wang, D. and Li, Q. (2009) Experimental study of mobile RFID performance, in *International Conference on Wireless Algorithms Systems and Applications*, Boston, MA, pp. 12–20.
[20] Jo, M., Youn, H. Y., Cha, S.-H. and Choo, H. (2009) Mobile RFID tag detection influence factors and prediction of tag detectability, *IEEE Sensor Journal*, **9**(2): 112–119.
[21] Schneegans, S. Vorst, P. and Zell, A. (2007) Using RFID snapshots for mobile robot self-localization, in *Proceedings of the 3rd European Conference on Mobile Robots (ECMR 2007)*, Freiburg, Germany, pp. 241–246.
[22] Tanaka, T. and Sasase, I. (2007) Interference avoidance algorithms for passive RFID systems using contention-based transmit abortion, *IEICE Transactions on Communications*, **E90-B**(11): 3170–3180.
[23] Bhuptani, M. and Moradpour, S. (2005) *RFID Field Guide: Deploying Radio Frequency Identification Systems*, Sun Microsystems Press, Prentice Hall.
[24] Ahson, S. A. and Ilyas, M. Eds., (2008) *RFID Handbook: Applications, Technology, Security and Privacy*. Boca Raton, FL: CRC Press.
[25] Krishna, P. and Husalc, D. (2007) RFID infrastructure, *IEEE Communications Magazine*, **45**(9): 4–10.

2

Performance Metrics and Operational Parameters of RFID Systems

Raj Bridelall[1] and Abhiman Hande[2]

[1]*Axcess International, Inc.*

[2]*Texas Micropower, Inc.*

2.1 Overview

Automatic Identification and Data Capture (AIDC) technologies come in a wide variety of functionality, all targeted towards quickly linking a physical item with associated data contained within an information-technology (IT) system. Unlike other forms of AIDC technology such as magnetic stripe and barcodes, end-users often select RFID technology for its transparent ability to automatically locate and monitor the condition of physical assets and personnel using minimal or no manual intervention. Tagged items are generally associated with high value, high liability, or both. Examples of high value assets include laboratory equipment, medical instrumentation, pre-fabricated construction material, heavy-duty tools, and controlled medical substances. Personnel tracking applications include unattended contractor time and attendance logging for automatic invoicing, locating miners and first responders for safety, and the automatic accounting of chemical and nuclear plant employee whereabouts during an emergency. Organizations spend a large sum of money to deploy RFID technology because they expect significant improvements in operational efficiency, personnel safety, and exposure to liability.

It is widely understood that a supplier's failure to map the true capabilities of the technology to actual application requirements can lead to failed pilots and significant economic losses. Over a period of several decades, innovators have introduced numerous

RFID Systems: Research Trends and Challenges Edited by Miodrag Bolić, David Simplot-Ryl, and Ivan Stojmenović

types of homogeneous RFID technologies to address a broad range of requirements for unattended data capture across various physical asset tracking and personnel location applications. However, it has become evident after a period of trial and error that not all RFID technology types can be successfully configured for a given application, nor does a single technology type exist that can be applied across all application categories.

This chapter examines key operational parameters that make up the various technology categories and how these map to both performance and application requirements. The second section examines the technical trade-off required to achieve the desired range, throughput, omnidirectionality, localization, environmental compatibility, security, and standards compliance. Parameters that affect operating distance are covered in detail, including the amount of power that the interrogator radiates, interrogator and tag sensitivity and others. Next, system throughput is evaluated by analyzing relevant parameters such as the bandwidth occupied, data rate, modulation scheme, signal to noise ratio and channel sharing mechanisms. In addition, interrogator and tag interferences are discussed briefly. The major application of RFID systems is to detect the presence of tagged objects and/or people. Another important application is to provide the location of the tagged objects or people. Discussion on localization accuracy starts with the basic algorithms for estimating positions followed by the impact of multi-path and omni-directionality on localization accuracy. Impact of materials such as water or metal within proximity of interrogators and tags are examined. Other factors considered include reliability of RFID systems, size and thickness of RFID tags, health and safety aspects of RFID systems, security, and total cost of ownership.

In the third section, several classifications of RFID systems are discussed. Based on the type of EM link formed with the interrogator, we distinguish between near-field versus far-field RFID systems. Commercial near-field RFID systems that operate at low, high and ultra-high frequencies are described and compared. Then, basic features of narrow-band, wide-band and ultra wide-band far-field RFID systems are presented. Next, classification is considered based on the way the tags obtain their power. We then describe basic characteristics of active, passive and semi-passive RFID systems. The final section presents concluding remarks and some research directions.

2.2 Key Operational Parameters

This section will identify the key operational parameters that affect RFID performance in terms of range and throughput. Later sub-sections will also touch on factors that influence design choices for size, scalability and security.

The key parameters that affect range and data rate are interdependent as shown in the dependency graph of Figure 2.1. In this figure, the parameters and decision choices that the designers/users of RFID system have control over are shown within the hexagons. These parameters include operating frequency, transmit power, bandwidth, digital modulation encoding, and maximum tolerable bit error rate (BER). The key operational parameters that are affected by those design parameters include operating distance (range) and system throughput (proportional to data rate). These operational parameters are shown within the rectangles of Figure 2.1. The oval objects in Figure 2.1 are intermediate factors that are influenced by one or more design parameters. They include signal-to-noise ratio (SNR), noise level, receiver sensitivity, and noise figure (NF). For example, the amount of "noise"

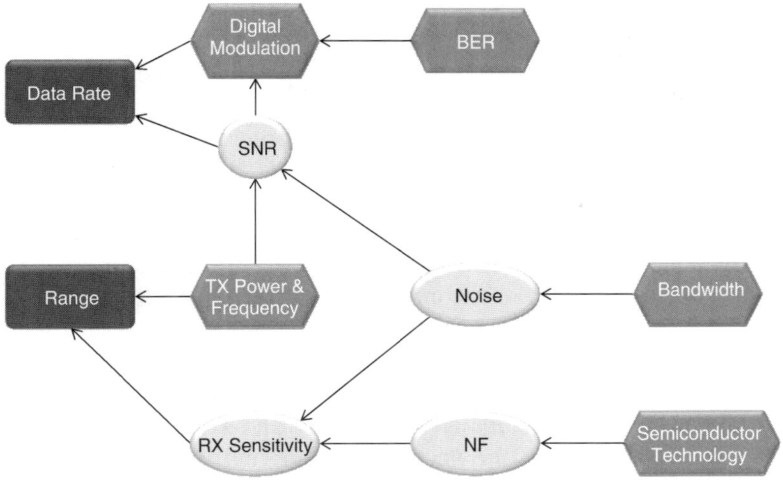

Figure 2.1 Key operational parameters and design trade-off.

at the input of a receiver is strongly dependent on the "bandwidth" of the receiver. Noise figure is a measure of degradation of the signal to noise ratio, caused by components in the RF signal chain. NF depends on the type of semiconductor technology and determines how much the noise at the input of the receiver is amplified relative to the signal before arriving at the demodulator. We would like to point out that some parameters and some possible dependencies are omitted in Figure 2.1 for reasons of clarity.

The specific parameter combination and their values ultimately determine the type of tag. Although there are a large number of possible parameter permutations, the type of tag can be categorized by the type of communications link (near-field or far-field), the method of tag transmission (emission or reflection), and the type of power supply (battery or energy harvesting). The third section covers the benefits and deficiencies of each tag type. Two of the most widely deployed tag types for long-distance operation are *passive* and *active* operating in the ultra-high frequency (UHF) bands. The impact from adjusting any of the key operational parameters of just these two tag types adequately illustrates RFID system behaviour without much loss of generality.

The far-field begins where the wave impedance quickly settles towards the free-space impedance value of 377 ohms [1]. For electrically small antennas, this distance is

$$R_{NF} = \frac{\lambda}{2\pi} \tag{2.1}$$

This distance is essentially the radius of the near-field region around an antenna, and is less than one foot for popular commercial UHF RFID systems that operate in the 915 MHz band.

Passive tags transmit data by reflecting power from the interrogator. This is also referred to as *backscatter modulation* for systems that operate in the far-field, and *load modulation* for systems that operate in the near-field. Passive tags harvest energy from the interrogator instead of using a battery. Active UHF tags, on the other hand, utilize an on-board battery for the power needed to emit a modulated RF signal that encodes the data.

The most commonly utilized frequencies for both active and passive RFID operation are at or near 125 kHz, 13.65 MHz, 315 MHz, 433.92 MHz, 915 MHz, 2.45 GHz, and 5.8 GHz. Systems operating in bands below 70 MHz tend to operate in the near-field with inductively coupled coil wire antennas because far-field antennas at those frequencies would be impractically large. The majority of standards prescribe passive UHF RFID operation in the region between 860 MHz and 960 MHz because of its favorable output power and RF energy harvesting characteristics [2]. Unfortunately, this is the only spectral region without uniformly allocated frequency bands around the world. Consequently, international standards are divided between the European allocations from 865–868 MHz and the North American allocations from 902–928 MHz [3, 4]. Regulations outside of Europe and the Americas have also opted for a blend of the European and American regulatory allowances. Figure 2.2 shows the distribution of countries that follow the FCC allocations for Effective Isotropically Radiated Power (EIRP), and countries that follow the EU allocations for Effective Radiated Power (ERP). Countries that adopt a blend of the FCC and EU regulations are shown in detail, and the EIRP amount is shown unless otherwise indicated as an ERP amount. The difference between EIRP and ERP amounts relates to the type of antenna assumed, and this is further addressed in a later section. Figure 2.2 also indicates those countries that use Frequency Hopping Spread Spectrum (FHSS), and whether the specification is for indoor (InDr) or outdoor (OutDr) environment, and if the band is licensed (Lic). These allocations are still evolving as of the time of this writing. Nevertheless, regardless of the operating frequency selected, performance will be constrained by the amount of signal power allowed, the amount of noise or interference in the system, and the available bandwidth.

2.2.1 Operating Distance

Operating distance is defined as a maximum distance between the interrogator's antenna and the tag at which the interrogator can reliably read from or write to the tag. In passive RFID systems, maximum operating distances for reading and writing operations are not always the same. The reason is that when writing to its memory, most tags require more power and, therefore, the operating distance is reduced. The maximum distance for reading operation is called *read range* while the corresponding distance for the writing operation is called *write range*.

All things being equal, the greater the power that the interrogator is allowed to radiate, the greater the distance at which a tag will be able to receive and interpret commands. Reciprocally, the more power that a tag can direct towards the interrogator, the greater the distance at which the interrogator will be able to receive and interpret tag responses. Power is often listed in decibels (dB) of milliwatts (mW) or dBm, and generally includes the power gain of the antenna supplied with the unit. Receiver sensitivity and interference are the next dominant parameters that affect range. However, these tend to be a strong function of bandwidth and another section addresses this relationship.

2.2.1.1 Radiated Power in the Americas

Most interrogators designed for sale in FCC regulated environments can radiate up to four Watts of power in the 915 MHz band. This includes the antenna gain and any losses

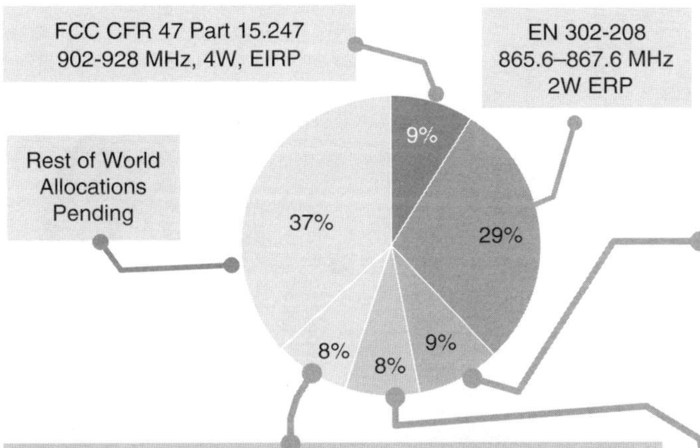

Similar Range to FCC

	Start	End	Watts Sharing
Australia	920.00	926.00	4.00
Brazil-A	902.00	907.50	4.00
Brazil-B	915.00	928.00	4.00
Hong Kong B	920.00	925.00	4.00
Japan Lic.	952.00	954.00	4.00 LBT
Korea A	908.50	910.00	4.00 LBT
Korea B	910.00	914.00	4.00 FHSS
New Zealand	864.00	868.00	4.00
South Africa B	915.40	919.00	4.00 FHSS
South Africa C	919.20	921.00	4.00 CW
Thailand A	920.00	925.00	4.00 FHSS-Lic.

Taiwan A	922.00	928.00	1.00 ERP FHSS InDr
Armenia	865.60	867.00	0.50 ERP
Philippines	918.00	920.00	0.50 ERP
Singapore A	866.00	869.00	0.50 ERP
Taiwan B	922.00	928.00	0.50 ERP FHSS OutDr
Thailand B	920.00	925.00	0.50 ERP FHSS-NoLic
Vietnam A	866.00	869.00	0.50 ERP
Vietnam C	920.00	925.00	0.50 No Lic
Malaysia A	866.00	869.00	0.05
Japan UnLic.	952.00	955.00	0.02 LBT
Indonesia	923.00	925.00	0.00

Lowest Performance Regions

India	865.00	867.00	4.00 ERP
Israel	915.00	917.00	2.00
Belgium	865.60	867.60	2.00 ERP
China-A	840.50	844.50	2.00 ERP
China-B	920.50	924.50	2.00 ERP
Hong Kong A	865.00	868.00	2.00 ERP
Iran	865.00	868.00	2.00 ERP
Macedonia	865.60	867.60	2.00 ERP
Malaysia B	919.00	923.00	2.00 ERP
Singapore B	920.00	925.00	2.00 ERP
Vietnam B	920.00	925.00	2.00 ERP Lic.

Similar Range to Europe

Figure 2.2 Worldwide regulatory allowances for UHF RFID operation as of January 2009.

in the system. Specifically, the FCC limits the power delivered into the antenna to 1 Watt and the antenna gain to 6 dBi. The norm for antenna gain specification is with respect to an Effective Isotropic Radiated Power (EIRP) as denoted by the small 'i' next to the 'dB' unit. The isotropic radiator is a theoretical antenna that radiates with zero dB gain in all directions. Such an antenna is, however, not physically possible. In practice, adding antenna gain adds directionality to the radiated pattern. For example, a theoretical half-wave dipole antenna has a gain of about 2.15 dBi [1]. The FCC also allows for antenna gains beyond 6 dBi in the 915 MHz band, but with a corresponding reduction of power to the antenna so that the radiated power in the direction of maximum gain does not exceed 4 Watts. If a circularly polarized antenna is used, then the antenna gain can actually double. For example, commercially available systems deliver either one Watt into a 6 dBi *linearly* polarized antenna or 1 Watt into a 9 dBi *circularly* polarized antenna. This is possible because the FCC measures antenna gain with respect to a linearly polarized receiving element that effectively captures half (−3 dB) of the incident circularly polarized power.

2.2.1.2 Radiated Power in Europe

The European specifications cite the maximum output power in terms of the Effective Radiated Power (ERP), which is different from EIRP. The ERP specification is with reference to a theoretical half-wave dipole antenna. As an example, the European Union (EU) standard EN302-208 establishes the maximum radiated power at 2 Watts, ERP [4]. System manufacturers interpret this as 2 Watts radiated by a theoretical half-wave dipole in the direction of highest gain. Figure 2.3 illustrates the difference between these two interpretations.

European regulations for UHF RFID specify 15 channels of 200 kHz bandwidth each, but only ten of those allow a maximum of 2 Watts ERP [4]. With everything else being equal, this power level limits the maximum unobstructed theoretical read range to nearly two-thirds that of systems operating within FCC regulations. A lower UHF operating frequency, nevertheless, is preferred because of the more favorable energy capture characteristics of a longer wavelength. The *Friis* far-field transmission formula for unobstructed

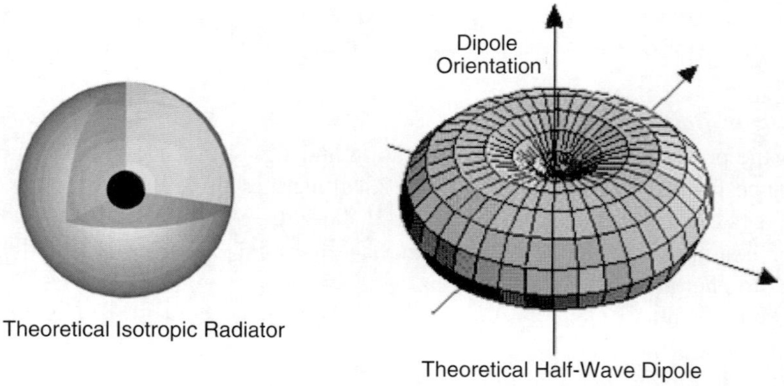

Figure 2.3 Radiated power relative to EIRP and ERP.

narrow-band propagation shows this effect [5].

$$P_{i\_tag} = P_s \left(\frac{\lambda}{4\pi r} \right)^2 \Psi_T \psi_{ri} \tag{2.2}$$

That is, the amount of signal power that the tag receives, $\mathbf{P_{i\_tag}}$ is proportional to the wavelength $\boldsymbol{\lambda}$ squared. $\mathbf{P_s}$ is the power supplied by the interrogator, **r** is the operating distance, $\boldsymbol{\Psi_T}$ is the interrogator antenna gain, and $\boldsymbol{\psi_{ri}}$ is the tag's antenna sub-system gain when it is absorbing energy.

2.2.1.3 Radiated Power in the Rest of the World

Adoption trends show that most countries are either following UHF RFID regulations similar to the Americas, the European Union, or a blend of both. Countries that cannot allow as much power in the UHF frequency bands may achieve at least equivalent or better performance with active RFID even when the permitted power levels are only a few microwatts. For example, technology providers have successfully deployed approved products that transmit no more than 2 microwatts in the 315 MHz and 433 MHz bands to achieve performance significantly exceeding that of passive systems operating at the maximum power levels in the 915 MHz FCC bands [6].

2.2.1.4 Interrogator and Tag Sensitivity

Receiver sensitivity is the minimum signal power that is required to produce a specified output signal. The interrogator and tag receivers will decode a command if the average required signal power is greater than their respective receiver sensitivity levels. The amount by which this signal power must exceed the receiver input noise power is the minimum signal-to-noise ratio (SNR) required to recover data of a specified digital modulation scheme and maximum tolerable bit error rate (BER). Digital data communications systems typically accommodate a BER of one bit per million received (10^{-6}). Applications such as digitized voice transmissions can tolerate a higher BER. In general, a higher SNR reduces the BER. However, increasing the SNR for a specified range implies increasing the transmitted signal power or reducing the receiver input noise power. Figure 2.1 illustrates the various dependencies on SNR.

Constraints on Passive Tag Sensitivity

The lower the power a tag requires for activation and signal decoding, the greater its operating distance from the interrogator. Figure 2.4 illustrates this relationship for a selected sub-set of system types that operate under different regulatory allocations. Figure 2.4 shows, for example, that an EU compliant tag operating at 868 MHz that requires 50 microwatts to energize and decode a signal will respond from an unobstructed distance of about 3.25 meters, under ideal conditions. The same tag can respond from an unobstructed distance of about 5 meters if the tag sensitivity improves to 20 microwatts. Signal power diminishes as the inverse square of the distance ($1/\mathbf{r}^2$) in the far-field, and as the inverse sixth power of the distance ($1/\mathbf{r}^6$) in the near-field [5]. Therefore, near-field tags will operate at a significantly reduced distance from far-field tags of the same sensitivity.

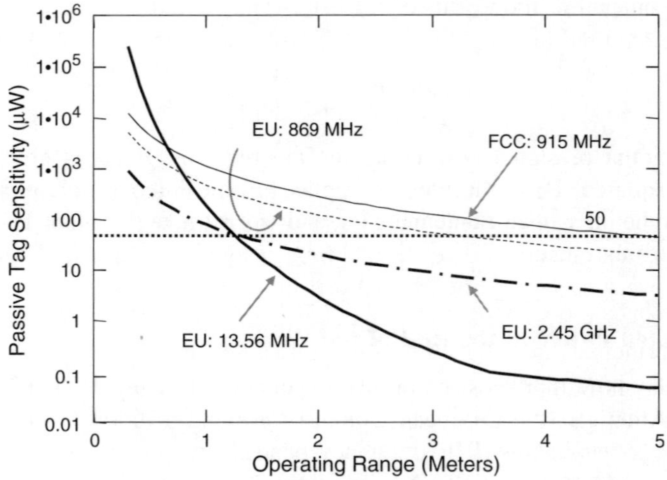

Figure 2.4 Effect of sensitivity on range.

Constraints on Active Tag Sensitivity

Technology providers optimize passive tags for RF energy harvesting. Therefore, their receivers are inherently less sensitive than those of active tags. Passive tag receivers typically consist of signal envelope detectors constructed from passive diodes or an equivalent construction [7]. They detect a signal only when the input signal is sufficiently strong to overcome the passive diode detector forward bias threshold. Active tags utilize the on-board battery to bias active rectification circuits, which substantially lowers this forward bias threshold. Consequently, the sensitivity difference can be significantly greater than 100 dB [7].

Constraints on Interrogator Sensitivity for Passive UHF Tags

The interrogator signal received from a backscattering passive UHF tag decreases with free-space distance as follows [5]:

$$P_B = P_s \left(\frac{\lambda}{4\pi r} \right)^4 \Psi_T^2 \psi_{ro} \tag{2.3}$$

That is, the amount of signal power $\mathbf{P_B}$ received by the interrogator decreases at the rate of $1/\mathbf{r}^4$ for passive UHF systems. $\mathbf{P_s}$ is the power generated from the interrogator and delivered to the antenna, $\mathbf{r}$ is the operating range, $\mathbf{\Psi_T}$ is the interrogator antenna gain, and $\boldsymbol{\psi}_{\mathbf{ro}}$ is the tag antenna subsystem gain when it is reflecting energy, including any transmission losses. In an optimized design, the interrogator decoding sensitivity should be at least $\mathbf{P_B}$ when the tag is at its maximum operating distance. Solving for $\mathbf{r}$ in Equation 2.2 and setting $P_{tag} = P_{tag\_sens}$ gives the maximum powering range $\mathbf{r_{max}}$ as a function of the tag's decoding sensitivity,

$$r_{\max} = \frac{\lambda}{4\pi} \sqrt{\frac{P_s \Psi_T \psi_{ri}}{P_{tag\_sens}}} \tag{2.4}$$

Substituting Equation 2.4 into Equation 2.3 and setting $P_B = P_{I\_sens}$ to maintain maximum range gives the required interrogator sensitivity as a function of the tag sensitivity,

$$P_{I\_sens} = \frac{P_{tag\_sens}^2}{P_s} \frac{\psi_{ro}}{\psi_{ri}^2} \tag{2.5}$$

Intuitively, for fixed tag sensitivity, the greater the power ($\mathbf{P_s \psi_{ri}}$) transferred to the passive tag, the further out it can respond with a valid signal, and the weaker the signal that the interrogator must accommodate. In addition, a lower tag antenna gain in reflection mode, $\boldsymbol{\psi_{ro}}$ will weaken the reflected signal that the interrogator must accommodate. This relationship is independent of the interrogator antenna gain ($\mathbf{\Psi_T}$) as anticipated from the RF reciprocity principle that balances the outbound gain with an identical inbound gain. As a final insight, the required interrogator sensitivity under ideal propagation circumstances is at most twice that of the passive tag's sensitivity on a logarithmic scale because of the round trip that the same signal must travel.

Constraints on Interrogator Sensitivity for Active UHF Tags

Since active UHF tags transmit by generating a signal rather than backscattering, the interrogator will receive signal power,

$$P_I = P_{s\_tag} \left(\frac{\lambda}{4\pi r} \right)^2 \Psi_T \psi_r \tag{2.6}$$

Even though interrogators can radiate a relatively large signal for powering passive tags, the available backscatter signal ($\mathbf{P_B}$) from passive tags is still orders of magnitude less than a low-power signal ($\mathbf{P_I}$) arriving from a typical active tag from the same distance. For example, from equation 2.2, a four Watt EIRP (~36 dBm) signal transmitted from an FCC compliant interrogator operating at 915 MHz will become approximately 3 microwatts (−25.27 dBm) after traveling a distance of 30 meters in free space. A battery-powered backscatter tag will reflect a portion of the three microwatts towards the interrogator. In comparison, a low power active tag in the same band will typically transmit 1 milliwatt (0 dBm) of signal, which is over 300 times more powerful than the signal backscattered from a passive tag. This means that an interrogator with the same sensitivity can decode the Active tag's response from a distance that is a factor of approximately 17 (that is $\sqrt{300}$) further away.

With all other parameters unchanged, increasing range means either increasing the average signal power transmitted or increasing the receiver sensitivity. However, regulatory rules limit the transmitted power, and the receiver noise figure bounds its sensitivity. The theoretical noise floor is a function of bandwidth and temperature. For example, at room temperature, the theoretical minimum noise spectral density is −174 dBm/Hz [8].

Other Design Considerations

The operating distance depends on many other factors including: the environment, the orientation of the antennas, the directionality of the antennas, the modulation scheme, the type of RFID system used, the antenna gains, and many other factors. Some of the key environmental factors include the proximity of metal to the tag or reader antennas, the presence of in-band noise sources, and operation indoors or outdoors. Relative antenna orientations will affect the amount of received power incident at the tag and interrogator antennas. For a given radiated power, greater operating distance can be achieved with directional antennas; however, this would limit readable locations. The type of RFID system used significantly affects operating distance. For instance, mobile interrogators are designed to have relatively small form factor and long battery life. Therefore, large operating distance is not generally available with hand-held and mobile devices. The modulation scheme can affect the operating distance since different modulation types provide different levels of noise resistance.

For practical implementations, transmission power, noise figure, and bandwidth affect key application level considerations such as power consumption, cost, and size. Increasing antenna gain in order to increase the received signal strength tends to increase the size of the antenna. Higher antenna gain also favors transmission or reception in one direction over others. On the other hand, lower gain antennas must be used if the application requires omni-directional performance. A lower noise figure amplifier increases sensitivity but also tends to demand more power or require a more expensive semi-conductor technology [9]. A higher output power interrogator amplifier uses more power and tends to require additional heat sinking elements that also increase size and bulk. This trade-off between range, bandwidth, power consumption, cost, and size are some of the most important RFID systems design considerations.

2.2.2 System Throughput

Given a data exchange protocol, the amount of bandwidth per channel that an interrogator can legally transmit is directly proportional to the maximum achievable data rate of a specified modulation scheme. Interrogators select an available channel from the sub-divided frequency band by utilizing traditional narrow-band multiple access techniques such as frequency hopping or carrier sensing. Therefore system throughput is evaluated by analyzing relevant parameters such as the bandwidth occupied, data rate, modulation scheme, signal to noise ratio and channel sharing mechanisms. Although interrogators can simultaneously operate in separate channels, increasing the number of available channels does not necessarily lead to an overall increase in system throughput. Interrogator interference and Tag interference are two phenomena that account for this barrier as described in Sections 2.2.2.6 and 2.2.2.7 respectively.

2.2.2.1 Occupied Bandwidth

Once the interrogator selects a channel, it may modulate its carrier in order to send data to the tag. Carrier modulation generates a power spectral density (PSD) that the regulatory bodies measure. The PSD is essentially an electromagnetic (EM) energy footprint that the interrogator creates across the occupied channel. The interrogator generally centers the carrier within the channel and the data energy is distributed on either side of the carrier. For a selected modulation scheme, higher data rates result in spreading the data energy further away from the carrier and potentially into adjacent channels. Therefore, the specified channel bandwidth ultimately constrains the data rate.

The majority of low-power RFID systems utilize a form of ASK modulation for both the forward and return links. Wireless network standards such as IEEE 802.11b (Wi-Fi)

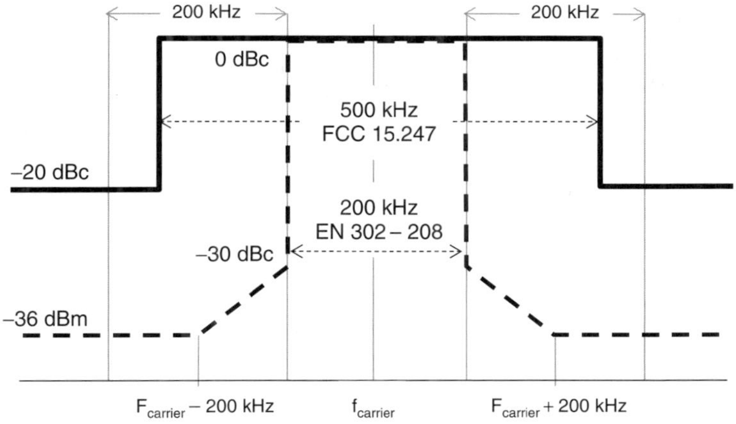

Figure 2.5 Comparison of FCC and EU spectral masks.

and IEEE 802.15.4 use other digital modulation schemes such as FSK, PSK, and QPSK to transmit a higher data rate within the same channel bandwidth. In general, more complex modulation schemes require higher SNR, as well as increased power demands from the transceiver architecture. The simplicity of ASK modulation is often the primary driver for minimizing cost, power consumption, battery size and overall tag footprint.

Each standard specifies a spectral mask within which the transmitted signal must be contained. Figure 2.5 compares the UHF RFID spectral masks of the FCC and EU regulatory domains. The main lobe of the EU mask is not only 200 kHz narrower, but also imposes a more stringent limit on sidebands and spurious emissions outside of the band. Unlike the FCC, the EU standards also regulate the backscatter modulation from tags. Figure 2.6 shows the backscatter spectral mask prescribed by the EN 302-208 standard. The center frequency of the interrogator carrier f_c and its value depend on the channel that the interrogator selects for transmitting. The modulated backscatter energy is constrained relative to this center frequency. Some exceptions are possible, as noted in Figure 2.6, where the standard limits any spectral response above 870 MHz to −36 dBm regardless of the channel the interrogator selects.

2.2.2.2 Protocol Throughput

Throughput in terms of the number of tags read per second depends on many factors including:

- the data rate of the forward link (interrogator commands to the tag);
- the data rate of the return link (tag responses to the interrogator);
- the Media Access Control (MAC) protocol;
- the operating environment.

The air interface and MAC protocols of the selected standard determines the number of bits transferred in each direction before a single tag read can be completed. Some protocols

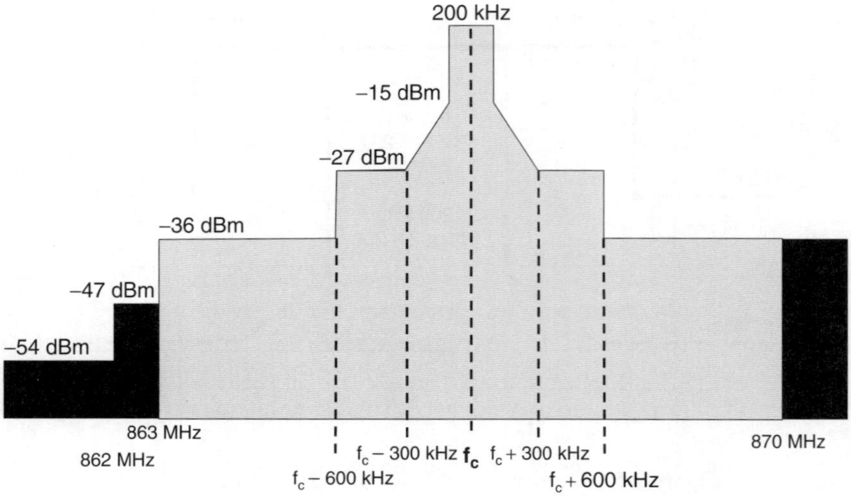

Figure 2.6 EN 302-208 backscatter spectral mask.

do not require a forward link message in order to read tags and result in significantly faster throughput. However, the trade-off is an inability to command tags to configure their behavior in real-time before they automatically respond, or to write data to their on-board memories. Standards often refer to these unidirectional protocols as Tag-talk-first (TTF). Reader-talk-first (RTF) protocols are more prevalent, however, because of the added degree of control over the link characteristics. For example, in the popular EPC Class I Generation 2 (EPC C1G2) or the ISO 18000-6c standards, the interrogator initiates the data transfer protocol and configures tags to respond at data rates between 40 kbps and 640 kbps [10]. This provides for some degree of adaptability depending on the EM properties of the application environment.

The MAC protocol used affects how collisions in tag-to-reader communications are handled. The RF environment can also affect the protocol throughput since the number of retransmissions is partially dependent on the environmental conditions.

2.2.2.3 Modulation Scheme

The achievable bit rate is a direct function of bandwidth occupied, bit encoding scheme, and SNR [11]. Once allotted a fixed bandwidth, the maximum bit rate achievable depends on the spectral efficiency of the selected bit-encoding scheme. For example, non-coherent ASK modulation requires a bandwidth that is twice the bit rate and is often described as having a spectral efficiency of 0.5 bits-per-second-per Hertz (bits/sec/Hz). Similarly, binary phase shift keyed modulation (BPSK) theoretically provides up to 1 bit/sec/Hz, whereas quadrature phase shift keyed modulation theoretically provides up to 2 bits/sec/Hz [11]. The actual data rate and BER achieved depends on the available SNR.

Nearly all passive and most active RFID technologies utilize non-coherent ASK modulation and demodulation because their implementation inherently requires low-power and is low-cost. It is also possible to utilize a sub-carrier modulation type such as FSK and PSK

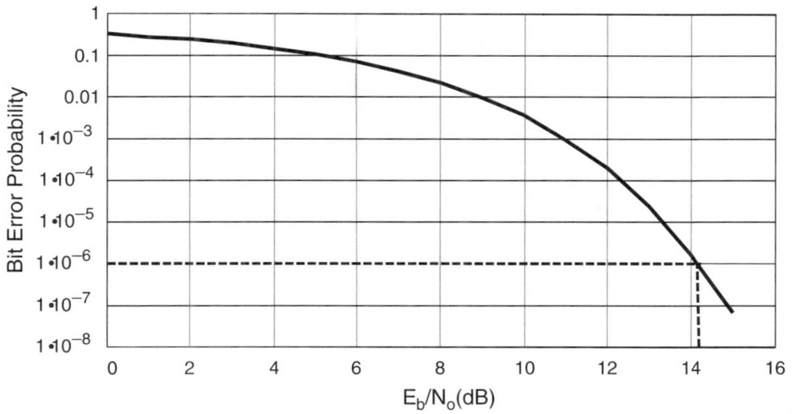

Figure 2.7 Signal quality for non-coherent ASK demodulation.

along with ASK. For example, FSK sub-carrier modulation is achieved by manipulating the ASK modulation speed in a data stream to encode binary digits.

The bit error probability (**p_e**) for non-coherent ASK demodulation is

$$p_e = \frac{1}{2} e^{\frac{1}{2}\left(\frac{E_b}{N_0}\right)} \tag{2.7}$$

$\boldsymbol{E_b}$ is the energy per bit in Joules, and $\boldsymbol{N_o}$ is the single-sided noise power spectral density in Watts per Hertz [11]. This is plotted in Figure 2.7 where it is observed that slightly more than 14 dB of bit energy per unit of noise power spectral density is required to achieve a lower bit error rate than one bit per million received. In general, the greater the spectral efficiency of the modulation scheme, the greater the SNR required for an acceptable bit error probability.

2.2.2.4 Signal to Noise Ratio

The maximum achievable bit rate for a UHF backscatter system is

$$R_{bit}(r, p_e) = P_s \Psi_T^2 \Psi_{ro} \times \left(\frac{\lambda}{4\pi r}\right)^4 \times M(r, R_o, N_B) \times \left(\frac{1}{k_B T_0 f_r}\right) \times \frac{1}{2 \ln\left(\dfrac{1}{2p_e}\right)} \tag{2.8}$$

The parameters **k_B** and **T_o** are the reference temperature and Boltzmann constants respectively [5]. The signal attenuation factor M (r, R_o, N_B) accounts for non-line-of-sight conditions using a model for signal propagation where

$$M(r, R_o, N_B) = \frac{1}{\left(1 + \dfrac{r}{R_0}\right)^{2(N_B - 2)}} \tag{2.9}$$

The parameters **R_o** and **N_B** are the free-space equivalent breakpoint distance and the environmental attenuation factors respectively. For free space, **N_B** $= 2$ and the factor becomes unity. A receiver manufacturer typically specifies the amplifier's noise factor, which is

$$f_r = \frac{SNR_{in}}{SNR_{out}} \tag{2.10}$$

This is the same as a noise figure which is typically specified with a dB value. From Equations 2.3 and 2.8, the SNR can be written as

$$SNR = \left(\frac{P_B M(r, R_o)}{k_B T_0 f_r} \right) = R_{bit} \times 2 \ln \left(\frac{1}{2p_e} \right) \tag{2.11}$$

Solving for E_b/N_o in Equation 2.7 and substituting the result into Equation 2.11 gives the minimum required SNR as

$$SNR_{\min} = R_{bit} \times \left. \frac{E_b}{N_0} \right|_{p_e} \tag{2.12}$$

That is, the minimum SNR needed to decode a backscattered signal with maximum tolerated probability of bit error (**p_e**) is directly proportional to the bit rate. Equation 2.8 proves that a higher SNR is required if the application requires more range (**r**), a lower bit error probability (**p_e**), or a higher throughput from increasing the bit rate (**R_{bit}**).

2.2.2.5 Channel Sharing

Interrogators designed for use with passive tags must transmit sufficient power to energize tags and consequently do not utilize multiple access techniques such as Direct-Sequence-Spread-Spectrum (DSSS) or Ultra-Wide-Band (UWB) that spread energy across a wide spectral region using relatively short time period signals. Rather, multiple access techniques that require channel hopping are used because the tag can collect energy from a non-modulated carrier using a simple diode rectifier front-end. Consequently, interrogators for passive tags are limited to time-sharing or activity-based media access (MAC) protocols. They either randomly select a channel and occupy it for a fixed duration before moving on to another, or wait until a channel becomes vacant before transmitting. Standards have dubbed the former method as frequency hopping (FH) and the latter a Listen-Before-Talk (LBT). FH with a randomized hopping pattern is a requirement for UHF transmission levels greater than 125 milliwatts under FCC regulated domains. The intention is to spread the power averaged across the band over a specified duration. FCC compliant UHF interrogators that transmit the maximum allowed power must frequency hop randomly across 50 of the 52 channels allocated within the 902 MHz to 928 MHz band. Each channel is 500 kHz wide with spectral mask shown in Figure 2.5. Interrogators must also hop at a rate such that the average time transmitting in any one channel is less than 400 milliseconds over any 20-second period.

The European regulations prescribe an LBT/FH channel sharing method since there are far fewer channels for FHSS to be effective. Figure 2.8 illustrates the European channelization for UHF RFID. The EN 302-208 standard currently defines channel occupancy as any carrier signal that can be detected above −96 dBm. The link margin for an EU compliant interrogator transmitting a signal at full power (2 Watts ERP) is

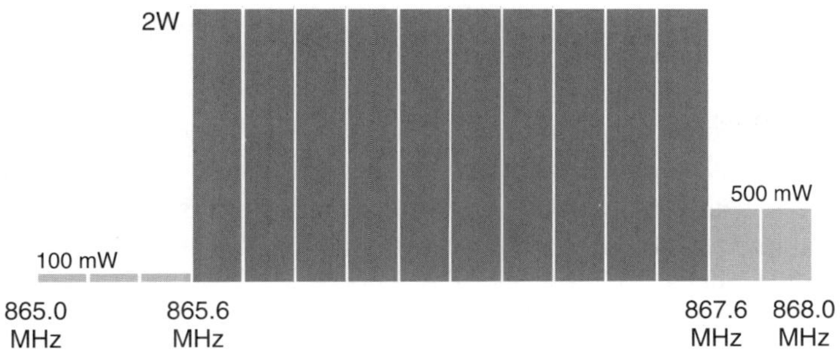

Figure 2.8 EU channelization under EN 302-208.

33 dBm – (−96 dBm) = 129 dB. From the Friis equation, this equates to several miles. Therefore, it is possible for EU compliant systems to become sluggish as the number of interrogators within a radius of several miles exceeds the number of available channels. A typical warehouse can utilize dozens of interrogators, including those fixed at dock doors, conveyor belts, and storage rooms, as well as mobile and hand-held interrogators brought within close proximity to other interrogators. Therefore, it is likely that the EU regulations will continue to undergo modifications to mitigate such constraints.

2.2.2.6 Interrogator Interference

When interrogators operate simultaneously within a few channels of each other, they produce interfering signals that affect the operation of both tags and interrogators within range. The impact on broadband backscatter tags is more significant than for narrow-band active tags. Another section of this chapter covers the performance impact from this type of tag interference phenomenon, which is different from MAC-related tag interference also known as tag collision.

The mixing of a carrier from the desired interrogator with a carrier from another nearby interrogator causes interference. Nonlinearities from practical RF amplifiers and mixers produce signal components called inter-modulation products (IM) that appear within the receiver bandwidth and can distort the tag data signal. The relative frequency position of the individual spectral energy components received are illustrated in Figure 2.9, namely the spectrum of the carrier signal plus the backscattered data (grey arrow and lines), the

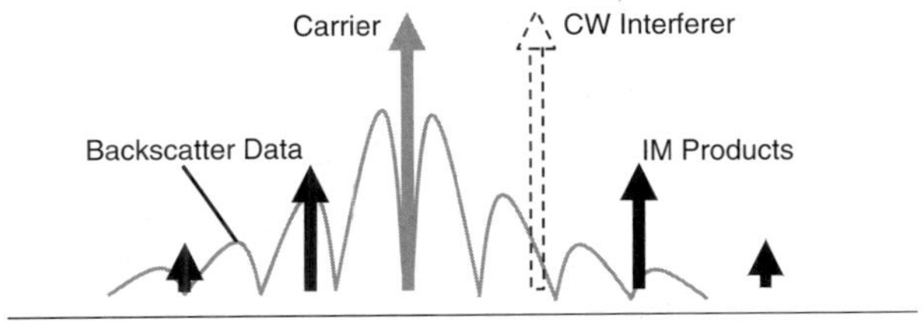

Figure 2.9 Interrogator interference.

continuous waveform (CW) interferer (dotted arrow), and multiple IM products components (black arrows). Demodulators can recover the data signal contained within the upper side band or the lower sideband of the carrier, but not when unwanted signal components overlap the data sidebands as shown. The C1G2 standard defines a “dense reader environment” as one where as many interrogators are operating as there are channels allocated in the band [10]. The net effect from any interference is reduced system throughput. That is, adding a second interrogator does not necessarily double the throughput of tags identified per unit interval of time.

2.2.2.7 Tag Interference

Increasing the number of available channels in a band may increase aggregate throughput for active tags but not necessarily for passive tags. The reason for this is that passive tags have fixed broadband receivers that listen to the entire band. Unlike active tags, passive tags cannot dynamically hop to another vacant channel to establish an isolated link with an interrogator. They simultaneously receive and integrate signals within range from all interrogators transmitting anywhere in the band. Figure 2.10 illustrates how multiple interrogator signals can add destructively at the tag to distort the received signal. The figure shows interference between the modulated and unmodulated signals in two different channels. Other equipment that shares the band such as cordless phones and wireless local area networks can also add to this distortion. Under certain circumstances, adding a low pass filter to the tag’s baseband receiver circuit may partially mitigate this issue [12].

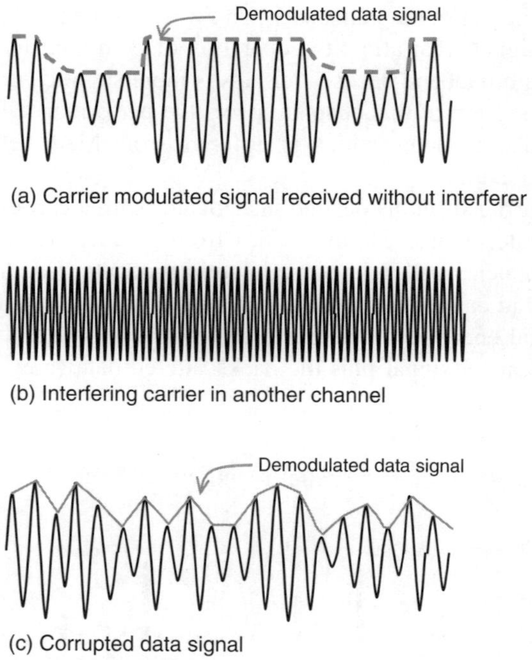

Figure 2.10 Signal from tag interference.

Table 2.1 Common RTLS techniques for indoor positioning.

Number of antennas	Location method	Positional Information
1	Proximity (near-field)	In or out of a magnetic field zone
	Proximity (far-field)	Radial proximity to antenna
2	Triangulation	A 2D position between antennas
3 or more	Trilateration	A position in 3D space

2.2.3 Localization

In addition to automatic identification, advanced RFID implementations can also provide information about a tagged item's position, speed, and direction of travel. These Real-Time Location System (RTLS) techniques range from simple proximity to computational positioning using a local or a global positioning system (GPS). Table 2.1 summarizes common techniques for indoor positioning solutions. Localization accuracy depends on operational parameters as well as the effectiveness of the signal processing algorithms used. In this section, we first describe several basic algorithms for localization. They include proximity detection (which only determines that the tag within the reading range of the reader), as well as algorithms for determining 2D and 3D position of the tagged objects. These algorithms rely on estimating the range between the interrogator's antenna and the tag. Range estimation algorithms calculate the range by correlating the distance with either the time of flight (TOF) of a signal, the Received Signal Strength (RSS) or with the ratio of phase versus frequency variations. The ranging accuracy is affected by many factors. Multi-path conditions depend on the indoor layout structure, line-of-sight conditions, mobility of objects, height of the tag's antenna and others. Next, better time resolution (or proportionally larger available bandwidth) results in better ranging accuracy. Hence, in the UHF range, higher localization accuracy is expected in North America than in Europe due to wider available bandwidth. The level of SNR is important since low levels of SNR can affect the accuracy of the signal processing algorithm applied for range estimation. Tag antenna types as well as their orientation are important factors in determining localization accuracy. RTLS usually requires tags that are orientation insensitive. Proximity of metals and liquids can cause detuning which will degrade either powering range for passive RFID systems or SNR for active RFID systems, which then affects RSS and Angle-of-Arrival (AoA) estimates.

2.2.3.1 Proximity

Presence determination at specified spatial regions may be the lowest cost RTLS method because fewer antennas are required, and the amount of computation is limited to determining whether or not the tag is within range of the antenna coverage pattern. Near-field antennas provide the highest degree of control when the coverage area must be relatively small and non-overlapping. These include transitional and restricted areas such as doorways and work stations. Far-field antennas cover large areas with less precision because of signal reflections. They are adequate when the zones are large and far enough apart from each other, for example, detached buildings across a campus.

The near-field zone is also called a control point when the presence of a tag in the near-field is utilized to control an apparatus such as an automatic gate motor, a door latch, a camera, a buzzer, or some other actuating device. Control points are less reliable with far-field signaling because of signal reflections from multi-path propagation.

2.2.3.2 Positional Computation

Triangulation and trilateration techniques are used to determine 2D or 3D position of a stationary tag using stationary interrogators. These techniques require additional signal processing, computation, and data exchange between interrogators in order to determine the velocity of a tagged item. Trilateration algorithms rely on an estimate of the radial distance between the tag and the interrogator. RSS and TOF algorithms can provide a distance estimate with varying degrees of accuracy depending on multi-path conditions, bandwidth, and SNR [13]. Triangulation algorithms determine a tag's velocity by successively computing the AoA from the average RSS.

Triangulation Method

Given two antennas positioned a known distance **D** apart as shown in Figure 2.11, with each having sub-systems capable of estimating the tag's AoA, the perpendicular distance to the tag **r** can be estimated from trigonometry using the law of sines [14],

$$\hat{r} = D\frac{\sin(\hat{\alpha})\sin(\hat{\beta})}{\sin(\hat{\alpha}+\hat{\beta})} \tag{2.13}$$

The accuracy of this estimate depends upon the SNR, multi-path signal propagation conditions, and the accuracy with which each interrogator is capable of estimating the two angles of arrival, $\boldsymbol{\alpha}$ and $\boldsymbol{\beta}$. interrogators can also conceivably estimate the AoA with phased-array antennas capable of adaptively steering the beam to search for the maximum likelihood source direction [16].

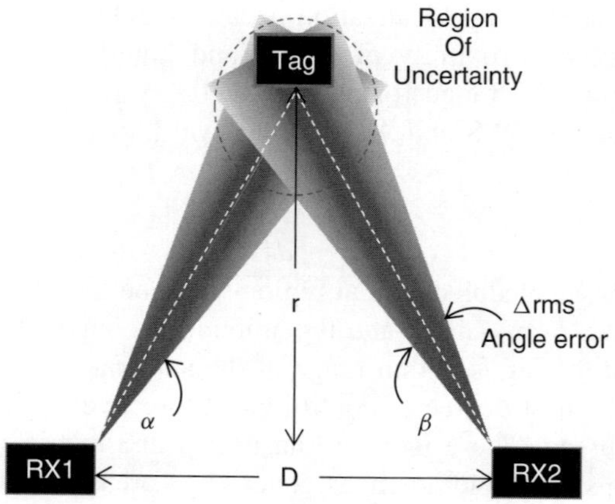

Figure 2.11 Triangulation.

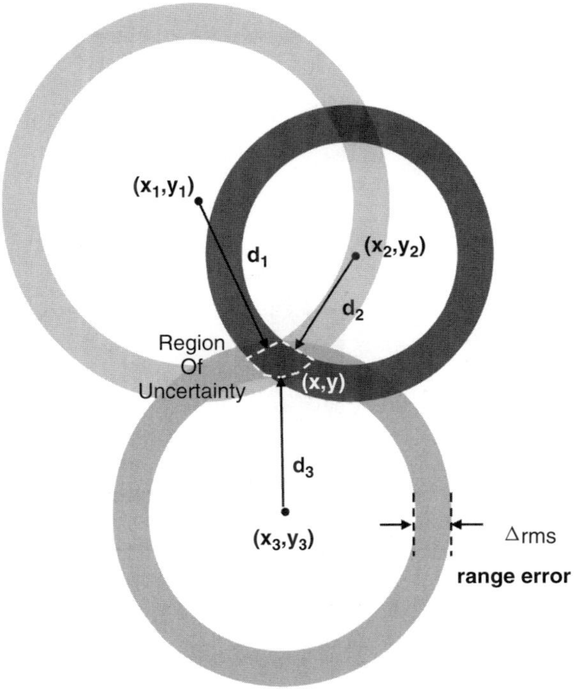

Figure 2.12 Trilateration.

Trilateration Method

With three or more antennas, it is possible to approximate the 3D coordinates of the tagged item. Each antenna sub-system first estimates the tag's radial distance and then sends this data to a common computer that can solve for the intersecting area illustrated in Figure 2.12. This technique requires that every antenna cover the same target area. The position estimate of $(\mathbf{x}, \mathbf{y}, \mathbf{z})$ is determined from estimates of each radial distance $\{\mathbf{d_1}, \mathbf{d_2}, \mathbf{d_3}\}$ and the known coordinates of each fixed position antenna $(\mathbf{x_n}, \mathbf{y_n}, \mathbf{z_n})$. The uncertainty of each distance estimate translates into a tag position uncertainty.

As an example, by assigning $(\mathbf{x_1}, \mathbf{y_1})$ as the origin in the layout shown in Figure 2.13, and simultaneously solving the set of circle equations for a common solution, the position coordinate estimate becomes,

$$\hat{x} = \frac{\hat{d}_1^2 - \hat{d}_2^2 + x_2^2}{2x_2} \tag{2.14}$$

and

$$\hat{y} = \frac{\hat{d}_1^2 - \hat{d}_3^2 + x_3^2 + y_3^2}{2y_3} - \frac{x_3}{y_3}\hat{x} \tag{2.15}$$

A 3D solution is determined similarly by simultaneously solving the three equivalent equations for a solid geometric sphere.

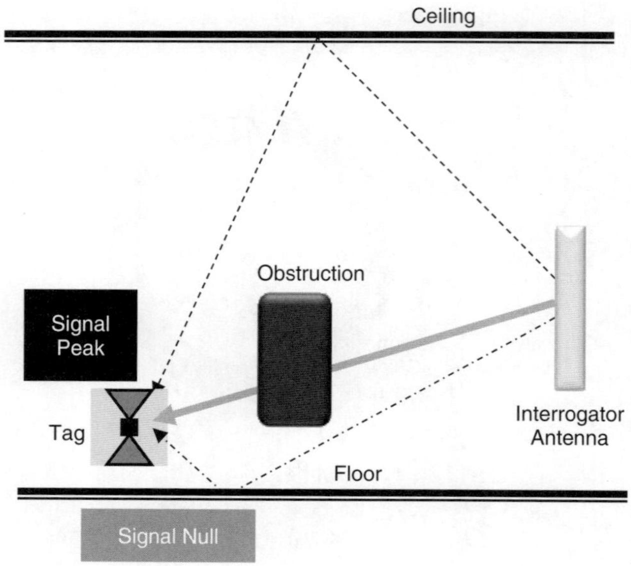

Figure 2.13 Multi-path interference.

2.2.3.3 Multi-Path Impact

The signal arriving at the tag or the interrogator is a combination of signals reflected from materials in the environment plus the direct path transmission. The theoretical impulse response for a multi-path signal is,

$$h(t) = \sum_{i=1}^{L} \alpha_i e^{j\phi_i} \delta(t - \tau_i) \tag{2.16}$$

The amplitude and phase of each signal component i are α_i and ϕ_i respectively. The TOF for the ith component is τ_i. L is the total number of received signal components. Since time of flight is directly proportional to the distance and the direct path signal is the first to arrive, the goal of the signal processing algorithms is to determine the direct path signal from the received composite signal.

The RSS is the vector sum of all signal components. For non-line-of-sight conditions, the direct path signal may be much weaker than it would have otherwise been. The reflected signal components will combine constructively or destructively as illustrated in Figure 2.13. Therefore, RTLS engines average the RSS over several transmissions in order to improve the radial distance estimate. This implies that tags must transmit more frequently at each position in order to maintain the same velocity and positional update latency. Therefore, the trade-off for higher speed and accuracy is higher power consumption.

In general, better localization accuracy can be achieved by collecting more diversified information that includes time, frequency, space, and polarization. For example, a Direct-Sequence Spread Spectrum (DSSS) technique can be used to increase the time resolution of the signal in order to estimate TOF more accurately. With such an approach, the receiver correlates a known pseudo-random number (PN) sequence with the composite signal to resolve each signal component. The algorithm measures the number of chips between the transmitted and the backscattered signal. The TOF estimate and the ability to resolve the first arrival signal improve with PN sequence length. However, the chipping sequence rate must also increase in order to maintain the same data rate [17]. A faster chipping rate also requires more bandwidth, and this may not be available. Increasing the bandwidth raises the noise floor and consequently reduces the receiver sensitivity, which ultimately reduces link margin and range.

TOF of the first arrival signal can also be determined from UWB transmissions, which consist of nanosecond wide pulses. The time duration between pulses encodes the data. Signal absorption and multi-path conditions can cause inter-symbol interference making it nearly impossible to distinguish between pulses. Shortening the pulses and improving the SNR can improve the TOF estimate, but shorter pulses also occupy more bandwidth. UWB transmissions occupy several gigahertz of bandwidth and few regulatory bodies currently allow them at power levels below that of non-intentional radiators [3].

2.2.3.4 Omni-Directionality Impact

RTLS assumes that each interrogator can simultaneously read the tag throughout the common coverage area regardless of its orientation with respect to the interrogator. However, it is likely that an RF opaque material that the tag is attached to will block its signal from reaching more than one interrogator. The degree of RF transparency depends on the type of material and the operating frequency as described in Section 2.2.4. While omnidirectional antennas and RF transparent materials are likely to provide the best results, RF transparency is not a guarantee.

2.2.4 Impact of Materials

RF reflective materials in the environment contribute to multipath conditions that reduce localization accuracy, read rates and operating distance of RFID systems. When these types of materials come into close proximity with either the tag or the interrogator, they could cause an adverse change in the receiver's impedance characteristics that reduce the amount of energy coupled. The type of material and the selected operating frequency are the two most important parameters that determine the impact on receiver sensitivity.

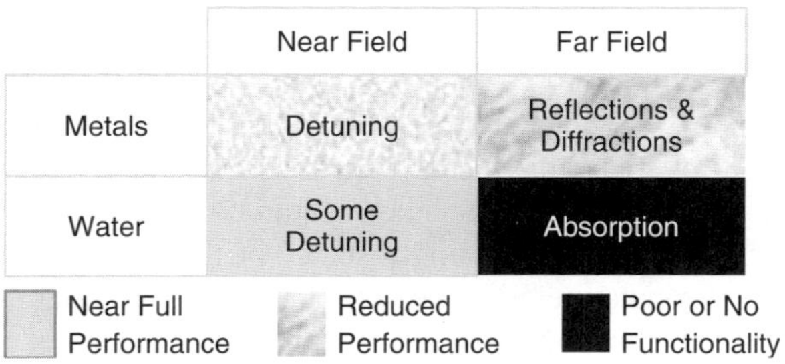

Figure 2.14 Effect of materials.

Figure 2.14 summarizes the performance impact of materials on each system type. Antenna detuning or signal absorption occurs when material with dielectric properties substantially different from air comes within proximity of either the interrogator or tag antennas to cause an impedance mismatch. Any change in the impedance from a conjugate match with the propagating medium will decrease the amount of power transferred to the tag. A passive tag will receive less energy for activation, and a battery-based tag will experience a lower SNR. The net result is a decrease in operating distance and read rates for either type of tag. The degree to which these materials will degrade performance depends on the type of electromagnetic link and the material as shown.

2.2.4.1 Metals

When attached to material with high metallic or water content, the electrical coupling between the tag's antenna and the material causes an impedance change that reduces the power coupled into the tag. Manufacturers seemingly optimize planar or printed antennas either for free-space impedance characteristics or for the type of material intended for tagging. The electric field intensity near the antenna-material boundary appears to dominate a designer's ability to practically implement the required impedance matching conditions [18]. Therefore, designers often add dielectric materials to the antenna to alleviate a trade-off between free-space and direct attach performance. However, dielectrics form an antenna spacing that forces non-planar and thicker antenna designs.

2.2.4.2 Water

Microwave ovens work because water molecules in the food absorb the microwave energy. The H_2O molecules behave like tiny dipole antennas that attempt to align themselves with the oscillating EM fields. The water molecules generate heat in the process, much in the same way that a mechanical piston does when it absorbs energy to smooth out the vibrations. Any UHF tag antenna that is within close enough proximity to water will lose energy to the vibrating H_2O molecules because the molecules effectively form a lower impedance path for absorbing the EM energy. Water also acts like a dielectric that could detune the antenna. The degree of impact depends on whether the tag is completely immersed or attached to a water container.

2.2.5 Other Factors Considered

2.2.5.1 Read Rates

One of the most important problems facing the RFID industry nowadays is that RFID systems are not 100% reliable. This means that if 100 tags are in the operating distance of the interrogator, very rarely will the interrogator be able to read all 100 tags. The percentage of tags that are in the operating distance of the interrogator and that are read (identified) by the interrogator relative to the total number of tags that are within the operating distance of the same interrogator is called the read rate. Low read rate can occur due to poor positioning of RFID tags so that their orientation relative to the reader antenna is unfavorable, close proximity of tags to one another, proximity of tags to water

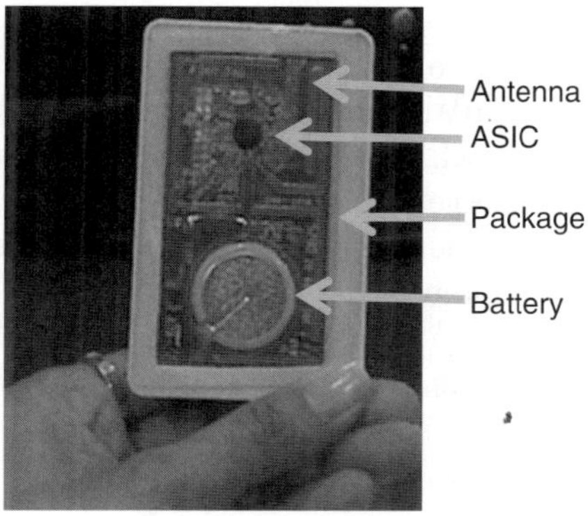

Figure 2.15 Packaged active tag. Courtesy Axcess International, Inc.

or metal, the number of tags within the operating distance, relative speed between the tagged objects and the interrogator, and many other factors.

2.2.5.2 Limitations on Size and Thickness

The most basic passive tag consists of a substrate containing the electronics and antenna. An active tag also includes a battery and may incorporate additional components such as actuators, sensors, and indicators requiring additional electronics. Figure 2.15 shows an active tag packaged into a credit card form factor. Single chip implementations for active tags are now common [19]. Given the ability to reduce the size of the electronics, the size of the antenna and the battery remain the two dominant factors that constrain the tag's form factor.

Size Impact from Antennas

The antenna size is a function of wavelength, gain, radiation pattern, bandwidth, and the required electrical impedance characteristics. Without the ability to further increase source power or sensitivity for a selected operating frequency, enhanced range is still possible by increasing the receive antenna gain if the required impedance matching characteristics are maintainable. The concept of an effective aperture (A_{eff}) relates the wavelength (λ) and antenna gain (Ψ) for far-field antennas where

$$A_{eff} = \frac{\lambda^2}{4\pi}\Psi \tag{2.17}$$

The ratio of the effective aperture to the physical antenna size A_{phy} is the *aperture efficiency*, which is typically about 80% for the best designs. The actual value depends on the type of antenna construction [1]. The interrogator's antenna size is comparable to the wavelength in the far-field region. The antenna size decreases with increasing frequency and, therefore, there is a tendency to increase the operating frequency of RFID systems.

In the near-field region, interrogator antennas are considered electrically small even though they can be several meters in diameter [20]. Near field RFID tag and interrogator antennas are basically loops of conductive material such as copper wires.

Size Impact from Batteries

The battery size depends on the volumetric energy density of the battery type. Lithium coin-cell batteries store near 90 μW-Year of energy per cubic centimeter. Therefore, the size of the battery depends on both the tag's power consumption and the desired operational life before needing a recharge or a replacement. A 'low-power' narrow-band active tag consumes an average of 30 microwatts under typical usage scenarios [1]. Therefore, a two cubic centimeter size Lithium battery will last,

$$90\,\mu W \cdot year/cm^3 \times 2 \cdot cm^3 \times \frac{1}{30\,\mu W} = 6\ years \tag{2.18}$$

The actual battery capacity should be degraded based on its leakage current or shelf-life rating. Active tags use different solutions to preserve battery life including:

- low-power motion sensors integrated with tags so that tags do not report when they are stationary;
- incorporating short-range radio in the long-range active tag and using short-range communication as an activation (waking-up) point so that the tag is awakened after it enters the area of interest [6].

By using these techniques, tags can reduce their beaconing frequency and reduce power consumption. For example, consider an application that requires detecting the presence or condition of tagged assets in storage. When using dual-frequency active tags, for example, those from Access International, Inc., the presence of these assets can be detected at the activation point that wakes up the tags as they enter the storage area. After that, tags will periodically report their condition and status to the interrogator. A very low beaconing frequency such as once per hour suffices for this typical scenario. In this way, the battery life can be significantly prolonged as pointed out above (about six years for this scenario).

2.2.5.3 Health and Safety

In 1995, the European Telecommunications Standardization Institute (ETSI) subcommittee, European Committee for Electro Technical Standardization (CENELEC) published the pre-standard ENV50166-2 advising that uncontrolled emissions should not exceed 10 watts per square meter when averaged over any six-minute interval, and within about 20 centimeters of the antenna. However, for a controlled environment, the exposure limits are frequency dependent. Exposure limits are much more stringent for microwave frequencies starting at 1 GHz [21]. The implication is that designers should consider the frequency dependent Maximum Permitted Exposure (MPE) when installing antennas in people-occupied facilities.

2.2.5.4 Security

Secure tags are capable of disguising their over-the-air data exchange. The level of security is proportional to the difficulty or cost to infer, reveal, or capture the actual data exchanged. The ease or cost with which the exchanged data is recoverable and applied to expose a person's identity or activity relates to the privacy factor. Cryptography schemes can encrypt and decrypt data using secret digital keys and passwords. The strength of the cryptographic scheme is highly correlated to cost and power consumption. Therefore, adding security features will likely reduce the tag's range and throughput performance.

2.2.5.5 Total Cost of Ownership

Tag cost is often a key point of comparison among the various technology types. However, a complete RFID system also requires interrogators, label printers, software, and networking technologies to provide the value sought. Other cost components also include the manual labor to install, integrate, apply, and use the technology. The sum of all related RFID expenditures is the total cost of ownership (TCO). Tag price tends to be more sensitive in applications such as the retail supply chain where the tag is likely to be disposed of once read. Conversely, reusable tag embodiments and higher order functionality such as RTLS, tamper alert, and conditional sensing, tend to increase tag price. Therefore, return-on-investment (ROI) analysis for active tags often considers amortizing the tag price over its useful life.

2.3 Classification of Commercially Available Products

Three RFID characteristics form the orthogonal bases for their categorization. These are: the way they obtain power, the way they transmit data, and the type of EM link formed with the interrogator. The four near-field and four far-field combinations form the eight categories shown in Figure 2.16 together with the major applications of RFID systems for each combination. The type of EM link impacts performance near liquids and metals as described in Section 2.2.4. The two most popular categories are active

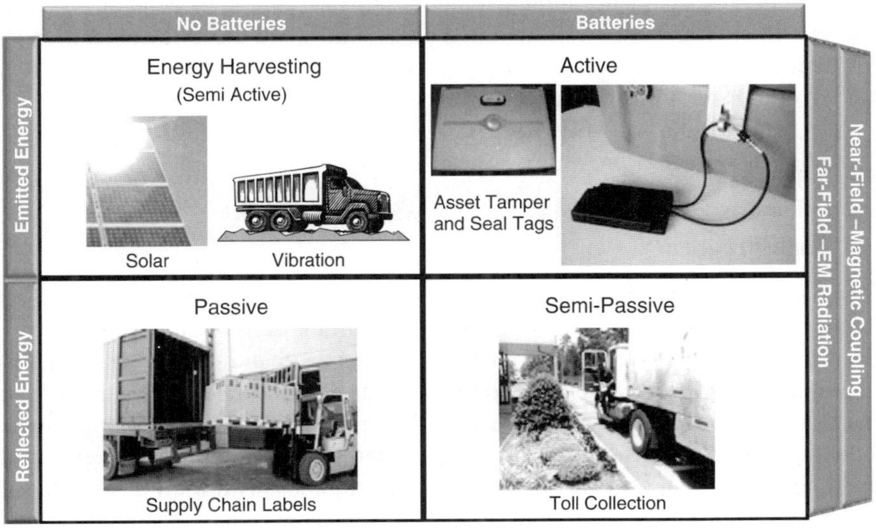

Figure 2.16 Tag technology categories.

and passive, with their key performance attributes described earlier. Conceptually, it is convenient to consider *semi-passive* as passive with a battery, and *semi-active* as active without a battery. Batteries can improve both the sensitivity and functionality of a passive architecture. The absence of a battery forces reliance on other energy sources, for example, solar, thermal, and vibration, which may not be readily available. Figure 2.17 summarizes the trade-off between each of the remaining categories focused on the link between the tag and the interrogator. The highlighted portions of Figure 2.17 imply that products that integrate near-field and far-field capabilities can provide the combined benefits shown. Some of these architectures will be covered in Chapter 18. Figure 2.18 is a qualitative categorization of the two key performance requirements of range and throughput with respect to representative applications and tag types.

2.3.1 Near-Field Coupled Systems

The most significant feature of near-field coupled systems is their resilience to impedance perturbation from the proximity of liquids and metals as described in Section 2.2.4. This near-field characteristic allows for covert antenna installations, such as burying them in the ground beneath control points such as gates and doorways. Control point boundaries can also be sharper because the near-field power declines three times faster on a logarithmic scale than in the far-field. However, this characteristic becomes a deficiency when the application also calls for long-range communications and higher throughput. By extension, both capabilities are obtainable by producing a combined near-field and far-field tag.

2.3.1.1 Low-Frequency

Tightly wound low frequency antennas operating in the 130 kHz region have made it possible to construct capsule-size tags. Prevalent applications are animal tracking in areas such as wildlife management, animal rearing, and pet safety. In general, systems that must operate in environments with high humidity, fluids, and metallic content tend to use low-frequency near-field communications. A low frequency near-field antenna typically requires several dozen turns of a copper wire to provide sufficient operating range. However, these types of antennas often break from bending when implemented in a planar form factor such as within proximity access control cards.

2.3.1.2 High-Frequency

In the higher frequency band such as 13.56 MHz, the antenna requires fewer turns for equivalent form-factor LF tags that can operate within the same coupling distance. Fewer turns make them less expensive to print or etch antennas than their LF counterparts. The higher frequency also facilitates more complex modulation schemes that can increase data rate and throughput. HF designs have become the preferred approach for contactless smart card, electronic transit pass, and electronic passport applications. Their accommodating near-field characteristics near liquids and metals combined with the higher data rates as compared with LF have increased their utility in short range supply chain operations such as automatic item identification in the pharmaceutical supply chain.

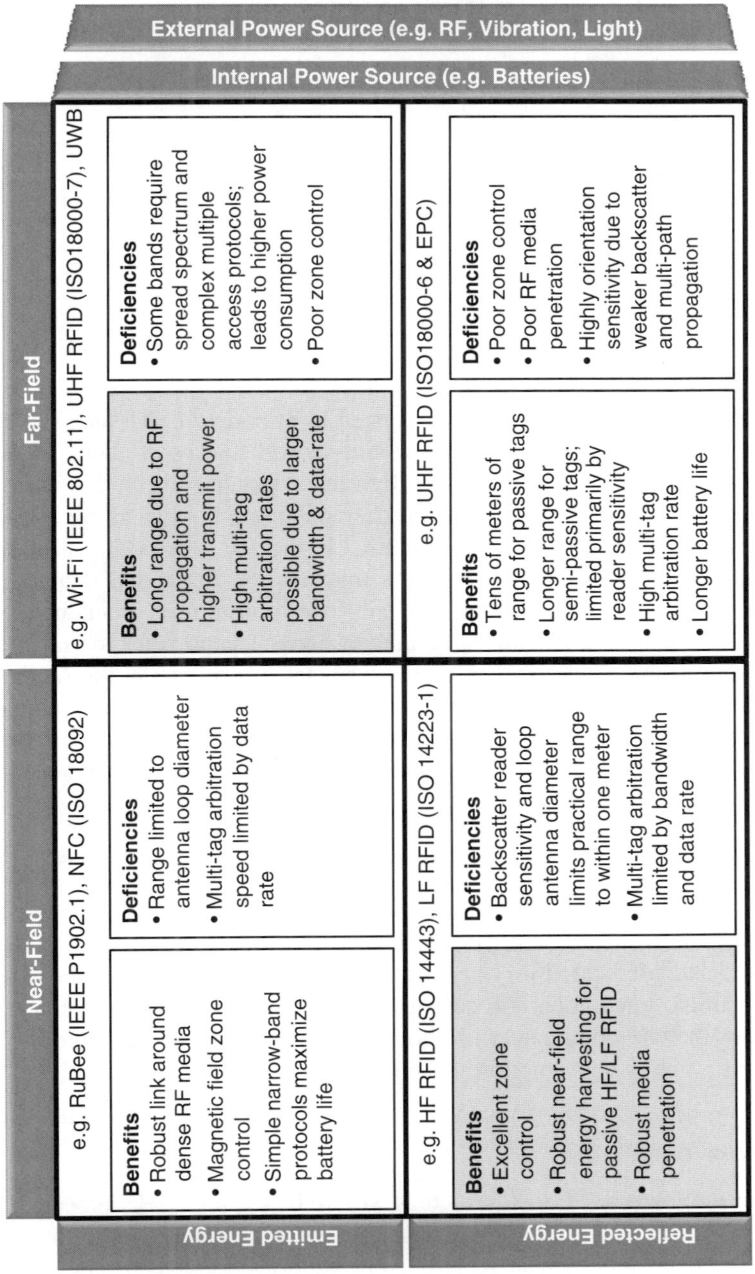

Figure 2.17 Performance and feature trade-off between technology categories.

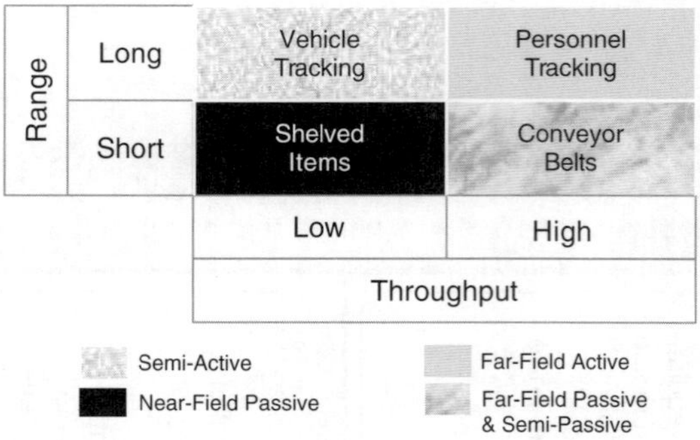

Figure 2.18 Application vs. technology.

2.3.1.3 Ultra High Frequency

Near-field UHF RFID technology was introduced to provide an alternative to HF but at a much higher data rate. Tags can be read within short distances of about 1 ft with UHF antennas designed for near-field operation. High read rates have been demonstrated with near-field UHF tags placed on metals, on bottles containing liquids, as well as immersed in liquids. The advantages in comparison with LF and HF tags is that near-field UHF technology provides significantly higher data transmission rates, and that tag antennas can be relatively smaller since they require only one loop turn of a conductive trace or wire. The tags use loop-like antenna structures of about 1 centimeter in diameter, which may also be coupled to dipole-like structures for better sensitivity in the far-field region [22]. Tags that support both near-field and far-field UHF antennas can communicate with the interrogator using either near-field or far-field or both, depending on the type of interrogator antenna, and the power transmitted from the interrogator antenna.

2.3.2 Far-Field Propagating Systems

Early far-field RFID systems utilized narrow-band communications in the lower UHF bands. However, the proliferation of commercial wireless standards in the 2.45 GHz Industrial, Scientific, and Medical (ISM) band, for example, Wi-Fi® and Bluetooth®, drove significant electronic integration and cost reduction. This fueled the emergence of wide-band RFID tags that are compliant with those short range wireless technology standards, but not without a power consumption penalty.

2.3.2.1 Narrow-Band

Regulations in the lower UHF bands such as 315 MHz and 433 MHz allocate a few tens of kilohertz of bandwidth. This allocation suffices for a majority of traditional RFID applications. Implementations near 433 MHz are widely deployed for container security and military logistics. This application was partially a catalyst for the ratification of international narrow-band standards such as ISO18000-7 and ISO18185 [23, 24]. Regulatory regions that have no allocations near 433 MHz tend to allow operations near 315 MHz.

FCC CFR47 Part 15.231 limits the output power in these bands to approximately 6 microwatts [3]. Therefore, solutions in these bands are generally limited to active systems.

2.3.2.2 Wide-Band

Although RFID implementations require very little bandwidth to transmit their unique identification number, the wider bandwidth ISM bands at 915 MHz, 2.45 GHz, and 5.8 GHz have attracted many wireless technologies, including RFID. Wi-Fi, ZigBee®, and Bluetooth, are the most popular examples in this category [17, 25]. The number of active tag implementations using Wi-Fi, ZigBee, and Bluetooth are growing in the overall RFID market To share the band, these devices utilize spread spectrum with a relatively high data rate of at least one MBPS. In addition, active tags that operate in the ISM bands transmit at power levels that are at least an order of magnitude higher than narrow-band active tags. Consequently, they utilize proportionally larger batteries for equivalent longevity.

Wi-Fi and Bluetooth are now widely available in consumer products such as personal digital assistants and cellular phones. ZigBee specifies a suite of protocols that comply with the IEEE 802.15.4 standards [25]. This IEEE standard defines an ad-hoc mesh network where nodes within range of each other can exchange data. Nodes initiate communications by either continuously listening for requests from other nodes in the network, or listen only at pre-determined time-slot intervals. The former requires that their receivers remain continuously activated. This increases energy consumption and shortens battery life. Waking up at pre-determined time-slots requires clock synchronization. This tends to add both cost and power consumption to the architecture. Therefore, a key trade-off in mesh network architectures is power consumption for responsiveness. RFID tag implementations that are compliant with these standards will, therefore, suffer from these same constraints on power consumption and battery life. Consequently, commercially available tag products that are compliant with these short range wireless network standards typically provide for user replaceable batteries.

2.3.3 Ultra Wide-Band

The fundamental carrier signal for UWB systems is a nanosecond scale pulse width instead of a continuous wave (CW) frequency. The relative time separation between pulses encodes the data. While a relatively short time pulse improves the TOF estimate, it also spreads the occupied frequency across more than 500 MHz of bandwidth. Therefore, regulatory bodies severely restrict the output power to avoid interference with other equipment. For example, FCC CFR47 Part 15.517.c limits indoor UWB system power levels to −41.3 dBm EIRP, which is equivalent to about 74 nanowatts per MHz of resolution bandwidth [3]. This is less than 2% of the power density allowed for narrowband 433 MHz systems of equivalent bandwidth. UWB spectral allocation across the world is currently sparse.

The link budget, which is the difference between the transmitted power and the achievable receiver sensitivity at the required data bandwidth, provides a good indication of the fundamental range difference between a narrow-band RFID system and a RFID implementations based on UWB technologies of equivalent data rate.

2.3.4 Passive Solutions

Nearly all of the passive tag performance deficiencies are due to its reliance on the interrogator for energy and a carrier signal to reflect during backscatter communications. The tag must receive reliable power for sufficient time duration in order to maintain a stable logic state during the collision arbitration process. Regardless of the algorithm used, tags must either keep track of a time-slot, a bit position in their serialized identification number, or an acknowledged response. However, multi-path propagation, electromagnetic interference, and electrical impedance perturbation from material proximity often hamper a tag's ability to receive power for a sufficiently long period. Even when conditions allow the tag to obtain power from the interrogator for a sufficient period, or if a battery is available for powering the electronics, the significant loss in signal strength from backscattering becomes the next significant performance constraint as described in Section 2.2.2.4.

2.3.4.1 Silicon Based

The traditional passive RFID tag, regardless of operating frequency, utilizes a single silicon chip. A simple protocol such as ISO18000-6a will require a few thousand transistors, while a more complex protocol such as C1G2 will require tens of thousands of transistors [12].

2.3.4.2 Chipless RFID

Researchers continuously demonstrate low functionality tags implemented with surface acoustic wave (SAW) resonators, MEMS cantilever finger reflectors, and RF reflective materials. These technologies create a unique backscatter signature from the incident RF energy [26]. The signature is usually a function of the mechanical construction methodology so this restricts their use to applications where a unique identifier without any prescribed format suffices. Other limitations are typically an inability to remotely address a specific tag, to write data from a distance, and to implement a collision arbitration MAC protocol. Printed transistor and organic circuit technologies capable of implementing low complexity RFID tags are also now emerging [27]. However, practical implementations are still constrained by their relatively large transistors, high operating voltage, and low charge carrier mobility [27]. Printed transistor technology cannot yet achieve the density or electron mobility required to implement complex RFID protocols in practical form factors [28].

2.3.5 Semi-Passive Architectures

Also known as battery-assisted passive (BAP), semi-passive RFID is based on a hybrid architecture that reflects energy from the interrogator to transfer its data in the same manner as passive RFID, but utilize a battery for operating power. Therefore, their performance limitations stem from the same backscatter link constraints as passive tags. Although the battery facilitates improved receiver sensitivity through lower threshold voltage-biased rectifiers, the weak backscatter signal imposes greater sensitivity challenges on the interrogator. Equation 2.5 highlights that this challenge is exponential with improvements in tag sensitivity. Increasing the interrogator sensitivity will impose bandwidth restrictions, which ultimately reduces system throughput.

2.3.6 Far-Field Solutions

Semi-passive far-field tags utilize the same backscatter modulation schemes as passive far-field tags. Aside from power transfer, they will impose similar constraints on link performance. Compelling reasons for semi-passive far-field implementations are backwards compatibility with existing passive tag standards, the potential for longer operating range than passive tags, and support for sensors and actuators.

2.3.7 Near-Field Solutions

Semi-passive near-field architectures create controlled magnetic field disturbances that the interrogator senses and decodes. The benefits of semi-passive near-field solutions are similar to those of semi-passive far field solutions in that they are backwards compatible with passive systems but offer some performance and feature enhancements. Commercial examples are devices that comply with the RuBee® [29] and NFC [30] standards. At lower RF carrier frequencies, magnetically coupled near-field systems operate more reliably near liquids and metals and provide better zone control accuracy than far-field systems.

2.3.8 Active Architectures

Active tags use a battery for communications as well as other tag operations. Much higher operating distance can be achieved because of the higher receiver sensitivity possible with voltage biased front-end detectors. Several classifications of active tags can be made depending on the type of the standards used, their predominant application, wake-up mechanism, and frequency of operation. Active systems that utilize narrow-band standards such as ISO18000-7 or similar legacy protocols, are distinguished from those that are based on existing short range wireless network standards such as Wi-Fi.

Active RFID solutions span the same frequency bands and electromagnetic link types available for passive tags, plus additional frequency bands where passive RFID do not operate due to regulatory imposed transmit power limitations. For example, narrow band active RFID tags operate within the 315 MHz and 433.92 MHz bands for which passive RFID tags are not commercially available. The 433.92 MHz UHF band is more uniformly allocated world-wide, whereas UHF allocations for passive RFID are more fragmented. Although more uniformly allocated, regulatory bodies such as the FCC restrict transmitted power within the 433.92 MHz band to about 6 micro-watts. The interrogator transmission periodicity is also limited to one transmission every 10 seconds or less often for regular reporting. Active RFID architectures span single frequency (LF, HF, UHF, microwave, UWB) implementations as well as multi-frequency implementations that include LF wake-up and UHF transmission [19].

Architectures compatible with WLAN infrastructures tend to be deployed for RTLS applications that track high value assets. The tags are typically designed to use replaceable batteries because of higher power consumption demands for operational compliance with wireless network infrastructures. Tag architectures designed for compliance with ZigBee or BlueTooth tend to support some form of wireless mesh networking so that data can be moved through intermediate nodes and eventually to a network connected hub. This architecture requires that each node awaken often enough to maintain synchronization with the network timing parameters, and to relay data packets that are part of its route. Since each node can uniquely identify a tagged asset, these types of network compatible technologies are also considered a type of RFID solution. However, the trade-off is usually higher power consumption than low power, narrow-band implementations such as protocols that comply with or are similar to the ISO18000-7 standard.

2.4 Conclusion

In this chapter, the key operational parameters of an RFID system have been defined and the influence of other parameters, factors and design choices that affect them is elaborated. The key operational parameters are operating distance, system throughput and localization accuracy. This chapter provided detailed parameter and decision choices that the designers have control over including operating frequency, transmit power, bandwidth, digital modulation encoding, and maximum tolerable bit error rate. Factors that are influenced by one or more design choices include environmental characteristics based on deployment decisions such as signal-to-noise ratio (SNR), noise level, receiver sensitivity, and noise figure. The chapter also covers performance impact from materials such as water and metal on design considerations for read rates, size and thickness of RFID tags, health and safety aspects of RFID systems, security, and total cost of ownership. The available RFID technologies are then classified and qualitatively compared.

Although RFID performance is critically important for successful system implementation, technology selection guides that systematically address key operational parameters and performance metrics for proper deployment have not been sufficiently covered in available literature. Most of the previously published works on parameters of RFID systems address mainly performance of passive HF [31] or passive UHF systems [32], and as such they cannot be used to select the appropriate technology for a given application anywhere in the world. In [32], read rate and read range are defined as main parameters of passive UHF systems. In [31], RFID systems are specified by their cost, size and performance where performance is determined by read range, speed, integrity of communication and compatibility among the systems from different vendors. A second line of research is related towards quantifying performance of RFID systems experimentally [33]. A set of performance benchmarks is proposed mainly to determine read rates and ranges versus different parameters for passive UHF RFID systems. The key parameters that we selected in this chapter are different from the previous publications because the focus of this chapter is beyond passive and far-field RFID systems. This chapter represents a comprehensive survey of performance metrics and operational parameters of overall RFID systems.

Due to limited space, the factors that are also important but have not been treated in this book include the cost of RFID tags and readers, the effect of environmental factors such as temperature and humidity, the effect of the placement of the tags next to each other, unwanted reads, and several others. In addition, the analysis is restricted to RF and physical level parameters and do not include the effects of reader to tag protocols on performance and other high-level factors.

In this chapter mathematical expressions for quantifying the performance of several parameters are given. This represents a step towards addressing larger research challenges in which all the parameters can be quantified and cost functions defined when comparing different technologies for various applications. Possible research directions include modeling and RFID system, designing cost function and developing an application software that would compare different design choices for a specific set of applications. In conclusion, we foresee that this chapter can be useful for the classification of RFID systems and quantification of their performance.

Problems

1. A tag receives 1 microwatt of ASK-modulated signal at 1 kbps data rate. What receiver sensitivity is required to demodulate the signal with a BER no greater than 10^{-6}?
2. Consider a receiver with the sensitivity level solved from problem 1, and a corresponding FCC Part 15.231.e compliant transmitter operating in the 433.92 MHz band. If each transceiver in the system uses a 2.2 dB gain antenna, what is the link margin?
3. Given the same receiver sensitivity as problem 1, what is the link margin for an UWB system providing identical data rate and BER as the system in problem 2, but compliant with FCC 15.517.c using the 3.1 GHz to 10.6 GHz band?
4. Compare the theoretical unobstructed far-field range for each system of problems 2 and 3.

References

[1] Krause, J. (1950) *Antennas*, 2nd edn. New York: McGraw-Hill.
[2] Bridgelall, R. (1999) UHF Tags – the answer to the retail supply chain's prayers? *RF Innovations Magazine*, August.
[3] Federal Communications Commission (FCC). Code of Federal Regulations (CFR) Title 47, Part 15.
[4] European Telecommunications Standards Institute (ETSI). EN 302 208: Electromagnetic compatibility and radio spectrum matters (ERM) – Radio-frequency identification equipment operating in the band 865 MHz to 868 MHz with power levels up to 2 W, Part 2 – Harmonized EN under Article 3.2 of the R&TTE Directive.
[5] Bridgelall, R. (2002) Bluetooth/802.11 protocol adaptation for RFID tags, in *Proceedings of the 4th European Wireless Conference*, Feb. 28.
[6] Bridgelall, R. (2008) Introducing a micro-wireless architecture for business activity sensing, paper presented at IEEE International Conference on RFID, April 16.
[7] Mandal, S. and Sarpeshkar, R. (2007) Low-power CMOS rectifier design for RFID applications, *IEEE Transactions on Circuits and Systems*, **54**(6): 1177–1188.
[8] Smith, AA. Jr. (1998) *Radio Frequency Principles and Applications*. New York: I.E.E.E. Press.
[9] Gray, P.R., Hurst, P.J., Lewis, S.H., and Meyer, R.G. (2001) *Analysis and Design of Analog Integrated Circuits*, 4th edn. New York: John Wiley & Sons, Ltd, pp. 127–128.
[10] EPCglobal Inc. (2005) EPC™ Radio-Frequency Identification protocols Class-1 Generation-2 UHF RFID protocol for communications at 860 MHz–960 MHz. Ver. 1.1.0. Dec. 17.
[11] Skalar, B. (1988) *Digital Communications Fundamentals and Applications*. New Jersey: Prentice Hall.
[12] Karthaus, U. and Fischer, M. (2003) Fully integrated passive UHF RFID transponder IC with 16.7-μW minimum RF input power, *IEEE Journal of Solid-State Circuits*, **38**(10): 1602–1608.
[13] Assad, M.A. (2007) Master's thesis: A real-time laboratory test bed for evaluating localization performance of Wi-Fi RFID technologies, Worchester Polytechnic Institute, May 4.

[14] Thomas, G.B. and Finney, R.L. (1984) *Calculus and Analytic Geometry*, 6th edn. Reading, MA: Addison-Wesley Publishing Co., pp. 151–152.

[15] ISO/IEC 18000-6 (2004) Information technology automatic identification and data capture techniques – Radio frequency identification for item management air interface – Part 6: Parameters for air interface communications at 860–960 MHz.

[16] Tseng, J.D., Ko, R.J., and Wang, W.D. (2007) Switched beam antenna array for UHF band RFID system, *IEEE International Workshop on Anti-counterfeiting, Security, Identification*. Issue 16–18. April: 92–95.

[17] IEEE Std. 802.11. Specification is a basis for Wi-Fi™.

[18] Dobkin, D.M. and Weigand, S.M. (2005) Environmental effects on RFID tag antennas, in *IEEE International Microwave Symposium*, June.

[19] Axcess International Inc. (2007) Dot™ Micro-wireless technology for business activity monitoring. Press release, November.

[20] Sears, F.W., Zemansky, M.W., and Young, H.D. (1984) *University Physics – Part II*. 6th edn. Reading, MA: Addison-Wesley Publishing Co., pp. 616–617.

[21] IEEE C95.1-1991. Standard Safety Levels with Respect to Human Exposure to Radio Frequency Electromagnetic Fields (3 KHz–300 GHz).

[22] Nikitin, P.V., Rao, K.V.S., and Lazar, S. (2007) An overview of near field UHF RFID, IEEE RFID 2007 conference, *Grapevine*, TX, March 2007.

[23] ISO 18000-7 (2004) Parameters for active air interface communications at 433 MHz.

[24] ISO 18185-1 (2007) Freight Containers – Electronic Seals.

[25] IEEE Std. 802.15.4. Specification is a basis for ZigBee.

[26] Preradovic, S., Balbin, I., Karmakar, N.C., and Swiegers, G. (2008) A novel chipless RFID system based on planar multiresonators for barcode replacement, in *IEEE International Conference on RFID*, April 16, pp. 289–296.

[27] Subramanian, V. (2009) Printed electronic tags and sensors for smart packaging applications, in *Proceedings of the 8th Annual Conference on Flexible Electronics and Displays*, Feb. 2–5.

[28] Gowrisanker, S. (2009) Low temperature integration of CMOS devices on flexible substrates, in *Proceedings of the 8th Annual Conference on Flexible Electronics and Displays*, Feb. 2–5.

[29] IEEE Std. 1902.1. Specification is a basis for Rubee™.

[30] ISO 18092. Specification is a basis for Near-Field Communications at 13.56 MHz.

[31] Scharfeld, T.A. (2001) An analysis of the fundamental constraints on low cost passive radio-frequency identification system design, Master's thesis, Massachusetts Institute of Technology, Cambridge, MA, August.

[32] Nikitin, P.V. and Rao, K.V.S. (2006) Performance limitations of passive UHF RFID systems, in *Proceedings of IEEE Antennas and Propagation Symposium*, Albuquerque, NM, July, pp. 1011–1014.

[33] Ramakrishnan, K. and Deavours, D. (2006) Performance benchmarks for passive UHF RFID tags, in *13th GI/ITG Conference Measurement, Modeling, and Evaluation of Comp. and Comm. Systems*, Nürnberg, Germany, March 27–29, pp. 137–154.

3

UHF RFID Antennas

Daniel Deavours

University of Kansas, USA

Passive UHF RFID antennas are primarily based on a "printed" dipole. Figure 3.1 shows several commercial tags. At first pass, one can see that they are a combination of long and skinny dipoles, though some of them are fat (those tend to be more expensive and thus less popular commercially). They are certainly not always straight ribbon dipoles and include a number of wiggles, slots, curves, and other interesting geometries. The RFID IC seems to always be attached to some loop within the antenna structure. Some of these features are designed for aesthetics, but clearly some are functional, that is, they affect behavior of the antenna in some intentional way. What do all of these features do? How does size impact performance? How does the environment affect tag performance? What makes an antenna a "good" antenna? These are many of the topics that we will address in this chapter.

The chapter is set out as follows. In the first section, we review the behavior of a dipole antenna. Those who are familiar with antenna theory can safely skip this section, and those who are vaguely familiar or unfamiliar will find the most useful concepts and expressions there. The second section will focus on the T-match in its variants [1, 2]. The T-match is essentially the loop-like structure that we see in many of the RFID tags. For many of these analyses, we attempt to reduce the physical structure to circuit-equivalent approximations. Usually, some detail is lost in such simplifications, but they are also invaluable to providing intuition, insight, and design principles. In Section 3.3, we walk through a design process using a combination of equations and CAD tools to design an antenna.

In Section 3.4, we leave the dipole-based antennas and focus on microstrip antennas. Though the literature is now full of microstrip-based RFID antenna designs, we present here some theory and examples of how one can build microstrip RFID antennas using the same T-match used for dipoles. We also present some recent research that shows an interesting "hybrid" dipole/microstrip solution. We conclude with some remarks and references.

RFID Systems: Research Trends and Challenges Edited by Miodrag Bolić, David Simplot-Ryl, and Ivan Stojmenović

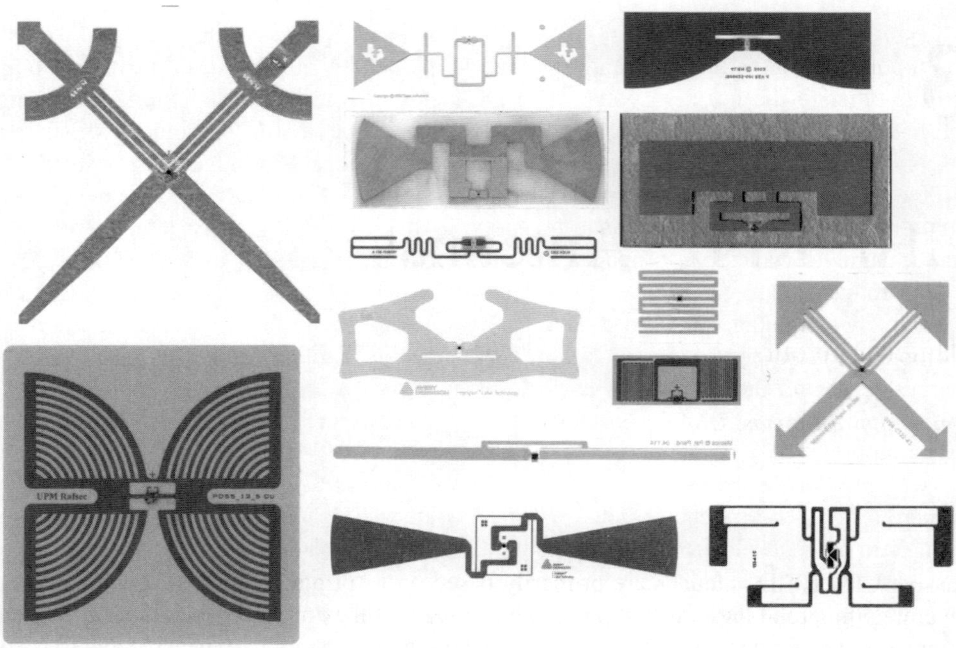

Figure 3.1 Variety of commercially available UHF RFID tags.

After reading this chapter, we hope you will have a better appreciation of the antenna designs. You should be able to look at a commercial antenna and identify the important aspects of the antenna. You should understand the fundamental limits that govern all printed antennas. We won't be able to tell you which tag is the right one for a particular application, but if you're ambitious, you might be able to design one yourself.

3.1 Dipoles and Relatives

The classic dipole antenna consists of two wires of equal length placed in line and attached to an AC source in the middle. To simplify the analysis, we often assume that the wires are hollow cylindrical tubes. RFID antennas are usually not made out of straight hollow wire tubes with the chip placed in the middle. However, understanding something about the basics of the simple dipole will translate well to the printed dipole. By "printed" we do not necessarily mean that it was literally printed (by a printing press or an ink jet printer, e.g.), though it certainly may be printed using conductive silver or other metallic inks. What we mean is that it is thin, flat, and all the metal residing in one plane. Cylinders are three-dimensional; printed antennas are (primarily) two-dimensional structures.

We will begin this section by reviewing the classic dipole antenna, then move on to printed dipoles, meandering dipoles, tip-loaded dipoles, T-match, bandwidth, gain, power transfer efficiency, and other important metrics.

3.1.1 Dipole

The dipole antenna is fed differentially, that is, one side is attached to a positive voltage and the other to the negative voltage. At DC, with such a structure, nothing interesting will happen. The structure will have some DC capacitance, that is, some positive charges will pile up in one of the poles, and an equal number of negative charges on the other end, and that's it. At AC, things begin to get interesting. Capacitors take time to charge, and moving currents along a wire create magnetic fields. Those moving currents also launch electromagnetic waves that propagate into the far field, which acts as a resistor.

Conceptually, one can think of the dipole as two inductive wire segments that are capacitively coupled to each other. Capacitance accumulates charge, which creates a force that opposes more charge accumulation. As charges accumulate in one of the poles, like charges repel and unlike charges attract, so the capacitance resists currents "piling up" in one pole or the other. Mathematically, this kind of force is similar to a spring that resists motion.

As current moves through one of the poles, it creates a magnetic field according to Ampere's law. The magnetic field serves as a kind of momentum, a force that resists changes in current: electrons in motion tend to stay in motion and electrons at rest tend to stay at rest. (This is partly due to physical momentum since electrons have mass, but the energy stored in the magnetic field is *much* stronger than the energy stored in electron momentum, so we usually ignore the momentum due to mass.) At some frequency, the two forces are equal and tend to cancel, causing resonance (Figure 3.2). Mathematically, this is no different than a child on a swing or a guitar string vibrating. A common analogy is a mass on a frictionless surface connected to a spring. At some frequency, the energy stored in the spring and in the momentum is the same and the two forces cancel themselves out in a kind of resonance, which is observed in a sinusoidal oscillation of the mass.

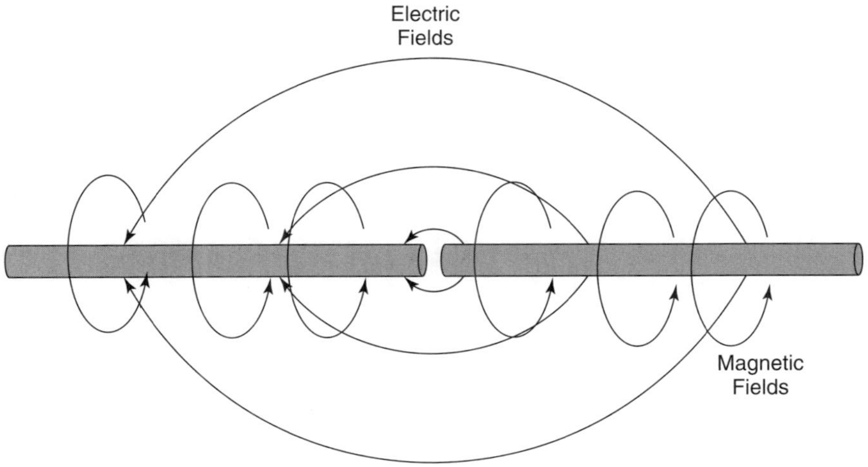

Figure 3.2 Electric and magnetic fields around a dipole antenna.

A very important concept with dipoles is that of impedance, which simply stated, is the ratio of voltage to current. Because the voltage and current of an AC system may not be in phase, the impedance may be a complex number, that is, having a magnitude and phase, or a real and imaginary part. Resonance can be defined as when the current and voltage is in phase, that is, the impedance has no imaginary part. This happens when the dipole is approximately one half of a wavelength in length.

Predicting the impedance looking into a dipole is a very fundamental problem, but also a complex problem. While a dipole has been around since Heinrich Hertz experimented with them in 1886, much has been learned (e.g. [3]), but unfortunately very little is simple. However, over a relatively short range of frequencies near resonance, there have been several circuit models that can moderately accurately predict the dipole impedance. Even those tend to be more complex than we would like to work with, so often we simplify the equivalent circuit to a series RLC circuit shown in Figure 3.3. It's important to note that this is only an engineering approximation, and only works moderately well over a short range of frequencies. The reactance does tend to follow the series LC circuit, so the model works very well when describing the reactance. However, the resistance is not constant. We will develop the resistance in more detail below. Again, over reasonably small ranges of frequencies, such as 3% or 5%, these assumptions hold fairly well.

3.1.2 Radiation

As with many things pertaining to a dipole, an accurate model is quite complex, but a reasonably good approximation is simple. Let's begin by looking at what happens with one very small segment of a dipole. Imagine chopping up the dipole into short segments, and on each segment assume the current is uniform. (These little dipoles are known as Hertzian dipoles.) Let I_0 be the current in the dipole, d the distance from the antenna (assumed to be more than a wavelength away), and θ the angle formed from the X axis as illustrated in Figure 3.4. Then the intensity of the electric field is given in Equation 3.1.

$$E = \frac{-jI_0 \sin\theta}{2\varepsilon_0 cd}\frac{L}{\lambda}e^{j(\omega t - kd)}. \tag{3.1}$$

Here, $j^2 = -1$, ε_0 is the free space permittivity, c is the speed of light, $k = 2\pi/\lambda$, and $\omega = 2\pi f$. The important point is that: (1) the power intensity is the square of the

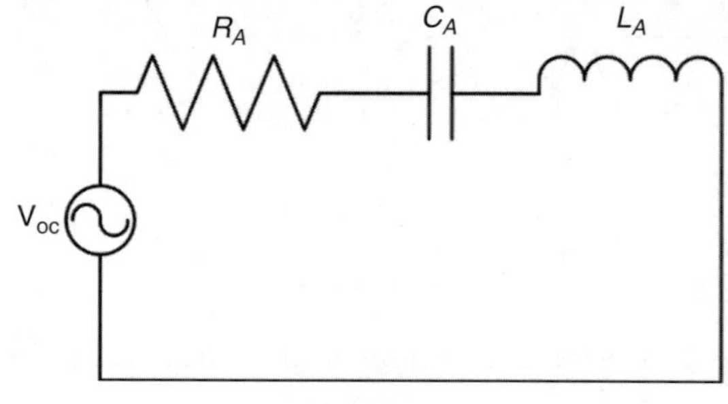

Figure 3.3 Simple circuit model of dipole antenna near resonance.

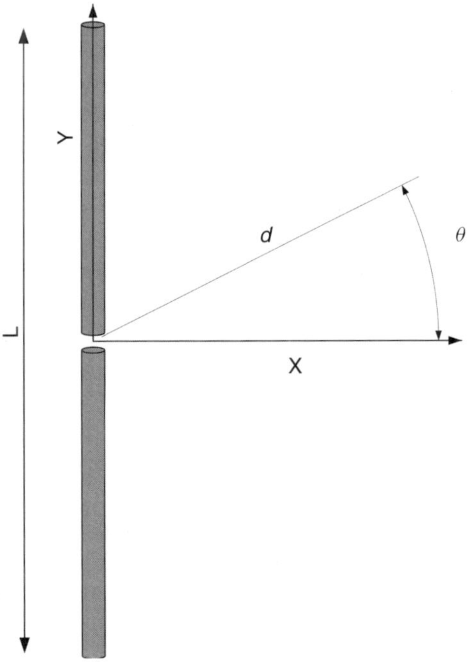

Figure 3.4 Radiation pattern around a dipole antenna.

E-field intensity, and the E-field drops off as $1/d$, so power drops off as $1/d^2$; and (2) the E-field intensity falls off with θ as $\sin\theta$. The radiation pattern is thus a toroid-like or "donut-shape" radiation pattern.

However, a half-wave dipole does not have a uniform current distribution over its length. At resonance, it is actually a half sine wave, so the current is larger in the center and falls off to zero at the ends. The resulting electric field intensity follows a pattern that is given in Equation 3.2.

$$\frac{\cos(\frac{\pi}{2}\cos\theta)}{\sin\theta} \approx \sin\theta \tag{3.2}$$

For many applications, the approximation of $\sin\theta$ is good enough. The bottom line is that a dipole radiation is at a maximum in the broadside direction, and goes to essentially zero in the direction of the poles. We'll see what that looks like when we consider a specific example in Section 3.3.

3.1.3 Impedance and Bandwidth

The first question we will ask is: what is the effect of the diameter of a wire to a wire dipole? The answer is that it affects a number of things, but most importantly, *bandwidth*. Informally, bandwidth is the range of frequencies over which the antenna provides a given level of performance. Usually, the limiting factor in antenna bandwidth is the antenna impedance, that is, the input impedance of the antenna changes (relatively) rapidly with frequency. For most antennas, such as the antenna you might have on a television, the intention is for the antenna to match the characteristic impedance of the transmission line connecting the antenna to the television. The characteristic impedance is essentially real, for example, 75 or 300 Ohms. This is also the case with RFID reader antennas that are commonly connected to the reader through a 50 Ohm coaxial cable.

With RFID tag antennas, the antenna is directly coupled to the RFID integrated circuit (IC). The IC usually has some kind of circuit that converts small voltages into usably large DC voltage to power the IC, usually using diodes or transistors in diode configuration. What that basically means is the IC impedance is not resistive, but contains a large reactive component. In this case, because diodes are used, the impedance tends to have a large capacitive component. For example, an IC may have a series resistive component of 10 Ohms and a reactive component of $-j150$ Ohms. If the IC has an impedance of $10 - j150$ Ohms, then to maximize the power delivered to the IC, the antenna should present an impedance of $10 + j150$ Ohms, that is, a conjugate pair. This follows from the maximum power theorem.

To measure the quality of the impedance match, we can use something called the *power transfer efficiency*, or *power transfer coefficient*, typically denoted in the literature by the Greek letter τ. If $\tau = 1$, then we know that the IC and antenna form a conjugate pair and the maximum power is transferred between the two. The astute reader might notice that only half of the power is transferred to the load; the other half is scattered (re-radiated) back into space by the antenna. This can't be helped.

The power transfer efficiency can be expressed in terms of the impedances [4].

$$\tau = \frac{4R_A R_C}{|Z_A + Z_C|^2} \tag{3.3}$$

In Equation 3.3, the subscript A represents the antenna impedance and resistance, and C the IC impedance and resistance. If you substitute $R_A = R_C$ and $Z_A = Z_C^*$, you'll see that $\tau = 1$ as expected.

Directly related to power transfer efficiency is return loss. Return loss is the fraction of available power not delivered to the load. There is a simple identity: power transfer efficiency + return loss = 1. To be specific, we can define the scattering wave s as shown in Equation 3.4.

$$s = \frac{Z_A - Z_C^*}{Z_A + Z_C} \tag{3.4}$$

Then the return loss is $|s|^2$, and the identity $\tau + |s|^2 = 1$. Return loss is usually specified on a decibel scale, that is, $RL = 20 \log_{10} |s|$. There are some convenient conversions to remember:

- a 10 dB return loss = 90% power transfer efficiency;
- a 20 dB return loss = 99% power transfer efficiency;
- a 3 dB return loss = 50% power transfer efficiency etc.

The IC impedance does change with frequency. We'll return to the IC impedance changing with frequency later on, for now, let's focus on the antenna. For the antenna, especially for small, short printed dipole antennas that are commonly used for RFID, the impedance changes rather rapidly with changes in frequency. As the impedance changes, τ also changes.

One way to quantify the change in impedance is through measuring the quality factor or Q of the antenna. The name may be rather misleading because a larger quality factor indicates an antenna with a small bandwidth, which is generally regarded as a lower quality antenna. The term "quality factor" was historically used to describe the quality of reactive elements such as inductors and capacitors. Hence, a high quality inductor is one with a large quality factor or Q. However, the concepts still apply to antennas, except that a large Q is often inversely proportional to the quality of the antenna. Generally,

$$Q = 2\pi \frac{\text{Energy Stored}}{\text{Average Power Dissipated}} \approx \frac{1}{\text{Bandwidth}} \tag{3.5}$$

So what does that mean? Remember that the dipole stores energy in the magnetic and electric fields. The more energy stored in those fields as a fraction of the power radiated, the larger the Q of the antenna. Power is dissipated by an antenna through radiation. (It can also be dissipated through resistive or Ohmic losses, dielectric (material) losses, but that negatively impacts efficiency. We'll return to this below.) So a very large Q antenna stores a lot of energy and radiates relatively little of it, and thus will have a very small impedance bandwidth. An antenna that has a large bandwidth stores relatively little energy and dissipates much of it.

How can you decrease the energy stored in a dipole? Simple: use thicker poles. A thin dipole stores a lot of its magnetic energy very close to the thin wire, while a thicker cylinder just stores a lot less magnetic energy. Interestingly, the thicker dipole has a larger surface area, which stores less capacitive energy, and as the diameter gets larger, the amount of energy is decreased in both the magnetic and capacitive fields in very close to the same amount. That means that a dipole basically resonates at the same frequency regardless of how thin it is. That assumption starts to break down when the dipole becomes fairly fat, but for thin wires, it holds remarkably well. Thus, we talk about the resonant frequency as (mostly) a function of its length, and the Q as a function of the length-to-radius ratio.

3.1.3.1 Printed Dipoles

Until now, we have considered a simple, cylindrical dipole, but nearly all RFID antennas are some variant of a printed dipole. A printed dipole isn't necessarily printed using, for example, an ink jet printer (though it could be); it simply means that the antenna is thin and flat. The term comes from the use of printed circuits such as a printed circuit board, which has been turned into the acronym PCB. This term comes from the old printing processes used to print a layer of resist before etching the pattern away using a corrosive solution. Some RFID tags are literally printed using silver inks. Regardless, the end result is an antenna that is fundamentally flat, not round.

What are the consequences of a flat dipole versus a round dipole? Bottom line, we see a loss of bandwidth. There is a rough equivalence between a wire dipole with radius r and a printed dipole with width W:$W = 4r$, which is shown in Figure 3.5.

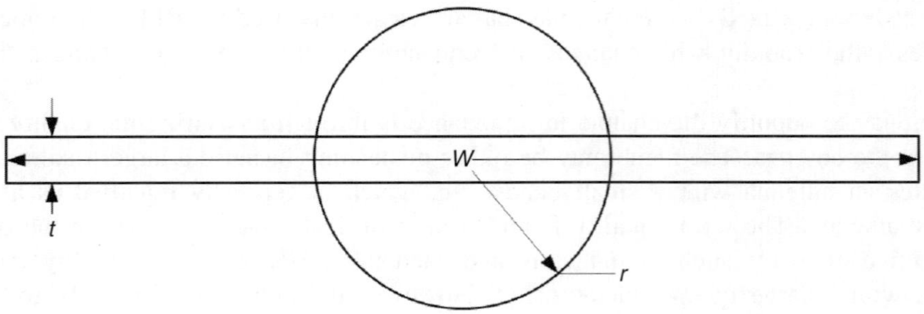

Figure 3.5 Relationship between cylindrical and ribbon dipoles.

Not surprisingly, wider printed dipoles have larger bandwidths. Unfortunately, wider dipoles take up a lot more space, and if the dipole is made out of silver ink, printing a wide dipole uses much more ink than a narrow dipole. If the antenna is etched from copper or some other subtractive process, then it makes sense to leave as much metal on the antenna as possible, but if the antenna is intended to be printed using expensive silver ink, then you may see designers trading off bandwidth for decreased cost. Because IC sensitivity (discussed below) has improved so much in recent years, it is now common for designers to trade off performance for reduced cost.

3.1.3.2 Meandering Dipoles

Unfortunately, much of the previous discussion has little to do with practical RFID antennas because most RFID tags are designed to comfortably fit within 50.8 mm by 101.6 mm label (2 by 4 inches). A large number of tags fall within the general size of 8 mm by 94 mm. Recall that we said that a resonant dipole is a little short of half a wavelength long, and at 915 MHz, that's about 150 mm, or more about six inches. So somehow we have to shrink the size of the dipole.

We basically have two options: (1) we can take the dipole and make it meander so that the full 150 mm of length will fit within 94 mm of space, or (2) we can just not use a resonant dipole. Most commercial tags use some combination of the two approaches.

What is the impact of shrinking the dipole from 150 mm to 90 mm? To demonstrate, we consider two antennas that are both 1 mm wide. The first is resonate length (151.8 mm). The second is a 1 mm line that is meandered within a 90 mm by 10 mm "box" so that it also resonates at approximately 915 MHz. See Figure 3.6 for an illustration of the tested antennas. The long, straight dipole was found to have $Q = 6.7$, which gives a

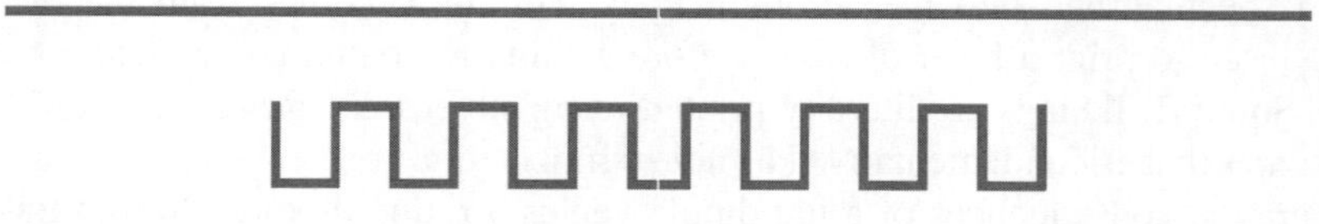

Figure 3.6 Straight vs. meandering dipole used to evaluate changes in Q for different dipole geometries.

3 dB bandwidth of approximately 130 MHz, which will cover the world-wide RFID band comfortably. The meandering dipole was found to have a $Q = 15.7$, or a 3 dB bandwidth of approximately 58 MHz.

What if the dipole is too short to resonate? That's actually OK, because we don't need the dipole impedance to be resistive. We actually want the dipole impedance to be inductive (positive reactance) because the IC typically has a negative reactance. We'll get into the details of the matching network in the next section. However, for now, we can say that a "short" dipole, or one operating below resonance, has a capacitive impedance (negative reactance). Somehow, we have to make that impedance inductive-looking (positive reactance), and we usually have to add some kind of inductor or inductor-like circuit. Remember, inductors store magnetic fields to operate. So, basically, we have to add reactance to the antenna, and remember that adding reactance increases the antenna Q.

We threw out a few examples about antennas, Q, and bandwidth, but it is very natural to ask a question about the fundamentals. Let's assume for a moment that we are not particularly clever about how we are building antennas. Perhaps there is something else out there, other than a dipole, which might have much better bandwidth. How can we quantify just how good these printed dipoles perform in terms of bandwidth?

This has been a topic of theoretical antenna work for a long time. Today, there is a generally accepted lower bound on the Q of an antenna that fits within a sphere of radius a. This is called the Chu limit, after Andrew Chu, who first published the concept in 1948 [5]. The limit is given here in Equation 3.6.

$$Q_{lb} = \eta_r \left(\frac{1}{(ka)^3} + \frac{1}{ka} \right) \tag{3.6}$$

Here, η_r is the relative free space impedance, which for most purposes, is unity. How is our dipole doing? Well, for one thing, we know that the printed dipole doesn't use space as efficiently as a cylindrical dipole, so we're giving up some bandwidth there. For a device that is 150 mm long, the theoretical smallest Q is 1.03. A Q of 6.7 is quite a bit worse. The 90 by 10 mm dipole? Its minimum Q is about 2.7, which would have more than 300 MHz of bandwidth, so a Q of 15.7 and bandwidth of 58 MHz is, again, considerably worse. So the bottom line is that printed dipoles are not at all bandwidth-efficient. However, they are very cheap to make and simple to design, and within the field of RFID antennas, performance almost always takes a back seat to cost.

Why are dipoles so bandwidth-inefficient? It's fairly simple. If you consider the bounding sphere, the dipole, and especially the printed dipole, uses very little of that volume. Other antennas that fill that volume more effectively will be able to store less energy, just as we saw thicker dipoles have smaller Q. A dipole built out of two hemispheres would have a very small Q and approach the Chu limit. But who wants to buy RFID tags out of two hemispheres?

3.1.4 Radiating Resistance

In the previous section, we focused a lot on the antenna reactance, Q, bandwidth, and resonance. Next, we will look at the radiating resistance. Simply put, if you were to break up the dipole into little segments of a dipole (perhaps "dipolettes"), each one would induce an electric field at some distance. If we integrated the square of the electric field (power) over all solid angles, we would find the total radiated power. That power lost to radiation is what we call the *radiating power*, and shows up in our circuit model as resistance, that is, the *radiating resistance* [3].

To cut to the chase and simplify, we've put the radiating resistance of a dipole in a convenient form:

$$R_A = 80\alpha^2 \left(\frac{\pi L}{\lambda}\right)^2 \tag{3.7}$$

Here, L is the total length of the dipole, and α is a term that is dependent on the current distribution along the dipole. If the current is uniform, then $\alpha = 1$. If the current is triangular, maximum at the center and zero at the ends, then $\alpha = 0.5$. If the current is distributed as a half sine wave, which is the case for resonant dipoles, then $\alpha = 0.62$. So, for example, our 150 mm dipole with a half sine wave distribution should have a radiating resistance of about 64 Ohms; we found 67 Ohms, so it was a pretty good estimate. The meandering 90 mm dipole would have a radiating resistance of about 23 Ohms (we saw 27.5 Ohms). If the antenna is very short, then the current tends to be more triangular, so the radiating resistance of a 90 mm straight (not meandering) dipole would fall to about 15 Ohms (about what we saw); another reason why short dipoles aren't as effective.

However, armed with this knowledge, we know that α^2 could be much larger than 0.38, which is the case for a resonant antenna with a half-sine wave current distribution, so there have been several "tricks" used to try to increase α. Remember that Q is a ratio of energy stored to power dissipated, and if you can increase the power dissipated without increasing the energy stored, then you can increase the bandwidth-efficiency of the dipole.

The main concept for most of these techniques is to try to make the currents as uniform as possible across as much length of the dipole as possible. Recall the 90 mm meandering dipole. Before, we had the meandering distributed evenly throughout the length of the antenna. Now, consider pushing all those meanderings to the edge of the antenna, and having a large section in the middle that is straight. Sometimes, instead of latitudinal meandering sections, people use longitudinal meandering, or what looks like a hat at the ends of the antenna.

Another way to increase α is to place a large capacitor at the ends. Obviously, adding a real capacitor would not be practical, but adding a very large metalized area, with enough area to store a significant amount of charges, can sometimes work. This is what we call capacitive tip loading. In some of our experiments, we found that the size of the capacitor needed to be impractically large for it to be an effective technique. However, on some of the smaller, wider tags, like the one shown below, the large metal area at the ends might be large enough to make a difference. Figure 3.7 shows some examples of commercially-available tags that use either capacitive or inductive ("hat") tip-loading technique.

The best way we found to increase α is to use a spiral inductor. Because of the reinforcing magnetic fields, the spiral inductor can create the same amount of inductance in a much smaller space than meandering lines [6]. Using a spiral inductor, we were able to get α close to 0.8, which is quite good. Unfortunately, the spiral inductor used to create that large α also stored a lot of magnetic as well as capacitive energy, and we ended up increasing the Q of the antenna, so tip-loading with a spiral inductor is definitely not a good design choice. Figure 3.8 shows a prototype of a spiral-loaded dipole that we designed.

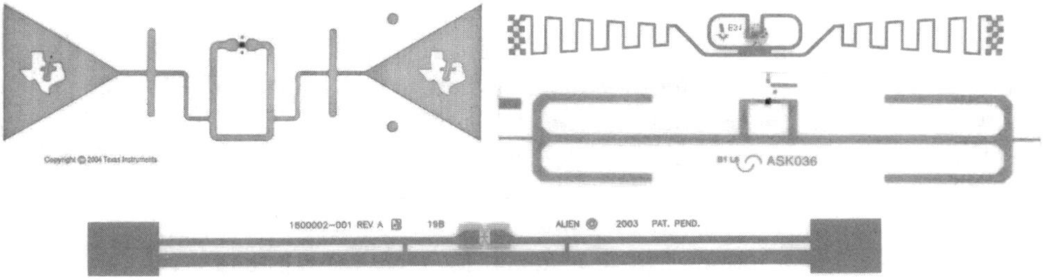

Figure 3.7 Examples of commercial tip-loaded dipole tags.

Figure 3.8 Example of a spiral-loaded tag. This tag gives excellent radiating resistance for its size, but poor Q.

From our experience, the simplest and most effective way to increase α and decrease Q is to use a meandering dipole where the meandering sections are pushed towards the end of the dipole. In doing that, we were able to decrease the Q from 15.7 to 13.8. Not a big difference, but it also didn't cost us anything in terms of extra size or material to get it. Another way to decrease Q is to use fatter traces, but there is an obvious limit to how fat you can make the traces and still fit within a confined area, plus there is the additional cost if one uses expensive metallic ink-based antennas.

Bottom line, printed dipoles have adequate, but not great bandwidth. It is very difficult (but possible) for an antenna to have such large Q and operate over a world-wide frequency range. However, below, we will explore some of the details of how one can use the antenna Q together with the reactance of the IC to further improve the bandwidth of the tag. Before we can do that, we have to understand one more part of the antenna: the matching circuit. But before we dive into the matching circuit, there is one more component to the antenna that is worth covering.

3.1.5 Polarization

A propagating electromagnetic wave is called a transverse electromagnetic (TEM) mode. In the TEM mode, the electric and magnetic fields form a 90 degree angle, and both are perpendicular to the direction of travel. Other modes exist, but they don't propagate in free space; only the TEM mode propagates efficiently through space. The right hand rule applies here: point your fingers of your right hand in the direction of the electric field vector, curl your fingers towards the direction of the magnetic field vector, and if you stick your thumb out, it will point in the direction of travel. To keep matters simple, we consider only the electric field component of the TEM wave.

Notice that for a particular direction of propagation, let's say in the Z direction, the E-field can be anywhere in the X–Y plane. Let's say you're standing and facing a tag

10 meters away. Let the X axis extend to your right out of the tag (the tag's left), the Y axis extends up from the tag, and the Z axis is pointing right at you. If the tag is a dipole antenna and the long dimension of the dipole is horizontal, that is, parallel to the ground, then we say that the dipole is *horizontally polarized*. The electric field will be oriented completely in the horizontal plane. If the dipole is perpendicular to the ground (straight up and down), then any transmitted or reflected wave will be *vertically polarized*, that is, the electric field will be oriented vertically.

Now consider a transmitting dipole antenna (let's say the tag) and a receiving dipole antenna that you're holding in your hand. What happens if the transmitting antenna is horizontally polarized and the receiving antenna is also horizontally polarized? The antennas work as expected, and power is transferred between the antennas as expected. But what happens if you rotate the receiving antenna by 90 degrees and hold the dipole vertically? The received power becomes essentially zero; there is a polarization mismatch between the transmit and receive antenna.

So far, the polarization we have considered is in dipoles, which are linearly polarized (LP), that is, the E-field is aligned linearly in one direction. Conceptually, imagine if you could rotate the antenna really, really fast, like 915 million rotations every second? (Which corresponds to one rotation per complete cycle of a wave.) This would result in an electric field vector that would corkscrew the air, which yields an electric field that we say is *circularly polarized* (CP). While it may not be realistically possible to spin the dipole at 915 million rotations per second, it is possible to rotate the electric field electrically through some other means, usually using some antenna other than a dipole. In fact, it's quite common for RFID reader antennas to be circularly polarized. It's possible for a wave to be partially circularly polarized and partially linearly polarized; to understand this, just imagine the superposition of the two. The resulting wave, we say, is *elliptically polarized*.

What happens when a LP antenna, such as a dipole, and a CP antenna, such as a CP patch antenna commonly used with RFID readers, meet? What happens is that there is a partial polarization mismatch, and exactly one half of the power is lost (compared to the case where there is a polarization match). This is commonly tolerated for a couple of reasons. First, it is common for tags to be oriented both vertically and horizontally. If we choose a vertically polarized reader antenna and some of the tags are oriented horizontally (or the reverse), the system performance will suffer severely. With a CP reader antenna, it doesn't matter if the tags are oriented horizontally or vertically or any angle in between – the performance will be the same, though degraded by 3 dB. Note that polarization is distinct from the radiation pattern discussed in Section 3.1.2.

There is a special class of RFID tag antennas that are called *dual dipole* antennas. (See Figure 3.9 for an example.) You can think of this antenna as two orthogonal dipoles. The dual-dipole antenna requires the use of a special, three-terminal IC, with the terminals labeled RF1, RF2, and GND (ground). The chip has a rectifier circuit so that it, basically, automatically selects which dipole it should use, and when helpful, can combine the signal from both dipoles. This allows the dual-dipole antenna to be polarization-matched to both linear and circularly polarized radiation, which is very convenient when using CP reader antennas. This is one reason why dual dipole antennas tend to perform so well. Another is because of the large size, the antenna bandwidth is excellent, and thus tends to be less

Figure 3.9 A dual dipole RFID tag.

affected by environmental parameters (discussed more in Section 3.4). The downside is the large size tends to significantly increase the cost of the tag.

3.2 T-Match and Relatives

3.2.1 The Classic T-Match

Remember that the IC has a significant (capacitive) reactive component to its impedance, and to have an efficient transfer of power between the antenna and the IC, we need to present an antenna with a conjugate impedance. We used an example of $10 - j150$ Ohms for a typical IC impedance. Thus, we would like the antenna to present $10 + j150$ Ohms. How do we do that? One way is to make the antenna short enough so that the radiating resistance is about 10 Ohms, and then meander the antenna so that it is electrically long and presents a large (inductive) reactance. While that will work, it's not a good, general design tool.

A variant of that design is to start with a meandering dipole, and make it sufficiently meandering so as to provide enough inductance. This may provide too large of a resistance, especially for antennas that are longer than resonant-length (for reasons we won't go into here), so one proposal is to use a "loading bar," which is a second, parallel conductive strip (dipole) nearby, which serves to effectively reduce α, which in turn reduces the radiating resistance. We don't recommend this approach because it requires lots of extra space for both the electrically long antenna and the loading bar, plus it intentionally decreases α, meaning the Q of the antenna is significantly increased in the process. The resulting antenna will of course work, but in our opinion, it does not make good use of space.

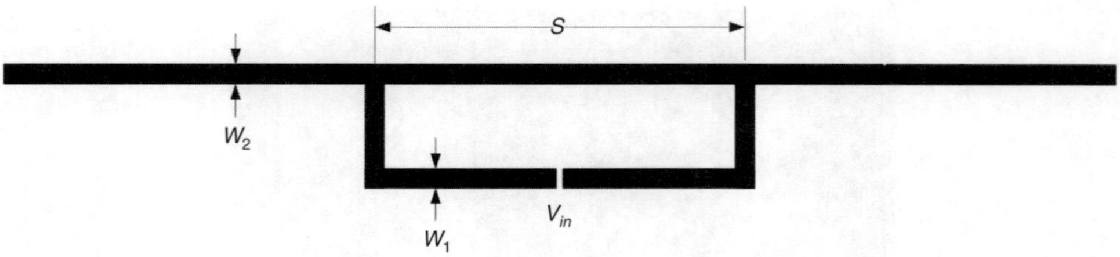

Figure 3.10 Dipole with T-match.

In practice, most commercial RFID antenna uses some kind of modification to the T-match. The T-match was a structure developed by Uda in the 1930s. The structure of the T-match is shown in Figure 3.10. You can see that there is a dipole component and a loop-like structure. If you extend that loop-like structure to the ends of the dipole, the T-match basically becomes the folded dipole. Hopefully, this starts to look a little bit familiar after examining the commercial antennas in Figure 3.1. The T-match basically has three critical parameters: the length of the T-match S, the width of the secondary trace W_1, and the distance between the two traces, see Figure 3.10.

A number of textbooks cover the T-match, so we will skip the analysis here and get to the results. The analysis follows a minor variation of the even/odd mode analysis with a second port added in the middle of the dipole (top trace). The equivalent circuit model is given below, where Z_d is the impedance (which is almost entirely reactive) seen from the ports when driving the two ports differentially, and Z_c is the impedance seen when driving the ports in common, or the common mode impedance, which is about that of the dipole impedance. Notice that Z_d is basically a short circuit transformed by a short length of transmission line, so if the line is relatively short and we consider a relatively narrow range of frequencies, we can model Z_d with an inductor. The α term here is a current splitting factor in the common mode, which acts as a transformer, and for equal diameter conductors, $\alpha = 1$. (When the loop becomes a half wavelength long, Z_d becomes an open circuit, and all we see is the dipole impedance scaled by $(1 + \alpha)^2$. This is the "folded dipole.")

The equation that governs the input impedance is given in Equation 3.8. This is the same as the circuit diagram shown in Figure 3.11.

$$Z_{in} = \frac{(1+\alpha)^2 Z_c Z_d}{(1+\alpha)^2 Z_c + Z_d} \tag{3.8}$$

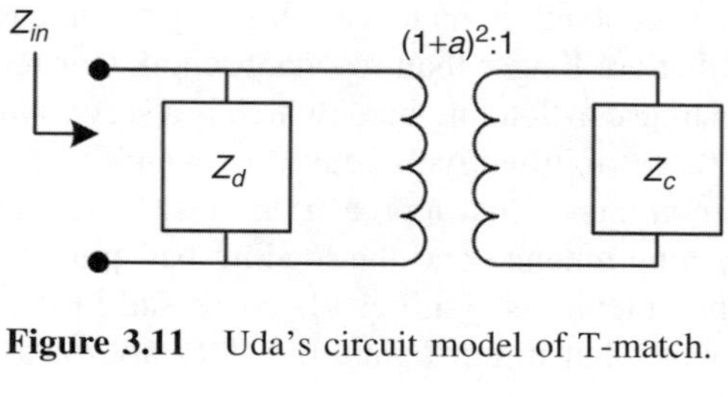

Figure 3.11 Uda's circuit model of T-match.

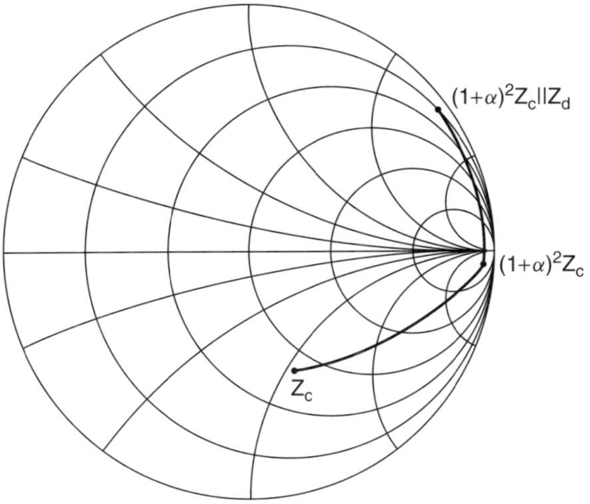

Figure 3.12 Smith chart view of impedance matching using the T-match.

To illustrate what the T-match does, we consider a dipole before and after a T-match. We use a Smith chart in Figure 3.12 to illustrate. If you're not familiar with a Smith chart, unfortunately there is not enough space to give a tutorial here. Basically, the Smith chart captures the positive complex plane through a Moebius transform of Z:

$$\hat{z} = \frac{Z_0 - Z}{Z_0 + Z}, \tag{3.9}$$

where Z_0 is commonly taken to be 50 Ohms for convenience, as it is here. The circles on the Smith chart represent lines of constant resistance; the radial arcs represent constant reactance; the horizontal line in the center is the real axis; the top region is inductive (positive reactance) and the bottom is capacitive (negative reactance); the left-most point is a short circuit or zero Ohms, and the right-most point is an open circuit.

The common mode is approximately that of a center-fed dipole of comparable size, and is plotted as Z_c in Figure 3.12. The splitting factor a provides an impedance scaling factor by $(1 + a)^2$. Finally, Z_d provides a shunt reactance. A short, shorted transmission line provides an inductive reactance, so following a line of constant susceptance, results in the final input impedance Z_{in}. Hopefully, if we designed the T-match correctly, then $Z_{in} = Z_{ic}^*$ (or whatever the desired criteria).

3.2.2 *The Modified T-Match*

If you look at a modern RFID tag, you will probably see something that looks like a T-match, but doesn't exactly fit the model of a T-match. In fact, only one of the antennas shown in Figure 3.1 uses a classic T-match. Two others use something we call the embedded T-match (the tag in the upper right corner of Figure 3.1 is an excellent example), where the T-match is inscribed into the dipole section, and behaves substantially like the classical T-match. The rest are some modification to the basic principle. Not surprisingly, we call these structures the modified T-match. In Figure 3.13, we show a

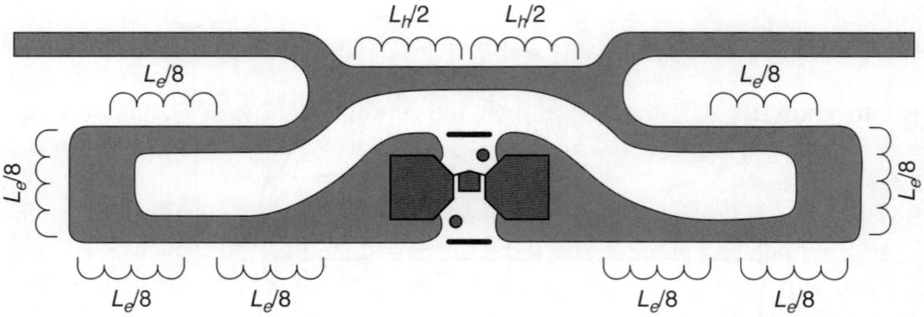

Figure 3.13 Intuitive inductor-based circuit model for the T-match. Both the series and shunt inductor are shown as distributed inductors.

commercial antenna's matching network, and interposed on the image is one way we propose that we think about the matching network. (The following is based roughly on the work of [7].) Imagine that each short trace acts like a small inductor [6]. Thus, we see distributed inductance all along that matching network. (Note that determining the actual inductance of straight wire segments in the presence of other conductors with currents causes mutual inductances, making exact calculations exceedingly difficult. This approach requires a little imagination, but it's not really that much of a stretch.) Again, imagine the inductances are distributed along that loop. For reasons that will become clear in a moment, we'll consider two parts of the loop: the "series" section, which we will call L_e, and the "shunt" portion, which we will call L_h.

We must emphasize that this model does *not* consider the coupling between components. When metal strips are in close proximity, as is shown in Figure 3.13, there is substantial coupling between the conductors, which is difficult to predict accurately without detailed computer models. However, as we will see later, while you may not be able to easily directly map the physical structure to literal inductor values, the qualitative circuit can be quite useful in guiding antenna design. So one can use a qualitative process and either experimentation or computer simulation to guide the design process. One alternative is to simplify the circuit and structure the matching circuit into simpler and largely uncoupled segments, which is not necessarily space-efficient, and therefore rarely used in practice.

Until now, we have neglected to describe the RFID IC's equivalent circuit. The circuit model that people use is determined by its purpose. We should mention that the RFID IC impedance is non-liner, that is, the impedance varies (sometimes considerably) with the input power. It is common to characterize the impedance of the IC at the smallest power setting, that is, the *turn-on power*, which will cause the IC to turn on and operate, that is, the *turn-on impedance*. It is reasonable to match the antenna at the IC turn-on impedance because that's where the IC needs the most efficient transfer of power; when there is excess power collected by the antenna, we can afford to lose efficiency in the power transfer. In fact, if there is too much power transferred to the IC, the sensitive circuits can be damaged, and so IC-makers build special overload circuits. IC design is outside the scope of this chapter, but it is important for us to understand how to model the IC impedance.

Sometimes, you may see the IC impedance specified as a complex number, that is, 20 $-j167$ Ohms. This number should be specified at a particular frequency, that is, 915 MHz, because to complicate matters, the IC impedance also changes with frequency. If you're only interested in designing a tag for a particular frequency or small range of frequencies, this might be fine, but if you're trying to design a tag for world-wide operation, then it's important to know the IC impedance over the world-wide band. Fortunately, most ICs can be modeled as a parallel RC circuit. This is not a perfect model, but it's accurate enough for most applications.

Putting together the dipole, which is approximately a series RLC circuit, the T-match, which is approximately a shunt-series inductor circuit, and the IC, which is approximately a parallel RC circuit, we end up with the circuit shown in Figure 3.14. Probably the worst assumption in this circuit is that the resistance of the dipole is constant. We know, for example, from [3] that the resistance is inversely proportional to the square of the frequency. So we know it's not perfect, but it is a good approximation, and it allows us to translate something as tricky as currents and fields into something as simple as lumped element circuits.

However, there is one more "trick" that we can use that will provide a lot more insight. Let's introduce a parameter $\beta = L_h/(L_e + L_h)$. Then it can be shown that the circuit shown in Figure 3.14 is identical to the figure shown in Figure 3.15. Here, $L_{se} = L_e L_h/(L_e + L_h)$, that is, the parallel combination of L_e and L_h. The proof [7] is not too difficult and can be performed by someone with a modest background in circuit theory.

There are a number of important results of the circuit in Figure 3.15. First, we can see how the ratio β can effectively scale the antenna (or IC) impedance. Also, note that β acts much like α in the T-match. In fact, one can draw an equivalence: $\beta^2 = 1/(1+\alpha)^2$. The classic T-match includes the inductors in the common mode impedance Z_c, while the distributed inductor model tries to separate the matching circuit from the antenna. This separation may be OK if the inductors are relatively small, but if they become too large, they become a part of the antenna and need to be explicitly included.

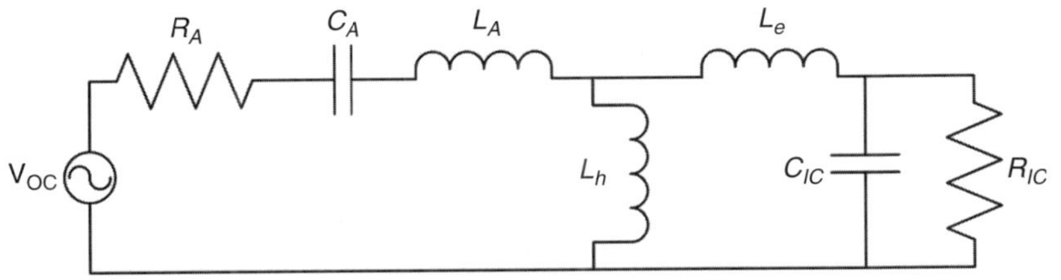

Figure 3.14 New inductor-based circuit model of the T-match dipole.

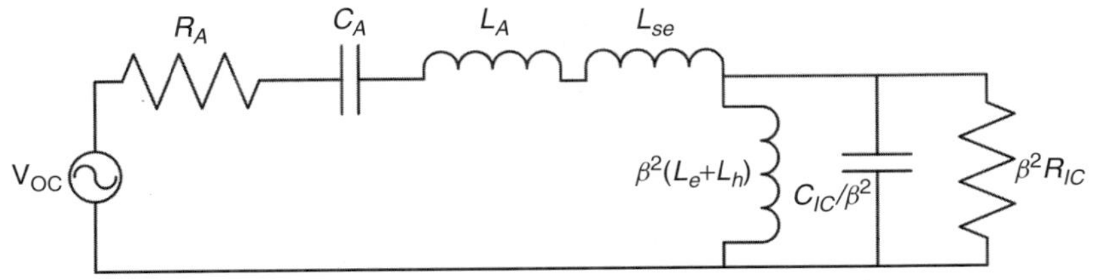

Figure 3.15 Transformed circuit model of RFID tag.

Second, an electrical engineer familiar with filters can spot this circuit and may recognize that the circuit in Figure 3.15 looks like a band pass filter, and a band pass filter can be used to extend the bandwidth of a circuit. Briefly, it works like this. Let the series and parallel RLC circuits have the same resonant frequency. At resonance, the series RLC circuit has a reactance with a positive slope. At resonance, the parallel RLC circuit has a reactance with a negative slope. Then, over some range of frequencies near resonance, the reactance will nicely cancel. If you're lucky (or clever, or diligent) enough so that the slopes are just right, you can actually increase the bandwidth out of your system by about a factor of three. For several practical examples, we've been able to achieve this three-fold increase in bandwidth, which is significant.

Unfortunately, this technique relies on the Q of the IC to be what you want it to be. Using an equal ripple (Tchebychev) filter, the Q of the antenna and IC should be identical. As we saw before, the Q of the antenna is largely determined by the size of the antenna, and the Q of the IC is usually set by the chip-maker. Fortunately, many antennas and chips have a Q around 10 to 15, which means that this technique can be used. For really small antennas where the antenna Q is very large, like 100, the bandwidth will be entirely dominated by the dipole Q and this technique won't offer any benefit.

Also unfortunate is that this technique requires both the series and parallel RLC circuits to resonate at the *same frequency*, which is also the (geometric) center frequency of the frequency range of interest. Recall from earlier, most commercial RFID antennas are not resonant antennas. In fact, it requires quite a bit of meandering to make an antenna resonate at around 900 MHz. (Look at the antennas in Figure 3.1 and guess which ones are resonant and which ones are not. We think only two of them are actually resonant antennas.)

3.3 Putting it Together: Building an RFID Tag

Next, let's walk through the process of designing an actual RFID tag using the lessons we've learned so far. We are using Ansoft Designer™ to do most of the "heavy lifting" for us, although Designer is by no means the only tool available for doing this kind of work. In general, any good method of moments (MoM) tool would be suitable.

Let's assume we're given the task of designing an RFID antenna to work on RF-friendly material (we assume air). The antenna must be cheap and therefore as small as possible. Initially, the antenna will be etched from copper but in the long term will be printed with silver, so it must be a thin-wire dipole. After consulting with a contract manufacturer, you learn that the manufacturer can reduce costs by US$0.003 if the maximum antenna length is 86 mm, and the antenna can be as wide as 12 mm. Any wider, and the antenna will cost a fraction of a penny more. The antenna should be designed to be used with a new, super-low-cost RFID chip that has a parallel resistance of 2200 Ohms and a parallel capacitance of 1.15 PF. The tag must give as much performance as possible in the European ETSI band and the FCC band in North America, that is, 865 to 930 MHz, and we can expect minimal detuning from the environment.

An 86 mm design is going to have an impact on the antenna Q, but we can make up for some of it with a 12 mm width. The line width may be a concern. 1 mm lines might be too costly, so we will try 0.5 mm lines and see if we can meet the objectives with that, and if not, try 0.75 mm lines.

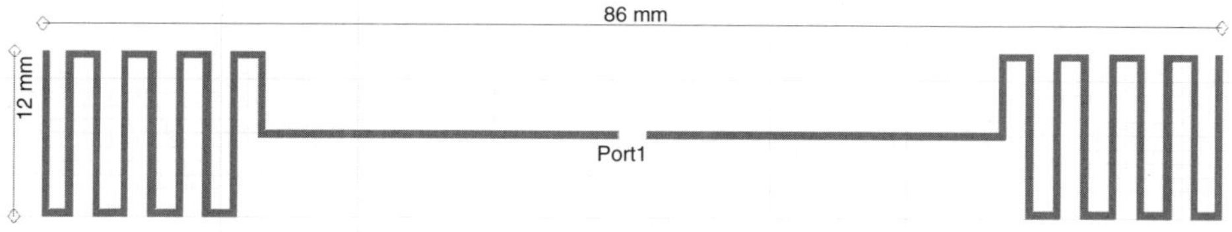

Figure 3.16 Meandering dipole used in the first cut.

First, we do a first rough pass at the antenna and see what kind of Q we can get out of it. The geometric mean of 865 and 930 MHz is 897 MHz, so we'll target that as the resonant frequency. We'll target an antenna with inductively loaded ends to get the smallest Q with the least amount of metal. Figure 3.16 shows a rough first cut. We adjusted the number of meanders so that it will resonate at about 897 MHz. It doesn't have to be exact; we'll tune it later. All we need now is a good estimate. We use the Equation 3.10 as a simple, reasonable estimate Q, which works well for a series or parallel resonant RLC circuit with resistance R:

$$Q = \frac{\omega}{2R}\frac{dX(\omega)}{d\omega}, \tag{3.10}$$

Estimating the Q, we see that it's about 17, which is probably good enough. We don't actually have to cover the *entire* band from 865 to 930 MHz; the region between 868 through 900 MHz is unused, and we may be able to use that to our advantage. We also notice that the resistance is about 35 Ohms, which is good for an 86 mm antenna; the extra width allows us to make longer and therefore fewer meandering segments, increasing α.

From this, can estimate a circuit shown in Figure 3.17. We know that that $(L_e + L_h)$ must resonate with the IC capacitance, so $(L_e + L_h) = 27.3$ nH. What remains is choosing β, that is, the ratio of L_e to L_h. One very useful rule of thumb is as shown in Equation 3.11:

$$\beta = \sqrt{2\frac{R_a}{R_c}}. \tag{3.11}$$

This estimate works well for choosing a 10 dB return loss (90% efficiency) over a broad band. So for this case, $\beta = \sqrt{2 \cdot 35/2200} = 0.1784$ is a good estimate. We decided to write a little computer program to vary β and see what happened. We found that we could

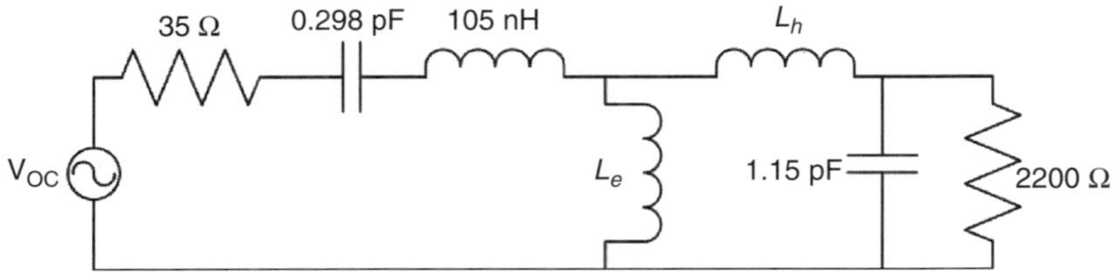

Figure 3.17 Circuit model of proposed antenna.

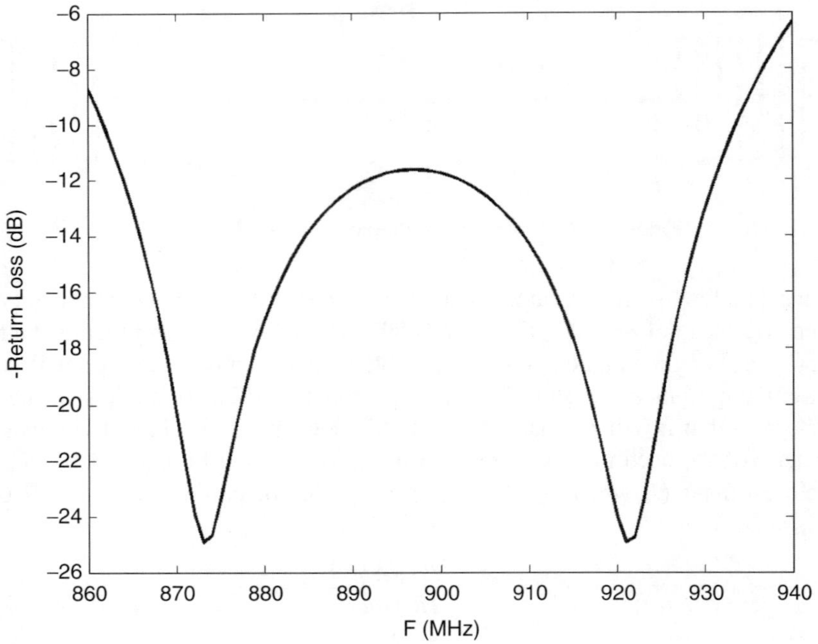

Figure 3.18 Return loss for circuit with $f_0 = 897$ MHz, $\beta = 0.165$.

actually give up some bandwidth to get a better return loss when we used $\beta = 0.165$. The resulting return loss is shown in Figure 3.18.

We see that we have a very nice result in which the return loss stays under 12 dB over the entire range from 865 to 930 MHz. However, we can note that the performance is best at about 875 MHz (with a very nice 99.5% efficiency), not at 866 MHz where we would like it to be, and again at 925 MHz, not 915 MHz. So let's try changing the resonant frequency. Recall that we set the center frequency to be 897 MHz. Next, we tried reducing the resonant frequency to 890 MHz, and then increased β a little bit so that $\beta = 0.17$. The resulting return loss is shown in Figure 3.19. The overall worst-case performance over band has little difference. We see that instead of 12 dB return loss at 930 MHz, it is only 10 dB, or still at 90% efficiency. However, the antenna operates at well over 99% efficiency at 865–868 MHz, and again has better than 99% efficiency at around 917 MHz and is better centered within the operating band. What we basically have done is chosen the geometric mean of the center of the two bands, 866 and 915 MHz. Then we adjusted β until we achieved a minimum return loss at those two frequencies (Figure 3.19).

Now that we have a return loss curve, the next step is to go back to the antenna and fit it with a T-match that will (hopefully) produce an antenna with the same return loss. The way we do this will require a few iterations. We know, for example, we can control the resonant frequency of the parallel RLC circuit by the size of the loop formed by the T-match, and we can control β by the ratio of L_e to L_h. The resonant frequency of the dipole antenna can be modified by the length of one of the meandering elements. So, armed with that intuition, we can start to construct a T-match to our desire. Because β is

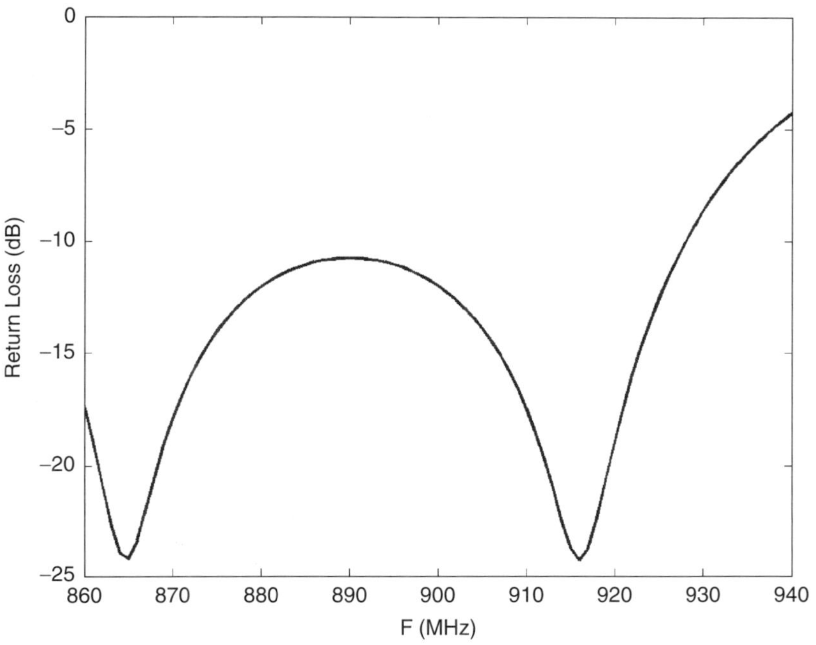

Figure 3.19 Return loss for circuit with $f_0 = 890$ MHz, $\beta = 0.17$.

small, we know that $L_h \approx 6L_e$, so that gives a good starting point on the proportions. For convenience, we plotted the impedance looking into the circuit-equivalent of the antenna from the IC in Figure 3.20. This is the impedance behavior that we want the simulated physical antenna results to duplicate.

It took us about 10–12 iterations, for maybe 20 minutes of work to get it just right, and we finally arrived at the modified T-match shown in Figure 3.21. The tag shown in its entirety is shown in Figure 3.22. As you can see, the design looks like a conventional RFID antenna design, except for the rather blocky design. (It's at this point in the design process where management will probably introduce a new requirement regarding aesthetics.) Next, let's look at the return loss of the simulated antenna versus the circuit model, which is plotted in Figure 3.23. We see that we are able to achieve remarkably close agreement between the simulated result and the circuit-based result. So the circuit is a very good guide for determining what is achievable. The reason for using a circuit is because it is much easier and faster to write a program or script to manipulate the circuit to achieve a desired level of performance than it is to manually change the physical layout of the antenna and run a full-wave simulation model over a range of frequencies. It also provides a great deal of insight and intuition as to the frequency response of tags.

We see that there is pretty good agreement, though it isn't perfect. Again, several of the assumptions that went into the circuit model, especially the assumption that the radiating resistance is constant over frequencies, are not accurate assumptions. But we can conclude that the circuit model was able to guide us to a solution that was able to meet the predicted performance remarkably well and with relatively little effort given the assumptions.

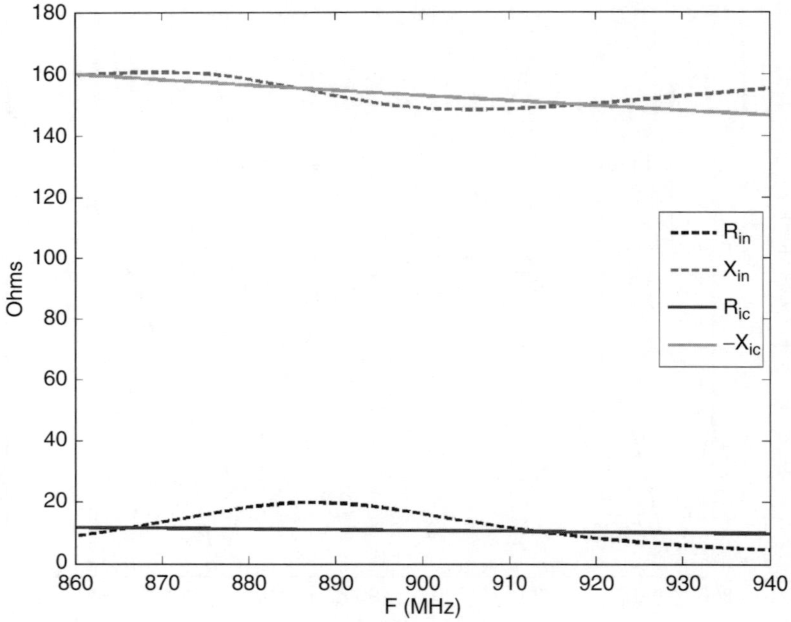

Figure 3.20 Circuit predicted impedance looking into antenna.

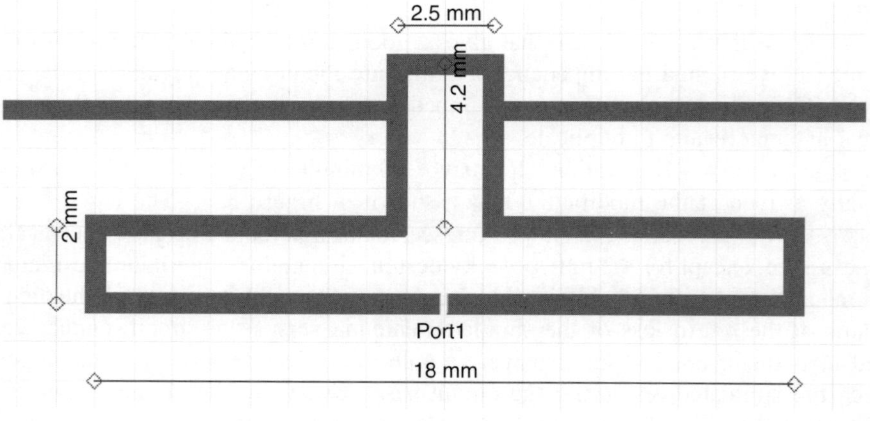

Figure 3.21 Geometry of matching circuit.

Next, we plotted the three-dimensional radiation pattern in Figure 3.24 using a dB scale. The long dimension of the antenna is oriented along the X axis. As we can see, the antenna exhibits the "normal" dipole radiation pattern where radiation is poor along the X axis and is at a maximum in the Y-Z plane.

So are we done with the tag design? Only if we're lucky! In reality, there are numerous ways in which reality can deviate from our nice circuit and simulation model. For one, the

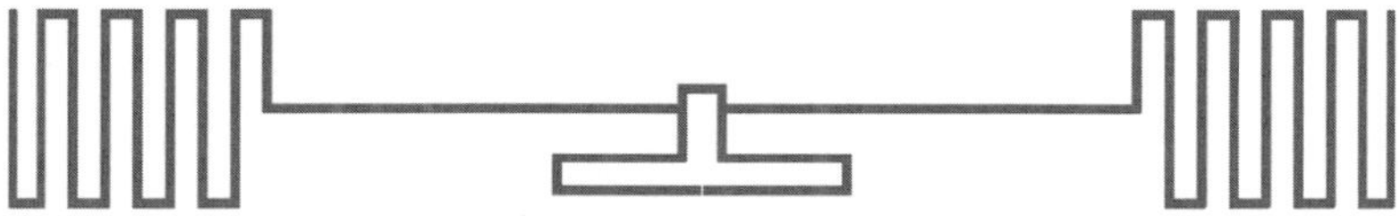

Figure 3.22 Geometry of the completed antenna.

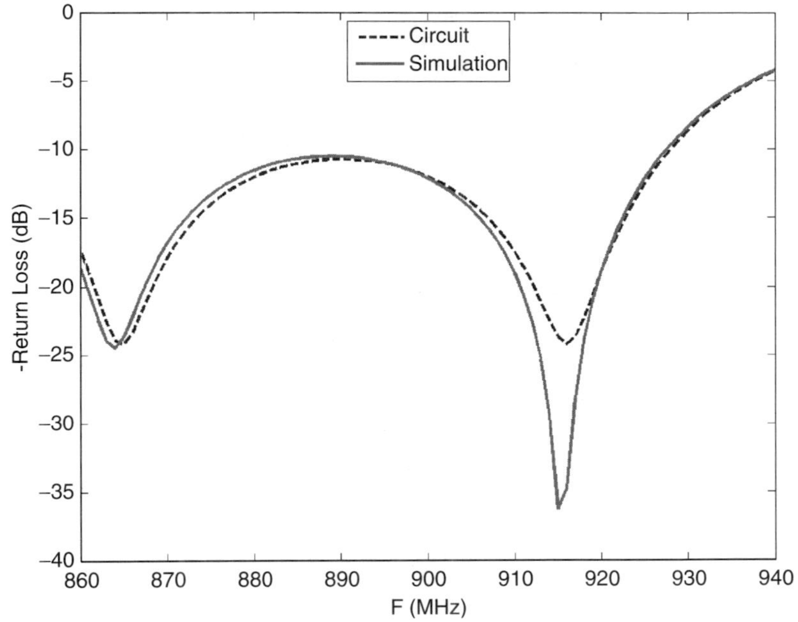

Figure 3.23 Return loss of circuit and simulated antenna.

data sheet of our fictitious IC may be accurate under assumptions that don't hold for us. For example, the method of attaching the IC to the antenna will introduce some parasitic reactance. What if we're using a different adhesive, or a different pressure setting when attaching the chip? Does the data sheet IC impedance assume a particular attachment? In reality, the chip attachment process introduces some "parasitics" that affect the impedance. Some data sheets give the probed IC impedance (with no attachment reactance), and you have to add about 0.1 to 0.15 pF of shunt capacitance to get values that you want to use. Add 0.15 pF of capacitance to the IC that we designed the antenna for and you'll see that our carefully designed antenna won't work nearly as well. One IC data-sheet that we had indicated the IC reactance should be about $j194$ Ohms, but we found through experimentation that after all the parasitics, we needed to match to $j127$ Ohms. That's one rare and rather extreme example, but a real one. Also, the impedance of these chips comes from analog RF front ends to the IC, and there are natural, random variations within the manufacturing process. Some chips will have different impedances than others. You may need to produce dozens of permutations and test dozens of tags with each permutation to get the design that you want that works well on average. Is this for a strap? Different

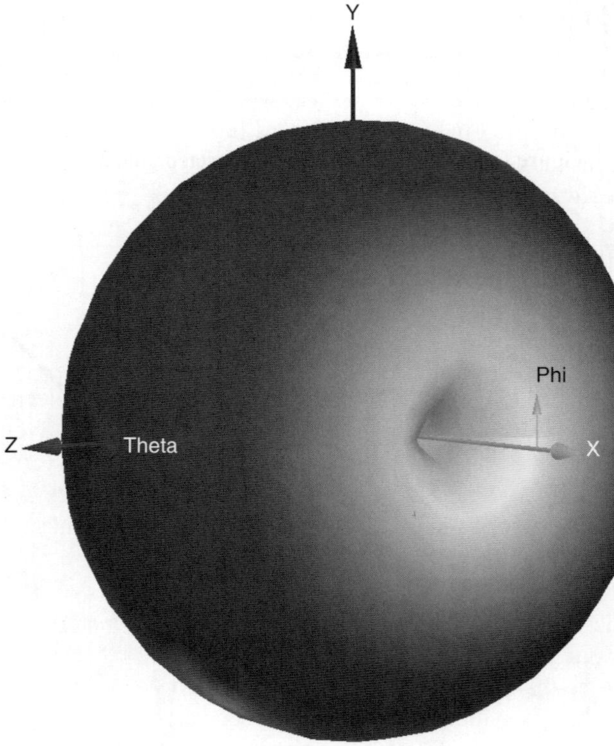

Figure 3.24 Radiation pattern of designed antenna.

straps introduce different parasitics. All this work was done putting blind faith into our simulation results to produce accurate results; did we configure the software properly? So don't be fooled into thinking that you can design to the data sheet and everything will magically work as expected. There's still a lot of hard work to be done before you can move from a prototype to production. There's no substitute for experience and good support from the chip vendor.

The aim of this chapter isn't to turn you into an expert in building RFID antennas. The aim is to introduce simple and practical ways of thinking about antennas that will give you insights into how the antennas work. We wanted to give enough information to really understand the mechanics in terms of simple circuit elements, and to give insights into the fundamentals of what can and cannot be achieved. The "trick" of using the tag and matching circuit as a band pass filter is still not well known in the industry, and if you look back to Figure 3.1, you can see some of the antennas on that page do not have enough meanders to be resonant dipoles, and thus cannot take advantage of the band pass filter. That might be fine for antennas that are used only in one region, but makes it hard to believe claims of "world wide operation" – unless the vendor's definition of *operation* is fairly loose.

3.4 The Environment

In the previous section, we pretended to be antenna designers for a little while and built an antenna that would work in these very nice, well-defined ways. What happens when we place the tag on a plastic bin? On a cardboard box? A bottle of water? Near metal? Unfortunately, the answers to those are complicated, and usually the results are that the tag degrades its performance one way or another. In this section, we'll examine the mechanisms of how the environment impacts the tag performance, and steps that can be done to mitigate that impact.

3.4.1 Dielectric Constant

To begin with, we can classify materials into three categories: dielectrics, conductors, and magnetic materials. Magnetic materials such as ferrites are relatively rare, so we'll ignore those. Dielectrics are basically electrical insulators; they have very large resistance to electrical current when a voltage is placed across them. Conductors, on the other hand, conduct electrical current very readily and offer very little resistance. At 900 MHz, no material is a perfect conductor or a perfect dielectric.

A dielectric material basically does one thing under the presence of an electric field: it resists the electric field. Under an electric field, a dielectric material produces its own electric field that opposes the applied electric field. Every material is made up of molecules and atoms. Two atoms that are bound in a molecule often do so because one atom is an electron-donor and the other atom is an electron-receiver, and the resulting molecule is slightly polar. Under an electric field, the polar molecule tries to align itself with that field, but the polar molecule produces its own, counteracting electric field. This alignment of polar molecules is the way that the material stores energy. Atoms themselves consist of a positively charged nucleus and negatively charged electrons in a surrounding electron cloud. Normally, the nucleus is in the center of the electron cloud, but when an external electric field is applied, the cloud can shift slightly to one direction and the nucleus the other direction to produce a slight polarization and an opposing electric field.

The amount of counteraction to the electric field is called the *dielectric constant*, or *permittivity*, and is represented by the symbol ε. Because it has units, we often talk about the *relative* dielectric constant, ε_r, which is relative to the dielectric constant of a vacuum, ε_0, that is, $\varepsilon = \varepsilon_r \varepsilon_0$. This relative dielectric constant has an important implication to the propagation of electromagnetic fields. We won't go into the details to derive this here, but we will simply state that the velocity of a propagating wave through a dielectric material is slowed down. Let c be the speed of light in a vacuum. Then the velocity of a propagating wave through a dielectric medium with permittivity ε_r is

$$v = \frac{c}{\sqrt{\varepsilon_r}}. \tag{3.12}$$

What does any of this have to do with RFID antennas? Remember that the antenna resonates when the electric field and magnetic fields essentially cancel. What happens if the electrical energy stored in the antenna is increased? The resonant frequency of the antenna decreases. Or, another way to think about it, if a half-wave dipole (where half a wavelength is measured in air) is completely embedded a dielectric material, the wavelength is reduced, so the antenna, relative to the wavelength in the dielectric, becomes larger, and it becomes larger by a factor of $\sqrt{\varepsilon_r}$. The new resonant frequency is decreased by a factor of $1/\sqrt{\varepsilon_r}$. If we know an antenna is going to be completely embedded in a dielectric material, we need to scale all the dimensions by a factor of $1/\sqrt{\varepsilon_r}$.

What frequently happens is that tags are placed on some object. If the object is relatively thick (comparable to the size of the antenna), then the tag is surrounded in part by the object (dielectric) and in part by air. See Figure 3.25. We can assume air has a relative dielectric constant of 1. In this case, we can estimate an *effective* dielectric constant. The effective dielectric constant is the dielectric constant of a (fictitious) homogeneous medium that would produce nearly-identical behavior to the nonhomogeneous environment. In this case, we can simply estimate

$$\varepsilon_{eff} \approx \frac{\varepsilon_r + 1}{2}. \tag{3.13}$$

This is only an estimate, which breaks down when ε_r becomes large, but it's a good estimate when ε_r is small, which is the case in a lot of instances. Then we can substitute ε_{eff} into the previous equations and estimate the change in resonant frequency, for example. Note: if the dielectric is thin, then you should *not* assume equal weighting; the actual weighting is really quite complex and involves complex mathematical equations, and all but the simplest structures require complex numerical calculations that are best done by computers.

A simple question to ask is then: what kind of materials gives what kind of dielectric constants? That's actually a hard thing to catalog because materials commonly found in the environment of RFID tags are rarely in their pure form. Cardboard is mostly air, except that it readily takes on water, depending on the humidity of the air, and can have a dielectric constant between 1.2 and 1.4. Plastics like ABS are a mixture of three (or more) materials in varying proportions, but 2.8 is typical. Polyethylene is about 2.2 or 2.3, polypropylene about 2.3, polystyrene between 2.4 or 2.5, and polycarbonate about 2.8. Nylons tend to be around 3.5, but can also vary considerably because nylons readily take on water from the environment. Glass for windshields and wine bottles are a challenge because glass can have a permittivity of 6 or larger. Water, we should mention, is about 80, and thus is one of the most challenging dielectric materials for RFID. Motor oil can vary considerably, but may be around 20 to 30. However, white petroleum jelly is about 2.1.

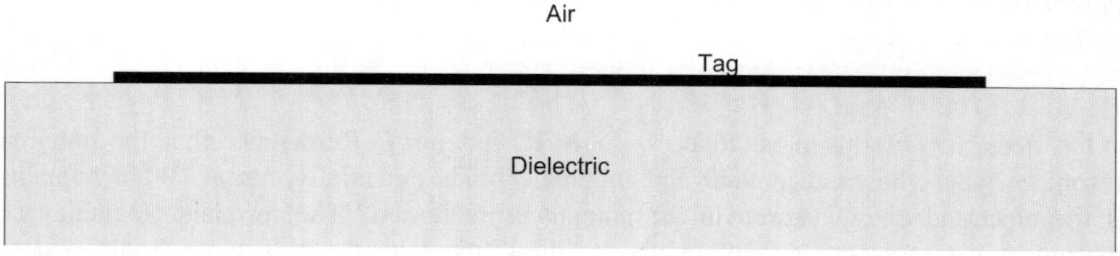

Figure 3.25 Tag on semi-infinite dielectric. This is a common model for a tag placed on an item.

3.4.2 Dielectric Loss

The other thing about dielectric materials is that in the presence of time-varying electric fields, the material may not align its electric field exactly at the same time as the electric field is being applied. You can imagine, a water molecule may not be able to spin around 915,000,000 times a second and keep up. In fact, it doesn't; its rotation lags the electric field a little bit. That lag introduces a complex component to the dielectric constant. The ratio of the imaginary to real is called the loss tangent, and is usually written $\tan\delta$. The larger that number, the more the material converts energy in the form of the electric field into heat, and it doesn't have to be very large at to start causing problems.

For RFID tags, this is not a good thing. We spent a lot of time looking at how an antenna can efficiently deliver power to the chip, but if there's a lossy dielectric material near the antenna, the antenna is going to lose some of its power to the dielectric. Just how bad is it? Again, a lot depends on the details. Rarely is a tag completely immersed in a dielectric. So we again use a theoretical homogeneous material with an effective loss tangent, determined in roughly the same way as we would estimate the effective dielectric constant.

Let $Q_d = 1/\tan\delta_{eff}$ be the quality factor of the effective dielectric homogeneous medium. Assume our antenna has a quality factor of Q_a. Then following the definition given in [5], we can find the efficiency of the antenna as shown in Equation 3.14:

$$\eta = \frac{Q_a Q_d}{Q_a + Q_d} \tag{3.14}$$

So it turns out that one way to combat dielectric loss is to have an antenna with a small Q_a. Unfortunately, as we saw earlier, a small Q_a requires a physically large antenna, which usually means more cost. But otherwise, if you place an antenna on a lossy material, you will expect to see a reduction in antenna performance.

Materials that have a lot of dielectric loss tend to be very polar molecules. So arcylates and vinyls tend to be very lossy. One of the most lossy materials that we've come across is phenolic resin, which is sometimes used to make hard, plastic dishes that you might have noticed gets very warm in the microwave. Phenolic resins have loss tangents between 0.2 and 0.4. Water is fairly common and pretty bad, and at around 900 MHz has a loss tangent of 0.05, or a Q of 20, which is about the same as acrylic plastic. But if you look at this from a purely efficiency standpoint, the antenna efficiency may only be 50%. The majority of the challenges around water come from the dielectric constant and the resulting change in the antenna impedance, and the modification to the radiation pattern (the antenna tends to radiate more energy into the water), which is why acrylic is a much easier material to work with for RFID than water even though it has a similar loss tangent. Most other plastics are fairly low loss, with the exception of Nylon (when hydrated) and most polyurethanes, which can be close to 0.1. Acrylic, as we stated is about 0.05, polyester or PET is about 0.02, and polycarbonate is about 0.012. The polyethylene, polypropylene, and polystyrene plastics are very low loss with a loss tangent of .001 or less, as is PTFE (Tefon™).

Now, let's briefly consider the combined effects of the dielectric constant and loss tangent on the performance of an RFID antenna. The dielectric constant will change the resonant frequency of the antenna, which has a similar affect to operating the antenna at higher frequency, higher by a factor of $\sqrt{\varepsilon_{eff}}$. Making a dipole assumption, this normally means the resistance of the antenna increases a little bit and the reactance of the antenna increases a lot, by a factor of approximately $2\sqrt{\varepsilon_{eff}} R_a Q_a$. That will normally result in a considerable impedance mismatch, which is the normal source of lost performance. Again, we see that Q is fundamental to the performance of tags and their sensitivity to dielectrics. In addition, any loss in the material will result in the antenna resistance decreasing and additional loss in performance as described by [15]. Note, however, that decreasing the resistance from dielectric loss might actually be a positive benefit. If reducing the resistance results in better efficiency in delivering power to the load, the antenna may lose some radiating efficiency but gain more power transfer efficiency and have a net increase in performance.

3.4.3 *Metals*

Metals produce a completely different set of challenges for RFID. Metals can very profoundly and negatively impact RFID tag performance [8]. The best way to explain this is using image theory. You know, for example, when you look in a mirror you see a mirror image of yourself. Everything in the mirror image is identical but backward, that is, left is right and right is left. At RF frequencies, a large, flat metal surface acts as a mirror for RF signals. So if an antenna is some distance h above a ground plane, we can replace the ground plane with a mirror image of the antenna a distance $2h$ away (see Figure 3.26).

Recall from earlier, a small dipole segment with a unit of current radiates that energy into space. If there is another small dipole segment very close by with a current exactly opposite, the electromagnetic fields produced by the two will nearly cancel. The only portion that will survive is due to the small offset in phase produced by the distance $2h$. So, we can surmise that when the dipole antenna is placed near a metal ground plane, we will see the radiating resistance reduce by a factor of roughly $\sin^2(2\pi h/\lambda)$, which, for small h, reduces to $(2\pi h/\lambda)^2$. (Note that if $2h = \lambda/4$, you have the condition in which the image significantly adds to the radiation in the direction normal to the ground plane.) If the tag is placed right on metal, h becomes so small that the resistance plunges to nearly zero. So placing a tag right on metal is a good way to prevent the antenna from working at all.

The reactance is much more difficult to analyze rigorously, so we'll take a more qualitative approach. The inductance of the wire antenna is slightly reduced by the presence of the ground plane, but since most of the energy is stored in the magnetic field very close to the wire, the affects are mild until the antenna gets very close. The capacitance of the antenna changes quite a bit. The antenna now stores energy in the electric field between the antenna and the ground plane. Recall from basic circuit theory that a series

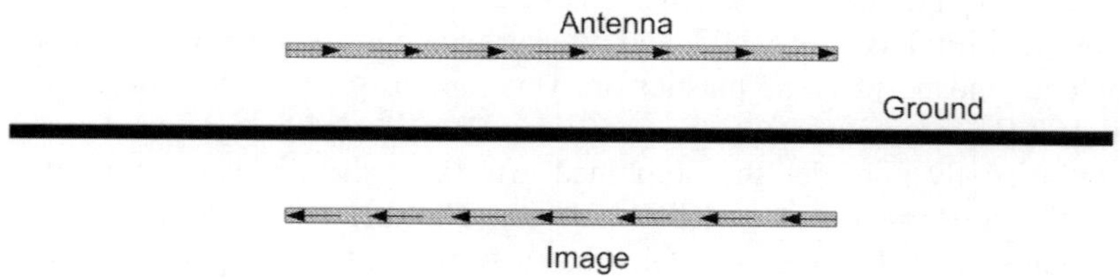

Figure 3.26 Tag operating above a ground plane, shown with image.

RLC circuit has the following resonant frequency.

$$f_0^2 = \frac{1}{2\pi LC} \tag{3.15}$$

With equal or a little less inductive energy and lot more capacitive energy, we expect the resonant frequency of the antenna to be related as follows. With C increasing and L the same or decreasing only slightly, we expect f_0 to decrease. That's exactly what happens. The resonant length of a dipole will decreases from about 0.46λ to about 0.38λ or less, for example; the actual value is dependent on h, the width of the dipole, and other factors. In Figure 3.27, we show the (simulated) input impedance of a dipole above a 300 mm^2 ground plane at varying heights on a Smith chart normalized to 50 Ohms. For heights less than 1/4 of a wavelength, we see both the input resistance and reactance decreasing with decreasing height. At about 6 mm, the effects of the ground plane affect the impedance in a significantly different way, causing both the resistance and reactance to move sharply to zero with decreased height. So a resonant dipole with impedance of about 70 Ohms in air, now with a 6 mm separation has an input impedance of 0.6 $-j20$ Ohms – a very significant change in impedance! This is why metal is considered one of the more challenging materials to work with for RFID.

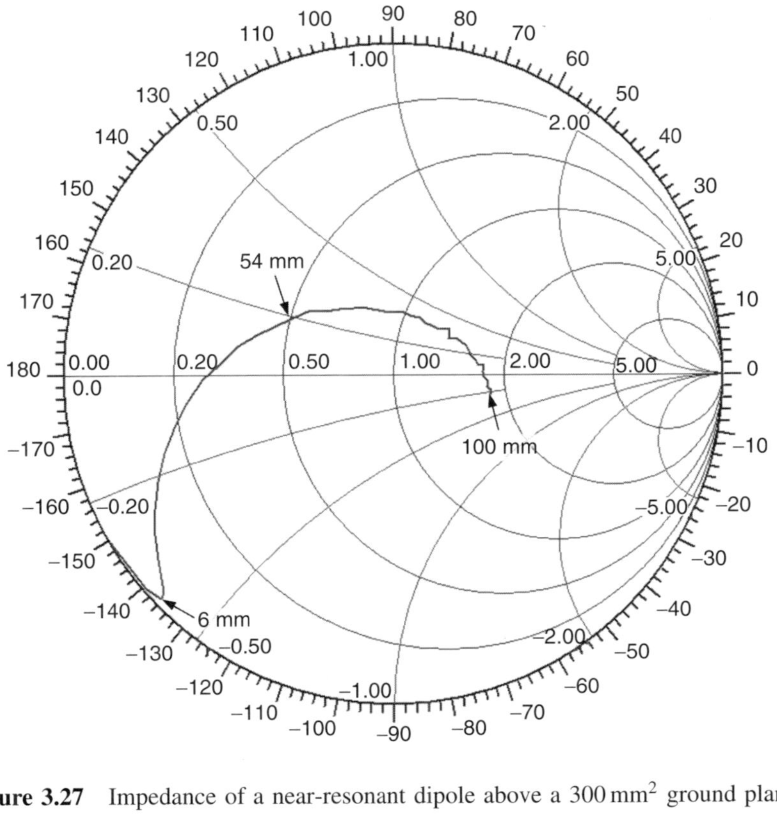

Figure 3.27 Impedance of a near-resonant dipole above a 300 mm^2 ground plane.

What can be done? Lots of things, but unfortunately, none of them are simple. In Section 3.4.5, we will investigate some alternatives. But before that, we will look at one more environmental issue affecting RFID tag performance.

3.4.4 Propagation

Until now, we've primarily focused on the impact that a nearby material has on the antenna. What about material interposed between the tag and the reader? Depending on the kind of material and size of material, it will have varying impact on the performance of the tag. For example, consider a tag that is on a pallet of cardboard boxes. The dielectric and metallic material inside the boxes can detune the material, but if the tag is surrounded by boxes, then those interposing boxes (and material inside the boxes) can pose additional obstacles to getting RF energy in and out of the pallet.

From the previous section, we can classify materials as dielectric or conductors. Let's consider conductors first. Conductors, naturally, readily conduct electricity. The phrase commonly used in the RFID industry is that "conductors reflect," and to a first approximation, that's correct. However, that's not quite the case. The term commonly used in the technical community is *scattering*. There are obvious differences in the connotations of the words *reflecting* and *scattering*, and those connotations are appropriate here. We usually think of mirrors as reflecting, that is, light bouncing off the surface of the mirror in a very predictable direction. Scattering is, well, bouncing off in a bunch of different directions at the same time, that is, all over the place. The description that is most apt depends on the size of the metallic object.

Recall that with electromagnetism, size is always relative to the wavelength. At 900 MHz, a wavelength is roughly a third of a meter (about 1 ft). If the metallic object is large relative to a third of a meter, like, say, a large metal wall, then the RF energy will reflect off the metal wall in much the same way that light reflects off a mirror. If the metallic object is small relative to a third of a meter, like, say, another RFID antenna, then the RF energy will scatter. In this example, a nearby tag will scatter RF energy in much the same way that it radiates RF energy, according to a radiation pattern (see Section 3.1.2). In fact, two RFID antennas can couple and become something like an antenna array with an entirely different radiation pattern, and is the subject of ongoing work within the research community.

So, if there is a large, intervening metallic wall, the wall will reflect RF energy in much the same way that a mirror reflects light. If there is a series of small metallic objects, some RF energy may be able to get through the space between metallic objects, and some of the RF energy may get through by scattering off of the metallic objects. Furthermore, other objects that are not intervening between the tag and reader may reflect or scatter RF energy. These multiple signals come together and may add (constructive interference) and serve to boost the signal, or cancel (destructive interference) and suppress the signal. In a complex environment, all of these factors are happening and the results can be very difficult to predict. This is one of several things that make RF and RFID behave somewhat randomly or magically to the initiated.

What about dielectrics? Recall that we can characterize dielectric material through two properties: the dielectric constant, and the loss tangent. Let us first consider the dielectric constant. Recall that one of the effects of the dielectric constant is that it causes the velocity of RF energy to slow down. This has two important consequences. First, any time there is an abrupt change in the dielectric constant in a medium, an electromagnetic wave incident on that transition undergoes both reflection and refraction, that is, some of the energy is reflected, and some it is transmitted through the medium but at a different angle. A set of equations called the Fresnel equations precisely describe what happens. Bottom line, at an air-dielectric boundary, the larger the dielectric constant of the material, the more energy is reflected. So, for example, trying to read through a stack of (full) milk cartons (dielectric constant about 80) will be very difficult because most of the RF energy will be reflected. But reading through paper or plastics can be relatively easy, with most of the energy being transmitted and little reflected.

Note that this is different from an antenna being placed very close to a dielectric material. One vendor has sort of a gimmicky trick for showing off their product. They place a tag within a fish tank (complete with water and fish) and a reader antenna very close to the tank. Most of the electromagnetic wave actually forms inside the fish tank, and thus does not experience the air–water dielectric boundary. If the vendor were to move the antenna back by half a meter or more, they would have a much more difficult time getting energy into (and out of) the fish tank.

Second, the loss tangent of a material causes RF energy to propagate poorly through the material. The loss tangent is a measure of how much energy is lost to the material per wavelength. Since water is common (and particularly problematic), consider a wave that is propagating through pure water with a dielectric constant of 80 and a loss tangent of 0.05. For a relatively small loss tangent, we can approximate the loss (in dB) after propagating a distance x as shown in Equation 3.16.

$$A_{dB}(x) = x8.686k_0\sqrt{\varepsilon_r}\ \tan\delta \tag{3.16}$$

Here, k_0 is the free space wave number, or about $19.16\,\text{m}^{-1}$ at 915 MHz. For water, this works out to 74.4 dB per meter. We can forget about using long-range RFID under water! 10 cm yields only 7.4 dB of loss, which is significant, but still possible to overcome. However, keep in mind that this component is only the propagation loss.

What does this mean? As an example, putting RFID tags on one liter milk cartons is probably not a good idea. The dielectric constant can change the antenna impedance significantly, creating power transfer inefficiencies. The dielectric loss of milk will result in a less efficient antenna. Trying to read through milk cartons will introduce a sizeable reflection loss at the air–milk boundary, and if milk has a larger loss tangent than water (which is very likely), then there will be significant propagation losses. Bottles of wine offer a similar challenge, but since wine bottles are generally cylindrical, there are possibilities of waveguide structures formed by the bottles, especially if foils are used in the bottle labels. A detailed analysis of tagging wine bottles and similar challenging RF product is the subject of ongoing research and the exchange at several research conferences.

3.4.5 Practical Steps to Overcome Environmental Challenges

In this section, we focus on overcoming some of the environmental challenges of RFID tags. Note that we do *not* address propagation challenges here; we focus only on getting the antenna to work well when placed on a challenging item.

The simplest way to overcome the environmental challenges is to physically separate the tag from the offending material. The further away the tag, the less the impact the material has on the tag. Again, *far* is defined in terms of a wavelength, or in this case, in terms of the length of the antenna. For metals and liquids, the separation has to be significant or else the tag will suffer from significant performance degradation. For example, for a 3.2 mm foam spacer from a metal plate, we observe that the typical tag degrades by about 16 dB. A 4.8 mm spacer degrades performance by about 13 dB, and a 6.4 mm spacer by about 11 dB. Tags vary by a few dB, but nearly all tags degrade pretty much the same. The exception to the rule are the exceptionally large, 90 mm^2 tags that perform significantly better, though they still degrade.

There are several solutions for solving the metal problem. The easiest thing to try is to provide some spacing from the metal object, usually using some thick foam backing. These are inexpensive, but tend to work poorly. When those don't work, people usually use a class of tags called "asset tags" or "metal tags" that are usually based on some kind of microstrip antenna, for example, [9]. These usually include a metal ground plane, a hardened case, and tend to be relatively expensive. On the positive side, they work well, sometimes better on metal than in air. On the negative side, they can be considerably more expensive. A second approach that has recently been developed is a hybrid approach: a special dipole antenna that can also work as a microstrip antenna. Below, we will go into more detail.

3.4.5.1 Separation

The simplest approach to dealing with metal is to separate the antenna from metal. Recall the results of dipole spacing from metal in Figure 3.27. A dipole near metal will experience very significant changes in input impedance, resulting in a substantial decrease in power transfer efficiency.

The larger the separation, the better the performance, though for complex reasons the relationship may not be monotonic for larger separation distances. Usually, people use a thin foam spacer made of polyethylene (PE) or polyurethane (PU). PE is popular because it tends to be the lowest cost, has a very low dielectric constant, and excellent resistance to water. PU is sometimes used because it can provide a nice, open cell "sponge" that can be easily compressed, which simplifies manufacturing, but unfortunately, sponges are also good at absorbing water. Typical thicknesses are 4.8 mm (3/16 inches), though 3.2 and 6.4 mm are sometimes used.

Suppose a tag can be read in free space at a distance of 32 feet, and with a 4.8 mm spacer, the tag suffers 18 dB of performance degradation, that is, it operates at 1.6% efficiency. The read distance will be reduced by a factor of 8, meaning the tag will be readable to about 4 feet. For a number of applications, this read distance is sufficient. So sometimes, nothing more need be done.

3.4.5.2 Microstrip Antennas

A microstrip antenna, see Figure 3.28, is the class of planar antennas that are designed to operate above a ground plane. Figure 3.28 illustrates a typical microstrip antenna with the associated fringing electric fields. The fringing fields are primarily where the radiation comes from. Energy is stored in the area between the antenna and the ground plane,

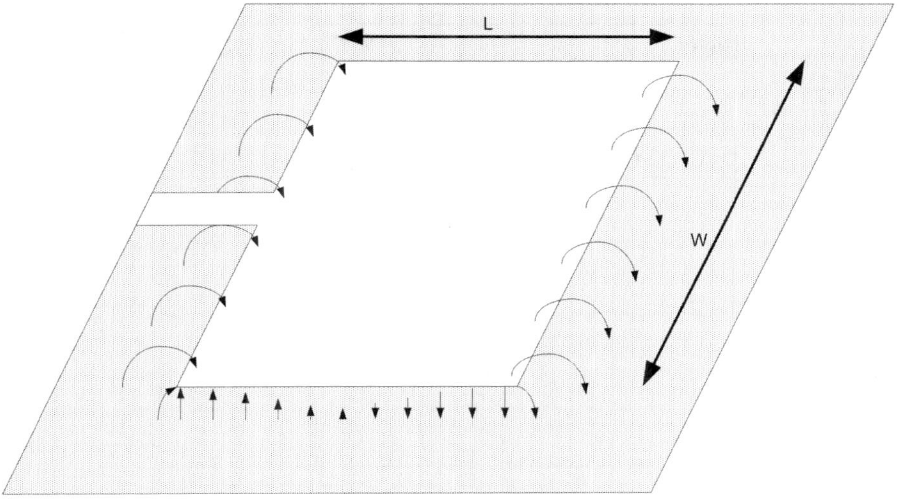

Figure 3.28 Microstrip antenna fed by a microstrip transmission line, shown with fringing electric fields.

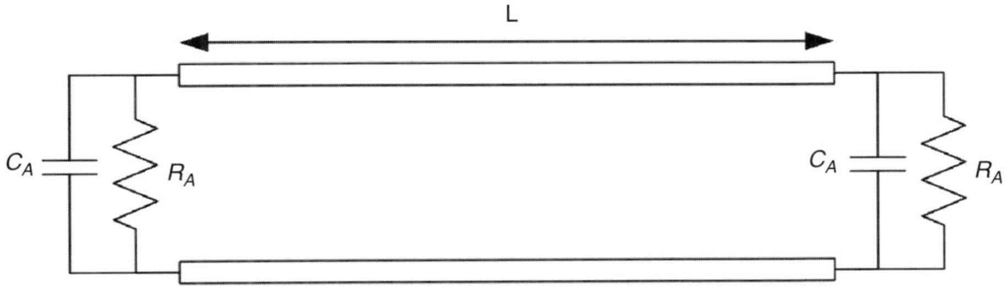

Figure 3.29 Transmission line model of a microstrip antenna.

similar to a parallel plate capacitor. Small, thin antennas will thus have a very large Q, while thick, wide antennas will have a smaller Q.

One especially simple but effective way to think of the microstrip antenna is to view it as a pair of RC circuit elements connected by a transmission line, illustrated in Figure 3.29. The resistances at the edges represent the (shunt) radiating resistance of the antenna, and the capacitors represent the energy stored in the fringing fields illustrated in Figure 3.28. Note that the resistance here is a shunt resistance in parallel with the capacitance, not the series resistance as in a dipole. Thus, the resistance represented in Figure 3.29 tends to be quite large, as opposed to small in the dipole case. We'll draw equivalence between the two views below.

The two circuit elements are connected by a microstrip transmission line of length L and has some characteristic impedance determined by the width, height, and dielectric constant of the substrate. Consider viewing the impedance of the antenna at the left radiating edge. The antenna will resonate when the impedance at the right edge, transformed by the transmission line of length L, presents the conjugate impedance to the RC element on

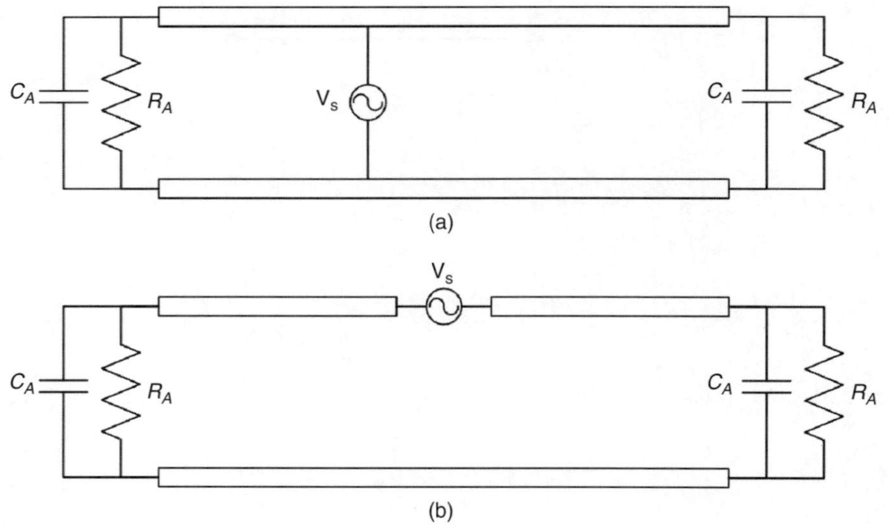

Figure 3.30 "Traditional" unbalanced way to feed the microstrip antenna (a) and "dipole" balanced way to feed the microstrip antenna (b).

the left. This happens when L is slightly less than half a wavelength. Typical resistance values may be 100 Ohms for a very wide, thick antenna, to 30,000 Ohms for a narrow, thin antenna.

Note that the circuit model of the microstrip antenna implicitly assumes that the way we are feeding the antenna is with respect to ground, as illustrated in Figure 3.30 top, hence a conceptual bias towards viewing the antenna as an unbalanced device, that is, a signal feed with reference to ground. If we split the transmission line exactly in half and feed one side with respect to the other, as illustrated in Figure 3.30 bottom, then we are back to something that resembles a dipole. In that case, the rather large edge resistance is transformed by a (roughly) quarter wave transmission line, which will result in a small series resistance. In fact, the parallel RC circuit will be transformed to a series RC circuit. The circuit equivalent of the frequency-dependent behavior of a resonant antenna shown in Figure 3.30(a) will be that of a *parallel* RLC circuit, while same antenna fed as shown in Figure 3.30(b) will behave as a *series* RLC circuit, or qualitatively the same impedance behavior as a free-space dipole. The only difference is that the series resistance can be small, for example, less than an Ohm.

The traditional challenge with microstrip antennas is that microstrip antennas are generally thought of as unbalanced devices, that is, you feed them with a signal (in Figure 3.30, a microstrip transmission line), which is referenced to the ground plane. The question has been: how do you attach the chip to the antenna? One simple way is to attach one terminal of the IC to the signal source and the second to the ground through a "via" or through-hole connection. Drilling a hole or wrapping a conductor is an extra step that can be cumbersome and adds to the cost of the tag.

It may be surprising (or not) that the T-match structure, or the modified T-match, can be used to transform the antenna/IC impedance to something that is usable. Recall in

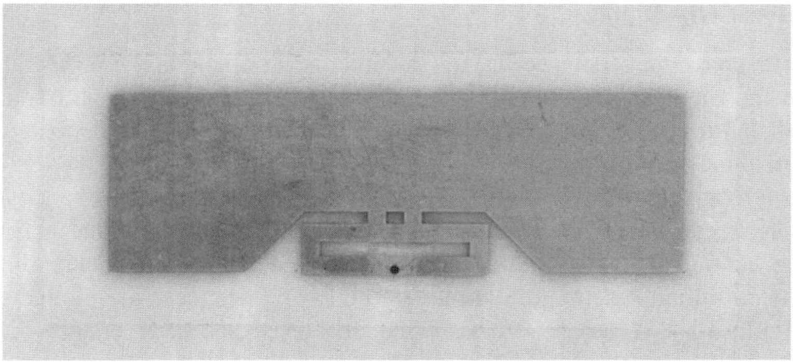

Figure 3.31 Narrowband microstrip antenna utilizing a modification to the modified T-match.

the modified T-match, the term β is used to transform the IC resistance to the dipole resistance, reducing the IC resistance by a factor of β^2. There is no theoretical limit to how small we can make β, although there are practical limits. If $\beta = .01$, for example, then $\beta^2 = 10^{-4}$, which can transform a 2200 Ohm resistor to 0.22 Ohms if necessary. If that's not sufficient, we can repeatedly apply the T-match. In Figure 3.31 we show an antenna that is designed to produce an impedance match for an antenna separated from the metal ground plane by 1.6 mm of HDPE. The antenna size is approximately 94 mm long and 30 mm wide. The matching network is can be viewed as two applications of the modified T-match. Because of its size, the bandwidth of this antenna is very small. At another extreme, the antenna shown in Figure 3.32 uses a very simple T-match. It is physically large, measuring 100 mm wide and 138 mm long, and a 15 mm expanded polystyrene substrate. This antenna was designed using the filter theory described above to give 90% power transfer efficiency from 865 to 955 MHz.

Figure 3.32 Large, wideband microstrip antenna utilizing classic T-match.

3.4.5.3 Microstrip Dipole

Recall that a convenient perspective on the microstrip antenna is to view the antenna as a center-fed dipole with a small radiating resistance, that is, a small fraction of an Ohm (recall Figure 3.27). Remarkably, the unmodified T-match is adequate to perform an impedance match.

Consider a rectangular dipole antenna with a width of about 15 mm. We vary the length of the dipole from 70 mm (well below resonance) to 320 mm (past the anti-resonance) and observe the impedance on a Smith chart. Second, we observe the same dipole placed 3.2 mm from an infinite metal ground plane. The two impedances are plotted on a Smith chart in Figure 3.33. As one can clearly see, the resistance at resonance is drastically reduced to a little more than a tenth of an Ohm.

Next, consider the permissible region of the matching circuit for a T-match circuit. While α can take on any positive number in theory, there are practical limits. For simplicity, we assume that α is bound by 10. Recall that the T-match first transforms the common mode impedance by a factor of $(1 + \alpha)^2$, and second transforms the impedance by a shunt reactance Z_d, resulting in the circuit shown in Figure 3.11. Figure 3.34 illustrates the permissible region, that is, the region in the Smith chart in which the common mode impedance can be matched to the conjugate IC impedance using the T-match. Notice that the dipole impedance shown in Figure 3.33 will have considerable overlap within the permissible region.

Next, consider the impedance of the microstrip dipole shown in Figure 3.33. Clearly, there is a small region close to the series resonance of the microstrip dipole that crosses into the permissible region. Depending on the particular IC and whether the radiating resistance is large enough, there may be a small region of overlap when the antenna is slightly longer than resonant, but that doesn't happen for the example antenna and IC we show here.

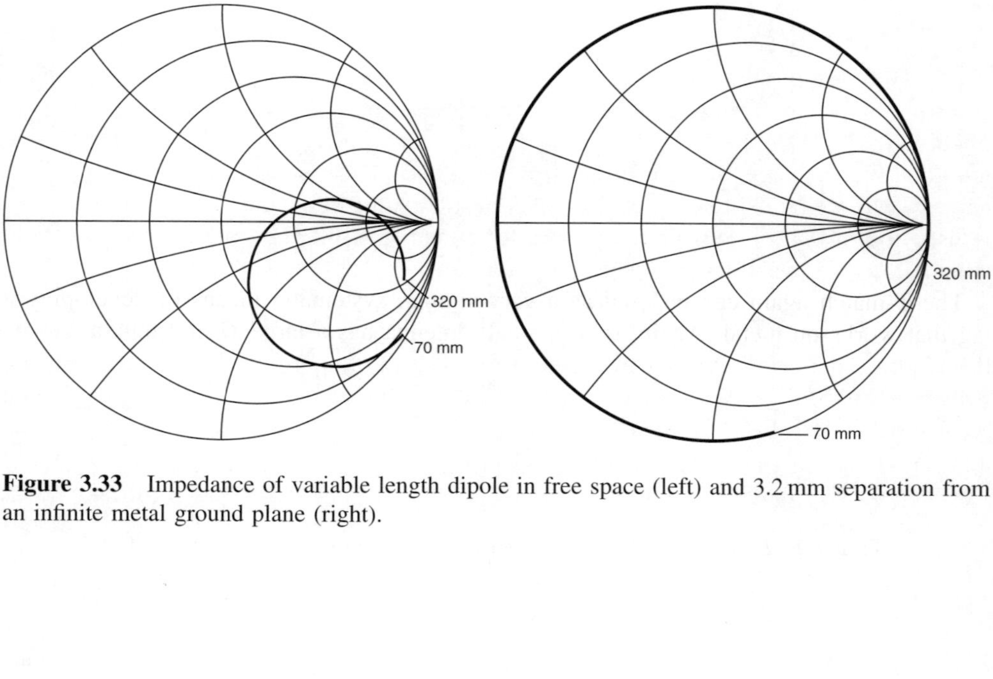

Figure 3.33 Impedance of variable length dipole in free space (left) and 3.2 mm separation from an infinite metal ground plane (right).

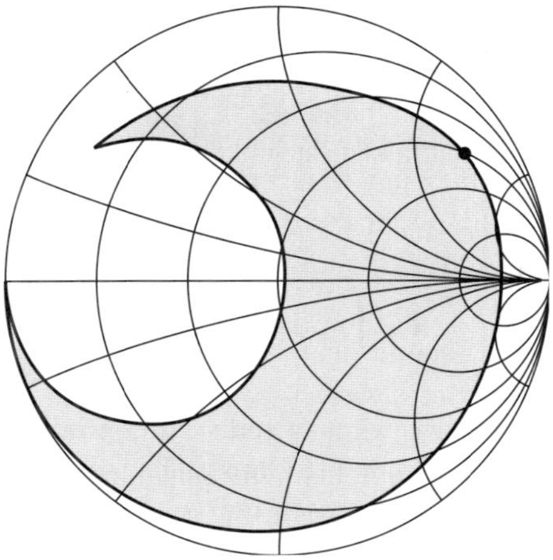

Figure 3.34 Permissible region. If Z_C lies in the shaded area, then the T-match can provide a conjugate impedance match.

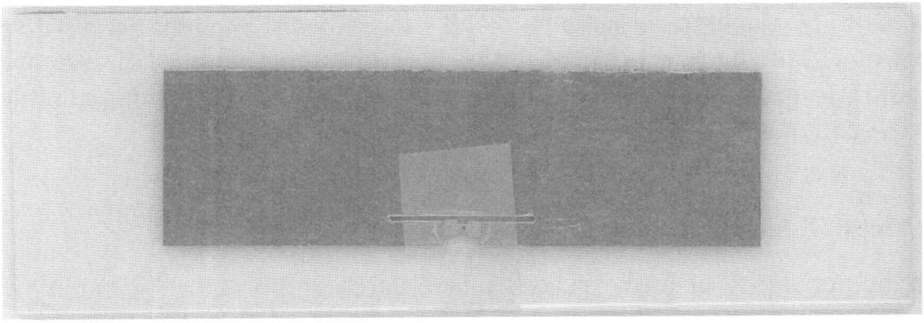

Figure 3.35 A prototype microstrip antenna using an embedded T-match.

It may be quite surprising that with a modest value of α and Z_D, one can find a conjugate impedance match for a microstrip dipole using only the T-match. A prototype of this kind of tag antenna is shown in Figure 3.35. The resulting antenna is 100 mm long and 30 mm wide, and constructed on a 3.2 mm HDPE substrate. The width of the narrow strip region is used to determine the desired value of α, and the length of the slot is used to control Z_D. Small conductive losses cause Z_D to become slightly resistive, which affects Z_{in}, and must be accounted for.

The T-match again can be used as a very simple, systematic means of developing a tag that works on metal. Again, such a small antenna has a large Q and thus a narrow bandwidth. The antenna shown in Figure 3.35 has a Q of approximately 100, and thus a 3 dB bandwidth of approximately 9 MHz. At the edge of the bands, the performance can be as much as 10 dB above the performance in the center of the band. Depending on the application, this may or may not be acceptable.

For some applications, it is more important for the tag to have more uniform sensitivity across the band. Because we're starting with an antenna with a Q of 100, there isn't much we can do to improve on the situation without sacrificing performance. One way to achieve broadband performance with a small, microstrip antenna is to purposely add loss to the system. One way to add loss is to use a dielectric with higher loss, or a less conductive material for the antenna. Or, we can add loss in the power transfer efficiency.

To purposely decrease the efficiency to a certain level, we can rearrange the expression for power transfer efficiency (3) in terms of the antenna resistance:

$$R_A = R_{IC}\left(\frac{2-\tau}{\tau} \pm 2\sqrt{\frac{1}{\tau^2} - \frac{1}{\tau}}\right) \tag{3.17}$$

Assuming $R_{IC} = 20$ Ohms and we want to reduce the power transfer efficiency to 0.5, we find that $R_A = 3.43$ Ohms or $R_A = 116$ Ohms. The smaller resistance value decreases the slope of the reactance, which increases the bandwidth, while the larger resistance increases the slope of the reactance, which decreases the bandwidth. Clearly we want to use the smaller resistance. By matching to $R_A = 3.43$ Ohms, we will reduce the slope of the reactance by a factor of about 5.8, which is significant. It is difficult to predict the resulting bandwidth since there are several other factors involved, including the resonance of various circuits. However, as a rule of thumb, by reducing the efficiency by a factor of 2, we lower the Q of a high- Q antenna by a factor of 2 and increase the 3-dB bandwidth by a factor of 2. This is only a rule of thumb and relies on a number of assumptions that may not hold. However, with antennas that have very large Q, the technique of decreasing the Q of the antenna by reducing R_A can be effective.

3.4.5.4 Metal-Tolerant Antennas

Owing to the popularity of foam spacers, our research group at the University of Kansas started investigating whether there was something that could be done to make tags using foam spacers to work better [10]. Intuitively, we know that there is considerable design freedom to pick a particular set of antenna geometries, electrical lengths, and T-match sections. Given that, it seemed obvious that certain choices would be better than others. We suspect that most commercial antennas were not designed with near-metal performance in mind, and so it seemed likely that there was considerable room for improvement.

Recall that a number of things happen when a dipole antenna is placed near metal, namely that the radiating resistance decreases drastically, the resonant frequency decreases modestly, and that the antenna Q increases substantially. Also, keep in mind that both the antenna and the matching network are affected by the presence of metal, and that in the previous section we showed that it is possible to use the T-match to perform impedance matching of a microstrip dipole. It is at least conceivable, then, that one may be able to find an antenna design that behaves as a dipole with a proper T-match in air, and as a microstrip dipole with the appropriate T-match when near metal (with a HDPE foam substrate, e.g.). Figure 3.36 shows such a tag. The tag is mounted on a 3.18 mm HDPE foam separator.

We have not yet fully devised a simple design methodology for constructing such a tag, but we can present an analysis of the tag here. Because it uses a loop-based T-match, we use the distributed circuit model to describe the antenna and matching circuit behavior both in air and when on metal with a 3.2 mm HDPE foam substrate ($\varepsilon_r = 1.095$). The

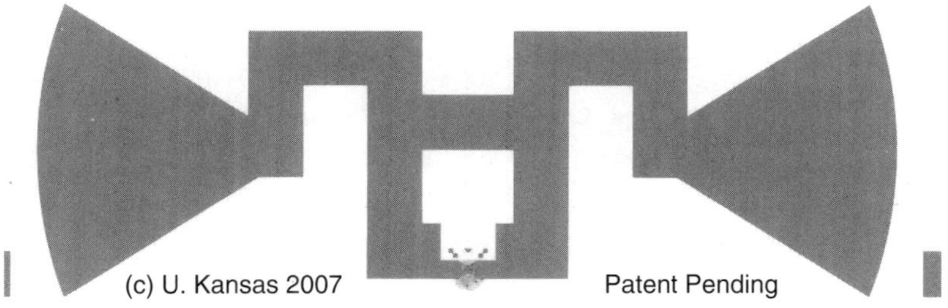

Figure 3.36 A combined dipole/microstrip antenna.

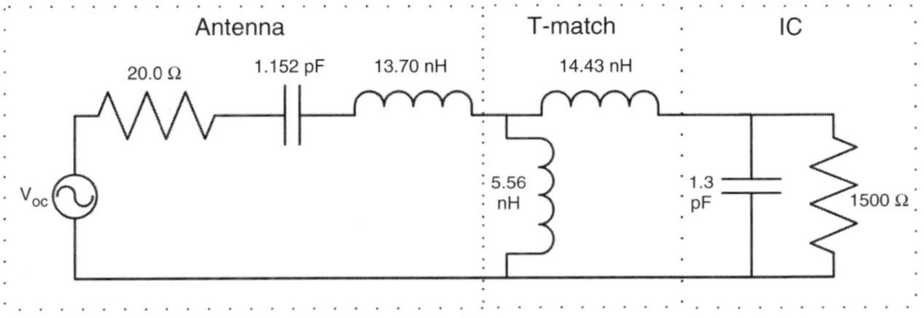

Figure 3.37 Circuit model of combined dipole/microstrip antenna functioning in air.

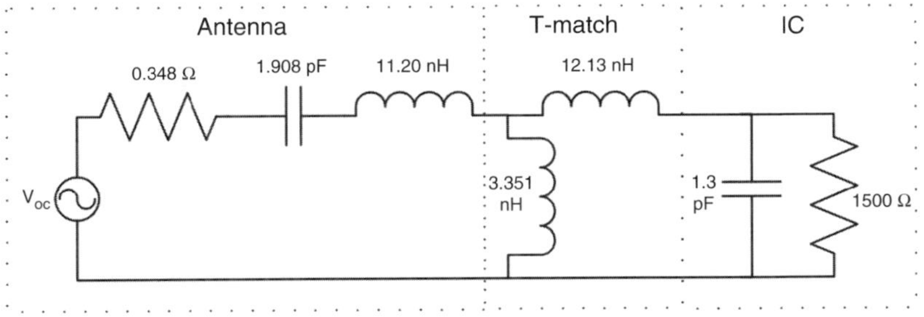

Figure 3.38 Circuit model of combined dipole/microstrip antenna functioning on metal.

circuit-equivalent models for the antenna in air and on an infinite ground plane are shown in Figure 3.37 and Figure 3.38 respectively. Note that the resistance in air of 20 Ohms reduces considerably to a fraction of an Ohm on metal. For the matching circuit, we note that both the shunt and series inductor reduces when placed on metal; the inductance per unit length is partially cancelled by the presence of the ground plane, turning the traces into transmission lines.

Next, let's look at the input impedance as a function of frequency. In air, the impedance is given in Figure 3.39(a), along with the conjugate IC impedance. We see that the antenna doesn't quite achieve a conjugate impedance; the reactance lines cross at about

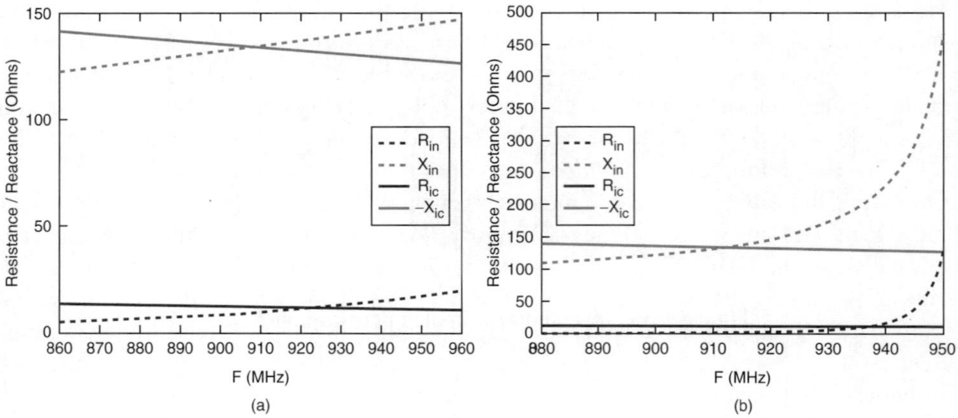

Figure 3.39 Impedance of combined dipole/microstrip antenna in air (a) and on metal (b).

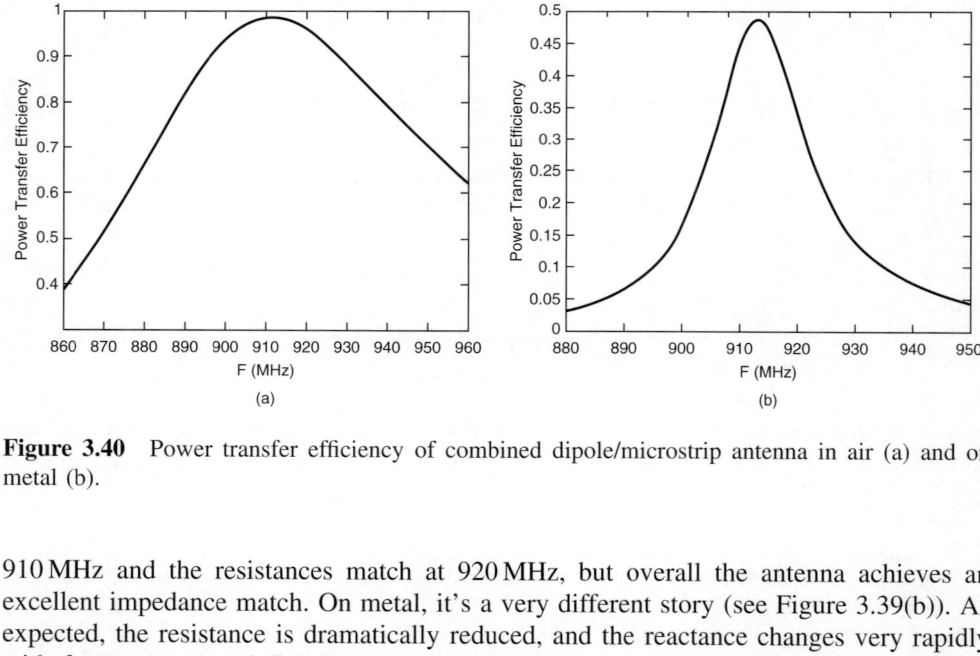

Figure 3.40 Power transfer efficiency of combined dipole/microstrip antenna in air (a) and on metal (b).

910 MHz and the resistances match at 920 MHz, but overall the antenna achieves an excellent impedance match. On metal, it's a very different story (see Figure 3.39(b)). As expected, the resistance is dramatically reduced, and the reactance changes very rapidly with frequency, especially above resonance. Clearly, we have a very large Q antenna. Indeed, from the circuit values we estimate a Q of about 250.

Next, let's translate the impedance information into power transfer efficiency. The efficiency is plotted in air and on metal in Figure 3.40(a) and (b). We see excellent performance in air, though not perfect, and certainly not wideband. On metal, however, we see the efficiency peak at about 50% efficiency, but the bandwidth is extremely narrow. This should be expected with such a large Q, and is about the best one can do. Achieving a higher peak performance would result in an even narrower bandwidth. (Indeed, performance and bandwidth are often traded off.)

How does the tag perform? The data sheet for the IC indicates a minimum power requirement of −14 dBm (39.8 milliwatts). For a reader at full power (30 dBm and 6 dBi transmit antenna) and a 3 dB polarization loss (assuming the reader is transmitting a circularly-polarized wave and the tag is linearly polarized), both the reader and tag antenna are optimally aligned, the maximum free-space read distance for a 2 dBi tag is 7.36 meters (24.1 feet). Environmental effects may add or subtract from that number.

On metal, the directivity of the antenna is enhanced from 2 dBi in the dipole case to about 8 dBi. The antenna efficiency (primarily due to conductive losses) is reduced by about 6 dBi, so the resulting gain is back to 2 dBi. That is further reduced by about 3 dB by the peak power transfer efficiency, but at the edge of the bandwidth, can be reduced by 8 to 10 dB. So the effective gain of the tag (including power transfer efficiency) ranges between −1 and −11 dBi. Under similar assumptions, the free space read distance will vary between 5.2 meters (17.1 ft) and 1.6 meters (5.4 ft).

In the real world, such a narrow bandwidth antenna poses interesting behavior. Recall that the EPC protocol (and most country regulations) require the RFID reader to frequency hop. Every 0.4 seconds, the reader will hop to a new frequency according to a pseudo-random hopping sequence. In the 915 MHz band in North America, there are 50 channels, so it takes 20 seconds for the reader to cycle through all 50 channels. A narrow-band RFID tag will be seen with decreasing frequency at increased distance. The reader may take 7 seconds on average to see the tag at the furthest read distance, may take only one or two seconds at half the furthest read distance, and will always see the tag at about one fourth of the furthest read distance. What is acceptable is entirely a function of the application.

For many asset-tracking applications, a read distance of 7 to 10 feet is sufficient, and the reader takes 2–3 seconds to scan the area, so a narrowband tag is quite acceptable, and in fact this hybrid dipole/microstrip antenna may be "overkill." If the tag is passing through a large portal with other losses (perhaps lossy dielectrics interposed) where the highest tag performance is necessary and at high speeds where the tag remains in the read field for only a fraction of a second, then the narrowband tag won't prove adequate. For those applications, a larger tag with a wider bandwidth response is necessary. In practice, those applications are relatively rare, and the market seems to readily adopt narrowband RFID tags for asset tracking.

3.5 Conclusions, Trends, and Challenges

We've covered a lot of topics in this chapter, starting with some of the basics about how antennas work, to advanced topics and leading-edge research. RFID antennas coupling to reactive RFID chips and using the T-match and modified T-match circuit offer a new host of possibilities of what can be done, if they are well understood.

For the majority of RFID applications, price is a dominant factor for the tag, so small, short, printed dipoles are the norm. We covered some of the basics about how energy is stored and dissipated in the antenna, and different ways to minimize the antenna Q. Next, we examined the T-match and modified T-match, the circuit equivalent, and how using the two and some simple filter theory we can develop wideband RFID tags. Next, we put all those concepts together and designed our own antenna for a set of hypothetical design requirements. Finally, we looked at the issues involved for RFID tags working in various environments. We looked at what the environment fundamentally does to those antennas, and what we can do to mitigate the impacts. We've seen that the future of RFID is likely to include tags that work much better near metal, and while we are not aware of any tags that work well near water, we see no reason why they could not be developed.

Hopefully, the reader has come away with an appreciation for how antennas are designed, what the various design elements are, what trade-offs are involved, how the design elements may be composed, and what some of the fundamental limits are.

References

[1] Marrocco, G. (2008) The art of UHF RFID antenna design: impedance-matching and size-reduction techniques, *IEEE Antennas and Propagation Magazine*, **50**(1): 66–79.

[2] Rao, K.V.S., Nikitin, P.V., and Lam, S.F. (2005) Antenna design for UHF RFID tags: A review and a practical application, *IEEE Transactions on Antennas and Propagation*, **53**(12): 3870–3876.

[3] Balanis, C. (2005) *Antenna Theory*, 3rd edn. Chichester: John Wiley & Sons, Ltd.

[4] Nikitin, P.V., *et al.* (2005) Power reflection coefficient analysis for complex impedances in RFID tag design, *IEEE Transactions on Microwave Theory and Technique*, **53**(9): 2712–2725.

[5] Chu, L.J. (1948) Physical limitations of omni-directional antennas, *Journal of Applied Physics*, **19**: 1163–1175.

[6] Wadell, B.C. (1991) *Transmission Line Design Handbook*. Wilmington, DE: Actech.

[7] Deavours, D.D. (2009) Analysis and design of wideband passive UHF RFID tags using a circuit model, in *IEEE RFID*, Orlando, FL.

[8] Aroor, S.R. and Deavours, D.D. (2007) Evaluation of the state of passive UHF RFID: An experimental approach, *IEEE Systems Journal*, **1**(2): 168–176.

[9] Ukkonen, L., *et al.* (2005) Reliability of passive RFID of multiple objects using folded microstrip patch-type tag antenna, in IEEE Antennas and Propagation Society International Symposium. 2005.

[10] Mohammed, N.A., Sivakumar, M., and Deavours, D.D. (2009) An RFID tag capable of free-space and on-metal operation, in IEEE Radio and Wireless Symposium, San Diego, CA.

[11] Dobkin, D.M. (2007) *The RF in RFID: Passive UHF RFID in Practice*. Oxford: Elsevier.

4

RFID Tag Chip Design

Na Yan[1], Wenyi Che[1], Yuqing Yang[1], and Qiang Li[2]

[1]*Fudan University, China*

[2]*Quanray, China*

This chapter will explore the fundamental theory of RFID tag chip design in detail, including RF/analog front end, baseband and non-volatile memories. The other focus of this chapter is to combine layers of a UHF RFID and analyze the energy/signal transmission theory applied to RFID. Some low-power and low-cost design techniques will be characterized.

4.1 Tag Architecture Systems

The tag system consists of RF/analog frontend, non-volatile memory and digital baseband sections. The functions of every module and design specifications of the system have been described from the standpoint of system architecture. High efficiency frontend design includes: (1) rectifier; (2) power (voltage) regulator; (3) demodulator; (4) clock extraction or generation; (5) backscattering; (6) power on reset; and (7) voltage (current) reference, etc. The emphasis of Section 4.1.1 is on rectifier design.

4.1.1 Tag Architecture

An RFID tag is a low-end wireless communication device, however, it has several features that distinguish it from other low-end wireless devices. The low fabrication cost is mandatory for applications such as logistics, tickets and personal identification purposes. Chip designers have to use a process comparable to the standard CMOS process to reduce extra masks and abolish almost all the complicated circuits to save the chip size. On the other hand, power supply for the tag system is harvested from the reader's interrogating

RFID Systems: Research Trends and Challenges Edited by Miodrag Bolić, David Simplot-Ryl, and Ivan Stojmenović

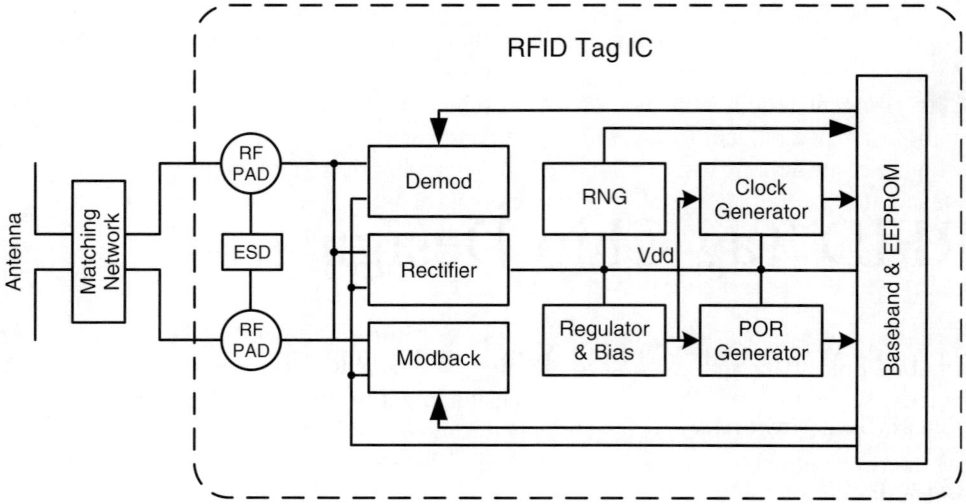

Figure 4.1 Tag system architecture.

electromagnetic (EM) field. Such characteristics limit the use of power-dissipative circuit modules, such as power amplifiers and high-speed high-precision A/D converters, frequency synthesizers, and so on. For uplink (R-T) communication, the tag changes its complex input impedance to backscatter the reader's interrogating power. For downlink (R-T), the envelope of the input RF carrier is extracted and converted to digital signals by an incoherent detector.

Figure 4.1 depicts a typical tag system. The matching network is to provide maximum power transmission from the antenna to the tag system. The modback circuit consists of several passive devices such as capacitors and switches. It performs the function of load modulation at the simplest circuit. The rectifier converts the input RF power to a DC voltage to power the whole system. Since the input RF power of the tag will vary a lot along with the change of the distance between reader and tag, a regulator is incorporated in the tag system to provide stable voltage supply. A bias circuit provides voltage and current references for the whole system. The clock signal for the digital baseband is generated by a low power oscillator, such as a ring oscillator. The demod module detects the input R-T signal [1]. A power-on reset (POR) circuit detects the rectified DC voltage level to estimate whether the total power is sufficient to supply the whole tag or not. The POR signal is used to set up the digital baseband which will process the input signal and control the logic operation of the whole tag system. EEPROM is also an essential part of the tag circuit, because it stores the data when the tag is not powered up. The passive operating manner of the tag makes it necessary to add a large storage capacitor C_S to smooth the supply voltage in case of the interrogating energy gap.

4.1.2 Design of High Efficiency Frontend Circuits

Power consumption is the primary limitation when designing the frontend circuitry of a tag. Let us take the Alien Higgs™-3 tag as an example. Its reported sensitivity is −18 dBm

[2], which actually means that the overall power consumption during the whole operation period should not exceed 16 μW. Even if the tag's rectifier reaches 50% PCE (power conversion efficiency), the tag's operating power needs to be lower than 8 μW. This is really stringent requirement for any IC designer. In order to design a tag with such a small power budget, many design methods have been proposed in recent years [3]. These technical contributions can be divided into two categories: (1) to lower down the power consumption of the tag's processing circuits; and (2) to increase the rectifier's PCE.

4.1.2.1 Low Power Frontend Circuit

There are two reasons for designing low power circuits: (1) to save the overall energy in a long time span; and (2) to decrease the transient power consumption. Most portable devices use the first reason because their design target is to prolong the battery life. However, low power RFID tag design follows the second reason so that the reader transmits EM energy continuously to power on the tag. Therefore, a sporadic large current is forbidden during the tag's operating period. Energy used to supply the tag system needs to be scattered over the whole time span.

Low voltage design in tag system has stimulated a great deal of interest in recent years. The sub-1 V technology trends [4, 5] have made it possible to enhance the system performance of RFID systems. Unlike the other IC design, there are two different supply voltage levels in the tag's frontend design. Figure 4.2 shows the supply voltage change during the "frame-sync" period of EPC Gen 2 protocol [6]. When the ASK modulation signal is at a high level, the tag senses the modulated carrier and the supply voltage increases. When the ASK modulation is at a low level, the opposite process takes place. Before time t_1, V_{dd} is at a level of V_{op}, which represents the operating voltage. During the period of DELIMITER, V_{dd} drops to a lower level labeled as $V_{\min}$, which should be higher than the tag's minimum supply voltage requirement. The I–V relationship in this period is:

$$C_s(V_{op} - V_{\min}) = \int_{t_1}^{t_2} I(t) \cdot dt \tag{4.1}$$

where $I(t)$ is the transient current of the whole tag system. Moreover, V_{op} is a function of the operating distance:

$$V_{op} \cdot I_{op} \propto 1/d^2 \tag{4.2}$$

Where, I_{op} is the operating current and d is the operating distance. Equations (4.1) and (4.2) indicate that a low $V_{\min}$ not only increases the operating distance, but also decreases the chip size by reducing the storage capacitor.

With a systematic view in designing the frontend circuitry, different building blocks have different $V_{\min}$ according to their operating speed. In order to have a global low voltage performance, multi-level supply voltage generation appears to be a good solution [7]. The essence of this method is to combine building blocks with similar supply voltage requirements together as a group, and to generate different voltage levels for different circuit modules. Each group has its optimal supply voltage level and operates independent of the others. Figure 4.3 presents a typical circuit architecture of a tag system using

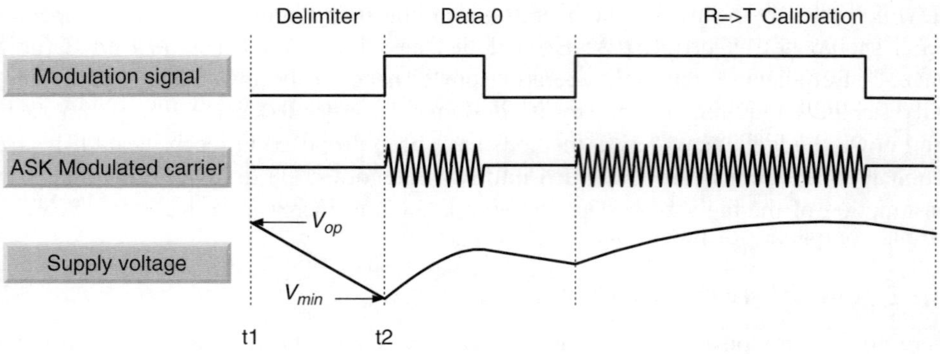

Figure 4.2 The "frame-sync" sequence of the Gen 2 protocol.

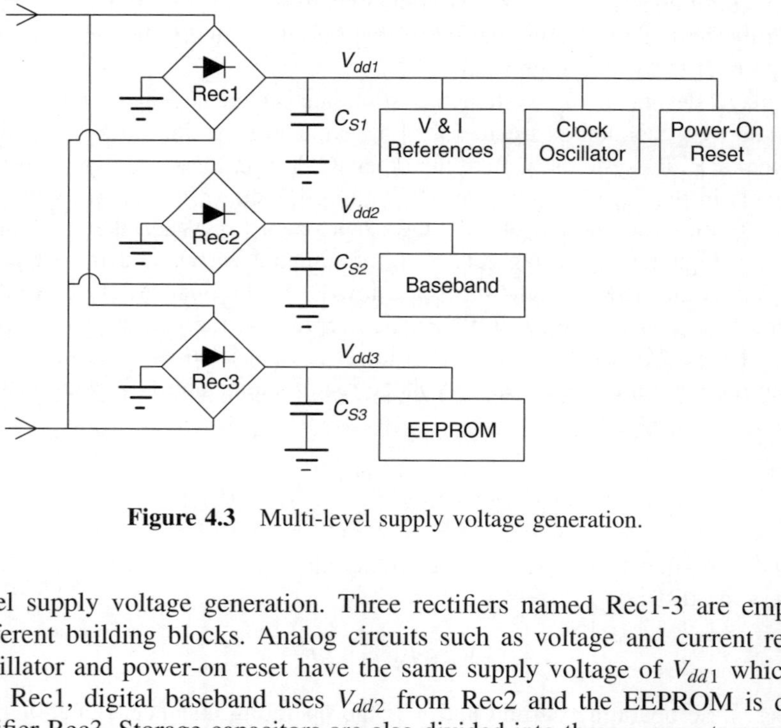

Figure 4.3 Multi-level supply voltage generation.

multi-level supply voltage generation. Three rectifiers named Rec1-3 are employed to drive different building blocks. Analog circuits such as voltage and current references, clock oscillator and power-on reset have the same supply voltage of V_{dd1} which is generated by Rec1, digital baseband uses V_{dd2} from Rec2 and the EEPROM is driven by third rectifier Rec3. Storage capacitors are also divided into three segments named C_{s1-3} for different supply voltages. In this way, each building block of the tag system can be optimized separately and the tag system has the lowest supply voltage requirement as a whole.

Another way to reduce power consumption is to use proper power management. Since all the functional blocks are not necessarily working in the whole operating period of the tag system, supply paths of each functional block can be turned on and off properly to have minimum power consumption averaged in time. For instance, when the tag is in the process of demodulating the signals, the modulation blocks might be shut down to save power and when the tag is transmitting signals, the demodulator might be shut down and vice versa. Some functional blocks like RNG (random number generator) can be turned on at the power-on period. This generates several bits of random number seeds before the whole digital baseband is set up. When the whole tag is power on, the RNG is turned off. The power management method saves total power consumption at the cost of more logic control devices. Besides, the power management circuit also consumes power and it has to be turned on all the time. Therefore, trade-offs need to be carried out to gain the optimal complexity of power management logic.

4.1.2.2 High Efficiency Rectifier

The rectifier is an essential component for all types of RFID tags. If PCE (η) can be improved from 17% [8] to 23.5% [9] with a constant load power of P_{load}, the necessary power threshold of tag $P_{tag,th}$ would be decreased by 38%. That is why increasing the rectifier's PCE has always been a hot topic in the RFID area. In this section, some common techniques, which enhance the efficiency of the rectifier, are introduced.

$$P_{tag,th} = P_{load}/\eta \tag{4.3}$$

The remotely powered characteristics of the RFID tag require the rectifier to be operated with limited input RF signal amplitude, which is often at the level of several hundred mV. In order to generate a 1 V supply voltage for the whole system, almost all the tag designers use voltage-boosting rectifier circuits [9]. An AC–DC charge pump is a common choice. Most of today's AC–DC charge pump circuits are based on the fundamentals of Dickson's structure [10]. Figure 4.4 shows the circuit architecture. From Figure 4.5, it is obvious that the PCE of Dickson's charge pump is determined by all the parameters such as stage number, or process related parameters such as the threshold voltage of the diode (in CMOS circuit design, the diodes are often replaced by diode-connected MOS transistors for the sake of the chip area and the small turn-on voltage), the loading current and the output voltage level.

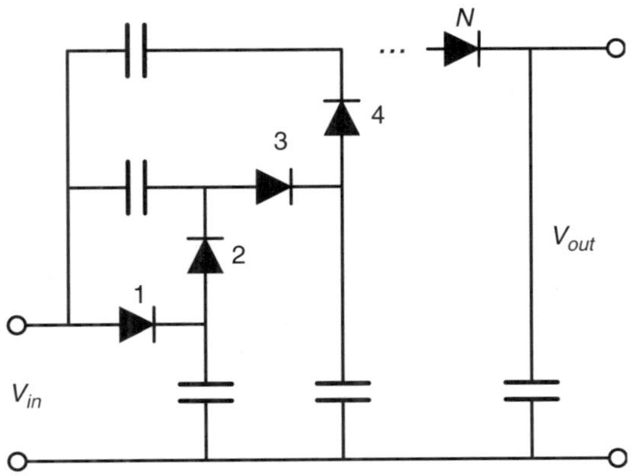

Figure 4.4 An N-stage Dickson charge pump.

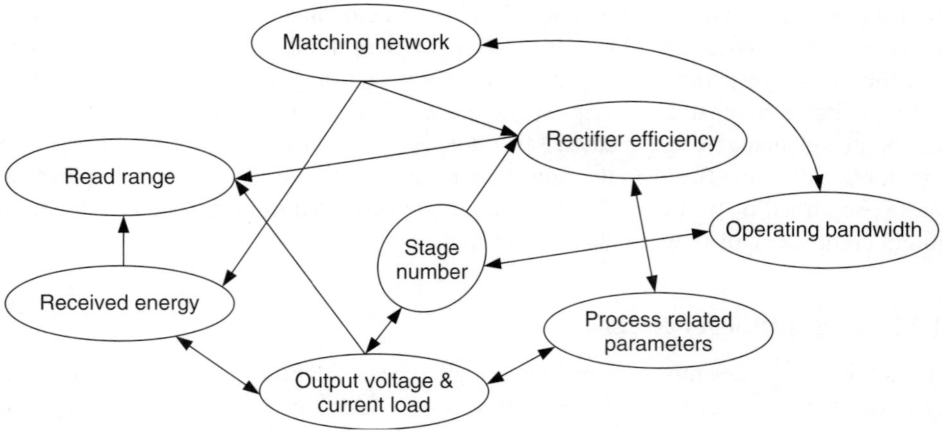

Figure 4.5 The factors affecting the power conversion efficiency of the charge pump.

Among all these factors determining PCE, the threshold voltage is found to be the crucial one. Therefore, Schottky diodes and zero thresholds MOS transistors are widely used in conventional charge pump circuit because their threshold voltages are much smaller than those of the common diode or MOS transistor [3, 11]. However, to integrate Schottky diodes in a tag, one needs to have more masks in the fabrication process which upsets the idea of low cost. On the other hand, zero threshold MOS transistors are sensitive to process variations and have the drawback of large leakage current. This situation promoted the invention of threshold compensation techniques with standard CMOS devices. Some recent works on charge pump design are illustrated below to explain the threshold compensation method.

Figure 4.6 illustrates an N stage self-threshold compensated charge pump circuit. A detailed circuit diagram of each stage is depicted in the cloud chart [12]. The gate of the NMOS transistor M_n is controlled by V_{outdc} which is the generated high DC voltage, and the gate of the PMOS transistor M_p is controlled by V_{indc} which is lower than the output. Such gate connections make both transistors operate in the linear region during the forward biasing period. The voltage drop on each transistor can be far less than its threshold. However, the constantly turned on manner of the transistors has also some drawbacks. Leakage current from V_{outdc} to V_{indc} is not negligible during the reverse biasing period of the circuit.

Figure 4.7 shows another topology of a threshold compensated charge pump circuit which has a balance in forward voltage drop and reverse leakage current [13]. The basic concept of this circuit is that it employs a constant current source to bias a diode-connected MOS transistor M_b and uses its gate-source voltage V_{GSb} to compensate the threshold of pass transistor M_i. The combined effect of forward voltage drop and reverse leakage current can be minimized by carefully setting the value of V_{GSb}. Another merit of this circuit is that it introduces two transistors, namely M_{s1} and M_{s2}, at each stage to shift the substrate voltage of M_i and thus eliminates the body effect which degrades the PCE in conventional charge pump circuit. It should be noted that in Figure 4.7, complementary blocks, namely Cell_P and Cell_N are applied for odd and even stages respectively. The

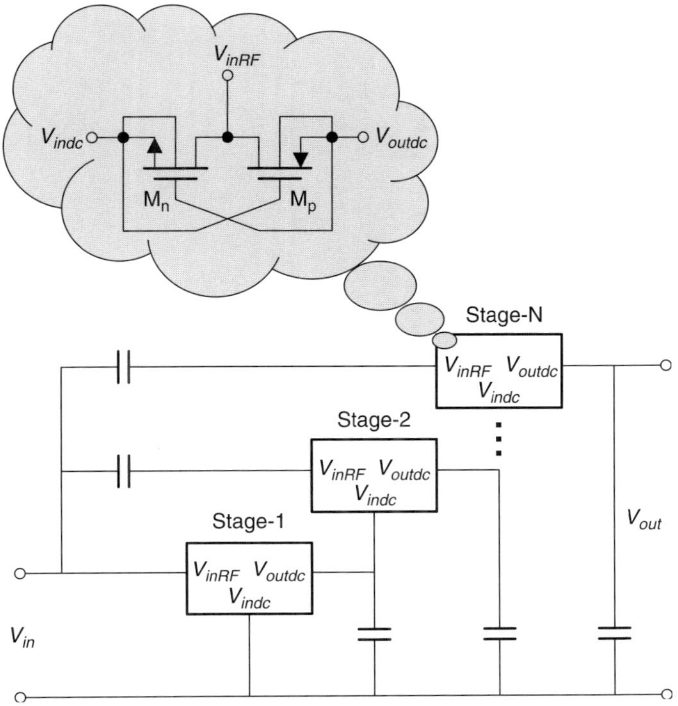

Figure 4.6 Charge pump circuit with a self-threshold compensation.

difference of Cell_N and Cell_P is that Cell_N uses NMOS transistor as M_i and M_b which are shown in the cloud chart.

Compared to the self-compensated and constantly compensated charge pump circuits which use a static voltage level to bias the gate of the pass transistor, Figure 4.8 shows a charge pump circuit which uses dynamic voltage levels for gate control [9, 14]. The symmetric structure of the circuit makes the differential RF inputs of each stage (denoted by V_{RF^+} and V_{RF^-}) have same amplitude and opposite phase from each other. In this way, during the operating period, only one pair of transistors is turned on at one time instant. For example, when $V_{RF^+} > (V_{indc} + V_{outdc})/2$ and $V_{RF^-} < (V_{indc} + V_{outdc})/2$, M_{n2} and M_{p1} are turned on and the other two transistors are turned off. The complementary on/off behavior of the transistor pairs minimizes the forward voltage drop on pass transistor and the reverse leakage current at the same time.

Table 4.1 lists some of the recent works on AC–DC charge pump circuits. Though the reported PCEs differ a lot from one another, almost all of these works utilized the threshold compensation method. It should be noted that all of the circuits with incident power less than −10 dBm used the differential structure, while the singled-ended structure was mostly chosen for incident power larger than −10 dBm. This is because that circuit using differential structure gains twice the incident RF voltage amplitude than those using single-ended structures. Such a method makes the differential charge pump structure a promising candidate for future tag design, which requires a better sensitivity performance than today's state of the art of −18 dBm.

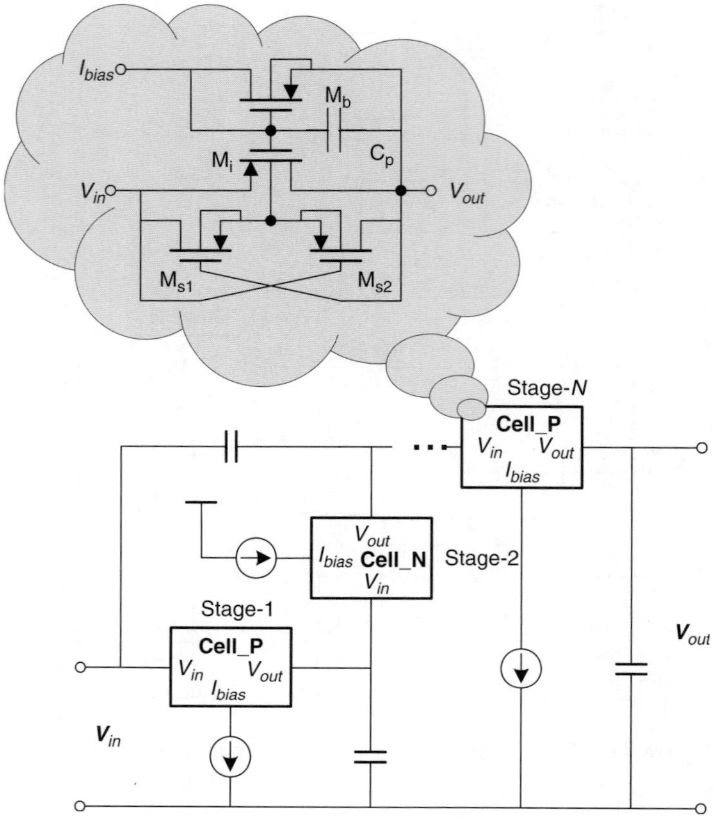

Figure 4.7 Charge pump circuit with constant threshold compensation and substrate shift.

4.1.2.3 Random Number Generator

Information security of RFID systems is now becoming a crucial factor in a number of applications such as personal identification and military cargo management. Therefore, a great deal of attention has been given to security-related research on cryptographic circuitries and algorithms. Since random numbers are employed in these algorithms as security keys, their randomicity performance determines the overall security level. There are several special requirements for RNG circuitries, which can be implemented in tag systems. These requirements and some real RNGs are discussed in the following section.

For the sake of flexibility and small chip size, pseudo random number generators (PRNGs) are widely used in conventional tag systems. However, PRNGs are mostly digital state machines which have repetitions of outputs with constant cycles. Such a feature makes them susceptible to attacks. For this reason, implementing a low power truly random number generator (TRNG) into a tag has become a new challenge for RFID.

According to the random number generation mechanism, there are three main categories of TRNGs hardware: (1) direct amplification of noise using a wideband high gain amplifier; (2) sampling of a high frequency oscillator with a jittered low frequency oscillator; and (3) discrete time chaos systems using analog signal processing technique. Though

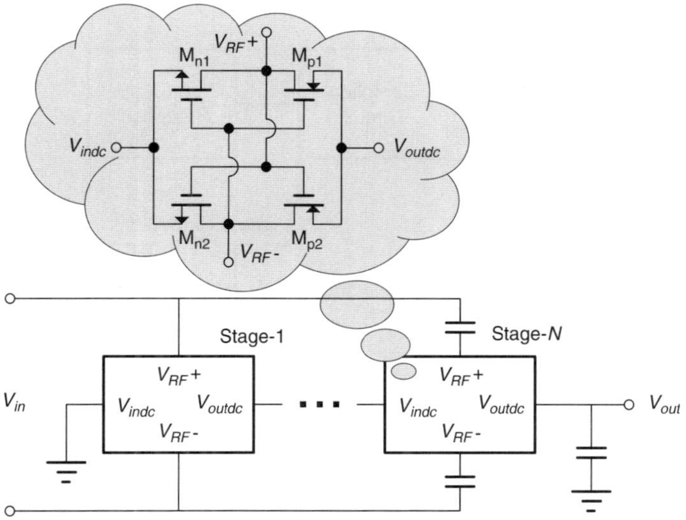

Figure 4.8 Symmetrical charge pump circuit with dynamic threshold compensation.

Table 4.1 Recent work on AC–DC charge pump circuits.

Authors	Structure	PCE (%)	Incident power	Process
[16]	Differential	36.6	−10 dBm	0.35 μm FeRAM CMOS
[9]	Differential	3.2	−12.1 dBm	0.5 μm standard CMOS
[9]	Differential	23.5	−20.7 dBm	0.18 μm standard CMOS
[9]	Differential	16.7	−22.2 dBm	0.18 μm standard CMOS
[12]	Single-ended	29	−9.9 dBm	0.35 μm standard CMOS
[13]	Single-ended	36	−9.5 dBm	0.18 μm standard CMOS
[14]	Differential	66	−12 dBm	0.18 μm standard CMOS

the first method is easy to realize, it requires a large gain and wideband amplifier which consumes lot of power. In the second category, the amplified noise will affect other parts of the circuit and thus brings the problem of physical isolation. The third method requires complicated analog circuitry with large current consumption [15]. Such drawbacks make these two methods unacceptable for tags.

With careful analysis and trade-offs among chip size, the second method of oscillator-based TRNG is now being accepted by researchers for its power consumption and output bit rate. It consumes less power, shows good quality against $1/f$ noise and insensitive to process variations [15]. Figure 4.9 is an ideal model to illustrate the operating principle of oscillator-based TRNG. In the circuit, white noise is deliberetely affixed to a low frequency clock to jitter its rising and falling edges. A fast oscillating clock is sampled by the jittery slow clock to generate serial random bits as seeds for post-digital processing. Since the edges of the jittery slow clock are unpredictable, the output of the circuit is truly random. There are several approaches nowadays to implement an oscillator-based TRNG into a tag and main difference is in the way to select the fast clock and noise source.

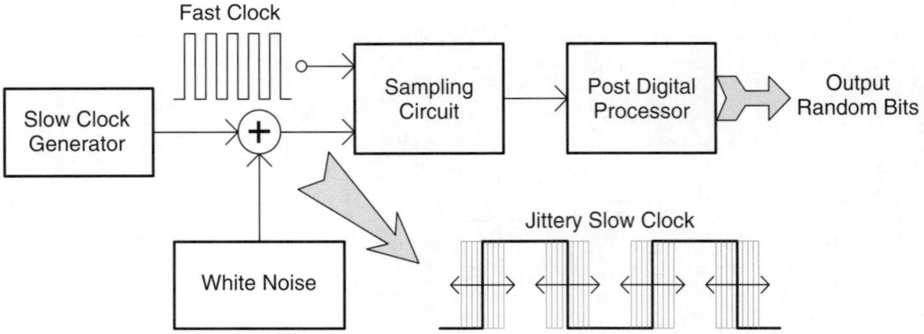

Figure 4.9 An oscillator-based TRNG.

Fast Clock

According to the basic mechanism of the circuit, the standard deviation of the slow clock's jitter $\sigma(T_{slow})$ needs to be several times of the period of fast clock T_{fast}. The higher the ratio between $\sigma(T_{slow})$ and T_{fast}, the higher the output randomicity will be. The optional source for a fast clock in a tag includes the input RF carrier and the system clock for baseband. To exploit the input RF carrier is a straightforward method and it is the signal with highest frequency in tag [15]. The circuit works well for tags operating in HF band but encounters some difficulties for UHF band. The 900 MHz UHF carrier presents difficulties in designing the sampling circuit. Extra power consumption and hardware overhead are required for accurate sampling [17]. Using the system clock as the fast clock is a feasible way for the UHF tag [18]. Since the prevalent system clock frequencies of today's UHF tags are from 1.28 MHz to 2.56 MHz [19], the sampling circuit can be realized by a T flip-flop, which is simple to be realized and consumes no static power.

Noise Source

Common noise sources in IC to jitter the slow clock are to use the channel noise of MOS transistors [17] and the thermal noise of resistors [15, 18]. Thermal noise of resistors is a good choice as its noise spectrum density is almost white in the entire frequency span. The ideal noise of resistors excludes the low frequency flicker noise which will contaminate the output randomicity. The drawback of using resistors as a noise source is that resistors occupy a larger chip area and cannot be fabricated in digital CMOS process. On the other hand, MOS transistors are smaller than resistors in size but their noise spectrum is not uniform. Such a feature makes the channel noise of MOS transistors a good option in designing TRNG for anti-collision purpose rather than security.

Table 4.2 lists some recent work on TRNG circuits for RFID tags. The level of randomness is represented by low, medium and high as specified by the pass rate performances of the NIST standards [21] and are concluded by randomicity levels of low, medium and high. As can be found in Table 4.2, all designs of TRNG circuits are making trade-offs in power consumption, output bit rate, chip size and randomicity performance. There still remains a big challenge for tag designers to design a low power high randomicity TRNG circuit with compact chip size.

Table 4.2 Recent work on TRNG for RFID tags.

Author	Power consumption	Bit rate	Chip size	Randomicity
[17]	0.528 μW	320 kb/s	0.0056 mm^2	low
[18]	1.04 μW	40 kb/s	0.05 mm^2	medium
[15]	2.3 mW	10 Mb/s	0.0016 mm^2	high
[20]	180 μW	50 kb/s	1.49 mm^2	high
[20]	2.92 μW	0.5 kb/s	0.031 mm^2	high

4.2 Memory in Standard CMOS Processes

In RFID tag chip, the embedded non-volatile memory (eNVM) is used to store the related essential information such as the electronic product code (EPC), the expiration date, the security passwords, the tag manufacturer information, and possibly user data. The memory organization of the tag defined in ISO18000-6C protocol is shown in Figure 4.10, which is redrawn from ISO 18000-6C. The capacity of the eNVM ranges from 96 bits to 2 Kbits. Comparable to the other memory applications, the capacity is very small and the memory area is still limited by peripheral circuits, such as the decoder, the latch, and the high voltage generator for the write/erase operation. In the past couple of years, conventional EEPROM memory has been widely used in RFID technology. However, the conventional EEPROM process requires additional mask steps and a long manufacturing cycle. So the small memory area results in high manufacturing costs. Obviously, it is necessary to find a low-cost, high-performance and reliable solution for embedded non-volatile memory.

4.2.1 Why Have a Standard CMOS eNVM?

Several memory solutions have been proposed recently to replace the EEPROM. Ferroelectric memory has been used as RFID tag memory [22, 23], because of the following advantages: no requirement for high voltage for the write operation, fast write cycles and 10-year data retention. Unfortunately, fabrication process modifications and additional mask steps are still required to produce FeRAM. Therefore, if one takes the cost and chip yield into consideration in a low-cost design, FeRAM is not the best solution.

One time programmable (OTP) non-volatile memory is compatible to standard CMOS process and does not require an additional mask layer [24, 25]. In the factory, the stored information is written into OTP memory one time, and then locked forever. The user can not rewrite/modify data of memory and therefore, the application of OTP memory in RFID systems is limited.

Resistive RAM (RRAM) stores information according to the electrical resistance of certain materials by applying an external voltage or current, instead of storing information in transistors (flash memory) or capacitors (DDRAM) [26]. RRAM is non-volatile, and its simple structure is ideal for future generations of CMOS chips. It requires no extra mask layers, but few process modifications. The type of memory has two states: high resistive and low resistive to be denoted as “1” and “0.” The low resistive state needs more current consumption than the high resistive state, which is still a challenge for low power tag chip design. The related researches and verifications are still under development and not mature.

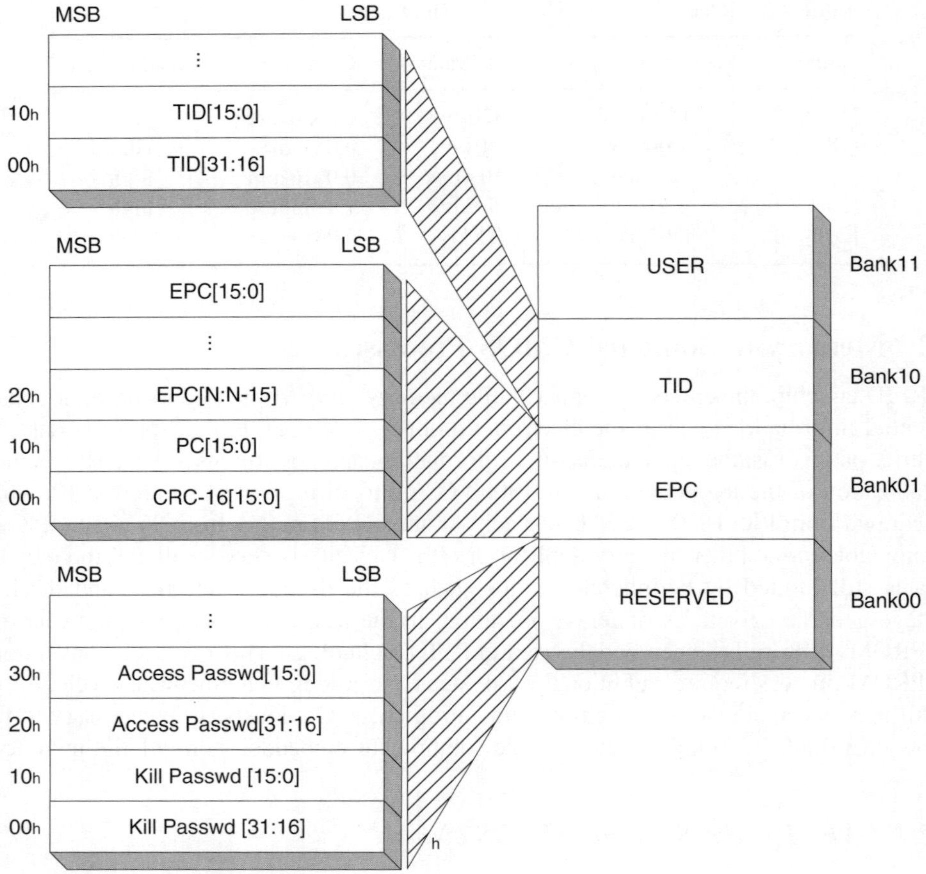

Figure 4.10 Logical memory map. Redrawn from ISO 18000-6C.

Along with the development of the standard CMOS process, the characteristic dimension of transistors is shrinking; the thick gate-oxide of transistors is becoming thinner and thinner. Consequently, the needed voltage for tunneling is decreasing. That is the reason why eNVM can be implemented in standard CMOS process and be a better choice in RFID tag chip design. Standard CMOS eNVM brings two advantages: (1) shorter time to market; and (2) cost advantage over traditional eNVM. Now it has already been used in RFID tag chip products (Impinj Monza™).

4.2.2 Basic Cell Structures and Operation Mechanisms

Aiming at low cost and low power, a single-poly nonvolatile memory (SPNVM) in standard 0.18 μm CMOS processes has been developed [27], which is compatible with the standard CMOS process.

A basic single-end SPNVM cell is shown in Figure 4.11. By storing a different amount of charges on a floating gate denoted as FG1, a different voltage can be established. As a

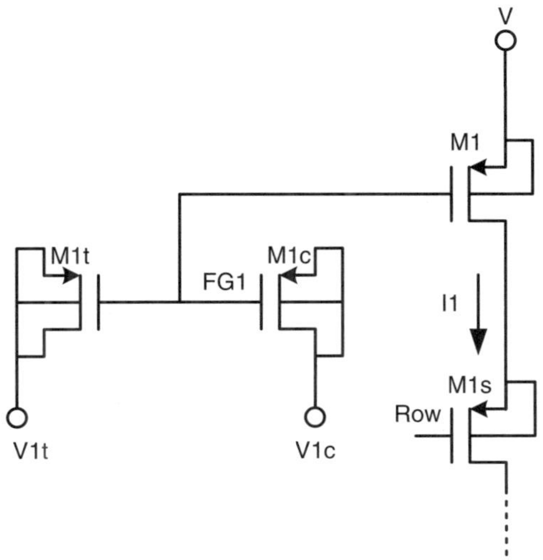

Figure 4.11 A basic single-end SPNVM cell structure.

result, it is feasible for the sense-amplifier to discern the logic value stored in the memory cell by comparing I_1 with reference current I_{ref}.

Figure 4.12(a) illustrates the key part of the SPNVM cell. It is composed of four transistors: a tunneling transistor M_{1t}, a coupling transistor M_{1c}, a reading transistor M_{1s}, and a select transistor M_{1s}. There is a common end between tunneling and coupling transistors, which act as a floating gate denoted as FG1. The physical size of coupling transistor M_{1c} is much larger than that of tunneling transistor M_{1t}. M_{1t} and M_{1c} are connected as capacitor transistors. Thanks to the capacitive divider, a large fraction of the voltage, which is enough to induce the FN tunneling at M_{1t}, will be applied. The coupling transistor and the tunneling transistor act as the control capacitor and the tunneling capacitor respectively.

Figure 4.12(b) shows the cross-section of the tunneling transistor and the coupling transistor. The source and drain area are connected to N-well. The gates must be connected together by poly-silicon, not by metal.

Table 4.3 illustrates a set of operation states and the applied voltages at corresponding nodes for the single-ended SPNVM. During the program operation, V_{1t} is pulled down to the ground while V_{1c} is connected to the high voltage V_{hh} generated from the charge pump, while the readout circuit is turned off ($V = 0V$, Row $= V_{dd}$). Due to the high electric field intensity established across the oxide dielectric of M_{1t}, electrons tunnel from M_{1t}'s bulk to the floating gate and decrease the floating gate voltage.

During the read operation, both V_{1t} and V_{1c} are pulled down to the ground, while the readout circuit is power-on ($V = V_{dd}$, Row $= 0V$). Then the readout current I_1 is generated and sensed by the sense-amplifier.

Let us compare the cost of a tag chip between the conventional EEPROM process and the standard CMOS process, based on SMIC 0.18 μm EEPROM/Standard process respectively, to exemplify the cost advantage of standard CMOS eNVM, as shown in Table 4.4. Suppose the two tag chips are the same, except the EEPROM circuit and the tag chip area is 0.5 mm^2, including the 512-bit memory cell. In the SMIC 0.18 μm

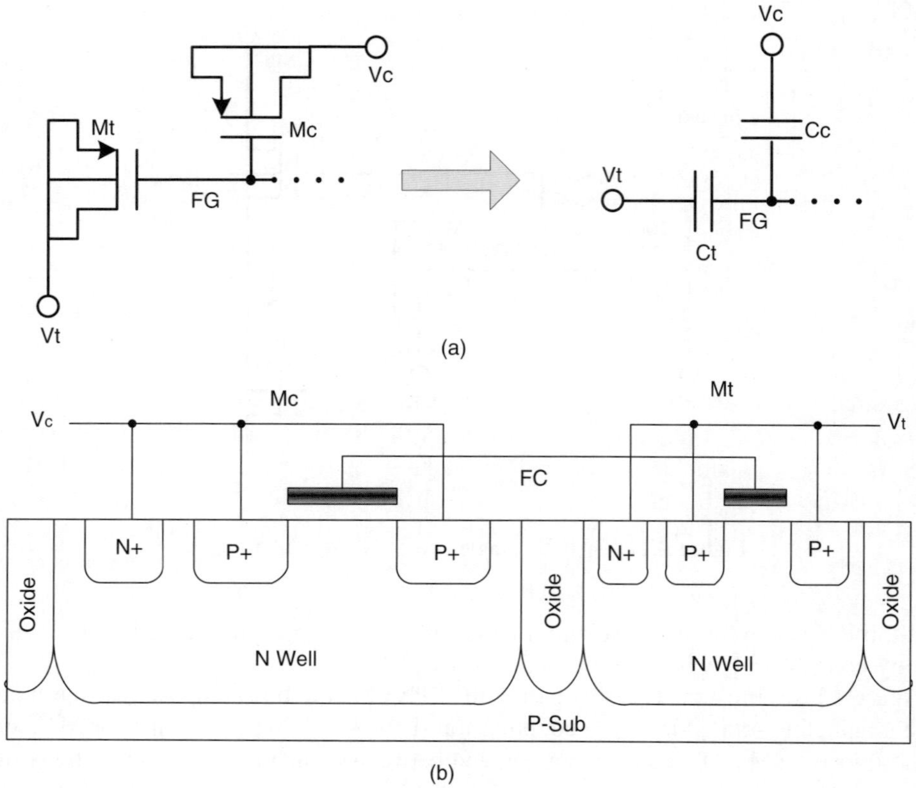

Figure 4.12 Schematic (a) control capacitor Mc and tunneling capacitor Mt; (b) cross-section.

Table 4.3 Operation states versus applied voltages.

Operation	V_{1t}(V)	V_{1c}(V)	V(V)	Row(V)
Program	0	Vhh	0	Vdd
Erase	Vhh	0	0	Vdd
Read	0	0	Vdd	0

EEPROM process, the EEPROM cell is 3.96 μm^2. If we define the proportion of control capacitance to tunneling capacitance as 9:1, then the maximum area of SPNVM cell is about 110 μm^2. The comparative results show that the total costs of the tag chip in the standard CMOS process reduce to 80.9% of the EEPROM process. The limitation of a standard CMOS NVM bit cell is a larger memory cell size compared with the conventional EEPROM bit cell size. However, it is not a severe constraint in RFID technology since the required data storage capacity is small and the capacity of 512-bit is enough in most cases.

Other attempts have been reported in this regard [28–31]. Pesavento *et al.* [28] reported an eNVM used in RFID chips in 0.25 μm standard CMOS processes by applying a

Table 4.4 Comparison of two tag chips with different processes.

Process (SMIC 0.18 μm)	EEPROM	CMOS
Area of memory bit cell	$3.96\,\mu m^2$	$110\,\mu m^2$
Number of mask layers	28	19
Number of photolithograph	29	19
Memory chip/tag in area	0.41%	11.3%
Cost increase due to area	0	10.89%
Cost increase due to additional masks	0	−30%
Total cost increase of tag chip	0	−19.1%

differential cell structure. Ohsaki *et al.* [29] used an nMOS transistor as a tunneling capacitor. However, it has been proved that a pMOS capacitor has better data retention than a nMOS capacitor due to its higher insulator energy barrier. In [30], each bit cell has its own localized sense amplifier and switching circuits. Although it offers better reliability, the bit cell is too complicated and thus is too large and consumes too much power. Kee-Yeol *et al.* [31] used a MIM capacitor as a coupling capacitor, but this reduces the memory retention time due to leakage from the poly contact and the metal interlayer dielectric.

4.2.3 Memory Architecture and Peripheral Circuits

Figure 4.13 presents the general architecture of eNVM. For eNVM memories with small capacity, the area is mainly occupied by peripheral circuits. The memory includes the following modules: (1) input/output buffers, which buffer address/data signals; (2) a sync module, in which the address/data signals are synchronized with the clock; (3) a decoder including a block/word/bit decoder, used to select the operated memory cell; (4) a charge pump, which generates the high voltage required for write operation; and (5) a sense amplifier that senses the state of a bit line and amplifies the signal. Considering the power consumption/read speed requirements for RFID applications and the intrinsic characteristics of the memory cell, eNVM memory is always designed to have parallel input and serial output. Namely there is only one sense-amplifier needed, since the data is read out one bit by one bit.

The sense amplifier (SA) is one of the most critical peripheral circuits in memory design and it function is related to the read access time. It is used to retrieve the stored data from memory, by amplifying small signal variations on the bit-lines. In general, increasing the memory capacity will result in increasing parasitic capacitance of the bit-line, consequently lowering the read speed of the sense amplifier and enhancing the average power consumption. There are mainly two kinds of architectures for SA, voltage-mode sense amplifier and current-mode sense amplifier shown in Figure 4.14(a) and Figure 4.14(b), respectively. In the case of low power supply, current-mode architecture is usually used for high speed and low power consumption because the parasitic capacitance on the bit-lines is very large. Most of the current-mode SAs are based on cross-transistors [32, 33].

On the other hand, high write voltage, for example, which is above 14 V in SMIC 0.18 μm EEPROM process, need to be generated on the chip for write operation. Most

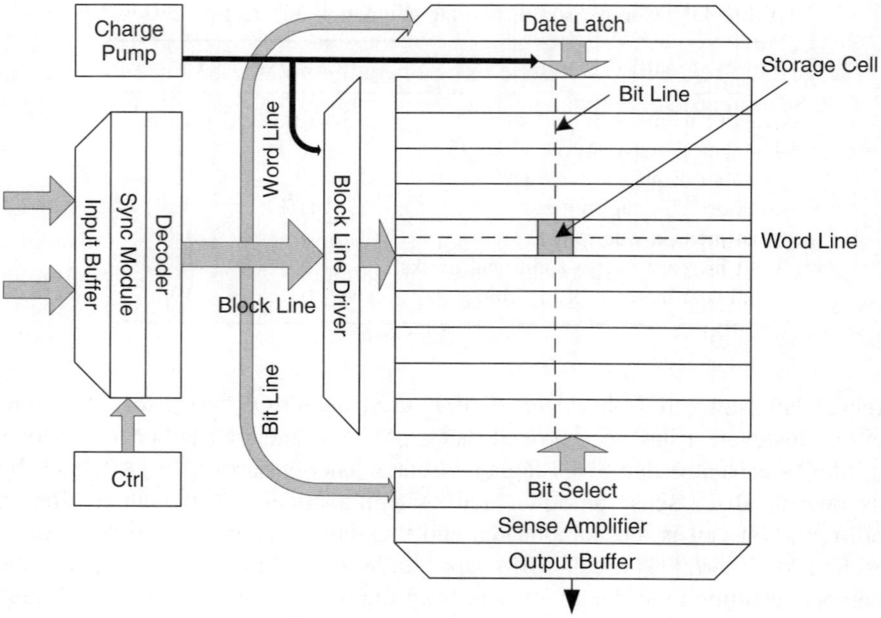

Figure 4.13 General standard CMOS eNVM architecture.

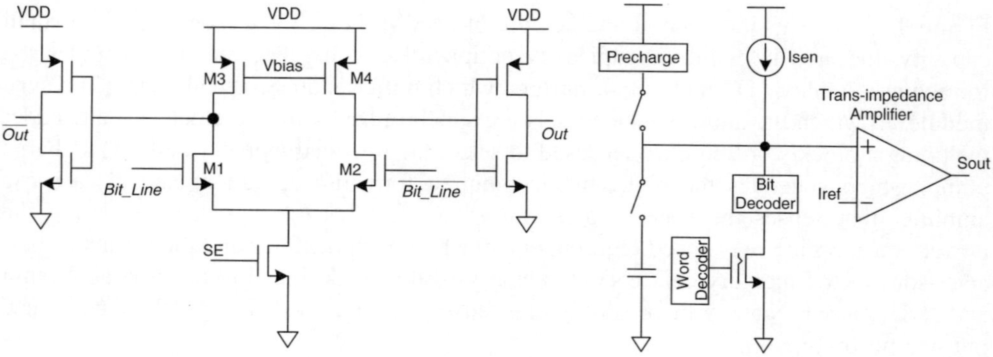

Figure 4.14 (a) Voltage-mode sense amplifier; (b) Current-mode sense amplifier.

charge pumps are based on the circuit proposed by Dickson [10]. However, due to the existence of the body effect, the pumping efficiency of Dickson charge pump has been degraded. In recent years, several attempts have been made to minimize the body effect of the charge transfer transistors to improve the pumping efficiency of the charge pump [34, 35]. A high performance ALL-PMOS charge pump for low voltage operations is proposed [36] and a maximum power efficiency of 87.1% has been reached at the load current of 2.5 mA for the power supply voltage of 1.5 V.

High voltage switch is another important part, which allows the voltage to shift from low voltage to high voltage. A high voltage device in standard CMOS processes is required here to ensure the reliability, since the write operating voltage is typically much higher than VDD. Several articles and patents have been published to address this issue [37–39].

4.2.4 Future Challenges

As one of the leading innovators in developing RFID products, Impinj Inc. has invented AEON/MTP NVM which can be produced in standard CMOS processes. Impinj had already completed the qualification of its AEON Memory. Available in 0.18 μm and 0.25 μm TSMC CMOS processes, AEON Memory has attracted a lot of attention from the NVM industry. Impinj Inc. [40] shows the specifications for a 4 K-bit configuration with integrated high voltage on TSMC's 90 nm LP process.

It seems that standard CMOS eNVM is a perfect solution when only small amount of NVM is needed. However, with the opportunities provided by the standard CMOS eNVM technology, designers still have to face some challenges in order to make the technology competitive, including gate oxide scaling, high voltage technology, technology transfer and process control monitoring (PCM).

Some tests should be conducted to find out the failure rate dependence on oxide thickness. In addition, experiments need to be done to improve the traditional HV device, like LDMOS (laterally diffused MOS), in order to ensure reliability. Some specific design rules and PCM strategies should be optimized to ensure manufacturability and reliability at future technology nodes. In spite of the challenges it faces, eNVM in standard CMOS processes is still one of the best solutions for RFID applications due to its better data retention, lower cost and higher yield.

4.3 Baseband of RFID Tag

A conventional RFID tag contains three essential components: (1) the front end; (2) the baseband; and (3) the non-volatile memory. The front end obtains energy from radio waves, receives and transmits signals from and to interrogators. The non-volatile memory stores information, and the baseband executes operations required by protocols. In this section, the attention is focused on the low-power design of RFID tag baseband. Initially, an introduction to baseband including its function and blocks is given, and later low power design techniques, attempting to reduce both clock rate and supply voltage, are discussed in detail.

4.3.1 Introduction

The baseband of an RFID tag should manage to execute the following operations:

1. **Waveform coding and decoding**: EPC C1G2 protocol defines the interrogator-to-tag link uses PIE (Pulse Interval Encoding) code, and the tag-to-interrogator link uses FM0 baseband signal or Miller (M = 2, 4, or 8) modulation of a subcarrier signal to encode the backscattered data. The baseband should have the ability to decode the PIE coded command from the interrogator and to encode data to FM0 or Miller waveform.

2. **Command decoding and tag state transmission**: The baseband should recognize commands from interrogator, reply (or not) to interrogator and change current state according to EPC C1G2 protocol. The baseband should keep its current state until the next command arrives while powered up.
3. **Link timing**: The EPC C1G2 protocol specifies the link timing requirements such as T_1 (time from interrogator transmission to tag response) and T_2 (interrogator response time required if a tag is to demodulate the interrogator signal, measured from the end of the last bit of the tag response to the first falling edge of the interrogator transmission). The baseband should provide the accuracy to meet timing requirements.
4. **Non-volatile memory interface**: The interrogator and the tag have the ability to exchange data. The baseband should incorporate an interface with the non-volatile memory, through which reading from and writing to the memory are performed.
5. **Pseudo-random number generator (PRNG)**: EPC C1G2 requires the use of a random or pseudo-random number. To achieve the balance between power consumption and randomness, a PRNG is usually implemented in the baseband. The seed of PRNG may be generated by a True-Random Number Generator (TRNG), which is an analog circuit and measures a physical noise such as clock jitter.

Figure 4.15 shows the block diagram of a baseband in a RFID tag chip. Since the signal of the interrogator-to-tag link is strong and with high SNR, non-coherent demodulation is widely used in the tag when it comes to PIE demodulating. According to the EPC C1G2 protocol, the length of RTcal is equal to the length of symbol data-0 plus that of symbol data-1, and the length of symbol data-1 should be 1.5 to two times of that of symbol data-0. The demodulator of a tag measures the length of RTcal and divides it by two, the result of which is called pivot, and is considered as the decision threshold. Any subsequent PIE symbol shorter than a pivot will be decided as data-0, otherwise data-1. The performance of non-coherent demodulation is affected mainly by the frequency of the clock, and how to determine the clock frequency is discussed in the subsequent section.

4.3.2 Low Power Baseband Design

Power and cost are the two permanently hot topics for RFID tag chips. Power consumption in CMOS circuits consists of two parts: static power and dynamic power. Reduction of

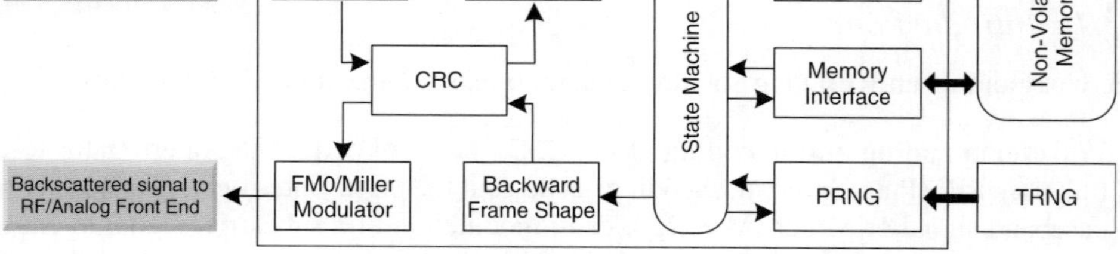

Figure 4.15 Block diagram of an RFID tag baseband.

the former in a digital circuit depends mainly on the advancement of manufacturing aspects. We will focus here on reducing dynamic power consumption. Dynamic power is comprised of short-circuit power and switching power. Conventionally dynamic power can be approximated by:

$$P_{dyn} \propto \alpha_{0\rightarrow 1} V_{dd}^2 C_L f_{clk} \quad (4.4)$$

Here V_{dd} is the supply voltage, C_L is the equivalent capacitance load, f_{clk} is clock rate, and $\alpha_{0\rightarrow 1}$ is the activity factor. From Equation (4.4) it makes sense that reducing clock rate and supply voltage are common techniques to achieve low power in RFID tag chips.

First, we will discuss the minimum clock rate required by EPC C1G2 protocol. The clock rate is mainly restricted by the requirement of PIE decoding accuracy and backscatter link frequency (BLF) tolerance. Second, methods to balance clock rate and power performance are introduced. The method called clock management ensures that function blocks operate at their lowest clock rate. A clock gating method can reduce short-circuit power consumption. In addition, adopting dual edge triggered (DET) flip-flops at the circuits that require the highest clock can help cut the highest clock rate to half, which can reduce the power of the oscillator significantly.

The sub-threshold technique and the adiabatic technique, which are methods aimed at low-power low-speed application, such as RFID tag chip, are also introduced in this section.

4.3.3 Clock Rate

The maximum clock rate of the baseband is restricted by two factors: power requirement, and time resolution during non-coherent demodulation of PIE codes. The effect of the former factor is somewhat clear because lower frequency leads to both a lower dynamic power of digital circuit and a lower current consumption of oscillator. The influences of time resolution lies in ambiguity between symbols data-0 and data-1, and BLF offset. As the relationship between power consumption and clock rate can easily be found in any books and papers which talk about low power design, here more concern is focused on analyzing how the EPC C1G2 protocol restricts the clock rate to produce lower. In general, the clock rate of the tag is an integer multiple of the maximum BLF, which is 640 KHz defined in EPC C1G2 protocol. Impinj Inc. [41] first performed research on this problem and concluded that 1.92 MHz is more reliable than 1.28 MHz when demodulating clock rate is taken into consideration.

As mentioned above, the tag prefers non-coherent demodulation of PIE encoded waveform from the interrogator. The tag measures the length of RTcal (which should be equal to the sum of length of data-0 and data-1) and each data symbol, and compares the length of data symbol to half of that of RTcal (the so-called Pilot). The difference value between the pilot and the worst data symbol (the minimum length of data-1 or the maximum length of data-0) is the margin for decoding, the so-called decode margin. However, quantization errors introduced while measuring the length can inevitably deteriorate the decode margin. When the decode margin is zero, the demodulator cannot distinguish data-0 and data-1 from each other.

Figures 4.16 and 4.17 show the decode margin deteriorated by quantization error when the clock rate is 1.28 MHz and 1.92 MHz. They show that it is safe to use 1.92 MHz as

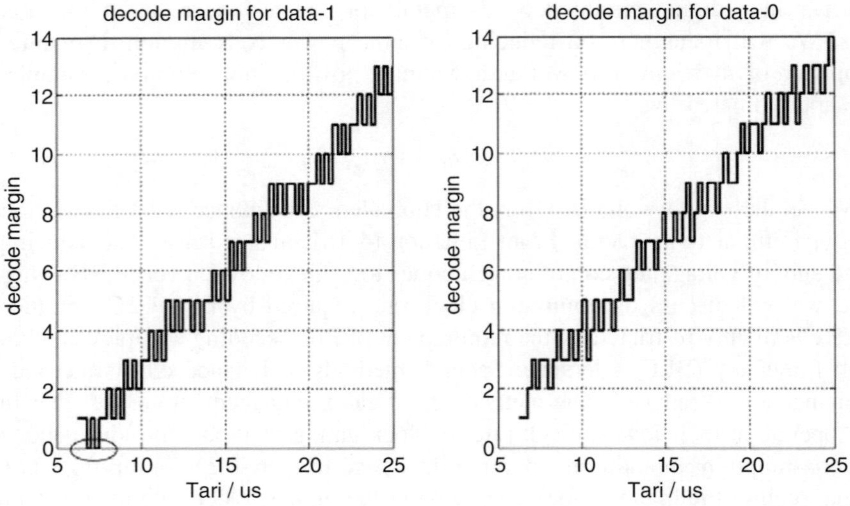

Figure 4.16 Decode margin for 1.28 MHz with 5% clock uncertainty (violations marked with the circle).

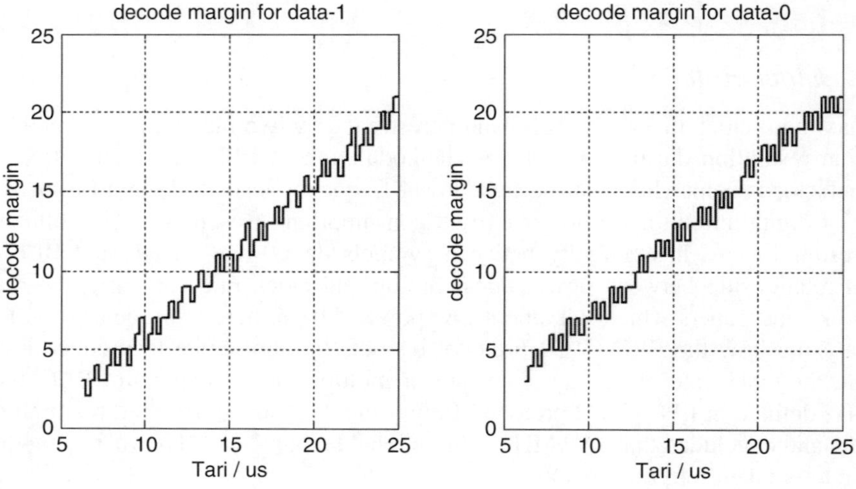

Figure 4.17 Decode margin for 1.92 MHz with 5% clock uncertainty.

the demodulating clock rate, because the decode margin may be zero when the clock rate is 1.28 MHz (marked with the circle).

According to the EPC C1G2 protocol, tag backscatters its data at a rate specified by *TRcal* and a parameter called DR (Divide Ratio), both of which are sent by interrogator in a Query command. The tag computes its BLF by the following equation:

$$BLF = \frac{DR}{TRcal} \tag{4.5}$$

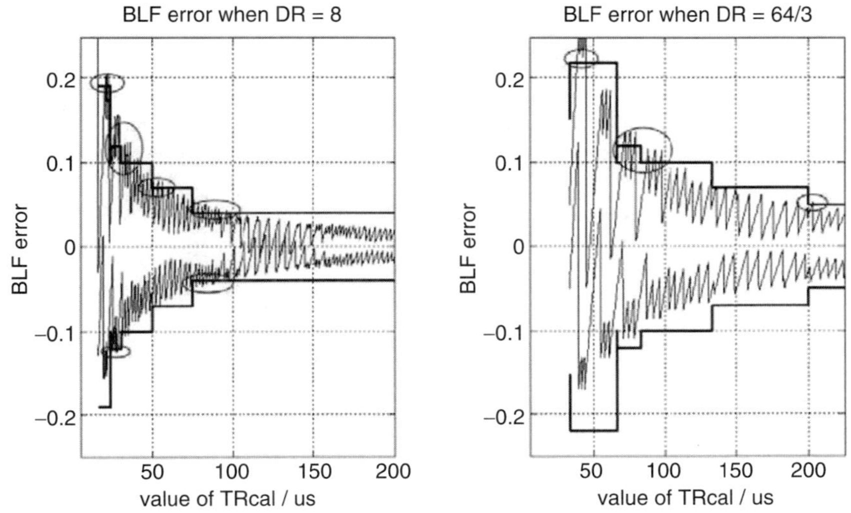

Figure 4.18 BLF error when the clock rate is 1.28 MHz (violations marked with the circle).

Here $DR = 8$ or $DR = 64/3$ as specified by the interrogator and $TRcal$ denotes the length of $TRcal$.

The BLF frequency offset occurs since the length of $TRcal$ is quantized to the integer times of the period of sampling clock while the tag measures $TRcal$ by sampling the PIE coded signal at a sampling rate of f_{clk}. Then the measured $TRcal$ ($\widetilde{TRcal}$) can be represented by:

$$\widetilde{TRcal} = nT_{clk} = TRcal + e_{TR} \tag{4.6}$$

where T_{clk} is the reciprocal of f_{clk}, $n = \frac{TRcal}{T_{clk}}$ or $n = \frac{TRcal}{T_{clk}} + 1$, where x denotes the largest integer not greater than x and e_{TR} is the quantization error of $TRcal$.

Figures 4.18 and 4.19 show the maximum possible BLF error due to quantization error when clock rate is 1.28 MHz and 1.92 MHz respectively. The bolder line is the error limit defined in the EPC C1G2 protocol. All violations are marked with the circle. They show that when clock rate is 1.28 MHz, BLF error may outrun the bound specified by EPC C1G2 protocol.

4.3.4 Clock-Related Low-Power Techniques

Reduction of the clock rate is a widely used low power technique. According to the above analysis, the baseband contains three clock domains: the sampling clock domain, the backscattering clock domain and the controlling clock domain. To obtain adequate time resolution, the demodulator and timing control module should operate at the clock rate computed in Section 4.3.3. This can easily be taken from the waveform of FM0 and Miller that the modulator module should operate at a frequency which is twice of BLF. Other modules, such as the finite state machine and the non-volatile memory interface

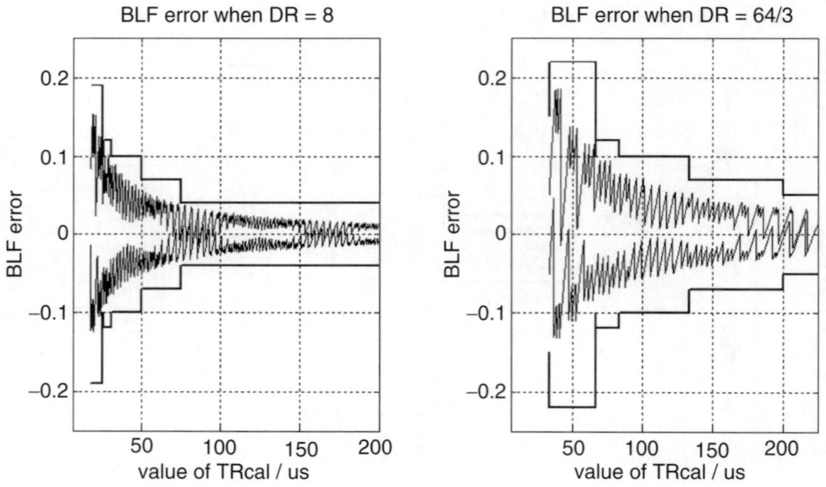

Figure 4.19 BLF error when the clock rate is 1.92 MHz.

can operate at the lowest frequency. Through grouping clocks, power and performance can be well balanced.

Not all registers work at every clock cycle. The technique of disabling the clock of registers when the values are not changed, known as clock gating, is effective in lowering power consumption of digital circuits. Figure 4.20 illustrates clock gating. When the signal EN is inactive (low), no clock pulse can be propagated to the clock pin of registers. Compared to its equivalent implementation using the multiplexer (left), clock gating outperforms it at power and area cost. The latch (sometimes the register) placed before the AND gate is used to eliminate glitches. Clock gating can be applied at every level of hierarchy, and it can be inserted both manually by RTL coding and by EDA tools such as Design Compiler of Synopsys during synthesis [41].

Another low power technique is based on DET FF (dual edge triggered flip-flop). Using DET FF in the demodulator and timing module can release the highest clock rate to a half. Though this will introduce more dynamic power, it can significantly reduce the power of oscillator in the analog front end and contribute to the overall decrease of power consumption.

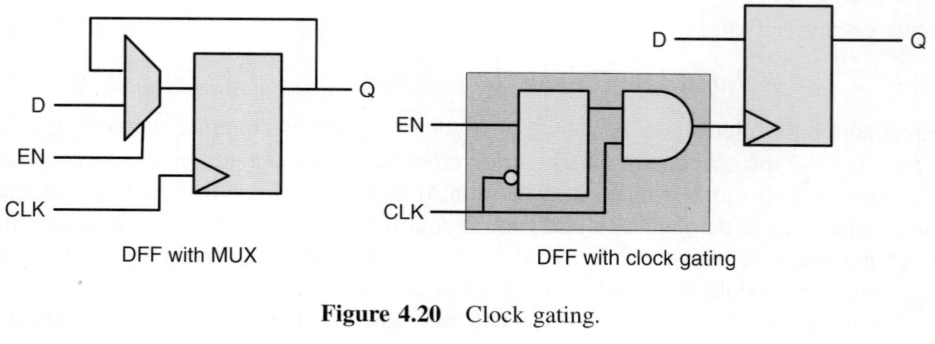

Figure 4.20 Clock gating.

4.3.5 Sub-Threshold Digital Circuit

Standard CMOS logic operates less than a voltage supply greater than the sum of threshold hold voltages of PMOS and NMOS FET. The sub-threshold digital circuit is a technique that lets the MOS FET operate at a voltage slightly higher (100 mV greater) or no higher than the threshold voltage. In this case, the current in the sub-threshold region is:

$$I_{Dsub} = \frac{W}{L}\mu_n C_{ox} n \left(\frac{KT}{q}\right)^2 e^{\frac{V_{GS}-V_{T,g}}{kT/q}} \left(1 - e^{-\frac{V_{DS}}{kT/q}}\right) \tag{4.7}$$

where $V_{T,g} = V_T + nkT/q$, V_T is the threshold voltage, n is empirical parameters, with $n \geq 1$. In sub-threshold technology, circuits are derived by the I_{Dsub}. The propagation delay of an inverter is:

$$T_d = \frac{KC_g V_{dd}}{I_{0,g} e^{\frac{V_{GS}-V_{T,g}}{nV_{th}}}} \tag{4.8}$$

where, C_g is the effective capacitance load. For a digital circuit, whose critical path delay is ξ times of inverter propagation delay, the maximum operating frequency is

$$f_{\max} = \frac{1}{\xi \cdot T_d} = \frac{I_{0,g} e^{\frac{V_{GS}-V_{T,g}}{nV_{th}}}}{\xi KC_g V_{dd}} \tag{4.9}$$

Sub-threshold technology is more than just decreasing the power supply voltage. Circuits operating in sub-threshold region should be carefully designed. Wang [42] gives an example of how to balance the leakage current by modifying the logic architecture.

In contrast to sub-threshold analog circuit design, the difficulty of sub-threshold digital circuit design is integrating it into existing EDA flows. The key point of being compatible with existing EDA flow is to design sub-threshold libraries for synthesis, automatic placing, routing and static timing analysis.

4.3.6 Adiabatic Circuit

The adiabatic circuit is also called the energy-recovery circuit. It achieves low power consumption by restricting the currents to flow across devices with low voltage drop and by recycling the energy stored in their node capacitors using an AC-type power supply rather than a DC type [43]. The adiabatic circuit must follow the following key rules:

1. A device can be turned on only when the source-drain voltage is zero.
2. The source-drain voltage can be changed only when the device is off.
3. Any voltage change must be done gradually.

An example of adiabatic CMOS logic circuits is shown in Figure 4.21. The equivalent diagram of charging and discharging process is shown in Figure 4.22. Instead of holding a constant supply voltage, the power source of adiabatic circuit rises linearly during the charging process. The voltage drop on the effective resistance of the driven device is

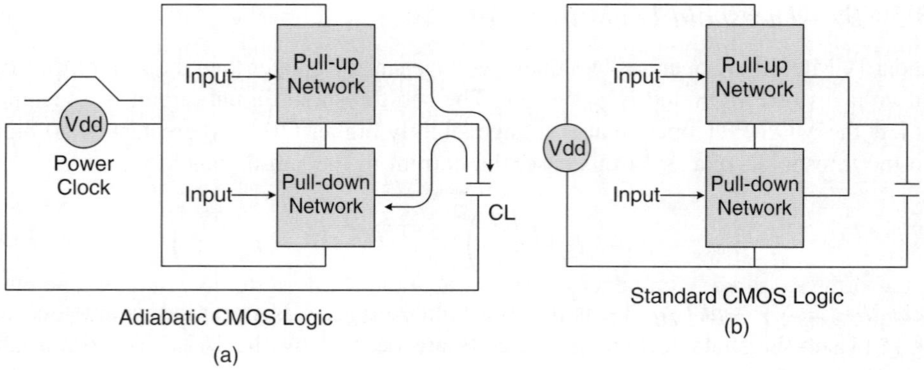

Figure 4.21 Example of adiabatic CMOS circuit (a) versus standard CMOS Logic (b).

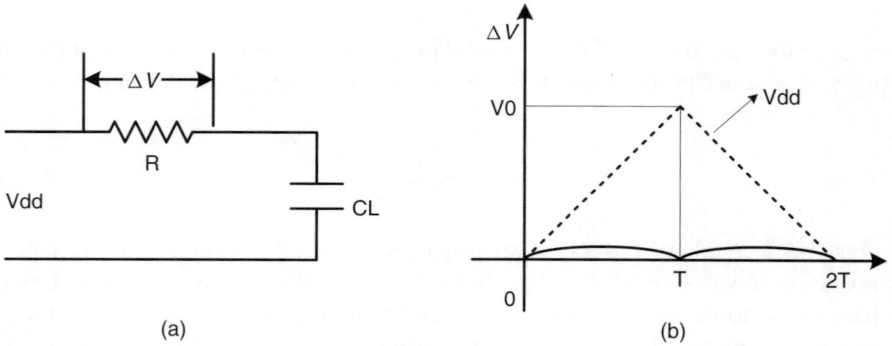

Figure 4.22 Energy dissipation in the adiabatic circuit.

small; energy dissipation on it, therefore, is small too. The overall energy dissipated in the transition is reduced to:

$$E \approx 2\frac{RC}{T}CV^2 \tag{4.10}$$

When T is sufficiently larger than RC, energy dissipation during the transition can become almost negligible.

The EDA for adiabatic digital circuits design encounters a number of challenges. Because the power clock differs from conventional concept of clocks, there is still a long way to make it compatible with current timing-driven synthesis and placing and route EDA tools.

4.4 RFID Tag Performance Optimization

In terms of the applications of RFID, high performance and low cost are two equally important aspects that impose trade-offs in the design of an RFID tag. New technologies need to be explored to improve the RFID tag performance. This section mainly summarizes the corresponding optimization technologies, and mentions the possible future technologies in the RFID tag design. Finally, several products are introduced and comparative results of their features are presented, showing the level of development of tag design.

4.4.1 Low Power

Tag performance directly influences the success of the whole system, including read range, singulation speed, read rate, etc. [44]. Low power design is always an important issue in backscattering RFID system. It can improve the tag's performance, including communication reliability and read/write range. However, accounting for decreasing power consumption and battery-assisted/semi-passive system, the read range is not only determined by the tag's power consumption, but also by the reader's sensitivity. Cooperative design between the reader and the tag in RFID system is very important. At the same time, the equivalent impedance of antenna is influenced by the environmental temperature, the surrounding objects, and so on. The impedance matching between tag antenna and chip will be altered so that the read range is shortened. An automatic impedance matching system (AIMS) may be considered a good solution to this problem. AIMS includes an impedance mismatch detector, a tunable matching network, a control unit, etc., which are always used for transceiver antenna impedance matching. Recently, the technology has been developed in RFID NFC system [45]. An on-chip tuner should be implemented to adjust for variations in antenna impedance. The tag system measures the output voltage of the rectifier to estimate the impedance mismatching, then dynamically adjusts the capacitor array according to the negative feedback network, and finally it provides the optimal matching state and increases the read range.

The singulating speed is determined by different anti-collision algorithms, the communication speed, the decoding arithmetic, etc. A detailed analysis has been given in [44]. Now the singulating speed is calculated based on tag-read operation. But for some extra applications, a high tag write speed is very important. For example, in the WULIANGYE application, which is a famous wine in China, RFID tags on wine bottles need to be encoded quickly when they move along a bottling line, so that the singulating write speed becomes a key factor.

The read rate is defined as a percentage of successful read operation. In real applications, RFID readers are not always able to access tags on a 100% basis. Currently, research is being conducted to resolve this problem, and analysis and related improvement methods have been given in [44]. These methods include the Swiss cheese effect, power management and state recovery technology, and automatic impedance matching.

4.4.2 Low Cost

Reducing the RFID tag cost is the key element to widen the applications of RFID technology. Now, the average price of passive UHF RFID tag is several cents, and printed RFID tags have a lower cost. In general, the tag chip accounts for 55% of the total tag cost, and antenna manufacture and package costs occupy 45%.

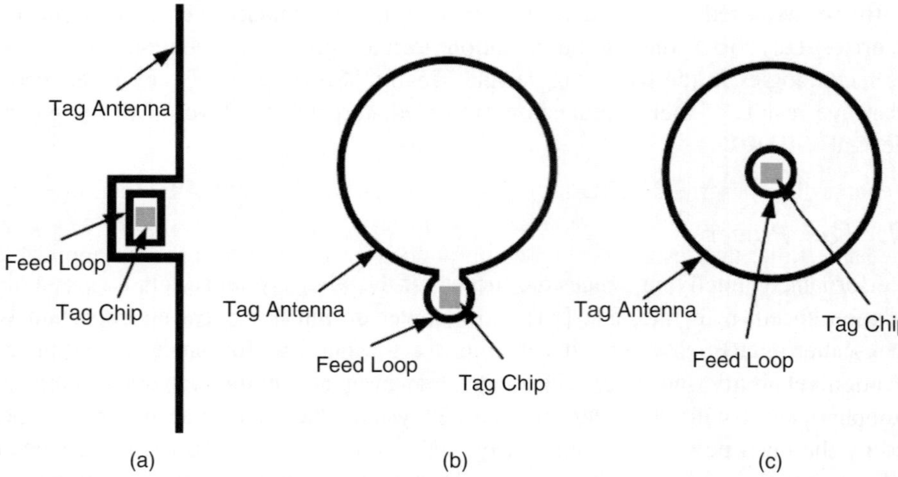

Figure 4.23 The different packages of on-chip antenna and tag chip.

Besides reducing the chip area, low cost can be achieved by using:

- standard CMOS process including eNVM memory and Schottky diode always used in rectifier circuit design;
- an on chip antenna (OCA).

RFID tags with on chip antenna (OCA) have several advantages including: reducing the assembly cost of RFID tags, providing more universal applications, especially item-level tagging (ILT) which puts a strict demand on tag size and cost [46]. Figure 4.23 presents the different packages of on-chip antenna and tag chip for the different applications. The detailed design technology is related to antenna theory, and is beyond the scope of this chapter.

Next, we compare several UHF tag (chip) features in Table 4.5. All five tag chips are the most up-to-date products from Impinj, Texas Instruments, Philips, Alien, and Quanray (a rising RFID design house in China). Different characteristics are compared including read/write sensitivity, programmable non-volatile memory capacity and the chip area. From the view of power consumption, Higgs™-3 is the lowest one, which is only

Table 4.5 Comparison of several UHF tag chip products.

Manufacturer	Frequency range	Read/write sensitivity	Programmable NVM	Area (μm^2)
[47]	860–960 MHz	−15 dBm/−12 dBm	256-bit	720 × 720
[48]	860–960 MHz	−13 dBm/−9 dBm	192-bit	–
[40]	860–960 MHz	7 m, US 6.6 m, Europe	512-bit	450 × 450
[2]	860–960 MHz	−18 dBm	800-bit	650 × 650
[50]	860–960 MHz	−14 dBm	2560-bit	720 × 720

−18 dBm. If we suppose that the conversion efficiency of the rectifier is about 30%, and the power supply is 1.5 V, then the total read current is 3.17 μA.

4.5 Conclusion

This chapter has described RFID tag chip design. Design aspects of the tag architecture and of every module are given in detail, and some design considerations are also explained. At the same time, the design techniques introduced in this chapter are also applicable to semi-passive and active RFID tag chip design. Low-power and low cost design techniques are always the most important research directions in order to widen the applications of RFID systems. Since some design techniques are immature, such as eNVM compatible with the standard CMOS process and adiabatic circuit technology, the tag chip using these techniques is still under development. However, the authors believe that, with the advancements in research, these techniques will be used in tag chip products, and more new design technologies will be developed for the application in RFID systems.

Problems

1. How to set parameters *TRcal* and *DR* of Query command to let tag replay at a BLF equal to:
   640 KHz
   320 KHz
   160 KHz
   80 KHz
2. Calculate CRC16 word for the following data (in Hex);
   0000
   0800_1111
   1000_1111_2222
   1800_1111_2222_3333
3. Explain why there should be a latch in clock-gated architecture.
4. Try to construct a DET using a positive edge DFF and a negative edge DFF.
5. List the advantages and disadvantages of the Miller Code.
6. Please list three main factors which affect the power of tag baseband, and the corresponding low-power techniques.
7. List the methods to round a number to an integer. What is the effect of employing different rounding methods in tag baseband?
8. The power threshold for tag Ptag,th is −14 dBm, and power consumption of the circuits Pload as a whole is 12 μW. What is the PCE η of the rectifier?
9. Please check the system architecture of Figure 4.1. A multi-level supply voltage generation method is to be applied in the design, and supply voltage requirements for these building blocks are 0.65 V for Demod, 0.8 V for RNG, 0.8 V for POR generator,

0.85 V for bias and regulator, 0.6 V for clock generator, 0.6 V for baseband, and 1.5 V for EEPROM. How to group the building blocks in order to get the systematic low voltage performance?

10. What is the operation region of biasing transistor Mb in the cloud chart of Figure 4.6, and why?

11. If a 1.28 MHz system clock is applied to the circuit in Figure 4.8 as fast clock, why should the sampling circuit be a T flip-flop instead of D flip-flop?

NB Solutions to the problems are provided on the book's website.

References

[1] Zhu, Z. (2004) RFID analog front end design tutorial. Available at: http://autoidlab.eleceng.adelaide.edu.au/Tutorial.html.

[2] Alien (n.d.) Datasheet: Higgs™-3 EPC Class 1 Gen 2 RFID Tag IC.

[3] Barnett, R., Balachandran, G., Lazar, S., Kramer, B., Konnail, G., Rajasekhar, S., and Drobny, V. (2007) A passive UHF RFID transponder for EPC Gen 2 with −14 dBm sensitivity in 0.13 μm CMOS, *IEEE International Solid-State Circuits Conference, Digest of Technical Papers*, pp. 582–583.

[4] Giustolisi, G., Palumbo, G., Criscione, M., and Cutrì, F. (2003) A low-voltage low-power voltage reference based on subthreshold MOSFETs, *IEEE Journal of Solid-State Circuits*, **38**(1): 151–154.

[5] Serra-Graells, F. and Huertas, J. (2003) Sub-1-V CMOS proportional-to-absolute temperature reference, *IEEE Journal of Solid-State Circuits*, **38**(1): 84–88.

[6] Che, W., Yan, N., Yang, Y., and Min, H. (2008) A low voltage low power RF/analog front-end circuit for passive UHF RFID tag, *Chinese Journal of Semiconductors*, **29**(3): 434–437.

[7] Hu, J. and Min, H. (2005) A low power and high performance analog front end for passive RFID, in *IEEE Workshop on Automatic Identification Advanced Technologies*, pp. 199–204.

[8] Facen, A. and Boni, A. (2006) A CMOS analog frontend for a passive UHF RFID tag, in *International Symposium on Low Power Electronics and Design*, pp. 280–285.

[9] Mandal, S. and Sarpeshkar, R. (2007) Low-power CMOS recti?er design for RFID applications, *IEEE Transaction on Circuits and Systems – I: Regular Papers*, **54**(6): 1177–1188.

[10] Dickson, J.F. (1976) On-chip high-voltage generation MNOS integrated circuits using an improved voltage multiplier technique, *IEEE Journal of Solid State Circuits*, **11**: 374–378.

[11] Karthaus, U. and Fischer, M. (2003) Fully integrated passive UHF RFID transponder IC with 16.7-μW minimum RF input power, *IEEE Journal of Solid State Circuits*, **38**(10): 1602–1608.

[12] Kotani, K. and Ito, T. (2007) High efficiency CMOS rectifier circuit with self-vth-cancellation and power regulation functions for UHF RFIDs, *IEEE Asian Solid-State Circuits Conference*, pp. 119–122.

[13] Wang, X., Jiang, B., Che, W., Yan, N., and Min, H. (2007) A high efficiency AC-DC charge pump using feedback compensation technique, in *IEEE Asian Solid-State Circuits Conference*, pp. 253–255.

[14] Sasaki, A. and Ito, T. (2008) Differential-drive CMOS rectifier for UHF RFIDs with 66% PCE at −12 dBm input, in *IEEE Asian Solid-State Circuits Conference*, pp. 105–108.

[15] Bucci, M., Germani, L., Luzzi, R., Trifiletti, A., and Varanonuovo, M. (2003) A high-speed oscillator-based truly random number source for cryptographic applications on a smart card IC, *IEEE Transactions on Computers*, **52**(4): 403–409.

[16] Nakamoto, H., Yamazaki, D., Yamamoto, T., Kurata, H., Yamad, S., Mukaida, K., Ninomiya, T., Ohkawa, T., Masui, S., and Gotoh, K. (2006) A passive UHF RFID tag LSI with 36.6% efficiency CMOS-only rectifier and current-mode demodulator in 0.35 μm FeRAM technology, in *IEEE International Solid State Circuits Conference*, pp. 1201–1210.

[17] Balachandran, G. and Barnett, R. (2008) A 440-nA true random number generator for passive RFID tags, *IEEE Transactions on Circuits and on Circuits and Systems – I: Regular Papers*, **55**(11): 3723–3732.

[18] Chen, W., Che, W., Bi, Z., Wang, J., Yan, N., Tan, X., Wang, J., and Min, H. (2009) A 1.04 μW truly random number generator for Gen2 RFID tag, in IEEE Asian Solid State Circuits Conference, accepted.

[19] Luo, Q., Guo, L., Li, Q., Zhang, G., and Wang, J. (2009) A low-power dual-clock strategy for digital circuits of EPC Gen2 RFID Tag, in *IEEE International Conference on RFID*, pp. 7–14.

[20] Holleman, J., Otis, B., Bridges, S., Mitros, A., and Diorio, C. (2006) A 2.92 µW hardware random number generator, in *Proceedings of the 32nd European Solid State Circuits Conference*, pp. 134–137.

[21] Ruhkin, A., Soto, J., Nechvatal, J., Smid, M., and Barker, E. (2001) A statistical test suite for random and pseudorandom number generators for cryptographic applications, NIST, Tech. Rep. 800–22.

[22] Nakamoto, H., Yamazaki, D., Yamamoto, T., Kurata, H., Yamada, S. Mukaida, K., Ninomiya, T., Ohkawa, T., Masui, S., and Gotoh, K. (2007) A passive UHF RF identification CMOS tag IC using ferroelectric RAM in 0.35-um technology, *IEEE Journal of Solid State Circuits*, **42**: 101–110.

[23] Derbenwick, G.F. (2008) Invited paper FR008 embedded ferroelectric memory for RFID tag applications, *17th IEEE International Symposium on the Applications of Ferroelectrics*, vol. 2.

[24] Lee, M.C., Barsatan, R., and Chan, M. (2007) OTP memory for low cost passive RFID tags, in *IEEE Conference on Electron Devices and Solid State Circuits*. pp. 633–636.

[25] Barsatan, R., Man, T.Y., and Chan, M. (2006) A zero-mask one-time programmable memory array for RFID applications, in *IEEE International Symposium on Circuits and Systems*, pp. 975–978.

[26] *Science News* (2007) Next-generation RAM: Remembering the future, *ScienceDaily*. Available at: http://www.sciencedaily.com/releases/2007/12/071221174912.htm.

[27] Dixian, Z., Na, Y., Wen, X., Liwu, Y., Junyu, W., and Hao, M. (2008) A low-power single-poly non-volatile memory for passive RFID tags, *Journal of Semiconductors*, **29**(1): 99–104.

[28] Pesavento, A. and Hyde, J.D. (2007) PFET nonvolatile memory, United States Patent, Patent No.: US, 7,221,596B2.

[29] Ohsaki, K., Asamoto, N., and Takagaki, S. (1994) A single poly EEPROM cell structure for use in standard CMOS processes, *IEEE Journal of Solid State Circuits*, **29**: 311–316.

[30] Raszka, J., Advani, M., Tiwari, V., Varisco, L., Hacobian, N.D., Mittal, A., Han, M., Shirdel, A., and Shubat, A. (2004) Embedded flash memory for security applications in a 0.13 µm CMOS logic process, in *IEEE International Solid-State Circuits Conference, Digest of Technical Papers*, vol.1.

[31] Kee-Yeol, N. and Kim, Y.-S. (2006) High-performance single polysilicon EEPROM with stacked MIM capacitor, *IEEE Electron Device Letters*, **27**: 294–296.

[32] Schuster, S.E. and Matick, R.E. (2009) Fast low power Edram hierarchical differential sense amplifier, *IEEE Journal of Solid State Circuits*, **44**: 631–641.

[33] Makosiej, A., Nasalski, P., Giraud, B., Vladimirescu, A., and Amara, A. (2008) Double-gate sub-32nm CMOS SRAM current and voltage sense amplifiers, insensitive to process variations and transistor mismatch, *IEEE International SOI Conference*, pp. 63–64.

[34] Li, N., Huang, Z., Jiang, M., and Inoue, Y. (2008) High efficiency four-phase all PMOS charge pump without body effects, in *International Conference on Communications, Circuits and Systems*, pp. 1083–1087.

[35] Hsu, C.-P. and Lin, H. (2007) Analysis of power efficiency for four-phase negative charge pumps with body potential control, in *IEEE International Conference on Integrated Circuit Design and Technology*, pp. 1–4.

[36] Yan, N., and Min, H. (2006) High efficiency all-PMOS charge pump for low-voltage operations, *IET Electronics Letters*, **42**(5).

[37] Bin, W. (2006) Graded-junction high-voltage MOSFET in standard logic CMOS. United States Patent, Patent No.: US, 7145203.

[38] Bianchi, R.A., Monsieur, F., Blanchet, F., Raynaud, C., and Noblanc, O. (2008) High voltage devices integration into advanced CMOS technologies, in *IEEE International Electron Devices Meeting*, pp. 1–4.

[39] Sun, W.F. and Shi, L.X. (2003) High reliability HV-CMOS transistors in standard CMOS technology, in *Proceedings of the 10th International Symposium on the Physical and Failure Analysis of Integrated Circuits*, pp. 25–28.

[40] Impinj, Inc. (2008) PRODUCT BRIEF AEON® MTP EEPROM Architecture. Gen 2 Tag clock rate – what you need to know, Impinj White Paper. Available at: www.impinj.com/WorkArea/downloadasset.aspx?id = 2541.

[41] Impinj, Inc. (2008) Clock Gating Methodology for Power and CTS QoR, Synopsis White Paper.

[42] Wang, A. and Chandrakasan, A. (2005) A 180-mV Subthreshold FFT processor using a minimum energy design methodology, *IEEE Journal of Solid-State Circuits*, **40**(1): 310–319.

[43] He, Y. and Min, H. (2007) Adiabatic circuit applied for LF tag, Auto-ID Labs White Paper WP-HARDWARE-041.

[44] Miles, S.B., Sarma, S.E., and Williams, J.R. (2008) *RFID Technology and Applications*. Cambridge: Cambridge University Press.

[45] Roland, M., Witschnig, H., Merlin, E., and Saminger, C. (2008) Automatic impedance matching for 13.56 MHz NFC antennas, *Communication Systems, Networks and Digital Signal Process*, CNSDSP 2008. 6th International Symposium on, pp. 288–291.

[46] Jingtian, X., Na, Y., Wenyi, C., Xiao, W., Hongyan, J., and Hao, M. (2009) On-chip antenna design for UHF RFID, *Electronics Letters*, **45**(1).

[47] Impinj (n.d.) Datasheet: EPCglobal™ Generation 2 RFID Monza™.

[48] Texas Instruments (n.d.) Datasheet: UHF Gen2 STARP RI-UHF-STRAP-08.

[49] Philips (n.d.) Datasheet: SL3 ICS 10, UCODE EPC G2.

[50] Quanray (n.d.) Datasheet: Qstar EPC C1G2 RFID Tag IC.

5

Design of Passive Tag RFID Readers

Scott Chiu

Intel Corporation

5.1 Overview

With the recent advances in global standardization, RFID has become an important tool for supply chain management [1] while expanding into many possible consumer applications through initiatives like Mobile RFID [2].

Similar to the evolution of other innovative technologies, many systems exist in the field which can be characterized by:

1. **The operating frequency band**: In general, RFID relies on unlicensed bands for communications. For example, ISM[1] bands in the United States, HF (13.5 MHz), VHF (40 MHz), or UHF (860–960 MHz, or 2.4 GHz). Currently HF is more prevalent in applications while UHF has been gaining popularity with lower cost tags, and the flexibility of a longer reach.
2. **How the tag and the communication link are powered**: Passive tags are powered by harvesting the RF energy transmitted from the reader, and the tag-to-reader communication is maintained by modulating (backscattering) the reader-transmitted RF energy; active tags are battery-powered, and are capable of sending signals back to the reader; semi-active tags fall in between – these tags are battery-powered but rely on backscattering for uplink communications. The trade-off here is cost versus reach. With on-tag power source, the active or semi-active tags have a longer range, but with a higher cost and a shorter life-span (dependent on how the battery power is consumed).

[1] Industrial, Scientific, and Medical bands.

RFID Systems: Research Trends and Challenges Edited by Miodrag Bolić, David Simplot-Ryl, and Ivan Stojmenović

3. **The coupling mechanism**: Near-field with inductive or electrical field coupling, or far-field with electro-magnetic wave coupling.

This chapter will focus on the emerging UHF (860–960 MHz) ISO 18000-6C compliant passive tag RFID reader design. Readers with an interest in the RFID evolutions should refer to [1] for detailed discussions of legacy systems.

In the following, Section 5.2 offers an introduction to passive RFID operations using the ISO 18000-6C [3] air interface. The emphasis is to make a link from the issues encountered in the field to the mitigation specified in air interface. An example of an inventory round is then used to demonstrate the tag access. Section 5.3 describes how the application impacts the design decisions, and the approach to derive the design specifications. A generic implementation is used to demonstrate the challenges of reader design, and methods to mitigate the non-idealities. Section 5.4 discusses more advanced design topics of the reader designs, and Section 5.5 closes with a conclusion.

5.2 Basics of Passive RFID Operation

The RFID system is an ad hoc network formed between a reader and the tags covered by its RF energy field for a duration controlled by the reader. The underlying reader (interrogator) selects the target tag to establish a link and the start/end time of the communication – it serves as the master in the system and co-exists with other reader systems in the same physical location.

A passive tag is powered by energy harvested from the reader transmitted carrier power [4–6]. As shown in Figure 5.1, the reader initiates an inventory round (the period

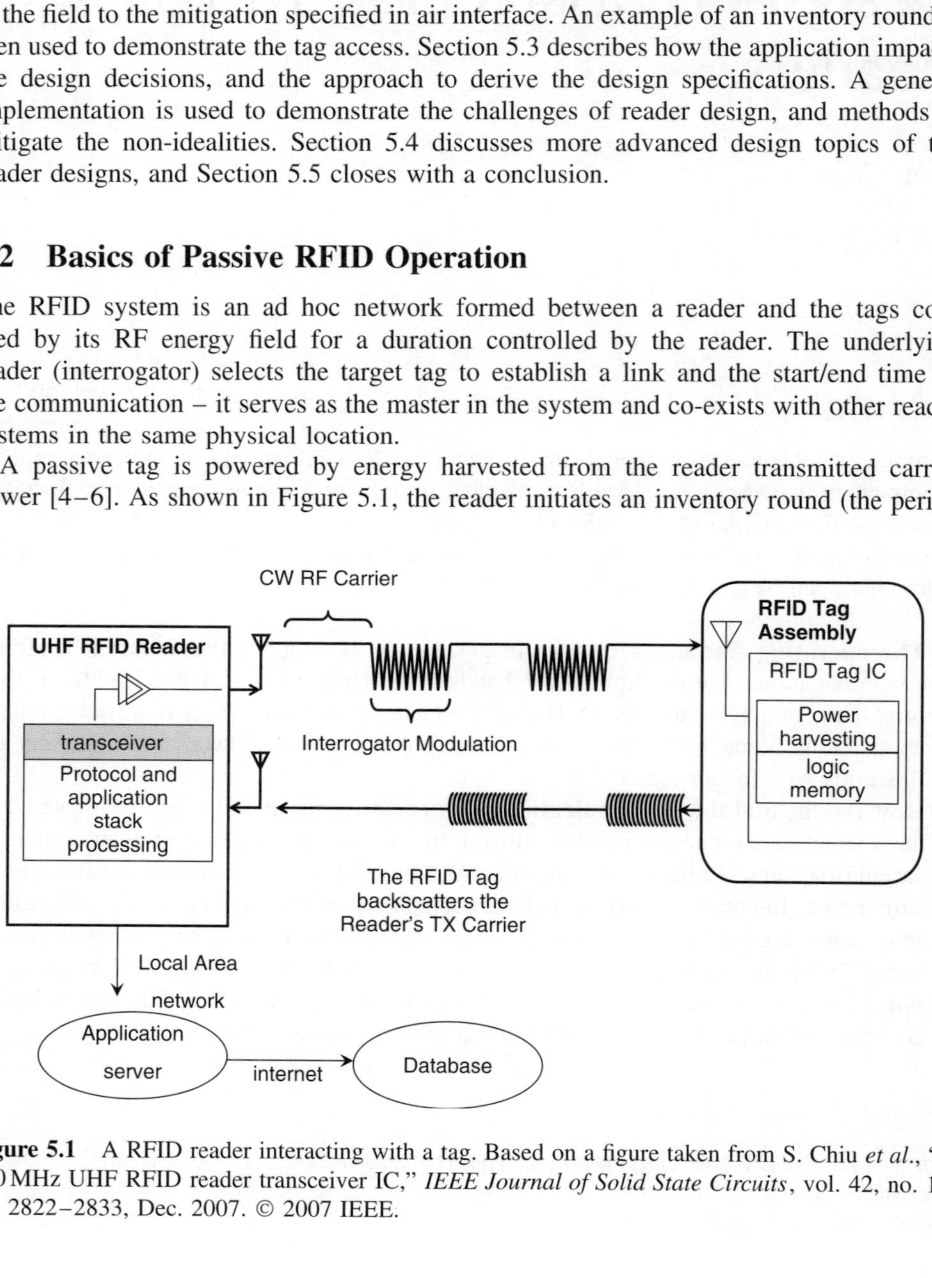

Figure 5.1 A RFID reader interacting with a tag. Based on a figure taken from S. Chiu *et al.*, "A 900 MHz UHF RFID reader transceiver IC," *IEEE Journal of Solid State Circuits*, vol. 42, no. 12, pp. 2822–2833, Dec. 2007. © 2007 IEEE.

needed to read all tags within its field of view) by sending an unmodulated carrier signal to "energize" the tags. The unmodulated carrier (CW) is sent throughout the inventory round, except during the period the reader talks, to keep the tags alive. Once the tags are powered up, the reader enters "talk" mode to initiate the communication by modulating its carrier to issue a command. After sending a command, the reader stops modulation to enter "listen" mode. During this period, the reader sends a CW while turning on the receiver to read tag responses. The tag of interest responds by backscattering the reader's CW signal. The backscattering is achieved by reflecting the incoming carrier power by changing the state of the tag antenna impedance. Note that the use of "talk" and "listen" is to emphasize the asymmetric nature of this half-duplex link.

5.2.1 An Introduction to ISO 18000-6C Air Interface

All RFID readers must abide by two types of rules for interacting with the tags: (1) local regulations established by a governing organization, or a government entity; (2) an air interface standard defining the protocols for the reader and tags to interact.

The aim of the local regulation is to ensure that the RFID reader and tag operation does not interfere with existing public or commercial services. In the United States, UHF RFID readers operate at the 902–928 MHz ISM (Industrial, Scientific, and Medical) band as defined by FCC (Federal Communications Commission) title 47 part 15 [7]. In Europe, the European Telecommunication Standard Institute (ETSI) set the guideline with ETSI 302 208 [8] in the 865–868 MHz band. These publications define the spectrum mask and maximum out-of-band emissions – for example, an ETSI compliant device can emit up to 33 dBm of power within 865.6 to 867.6 MHz with spectrum mask as shown in Figure 5.2. In terms of reader implementation, the in-band spectrum mask performance is usually dominated by the baseband filtering and transmitter linearity; the out-of-band emission is mostly determined by the inter-modulation of the carrier (which is at high power), and by spurious tones coupling into the frequency generation and transmit paths. The latter effect is often neglected in the early design process, and results in great pain late in a product development. Additionally, the adjacent and alternate channel spectrum masks are defined in reference to an absolute power level. This means the linearity requirement can be substantially relaxed for readers operating at a lower transmit power level.

The air interface consists of a set of rules governing the communication between the reader and the tags. There are two active international standard bodies for air interface definitions: the International Standard Organization (ISO) which defines RFID specifications including ISO 18000-6A, B, and C types of tags [9], and EPC global which is behind the EPCglobal Class 0 [10], Class 1 Generation 1 (C1G1) [11], and Class 1 Generation 2 (C1G2) [12]. The standards of ISO 18000-6C and EPC C1G2 are harmonized to reach a global standard, and are referred to as ISO 18000-6C in this chapter.

ISO 18000-6C consists of the physical layer interface and the command sets for implementing the protocol. The physical interface defines the modulation (waveform and spectrum mask), encoding scheme, data rate, link timing, and message constructs (preamble type, length, and message body). Table 5.1 summarizes the encoding and modulation types – the difference between reader to tag and tag to reader encoding schemes shows the asymmetric nature of the RFID system.

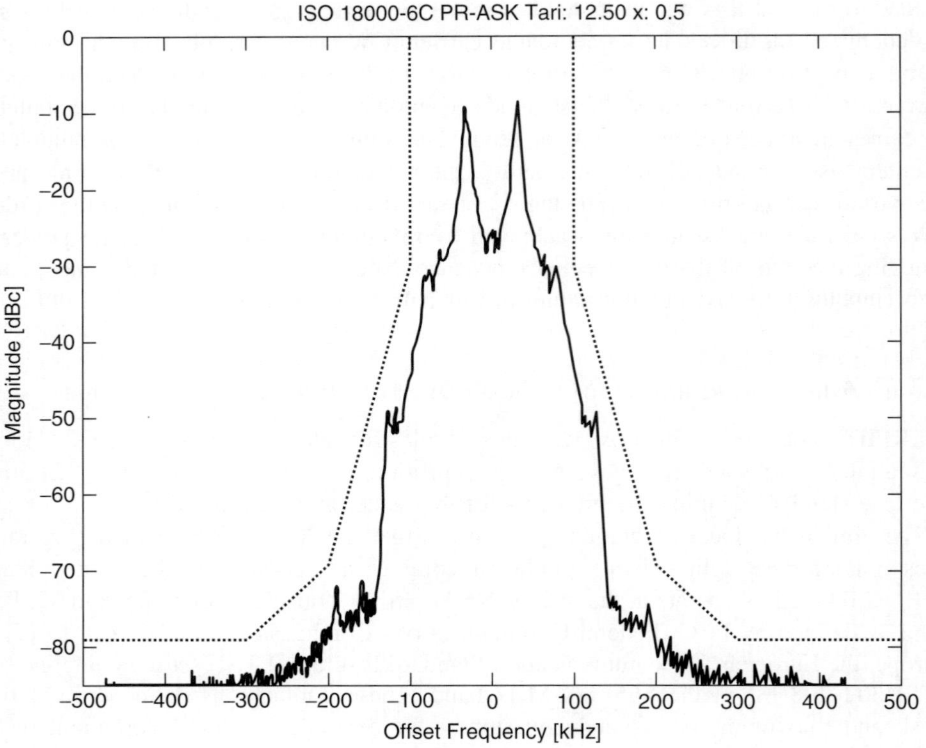

Figure 5.2 Transmit spectrum mask example: PR-ASK modulated reader with an ETSI 302-208 mask (Tari = 12.5 µs, with data '1' 0.5 Tari longer than data '0').

Table 5.1 Summary of ISO 18000 6-C encoding and modulation specifications.

Reader to tag		Tag to reader	
Encoding	Modulation	Encoding	Modulation
PIE (Pulse interval encoding)	DSB-ASK PR-ASK SSB ASK	FM0, Miller subcarrier	DSB-ASK PSK

For the downlink, all three reader-to-tag modulation schemes are forms of Amplitude Shift Keying (ASK) for the ease of tag demodulation. Double Side Band ASK (DSB-ASK) is a real-value modulation which offers the easiest implementation and the possibility of supply modulation [13]. Single-Side Band ASK (SSB-ASK) and Phase-Reversal ASK (PR-ASK) offer better spectrum efficiency with an increase in modulation complexity (more components in the transmit path). Based on the air interface standard, the bandwidth required is not absolute since the reader can lower the signaling rate to reduce the spectrum occupied (note that the signaling rate range is specified to be between 6.25 and

25 μs, and is controlled by reader through reader to tag preamble). This allows DSB-ASK to operate in the allocated channel except at a reduced data rate. Pulse Interval Encoding (PIE) is used for transmit coding since (1) the reader maintains the carrier wave power except for ½ Tari time slot at the end of a symbol. Tari is the basic reader to tag signaling time unit which is equal to length of a symbol "0." This allows the tag to harvest the maximum amount of power during an inventory cycle; (2) it allows the tags to decode by using its internal clock to differentiate an encoded "one" from a "zero" symbol through counting.

Since the tag demodulates the signal by tracking amplitude variations, excessive ripples can trigger false transitions while slow edges can blur the difference between a "1" to a "0." Thus the air interface defines a waveform mask in additional to the spectrum mask. Figure 5.3 shows the coding and modulation using a PR-ASK modulated reader to tag packet as an example. As depicted in Figure 5.3 a packet is composed of a preamble and the succeeding data symbols. The preamble of the packet consists of a delimiter, a "0" symbol, and two special PIE symbols RTcal and a TRcal. RTcal is equivalent in length to a "0" plus a "1" symbol for setting the tag symbol slicing threshold. TRcal defines the tag-to-reader link signaling rate with a length equal to the inverse of the link frequency.

For the uplink, both Amplitude Shift Keying (ASK) and Phase Shift Keying (PSK) are allowed in tag implementations. When the tag backscatters through changing the real portion of the antenna load, the modulation appears as ASK. When the scattering is done with changing the imaginary portion of the antenna load, the modulation appears as PSK.

There are two tag encoding schemes specified in ISO 18000-6C: FM0 or Miller modulated subcarrier (an example is shown in Figure 5.4 with various signaling rates). Use of the tag encoding scheme is chosen by the reader initiating the session. Both encoding schemes avoid signal contents at DC. FM0 is a baseband encoding which spreads the signal power around the carrier frequency. Miller uses frequency shifts to offset the signal

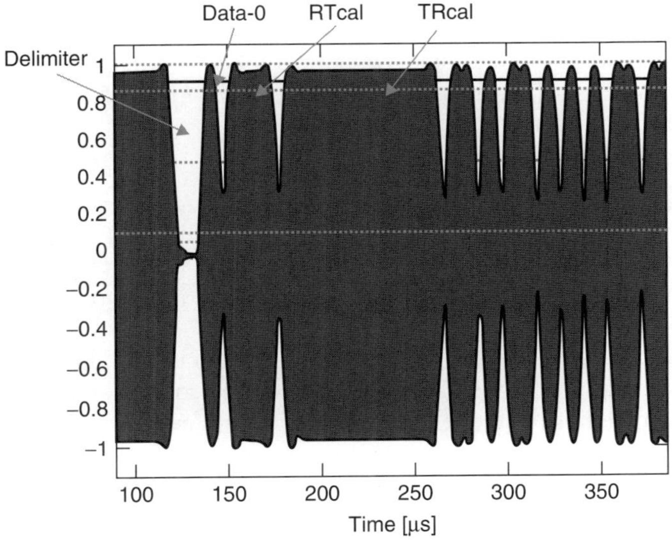

Figure 5.3 An example of a RFID reader-to-tag communication packet.

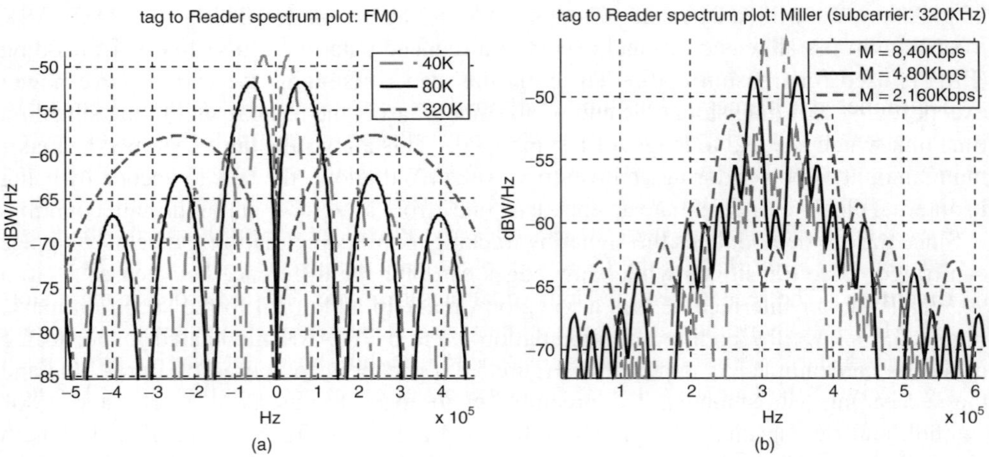

Figure 5.4 ISO 18000-6C tag modulation spectrum plot in reference to the carrier frequency: FM0 modulated (a); Miller subcarrier modulated (b), the subcarrier is shifted to 0 Hz.

power into a sub-carrier frequency. For example, for an 80-Kbps data with M = 4 Miller subcarrier modulation, the center of the signal spectrum is 320 KHz offset from the actual carrier. Miller modulated subcarrier encoding is especially useful in a dense reader environment due to the following: (1) offset tag responses to the subcarrier frequency to avoid the noise coupled from the reader transmission. To keep the tag powered throughout the inventory round, the reader is transmitting CW (carrier) at full power while receiving tag responses. Since the received tag power can be 90 dB[2] smaller than the transmitted power, the close-in noise of the transmitter dominates the receiver noise floor even considering a very good transmit-to-receive isolation. Moving the tag response away from the carrier mitigates this effect on tag sensitivity, especially in a dense-reader environment where other readers may be transmitting carrier at the same time; and (2) the subcarrier encoding provides more transitions in the same symbol period, which helps the receiver track the tag data rate variation at link frequency of 300 KHz for up to 15% initial tolerance and 2.5% drift within one data packet.

5.2.2 Tag Singulation and Access

After understanding the basic means of establishing a link between one reader and one tag, this section illustrates the case of communicating with many tags using the interface protocol.

Before the reader powers up the passive tags in its RF range, it does not have any prior knowledge of the tag population. The air interface protocol thus defines three sequential steps to establish the links with the targeted tags: select, inventory, and access, each with associated commands for managing a tag population.

[2] Assuming the transmitted carrier is at 30 dBm, with a tag backscattered power at −75 dBm, and 15 dB of transmit to receive isolation (in the case of using a bi-static configuration; to be discussed later in the chapter). The transmitter power is 90 dB higher than the tag reflection. However, depending on the targeted performance, this value can be high or lower in a specific system.

1. **Select**: During this phase, the reader commands the tag population to identify the group included into an inventory round. The Select command has comparison criteria specified by an array of four entries [MemBank, Pointer, Length, Mask]. MemBank determines the memory block under comparison – it can be UII (Unique Item Identifier) memory which contains a CRC-16 generated at power up, the protocol control word, and EPC (Electronic Product Code), TID (Tag Identification) memory for tag class information, or the user memory. CRC-16 is a 16-bit cyclic redundant check code computed to ensure the UII data integrity. Pointer and Length specify the target location and size; Mask is the bit string to compare against.
2. **Inventory**: The reader identifies individual tags from the selected group using a slotted ALOHA algorithm with (reader) query and (tag) acknowledgement. The Query command starts a new inventory round with a 16-bit slot count parameter Q. Each tag receiving the Query command generates a slot count between 0 to 2^Q-1. The tags with slot count 0 reply with its RN16 (a 16-bit random number from the tag pseudo-random number generator). To continue on the same inventory round, the reader issues QueryRep to decrement tags' slot counter by 1, and listens to the reply from tags with slot count 0. Selection of Q determines the statistical performance of this operation. Choosing a large Q avoids tag collisions (many tags reply in the same slot) at the expense of many unused slots; choosing too small the Q with respect to the tag population increases collisions and retries, which exhibits a degraded performance. Unfortunately, the tag population is not known before an inventory round. The air interface provides a command QueryAdjust to allow the reader to increment or decrement Q value in the current inventory round. This allows adaptive changes of Q during this phase based on heuristics (for example, from the collision statistics).
3. **Access**: Once a tag is uniquely indentified, the reader can choose to perform different operations ranging from read/write to the specified memory bank using a Read or a Write command, to lock the tag for secure access (using Lock command to prevent or allow access to the RFID password or a specific memory band), or even permanently disable the tag (using Kill command).

A possible sequence of reader and tag interactions is shown in Figure 5.5. In this example, the reader starts by picking a selected group of tags in a tag population, and begins an inventory round. The first query attempt results in collisions with multiple tags responding at the same time. In practice, the reader will follow an adaptive procedure to determine whether to collect more statistics or to change the Q – this example is a simplified case – the reader then sends a QueryAdjust command with a larger Q, and gets a tag reply RN16 (a 16-bit random number generated by the tag) successfully. The reader sends an ACK command with the same random number attached for tag to validate the data integrity. After verifying the RN16 is a match, the tag sends its unique item identifier with CRC-16.

At this point, the reader can decide to continue on the inventory process, or access the identified tag. In this example, the reader begins the access by issuing a req_RN command to "open" a tag. The tag replies a different RN16 (RN16_2 as seen in Figure 5.5) to be used as a key for the following access operations. The reader asks for the contents by issuing a read command with a specific memory bank in tag memory. After the tag replies with its stored data, the reader "closes" the selected tag, and resumes the inventory round with a QueryRep.

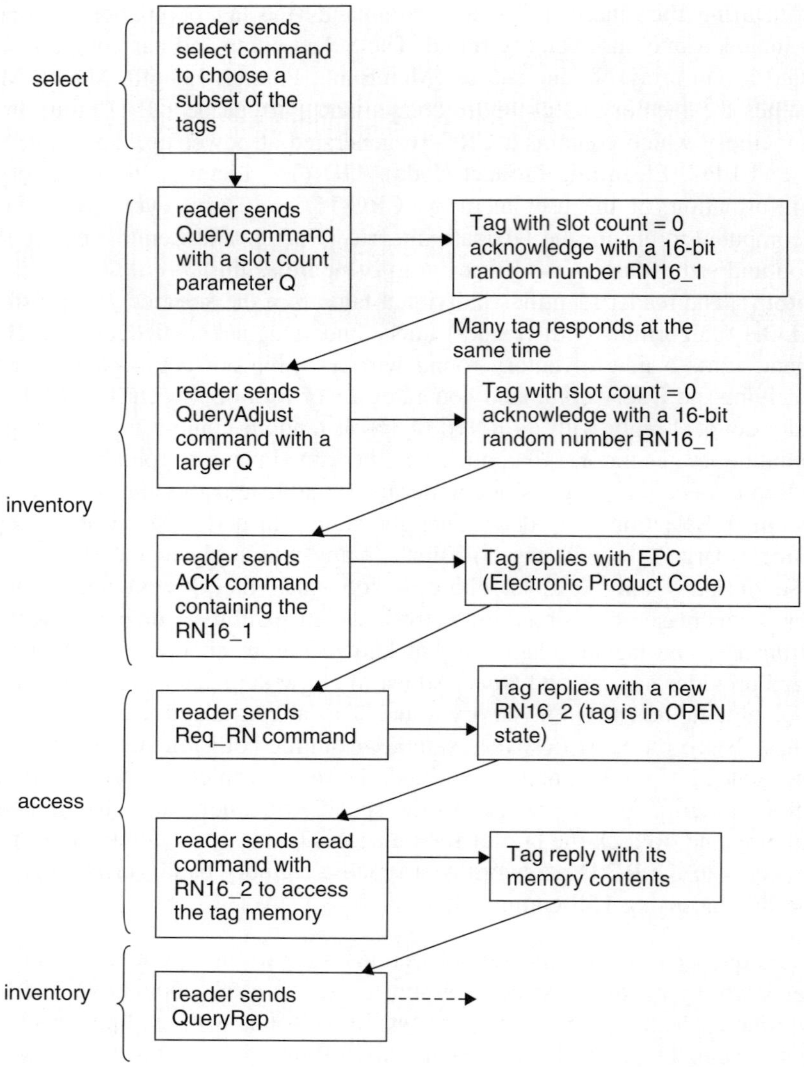

Figure 5.5 Example of a protocol followed between a reader and a tag.

5.3 Passive RFID Reader Designs

Depending on applications, commercially available UHF RFID readers have form factors ranging from the size of a notebook to palm top as illustrated in Figure 5.6. In this section, we will discuss the difference in design from a technical perspective by discussing the system specifications. A generic RFID reader implementation is then provided as an example to provide a link between the requirement and the circuit to support the functionality.

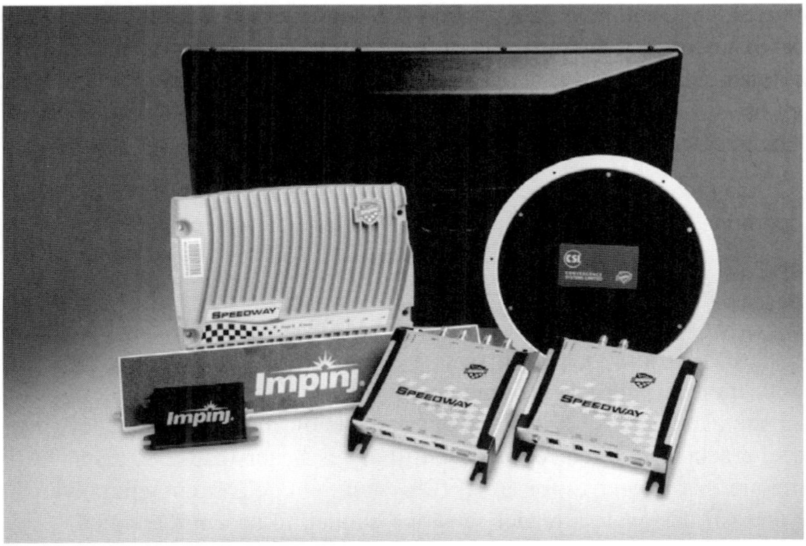

Figure 5.6 A few example implementations of UHF ISO 18000-6C compliant RFID readers. Reproduced by permission of © Impinj Inc.

5.3.1 RFID Reader Read Range and Transmitted Power

In principle, creating an UHF RFID reader specification is a practice no different from that of other wireless systems. Guided by a requirement document derived from standards, local regulations, and specific customer applications, the specification should encompass the parameters on dynamic range, linearity, and phase noise, etc. [14]. In the following, we will focus on deriving the relationship between the required reader sensitivity, read range, and transmitter power, which is specific to passive tags.

The performance of an RFID reader is characterized by its read range – unlike the conventional wireless communication devices, the read range of a passive RFID system is affected by three factors: (1) the power delivered and rectified to wake up the tag; (2) the power reflected from the tag; and (3) the reader sensitivity.

$$\mathrm{Pr} = \frac{Pt \cdot Gt \cdot Gr}{\left(4\pi \frac{d}{\lambda}\right)^2} \tag{5.1}$$

Equation 5.1 shows a general range equation describing the relationship of the received power at the receiver (Pr) to the transmitted power (Pt) referenced to an isotropic radiator, the transmitter antenna gain[3] (Gt), the receiver antenna gain (Gr), and the path loss (which is the term in the denominator based on the distance between them: d, and the radio wavelength: λ). This equation defines the power delivered to the tag.

[3] Antenna gain is to be discussed further in Section 5.3.2.1.

Referring back to Figure 5.1, the path from a reader to a tag is traversed twice between transmit and receive: once with the carrier sent from the reader, and again with the received power reflected from the tag; thus the pass loss doubles in dB comparing to other wireless systems. In addition, a loss term L_{tag} is added to account for the limitation in its antenna construction, polarization loss, and matching:

$$\mathrm{Pr} = \frac{Pt * Gt * Gr * L_{tag}}{\left(4\pi \frac{d}{\lambda}\right)^4} \tag{5.2}$$

For example, for an RFID system operates at 900 MHz UHF band, L_{tag} loss of 15 dB (3 dB of polarization loss and 12 dB of reflection loss for Raleigh range $\sim \lambda^{-4}$[1]), 3 dB gain on both transmit and receive antenna, 30 dBm transmitter, and the tags consuming 3.5 µW on the average with a 20% power conversion (harvest) efficiency, a read range of 9 m can be achieved when receiver sensitivity is better than −80 dBm. In the above, two criteria are met to budget for an RFID system: (1) deliver enough power to keep the tag alive; and (2) designate a reader sensitive enough to accept the signal.

Figure 5.7 depicts the read range versus receiver sensitivity and available power to the tag with the above assumptions. Note that the tag power consumption is a major part of the equation – if the tag consumes more power to stay alive (implementation needs more power or the conversion efficiency is lower), the link becomes downlink dominated – further improvement to receiver sensitivity will not yield any range increase. Though the range is limited by the tag on downlink currently, the RFID reader design should accommodate the improvement in tag performance over time.

Also implied in the equations is that the transmit power can be raised to improve the reader range. Higher transmit power requires much linear transmitter – a 1 dB increase in power means 3 dB increase in third order linearity requirement. This approach results in a less efficient power amplifier plus a larger case and board to ease thermal considerations.

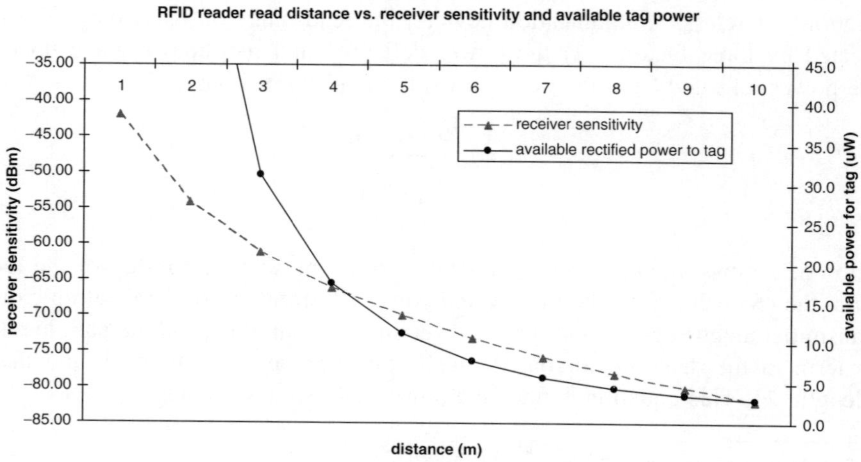

Figure 5.7 Sensitivity and available tag power versus reader range (from the example).

The other way to improve the system performance is to increase the reader or tag antenna gain. The reader antenna can be optimized based on size, cost, and applications. The fixed reader on the dock door usually has a larger size comparing to the lower performing hand-helds. In addition, tags are not created equal – at the same operating band, larger tags with more sophisticated antenna design tend to have better antenna efficiency. Therefore, it is important to select the tags for a given application.

5.3.2 RFID Reader Implementation

An UHF RFID reader is composed of four basic building blocks as depicted in Figure 5.8:

1. *Antenna interface* is where the RFID reader converts the electrical signal into electromagnetic waves and vice versa. Depending on the application, several antenna elements can be connected to a single reader through an antenna multiplexer. As an example, Figure 5.8 shows two antennas covering different locations accessing items on a conveyer belt.
2. *Transceiver* is the physical front-end translating bits into radio frequency transmission, or recovering backscattered tag responses into bits. To keep track of the operation status, a radio controller is used to: (1) monitor the reflected power level due to antenna impedance mismatch (in the case of someone disconnecting the transmit antenna); (2) transmit the power level for closed loop power control; (3) initialize the radio to operate at a target channel; and (4) ensure link timing conformance to the standard.
3. *Protocol and application processor* issues commands to the radio transceiver for tag inventory and read/write accesses described in Section 5.2.2. By embedding smartness to achieve performance on top of what is suggested in the air interface standards, the manufacturers differentiate their products in protocol processing, for example, using an improved Q algorithm to achieve faster inventory rounds [15]. In addition, one or more layers of the RFID system software stacks reside on the RFID reader with the rest located remotely based on applications. An example is the RFID system software architecture proposed by Microsoft [16] (we will leave a detailed description of the system software stack implementation to other chapters of this volume).

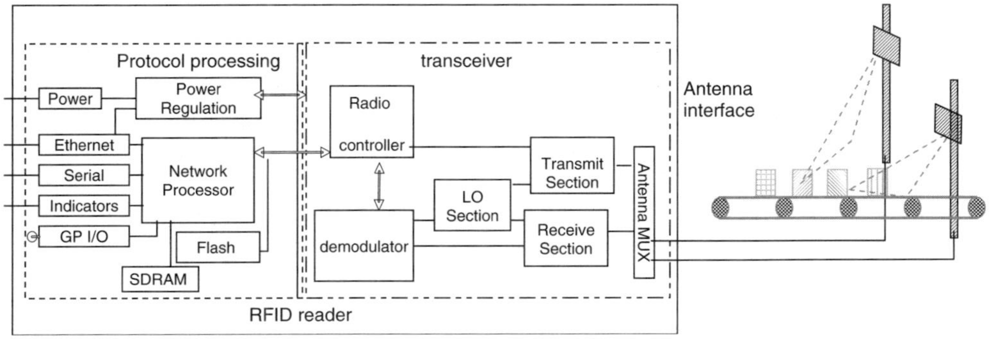

Figure 5.8 Block diagram of a generic RFID reader implementation. Reproduced by permission of © Intel Corporation.

4. *Network IO interface* is the link between the reader and the server(s) processing a higher level of the software stacks. Depending on the applications, one or more of the standard interfaces are offered – Ethernet is the predominant choice, especially with the possible use of power over Ethernet for supply generation for fixed-location readers. USB or PCMCIA are the choices for attaching an RFID reader to a portable device.

In the following, we will introduce the RFID specific hardware implementation of these components.

5.3.2.1 Antenna Interface

The antenna is the physical interface converting between electrical signals and electromagnetic waves. Depending on the application environment, the RFID reader designer need to prioritize different considerations based on cost, material, package, allowed space or geometric shapes. This multi-dimensional optimization makes antenna selection often determined by "engineering judgments." In this section, we will instead provide an introduction to the critical parameters with the aim of understanding the antenna used in RFID test/validations, followed by a discussion of antenna diversity techniques commonly used in current RFID readers.

The important parameters for an RFID reader antenna are:

- *Radiation pattern and directivity:* Isotropic radiation (the radio power emits evenly like a globe) exists only in theory as a reference. In real life, power emission varies depending on the direction looking into the radiator. The variation of the power as a function of the direction away from the antenna observed from the far-field is defined as the *radiation pattern*. To visualize the three-dimensional pattern in a two-dimensional plot, it is specified or plotted with respect to elevation and azimuth angles. Figure 5.9 depicts two antenna pattern plots – the narrow one is from a 10-element Yagi antenna, and the wider one is from a 3-element directional antenna. *Directivity* is a measure of how "directional" an antenna's radiation pattern is in reference to an isotropic radiator. 0 dB directivity means an isotropic antenna, and 3 dB equals a peak radiation at twice the power level as compared to an isotropic radiator. The "peak directivity" is usually specified for antenna.
- *Beam width:* The angle between two points (in the radiation pattern plot) where the magnitude of the radiation decreases by 50% (3 dB) from the peak of the main lobe. Using the same example from Figure 5.9, the 3-element antenna has a beam width of 70 degrees, while the Yagi's beam width is close to 40 degrees. In ETSI 302 208 [8], the antenna needs to have a beam width $\leq 70°$ in the horizontal direction for transmission between 500 mW to 2 W in a channel.
- *Efficiency:* The efficiency is the ratio between the power emitted from and the power delivered to the antenna. The loss is due to the resistive loss of the antenna or the impedance mismatches. For high power, high performance RFID systems, the efficiency of the antenna has a large impact on the overall power consumption.
- *Gain:* This is the ratio between the power transmitted in the radiation direction and an isotropic radiator (as a reference). The "peak" antenna gain is usually specified.

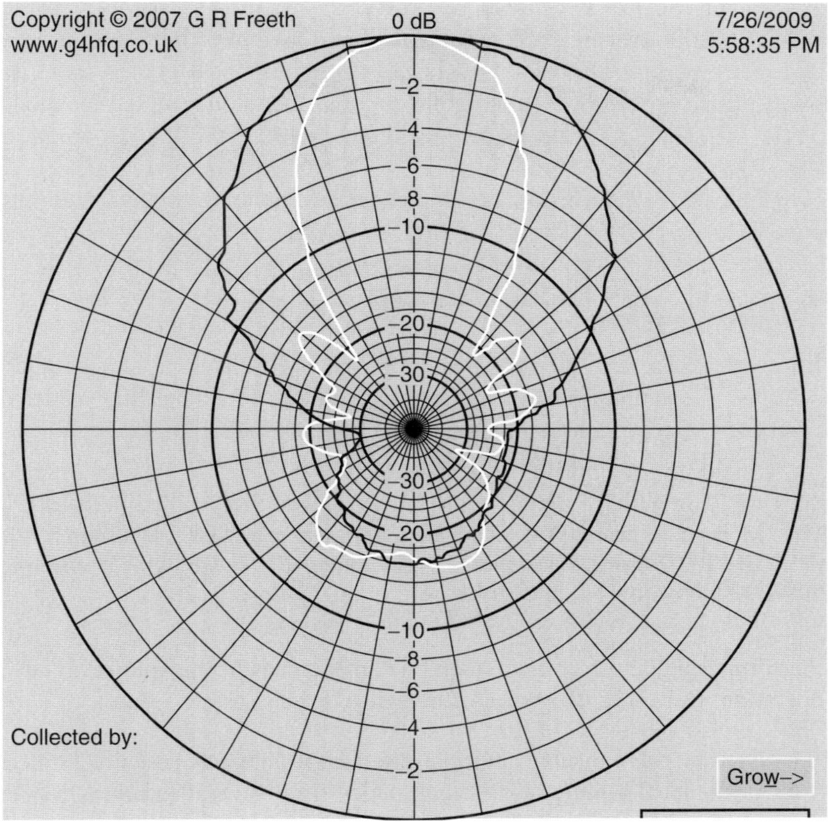

Figure 5.9 Illustration of antenna radiation pattern. Reproduced from PolarPlot by permission of © Bob Freeth.

- *Polarization loss factor:* An antenna is *linearly polarized* if the antenna radiation pattern follows a specific plane of emission. When two orthogonal radiators are combined for the transmission, the polarization rotates either counter-clockwise or clockwise ("right hand" or "left hand" *circularly polarized*). Since no excitation can happen between two linearly polarized antenna placed orthogonally to each other, a linearly polarized reader antenna may run the risk of losing tags with linearly polarized tag antenna. The polarization factor is used to account for the potential loss of signal power due to the different orientation of the tags from the reader. Given a reader with a circular polarized antenna and a tag with linearly polarized antenna, the polarization loss factor is 3 dB in the worst case.

The key point here is that each antenna design has a specific pattern of coverage – by understanding these parameters, we can specify the right type based on cost, coverage area, and size. A common practice in RFID system design is to place the complexity in the reader antenna design to trade off the cost of the tag. In addition to the individual

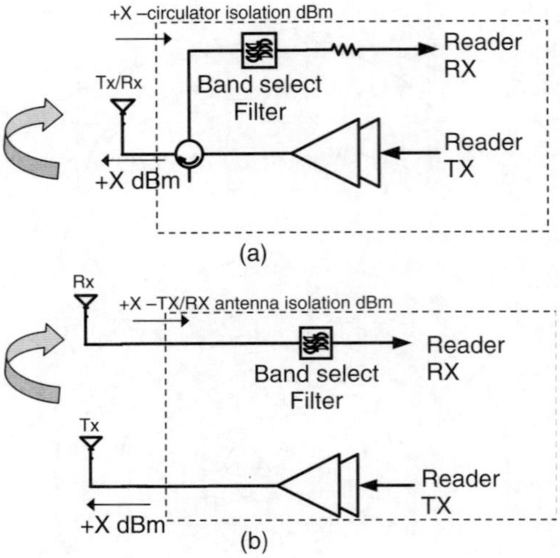

Figure 5.10 Mono-static and bi-static configuration. Based on a figure taken from S. Chiu *et al.*, "A 900 MHz UHF RFID reader transceiver IC," *IEEE Journal of Solid State Circuits*, vol. 42, no. 12, pp. 2822–2833, Dec. 2007.© 2007 IEEE.

antenna performance, the use of more than one antenna has been explored to advance the system operation. Currently two types of diversity techniques are in use:

1. **Spatial diversity**: By placing multiple directional antennas (usually limited to less than four per reader) aiming at different angles from several locations, for example, of a dock door. This approach enhances the coverage of the reader, especially when tags are located in different spots in a pallet, or being obstructed by a metal shield in some direction.
2. **Transmit and receive diversity**: This is shown in Figure 5.10(b) as a bi-static configuration with a transmit antenna and a receive antenna, comparing to a mono-static configuration (Figure 5.10(a)) which uses a single antenna. In the case of a mono-static configuration, a circulator (an expensive cost adder) or an attenuator (a reduction in sensitivity) is added in the receive path to avoid saturation from transmit to receive coupling. The bi-static configuration provides much better isolation to enable higher performance by reducing the large signal compression of the receiver frontend and the transmit noise coupled into the receiver. If the transmit/receive antenna produces 25 dB of isolation, this example shows at least 10 dB difference compared to using a circulator of 15 dB isolation. In practice, the difference can be larger due to the component margin and board impedance mismatches. However, transmit/receive diversity doubles the antenna cost and size. Currently bi-static configuration is limited to high performance systems, and mono-static configuration dominates less demanding applications.

5.3.2.2 RFID Transceiver

Serving as the physical layer interface, the RFID transceiver consists of three main functional blocks: the transmit path, the receive path, and the frequency generation section, as

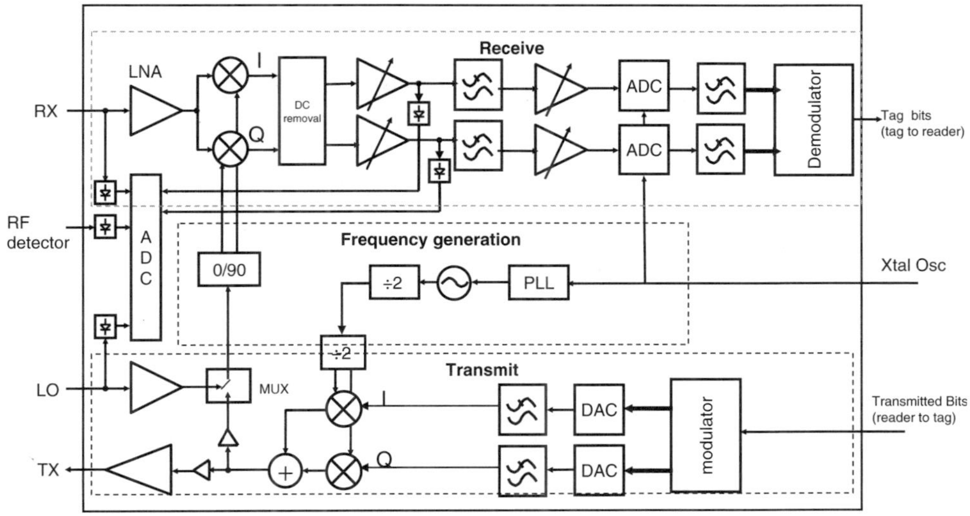

Figure 5.11 A RFID transceiver implementation. Based on a figure taken from S. Chiu *et al.*, "A 900 MHz UHF RFID reader transceiver IC," *IEEE Journal of Solid State Circuits*, vol. 42, no. 12, pp. 2822–2833, Dec. 2007. © 2007 IEEE.

depicted in Figure 5.11 using a direct conversion implementation. The transmit path carries out the task of baseband modulation and passband signaling by frequency up-conversion; the receive path performs frequency down-conversion and demodulation; the frequency generation section generates and distributes the carrier. These functions have been discussed extensively in wireless transceiver literatures [17, 18], and will not be repeated here. In the following, we will focus on the implementation differences originated from the unique attributes pertaining to a RFID system.

Receive Path

The biggest differentiation is the requirement to handle a large transmitter leakage during tag reception. Depending on the transmit to receive isolation, and the transmitter power level, the receiver typically needs to tolerate transmitter leakage signals (self-jammer) greater than 5 dBm (assuming a 30 dBm transmitter with 25 dB transmit to receive isolation). This requires a receiver front-end with high compression point, and limits the amount of front-end gain to less than 0 dB. Since there is no effective gain before the received signal reaches baseband, any noise added before the baseband amplifier impacts the receiver noise figure. Thus the receiver front-end implementation requires high compression point and low noise.

Due to the large self-jamming signal residing at the carrier frequency, direct conversion architecture is commonly employed to remove the self-jammer by mixing down it to DC. There are four types of DC offset cancellation methods used in practice [19] as illustrated in Figure 5.12: (a) using a simple RC-high pass filter; (b) using a high pass filter with sample and hold to memorize the low frequency signal contents; (c) using active cancellation with DC estimated by a low-pass filter; and (d) cancel the DC after converting the signal to digital. The first method creates a conflict between passing the

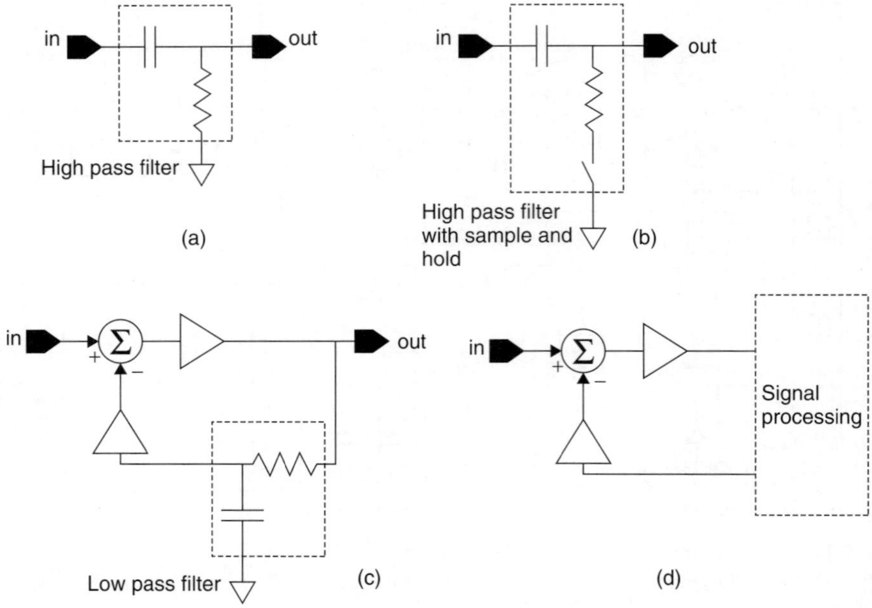

Figure 5.12 DC offset correction. Based on a figure taken from S. Zhou and M.C.F. Chang, "A CMOS passive mixer with low flicker noise for low-power direct-conversion receiver," *IEEE Journal of Solid State Circuits*, vol. 40, no. 5, pp. 1084–1093. May 2005. © 2005 IEEE.

signal contents and responding to DC changes from transmit power fluctuations. Since the tag response can be very close to DC due to the modulation schemes used, the high pass filtering corner needs to be set low, and incurs a long transient whenever the self-jammer level changes. For example, when the carrier goes from modulated to unmodulated, and vice versa. The second method eliminates this issue with an additional sample and hold control. The DC level during receive is sampled and held during transmit (when the modulated transmit changes the DC level). This approach allows a quick turnaround time at the expense of a large sampling capacitor to minimize the charge leakage impact. The third method removes DC (down converted self-jammer) at the mixer output, and reduces the mixer output compression requirement. However, the active circuitry adds noise into the system before the received signal can be amplified. This impacts the receiver noise floor, and lowers the receiver sensitivity. The last approach is not feasible for RFID receiver due to the large dynamic range required with a high power self-jammer.

Besides the risk of driving the receiver front-end into compression, the noise of the self-jammer can substantially degrade the receiver sensitivity. For example, the equivalent receiver noise figure will be 60 dB under a 6 dBm self-jammer with −120 dBc/Hz phase noise ($6 + (-120) - (-174) = 60$ dB at room temperature). To mitigate this effect, the same LO (local oscillator) source is used for both transmit up-conversion and receive down-conversion as shown in Figure 5.11. This provides the basis of correlating the phase noise down to DC[4] at the receiver mixer output. Furthermore, to capture the phase

[4] Multiplying a sinusoid with itself produces a signal component at DC and one at twice the frequency. It is assumed that the high frequency content is removed by low pass filtering in the signal conditioning path.

noise added by the components after the transmit up-mixer, an external LO (as shown in Figure 5.11) can be used to couple the transmitter power amplifier output for use as the receiver LO.

Tags are designed for low cost with a loose margin to account for process and device variations. For example, ISO 18000-6C specifies a Miller modulated tag to respond within ±15% of nominal data rate and ±2.5% drift over a tag response with 250 KHz link frequency.

The receiver demodulator needs to handle these variations by: (1) keeping the baseband filter bandwidth wide to cover all the tag data rate variations while narrow enough to reject neighboring channel interferences; this requires a balance between filter complexity and receiver selectivity to cover the tag reply power spectrum; and (2) tracking the data rate changes using a continuous timing recovery circuitry in the demodulator.

Frequency Generation

The carrier generation specification is determined by analyzing air interface standards and applicable local regulations. For ETSI applications [8], the most stringent phase noise requirement is −116 dBc/Hz at 200 kHz offset[5] originated from adjacent channel interferer cross-mixing, and −144 dBc/Hz at 3.6 MHz from out-of-band spurious emission regulation. This is tough to meet especially for highly integrated RFID transceivers.

The good news is: settling time is not explicitly defined by any local standards, though it is desirable to keep it to within a few hundred μsec to enable fast monitoring of available spectra. Based on these requirements, an integer-N PLL can be used for the best spectral performance and implementation simplicity; though a low noise Fractional-N PLL is applicable in less demanding applications. In addition, the VCO frequency should be selected to be two or four times the carrier frequency to avoid high power transmit output coupling back to the VCO. If this is not possible, extra attention should be paid to isolate the VCO from noise coupling at the carrier frequency.

Transmit Path

Direct conversion is commonly used to lower implementation costs. The data bits are streamed in from the protocol processor, encoded in pulse interval encoding, and shaped to limit the spectrum of the transmission. The baseband signal is then up-converted to passband through IQ mixing and amplified through a power amplifier to the antenna interface. There are several issues to be noted for a RFID transmitter:

1. **Receive to transmit link timing**: Specified as the time interval between the reader responds and the reply from tags ends, this includes the group delay through the receiver path as well as that of the transmit path. Referred to as "T2" time in ISO 18000-6C link timing, this is defined with respect to the tag modulation symbol period. As an asymmetrical communication system, the tag response data rate can be high while the reader transmit data rate is low (for example, with Tari = 25 μsec) to limit spectrum usage. Since the transmitter group delay scales with the transmit symbol

[5] Given −116 dBc/Hz phase noise, the noise power within 200 KHz (53 dB) with a −35 -dBm adjacent channel interferer is −98 dBm. Assuming 12 dB is required to demodulate FM0 , and 1 dB implementation loss, the receiver can maintain a target sensitivity at −85 dBm.

time basis (Tari), "T2" is increasingly hard to hit with higher tag data rate. One way to get round this problem is pre-empt transmission which attempts to send the transmit preamble before fully decoding the tag reply.

2. **Transmit linearity**: Incurred by a modulated signal passing through a nonlinear circuit, spectral re-growth increases the original baseband signal bandwidth by mixing among its different frequency components. For RFID readers, −40 dBc IM3 (third order inter-modulation) of the transmitter front-end [13] is required to meet various spectrum mask requirements at 30 dBm output level. To meet the stringent linearity requirement, a low efficiency Class A or Class AB amplifier ($\sim$at 10% power added efficiency or below) is used due to the large back-off needed. Another option is to explore the drain modulation with a non-linear switch mode power amplifier [20]. There are two common design challenges: (1) the signal bandwidth of drain modulation is roughly 10 times larger comparing to that of I/Q modulation. This poses as a limitation on wide bandwidth protocols, but presents a lesser challenge for the narrow band RFID modulations; and (2) non-linearity in the modulation path – using the power amplifier linearization techniques developed can mitigate the problem. Detailed treatment of this topic is beyond the scope of this chapter (the interested reader can refer to [21]).
3. **Transmitter AM noise**: Even when the receiver front-end is not compressed, the receiver sensitivity can still be impacted by the phase and amplitude noise of the self-jammer (transmit to receive coupling). As an example, assuming a 6 dBm self-jammer carries $-120\,\text{dBc/Hz}$ phase noise, and $-140\,\text{dBc/Hz}$ amplitude noise, the equivalent receiver noise figure caused by the phase noise will be 60 dB, and by the amplitude noise 40 dB, respectively. In the case of phase noise, using a correlated frequency source for receive LO can attenuate the phase noise substantially. However, since the amplitude information is not maintained through the receive mixer LO, the coupled AM noise does not get attenuated, and becomes the next obstacle to achieve good receiver sensitivity. Aside from reducing transmit AM noise though minimizing transmit mixer noise (for example, increasing the bias current) using self-jammer cancellation offers another option which will be discussed in the next section.

5.4 Advanced Topics on RFID Reader Design

This section discusses the current challenges of RFID reader design, and the approaches undertaken to address these issues. It should be recognized that the UHF RFID development is relatively new so far, and new problems in addition to the ones reviewed may occur as the deployments grow.

5.4.1 Integrated Transceiver

Due to the stringent performance requirements, the reader market used to be dominated by readers constructed with discrete components. With the introduction of fully integrated RFID reader transceivers [13, 22, 23], the reader has benefited from the reduction in size, power, and cost. It is expected that the trend of integration will continue and move towards CMOS implementations.

To obtain a high performance CMOS implementation, two RFID specific issues should be resolved:

1. **Low frequency signal contents of the tag reply overlaps with the CMOS 1/f noise range**: Since the transceiver can provide very limited RF gain, the device 1/f noise of the mixer weighs heavily on the receiver noise figure within a few hundred KHz for Miller encoded and a few tens KHz for FM encoded tags. There are many developed architecture and circuit techniques to deal with 1/f noise – of these options, the passive mixer [19, 24] seems the most promising for RFID transceiver implementation.
2. **Low breakdown voltage of the CMOS device at advanced CMOS process nodes**: For RFID applications with transmit power up to 30 dBm, the transistor drain stress limits the conventional power amplifier design to high-voltage tolerant RF processes. To increase the possibility of integration, the use of Distributed Active Transformer (DAT) [25] by power combining is a possible option for further study. On the receiver end, the stress on the transistor bounds the mixer input and output compression. This is handled by reducing receiver RF front-end gain and/or attenuating the signal before it reaches the receiver. However, a better alternative is to actively cancel the self-jammer which will be discussed in the following section.

5.4.2 Cancellation of Transmitted Carrier Leakage

For a mono-static configuration, a direct coupling path exists from transmit to receive by sharing the same antenna. The receiver is kept from saturation by adding either a circulator or an attenuator in the signal path. Using an attenuator is less expensive, but at the cost of reducing the received signal strength by the same amount. With bi-static configuration, the leakage is minimized by antenna isolation (for example, pointing the antenna in different directions). Depending on the application scenario, however, the antenna isolation may be substantially less than what can be achieved. Going forward, this issue is expected to become more serious with shrinking reader sizes and converting towards mono-static design for cost reduction.

Since the coupling source is known, an alternative is to cancel the transmitter leakage with a replica from the source, the transmitted carrier. This is a simplified case of the general carrier cancellation being explored for amplifier linearization and noise reduction [26]. The formulation of this problem into an optimization loop is depicted in Figure 5.13 with four stages: (1) picking the replica source; (2) devising the method to vary the phase and amplitude of the replica; (3) deriving the error detection scheme; and (4) finding the search mechanism for an optimal solution. Without going into details of every possible implementation, the following discussion focuses on important points specific to RFID reader designs:

- The replica's phase can be changed by a phase shifter or by varying the relative I/Q amplitude in the following equation: $\cos(wt - w0) = \cos(wt)\cos(w_0) + \sin(wt)\sin(w_0)$. The angle shifted is represented by the ratio between $\sin(w_0)$ and $\cos(w_0)$. The latter method is more suitable for integrated circuit implementation with a digital-to-analog converter.

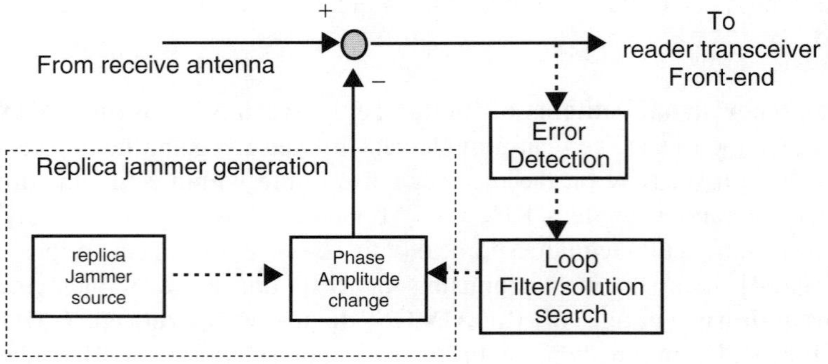

Figure 5.13 Self-jammer cancellation problem formulation.

- Error detection can be performed by observing the DC residue after the frequency down-conversion (assuming a direct conversion receiver). However, co-channel interferers, located at the same frequency as the self-jammer, can confuse the calibration circuit during initialization. This can limit the cancellation performance especially in a dense reader environment. Instead of relying on detecting the residual carrier by measuring the DC amplitude, more sophisticated modulation can be applied to the transmitted carrier during calibration phase to offset the residual power away from DC to avoid detection interferences.

Two implementation examples are shown in Figure 5.14 to illustrate the cancellation procedure. In Figure 5.14(a), the replica jammer is tapped from the transmitter output using a coupler. Two orthogonal vector signals are generated by an IQ phase splitter. The phase/amplitude change is implemented with two programmable variable attenuators commanded by the radio controller. The amplitude change is made by varying the amplifier gain on the I/Q paths in tandem, while the phase change is done by changing the gain of I relative to that of Q path. A power combiner is added before the received signal hits the received input to cancel out the self-jammer. The algorithm for detecting the error and close the control loop is implemented in software running on the radio controller by checking the DC amplitude level from the receive I and Q mixer outputs. A similar implementation for cancelling jammer with polar (phase and amplitude) control is illustrated in Figure 5.14(b). Rather than transforming the replica into orthogonal signals, the amplitude and phase are directly operated on using an amplifier and a phase shifter. There is no fundamental difference between these two implementations, except that care must be taken to ensure the replica signal generation path must have much lower noise than the receiver path. Otherwise, the cancellation mechanism can potentially add more noise to the system.

5.4.3 Dense Reader Operations

It is easy to conceive the coexistence of many readers in one physical location as RFID starts to be integrated into our daily life. When many readers establish concurrent links

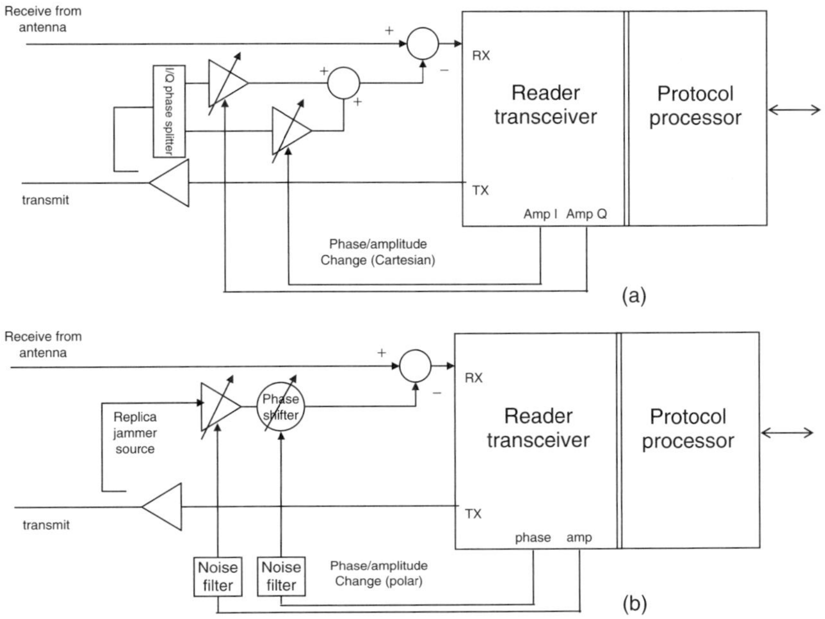

Figure 5.14 Transmitter carrier leakage cancellation implementation.

with the same tag population, we have to build the reader with smart intelligence to handle the following scenarios:

- *Reader transmission interferes with tag responses*: Reader transmission is usually much higher than tag responses – the response can be easily overwhelmed even when the reader is not talking (readers still transmit carrier waves). During the dense reader operating mode, ISO 18000-6C air interface separates the tag response from the reader carrier spectrally. In the US, the tag response uses Miller subcarrier coding $M = 4$, 62.5 Kbps to focus the backscattering energy right in between the center of two channels. In Europe, the standard recommends using only channels 1, 4, 7, and 10 with Miller subcarrier coding $M = 4$, 75 Kbps to avoid the reader and tag operating in the same channel.
- *Reader-to-reader interference*: When there are many readers working at the same time, two readers can choose the same channel, called co-channel interferers, or right at the next channel, called adjacent channel interferers. When two co-channel readers are talking at the same time, the tag population cannot differentiate where the commands are from unless they are physically separated by the channel reuse range of a cell. Since there is no base station in the system, the interference is mitigated through some combination of antenna isolation [27], synchronization [28], and interference avoidance. Synchronization describes the agreement (1) in timing for either reader talking or tag responding so that reader and tag messaging do not overlap in time; or (2) in coordinated channel usage to avoid the same channel being reused by closely located readers. This

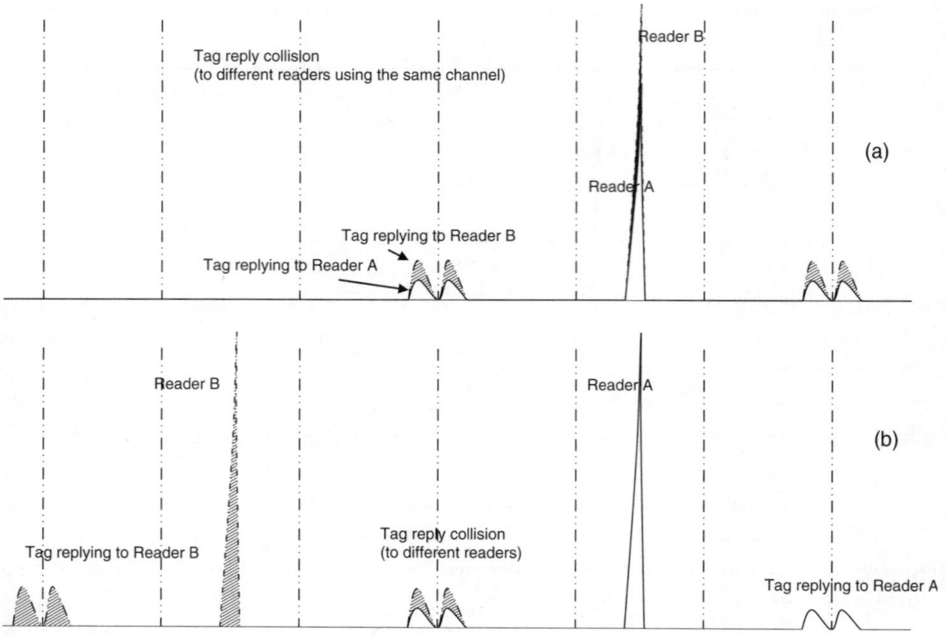

Figure 5.15 Tag reply collisions.

is done either with a central controller if all readers on the same network, or through side-band communications. For interference avoidance, for example, in the US, the local regulation requires the reader to randomly hop within the allocated channels from 902–928 MHz. Before hopping into the next channel and starting to power up tags, the reader can listen to the target channel to decide whether there is a possibility for collision or adjacent channel interference.

- *Tag to tag interference*: Depending on the spectral location, we can determine two different types of interferences: (1) the tag responses come from responding to two readers occupying in the same channel (depicted in Figure 5.15(a)). From the reader perspective, it is similar to tag collision as described before. (2) The tag responses come from two readers occupying in adjacent channels (as shown in Figure 5.15(b)). As can be seen from Figure 5.15, the collision is one-sided for the tag responses. A reader can thus deduce the tag signals by throwing away the collided portion with filtering.

5.5 Conclusion

This chapter provides an overview of UHF passive reader hardware design with a focus on ISO 18000-6C. Similar to the spread of the bar code technology, the proliferation of passive RFID technology would depend on continuing cost reduction while building up the infrastructure. Two of the most prominent issues encountered today are: (1) the active power consumption of the reader is still too high to enable general portable and embedded usage models. This chapter has introduced a few problems and approaches, but further progress in technology is necessary to enable the applications; (2) the integration

Table 5.2 Spurious emission requirement for ETSI 302 208 during reader operation.

	47–74 MHz, 87.5–118 MHz, 174–230 MHz, 470–862 MHz	Other frequencies below 1 GHz	Frequencies above 1 GHz
Emission limit	4nW	250 nW	1 μW

of RFID into the application environment – progress is being made both on software and hardware, but the application experience is still limited for this new technology. Despite the challenges, with the key technical/application problems being addressed and resolved over the last few years, it is expected that we will see reader implementation reach maturity to enable an explosive uptake of adoption in the near future.

Problems

1. Spurious emissions are transceiver spectral emissions at frequencies other than the intended carrier frequency. For ETSI 302 208, the spurious emission requirement is partially specified in Table 5.2. (a) What is the spurious emission represented in dBc when the reader is transmitting 30 dBm? (b) There are four proposed RFID reader channels (865.7 mHz, 866.3 MHz, 866.9 MHz, and 867.5 MHz). What is the minimum phase noise requirement for the RFID reader at 3.7 MHz from the carrier? (c) Does the transmit noise requirement change if the reader is operating at 20 dBm?

2. Receiver sensitivity is defined as the minimum detectable signal power level; given Eb/No = 15 dB is required to demodulate the incoming tag signal at a satisfactory packet error rate (for example, 10^{-3}), the data rate is 75K bits per second, with a receiver noise figure of 15 dB, what will be the receiver sensitivity under a white Gaussian noise channel?

3. An RFID reader operates at 2.4 GHz with transmit power of 20 dBm, and reader transmit antenna gain of 3 dB. (a) Using the range equation (Equation 5.1) what is the reader-to-tag path loss when the distance between tag and reader is 1 m and 2 m? (b) What is the available power (assuming tag antenna gain = 3 dB, and a 3 dB polarization loss) for the tag to harvest when the distance between tag and reader is 1 m and 2 m? (c) With the range equation (Equation 5.2), and a total of 19 dB of tag loss (L_{tag}), what is the minimum receiver sensitivity of an RFID reader operating at 2.4 GHz at 1m away? At 2m away? (d) What can you do to improve the operating range given that the reader transmit power is limited to 20 dBm? Please assume two different scenarios (1) the tag does not have enough power to activate; (2) the backscattered power form the tag is too small to demodulate.

References

[1] Finkenzeller, K. (2003) *RFID Handbook: Fundamentals and Applications in Contactless Smart Cards and Identification*, 2nd edition, Chichester: John Wiley & Sons, Ltd.

[2] Park, J.S. (2007) Trends for mobile RFID reader SoCs, developed by Korean ASIC companies, RFID Workshop, IEEE RFIC Symposium.
[3] Information technology-Radio frequency identification for item management, Part 6: Parameters for air interface communications at 860 MHz to 960 Mhz, ISO-IEC CD 18000-6C, 2007.
[4] Balachandran, G.K. and Barnett, R.E. (2006) A 110 nA voltage regulator system with dynamic bandwidth boosting for RFID systems, *IEEE Journal of Solid State Circuits*, **41**(9): 2019–2028.
[5] Karthaus, U. and Fischer, M. (2003) Fully integrated passive UHF RFID transponder IC with 16.7-μW minimum RF input power, *IEEE Journal of Solid State Circuits*, **38**(10): 1602–1608.
[6] Pillai, V., Heinrich, H., Dieska, D., Nikitin, P.V., Martinez, R., and Rao, K. V. S. (2007) An ultra-low-power long range battery/passive RFID tag for UHF and microwave bands with a current consumption of 700 nA at 1.5 V, *IEEE Transactions on Circuits and Systems* I: Regular Papers, **54**(7): 1500–1512, July.
[7] Operation within the bands 902–928 Mhz, 2435–2465 MHz, 5785–5815 MHz, 10500–10550 MHz, and 24075–24175 MHz, FCC title 47 part 15.
[8] Electromagnetic compatibility and radio spectrum matters (ERM): Radio frequency identification equipment operating in the band 865 MHz to 868 MHz with power levels up to 2 W. Part 1: Technical requirements and methods of measurement, ETSI EN 302-208-1, 2007.
[9] Information technology-Radio frequency identification for item management, Part 6: Parameters for air interface communications at 860 MHz to 960 MHz, Amendment 1: Extension with Type C and Update of Types A and B, ISO-IEC CD 18000-6, 2004.
[10] Protocol specification for a 900 MHz class 0 radio frequency identification tag. MIT Auto-ID Center, Feb. 2003.
[11] 860 MHz-930 MHz class I radio frequency identification tag radio frequency & logical communication interface specification candidate recommendation. ver. 1.0.1, MIT Auto-ID Center, 2003.
[12] EPC UHF radio frequency identification protocols: Class 1 generation 2 UHF RFID. Ver. 1.2.0, EPCglobal, 2007.
[13] Chiu, S. *et al.* (2007) A 900 MHz UHF RFID Reader Transceiver IC, *IEEE Journal of Solid State Circuits*, **42**(12): 2822–2833.
[14] Sklar, B. (2001) *Digital Communications: Fundamentals and Applications*, 2nd edn. Englewood Cliffs, NJ: Prentice-Hall.
[15] Maguire, Y. and Pappu, R. (2009) An optimal Q-algorithm for the ISO 18000-6C RFID protocol, *IEEE Transactions on Automation Science and Engineering*, **6**(1): 16–24.
[16] MSDN Architecture Center. Available at: http://msdn.microsoft.com/en-us/library/aa479362.aspx#rfidtechover_topic5.
[17] Lee, T. (2003) *The Design of CMOS Radio-Frequency Integrated Circuits*, 2nd edn. Cambridge: Cambridge University Press.
[18] Razavi, B. (1997) *RF Microelectronics*. Englewood Cliffs, NJ: Prentice Hall.
[19] Zhou, S. and Chang, M.-C.F. (2005) A CMOS passive mixer with low flicker noise for low-power direct-conversion receiver, *IEEE Journal of Solid State Circuits*, **40**(5): 1084–1093.
[20] Berglund, B., Johansson, J., and Lejon, T. (2006) High efficiency power amplifiers, *Ericsson Review*, **3**: 92–96.
[21] Raab, F. *et al.* (2003) RF and microwave power amplifier and transmitter technologies – Part 4, *High Frequency Electronics*, Nov. pp. 38–49.
[22] Khannur, P.B. *et al.* (2008) A universal UHF RFID reader IC in 0.18-μm CMOS technology, *IEEE Journal of Solid State Circuits*, **43**(5): 1146–1155.
[23] Austria Microsystems. Available at: http://www.austriamicrosystems.com/eng/content/view/full/7540.
[24] Valla, M., Montagna, G., Castello, R. Tonietto, R., and Bietti, I. (2005) A 72-mW CMOS 802.11a direct conversion front-end with 3.5-dBNF and 200-kHz 1/f noise corner, *IEEE Journal of Solid State Circuits*, **40**(4): 970–977.
[25] Aoki, I., Kee, S.D., Rutledge, D.B., and Hajimiri, A. (2002) Fully integrated CMOS power amplifier design using the distributed active-transformer architecture, *IEEE Journal of Solid State Circuits*, **37**(3): 371–383.
[26] McNeilage, C., Ivanov, E.N., Stockwell, P.R., and Searls, J.H. (1998) Review of feedback and feedforward noise reduction techniques, in *Proceedings of the 1998 IEEE International Frequency Control Symposium*, May, pp. 146–155.

[27] Leong, K.S., Ng, M.L., and Cole, P.H. (2006) Positioning analysis of multiple antennas in a dense RFID reader environment, in *International Symposium on Applications and the Internet Workshops*, Jan., pp. 56–59.
[28] Leong, K.S., Ng, M.L., Grasso, A.R., and Cole, P.H. (2006) Synchronization of RFID readers for dense RFID reader environments, in *International Symposium on Applications and the Internet Workshops*, Jan, pp. 48–51.

6

RFID Middleware: Concepts and Architecture

Nathalie Mitton, Loïc Schmidt, and David Simplot-Ryl

INRIA Lille-Nord Europe, France

6.1 Introduction

The RFID middleware is a central point in the integration process of any RFID solution [1–3]. There are several kinds of RFID tags and consequently several kinds of readers. If the end-user application is directly connected to readers, the management of all of them can be a hard task. Furthermore, this application would have to deal with a larger amount of data than necessary. Indeed, the RFID readers have to regularly scan their environment to know which tags are present. A middleware would enable data selection, that is, the application would only receive the information it requires. The RFID middleware is a set of components which aims to manage RFID readers, deals with RFID-events and data, and is connected to end-user applications. This chapter describes the general architecture of such a middleware.

We will start by presenting the management of readers. This means reader protocol/interface, monitoring, and settings. The reader protocol deals with interactions between readers and controllers. It offers a way to insulate the host from the technical aspects of how a reader communicates with tags. There are two main ideas in such a protocol. Firstly, it has to provide an abstract syntax for messages between the controller and the reader. Secondly, it must implement different transport layers (Bluetooth, wifi, serial, TCP/IP, . . .). An issue here is how to define a protocol which can interface with multiple kinds of readers, not only RFID, but also sensors, bar-codes, etc, in a transparent way for upper layers [4].

RFID Systems: Research Trends and Challenges Edited by Miodrag Bolić, David Simplot-Ryl, and Ivan Stojmenović

The next point is the data management. Indeed, readers can be configured to be aware of everything happening in a warehouse. This results in a large amount of data, which must be treated and stored intelligently in a database. The important point here is the data "filtering and collection." It means transforming a tag reading flow in a comprehensive "what, when and where" record. The way to retrieve data has a direct impact on the ease of developing and connecting the system with end-user applications. This is the role of the ALE standard ratified by EPCglobal Inc. [5]. The application level events (ALE) is an interface between business applications and the middleware data.

Finally, we will present how data are stored in the system and shared between multiple sites, multiple partners. The idea is to provide databases (EPCIS) [6] and a way to retrieve and query the right one. The main component here is the object name server (ONS) [7]. This component provides a mechanism to retrieve information about an object from its identification number. When an application asks the middleware more details about an object, the system first checks its local database. The EPCIS acts like a database, storing EPC-related events. If the object requested is not registered in it, the system then forwards the request to the ONS owning the database which contains this information. The ONS acts like the well-known Internet domain name server (DNS) [8], with an ONS root. Having a unique ONS root is both a technical and political problem. The challenge here is to define a multi-root structure for ONS providing load-balancing and scalability for this service, but also a shared governance for the political concern. Another way of retrieving information is the discovery service. It has yet no standards defining this service, but it aims to offer the ability to find all servers sharing RFID events about a product.

6.2 Overview of an RFID Middleware Architecture

6.2.1 The Need for a Middleware

A middleware is an intermediate software layer between the applications and the network, allowing the dialog between heterogeneous applications. It aims to accomplish technical tasks for business applications connection and data exchange. In RFID systems, such a middleware has to manage readers, to deal with RFID events and to be connected to end-user applications. In some cases, there is no need for a 'middleware', as in small and unique applications like "count number of tags read", or "list the tags in the scope of a reader." But when the need to share information coming from readers between several applications is rising, then messages from readers need to be application specific. For example, a conveyor application does not just need the list of tags read, it also needs to know which of these read tags is from the pallet and which are tags from products. In most cases, physical details are not useful for business logic. Therefore, a middleware becomes mandatory when the RFID applications become more complex. It is a unique access point for collected data. It associates read items with an activity and a location. It homogenizes the access of the data via a standardized interface which allows:

- Reader management (decoupling physical and logical readers).
- Data management (formatting, aggregation, filtering).
- Call to application functions or results sending.

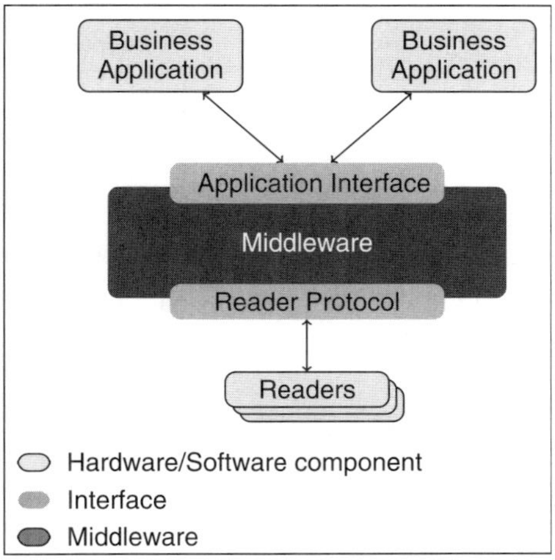

Figure 6.1 Outside a RFID middleware.

- Rules and exceptions management.
- Data collection and treatment in real time or in asynchronous mode.

RFID middleware is considered one essential intelligence-added component of any RFID system and could be linked with other company information systems such as external databases, business partner information systems or warehouse management systems, etc. Figure 6.1 shows the position of a middleware in an RFID system as well as two key concepts from an external point of view: the reader protocol and the application interface. Indeed, the middleware is the link between the readers and the business applications. Let's now have a look at this middleware black box to see other key concepts of a RFID middleware and how they are used to perform the tasks of the middleware.

The middleware specification and challenges are nowadays partially addressed in the scope of the Architectural Framework of EPCglobal [9]. EPCglobal[1] is a subscriber-driven organization comprised of industry leaders and organizations focused on creating global standards for the EPCglobal Network. EPCglobal has developed a collection of interrelated standards for hardware, software, and data interfaces, together with core services that are operated by its delegates. In this chapter, we will focus on RFID middleware functionalities and services and will mostly refer to the standards developed by EPCglobal.

6.2.2 Architecture

Connecting readers directly to applications raises several main problems:

- Applications are reader-dependent. This means that if the need to change the readers appears and if the new readers do not use the same communication protocols as the former ones, part of the application also needs to be recoded.

[1] http://www.epcglobalinc.org

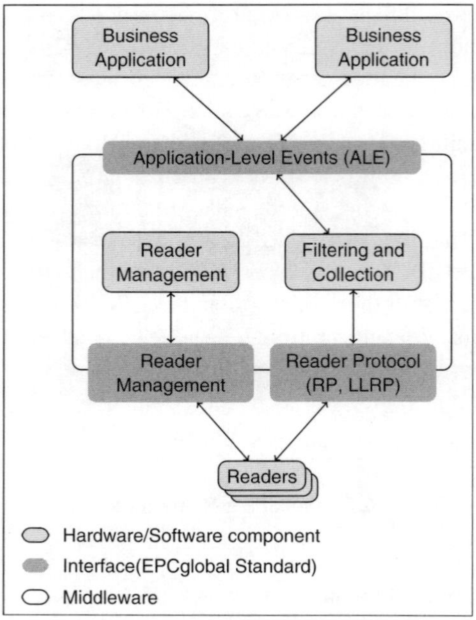

Figure 6.2 Inside a RFID middleware.

- Every application connected to the readers receives the same information about the tags read. Moreover, this information is raw (it is a list of tag identifiers). Nevertheless, it may need only a part of these information, which may differ from one application to the next. Applications are thus flooded with too much information.
- Finally, if there is a great amount of readers or/and applications, the system is likely to fail since it is saturated and overloaded.

Thus, connecting readers directly to applications is not scalable and wastes resources uselessly. Using a middleware between readers and applications solves these problems. To do so, the middleware is divided into several blocks, as shown in Figure 6.2. This figure shows a part of the EPCglobal Architecture Framework [9]. The reader protocol interface allows the first problem regarding the heterogeneity of the readers to be solved. This interface provides a way to monitor different readers, can configure them, check their status (reading, idle, etc.), possibly correct any operational problem, etc. Reader protocols will be detailed in Section 6.3.

The two other problems are managed through both the Application-Level Events (ALE) interface and the Filtering and Collection module. This latter module can also be seen as the ALE engine, that is, it allows operations on data, based on the events defined by applications through the ALE interface. These modules (ALE interface and Filtering and Collection) will be described in Section 6.4. Indeed, thanks to the middleware, information about tags is sent to applications only if the latter have emitted a data request through the ALE Interface. Then, instead of forwarding raw information, this is filtered, grouped, aggregated by the ALE engine before being sent to applications. In this way, applications directly receive exploitable data.

Then, information retrieved from tags through the middleware may need to be persistent, that is, one should be able to access the data afterwards. This is, for instance, the case for traceability business applications. Indeed, we need to record the data for later use, to translate ALE events into business events, that is, to give them a business signification. For instance, a tag's reading states that Object *A* has been packed with Object *B*, which can be translated as "Order number 10 is ready for expedition from date 2009-03-12 at 12:28." This event needs to be stored. This illustrates the role of the EPCIS (EPC Information Services). The goal of EPCIS is to enable disparate applications to leverage Electronic Product Code (EPC) data via EPC-related data sharing, both within and across companies. In order to perform these operations, the EPCIS provides interfaces to capture (store) and to query (retrieve) information.

Finally, a business application may need to retrieve a product name or manufacturer from its EPC identifier that is not stored in the company EPCIS. To do this, it needs to query an Object Name Service (ONS), which, in a DNS fashion, will answer the request.

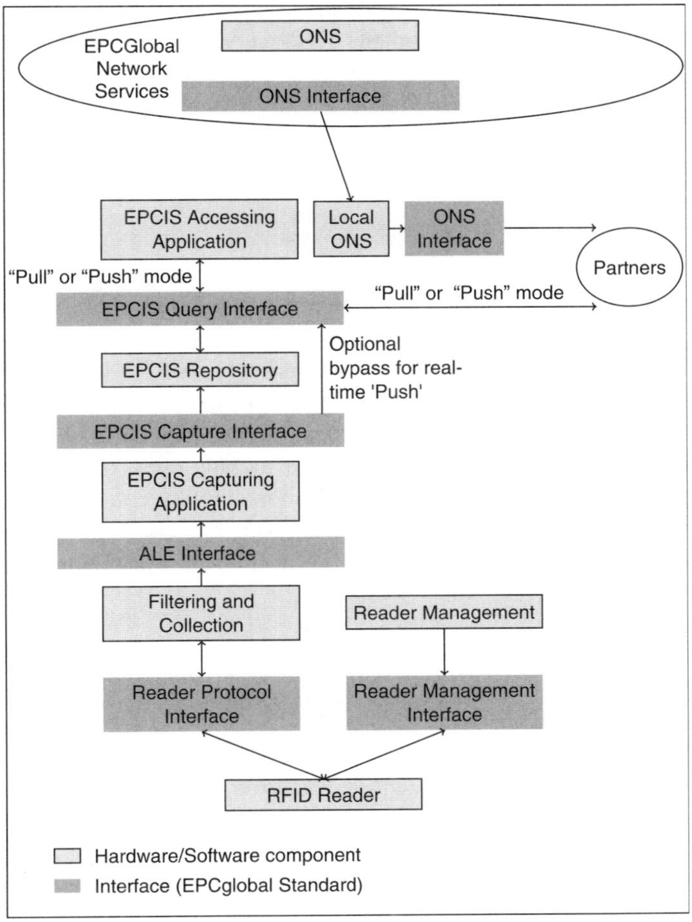

Figure 6.3 EPCglobal Network Architecture.

All these modules such as EPCIS or ONS have to be connected, directly or not, to the middleware as shown in Figure 6.3. We detail these modules and their interaction with the middleware in Section 6.5.

6.3 Readers Management

The aim of the reader protocol and of the reader management are respectively to provide:

- a uniform communication interface between different readers and middleware so that the application can function independently of the hardware;
- a way to manage and monitor readers.

In this section, we firstly present the role of the reader protocol and the two standards ratified by EPCglobal, the Reader Protocol and the Low-Level Reader Protocol. The second part describes the management of readers: it exposes the need of a management tool for readers and presents EPCglobal Standards which deal with the reader management (the Reader Management Standard and the Discovery, Configuration and Initialization Standard).

Reader management should also include the management of interference between readers. Techniques like reader anti-collision protocols or activity scheduling can be applied [10, 11]. These techniques are described later in this book.

6.3.1 Reader Protocol/Interface

The reader protocol defines interactions between readers and application softwares (or hosts). Figure 6.4 shows the links between the host, a reader and tags. There are a lot of parameters to give to readers in order to work efficiently, such as the number of antennas connected to it, the transmitter power, etc. The role of the protocol is to provide an interface, or API, to control and command the reader to launch reading actions (tag ID or tag additional data), but also writing, killing or locking ones, etc.

The most common case is that reader's vendors provide their own protocol, their own API to communicate specifically with their own readers. The problem here is the difference between these hardware specific control interfaces. Different architectures,

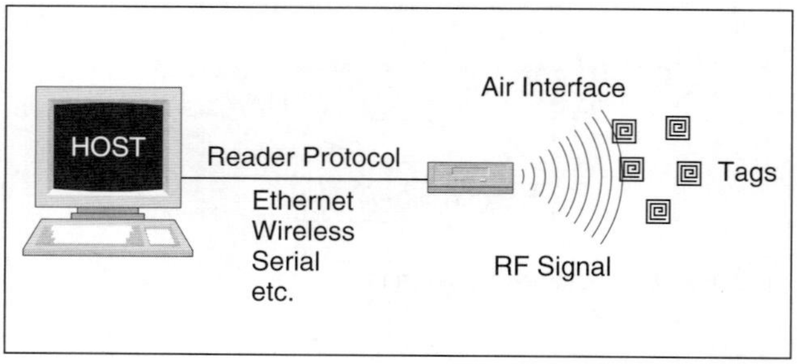

Figure 6.4 A host-to-reader connection through reader protocol.

different protocol options, etc. increase the maintenance task cost when new drivers or new hardwares need an application modification. This cost is also significantly increased when large number of readers have to be operated by various application softwares. The role of the middleware here is thus to provide a Hardware Abstraction Layer (HAL) between readers and host. The HAL goal is to provide an API, offering a way to send commands to readers independently of the reader specific API commands. An underlayer will then translate these commands into reader-readable commands and send them through a specific transport protocol to the reader.

EPCglobal defines two reader protocols in order to insulate the host from details of the reader and tags communication. The reader protocol deals with the RFID reader's commands such as inventory tags, read, write, kill or lock tags. The low-level reader protocol is low-level because of its awareness of the air protocol between readers and tags. This section describes these protocols in more details.

6.3.1.1 Reader Protocol

The EPCglobal reader protocol [12] standard defines three layers (see Figure 6.5):

- **Reader Layer** defines the abstract syntax used in messages between the readers and host.
- **Messaging Layer** specifies how the message is encompassed in the network transport dependent message.
- **Transport Layer** corresponds to the networking facilities provided by the operating system or equivalent.

Defined from a conceptual object view, the RP specifies commands in the Reader Layer that readers may perform. In the specification, the object ReaderDevice implements the command getEPC which returns the EPC of the reader. No matter how it is implemented by manufacturer, the protocol knows that this command performs this operation. A messaging/transport binding (MTB) is composed by one messaging layer and one transport layer, providing different ways of transport (e.g. TCP/IP versus Bluetooth versus serial). The role of a MTB is to transport the RP command between host and readers. It defines transformations of messages from a Reader Layer before sending it to the Transport Layer.

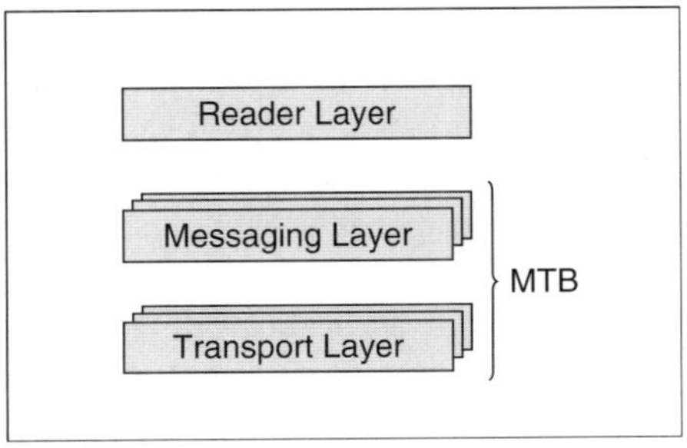

Figure 6.5 Layers of the EPCglobal Reader Protocol.

It also specifies which Transport Layer to use and how to set up connections. The reader protocol provides different MTBs but others can be defined.

6.3.1.2 Low-Level Reader Protocol

In contrast, the Low-Level Reader Protocol (LLRP, also defined by EPCglobal [13] is sensitive to the details of the air-interface protocol (between readers and tags) and provides specific parameters and controls in order to reveal the command and timing parameters of the RFID air protocol to the controller. The control of readers is based on specifications (called ROSpecs for Reader Operation Specifications). These specifications are sent to and stored in the reader, so readers can perform operations without overloading the network with command and status data. In general, a reader can store more than one specification. This implies a significant cost to the reader because of the amount of computational and memory resources needed by the ROSpecs.

The ROSpec describes a reader configuration, including the reader's ID, the antennas to be used and the air protocol at each antenna, start and stop triggers for inventory action, transmit power and receiver sensitivity, and a priority. When a trigger condition is met, the reader performs operations of the corresponding ROSpecs. Thanks to the priority level, some of these specifications can be pre-empted by a higher priority ROSpecs.

The LLRP standard defines parameters and controls for the UHF Gen 2 air protocol, but it also provides mechanisms for additional commands or additional parameters. Readers can ignore such kind of extensions, but they have to return errors if these extensions are not supported.

To conclude this part on reader protocols, we can say that events generated by readers through these protocols are of the form "Reader A saw EPC X at time T" each time a tag is read. This may seem meaningless in a business context, and may overload the network with useless messages to end-user applications. Therefore, filtering and data aggregation are needed in order to provide application-specific events. This part of the middleware is described in Section 6.4.

6.3.2 Manage and Monitor

In systems with a lot of readers, having a way to manage them and to monitor them can be very useful. Indeed, there is a lot of parameters to monitor in a reader. Therefore, we need a reader management tool that offers a way to monitor readers in a RFID middleware. It permits retrieval of information from readers such as their identity, their number of antennas, the air protocol configured, the number of tags read, the status of the communication channels, etc. The reader can also send some alerts for an operational problem.

In large RFID deployments, interference between readers can occur when readers placed in the same area attempt to send messages to RFID tags. Regarding this issue, the communication channel's configuration and monitoring of readers is important. This manages reader transmission simultaneously in order to prevent interferences for example.

Due to the number of different vendors of different RFID readers, the managing and monitoring task is very complex. Indeed, readers do not have same hardware capabilities or configuration, control and monitor parameters. A hardware abstraction layer is also required more here.

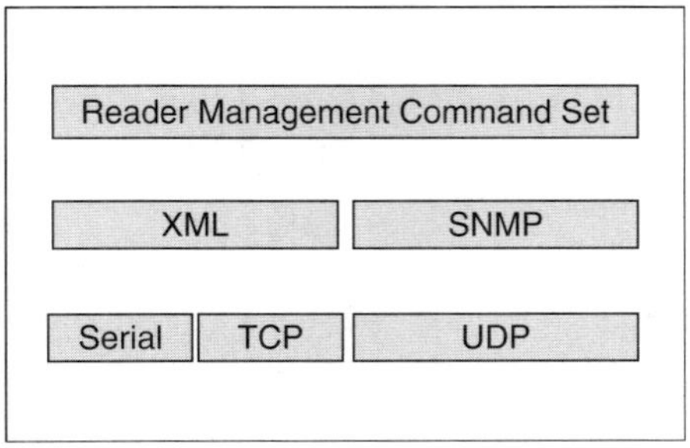

Figure 6.6 Protocol layers mapping.

The EPCglobal Reader Management standard [14] (RM) defines management protocol specifications. Based on the same structure as the reader protocol (Figure 6.5), this standard defines MTBs for the management process of reader health. Figure 6.6 shows two MTB examples.

The command set described in the RM specification allows the host to set or get a description from the reader, the location description, the operational status, etc. It also defines a lot of commands for the management and monitoring of readers, antennas, notifications, alarms, etc. All these operations can be categorized into three types of operations according to the task they perform:

- **DO-type** The reader performs an action.
- **SET-type** The reader changes the internal variable state.
- **GET-type** The reader sends the internal variable state to the host.

Some of these operations require special communication patterns, such as request/response, or asynchronous messages. The specifications of the RM Protocol define three kinds of communication channels with their own behavior. Used for communication between the Reader Layer and the Message Layer, channels are independent. To be compliant with these specifications, readers may support one or more of each of the following channels:

- **Command** Following the request/response pattern, this channel is used for requests from the host to the reader, and for responses from readers.
- **Alarm** This channel allows asynchronous messages from the reader to the host only.
- **Notification** Notification channels are used for delivering tag data.

The RM standard aims to provide mechanisms to monitor reader's operational status and to notify the host of potential operational problems. Another standard used for reader management is the Discovery, Configuration and Initialization [15] (DCI). Using the Control and Provisioning of Wireless Access Points (CAPWAP) protocol [16], the DCI specifies the device called the Access Controller. This device can perform some initial operations such as reader's identification, configuration, and network connectivity management. The aim is to provide a way for readers to discover hosts and hosts to discover readers. The Access Controller can also configure and initialize readers.

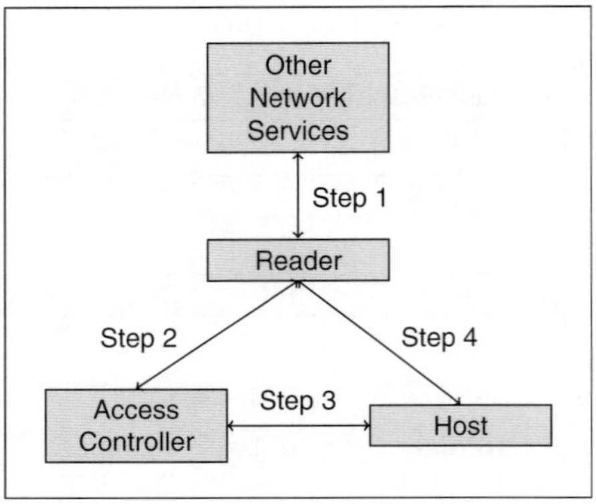

Figure 6.7 DCI overview.

Configuration includes of course configuration of the reader, but also updates the software and the firmware of readers. The initialization step provides parameters to the reader in order to begin the operation. Figure 6.7 describes the steps of DCI.

The first step is the reader network addresses assignment and the access controller addresses determination. Step 2 represents the discovery of host and reader initialization. Thirdly, the access controller gives indication to the host, and finally, the communication starts (Step 4).

Now we can manage readers, plug them and retrieve data from tags through a reader protocol and reader management protocol. The next important component in a RFID system is a component that filters and aggregates all tag data collected from readers.

6.4 Data Management and Application-Level Events

In a middleware RFID, we find the ALE Interface and the ALE Engine, also called *Filtering and Collection*. The ALE specification [5] is a software specification indicating required functionality and behavior, as well as a common API expressed through XML Schema Definition (XSD) and Web Services Description Language (WSDL). The ALE is thus an interface and provides all the functionalities this implies. Through this interface, clients may interact with filtered, consolidated EPC data and related data from a variety of sources. Therefore, the ALE interface cannot be decoupled from the data management held by the ALE engine. The goal of the ALE (interface and engine) is to reduce the volume of data that comes directly from EPC data sources such as RFID readers into coarser "events" of interest to applications. In this layer, common optimizations are efficient RFID data management and load-balancing methods [17, 18], indexing continuous queries [19], enhancement and contextualization of information [20, 21] .

In this section, we focus on the way the data can be handled and the Application Level Event or ALE, a standard created by EPCglobal. We first highlight the role of the ALE as well as the functionalities it provides. Then, we detail the specification standard of the ALE.

Note that the standard gives only the specification of the ALE and the functionalities it has to offer. It does not define the way these functionalities should be carried out.

6.4.1 Data Management and ALE Functionalities

The role of the ALE is to provide independence between the infrastructure components that acquire the raw data (like RFID readers, for instance), the architectural component(s) that filter and count that data, and the applications that use the data. This allows changes in one without requiring changes in the other, offering significant benefits to both the technology provider and the end-user.

ALE standard specifies an interface (not an implementation) which involves:

- receiving data from one or more data sources (readers);
- accumulating data over intervals of time;
- filtering to eliminate duplicated data or data with no interest;
- counting and grouping.

ALE allows aggregation and filtering of tag data over a period of time. An ALE server specifies when to start collecting data, when to stop collecting data, how to organize and sort the data and when to send the data to interested parties. An ALE client allows communication with any compatible ALE server to define data requirements and receive reports.

To illustrate the benefit of the ALE, let's assume a smart shelf as depicted in Figure 6.8(a). It is equipped with three RFID readers which all have two antennas (each antenna reads a unique board of the shelf). We assume that there are objects lying on the shelf, all equipped with a RFID tag. If there is no middleware or ALE, every reader of the shelf communicates continuously the list of tags/objects it has on it. This information is clearly neither useful not exploitable. Indeed, if an application is not written in a manner to handle this large throughput of data, it can suffer from severe scalability problems, perform poorly, or at worst, crash. This is where the ALE specification comes in. It defines a configurable set of data gathering techniques that higher order business

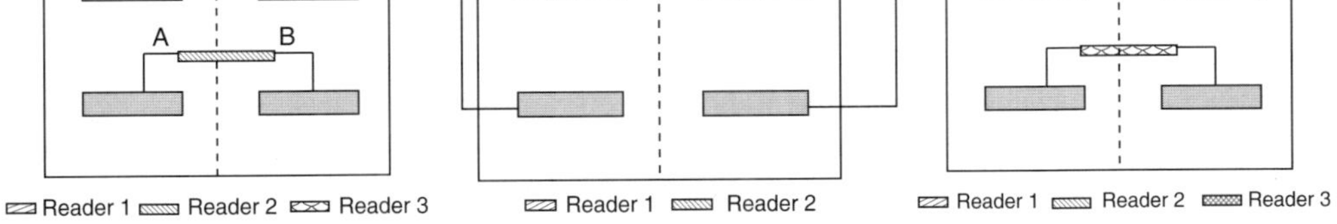

(a) Physical view. (b) Logical View 1. (c) Logical View 2.

Figure 6.8 Smart shelf.

applications can then use to receive more specialized set of tags to process. ALE provides a standard way to describe and define these data gathering techniques to help decouple a business application from the source of its RFID data.

First, ALE configures logical readers, which are discorrelated from the physical readers and antennas. This can be traduced by the fact that according to the application, we can configure logical readers in several ways. For instance, we may need to read and manage independently the right and left sizes of the shelf as one can see on Figure 6.8(b). In such a case, we configure two logical readers on the shelf: one reading the left side, composed by Reader 1 and Antenna *A* of physical Reader 2 and one monitoring the right side of the shelf, composed by physical Reader 3 and Antenna *B* of physical Reader 2. Another application having access to the same shelf may need to manage the object of the shelf according to height of the board on which they are. In this latter case, we would configure three logical readers as shown by Figure 6.8(c). The first logical reader is thus composed of Antenna *A* of both physical readers 1 and 3, the second is composed by Antenna *B* of both physical readers 1 and 3, the third logical reader is physical Reader 2 itself. This then allows a clearer view of the shelf by the application and allows it to query only one logical reader at once.

Second, thanks to the ALE, we will be able to configure, aggregate and filter the events we need. For instance, we may need one report every minute. In this case, the ALE aggregates data over time and over all readers and antennas and filters duplicated data. ALE sends a report every minute instead of continuously.

We may also need to know only which objects which have been removed or/and added in a period of time. In such a case, ALE filters every parasite data and reports only objects that have (dis)appeared during the last period of time.

Finally, ALE allows grouping and counting objects. For instance, a request would be to know how many objects of each kind are available on the shelf. In this case, ALE groups the objects kind by kind and counts them. The shelf then only reports a line by kind of objects (for instance, at time T, shelf had 3 books and 10 pens).

All this reduces the volume of data as soon as possible (reader or middleware) and makes data coming from RFID systems understandable by the information services.

6.4.2 *Specs and Reports*

In order to provide every functionality mentioned, several aspects have to be specified. Indeed, to allow the middleware to manage the data (filtering, aggregating, etc.), we first need to define a cycle during which data are acquired and managed.

For it, we first define a *reader cycle*. A reader cycle is the smallest unit of interaction between the application and the (logical) readers. Nevertheless, this unit is highly reader-dependent.

For example, for an ALE implementation directly embedded in an RFID reader device, the reader is here represented by the communication pathway between the ALE subsystem and the RF protocol subsystem. In such a case, a reader cycle might represent one iteration of the RF protocol used to communicate with RFID tags. But if the ALE implementation is provided by a middleware which communicates with an outboard RFID reader device, a reader cycle is a unit of interaction defined by the protocol used to interact the middleware and the reader. This latter may correspond to one or several RF protocol iterations.

Then, we define the *event cycle* (for reading API) or *command cycle* (for writing API). An Event Cycle (resp. Command Cycle) consists of a boundary which indicates when to start collecting data (resp. writing data) and when to stop collecting data (resp. writing data on tags). This start and stop conditions can be time, stability or trigger-based. A time-based condition is simply a number of milliseconds to wait before starting or stopping the boundary. A stability-based condition is only valid for stopping a cycle and is used to indicate stability in the set of tags detected (no new tags added or removed). A trigger-based condition allows an external controlling process to start or stop the cycle. Note that triggers are an expected extension point in the ALE specification as the ALE specification does not define any standard trigger mechanisms.

If we take the example of the smart shelf again, a time-based condition event cycle example would be "Read from Reader A for five seconds and Report every tag whose EPC belong to the A class." The event cycle in such a case is 5 seconds.

At the conclusion of an event cycle (resp. a Command Cycle), data that has been collected (resp. tags that have been operating as well as the operation result) during the cycle is filtered and grouped. A set of reports are generated and sent to the ALE client.

The ALE standard defines *specs* and *reports* linked to these cycles. "Specs" define a *language*, a format to send the client request, to make it understandable by the system. For instance, we will use *spec* definition to send "Read from Reader A for five seconds and Report every tag whose EPC belong to the A class." "Reports" define the way the result of the request should be sent. These Specs and Reports are translated in a XML language Specification Definition [22–24]. There are four specs and reports:

- Event Cycle Spec (ECSpec): ALE Client request in Reading API.
- Event Cycle Report (ECReport): ALE response to an ECSpec.
- Command Cycle Spec (CCSpec): ALE Client request in Writing API.
- Command Cycle Report (CCSReport): ALE response to a CCSpec.

6.4.2.1 Event Cycle Specs and Reports

An ALE Client's Declarative Specification (ECSpec) includes three fields:

- *Location (where)*: Specifies the logical reader(s) required.
- *Boundaries (when)*: Specifies the event cycle type (Continuous, Interval or Trigger) and boundary (Duration, Field status or Trigger). This defines when to start and stop collecting data.
- *Report contents (what)*: Specifies what kind of events should be reported.

The report content is driven by the following options:

- *Output Set*: Specifies what tags should be reported from the last reading: only current tags, new tags (addition) or missing tags (deletions).
- *Filtering*: Applies a set of patterns to the tags being reported to filter them in or out. Filtering is optional and more than one or both types can be used.
- *Grouping*: Applies a set of patterns to the tags being reported to group logical units together. For instance, you could group all products by a certain company or all products of a certain type together.

- *Data Format*: Indicates how the data should be reported: only a count of data (very useful for grouping), or the actual tag ID values in one of their supported formats (as defined by the EPC Tag Data Standard).
- *If empty*: Specifies what to report if there is no EPC to send.

A report (ECreport or CCReport) is then composed of two fields: the header and the output of the report. This latter is the answer content to a given ECSpec or CCSpec. The former includes several fields of which the most important are the ECSpec or CCSpec it answers, the start and stop condition and time of the event cycle, the readers handled and if any, the readers that ALE failed to question.

To illustrate this, let's go back to our smart shelf. Let's assume that we have the following request: "Report once per minute about things added and removed from Shelf Reader number 1." The ECSpec to be built for such a request should thus contains two different reports, one for new objects and one for objects taken away. Note that spec report names are arbitrary and are only used to identify different reports. The final ECSpec will thus be:

Logical Readers	Reader 1		
Time Boundaries	Start Continuous		
	Stop Duration = 60 seconds		
Output	Report Spec 1 "New"	Output set	Additions
		Content	Field epc (epc-uri)
		Report if empty	False
	Report Spec 2 "Taken away"	Output set	Deletions
		Content	Field epc (epc-uri)
		Report if empty	False

This ECSpec will be translated in a XML language into:

```
<ale:ECSpec>
 <logicalReaders>
  <logicalReader>Reader1</logicalReader>
  </logicalReaders>
  <boundaries>
   <duration unit=”MS>60000</duration>
  </boundaries>
  <reportSpecs>
   <reportSpec name=New Stuff>
    <reportSet set=ADDITIONS/>
    <output includeEPC=true/>
   </reportSpec>
  <reportSpec name=Stuff Taken Away>
    <reportSet set=DELETIONS/>
    <output includeEPC=true/>
  </reportSpec>
  </reportSpecs>
</ale:ECSpec>
```

To which the shelf will answer with a ECReport:

	Reader	1	
Header	Start time	Timestamp = 2009-03-11T16:37:00Z	
	Duration	60000 ms	
	Report 1		
	Tag1	ID	10210
Output		Designation	Cookies
	Tag2	ID	31210
		Designation	Chocolate

An ALE Event Cycle Spec is presented to the API in one of the three following modes:

- *Subscribe Mode*: Defines the ECSpec ahead of time like Poll. It does not require the ALE client to continually request data from the server. Instead the ALE client adds a subscription to the ECSpec that the ALE server communicates. One example of an ALE subscription would be an ALE client that opens a TCP port that the server can connect to and send reports through it.
- *Immediate Mode*: Allows the client to send an ECSpec and have it be processed and fulfilled immediately, returning the reports when the ECSpecs first Event Cycle Boundary completes. This option requires the client to continually send the ECSpec definition to the server with every request.
- *Poll Mode*: The Poll mode of interaction with an ALE server is similar to Immediate, except the ALE client defines the ECSpec on the server ahead of time. It gives the ECSpec a name which allows the ALE client to request data from it using the name instead of sending the ECSpec every time.

6.4.2.2 Command Cycle Specs and Reports

Command Cycle Specs and Reports concern writing API on tags. With a CCSpec, an ALE client describes which operations to perform on tags during a command cycle. Each active CCSpec gets exclusive access to the tags concerned. As for the ECSpecs, a CCSpec defines the location (which reader(s)?) and the start and stop conditions of the command cycle. It also defines some specifications, such as the filters to apply (if any), the kind of report if no tag to handle and a list of operations. Each operation to be performed specifies the operation type, the field to operate upon and the data input format. These writing operations are the following:

- *Initialize*: Prepare a bank for using variable fields.
- *Read*: Read contents of a field.
- *Write*: Write contents of a field.
- *Add*: Add a new field.
- *Delete*: Remove field.
- *Lock*: Change access permission.
- *Kill*: Kill tag.
- *Password*: Provide password to enable subsequent ops.
- *Check*: Check memory state consistency.

Then, every operation is performed in the same way as the event cycles. For instance, we need to write in a tagged product which enters the warehouse. The ALE CCSpec sent will be:

Logical Readers	Reader 1				
Time Boundaries	Stop	Tag count = 1			
			Filter	None	
				INIT PRODUCT	afi = 0xC1
Commands	Command Spec 1		Commands	ADD FIELD	Warehouse = #1
				ADD FIELD	Date = $2009-03-12T12:57:00Z$

This CCSpec will also be translated into a XSD standard. It is then followed by a CCReport as following:

	Reader	1	
Header	Start time	Timestamp = 2009-03-12T12:57:00Z	
	Duration	$135\,ms$	
	Report 1		
Output		Op1	SUCCESS
	Tag1	Op2	SUCCESS value = 1
		Op2	SUCCESS value = 2009-03-12*T*12:57:00*Z*

6.4.3 Research Challenges

The ALE may be connected to several readers and these readers may read several tags. In such a case, the ALE may have too many tags to filter and aggregate. In order to provide a scalable middleware, the task of filtering and aggregation should be distributed over different entities. This would allow the ALE engine to manage more readers. Nevertheless, this is a challenging task as it has to remain transparent to the application, that is, the latter should not have to change its behavior when using this distributed ALE engine.

A simple solution would be to duplicate the ALE engine to manage readers, but it is not transparent for the application since it has to send one ECSpec by ALE. In addition, this solution would not be scalable. Another drawback is that the application has to handle the fact that a tag may be read by two readers belonging to different ALEs. The tag ID will appear on both ECReports and the application has to work on the reports to delete duplicated tags.

In [25], the authors propose a Global-ALE which upon receiving an ECSpec from the application splits it and distributes it to the sub-ALEs. The Global-ALE receives ECReports from sub-ALEs and builds one global ECReport for the application. This solution provides transparency regarding the application. It also limits the overload of readers events, but another bottleneck may appear between sub-ALEs and the Global-ALE.

An efficient distributed ALE will contain both solutions. On one hand, it must be transparent for the application, that is, the application sends only one ECSpecs and receives only one ECReport. On the other hand it should be distributed in a peer-to-peer manner so that it minimizes the bottleneck problem. This is one of the hot issues on which researchers are currently working.

6.5 Store and Share Data

In order to store and share data between different sites of the same company or between companies, EPCglobal defines three standards (two of them are ratified, one is in development). The first one, called EPC Information Services [6], provides interfaces between applications that capture data and those that need it and can store the data captured. The second standard defines an Object Naming Service [7] that performs a simple operation: return an EPC related ressource (the EPCIS address which contains data of a given EPC, html link, web services, . . .). The last standard (in development) describes the EPC Discovery Services [26], which search and return locations storing EPC-related data.

6.5.1 EPC Information Services

The EPCIS standard is organized into three layers:

- *Abstract data model layer*: specifies the generic structure of EPCIS data with general requirements for creating data definitions.
- *Data definition layer*: specifies which data is exchanged via EPCIS.
- *Service layer*: specifies service interfaces: capture and query.

The EPC Information Services (EPCIS) standard defines three interfaces, the *EPCIS Query Callback Interface*, the *EPCIS Query Control Interface* (both part of the *EPCIS Query Interface*) and the *EPCIS Capture Interface*, which interact with two modules: the *EPCIS repository* and the *EPCIS Capturing Application*. Figure 6.9 positions these interfaces between components. We explain these five new notions in this section (the ALE Interface was described in Section 6.4).

The *EPCIS Capturing Application* aims to process ALE events with business information. This is the component that can check for business context problems and react

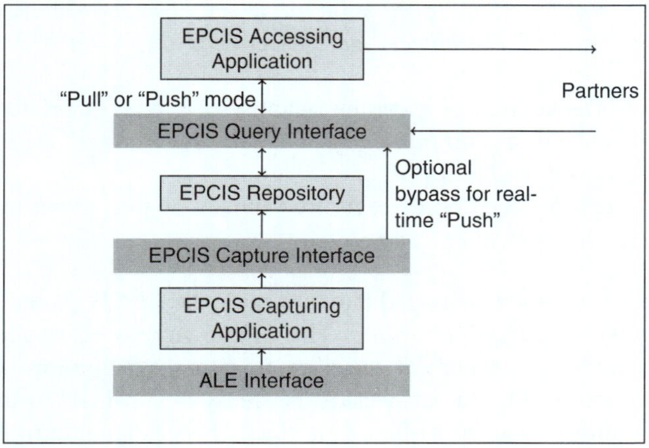

Figure 6.9 EPCIS interfaces.

to them. For example, it can divert a bad case in the network area in a conveyor system. It captures ALE events ECReports and CCReports (see Section 6.4) and rearranges them, that is, it translates ALE events into EPCIS events. It adds some business context and creates one or more business events. It allows the multiple ALE events to continue. Moreover, it allows the aggregation of other data sources that the ones coming from the filtering interface like user inputs or business information systems. It seems to have the same role as the ALE Engine except that the *EPCIS Capturing Applications* knows about the business steps context.

The *EPCIS Capture Interface* is the interface that defines how EPCIS events are delivered to the higher levels of the system (to the EPCIS Repository, to EPCIS Accessing Applications, or to a partner). It consists of one method *capture* with a list of EPCISEvent for single argument. This method returns nothing or void. The standard from EPCglobal defines a binding via an HTTP URL to send events from the capturing application to the higher levels through the capture interface.

The *EPCIS Repository* stores EPCIS events in order to offer persistence to EPC-related data. This data can be accessed through an EPCIS Query Interface by EPCIS Accessing Application or partners. It acts as a database and can be implemented in different ways using known database systems such as MySQL, Oracle,... An EPCIS event is a registering of something that occurs in the real world. Most of the time, but not necessarily, it is triggered by the reading of a RFID tag. An event has four dimensions:

- *What*: what object is concerned?
- *When*: when did it happen? (timestamp)
- *Where*: where the event happened? (location identifier)
- *Why*: from what business process is it issued?

An event is either:

- *Object Events*: Observation of a list of data tag during a business step in a given place and time *These objects were read at the distribution center #9 at 10:01AM, in receiving mode*.
- *Aggregation Events*: Physical association of a set of tags with a parent one at a given place and time. *These objects were assembled on Pallet #12 at Pallettizer #4 at 12:32PM*.
- *Quantity Events*: Declaration of a certain quantity of a type of products (object class) at a given time and place. *200 bottles of "Chateau La Pompe" were identified in the stock of shop #234 today at 3:20PM* or
- *Transaction Events*: Association of registered objects with a given transaction. *Order #123 was set up with objects x, y and z*.

The *EPCIS Query Interface* is divided into two parts: the *EPCIS Query Control Interface* and the *EPCIS Query Callback Interface*. The *EPCIS Accessing Applications* retrieves EPC-related data through the control interface in either synchronous or asynchronous mode. In asynchronous mode, the answer of the query is "pushed" to the corresponding Accessing Application(s) via the callback interface. This latter interface is also used to send data immediately from Capturing Application(s) to Accessing Application(s) (the "optional bypass" arrow on Figure 6.9).

The *EPCIS Accessing Application* is the business application that uses EPC-related data. It is the module which generates and sends ECSpec and CCSpec (see Section 6.4). For instance, it can be the bill module of a company which checks whether an order has been expedited before editing the bill. An EPCIS Accessing Application can be either a part of the enterprise system or a partner application. For instance, this may be the system of a delivery company which checks whether an order is ready before coming to retrieve it. A partner can access the EPC-related data of another enterprise through the EPCIS Query Interface. But partners have to know the EPCIS Repository address to query it. There are two EPCglobal standards (one ratified and one in development) in order to retrieve partners' EPCIS Repository address. These two standards are presented below.

6.5.2 Object Naming Service

The Object Naming Service (ONS) is a central look-up service of the EPCglobal Network [7]. Its main function is the address retrieval of manufacturer information services for a given Electronic Product Code (EPC) identifier. It locates the EPCIS of the issuing authority for the EPC (usually a manufacturer).

Today, the ONS is built on top of the Domain Name System (DNS) [8], that is, it is a hierarchical organization as shown in Figure 6.10. To use it, the EPC is first translated into an URI format, equivalent to a URL address. If the local root does not know this address, it forwards it to the ONS Root. The ONS Root contains the address of the local ONS for each EPC manager number. For instance, it knows the IP address of the local ONS hosting every EPC starting with 061414. Thanks to the DSN mechanisms, the requested is forwarded to the local ONS containing the EPC requested. In its turn, based on the same mechanism, the local ONS routes the request towards the proper EPCIS.

A local ONS contains the address (URL) of EPCIS of every EPC (e.g. SGTIN, SSCC, ...[2]). Therefore, the local ONS is able to find the right EPCIS thanks to the DNS and to redirect the request to it. Finally, the EPCIS contains data about a specific EPC.

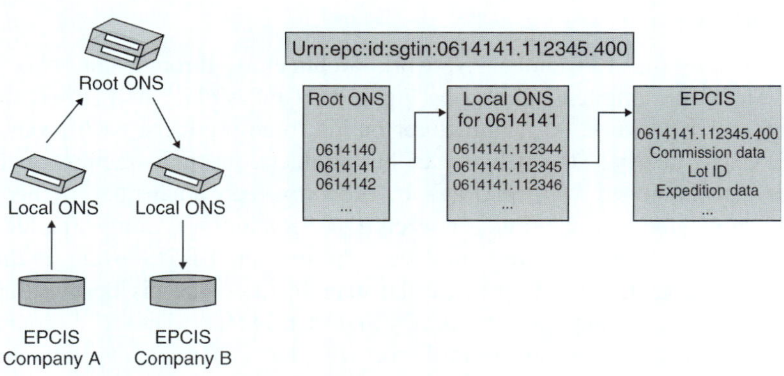

Figure 6.10 ONS look-up service.

[2] Two standards explain the different Electronic Product Codes [29, 30].

Nowadays, the number of requests to an ONS remains low due to the relatively low number of RFID systems currently deployed. Therefore, till April 2008, there existed only six *duplicated* ONS roots in the world operated by Verisign. This solution is not practicable or scalable as the overhead for preserving data consistency between roots is high together with memory overhead. Furthermore, this solution is not compliant with every country's systems.

Since the number of RFID systems will increase in the coming years, the EPCglobal network needs several other ONS roots in other countries, but the standard actually states the existence of a unique central root directory. Therefore, at the GS1 France initiative, EPCglobal will focus on designing and evaluating a multi-roots ONS system that will take into account security, stability, performance as well as interaction with DS (Discovery Services). The aim is to show that several ONS roots can work together and safely share the management and the governance of the network.

6.5.3 Discovery Services

Discovery Services simplify the data exchange process by offering a service that links information about RFID-enabled products as they move through the supply chain. The addition of Discovery Services to the EPCglobal Network offers trading partners the ability to find all parties who have possession of a given product and to share RFID events about that product.

The main difference with the ONS is that each EPCIS having information about a specific EPC register themselves on the DS. The company can query the DS which will return the list of all previously registered EPCIS containing information on this EPC. Then the company can retrieve all the knowledge of the entire network for this EPC by querying these EPCISs one after the other. Nevertheless, nowadays, there exists no standards or norms. This is in the scope of EPCglobal who defines Discovery Services as a standard in development.[3] Today, several protocols are proposed to feed the standard by companies such as IBM [27] or Afilias [28] but the standard is still not in production.

6.6 Example

We can illustrate the EPCGlobal Network architecture through the two following examples. The first example displayed by Figure 6.11 takes place in a closed architecture of a manufacturer network. A manufacturer builds some products, each product labeled with an EPC. To facilitate the delivery of his products, the manufacturer gathers them in a pallet also equipped with an EPC. In order to keep an history of his products, the manufacturer records the link between the "containing" entity (pallet) and the "contained" objects (manufactured products) through an EPCIS event in his EPCIS repository (called *AggregationEvent*). So the manufacturer, via his business application sends the ECSpecs to configure the ALE (Figure 6.11 ①). The ALE configures the readers (Figure 6.11 ②). When the ALE receives the readers events with the relevant EPC (Figure 6.11 ③), it sends the ECReport to the EPCIS through the Capture Interface and Capture Application (Figure 6.11 ④).

[3] http://www.epcglobalinc.org/standards/discovery

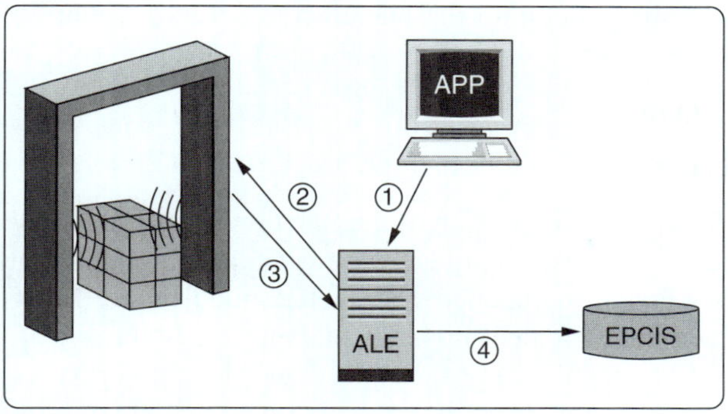

Figure 6.11 Manufacturer network.

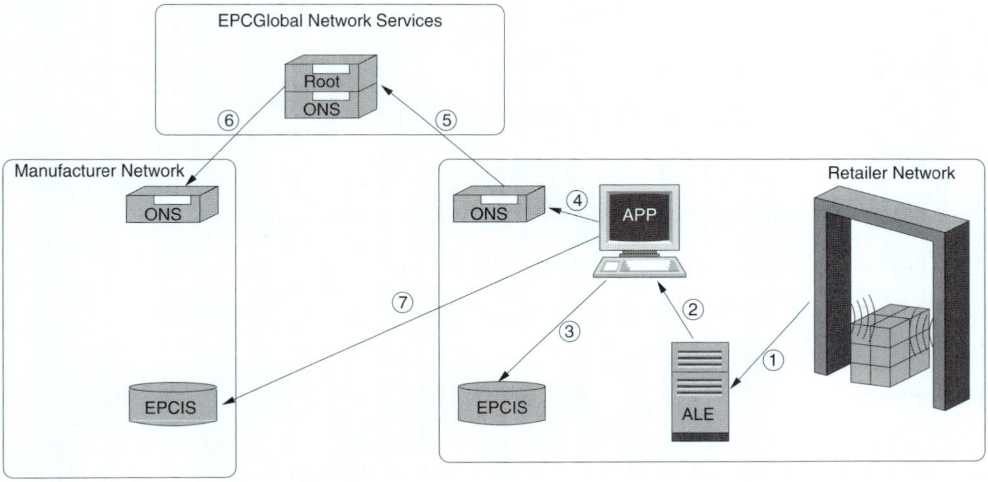

Figure 6.12 Across networks.

The second example plotted by Figure 6.12 shows how and when data can be exchanged and shared, across enterprises. The same manufacturer as above sends his pallet to the retailer. When the retailer receives the pallet, its EPC is read with or without the contained EPC-labeled objects (Figure 6.12 ①) and sent to the application (Figure 6.12 ②). The retailer needs to check the content of the pallet. His application will first query the local retailer EPCIS "Have you recorded an event based on this EPC?" (Figure 6.12 ③) but as we saw in the first example, the correlated AggregationEvent is indeed stored in the manufacturer EPCIS. Then, two possible options: (i) either the retailer knows the entry for the Query Interface of the manufacturer's EPCIS, so the application can directly query this EPCIS (Figure 6.12 ⑦); or (ii) the retailer's application can ask the ONS "Give me the address of the local ONS requested by this EPC" (Figure 6.12 ④ and ⑤), then the request is directed to the manufacturer's local ONS (Figure 6.12 ⑥). Finally, the retailer can retrieve the EPCIS address for the event record. Now he can compare the list of objects contained by the pallet with the manufacturer record (Figure 6.12 ⑦) and check the consistency of his order.

6.7 Conclusion

This chapter has drawn up a description of the functionalities offered by a middleware and highlighted the necessity for such a middleware when the amount of data and applications dealt with increase. For all components of such a middleware, EPCGlobal gives some standards but the way to implement it is still challenging. In the actual economic context, RFID applications have been finding their place and the coming years will witness their development. Researchers are currently proposing new models to implement every brick of the middleware in a still easier, faster and simpler way that will appear very soon. With the support of more and more companies and researchers, EPCGlobal is developing the standards, to give access to the largest number of companies.

Problems

1. Who is the entity which produces RFID standards?
2. What is a RFID middleware and what is its purpose?
3. What are the three possible modes for querying a source?
4. What is the entity in charge of reader management?
5. What principles do the current ONS rely on?

References

[1] Liu, S., Wang, F. and Liu, P. (2006) Integrated RFID data modeling: An approach for querying physical objects in pervasive computing, in *Proc. 15th ACM Int. Conf. Information and Knowledge Management (CIKM'06)*, Arlington, Virginia, USA, poster. pp. 822–823.

[2] Wang, F., Liu, S., Liu, P. and Bai, Y. (2006) Bridging physical and virtual worlds: Complex event processing for RFID data streams, in *Proc. 10th Int. Conf. Extending Database Technology (EDBT'06)*, pp. 588–607.

[3] Weinstein, R. (2005) RFID: A technical overview and its application to the enterprise, *IT Pro* pp. 27–33.

[4] Schmidt, L., Mitton, N. and Simplot-Ryl, D. (2009) Towards unified tag data translation in the internet of things, in *Proc. 1st Wireless Communication Society, Vehicular Technology, Information Theory and Aerospace & Electronics Systems Technology (VITAE'09)*, Aalborg, Denmark.

[5] EPCglobal (2008a) The Application Level Events (ALE) specification.

[6] EPCglobal (2007a) The EPC Information Services (EPCIS) standard, version 1.0.1.

[7] EPCglobal (2008c) Object Naming Service (ONS) Standard, version 1.0.1.

[8] Mockapetris, P. (1987) Domain names – concepts and facilities. *IETF RFC* 1034.

[9] EPCglobal (2007b) Epcglobal Architectural Framework, version 1.2.

[10] Eom, J. B. and Lee, T. J. (2008) RFID reader anti-collision algorithm using a server and mobile readers based on conflict-free multiple access. In *Proc. IEEE Int. Performance, Computing and Communications Conference (IPCCC 2008)*, pp. 395–399, Austin, Texas, USA.

[11] Joshi, G., Mamum, K. A. and Kim, S. (2009) A reader anti-collision MAC protocol for dense reader RFID system, in *Proc. WRI Int. Conf. Communications and Mobile Computing (CMC'09)*, Kunming, Yunnan, China. pp. 313–316.

[12] EPCglobal (2006) The Reader Protocol (RP) standard.

[13] EPCglobal (2007c) The Low-Level Reader Protocol (LLRP) standard.
[14] EPCglobal (2007d) The Reader Management (RM) standard.
[15] EPCglobal (2009a) The Discovery Configuration and Initialization (DCI) standard for reader operations.
[16] IETF (2009) rfc5415 Control and Provisioning of Wireless Access Points (CAPWAP).
[17] Park, S. M., Song, J. H., Kim, C. S. and Kim, J. J. (2007b) Load balancing method using connection pool in RFID middleware, in *Proc. 5th ACIS Int. Conf. Software Engineering Research, Management & Applications (SERA 2007)*, pp. 132–137.
[18] Wu, J., Wang, D. and Sheng, H. (2007) ECA rule-based RFID data management, in *Proc. 1st Annual RFID Eurasia*, pp. 1–5.
[19] Park, J., Hong, B. and Ban, C. (2007a) Efficient transformation scheme for indexing continuous queries on RFID streaming data, in *Proc. 2nd Int. Conf. Systems and Networks Communications (ICSNC 2007)*, Cap Esterel, France. pp. 41–46.
[20] Moon, M., Kim, Y. and Yeom, K. (2006) Contextual events framework in RFID system, in *Proc. 3rd Int. Conf. Information Technology: New Generations (ITNG'06)*, Las Vegas, Nevada, USA, pp. 586–587.
[21] Wang, F. and Liu, P. (2005) Temporal management of RFID data, in *Proc. 31st Int. Conf. Very Large Data Bases (VLDB 2005)*, Trondheim, Norway, pp. 1128–1139.
[22] Biron, P. and Malhotra, A. (2001) XML shema part 2. W3C recommendation.
[23] Bray, T., Paoli, J., Sperberg-McQueen, C. M., Maler, E. and Yergeau, F. (2004) EXtensible Markup Language (XML) 1.0 (third edition), W3C recommendation.
[24] Thompson, H., Beech, D., Maloney, M. and Mendelsohn, N. (2001) XML Shema part 1. W3C recommendation.
[25] Liu, F., Jie, Y. and Hu, W. (2008) Distributed ALE in RFID middleware, in *Wireless Communications, Networking and Mobile Computing, 2008. WiCOM '08. 4th International Conference*, pp. 1–5.
[26] EPCglobal n.d. The Reader Management (RM) standard (to appear).
[27] Rantzau, R., Kailing, K., Beier, S. and Grandison, T. (2006) Discovery services enabling RFID traceability in EPCglobal networks, in *Proc. of the 13th International Conference on Management of Data (COMAD)*, Delhi, India.
[28] Afilias (n.d.) Afilias provides free discovery services for RFID pilots.
[29] EPCglobal (2008b) Epcglobal Tag Data Standard (TDS).
[30] EPCglobal (2009b) Epcglobal Tag Data Translation (TDT) standard.

Part Two

Tag Identification Protocols

7

Aloha-Based Protocols

Kwan-Wu Chin and Dheeraj Klair

University of Wollongong, Australia

RFID readers aim to identify tags quickly. The key constraint, however, is that a tag can only be read one at a time. Hence, in scenarios with multiple tags, for example, tagged items on a pallet, reading becomes problematic when multiple tags reply simultaneously because their respective signals collide and corrupt one another. RFID readers, therefore, run an anti-collision protocol to arbitrate tag replies so that they experience minimal or no collision, and hence achieve fast identification.

Unfortunately, anti-collision protocols face a key challenge. That is, they are unaware of the identity (ID) or the number of tags in a reader's interrogation zone. If this information is available, then an anti-collision protocol can easily instruct each tag to transmit at a time that does not overlap with other tags' transmission. For example, if there are two tags with ID X and Y, an anti-collision protocol can inform the tags to transmit at time t_1 and t_2 respectively, where t_2 occurs immediately after the reader has acknowledged receiving tag X's transmission. However, in practice, both these bits of information are unknown. As a result, anti-collision protocols have difficulty ensuring tag replies are collision-free. Moreover, an anti-collision protocol needs to ensure tags' replies are consecutive in order to maximize reading rate. This also means the reader does not wait idly for a tag reply. Otherwise, it will experience prolonged identification delays, and also energy and bandwidth wastage.

To date, researchers have proposed two categories of anti-collision protocols: Aloha and tree. Each protocol category comes with its own advantages and disadvantages. In general, tree protocols promise deterministic identification but are complex, incur significant memory overheads, and require complex hardware [1, 2]. In contrast, Aloha protocols have simpler reader designs, lower protocol complexity and bandwidth requirements, smaller number of reader to tag commands, and are able to adapt dynamically to varying tag population.

RFID Systems: Research Trends and Challenges Edited by Miodrag Bolić, David Simplot-Ryl, and Ivan Stojmenović

Table 7.1 RFID standards and products that use Aloha-based tag reading protocols.

Standards	Protocol
ISO 18000-3 "MODE 1"	Pure Aloha or Framed Slotted Aloha
ISO 18000-3 "MODE 2"	Tags select from eight channels and use Slotted Aloha during transmission
ISO 14443-3 Type B	Dynamic Framed Slotted Aloha
ISO 18000-6A	Frame Slotted Aloha with muting and early-end
EPCGlobal Class 1	Basic Framed Slotted Aloha with early-end
Philips I Code	Dynamic Framed Slotted Aloha

Table 7.1 presents the standards that use Aloha-based protocols. These standards are managed by two bodies: EPCglobal and the International Standards Organization or ISO. The former develops standards targeted at supply chain networks whereas the latter specifies the air interface for tracking cattle, payment systems and smart cards. Apart from these organizations, there are also propriety standards by Philips. In general, standards over low bandwidth air interface rely on Aloha variants. Otherwise, systems have the flexibility to either choose tree or Aloha variants.

Henceforth, this chapter presents a comprehensive review of Aloha-based anti-collision protocols. In Section 7.1, we will present protocols based on Pure Aloha (PA). Then, in Section 7.2, we introduce Slotted Aloha (SA) variants and show how they effectively address the partial collision problem suffered by PA variants. This is then followed by an analysis comparing all Pure and Slotted Aloha variants in Section 7.2.1. This is followed by an introduction to Framed Slotted Aloha (FSA) protocols in Section 7.3, where tags are read using frames with either a fixed or varying number of slots. Moreover, we show how each frame is sized using a tag estimation function. Besides that, in Section 7.3.3, we review protocols that combine the advantages of both Aloha and tree algorithms. Finally, Section 7.4 presents future works and concludes the chapter.

7.1 Pure Aloha

Figure 7.1 shows three tags in a reader's interrogation zone. Upon being energized, each tag sets itself to respond randomly after receiving a read request from the reader. If the reader receives a tag response successfully, it transmits an acknowledgment (ACK). On the other hand, if there is a collision, the reader transmits a negative ACK (NACK), which causes a tag to retransmit its ID after a random delay.

RFID systems based on Pure Aloha, however, suffer from the well-known partial collision problem that limits their throughput to 18% [3]. To improve the performance of Pure Aloha, researchers have proposed various optimizations to reduce collision and increase its tag identification rate [4]:

- *Pure Aloha with Muting.* In this variant, the reader sends a mute command after identifying a tag. As a consequence of receiving the mute command, the identified tag stops responding to future read requests. This means every successful tag identification

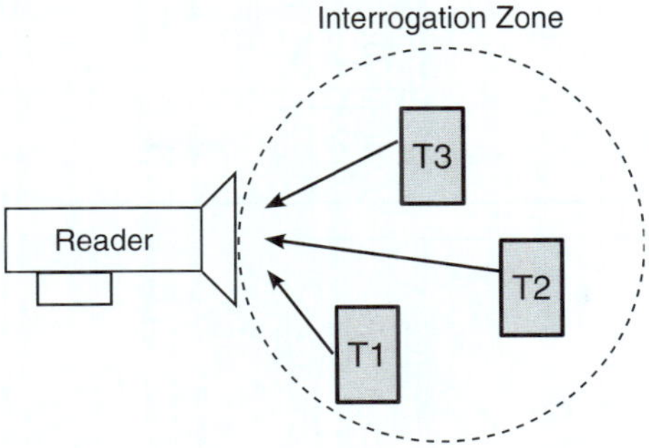

Figure 7.1 Reader and tags interactions.

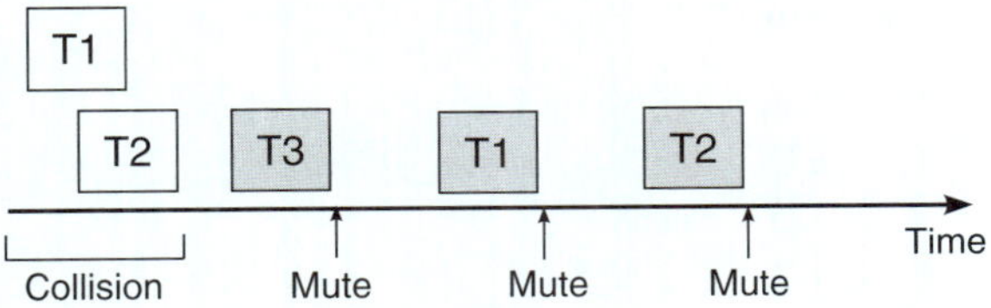

Figure 7.2 Pure Aloha with muting.

reduces the number of tags and the reader's offered load. Figure 7.2 shows the behavior of Pure Aloha with muting. Initially, the transmission from tags 1 and 2 partially overlaps, which results in a collision. These tags then retransmit after a random delay. After receiving a tag ID successfully, for example, tag 3, the reader sends a "mute" command to the tag.

- *Pure Aloha with Slow Down.* In this variant, once a tag is identified, it is instructed to increase its back-off time using a slowdown command. In Figure 7.3, if slowdown is not used, we see tag 3's reply interfering with the transmission of tag 1 and 2. However, with slowdown, the reader instructs tag 3 to back off for a longer period of time. Consequently, the reader has a higher probability of identifying tags 1 and 2 successfully.
- *Pure Aloha with Fast Mode.* The reader transmits a "silence" command after detecting the start of a tag transmission. This command has the effect of muting other tags. Tags are allowed to transmit again after the reader has sent an ACK command or until their waiting timer expires. Figure 7.4 shows Pure Aloha with fast mode. Once the reader detects a transmission from tag 2, tag 1 and tag 3 are silenced until the tag finishes its transmission. Otherwise, these tags would have collided with the reply from tag 2.
- *Hybrids.* Lastly, it is possible to combine the above features to create other Pure Aloha variants. Namely, Pure Aloha with fast mode and muting, and Pure Aloha with fast mode and slowdown. In the former variant, tags are temporarily silenced whenever a tag has started its transmission. A tag is then muted after identification. On the other hand, instead of muting tags, the latter variant slows the reply of identified tags so that their replies are less likely to collide with those from unidentified tags.

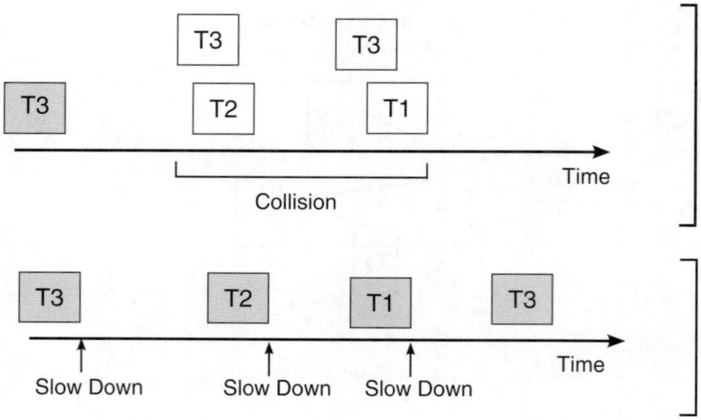

Figure 7.3 Pure Aloha with slowdown.

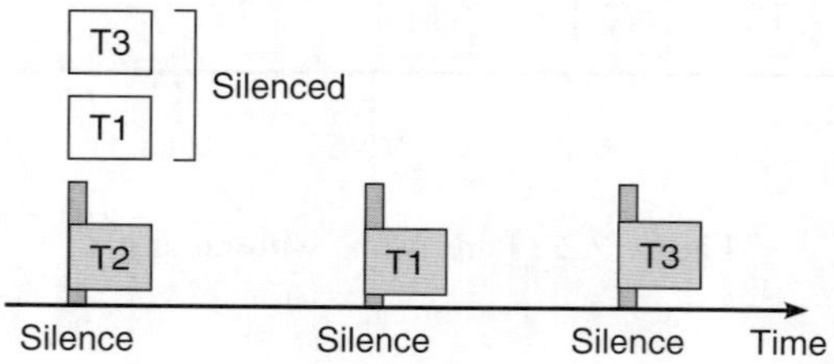

Figure 7.4 Pure Aloha with fast mode.

7.2 Slotted Aloha

The main problem that limits the reading rate of Pure Aloha systems is partial collisions. To overcome this fundamental problem, RFID systems can employ Slotted Aloha, which has a maximum throughput of around 36% [3]; viz. doubles that of Pure Aloha. Instead of replying on a continuous timeline, tags are now required to respond at pre-defined slots. This means the reader and tags are tightly synchronized and tags are only allowed to transmit at the beginning of a slot. If there is a collision, tags wait for a random number of slots before retransmitting again. The key performance gain obtained by Slotted Aloha is due to the fact that collisions only occur at the start of each slot as opposed to any time in Pure Aloha.

Similar to Pure Aloha, there are numerous variants:

1. *Slotted Aloha with Muting or Slowdown*. This variant has the same operating principle as Pure Aloha with muting or slowdown, but instead operates in a slotted manner.
2. *Slotted Aloha with Early End*. A reader closes a slot if it does not detect any transmissions at the beginning of a slot. Two commands are used: start-of-frame (SOF) and end-of-frame (EOF). The former is used to start a reading cycle, and the latter is used by the reader to close an idle slot early. Figure 7.5 depicts how early end is used to terminate two idle slots. As a result, tags can transmit sooner, leading to a higher

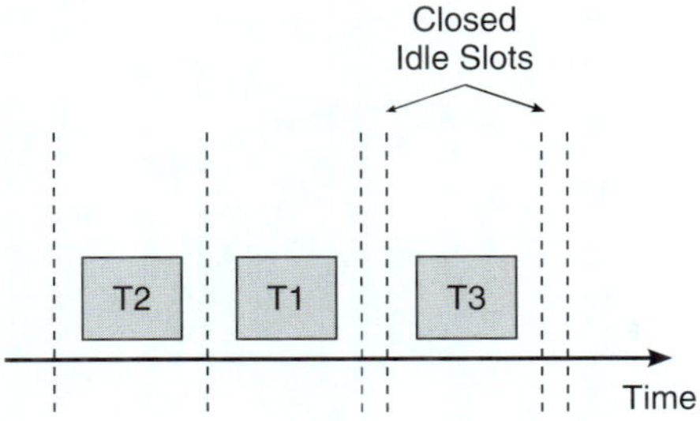

Figure 7.5 Slotted Aloha with early end.

reading rate. Moreover, a reader is able to conserve energy as it can start receiving the next reply sooner [5].

3. *Slotted Aloha with Early End and Muting*. After identifying a tag, the reader sends a mute command to the tag, thereby removing the tag from contending in subsequent slots. In addition, as required by the early end feature, idle slots are closed early using the EOF command.
4. *Slotted Aloha with Slow Down and Early End*: This variant combines slowdown with the early end feature. In other words, as well as closing idle slots, identified tags are instructed to slow their replies.

7.2.1 Pure versus Slotted Aloha Variants

Klair *et al.* [6] and Rivera *et al.* [7] have conducted extensive analytical and simulation studies of Pure and Slotted Aloha variants. The key simulation parameters used are as follows. The tag's data rate is 26 kbps as per ISO 15693. They modeled the communication between a tag and the reader as a Poisson process. Specifically, the mean arrival time between tag responses is $\frac{1}{\lambda}$, where λ is the average duty cycle of tags. Each tag ID is 112 bits in size. They bound the retransmission delay to K random slots; Schwartz [3] found $K = 5$ to be optimal as higher values prolonged identification delays.

From Figure 7.6, we see that the throughput for Pure Aloha with fast mode is as high as Slotted Aloha variants – throughput is defined as the number of tags identified per second. In addition, Pure Aloha protocols using the fast mode feature approach the maximum throughput when there is a large number of tags. Therefore, when protocols are operating at their maximum system efficiency, fast mode variants of Pure Aloha operate below their maximum system throughput. This means, given the same number of tags, Pure Aloha variants using fast mode experience more free channel time as compared to other protocols, hence tags' responses are less likely to collide. Note, the performance of Pure and Slotted Aloha protocols are limited by collisions. To clarify, Pure Aloha protocols suffer from partial collisions, which limit their performance to 18%. On the other hand, Slotted Aloha protocols have a maximum throughput of 36% because collisions can only occur at the start of a slot [3].

Figure 7.7 plots the average number of collisions when reading n tags. Among Pure Aloha variants, Pure Aloha has the highest collision count because of its vulnerability

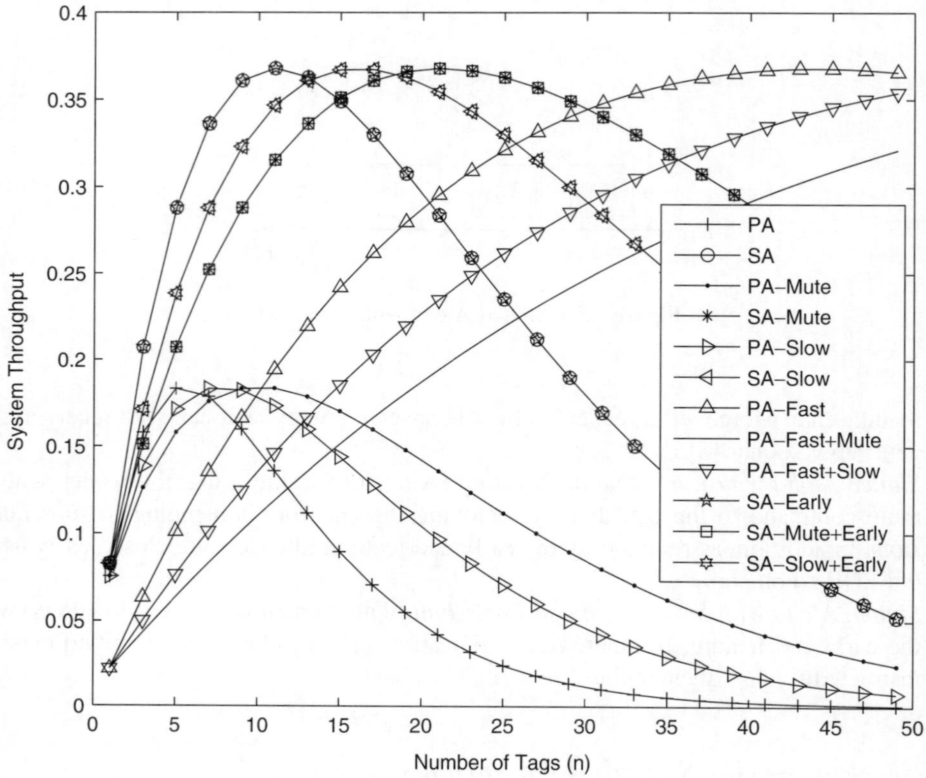

Figure 7.6 $\lambda = 20$, $K = 5$, where λ is the offered load and K is the maximum retransmission delay [6]. *Computer Communications*. Reproduced by permission of © 2008 Elsevier B.V. All rights reserved.

period of 2T, where T is the transmission time of a tag ID. On the other hand, Pure Aloha with fast mode and muting has the lowest number of collisions. This is because both fast mode and muting reduce collisions. Among slotted Aloha variants, conventional slotted Aloha has the highest number of collisions. On the other hand, slotted Aloha with muting achieves the lowest collision count. Note that, early end has no effect on the number of collisions. Overall, Pure Aloha fast mode variants have the best performance, that is, they outperform all slotted Aloha variants. The trade-off, however, is the increased system complexity because a separate channel is used to transmit the fast mode command.

Figure 7.8 plots the number of idle slots that occur when reading n tags. We can see that for Pure Aloha, Pure Aloha with muting, and Pure Aloha with slowdown, with increasing tag population, the number of idle slots reduces proportionally. This is because the probability that at least one tag choosing a given slot increases with tag numbers. For Pure Aloha, in particular fast mode variants, the number of idle slots increases initially, and then begin to reduce when the tag population exceeds 35. This is because with fast mode, collision reduces, and therefore, there are fewer tag retransmissions, which explains the initial small number of idle slots. However, as expected, when we increase the number of tags, the offered load to the reader becomes higher, which reduces the probability of idle slots.

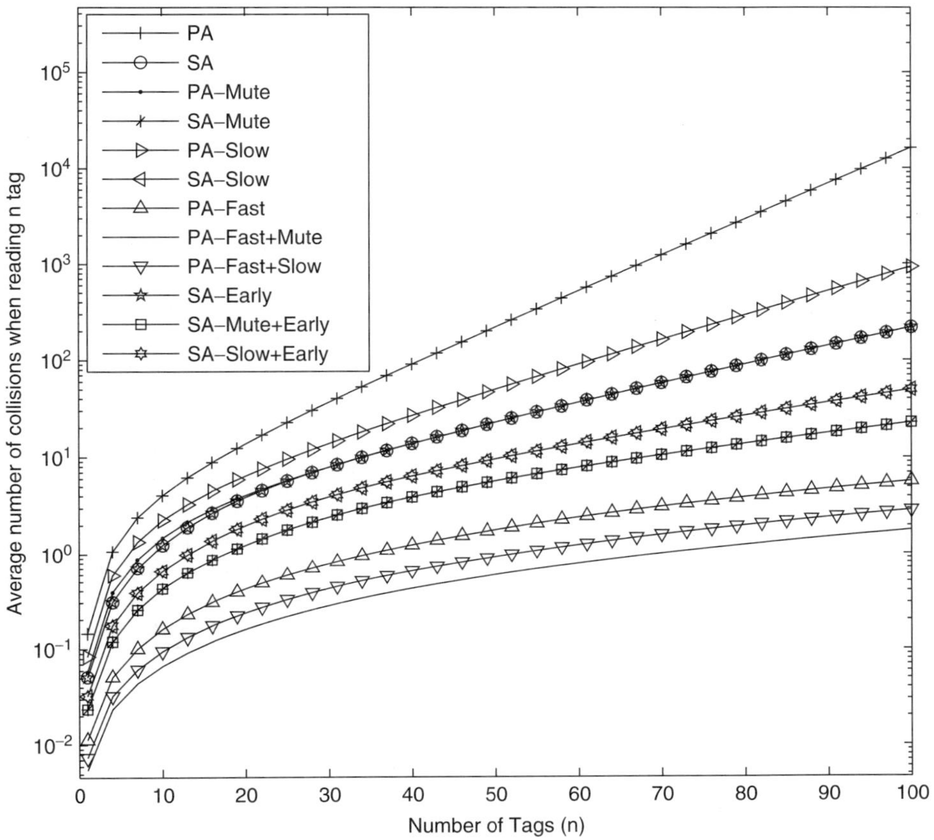

Figure 7.7 Number of collision to read n tags.

Similar to Pure Aloha variants, slotted Aloha variants initially observe a lower number of idle slots before peaking, and gradually dropping with increasing number of tags.

In summary, the key results from [6] and [7] are:

- Fast mode variants experience the smallest number of collisions. This means, a reader using these variants will be able to identify the highest number of tags quickly.
- Early end variants reduce energy consumption from idle listening significantly. However, its use does not contribute to any reduction in collisions. Apart from that, as the number of tags increases, idle listening reduces, and hence it does not provide any benefits in terms of energy savings.
- Pure Aloha with fast mode and muting can read the highest number of tags. Overall, this variant experiences the smallest number of collisions, and hence the highest read rate.

7.3 Framed Slotted Aloha

A major problem with Pure and Slotted Aloha variants is that tags reply at least once in a reading cycle. To address this problem, framed slotted Aloha (FSA) variants restrict

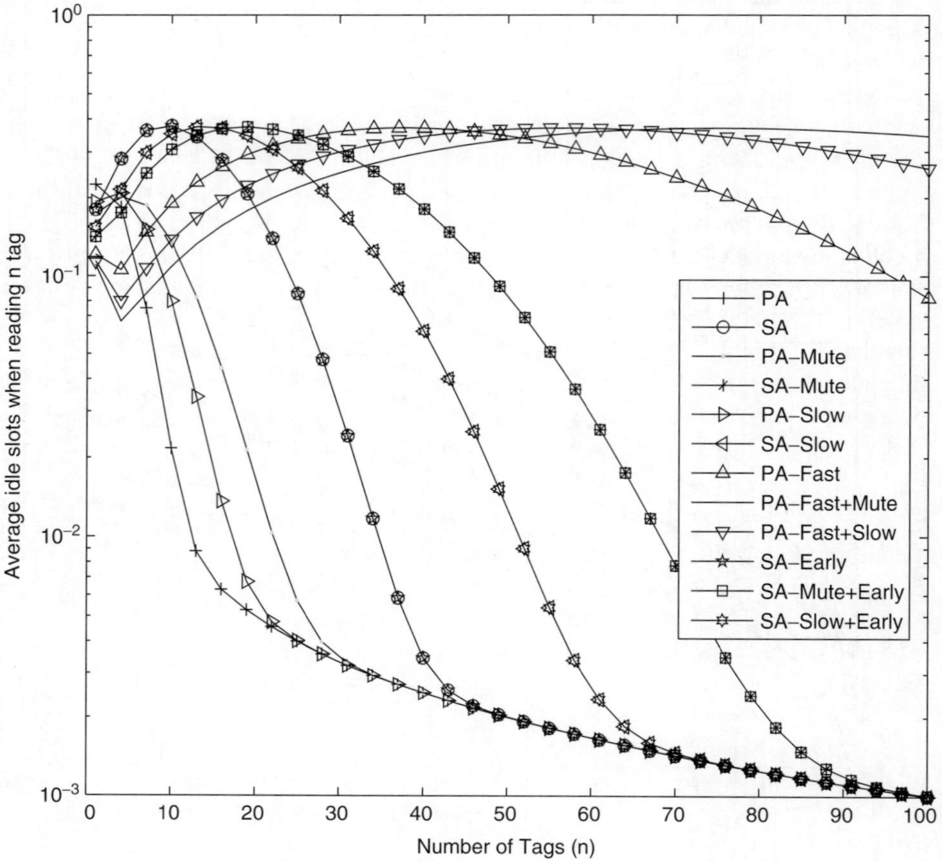

Figure 7.8 Number of idle slots encountered when reading *n* tags.

tag reply to only once in each frame. As a result, the offered load to the reader reduces, and consequently so do collisions. Moreover, it allows a reader to estimate the number of tags, and consequently, the number of slots or frame size required to minimize collisions and idle slots.

To date, there are three categories of FSA-based anti-collision protocols, as determined by a reader's ability to adjust the frame size used to read tags. The following sections review readers that use (i) a fixed/basic frame, where the same frame size is used in each read cycle; (ii) a dynamic frame, where the size of the frame is determined by a tag estimation function; and (iii) an enhanced/hybrid, where a reader adjusts its frame size dynamically as well as segregates tags into small groups.

7.3.1 Basic

Basic FSA (BFSA) anti-collision protocols can be classified according to their use of the following features: muting and early end. In other words, (i) BFSA-no-muting,

(ii) BFSA-with-muting, (iii) BFSA-no-muting-with-early-end, and (iv) BFSA-muting-early-end. These variants work similarly to those in Section 7.2 but operate on a frame-by-frame basis.

A fundamental problem with BFSA-no-muting is that a reader's identification delay increases exponentially when the number of tags is significantly bigger than the frame size used to read them. To address this problem, Hwang *et al.* [8] propose segregating tags into small groups so that the number of tags replying is smaller than the frame size. To achieve this, the reader transmits a "comparison" bitstring whenever the ratio of slots with collisions to frame size exceeds 0.5. Tags then compare this bitstring against a part of their ID; for example, the 20th to 25th bit of their ID. Tags with a smaller result reply in the current frame. The disadvantage of Hwang *et al.*'s work is that it does not limit the number of tags in a group. Given that a reader does not know the tag IDs in its interrogation zone, the "comparison" string used may yield a group with no tags, or in the worst case, a group that encompasses all tags.

7.3.2 Dynamic

The main problem with BFSA variants is the use of a fixed frame size. A reader will experience many idle slots when the number of tags is small. Conversely, many collisions will occur when the tag population increases beyond the frame size. Therefore, there is a need to adjust the frame size dynamically to match the number of tags.

The Q algorithm [4] uses a simple frame adjustment technique. A reader initially broadcasts a slot counter Q and sets its frame size to 2^Q, where Q is an integer between zero and eight. Tags then choose a slot randomly from 0 to $2^Q - 1$. The reader then monitors each slot. An idle slot causes the reader to decrease Q by a constant c, whereas slots with collisions cause the reader to increase Q by the same constant. Here, $0.1 \leq c \leq 0.5$. Before the start of the next read cycle, Q is rounded and sent to tags. This technique is computationally inexpensive and quickly converges to the optimal frame size. Lee *et al.* [9], however, showed that this technique can be improved further as follows. First, to reduce the algorithm's signaling overheads, they propose to send the value of Q to tags only when it changes, that is, when rounding yields a new Q value. Secondly, a reader can use a different constant and scale to update the value of Q. Specifically, they propose to decrease Q by C_i for each idle slot, and to increase Q by C_c for slots with collisions. These constants are computed as follows. If the number of tags, m, is known,

$$C_c = -0.0491 \times ln(m) + 0.534 \tag{7.1}$$

$$C_i = (e - 2) \cdot C_c \tag{7.2}$$

Otherwise, the constants are calculated as:

$$C_c = \text{any value between 0.1 and 0.5} \tag{7.3}$$

$$C_i = (e - 2) \cdot C_c \tag{7.4}$$

where e is the mathematical constant 2.71828. ...

The frame size of most Dynamic FSA (DFSA) variants, however, is controlled by a tag estimation function. As shown by Schoute [10], the maximum throughput of FSA occurs when the frame size matches the number of tags. Hence, this function plays a critical role in the performance of FSA protocols.

A tag estimation function uses the status of slots in the last frame to compute a tag estimate. Specifically, the number of slots filled with zero (c_0), one (c_1), and multiple (c_k) tag responses. A tag estimation function then computes the optimal frame size for the next read round using c_0, c_1, and c_k as its parameters. Note, a *read round* is considered complete at the end of each frame.

To date, researchers have proposed a number of tag estimation functions. Vogt [11] presents two tag estimation functions, which we denote as Vogt-I and Vogt-II. The former sets the tag estimate to be $c_1 + 2c_k$ as it assumes at least two tags are involved with every collision. On the other hand, the latter is based on Chebyshev's inequality and aims to minimize the distance between the actual and theoretically computed result. In other words, justify

$$\varepsilon_{vd}\left(N, c_0, c_1, c_k\right) = \min_t \left| \begin{pmatrix} a_0^{N,t} \\ a_1^{N,t} \\ a_k^{N,t} \end{pmatrix} - \begin{pmatrix} c_0 \\ c_1 \\ c_k \end{pmatrix} \right| \tag{7.5}$$

where ε_{vd} corresponds to the error between an actual and theoretical read result as represented by vector $< c_0, c_1, c_k >$ and $< a_0^{N,t}, a_1^{N,t}, a_k^{N,t} >$ respectively. In other words, Vogt-II aims to determine a tag number estimate t that minimizes ε_{vd}. The elements of the vector $< a_0^{N,t}, a_1^{N,t}, a_k^{N,t} >$ correspond to the expected number of empty slots, slots filled with one tag, and slots with collisions, respectively. To calculate $a_k^{N,t}$, given a frame size of N, and t, the expected number of slots filled with k responding tags is given by,

$$a_k^{N,t} = N \times \binom{t}{k} \left(\frac{1}{N}\right)^k \left(1 - \frac{1}{N}\right)^{t-k} \tag{7.6}$$

Vogt also proposes frame sizes that yield low identification delays for a given tag range; see Table 7.2. For example, a frame size of 16 is considered optimal when the number of tags is in the range of one to nine.

Zhen *et al.* [12] propose a function that computes the expected number of tags replying in a collided slot. According to Zhen *et al.*, on average, 2.39 tags are involved in every collision. Thus, the number of estimated tags is $c_1 + 2.39c_k$. In addition, Zhen *et al.*

Table 7.2 Optimal frame sizes for a given tag range.

Frame Size (N)	Low (n)	High (n)
16	1	9
32	10	27
64	17	56
128	51	129
256	112	∞

[11]. Reproduced from © 2002 IEEE.

propose to over-estimate the tag population because doing so lowers identification delays. Based on their experimentations, they propose $1.4 \times (c_1 + 2.39c_k)$ as a tag estimate. On the other hand, for RFID systems that support muting, instead of using 1.4 as a scaling factor, they use 0.65.

Cha and Kim [13] present two tag estimation functions for muting based RFID environments. Cha-I estimates the number of tags n by solving the following equation,

$$C_{ratio} = 1 - \left(1 - \frac{1}{N}\right)^n \left(1 + \frac{n}{N-1}\right) \tag{7.7}$$

where C_{ratio} is computed after a read round as $C_{ratio} = \frac{c_k}{N}$. The tag estimate in Cha-II is simply $2.39c_k$.

The DFSA variant proposed by Khandelwal *et al.* [14] estimates the number of tags using,

$$n = \frac{log\left(\frac{c_0}{N}\right)}{log\left(1 - \frac{1}{N}\right)} \tag{7.8}$$

As before, N is the current frame size. Note, Equation 7.8 cannot be applied when $c_0 = 0$. When this happens, the tag estimate is $n = c_1 + 2c_k$. Lastly, Khandelwal *et al.* propose setting the frame size to $1.943 \times n$ times the estimated number of tags.

Floerkemeier and Wille [15] and Floerkemeier [16] present two estimation functions: Floerkemeier-I and Floerkemeier-II. These functions estimate tags based on the Bayesian transmission strategy proposed by Rivest [17]. In Floerkemeier-I, a reader not only considers read results in the current read round, but also records those in the last frame to determine the frame size to be used in the next read round. Specifically, Floerkemeier *et al.* select the frame length L to be used in the next round by maximizing the following equation,

$$\sum_{i=0}^{n_{\max}} U(\eta = i, N)P(\eta = i) \tag{7.9}$$

The function $U(\eta, N)$ determines the frame size N that yields the maximum throughput for a given number of tags η, and it is derived by [10] as,

$$U(\eta, N) = \frac{\eta}{N}\left(1 - \frac{1}{N}\right)^{\eta-1} \tag{7.10}$$

The function $P(\eta = i)$ returns the probability that there are i tags in the reader's interrogation zone. To derive $P(\eta = i)$, Floerkemeier *et al.* make use of information from the last t frames, and also the last j slots of the current frame. In addition, they adjust $P(\eta = i)$ to account for newly arriving and departing tags. The difference between the function Floerkemeier-I and Floerkemeier-II is that the latter calculates $P(\eta = i)$ on a slot-by-slot basis, as opposed to on a frame-by-frame basis. Moreover, Floerkemeier-II terminates and restarts a frame if it is found to be non-optimal. As mentioned by Floerkemeier [16], this scheme is computationally expensive given that the reader must iterate

$n_{\max}$ times every slot or frame to compute the optimal frame size. Hence, both functions are not suitable for resource constrained RFID readers.

Kodialam and Nandagopal [18] propose an estimation function that computes the expected number of idle and single response slots by inserting $k = 0$ and $k = 1$ in Equ. 7.6. The resulting equations are then used to derive two estimators, called zero estimator (ZE) and collision estimator (CE),

$$ZE = e^{-(n_0/N)} = \frac{c_0}{N} \tag{7.11}$$

$$CE = 1 - \left(1 + \frac{n_k}{N}\right) e^{-(n_k/N)} = \frac{c_k}{N} \tag{7.12}$$

In Equations 7.11 and 7.12, n_0 is the tag estimate obtained from ZE, and n_k is the tag estimate computed from CE, respectively. The value of c_0 and c_1 is obtained by observing the number of idle slots, and slots with a successful transmission in a given frame. They are then used to solve ZE for n_0 and CE for n_k. If $n_0 < n_k$, then the tag estimate is n_0, otherwise it is n_k. The authors assume that tag estimation occurs in a phase that precedes the identification phase. In addition, slots in the estimation phase are only 10-bits long.

Chen and Lin [19] introduce two estimation functions: Chen-I and Chen-II. In the former, the authors compute the probability of having exactly k tags in m slots as [20],

$$p(k, m) = \frac{(-1)^m N!n!}{m!N^n} \times \sum_{j=m}^{\min(N,(\lceil n/k \rceil))} (-1)^j \frac{(N-j)^{n-jk}}{(j-m)!(N-j)!(n-jk)!(k!)^j} \tag{7.13}$$

Using Equation 7.13, the authors calculate the probability of exactly m slots with zero tag responses, that is, $k = 0$. The actual value of m is c_0, which is obtained from the reader's feedback. The probability equation is then solved for the value of n, which is the tag estimate. The Chen-II function, however, computes the expected number of slots filled with zero and a single tag using Equation 7.6. The results, denoted as E and S, are then inserted into the following equation,

$$n = (N - E - 1)\frac{S}{E} \tag{7.14}$$

where N is the frame size. Equation 7.14 is then solved for the tag estimate n.

In addition to the above works, Wang *et al.* [21] propose preceding each reading or jump frame with a short collision detection (CD) frame; see Figure 7.9(a). Tags pick a random slot in the CD frame and transmit a 4-bit Manchester encoded packet. A reader then determines whether there are any collisions in a slot; see Figure 7.9(b). Tags that respond successfully are instructed to reply in the jump frame. Note, the number of slots in the jump frame is equal to the number of tags that successfully transmitted in the CD frame. The key advantage of this scheme is that collisions are short, and a reader does not waste much energy receiving bits corrupted by collisions. In addition, given that the CD frame is effectively a reservation frame, tags are able to transmit collision-free in the jump frame. Having said that, it remains important to dimension the CD frame appropriately to minimize collisions, as the number of slots in the jump frame is dependent only on the number of successful transmissions in the CD frame.

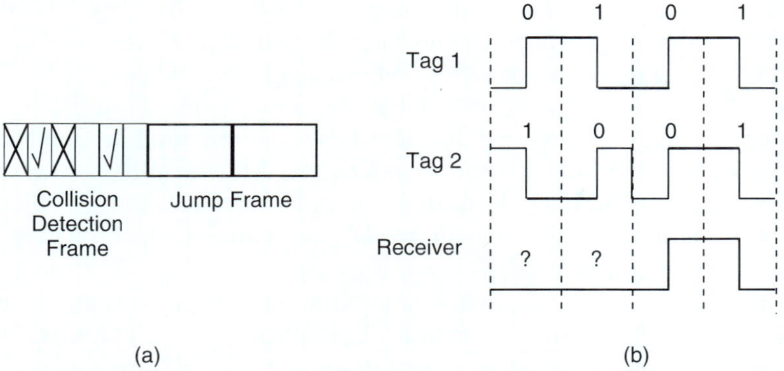

Figure 7.9 (a) A CD frame precedes each jump frame. The number of slots in each jump frame corresponds to the number of successful transmissions in the CD frame; in this example, only two tags successfully transmitted their ID. Tags that transmitted successfully in the CD frame then send their full ID in the jump frame. (b) Collision detection using Manchester encoding, where no transition in the middle of a bit indicates a collision.

7.3.3 Enhanced/Hybrid

A fundamental problem in all FSA protocols is that their frame size grows exponentially with increasing tag numbers. Hence, existing DFSA variants have a maximum frame size; for example, 256 [11] or 512 [22]. Unfortunately, persistent collisions occur when the tag population exceeds a variant's maximum frame size.

Lee *et al.* [23] address the aforementioned limitation by proposing enhanced-DFSA or EDFSA; an anti-collision protocol that grows the number of slots used to read tags linearly. The key idea is to determine whether the estimated number of tags would yield the maximum system efficiency given the current frame size. If not, tags are divided into M groups. Table 7.3 shows the value of M for a given tag range. Also shown are frame

Table 7.3 EDFSA frame sizes. n denotes the number of tags, N is the frame size, and M is the number of tag groups.

Number of tags (n)	Frame Size (N)	M
1–11	8	1
12–19	16	1
20–40	32	1
41–81	64	1
82–176	128	1
177–354	256	1
355–707	256	2
708–1416	256	4
1417–2831	256	8

[23]. Reproduced from © 2005 IEEE.

sizes for varying tag ranges that achieve maximum system efficiency. The reader then sends a number to the tags, which the tags then use with their ID in a modulo operation. Only tags that yield a remainder of zero are allowed to reply. Peng *et al.* [24] noticed two limitations with EDFSA. First, after one round of reading, there will be fewer unread tags in each group. Secondly, EDFSA will not adjust its frame size until the number of unread tags fall below a given threshold. They, therefore, conjecture that these observations lead to sub-optimal system efficiency. Henceforth, in order to restore system efficiency, they propose a protocol that re-adjusts the number of groups after the first round of reading to better match the current frame size.

To date, researchers have also proposed hybrid protocols that combine the advantages of both FSA and tree protocols. For example, Bonuccelli *et al.* [25] propose Tree Slotted Aloha (TSA), an enhanced FSA protocol that forms a tree-like structure during its reading cycles. TSA works by quickly segregating and identifying tags that have experienced collision in the following manner. A reader starts with an initial frame size l_0. Tags then select and record a random slot number, and transmit in the chosen slot. At the end of each frame, the reader broadcasts the slot numbers with collisions. Tags that transmitted in these slots are informed to reply in the upcoming frame. This process repeats at the end of each frame until a frame is collision-free. Bonuccelli *et al.* [25] compute the size of an upcoming frame as $\lfloor \frac{n-c_1}{c_k} \rfloor$, where n is a tag estimate obtained using Vogt's [26] function; recall that c_k and c_1 correspond to the number of slots with and without collision respectively.

In a different work, Klair and Chin [27] propose ResMon, an energy-efficient DFSA protocol that supports *both* tag identification and monitoring. ResMon is conceptually similar to Wang *et al.*'s [21] protocol, but adds an additional collision-free frame that is used solely for monitoring tags; see Figure 7.10. To clarify, ResMon uses three frames. The first frame, R_{frame}, is used by tags to reserve a collision free slot in the B_{frame}. Tags send a short, randomly generated bitstring to the reader. Collision is then detected as per Figure 7.9. If a tag's transmission is collision-free, the reader allocates the tag a slot in the upcoming B_{frame}, which the tag uses to transmit its full ID. As part of the reservation process, the reader also allocates tags a slot in the M_{frame}, which they then use to send the same 3-bit Manchester encoded bitstring to indicate they are still in the reader's interrogation zone. A reader transmits an acknowledgment if it receives this bitstring successfully. Otherwise, if there is an idle slot, the reader sends a negative acknowledgment (ACK) to compress the frame in order to remove the idle slot. This is

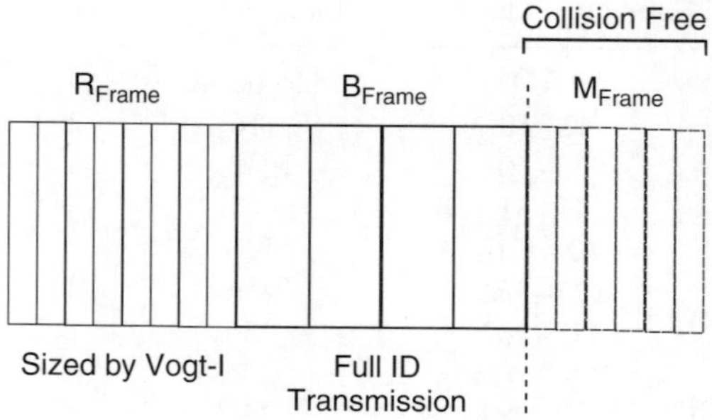

Figure 7.10 ResMon Frames.

carried out as follows. Assume there are four tags with transmission slot one, two, three, and four respectively. If the number two slot is idle, indicating a missing tag, the reader sends a negative ACK to inform all tags to reduce their slot number by one. As a result, in the next M_{frame}, these tags will transmit one slot sooner. Note, tags that have transmitted before the idle slot are not affected by the negative ACK as they are muted after transmission. Hence, only tags following an idle slot will re-adjust their transmission time.

The remaining two hybrid works make use of the Query Tree (QT) protocol [28]. Briefly, the QT protocol works as follows. Assume there two tags with ID 01 and 10 – each tag has a prefix matching circuitry. The reader first transmits a null string, which matches the prefix of both tags. As a result, both tags reply to the reader simultaneously and cause a collision. The reader then pushes the prefix "0" and "1" onto its stack. The next read cycle begins with the reader popping and transmitting the prefix "0". This causes tags with ID matching the prefix "0" to reply, that is, tag 01. After that, the reader transmits prefix "1". As only tags with prefix "1" can reply, the reader only receives a reply from tag 10. The main advantage of QT is its ability to systematically and deterministically identify tags. However, QT may require multiple iterations, that is, large tree depth, in order to identify tags. Specifically, the tree depth required to identify all tags may equal the length of tag ID. For example, as shown in Figure 7.11, the tree depth is four. Notice that if tags A and C or B and D are segregated into different groups, the resulting tree is shorter due to their unique prefix. To this end, researchers have devised various methods to reduce the probability of tags with the same prefix replying simultaneously.

Shin *et al.* [29] propose two algorithms that combine FSA protocols with the QT protocol. In the first protocol, called framed query tree, the reader transmits a set of frames, which tags then randomly choose one to reply in. As a result, there are fewer tags replying in each frame. Moreover, each of them may have a different prefix. The reader then identifies tags in each frame using the QT protocol. In the second protocol,

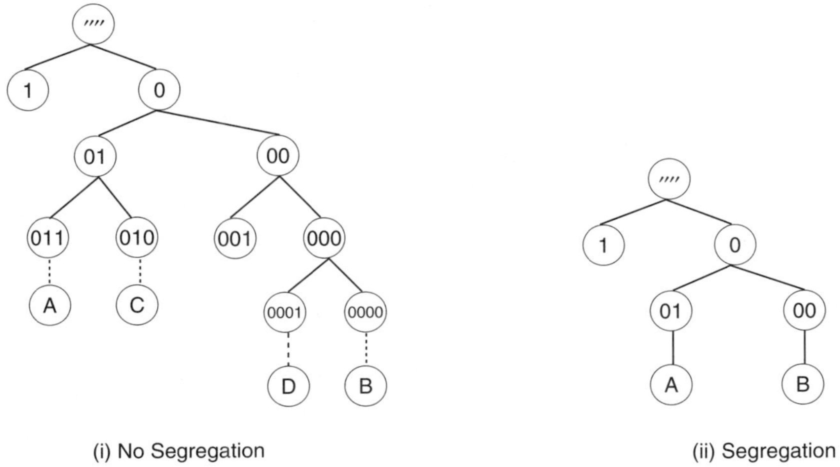

Figure 7.11 An example showing QT being used to identify the following tags: A (0111), B (0000), C (0101) and D (0001). In scenario (ii), only tags A and B contend with each other. Tag C and D will contend at a different time.

that is, query tree Aloha, the reader transmits a prefix, for example, "0", and a frame size. Tags with an ID that match the prefix randomly select a slot, and reply in the chosen slot. Upon detecting a collision, a new prefix is sent, for example, "01", to segregate the collided tags, and a new frame is then used to read tags. On the other hand, Namboodiri and Gao [30] introduce three anti-collision protocols that combine QT with DFSA. The key idea in all three protocols is to associate a frame or multiple slots with each QT query. In other words, with each query or a node on the tree, the reader sends a frame with multiple slots for tags to choose from. As a result, the reader is able to read tags with the same prefix. Thus, more tags can be read with the same number of queries as compared to conventional QT.

7.3.3.1 Performance

Klair *et al.* [31] have studied the performance of FSA variants that use the muting and early-end feature. Briefly, the simulation parameters used in their Matlab implementation are as follows. The tag to reader data rate is 26 kbps (ISO 15693). Hence, with a tag ID size of 96 bits, each slot is of duration $\frac{96}{26000}$. For non-muting experiments, the RFID reader is assumed to continue reading until all tags are read with 99% confidence level. On the other hand, when muting is enabled, it continues reading until there are no collisions in a given read cycle. To evaluate the performance of each tag reading protocol, they conducted experiments with increasing number of passive, static tags, and repeatedly read these tags for 100 simulation runs. They then recorded the mean delay, and the average number of idle and collision slots experienced by the reader when reading a given tag set. Lastly, they assume negligible propagation delay and channel error.

Figure 7.12 shows the delay incurred by each variant to read a given number of tags. We see that among non-muting variants, DFSA with early end has the lowest reading delay. This is because the early-end feature ensures the delay incurred by idle slots are minimized. Similarly, for muting variants, those that combine the early end yield the quickest reading time. The advantage of having a dynamic frame size is particularly critical for high tag numbers. For example, when there are more than 60 tags, the reading delay of BFSA variants grow much quicker than DFSA and EDFSA variants. The superiority of EDFSA variants is also evident. In particular, the reading delay of EDFSA with muting and early end is a few orders of magnitude lower than its DFSA counterpart.

Figure 7.13 shows the average number of collisions encountered when using framed Aloha variants. DFSA with muting has the lowest number of collisions as compared to other variants. On the other hand, BFSA non-muting variant has the highest number of collisions. Note, early end has no effect on collisions as it is used only to shorten idle slots.

Figure 7.14 shows the average number of idle slots that occur when reading n tags using framed Aloha variants. DFSA non-muting variants have the highest number of idle slots for a given number of tags. This is because of their varying frame size, and the use of muting – both of which reduce and increase the probability of collisions and idle slots respectively. On the other hand, BFSA muting variants have the lowest number of idle slots. This is because they use a fixed frame, and hence this causes a dramatic reduction in the number of idle slots.

A key issue with DFSA and EDFSA variants is the accuracy of their tag estimation function. In particular, an inaccurate estimate may lead to persistent collisions or a high

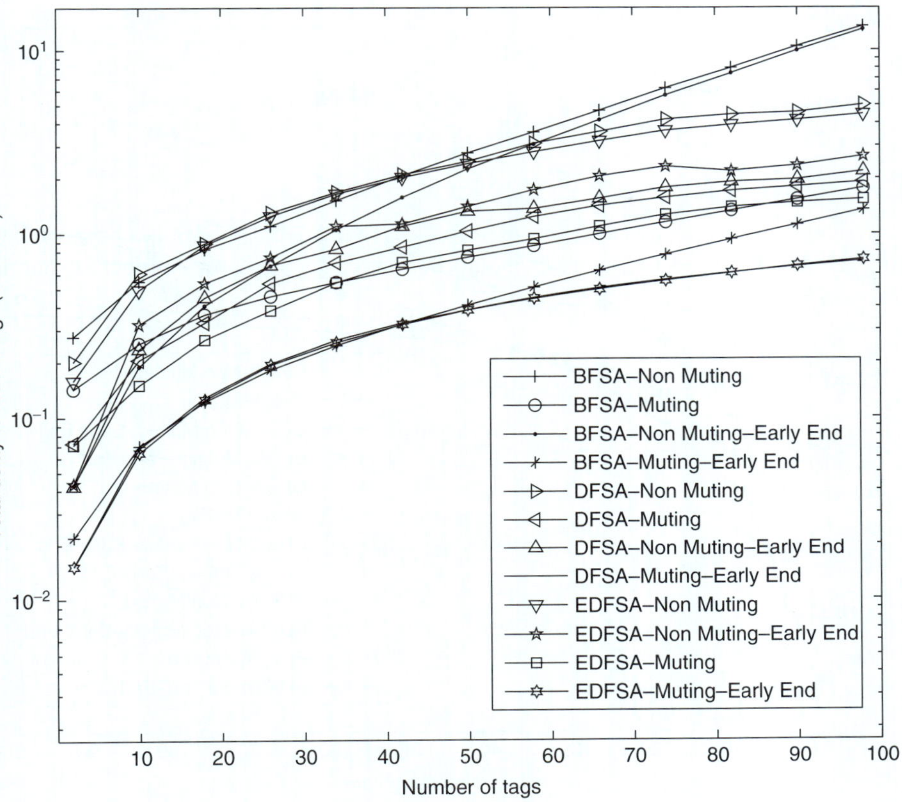

Figure 7.12 Reading delay of BFSA, DFSA, and EDFSA variants.

number of idle slots. To this end, Klair *et al.* [32] have compared the accuracy of the following functions: Vogt-I, Vogt-II, Cha-I, Cha-II, and Zhen *et al.* [12]. Their results show Vogt-II achieves the best accuracy for a wide range of tags. On the other hand, Cha-I is more accurate when the number of tags increases beyond the current frame size. In general, functions that derive tag estimates using probabilistic or statistical methods are more accurate with increasing tag numbers. These include Chen-I, Chen-II, Cha-II, Vogt-II, Floerkemeier-I, Floerkemeier-II, and the functions proposed by Kodialam and Nandagopal [18], and Khandelwal *et al.* [14]. Their better accuracy stems from not relying on a fixed multiple of c_k. Moreover, they are more robust in a wide range of tag population, whereas static estimation functions are suitable only for low tag ranges [11, 19, 26]. Apart from that, a critical observation is that frame sizes can only be a power of two. For example, when there are around 60 tags, a tag estimation function may set the frame size to 128. This reduces collisions, but at the expense of idle slots or system efficiency. Therefore, the early-end feature must be used to offset any frame size inaccuracies. Apart from that, tag estimation functions have varying computational requirements. Some involve only additions and multiplications; for example, Vogt-I, Cha-II, Q-algorithm, and those proposed by Zhen *et al.* [12]. On the other hand, Chen-II, Cha and Kim [13], Floerkemeier-I and Floerkemeier-II require the reader to compute

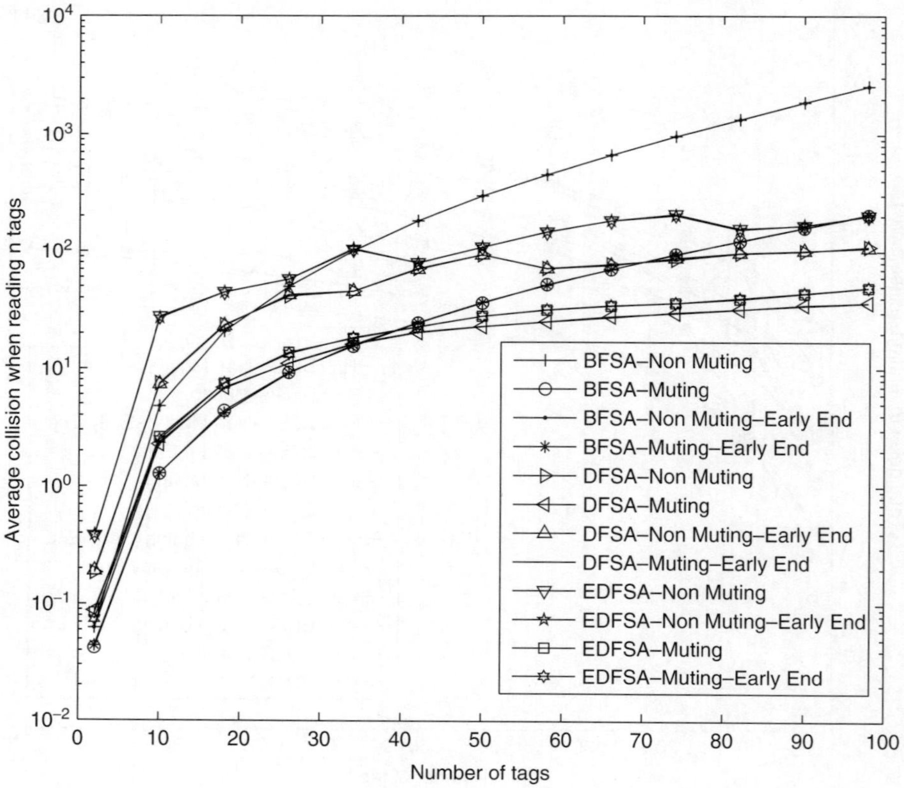

Figure 7.13 Average number of collisions encountered when reading *n* tags.

factorials and fractions, and hence they have a need for more computational power. Vogt-II, Cha-I and Chen-I have the highest requirement because all of them involve recursion.

To date, only two works have investigated the performance of anti-collision protocols in mobile tags scenarios, for example, tagged items on a conveyor belt. The key concern is whether an anti-collision protocol can quickly read all tags before they exit a reader's interrogation zone. In Lee *et al.* [33], the authors found DFSA protocols that use Zhen *et al.* and Cha-I with muting are able to quickly identify tags on a conveyor belt moving at varying speeds; their experiments involve, on average, 200 tags taking 0 to 30 milliseconds to transit a reader's interrogation zone. The key advantage of these algorithms is their ability to mute tags after identification. This prevents read tags from contributing to the offered load and colliding with unread tags. An important consideration when reading mobile tags is their varying signal over time, which causes some tags to be unreadable [34]. In this respect, Floerkemeier-I and Floerkemeier-II are designed to adapt to varying tag population as they re-adjust their frame size after each slot or frame.

Lastly, Namboodiri and Gao [30] conclude that using FSA with QT lowers the energy expenditure of a RFID reader. Similarly, Shin *et al.* [24] demonstrate that combining QT with FSA leads to low identification delays. Bonuccelli *et al.* [25] show that TSA

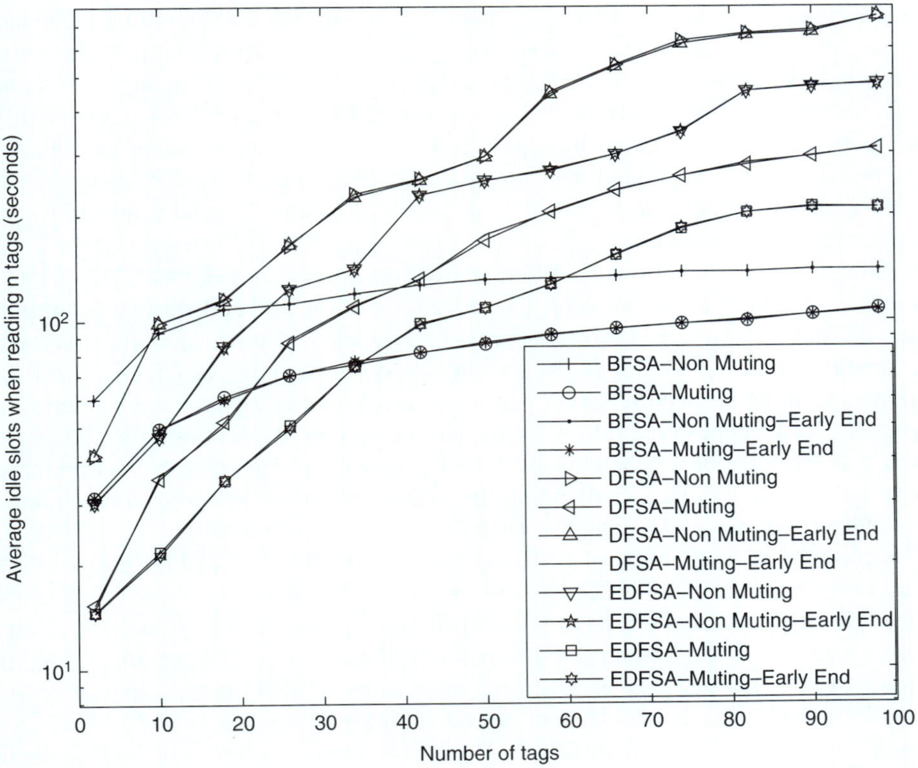

Figure 7.14 Average number of idle slots encountered when reading n tags.

achieves a higher system efficiency as compared to DFSA, EDFSA, QT, and QT with an aggressive enhancement when there are more than 60 tags. On the other hand, when the number of tags is below 50, QT with the aggressive enhancement has the highest system efficiency. The above results validate the advantages of hybrid protocols. Moreover, given the emergence of novel tree and Aloha variants as well as tag estimation functions, we expect researchers to propose better hybrid protocols in the future.

7.4 Conclusion

The performance of Aloha-based protocols increases as we move from Pure Aloha to DFSA variants. This, however, is at the expense of increased system cost due to the use of tag estimation functions that vary in complexity. Tag requirements also differ significantly. Pure Aloha protocols only require tags to have a timer, whereas Slotted Aloha and DFSA variants require tags to have a random number generator, synchronization circuits, and a timer. Having said that, DFSA variants are the most promising of all Aloha-based protocols because of their ability to read varying number of tags. This is particularly critical in practice because the vagaries of the wireless medium cause some tags to remain silent. In other words, the tag population will inevitably change with each read as the signal strength from the reader varies with each read. This is in addition to tag population changes caused by tags being muted and newly arrived tags. Other than that, the vagaries of the wireless channel also cause the capture or near-far effect, where tags with stronger signal strength are more likely to be read or detected by a reader. This is an interesting issue that has thus far received little attention. A key observation is that varying signal strength reduces tag population or contention level at a given reader. Moreover, if a reader supports muting, it will first identify tags with a strong signal followed by those with weaker signal strength.

Hybrid protocols that combine Aloha and tree protocols are becoming popular in the research community. These protocols take advantage of the probabilistic and deterministic nature of Aloha and tree protocols respectively to quickly identify tags. In particular, the key strength of tree protocols is their ability to systematically segregate collided tags, thereby preventing the occurrences of persistent collisions – a key problem in Aloha-based protocols. Consequently, hybrid protocols are very promising for future RFID systems. Given that there are only a handful of hybrid protocols to date, and the vast number of Aloha and tree protocols, a challenging research direction is therefore to determine the best combination that yields a high performing tag reading protocol.

The commercial availability of RFID readers such as SkyeTek's RFID reader [35] that mates with Crossbow's MICA2Dot sensor motes [36] has made it possible to create pervasive networks that can be used to track RFID tagged objects. A key research challenge in these networks is the limited resources on embedded devices. For example, according to Klair *et al.*'s [6] analysis of a sensor mote with an RFID reader, the energy consumed by a reader scanning 96-bits of tag ID is higher than a sensor node receiving and transmitting the same number of bits. Moreover, as a reader's scanning/reading duration increases, so does its energy consumption. Therefore, the commercial success of such RFID networks is dependent on having energy-aware, computationally inexpensive, low signaling anti-collision protocols that are capable of both identification and monitoring. In this respect, besides [27], few works have considered anti-collision protocols that can *both* identify and monitor tags efficiently. Accordingly, we believe the development of anti-collision protocols that operate efficiently in resource-constrained environments to be the next frontier in tag reading protocol research.

Problems

1. Show that pure Aloha's throughput is approximately 18%.

2. Show that slotted Aloha's throughput is approximately 36%.

3. Consider a reader using the Q algorithm. Assume an initial Q and c value of 4 and 0.5 respectively. Determine the frame size to be used in the next round if the reader experiences 4 collisions and 2 idle slots in the current round.

4. A reader, using a frame length of 8, estimates the vector $< c_0, c_1, c_k >$ after a read round to be $< 4, 2, 2 >$. Determine the tag estimate if the reader uses (i) Vogt-I, (ii) Cha-II, and the function by (iii) Zhen *et al.* [12], and (iv) Khandelwal *et al.* [14] with $N = 8$.

5. Use Vogt-II to estimate the number of tags in a reader's interrogation zone for the following vector: $< 4, 2, 2 >$. Assume the frame length is 8.

6. Repeat question five, but using Cha-I.

7. Draw the resulting query tree for tags with the following ID: 1111, 01010, 1010, 0001.

8. Consider a reader using ResMon that has allocated tag A, B, C and D slot 1, 2, 3, and 4 in the R_{frame}, respectively. Show how ResMon adjusts tag D's slot number after tag B and C leave.

9. Consider 16 tags with ID ranging from 0000 to 1111 in a reader's interrogation zone. Determine the number of tags competing in each frame if a reader using EDFSA transmits a modulo of two (0010).

NB Solutions will be provided on the book's website.

References

[1] Finkenzeller, K. (2003) *RFID Handbook: Fundamentals and Applications in Contactless Smart Cards and Identification*. Chichester: John Wiley and Sons Ltd.

[2] Hush, D.R. and Wood, C. (1998) Analysis of tree algorithms for RFID arbitration, in *The IEEE International Symposium on Information Theory*, pp. 107–114.

[3] Schwartz, M. (1988) *Telecommunication Networks Protocols, Modeling and Analysis*. New York: Addison-Wesley.

[4] EPCglobal (2005) Class 1 gen 2 RFID specifications. White paper. http://www.alientechnology.com/docs/AT_wp_EPCGlobal_WEB.pdf.

[5] Klair, D.K., Chin, K.-W., and Raad, R. (2007b) An investigation into the energy efficiency of pure and slotted Aloha based RFID anti-collision protocols, in *IEEE WoWMoM*, Helsinki, Finland.

[6] Klair, D., Chin, K.-W., and Raad, R. (2009) On the energy consumption of pure and slotted Aloha based RFID anti-collision protocols. *Computer Communications*, **32**(5): 961–973.

[7] Rivera, A., Klair, D., and Chin, K.-W. (2009) A simulation study on the energy efficiency of pure AJD slotted Aloha based RFID tag reading protocols, in *The 6th IEEE Consumer Communications and Networking Conference*, Las Vegas, USA.

[8] Hwang, T.-W., Lee, B.-G., Kim, Y.S., Suh, D.Y., and Kim, J.S. (2006) Improved anti-collision scheme for high speed identification in RFID system, in *Proceedings of the First International Conference on Innovative Computing, Information and Control*, pp. 449–452.

[9] Lee, D., Kim, K., and Lee, W. (2007) Q^+-algorithm: An enhanced RFID tag collision arbitration algorithm, in *The 4th International Conference on Ubiquitous Intelligence and Computing*, Hong Kong, China.

[10] Schoute, F.C. (1983) Dynamic frame length Aloha. *IEEE Transactions on Communications*, **31**(4), 565–568.

[11] Vogt, H. (2002b) Multiple object identification with passive RFID tags, in *The IEEE Intl. Conf. on Man and Cybernetics*, pp. 6–13.

[12] Zhen, B., Kobayashi, M., and Shimizu, M. (2005) Framed Aloha for multiple RFID objects identification, *IEICE Transactions on Communications*, **E88-B**, 991–999.

[13] Cha, J.-R. and Kim, J.-H. (2005) Novel anti-collision algorithms for fast object identification in RFID system, in *The 11th Intl. Conference on Parallel and Distributed Systems*, pp. 63–67.

[14] Khandelwal, G., Yener, A., Lee, K., and Serbetli, S. (2006) ASAP: a MAC protocol for dense and time constrained RFID systems, in *IEEE International Conference on Communications (ICC'06)*.

[15] Floerkemeier, C. and Wille, M. (2006) Comparison of transmission schemes for framed ALOHA based RFID protocols, in *Proceedings of the International Symposium on Applications on Internet Workshops*.

[16] Floerkemeier, C. (2007) Bayesian transmission strategy for framed ALOHA based RFID protocols, in *IEEE International Conference on RFID*, Grapevine, Texas, USA.

[17] Rivest, R. (1987) Network control by Bayesian broadcast, *IEEE Transactions on Information Theory*, **IT-33**(3), 323–328.

[18] Kodialam, M. and Nandagopal, T. (2006) Fast and reliable estimation schemes in RFID systems, in *SIGMOBILE: ACM Special Interest Group on Mobility of Systems, Users, Data and Computing*, pp. 322–333.

[19] Chen, W.-T. and Lin, G.-H. (2006) An efficient anti-collision method for RFID system, *IEICE Trans. Communications*, **E89**(B), 3386–3392.

[20] Feller, W. (1970) *An Introduction to Probability Theory and its Applications*. New York: Addison-Wesley.

[21] Wang, J., Zhao, Y., and Wang, D. (2007) A novel fast anti-collision algorithm for RFID systems, in *International Conference on Wireless Communications, Networking and Mobile Computing (WiCom 2007)*, pp. 2044–2047.

[22] Auto-ID Center (2003) 13.56MHz ISM band Class 1 radio frequency identification tag interface specification, version 1.0. HF RFID standard. http://www.epcglobalinc.org/standards/specs/.

[23] Lee, S.-R., Joo, S.-D., and Lee, C.-W. (2005) An enhanced dynamic framed slotted Aloha algorithm for RFID tag identification, in *The 2nd Intl. Annual Conference on Mobile and Ubiquitous Systems: Networking and Services*, pp. 166–172.

[24] Peng, Q., Zhang, M., and Wu, W. (2007) Variant enhanced dynamic frame slotted Aloha algorithm for fast object identification in RFID system, in *The IEEE International Workshop on Anti-Counterfeiting, Security and Identification*, China.

[25] Bonuccelli, M.A., Lonetti, F., and Martelli, F. (2006) Tree slotted Aloha: a new protocol for tag identification in RFID networks, in *International Symposium on World of Wireless, Mobile and Multimedia Networks*, pp. 603–608.

[26] Vogt, H. (2002a) Efficient object identification with passive RFID tags, in *The International Conference on Pervasive Computing (PerCom)*, Forth Worth, USA.

[27] Klair, D.K. and Chin, K.-W. (2008) A novel anti-collision protocol for energy efficient identification and monitoring in RFID-Enhanced WSNs, in *IEEE ICCCN*, St Thomas, US Virgin Islands, USA.

[28] Law, C., Lee, K., and Siu, K.-Y. (2000) Efficient memoryless protocol for tag identification (extended abstract), in *Proceedings of the 4th International Workshop on Discrete Algorithms and Methods for Mobile Computing and Communications*, pp. 75–84.

[29] Shin, J.-D., Yeo, S.-S., Kim, T.-H., and Kim, S.K. (2007) *Hybrid Tag Anti-Collision Algorithms in RFID Systems*. Berlin/Heidelberg: Springer.

[30] Namboodiri, V. and Gao, L. (2007) Energy-aware tag anti-collision protocols for RFID systems, in *The 5th Annual IEEE International Conference on Pervasive Computing and Communications (PerCom)*, pages 23–46, NY, USA.

[31] Klair, D.K., Chin, K.-W., and Raad, R. (2007c) On the suitability of framed Aloha based RFID anti-collision protocols for RFID-Enhanced WSNs, in *IEEE ICCCN*. Honolulu, Hawaii, USA.

[32] Klair, D., Chin, K.-W., and Raad, R. (2007a) On the accuracy of tag estimation functions, in *Sixth IEEE International Symposium on Communications and Information Technologies*, Sydney, Australia.

[33] Lee, W., Choi, J., and Lee, D. (2008) *RFID Handbook: Applications, Technology, Security and Privacy*, Chapter 9: Comparative Performance Analysis of Anti-Collision Protocols in RFID Networks, pp. 161–179. Boca Raton, FL: CRC Press.

[34] Floerkemeier, C. (2006) Transmission control scheme for fast RFID object identification, in *The 4th Annual Intl. Conference on Pervasive Computing and Communications Workshops*.

[35] SkyeTek (2009) Skyemodule M1-Mini. Datasheet. http://www.skyetek.com/Portals/0/SkyeModule_M1 Mini_060426.pdf.

[36] Crossbow (2009) The mica2 mote. Datasheet.

8

Tree-Based Anti-Collision Protocols for RFID Tags

Petar Popovski

Department of Electronic Systems, Aalborg University

8.1 Introduction

Radio Frequency Identification (RFID) systems are increasingly and ubiquitously digitalizing the physical environment by attaching tiny tags to objects and people [1]. They are expected to support a large volume/diversity of applications, ranging from logistics, transport, retail services, access control as well as many yet-to-be-imagined applications belonging to the "Internet of Things" paradigm [2].

The two key components of an RFID system are tags and readers. A tag is a small microchip equipped with antenna which is attached to the object or a person. A tag can be passive (not using a battery), semi-active (using a battery for sensing/processing, but not for communication), or active (battery-powered tag with very low power consumption). The readers, also known as interrogators, are transceivers that can communicate with the tags, by reading information from tags or writing information to them.

Although a variety of RFID systems has been used for a considerable time [3], in recent years RFID technology has exhibited immense growth [4]. There are several reasons for such growth. Tags and readers are becoming less expensive, therefore their number and ubiquity increase. Clearly, inexpensive tags will be present in much larger numbers relative to readers. In particular, RFID systems with passive tags that operate in the UHF (Ultra-High Frequency) band are gaining importance due to the increased read range. The potential of such systems has increased dramatically in recent years due to the

RFID Systems: Research Trends and Challenges Edited by Miodrag Bolić, David Simplot-Ryl, and Ivan Stojmenović

increased knowledge and applications of wireless networking. However, passive, battery-less tags pose unique challenges in the area of wireless networking due to the modest processing/storing capabilities of tags. Passive tags use the energy beamed from the reader to power their circuitry and also to transmit back to the reader using backscattering [3]. A tag can be attached to a sensor and act as a communication interface, thus creating a RFID sensor. One can easily conceive future systems comprising RFID tags and sensors that feature a combination of all types of tags; with passive tags far outnumbering active and semi-passive tags.

The communication mode in RFID systems can be described as follows: the reader sends a probe to a set of tags within its radio range. Based on the content of the probe, a tag decides whether to send its response by backscattering the power supplied by the reader. If more than one tag should respond to the probe, then multiple replies will simultaneously arrive at the reader, thus giving rise to the *tag collision* problem. If the replies of two or more tags collide, then there is very high probability that the signal will be corrupted and unreadable by the reader. Note that the tag collision problem occurs at the reader. If a tag is in the range of more than one reader, then a *reader collision* can occur at the tag [5]. This occurs when two or more readers transmit simultaneously, so that a tag that is in range of both readers cannot receive the probe sent by either readers (Figure 8.1).

Tag collision is resolved by running an anti-collision protocol (also called arbitration protocol or collision resolution protocol). The objective of an arbitration protocol is to make a transmission schedule for the tags, so that eventually each tag manages to send the reader a successful reply. Alternatively, instead of resolving the collision completely, the reader may need only partial resolution. For example, the reader may run a sequential decision process and gather data from the tags in order to carry out a statistical test of certain hypothesis and stop the arbitration process as soon as it has a sufficient set of responses. In general, the requirement of the arbitration protocol is determined by the higher-level task that is imposed on the reader network. Tag collision can be a particularly acute problem in UHF RFID systems, due to the larger reading range and thus the possibility of having larger number of tags in the reader's range.

Reader collision becomes a significant problem in environments with high reader density. Hence, the problem is to allocate communication resources (time, frequency) to the readers in order to minimize interference among them. For example, in Colorwave [6], the readers communicate with each other with the air interface that is also used to communicate with the tags and independently coordinate to minimize reader collisions using a

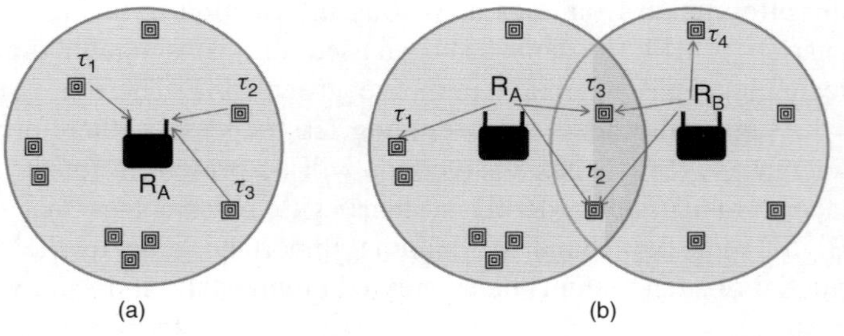

Figure 8.1 Illustration of (a) tag collision and (b) reader collision. Tag collision occurs at the reader R_A. The reader collision occurs at tags τ_2, τ_3, but not at τ_1, τ_4.

distributed time-division anti-collision algorithm. A comprehensive survey of the collision problems and collision resolution procedures in RFID systems is given in [7].

This chapter introduces the main ideas and perspectives of a class of anti-collision protocols termed *tree algorithms*. Tree protocols represent an important paradigm in the design of arbitration protocols for random access. These protocols have been introduced for a single channel with multiple access, where a single reader gathers replies from a population of tags that are in the range of that reader [8–10]. In essence, upon encountering collision, tree-based protocols recursively resolve the collision until all the tags initially involved in the collision have successfully sent their replies to the reader. Tree-based arbitration protocols represent one of the main approaches to resolve the tag collision problem in RFID systems [7, 11]. In recent years there have been a lot of studies that propose new variants of the tree algorithms [12–19]. The objective of this chapter is not to provide a detailed account of the different tree-based proposals and/or their quantitative comparison, but rather provide insight into the mechanisms used in tree-based approaches and possibilities to further optimize these algorithms in various scenarios.

This chapter is organized as follows. Section 8.2, where the principles of tree protocols are discussed in a generic setting, not strictly limited to RFID systems. This section first introduces several basic variants of the tree protocol, followed by techniques for improving the basic variants and finally an alternative arbitration framework is introduced which permits estimation of the tag population. Section 8.3 describes tree protocols that are used in specifications for UHF RFID systems. Section 8.4 presents practical issues that need to be considered, such as transmission errors and treatment of moving/late arriving tags. Section 8.5 discusses how tree protocols can be generalized to scenarios with multiple cooperative readers. The last section concludes the chapter.

8.2 Principles of Tree-Based Anti-Collision Protocols

This section introduces the principles and basic ideas used in tree protocols in a setting that is not necessarily limited to RFID systems. For example, we do not account for the computational/processing limitations of tags, error-prone communication, etc. The system model used in this section is idealized, but sufficient to introduce the mechanisms that are pertinent for understanding tree protocols. Relation to real-life systems and operation under non-idealized assumptions is given in Sections 8.3 and 8.4.

8.2.1 System Model

In this section we describe the context to present the basic ideas of tree-based anti-collision protocols. At the radio level, we assume that if a tag is within a distance D from the reader, then it can always receive probes from that reader without errors. Vice versa, if a single tag in the range of the reader transmits, then its packet is received successfully by the reader. We use the collision model for multiple access channel, which means that if two or more tags that are in the reader range transmit simultaneously, then the reader does not receive any of the transmitted packets. Note that these are rather strong assumptions for passive RFID systems, as many factors can cause transmission errors, such as fading, tag orientation, obstacles between the reader and the tag, etc. Nevertheless, this idealized MAC-layer model is sufficient to convey the concepts that are pertinent to the tree-based

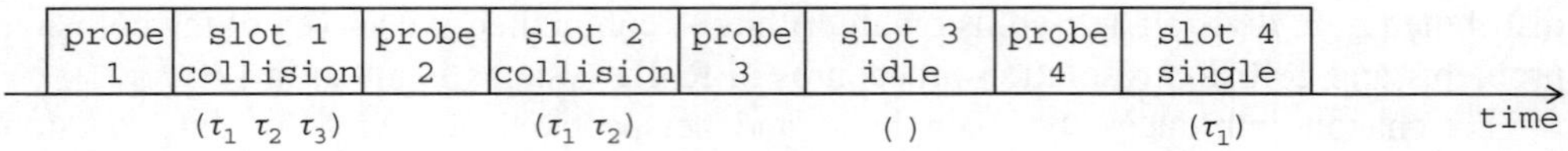

Figure 8.2 Time diagram that represents the channel over which the tag-reader arbitration and communication are done. Probes are sent by the reader, while in the other slots the reader receives. For example, in slot 1 tags τ_1, τ_2, τ_3 collide upon sending replies simultaneously.

protocols. In Section 8.4 we will relax the assumptions and discuss how various error types affect anti-collision protocols.

A tag sends a reply only if it is asked to do so by a probe sent by the reader. This is illustrated in Figure 8.2, where tags send replies in response to the probe. For example, upon receiving probe 1, tags τ_1, τ_2, τ_3 send their replies. By enabling replies from certain subsets of tags, the reader arbitrates the collisions on the channel. Note that the duration of the packet sent by each tag is constant and equal to a slot, so that if two tags transmit simultaneously their packets are completely overlapped. The reader then piggybacks feedback to tags in the next probe. For example, in Figure 8.2, `probe 2` sent by the reader carries feedback that informs the tags of the outcome (collision) in `slot 1`. When k tags transmit in the same slot, then the interrogator perceives the channel in that slot as:

- *Idle* (I) if $k = 0$ that is no tag replies.
- *Successful reception* (S) or *tag resolution* if $k = 1$.
- *Collision* (C) if $k \geq 2$.

In order to introduce tree-based arbitration protocols, we will assume that each tag is capable of producing random bits when asked to do so by the reader. This assumption will be revised in Section 8.4, where we will also discuss how the bits from the binary vector that describes tag's ID can be used for arbitration.

In the absence of errors, the efficiency of the arbitration protocols is measured both in terms of time and number of messages. If there are n tags, then we are interested in the average time $\bar{T}_n$ that the protocol consumes to identify all the tags. Let T_n be a random variable that stands for the time consumed when a particular instance of the arbitration protocol is executed. From Figure 8.2 it can be seen that T_n has two components: (1) the time used to send the probes; and (2) the time slots used for transmission from tags. We will assume that the duration of a probe sent by the reader is zero, and only focus on the time slots consumed for backscattering. With zero-length probe, the time efficiency of an arbitration protocol is defined as $\eta_n = \frac{n}{\bar{T}_n}$, see [20]. Note that another measure that is traditionally used to assess the performance of tree protocols is the average number of messages $\bar{M}_n$ sent during the arbitration process. This would be relevant if active tags were considered, as in that case each transmitted message consumes energy from the tag. On the other hand, when dealing with passive tags, $\bar{M}_n$ is rather irrelevant, as the tag energy is supplied by the reader and the average energy spent by the reader is linearly proportional to the average duration of the arbitration process.

8.2.2 Basic Tree Protocols

Tree-based protocols (also known as *splitting-tree protocols* or *tree-walking protocols*) have emerged as a solution to the multiple access problem over a shared medium. Several research groups have invented practically the same approach rather simultaneously [8–10]. The motivation for proposing these algorithms can be explained through a simple example with a single reader and two tags. The reader sends a probe requesting replies from the tags. Given that the reader does not know the tags' ID, this probe solicits an answer from unspecified tag and hence both tags are entitled to send a reply. After observing collision, the reader knows that there are at least two tags that have sent a reply. However, the number and the ID of these tags are unknown to the reader, giving rise to the symmetry problem, that is, all tags involved in the collision appear the same to the reader. In order to break this symmetry, the reader uses randomization. This means that after the collision, both tags toss a fair coin to generate a 0 or 1 value and in the next probe the reader asks only the tag that obtained 0 to send a reply. This procedure is repeated until the two tags have obtained different values for the random bits.

Figure 8.3 shows, through an example, how this basic idea can be generalized to more than two tags. In the example, the total number of tags τ_i is equal to $n = 8$. We define collision multiplicity (or conflict multiplicity) to be the number of tags that transmit when a collision is observed. The node labeled s_j refers to the outcome in the j-th slot. For example, the collision in the first slot s_1 has a multiplicity of eight, because all the tags sent their replies when responding to the initial probe. The level of a tree node is the path length from that node to the root of the tree. Each tree node is uniquely associated with a string called *address*. The address of a tree node is determined by the tossing outcomes for the tags belonging to that node. Using the tree representation, we can say that an

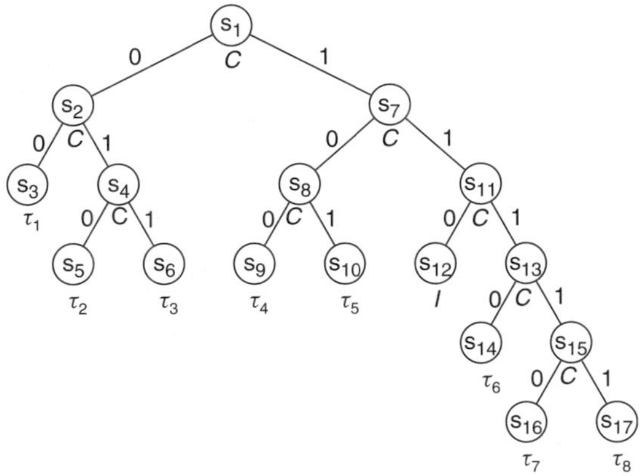

Figure 8.3 An instance of the binary tree algorithm for collision multiplicity of $N = 8$. Each vertex represents a slot that can be in one of the three states: Idle (I), Single (S) or Collision (C). For channel state "S," τ_i denotes the resolved or singulated tag.

address (a node) is enabled in slot s_i if the tags that belong to that node are allowed to transmit in s_i. At slot s_1 the tree root (level zero) is enabled, all eight tags transmit and the reader observes a collision. The address of the root node is empty, denoted by ϵ. Each tag tosses a fair coin and, in this instance of tree protocol, τ_1, τ_2 and τ_3 obtained value 0. At slot s_2, the left descendant of the root node is enabled and τ_1, τ_2, τ_3 are enabled to transmit. Note that the node enabled at slot s_2 is at level one of the tree and has address 0. The tags $\tau_4 - \tau_8$ that obtain coin value 1 after slot s_1 belong to the other node at level one, with the address 1. Having collision in slot s_2, the tags τ_1, τ_2 and τ_3 again toss fair coins. Only τ_1 obtained 0, while τ_2 and τ_3 obtained 1. Thus, when the node with address 00 (at level two of the tree) is enabled, only τ_1 sends a reply that is correctly received by the reader. Having transmission from a single tag, at slot s_4 another node at the same level is enabled, with address 01. Based on this explanation, the reader can go through the other steps of the example in Figure 8.3. Figure 8.5 shows the timing diagram of the tree algorithm from the example on Figure 8.3. It can be seen that the reader recursively resolves collisions until it obtains single (successful) answer from each tag.

Figure 8.4 shows a possible implementation of the basic variant of binary tree protocol by using a pseudocode. The code for the algorithm run at the reader represents one complete tree traversal. By setting **a** = *empty*, the reader can restart the traversal from the root of the tree. The pseudocode at the tag assumes that each probe carries the complete address. Note that the original tree algorithms proposed by [10] run in a completely distributed manner, without centralized probes, by using stack-based implementation.

```
a = empty; end=no;
while(end == no)
    transmit probe with address a;
    L = length(a);
// if a is not empty, then a = (a1a2 ··· aL)
    receive tag-reply;
    if tag-reply = IDLE
        transmit feedback;
    if tag-reply == COLLISION
        if L == 0
            a = (0);
        else
            a = (a1a2 ··· aL0)
    else
        if a == empty
            end = yes;
        else
            while aL == 1
                delete aL;
                L = L-1;
            if L == 0
                end = yes;
            else
                aL = 1;
```

(a) The algorithm run at the reader

```
a = empty ;end = no;
while (end == no)
    receive pr; %this is the address in the probe
    if a == pr;
        transmit;
        get feedback at the end of the slot;
        % current address is a = (a1a2...aL)
        if(collision)
            set b = randombit;
            set a = (a1...aLb);
        else
        if(single)
            set end = yes;
```

(b) The algorithm run at a tag

Figure 8.4 Pseudocode of the algorithms run at the reader and at a tag for the basic variant of binary tree protocol.

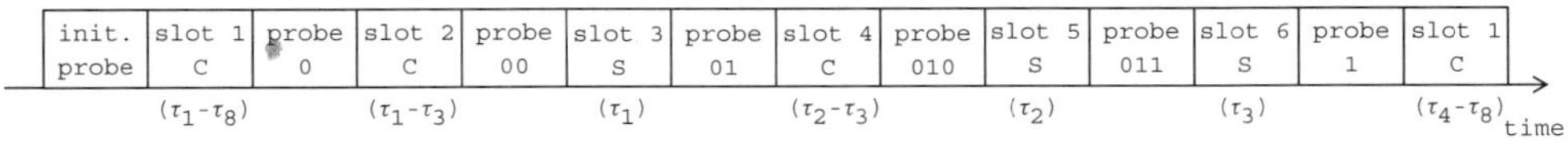

Figure 8.5 Timing diagram for part of the protocol execution described in Figure 8.3. The notation $(\tau_i - \tau_j)$ below a slot denotes the set of tags $\tau_i, \tau_{i+1}, \tau_{i+2} \dots \tau_j$ that transmits in that slot.

Tree algorithms belong to the class of collision resolution protocols, rather than collision avoidance protocols, where a representative of the latter are the protocols that use random backoff, such as the MAC protocol for IEEE 802.11 [21]. When there are no other errors except collisions, the tree algorithm is guaranteed to collect a reply from each tag that is in the range of the reader. An interesting dividend is that the algorithm creates ad hoc addresses for tags. Namely, after the collision resolution, the reader might need to communicate with a particular tag. Instead of using the full address of that tag, the reader can assign a short address to the tag, and this address is represented by the address of the tree node in which this tag has been successfully resolved or singulated. For example, in Figure 8.3 the short ad hoc address of the tag τ_3 is 011.

8.2.3 *Improvements to the Basic Tree Protocol*

In this section we describe several ideas that improve and generalize the basic tree protocol, described in the previous section.

At first, note that the described tree algorithm constructs a tree by binary branching as dictated by bit flipping. The branching can be generalized to M rather than two outcomes in the randomization process. Larger M decreases the probability of collision, but increases the probability of obtaining an idle slot. If the idle slot has the same duration as a single/colided slot, then it is clear that increasing M does not increase efficiency. However, an idle slot can have a shorter duration than a slot in which one or more tags are transmitting. This is because the reader, in principle, does not need to wait for the whole duration of the slot to detect that no tag has transmitted. Hence, the duration T_I of an idle slot can be lower than the duration of the slot with tag transmission, denoted by T_S. Such a model for collision resolution is termed a Carrier Sense Multiple Access (CSMA) model. If $T_I \ll T_S$, then an idle slot is very "cheap" in terms of duration and the tree protocol should be engineered in a way to produce idle slots with higher probability as compared to the probability of occurrence of collided slots. Nevertheless, in this text we will not specifically treat the CSMA model. Possible optimizations of the tree-based algorithms in CSMA setting are discussed in [22] and [18].

Massey in [23] and Tsybakov and Mikhailov in [10] proposed a simple way to improve the basic variant of the binary tree algorithm. They noticed that there are some tree nodes that certainly contain more than one tag and will thus certainly result in collision if the node is enabled. These nodes should be skipped during the traversal of the tree. For example, in Figure 8.3, after the collision in slot s_{11} and an idle slot s_{12}, it is clear that the enabling of the address 111 results in certain collision. Hence, after the idle slot s_{12}, the terminals belonging to node 111 toss a coin immediately and in slot s_{13} the enabled node is 1110. This algorithm will be referred to as Modified Binary Tree (MBT) algorithm.

However, as discussed in Section 8.4, this small optimization has a very high price if transmission errors occur with non-zero probability.

Early on, researchers observed that knowing the multiplicity (number) of tags involved in a collision can speed up the resolution of that collision. Capetanakis in [9] observed that binary tree algorithms are most efficient for conflicts of small multiplicity and applied this observation to devise a *dynamic tree algorithm* for scenarios with Poisson arrivals. The idea is to divide the initial set of n tags into smaller groups and then apply a tree protocol within each group. Multiplicity estimation to speed up tree protocols has been used in [24] and [20]. In both works, the proposal is a hybrid algorithm that consists of two phases. The first phase is devoted to the estimation of multiplicity. After obtaining an estimate $\hat{n}$, the second phase starts. The unresolved terminals are randomly split into approximately $\hat{n}$ groups and each group is resolved using the basic tree algorithm.

Note that both [24] and [20] have an explicit phase for estimating $\hat{n}$ and the accuracy of such an estimate is using algorithm-specific parameters that are chosen in advance. On the contrary, in [18] we have introduced a new framework for tree-based arbitration protocols where a running estimate $\hat{n}$ of the initial population is obtained, which becomes increasingly accurate as the collision resolution progresses. Before proceeding to the details of the framework from [18], we introduce the Clipped Binary Tree (CBT) algorithm, which is the main ingredient of tree algorithms with a running estimate.

CBT algorithm has been independently introduced by several authors in the context of random access for an infinite population of terminals that generates transmission requests with Poisson arrivals [25]. It is identical to the MBT algorithm except that it is stopped, that is, the tree is clipped whenever two consecutive successful transmissions follow a conflict. CBT resolves the batch conflict partially, since not necessarily all nodes of an initial batch are resolved during the execution of CBT. For example, referring to the tree in Figure 8.3, the first instance of the CBT algorithm starts in slot s_1 and terminates in slot s_6, resulting in three resolved tags. Now, the standard binary tree algorithm would continue to enable the node labeled with s_7, that is, all the tags that have flipped 1 after the initial collision. The key observation from [18] can be applied to this example as follows. After the termination of the first CBT algorithm, it is noted that 3 tags have flipped 0, so that the expected value of the number of tags that have initially flipped 1 is 3. This implies that the enabling of the node s_7, which belongs to the level 1 of the tree, will highly likely result in collision and therefore the next probe should directly enable a node that has a higher level in the tree. For example, if the node with address 10, from level 2, is enabled after the termination of the initial CBT algorithm, then one slot is saved by skipping the node s_7 at level 1. In such case, the next instance of the CBT algorithm would start by enabling the node with address 10 and terminate when tag τ_5 successfully sends its reply, that is, after the address 101 has been enabled. In the next section, we describe the framework that systematically estimates which tag subsets should be enabled for transmission after a single instance of CBT is finished.

8.2.4 General Arbitration Framework for Tree-Based Protocols

The key part of the framework proposed in [18] is the interpretation of the randomization procedures used in a tree protocol. Let us assume that, upon the transmission of the initial probe, each tag generates a random real number, uniformly distributed in the interval

$[0, 1)$. This random number is referred to as a token and let r_i denote the token generated by the tag τ_i. Let $\mathbf{x}_i = (x_{i1}x_{i2}x_{i3}\ldots)$ be the binary representation of fractional part of r_i. Then each $\mathbf{r}_i$ is an infinite string of 0s and 1s and:

$$r_i = r(\mathbf{x}_i) = \sum_{j=1}^{\infty} \frac{x_{ij}}{2^j} \tag{8.1}$$

The token r_i can be understood as an infinite reservoir for generating fair coin tosses, since each x_{ij} gets a value 0 or 1 with probability 0.5. For example, in Figure 8.3, after the initial collision, the node with address "0" is enabled, so that all the tags that have generated their tokens within the interval $[0, 0.5)$ are entitled to send a reply. This principle is generalized as follows. The act of enabling a tree node with a certain address from Figure 8.3 corresponds to the act of enabling transmission by tags that have generated their token in a certain sub-interval of $[0, 1)$. For example, when the tree node with address 10 is enabled in slot s_8, it corresponds to enabling transmission by tags that have their tokens generated in $[0.5, 0.75)$. More generally, if the node with address:

$$\mathbf{a} = a_1a_2\ldots a_K \tag{8.2}$$

is enabled, then this corresponds to enabling the interval $[b, c)$ where:

$$b = \sum_{i=1}^{K} \frac{a_i}{2^i} \qquad c = b + \frac{1}{2^K} \tag{8.3}$$

Hence, instead of using a binary tree, the arbitration process can now be represented using a sequence of enabled intervals. Figure 8.6 represents the example from Figure 8.3 using a sequence of enabled intervals. Such a representation brings a qualitatively new observation: after the termination of one instance of the CBT algorithm, we can obtain an estimate of $\hat{n}$ of the tag population size n and this estimate can be used to find out how many tags are still there to be resolved. For each slot on the abscissa of Figure 8.6, there is a rectangle that stands for the enabled interval. A shaded interval denotes that there is a collision in the slot, for example, when the enabled interval is $[0, 0.5)$ in the second slot. On the other hand, in slot 3 the rectangle is marked by τ_1 and it denotes that the enabled interval $[0, 0.25)$ contains only the tag τ_1. The enabled interval in slot 12 is empty.

We proceed to explain how the multiplicity estimation can be used in this framework. After the termination of the first CBT instance in slot 6, the reader observes 3 tags in the interval $[0, 0.5)$. If n is the total number of tags in $[0, 1)$, then the expected number of tags in $[0, 0.5)$ is $k = 0.5n$. Since $k = 3$, we can estimate $\hat{n} = \frac{k}{0.5} = 6$, so that the expected number of tags in the unresolved interval $[0.5, 1)$ is 3. The second instance of the CBT algorithm is finished when tag τ_5 is resolved, after which the reader observes that 5 tokens are located in the interval $[0, 0.75)$. Hence, a new estimate can be made of the initial tag multiplicity: $\hat{n} = \frac{5}{0.75} = 6.67$, while the number of tags left to be resolved in the interval $[0.75, 1)$ is estimated to be $\frac{5}{0.75}0.25 = 1.67$.

In general, assume that the arbitration is at the stage in which there are k resolved tags with tokens in $[0, p)$. Then there are $n - k$ unresolved tags (with n unknown) with their tokens in the interval $[p, 1)$. As stated before, knowing the multiplicity $n - k$ can speed

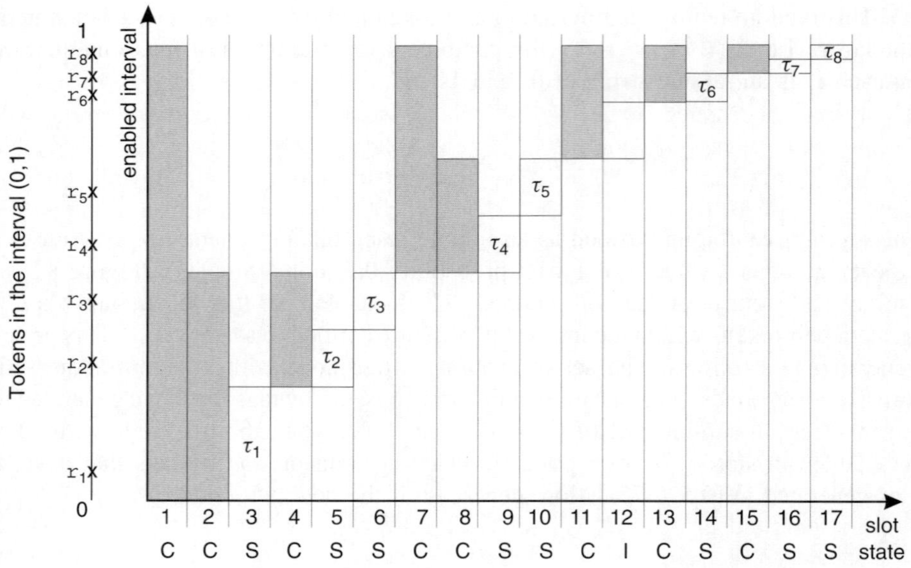

Figure 8.6 Binary tree from Figure 8.3 represented in the framework that uses tokens and sequence of enabled intervals. A rectangle denotes the interval enabled in a given slot. A shaded rectangle denotes collision; a rectangle labeled with τ_i denotes that only tag τ_i has its token in that particular enabled interval.

up the arbitration protocol for the remaining tokens. Considering the token generation process, the reader can make the following observation: *From n tokens that are uniformly picked from the interval* $[0, 1)$, *where n is unknown*, *k tokens have been picked from the interval* $[0, p)$. The probability of that event, conditioned on n, is

$$P(k|n) = \binom{n}{k} p^k (1-p)^{n-k} \tag{8.4}$$

For a given k, p, we use the maximum likelihood (ML) estimator:

$$\hat{n} = \frac{k}{p} \tag{8.5}$$

which is not necessarily an integer. The estimator is unbiased, since $E[\hat{n}|n] = n$, while its variance is:

$$\mathrm{Var}[\hat{n}|n] = \frac{n}{p} - n \tag{8.6}$$

As the arbitration protocol progresses, the value of p increases, which decreases the variance $\mathrm{Var}[\hat{n}|n]$ and therefore the obtained estimate becomes more accurate.

Using a framework based on token generation, the authors of [18] propose several arbitration protocols. Each of those protocols achieves faster collision resolution compared to basic tree protocols. In particular, the protocol termed Interval Estimation Conflict Resolution (IECR) has been reported to be the fastest known collision resolution protocol.

The algorithms in [18] operate by iteratively applying CBT. Let us assume that upon a termination of a CBT instance, the reader observes k resolved tokens in $[0, p)$. Based on the estimation of the conflict multiplicity, the next enabled interval is $[p, p + \Delta p)$. If a collision is detected, a new CBT instance starts and the process is repeated. If in $[p, p + \Delta p)$ there is a single token or no tokens, then the next slot is treated as a termination slot for that CBT instance and the operation of estimation/enabling new interval is reapplied. The speed of IECR is due to the optimized selection of the length of the enabled interval upon termination of a CBT instance. The authors in [18] show that the speed of IECR in resolving batch conflicts is asymptotically equivalent to the speed by which the collision resolution for Poisson arrivals is performed using the First Come First Serve (FCFS) tree algorithm [25].

In [18] it is shown that, when n is large, the length of the next enabled interval should be:

$$\frac{1.26p}{k} \tag{8.7}$$

where k is the number of tokens found in the interval $[0, p)$ after the termination of an CBT instance. In the same work the authors proved that such a selection of the enabled interval results in asymptotic time efficiency of the IECR algorithm that is equal to:

$$\lim_{n\to\infty} \tau_n = \lim_{n\to\infty} \frac{n}{T_n} = 0.487$$

which is identical to the efficiency of the FCFS algorithm [25] for Poisson arrivals.

The implementation of the optimized IECR algorithm in RFID setting is not straightforward, since the limits of an enabled interval $[a, b)$ are in general real numbers and each of them has a binary representation with infinite length. We say that a binary representation of a real number $x \in [0, 1)$, written as $x = 0.x_1x_2x_3\ldots$, has infinite length if there is no finite integer i so that for all $j > i$, the decimals are $x_j = 0$. The selection of an optimized length Δp may result in a sequence of enabled intervals that cannot be efficiently contained in the probe sent by the reader, as the interval is described by infinite-length numbers.

The observation in the previous paragraph makes another variant of the estimating tree algorithms, termed Estimating Binary Tree (EBT) [18], more suited for usage with RFID tags. In EBT, the length of the enabled interval is always equal to 2^{-l}, where l is an integer. The advantage of such a choice is that each enabled interval corresponds to a single node in the evolution of the binary tree, so that the enabled interval can be represented in a compact way, using the correspondence between the address in the tree and the actual interval, as described by Equations (8.2) and (8.3). The only thing that needs to be decided is how to select l, after a CBT instance is terminated. For example, a straightforward way is to select l in the way that it maximizes the probability of obtaining a single response:

$$l^* = \text{argmin}\left|2^{-l} - \frac{p}{k}\right| \tag{8.8}$$

Finally, *a priori* information about the size of the tag population can additionally improve the efficiency of tree protocols. For example, if it is known that the number of tags is at least $n_{\min} > 1$, then the initial probe does not need to enable the whole interval

Table 8.1 Time efficiency $\eta_n = \frac{n}{T_n}$ of various conflict resolution algorithms.

Number of tags n	MBT	EBT	EBT with $n_{\min} = 64$	IECR	IECR with $n_{\min} = 64$	Optimized IECR with known n
50	0.379	0.452	0.472	0.463	0.467	0.499
100	0.376	0.460	0.471	0.467	0.479	0.494
1000	0.375	0.463	0.465	0.483	0.486	0.488

$[0, 1)$, but it can start, for example, with the interval $\left[0, \frac{1.26}{n_{\min}}\right)$. Such an approach can avoid several of the collisions that occur in the initial stage, until the first CBT instance is terminated and the first estimation $\hat{n}$ is made.

8.2.5 Numerical Illustration

In this section we provide some performance figures to highlight the speed by which the described tree algorithms resolve the collisions in the cases without transmission errors. Table 8.1 illustrates the time efficiency, defined as $\eta_n = \frac{n}{T_n}$, where n is the number of tags and T_n is the average number of slots required to resolve all tags. The performance numbers are given for several algorithms. As a reference, we have also provided the time efficiency for the reference IECR version in which the number of tags n is known *a priori*. This reference algorithm operates analogously to the described IECR algorithm, except that it uses n instead of estimating $\hat{n}$, and after the termination of each CBT instance, it knows exactly the number of tags that are left in the unresolved part of the interval $[p, 1)$. It can be seen that estimating tree algorithms are superior to the Modified Binary Tree, which is in turn superior to the basic variant of binary tree protocol. If no *a priori* information is known about the tag population size, then the efficiency of EBT and IECR rises as the tag population increases. The required the number of resolved tags k that can produce a reliable estimate $\hat{n}$ grows slower than n. Therefore, tree protocols with running multiplicity estimation bring benefits already when a relatively low fraction of the tags is resolved. It is also seen that EBT pays the price in terms of time efficiency relative to IECR. For example, when $n = 50$, the average number of slots required to resolve the tags is $T_n = 110.62$, while it is $T_n = 106$ with IECR and around $T_n = 132$ with MBT. Having some or complete *a priori* knowledge about n improves performance. The results in Table 8.1 show that when n is known *a priori*, efficiency decreases as n increases. This is because for small n, the *a priori* information about n is more valuable compared to the estimation of the number of remaining tags in the unresolved interval. Such an advantage of the *a priori* knowledge diminishes as n increases.

8.3 Tree Protocols in the Existing RFID Specifications

Having introduced the main ideas used in tree protocols, in this section we turn to actual realizations of tree protocols in RFID systems. Although many of the principles are retained, there are changes that are necessitated by the limitations of passive UHF tags.

We describe two tree-based protocols used in EPCglobal Generation 1 tags: Class 0 and Class 1.[1]

In many applications, Generation 1 tags and the associated protocols have been displaced by Class 1 generation 2 tags [26]. Besides these two protocols, another standardized tree protocol for UHF tags is ISO 18000-6A [27].

8.3.1 Tree Protocol for EPCglobal Class 0

Thus far we have assumed that the tags can generate random numbers, which can in turn be used as tokens in anti-collision protocol. EPCglobal Class 0 tags ("Class 0 tags") do not use random bits, but arbitration relies on the unique ID of each tag. If the tag ID consists of L bits, then each ID $x_1 x_2 \ldots x_L$ is uniquely represented by a *token* in the interval [0, 1). The token is a real number that has a binary representation $0.x_1 x_2 \ldots x_L$, albeit not necessarily uniformly distributed in [0, 1). We can use again Figure 8.3 to exemplify arbitration based on tag ID. Tags τ_1, τ_2, τ_3 are allowed to transmit in slot s_2 since each of them has the first bit of the ID $x_1 = 0$, while the other tags $\tau_4 - \tau_8$ have $x_1 = 1$. Thus, from Figure 8.3 it can be inferred that the ID of τ_1 is $00\cdots$, the ID of τ_6 is $1110\cdots$, etc.

Tree protocol for Class 0 tags uses bit-by-bit query response rather than having probe packet from the reader followed by packets from tags. The reader starts arbitration by transmitting a special NULL sequence. Tags that receive this sequence backscatter the first bit x_1 of their ID. If x_1 is either 0 or 1 for all tags that backscatter, then the reader receives 0 or 1, respectively, after which it echoes that bit. On the other hand, if some tags transmit $x_1 = 0$ and other tags transmit $x_1 = 1$, then, due to the modulation format used for backscattering, the reader detects that both 0 and 1 have been sent. Note that this is different from the idealized collision model, used in Section 8.2.1. If the reader receives both 0 and 1, then it decides randomly to echo either 0 or 1. In the next step, tags that have x_1 equal to the bit echoed by the reader transmit the value of the bit x_2. On the other hand, tags that have x_1 different from the bit echoed by the reader are muted until the next transmission of a NULL sequence.

We can reuse the example from Figure 8.3 to illustrate the operation of the tree protocol used for Class 0 Generation 1 tags. The bit used to label each of the branches is the bit echoed by the reader. Assume that the tags have 5-bit addresses. The addresses of the tags are, for example, $X(\tau_1) = 00100$, $X(\tau_2 = 01010)$, $X(\tau_3) = 01100$, etc., where $X(\tau_i)$ stands for the address of τ_i. Each slot has duration equal to one bit. Assume that after initiating the tree at slot s_1, the reader chooses to echo 0, and after s_2 it chooses to echo 0. Here comes a change compared to the execution shown in Figure 8.3: τ_1 sends in sequence $x_3 = 1$, echoed by 1; it proceeds analogously for bits $x_4 = 0$ and $x_5 = 0$. Clearly, since τ_1 is the only tag that sends reply to the reader in these last three queries, each bit is received and echoed correctly by the reader, since we are neglecting noise-induced errors. After its fifth bit is being read, τ_1 goes into a dormant state. The reader sends again the NULL signal, receives 0 and 1, echoes 0, receives 1, echoes 1, receives 0 and 1, echoes 0, end proceeds to get the last two bits of τ_2. The resolution of the remaining tags proceeds in a

[1] At the time of writing, the specification for Class 1 Generation 1 tags was no longer available. Nevertheless, a "stable" reference for both protocols is [26].

similar way. One peculiar thing to be noted about this version of the tree protocol is that there are no slots with idle reply, such as slot s_{12} in Figure 8.3. The reason is that there is no collision, but instead a combination of 0 and 1, the reader always echoes bit value of an existing tag, so that there is always at least one tag that sends a reply.

Although Class 0 tree protocol is easy to implement and efficient in terms of transmitted symbols, it has several significant drawbacks. The drawbacks that stem from the physical-layer phenomena are treated in [26], here we outline some of the problems solely related to MAC-layer operation. The operation described above is rather a simplified version of the full protocol. In reality, each tag is in a default Dormant state and is wakened up by a Reset signal from the reader. After the Reset signal, the reader transmits a sequence of symbols to calibrate the tags, and only after such a calibration can a tag receive commands from the reader. Note that, besides the symbols 0, 1, and NULL, the specification defines other commands that can be sent by a reader. Note that the tags that arrive after the Reset/calibration commands have been sent, such as in a scenario with conveyor belt, cannot participate in the execution of tree protocol and should wait until another set of reset/calibration commands is issued. Hence, in order to be able to read late arriving tags, the reader should frequently issue calibration commands, which hinders the speed of the anti-collision protocol. Another remark regarding the tradeoff in time efficiency of the algorithms is in order. The protocol saves execution time by using short transmission with a symbol-level duration. On the other hand, short commands (0, 1, or NULL) used by the reader necessitate that the tree traversal is restarted from the root of the tree each time a tag is resolved. In general, short reader commands decrease the flexibility of the protocol to move within the tree and such flexibility is necessary to implement some of the time-efficient protocols, such as EBT.

8.3.2 Tree Protocol for EPCglobal Class 1

A serious drawback of Class 0 tags is that they are factory-written, such that the protocol does not make provisions for writing new data in the tags. This has been one of the main reasons for introducing EPCglobal Class 1 Generation 1 tags. Different from Class 0, in Class 1 the reader sends commands in packets and the tags reply to commands by whole packets. Each reader command has a synchronization sequence, such that, in principle, late arriving tags can join the arbitration protocol. Having reader commands with more content offers flexibility in traversing the tree, but the price is that the duration of a single protocol slot (containing reader command and tag reply) is much larger than the symbol duration. The Class 1 specification provides commands to support a tree-based protocol, but does not specify a particular way to traverse the tree [26]. The reader can use the PingID command, which is in fact a probe sent to a particular subset of tags. This command consists of: a pointer [PTR], which points to a location (or bit index) in the tag identifier, and [VALUE], which is a bit mask of length [LEN]. Tags that have a matching bit patters to VALUE in their ID at a location indicated by [PTR], are entitled to send reply to the probe. For example, if PTR = 1 and the bit mask is "11," then the tags that have ID $x11$, where x can be either 0 or 1, can send a reply to the probe. Interestingly, not all tags with a matching bit mask send a reply immediately. The reader delineates eight bins after the probe termination and the bins are labeled 000, 001, 010, ... 111. A tag that can match the required bit mask, uses the next three bits of its ID to determine in which bin to send its reply. For example, the tag with address $x11010\cdots$ will send a reply in the third bin after a probe with mask $x11$. Described in terms of arbitration framework with tokens, a probe simultaneously enables eight intervals. In addition, the set of bit masks that are used can be adapted to incorporate the ideas of multiplicity estimation, such as in EBT, and speed up tag resolution.

Both Class 0 and Class 1 tags from EPCglobal Generation 1, along with the associated arbitration protocols, have been gradually displaced by Class 1 Generation 2 tags [26], which use a variant of the ALOHA protocol. One of the main reasons for abandoning these tree protocols was the fact that they have difficulty in dealing with late arriving tags. Therefore, in Section 8.4.3 we will discuss which provisions can be made to create tree protocols capable of dealing with moving tags and late arriving tags.

8.4 Practical Issues and Transmission Errors

8.4.1 Token Generation

The two practical tree protocols for UHF RFID, described in Section 8.3 do not rely on random number generation at tags. Generation of random bits at a passive tag is not a problem and has been used for the ALOHA-based protocol for Class 1 Generation 2 tags. Hence, a future version of tree-based RFID protocol can rely on random bit generation rather than use the ID bits. This can be important from a security point of view, since the masks/addresses used in the probes sent by the reader are not correlated with the tags ID.

If ID bits are used for arbitration, then the tokens that correspond to the tags ID may not be uniformly distributed in $[0, 1)$, which poses problem for the estimation methods used in, for example, EBT. Note that EBT provides a better average performance than BT when the tokens are uniformly distributed in $[0, 1)$. If the ID of tags in a given set are not uniformly distributed on the interval $[0, 1)$, then the statistical estimation utilized is not valid. Thus, in such a case EBT may be even slower than the basic binary tree protocol. We illustrate this by an example. The set has five tags and their ID is in the interval $[0.25, 0.5)$. If the basic binary-tree algorithm is applied, then when all the tags are resolved, the next enabled interval is $[0.5, 1)$, which will result in an idle response and thus the algorithm will terminate. On the contrary, if EBT is applied, then after the tokens are resolved, the algorithm continues to estimate the cardinality of the tag set and enables intervals smaller than $[0.5, 1)$, so that it takes more time to terminate the arbitration protocol.

A way to randomize the identity of the tags is that the reader uses a predefined randomized permutation of the bits in the ID. This permutation can be used to scramble the addresses prior to the start of the EBT algorithm and thus randomize the prefixes used in collision resolution.

8.4.2 Transmission Errors

The described tree protocols operate under the assumption that there are no channel errors. The channel errors affect these protocols in different ways. In order to develop a systematic view on the error impact on tree protocols, we first divide the errors into errors that occur at a tag and reader.

8.4.2.1 Errors at the Tag

An error at a tag causes the tag not to send a reply back to the reader. We differentiate two error types at the MAC protocol layer: static and dynamic errors. *Static error* models the case in which, during a whole collision resolution session, the tag is in a "blind spot", so that none of the probes in the whole session is received by the tag. The effect of the blind spot is as if the tag is outside the range of the reader. Let p_s denote the probability that a tag experiences a static error during a given collision resolution session. On the other hand, a *dynamic error* occurs so that each probe sent by the reader is independently received with probability $(1 - p_d)$ and received incorrectly with probability p_d. Looking from a physical-layer perspective, static errors result from deep signal fade, while dynamic errors are caused by noise. Probability of static (dynamic) errors at tag τ_i is denoted by $p_{si}(p_{di})$. In general each $p_{si} \neq p_{sj}$ and $p_{di} \neq p_{dj}$ when $i \neq j$.

The impact of static errors on the protocol is simple – the tag does not participate in the reader session. One might think that a static error occurs when a tag is at a "blind spot" – a spot at a close distance to the reader that has a poor radio link to the reader. If a tag is at a blind spot in one session does not necessarily mean that the tag will be in blind spot in another reader session. For example, before the next session with the same reader, tags may be physically displaced, such that the radio link to the reader at the new location is uncorrelated with the link at the old location. Furthermore, if the next session is performed by another reader, its radio link to a tag can be also decorrelated, so that the tag is not at a blind spot for the second reader. This latter case can be represented by the same model used to represent readers with different reading ranges, as shown in Figure 8.1, the tag τ_4 is not in the range of reader R_A but is it in the range of reader R_B. The model in Figure 8.1 can be conceptually extended to account for tags that are in the blind spot of both readers, and hence they will be missed after both readers terminate their sessions. Reference [28] introduces several methods that rely on the described error model in order to approximately determine the probability of having missed tags after both readers terminate the sessions.

The impact of dynamic errors on the tree protocols is more involved. Assume that the reader is running the basic tree protocol, without any estimation or the optimization introduced by the MBT. Consider the example in Figure 8.6 and assume that tag τ_3 has not received the probe correctly in slot s_6, but it receives the probe in slot s_7 correctly. Accordingly, the reader observes an idle slot s_6. One can easily verify that until the end of the collision resolution session, this tag will not receive a probe that entitles it to send a reply. In terms of enabled intervals, starting from the slot s_7, the reader treats the interval $[0, 0.5)$ as resolved, while the token of the unresolved tag τ_3 lies in that interval. Note that, in the errorless case it is impossible for an unresolved tag to see a probe which informs the tag that the interval in which it belongs is resolved.

There are two approaches to mitigate the effects of dynamic errors. In the first approach, even after the interval $[0, p)$ is assumed to be resolved, the reader redundantly enables this interval or some of its subintervals in order to probe whether there is any missing tag left. In the second approach, if a tag that has a token in $[0, p)$ has not been singulated, while the reader sends out a signal to indicate that it considers the interval $[0, p)$ resolved, then the tag generates its token to be uniformly distributed within $[p, 1)$. The second approach implies that the estimates used in the framework with enabled intervals, such as for IECR or EBT, need to be changed, since the unresolved interval is statistically different from the resolved interval. Furthermore, if the resolution is purely identity-based, then the token re-generation may lead to unresolvable conflicts. For example, let τ_1 have identity $1x_2x_3\ldots$ and let τ_2 have $0x_2x_3\ldots$, that is, they differ only in the first bit. If tag τ_1 is missed after the interval [0,0.5), the tag can regenerate its token in [0.5, 1) by ignoring the first bit–however, doing so means it cannot be differentiated from τ_2 in the collision resolution process. Hence, it is more feasible to leverage on the approaches in which the reader re-enables some of the previously resolved intervals.

The basic binary tree algorithm is robust with respect to the errors at the tag, in a sense that these errors are not fatal to the algorithm operation. On the other hand, modified binary tree can experience a fatal error. Consider the situation with two tags τ_1, τ_2, with their respective tokens $r_1 = 0.1$ and $r_2 = 0.2$. Assume that the first probe, which enables $[0, 1)$ has been received correctly by both τ_1 and τ_2, so that the reader observes collision. Next, assume that the second probe, which enables $[0, 0.5)$ has not been received by either of the two tags. One can check that the MBT algorithm will enter an endless session, where the reader enables smaller and smaller subintervals from $[0.25, 0.5)$, getting only idle replies. On the other hand, if an estimating tree algorithm is applied, for example' EBT, without using the modification of the MBT, then the algorithm stays robust to errors, but the estimators will be less precise.

With static errors, the probability that a reader misses a tag is p_s. On the other hand, when there are dynamic errors that occur with probability p_d, the probability of missing a tag is not in a straightforward relation with p_d. We show through a simple example that the probability that a tag is missed depends on the other tags. Let there be two tags only, and let the token of tag τ_1 be in $[0, 0.5)$, while the token of τ_2 be in $[0.5, 1)$. Let μ_i denotes the event that τ_i is not read at the end of the reader session. Let us assume that the probability of dynamic error p_d is identical for both tags. Thus, $\Pr[\mu_1\bar{\mu}_2]$ denotes the probability that τ_1 is missing and τ_2 has been successfully read in the session. The probability that both tags will be missed $Pr[\mu_1\mu_2]$ is given as:

$$\Pr[\mu_1\mu_2] = p_d^2 + (1 - p_d)^2 p_d^2 \tag{8.9}$$

where the first p_d^2 is the probability that both tags do not receive the probe that enables $[0, 1)$. The second term comes from the probability that both tags receive the first probe, which is $(1 - p_d)^2$ and then each tag misses the probe that enables its subinterval – τ_1 misses the probe for $[0, 0.5)$ and τ_2 misses the probe that enables $[0.5, 1)$. In a similar manner, the following probabilities are determined:

$$\Pr[\mu_1\bar{\mu}_2] = p_d(1 - p_d) + (1 - p_d)^2 p_d(1 - p_d) \tag{8.10}$$

$$\Pr[\bar{\mu}_1\mu_2] = p_d(1 - p_d) + (1 - p_d)^2 p_d(1 - p_d) \tag{8.11}$$

$$\Pr[\bar{\mu}_1\bar{\mu}_2] = (1 - p_d)^4 \tag{8.12}$$

The marginal probability that τ_1 is missing is given by:

$$\Pr[\mu_1] = \Pr[\mu_1\mu_2] + \Pr[\mu_1\bar{\mu}_2] = p_d[1 + (1 - p_d)^2] = \Pr[\mu_2] \tag{8.13}$$

where the equality with $\Pr[\mu_2]$ is due to symmetry. We can now try to determine the conditional probability that τ_1 is missing given that τ_2 is missing:

$$\Pr[\mu_1|\mu_2] = \frac{\Pr[\mu_1\bar{\mu}_2]}{\Pr[\bar{\mu}_2]} = p_d \tag{8.14}$$

which is clearly different from $\Pr[\mu_1]$.

From this example, it is confirmed that the tag has less chance of being missed if its resolution requires the reader to go deeper in the tree, that is, towards smaller enabled intervals. This is because, for a deeper tree level, more probes are reaching a tag and thus the chance to respond is higher. In that sense, one can expect that tree protocols will more reliably cope with reading errors at the tag, as these protocols are intensively using probes, as opposed to the ALOHA protocol family, where relatively fewer probes are sent.

8.4.2.2 Errors at the Reader

The effect of non-ideal receptions at the reader cause erroneous interpretation of the channel state at the reader. For example, if only one tag transmits and the packet is not received correctly by the reader, it might be interpreted either as idle (not sufficient received power) or collision (transmissions are present, but are not decodable). Figure 8.7 shows the general error model for errors at the reader [23]. The value p_{uv} stands for the probability that the channel state u is perceived as channel state v. In the ideal model, assumed so far, we have $p_{uu} = 1$ and $p_{uv} = 0$ if $u \neq v$. Note that the model assumes that the probability of producing a valid single output when there is no transmission is $p_{is} = 0$. The actual values of the probabilities p_{uv} depend on the underlying physical phenomena, such as noise and fading.

It can be verified that the basic variant of the binary tree algorithm stays robust also with respect to the errors at the reader. Reference [23] describes an example in which MBT can again exhibit fatal behavior if $p_{sc} > 0$. Similarly to the case when the errors occur at the tag, the errors at the reader affect the correctness of the estimates made by the estimating tree algorithms. If the basic binary tree is used, with or without running multiplicity estimation, then when $p_{sc} > 0$, $p_{ic} > 0$, while all the other $p_{uv} = 0$, for $u \neq v$, the protocol will not produce missing tags. A tag can be missed if either of p_{si}, p_{ci}, p_{cs} is positive.

In practice, errors will occur both at tags and at the reader. In order to mitigate the problem of missed tags, a combination of the two approaches described above should

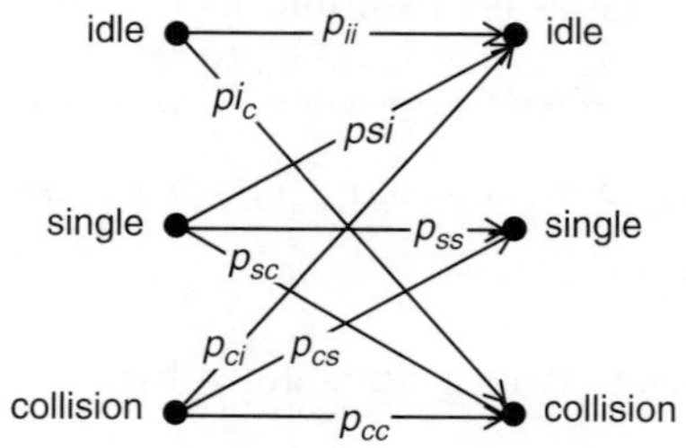

Figure 8.7 General model that represents the errors at the reader. p_{uv} stands for the probability that the channel state u is perceived as channel state v.

be used. The protocol designer might prefer to utilize approaches in which the reader re-enables some of the intervals that have already been enabled. This is because those techniques require minimal increase of the intelligence in the tag.

8.4.3 *Dealing with Moving Tags*

A particularly important scenario for RFID tags is one in which tags are physically moving in and/or out of the coverage region for a given reader. Such a case occurs when tagged items are put on a conveyor belt. This also makes conditions for protocol operation non-ideal, but in a way that is different from the impact that the static/dynamic errors have on the protocol. If we ignore other error types, then we can consider that the channel is ideal for the tags in range and no communication is possible with tags that are outside the reader's range.

Figure 8.8(a) depicts the situation in which the reader covers a total length L of the conveyor belt on which the tags are moving. It is again useful to represent the collision resolution process by using a two-dimensional arbitration space, depicted in Figure 8.8(b). When a given interval in the token space is enabled, a stripe of length L is enabled in the two-dimensional arbitration space, as depicted in Figure 8.8(b). If the tag density is uniform across the conveyor belt, then one stripe always contains the same expected number of tags, irrespective of the dynamics of the belt. The basic binary tree algorithm is, again, robust with respect to the tags coming in and out of range, since these effects are equivalent to the reading errors at the tag. Consider the example in Figure 8.9. At time t_1 the reader observes a collision due to transmission by τ_1 and τ_2 and starts to resolve this collision. The token of τ_1 is in $[0, 0.5)$, while the tokens of τ_2 and τ_3 are in $[0.5, 1)$. At the time t_2 it resolves τ_1. Unfortunately, at time t_3, when the reader enables the interval which would have resolved τ_2, the tag τ_2 is already out of range. Moreover, a new tag t_3 has arrived.

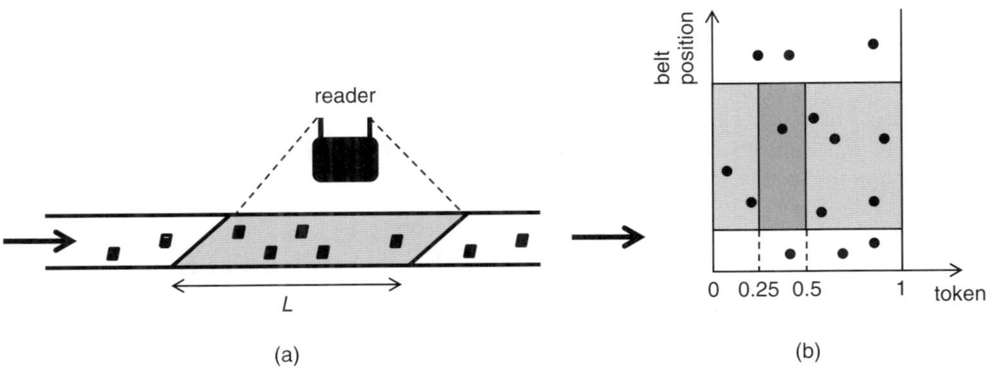

Figure 8.8 (a) RFID tags on a conveyor belt. The reader covers length L on the belt. (b) Equivalent representation in the two-dimensional arbitration space in one time snapshot. The abscissa represents the random token in $[0, 1)$. The light shaded stripe is the coverage area of the reader in that snapshot, while the dark shaded stripe is the area enabled when the tree protocol enables the interval $[0.25, 0.5)$.

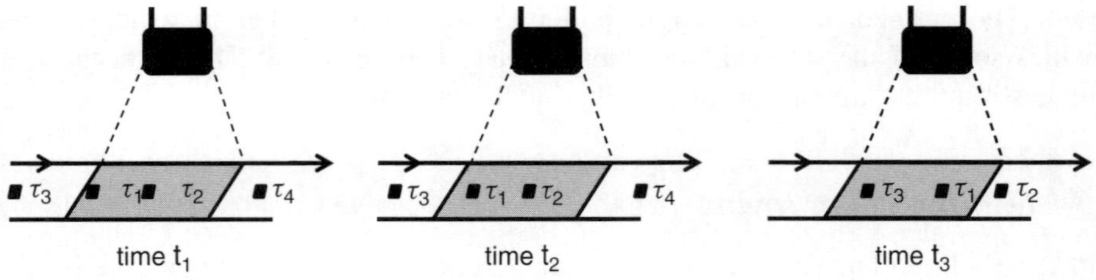

Figure 8.9 Movement of the tags on a conveyor belt, represented through three snapshots in time. The shaded area represents the reader coverage.

An important parameter for the performance of the collision resolution algorithms is the tag density in terms of number of tags per time unit that enter in the coverage area of the reader. Due to balance, this density is equal to the number of tags per time unit that exit the coverage area. When the tag density is small, the collisions are resolved quickly and there is a low probability that a tag will be missed. As the density increases, the average percentage of unresolved tags increases.

The setting is reminiscent of the original setting in which the tree algorithms have been proposed–the MAC protocol for an infinite population of terminals with Poisson packet arrivals [22]. In those scenarios, the maximal stable throughput of a given algorithm is represented by the average packet arrival rate which keeps the number of buffered packets finite. In the case of the conveyor belt, we again observe an infinite population of transmitting terminals, but if the packet is not delivered until a given deadline (i.e. until the tag leaves the coverage area of the reader), it is dropped. The packet deadline is determined as $\Delta = \frac{L}{v}$, where v is the speed of the conveyor belt. It is of interest to find the packet dropping rate as a function of the tag arrival density and the deadline Δ. Regarding estimating the tree algorithms, it is interesting to find how the estimate, obtained after a CBT instance is terminated, is related to the tag density.

But perhaps the most important parameter to determine by such an analysis is the density of outgoing tags that are not resolved. Namely, the problem of missing tags should be tackled by deploying multiple readers across the conveyor belt. The density of outgoing unresolved tags for reader R_A represents the input tag density for the next reader R_B. Thus, for a given number of tags per unit time on the conveyor belt, one can determine the required number of readers in order to keep the fraction of missing tags below a certain value.

8.5 Cooperative Readers and Generalized Arbitration Spaces

The described framework for tree protocols is suitable to define the *arbitration space* for a given random access protocol. The arbitration space is a set of points (real, integers, etc.), so that the collision resolution process can be represented via a sequence of enabled subsets from that arbitration space. In the approach described in the previous section, the arbitration space is the interval [0, 1) and is one-dimensional. In the following sections we will show that it is useful to generalize the ideas of arbitration space to higher dimensions.

8.5.1 Two-Dimensional Arbitration Space

The two-dimensional arbitration space can be conceptually introduced by considering two readers, as shown in Figure 8.1(b). The readers R_A and R_B can share information over a dedicated link, which uses a wire or radio interface that is different from the one used to communicate with tags. Alternatively, one can also think of the two readers as one common controller with distributed antennas. The readers cooperate with each other in the following ways: (a) they coordinate their transmissions in order to avoid reader collision and (b) they exchange information related to the tag reading process.

Let $\mathcal{T}_u$ denote the set of tags in the range of reader $u = A, B$. In general, the sets $\mathcal{T}_A$ and $\mathcal{T}_B$ are different. We assume that the density of the tags is uniform within the area covered by the two readers. This assumption is rather strong for RFID systems, where the tag density depends significantly on the usage scenario and the physical setup, such as a warehouse or shop. Nevertheless, such an assumption is very useful to introduce the principles of multi-dimensional arbitration spaces. Discussion on how the assumption can be relaxed is given in Section 8.5.2.

We consider the following arbitration algorithm. First R_A sends an initial probe, labeled A, and the tags from $\mathcal{T}_A$ reply. Next R_B sends another probe labeled B and the tags from $\mathcal{T}_B$ reply. We define the following sets: $\mathcal{S}_1 = \mathcal{T}_A \setminus \mathcal{T}_{AB}$, $\mathcal{S}_2 = \mathcal{T}_{AB}$, and $\mathcal{S}_3 = \mathcal{T}_B \setminus \mathcal{T}_{AB}$. These two initial probes are used to make each tag aware about the set $\mathcal{S}_j$ it belongs to. For example, a tag that received the probe from R_A, but not from R_B, belongs to $\mathcal{S}_1$. After the two initial probes, the arbitration should continue as three independent collision resolution processes, one for each set $\mathcal{S}_i$, respectively. The arbitration for the set $\mathcal{S}_2$ can be carried out by either of the readers. Let a session for the set $\mathcal{S}_i$ be the uninterrupted time during which the arbitration protocol is run only for the tags from the set S_i. There can be more than one session per set, such that τ_{ij} denotes the j-th session for the i–th set. For example, the readers can coordinate to run the following sequence of sessions: $\tau_{2,1}, \tau_{3,1}, \tau_{2,2}, \tau_{1,1}, \ldots$. Let the cardinality of set $\mathcal{S}_i$ be denoted by n_i and let k_{ij} be the number of resolved tokens during the session $\tau_{i,j}$.

The spatial division, introduced by the different coverage of each reader, is helpful for collision resolution as it inherently segments the tags into smaller groups. Such a segmentation suggests the following straightforward solution: run three sessions $\tau_{1,1}, \tau_{2,1}, \tau_{3,1}$ and in each session all tags from a given set are resolved, that is, $k_{i1} = n_i$. Such a solution features very limited cooperation between the readers and the estimation of the multiplicities n_1, n_2, n_3 is done independently. Nevertheless, if the spatial statistics of the tags is known, the readers can enhance their cooperation in estimating the population size.

The key idea of the two-dimensional arbitration space is, in addition to the dimension where random tokens are generated, another (spatial) dimension is utilized to resolve the collisions among tags. As an analogous notation to $\mathcal{S}_i$, let S_1 be the area covered only by R_A, S_2 be the area covered by R_A and R_B, and S_3 be the area covered only by R_B. Since the tags are uniformly distributed in the total area covered by the two readers, the probability that a tag belongs to $\mathcal{S}_i$ is $q_i = \frac{S_i}{S}$, where $S = S_1 + S_2 + S_3$. The considered readers are not mobile and they have already run a certain initialization process during which they have estimated the overlapping of their range and have thereby estimated q_i. Figure 8.10 depicts a representation of the two-dimensional arbitration space. The

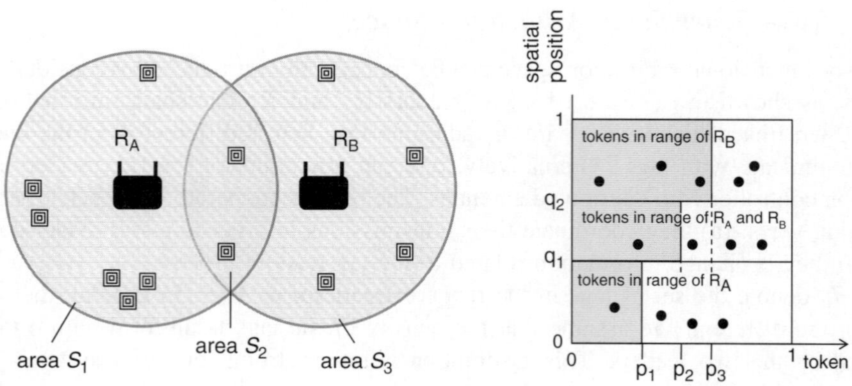

Figure 8.10 Representation of the compound random process: the tags are randomly distributed in a two-dimensional space and their tokens are randomly distributed in interval [0, 1). For a given tag, the random token represents its x-coordinate and the random placement its y-coordinate.

abscissa represents the dimension of random tokens, while the ordinate represents the spacial position of a tag. If two tags are in the same spatial set $\mathcal{S}_i$, then the only way they can be differentiated in the arbitration process is through usage of random tokens. On the other hand, if tag $\tau_1 \in \mathcal{S}_i$, $\tau_2 \in \mathcal{S}_j$ and $i \neq j$, then these two tags do not need to use random tokens to be differentiated. To facilitate the presentation, let us assume that initially, the system of readers runs independent instances of for example the IECR protocol for each set separately. For example, the collision resolution starts by having the reader R_A run the IECR protocol over the tags from the set $\mathcal{S}_1$. The reader has stopped when an instance of the CBT algorithm is terminated after k_1 tags have been resolved in the token interval $[0, p_1)$. Subsequent IECR instances are run for each of the $\mathcal{S}_i$, $i = 2, 3$, and an instance is stopped if after the termination of a CBT there are k_i tags within the interval $[0, p_i)$.

Instead of making separate estimates $\hat{n}_i = \frac{k_i}{p_i}$, the following approach can be used. Referring to Figure 8.10 and considering that the tags are uniformly distributed within the total coverage area of the two readers, the readers can share the information they have at observation time and one can obtain the following estimate for the total population of tags $n = n_1 + n_2 + n_3$:

$$\hat{n} = \frac{k_1 + k_2 + k_3}{p_1 q_1 + p_2 q_2 + p_3 q_3} \tag{8.15}$$

Using this estimation of the total population, an estimate can be obtained for the cardinality of each tag set $\mathcal{S}_i$ as follows:

$$\hat{n}_i^c = \hat{n} q_i \tag{8.16}$$

This estimate (using the superscript ^{c}) will be referred to as a cooperative estimate of n_i. In order to assess its correctness, its variance can be estimated as follows:[2]

[2] In the reference [19] the expression used for the variance $\mathrm{Var}[\hat{n}_i^c | n]$ is incorrect, this error is corrected by the expression in the present chapter.

$$\mathrm{Var}[\hat{n}_i^c|n] = nq_i \left(\frac{q_i(1-2p_i)}{p_1q_1 + p_2q_2 + p_3q_3} + 1 \right) \tag{8.17}$$

An alternative would be to use non-cooperative estimation, in which the cardinality of each set $\mathcal{S}_i$ is calculated separately:

$$n_i^o = \frac{k_i}{p_i} \tag{8.18}$$

and the variance of this estimate is determined as follows:

$$\mathrm{Var}[\hat{n}_i^o|n] = nq_i \left(\frac{1}{p_i} - q_i \right) \tag{8.19}$$

On can check that $\mathrm{Var}[\hat{n}_i^c|n] \leq \mathrm{Var}[\hat{n}_i^o|n]$, so that, on average, cooperative estimation provides more accurate estimates. It is important to point out that this is not the case, for example, if we fix n_1, n_2, n_3 (instead of fixing only n) and if $p_2 = 1$: then we clearly have $\hat{n}_1^o = n_1$.

Numerical illustration of the estimate variances is shown in Figure 8.11. The values of the resolved intervals at the moment of observation are $p_1 = 0.1$, $p_2 = 0.25$, $p_3 = 0.3$. The distance d between the readers controls the amount of overlap between coverage areas, and thereby the values of the probabilities q_i. Due to symmetry, $q_1 = q_3$ and $q_2 = 1 - 2q_1$. Each variance is linearly proportional to n, the total number of tags, and this figure plots the variances normalized with n. When the distance between the centers is $d = 0$, then $q_2 = 1$ and $q_1 = q_3 = 0$, hence the variance of $\hat{n}_2$ is maximal, while the variance of $\hat{n}_1$ is 0. When $d = 2R$, where R is the coverage radius, then $q_1 = q_3 = \frac{1}{2}$ and $q_2 = 0$. Figure 8.11 clearly shows that cooperative estimation significantly decreases estimate

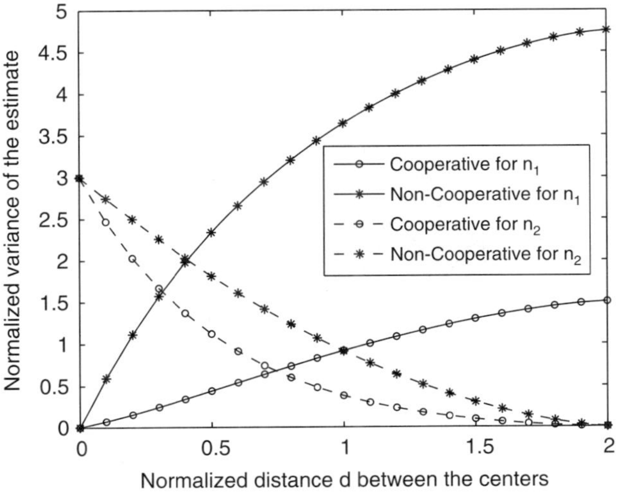

Figure 8.11 Different estimates of normalized distance by the total number of tags n.

variance of each set. As discussed before, such a correctness directly impacts the speed by which the tags within a certain subset are resolved.

To summarize, in the described method of arbitration, a CBT instance is run for one of the sets $\mathcal{S}_i$ at the time, while the estimation of the cardinality after each CBT instance is cooperative. From Figure 8.10 it can be seen that instead of enabled interval, during the arbitration for the set $\mathcal{S}_i$, there is an enabled area, rectangle of length p and height q_i.

8.5.2 Further Remarks and Multi-Dimensional Arbitration

In the previous example, the shape of a single enabled area has been restricted to a rectangle. In addition, we have assumed that the readers are unable to have an active role in determining the spatial division. If such a spatial control is possible, for example, by using multiple antennas and beamforming, then the enabled areas may have a more general shape. From the example shown in Figure 8.12, the multi-dimensional arbitration space can be utilized even with a single reader. In this example, the reader can control its transmission and create a beam that covers a sub-area of its total coverage area. Hence, the reader can spatially "address" smaller circles within its coverage area. The arbitration protocols for each subarea can be run in parallel (i.e. by interleaving the sessions in time) and the estimation can be made by aggregating the information from the different arbitration instances, as described in Section 8.5.1. In this particular case, it is suitable to use a three-dimensional arbitration space: two dimensions (x and y) are represented by the coverage area of the reader and the possibility to enable different spatial patterns within the coverage area, while the third dimension z is represented by the random tokens. Thus, if the possible spatial patterns are circles, then the enabled objects in the arbitration space are cylinders.

Additional acceleration of arbitration protocols can be achieved if the reader sessions take advantage of the spatial reuse and simultaneously inquire some of the tag subsets. For the example in Figure 8.10, the probing of set $\mathcal{S}_1$ and $\mathcal{S}_3$ can occur simultaneously. This can be generalized to multiple (>2) readers, where the sessions in the non-overlapping

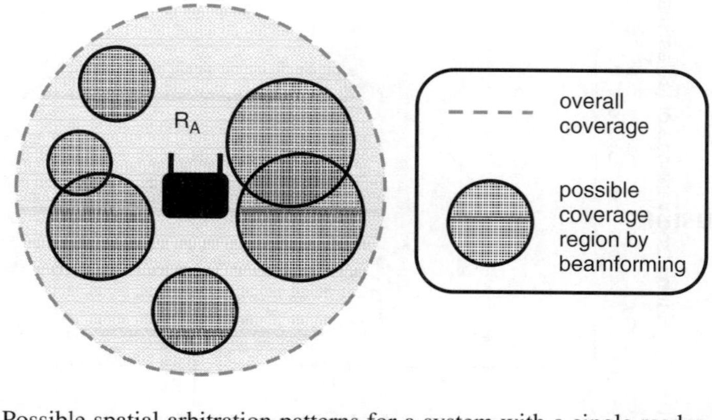

Figure 8.12 Possible spatial arbitration patterns for a system with a single reader and beamforming. The arbitration space is three-dimensional, one dimension being the random token and the other two dimensions are represented by the circle.

sets are scheduled to occur simultaneously. In the figure, the scheduling of the sessions is done in a round-robin manner, which in general is not optimal. In principle, a group of simultaneously scheduled sets that covers a larger area should be scheduled more frequently. Optimization of the arbitration speed achieved by the tree-based anti-collision protocols is difficult even in the case of one-dimensional arbitration space [18]. These difficulties are aggravated in the case of multi-dimensional arbitration. For example, in the case of two-dimensional arbitration, this optimization has two components: (1) selection of the length of each enabled interval in Figure 8.10; and (2) decision on the schedule used to query the three sets $\mathcal{S}_i$. A way to approach this optimization is to use Markov Decision Processes [29]: after a CBT instance is terminated, decide the next area to be enabled by the arbitration protocol, so as to minimize the expected duration of the protocol, while taking into account the conditions for spatial reuse and simultaneous transmissions.

It is interesting to note that, if the tag density in the coverage area of each reader is identical (or at least known), then the readers can cooperate even if their regions do not overlap. For the example in Figure 8.10, this is equivalent to setting $q_2 = 0$. This observation allows us to define a different problem: if tags are distributed uniformly in a certain area, we need to find which deployment of the readers produces the most efficient coverage of the area in terms of minimized expected duration of the collision resolution time.

Besides the described example of three-dimensional arbitration space for Figure 8.12, there are other ways to increase the dimension of the arbitration space. For example, in addition to the spatial dimension and the random token dimension, an additional feature of the RFID tags is taken into account when discriminating among the different tags. For example, tags may be attached to temperature sensors, so that the third dimension of the arbitration space is represented by the temperature that is read at a tag. The temperature range is mapped on the interval [0, 1) and the reading of each tag is a token distributed in this interval. This interval can be used as the third dimension for arbitration, in addition to the random token and the spatial position of tags. Now in the single step of the arbitration process, a 3-D region is enabled. Note that token distribution in the temperature dimension need not to be uniform, which should be reflected in determining the estimator of the tag population. Nevertheless, the task given to the reader network might require a partial resolution (e.g. estimate the average temperature or the temperature distribution with a sufficient accuracy) which should result in suitable selection of the sequence and shape of the enabled 3-D regions. In general, when there is an arbitration dimension across which the tags are not uniformly distributed, the width of the enabled intervals for that dimension should be adapted to the actual distribution of tags.

8.6 Conclusion

The problem of tag collision is fundamental in RFID systems and there are a number of protocols that deal with this problem. In this chapter we have described an important class of protocols for arbitration among collided tags, termed tree protocols. These protocols are based on a recursive solution and a query structure that forms a binary tree. We have introduced the basic principles of tree protocols, as well as several variants of tree-based algorithms. Furthermore, we have described an alternative framework for arbitration and representation of the tree-based algorithms from [18]. We have also described two variants of the tree protocol that have been subject to standardization by EPCglobal.

These practical RFID protocols show that there are many choices and approaches to implement the principles of tree-based protocols in real-life systems. Considering that the RFID systems are prone to transmission errors, we have discussed the impact that the transmission errors have on the tree-based arbitration protocols. Finally, we have discussed how tree protocols can be re-engineered in order to be adapted to a scenario with multiple cooperative readers. For that purpose, we introduced a generalized class of tree protocols via the notion of arbitration space. Such a generalization of the arbitration space offers a novel and versatile framework for developing future anti-collision protocols for RFID systems. By adjusting the arbitration space, these algorithms can be tailored to different applications, for example, in RFID sensor networks.

Problems

1. Let there be $n = 3$ tags in the range of the RFID reader. Assume that there are no transmission errors and the ideal collision model can be applied. Using the framework with enabled intervals, find the average duration of an anti-collision protocol if:
   (a) A basic binary tree algorithm is used.
   (b) A modified binary tree (MBT) algorithm is used.
   (c) Consider the situation in which the reader knows in advance that there are $n = 3$ tags. How can we use this information to optimize tree protocol? Note that since the reader knows that there are $n > 1$ tags, it will not enable the interval $[0, 1)$, as it will certainly result in a collision. Furthermore, assume that the algorithm is optimized in the following way: if the reader knows that in certain interval there are n tags, it enables $1/n$-th of that interval.
2. The pseudocode in Figure 8.4 assumes that the way in which the tree is traversed is decided by the reader, since it explicitly sends the address of the next enabled node. Let us assume that the "intelligence" of the tree protocol is transferred to the tags and the reader only provides: (a) the signal to initiate tree traversal; and (b) feedback to the tags. Write a pseudocode that describes the operation at the tag for this case.
3. In some models for collision resolution it is assumed that the multiple-access receiver, that is, the RFID reader, cannot distinguish between an idle slot and a collision slot. How should the binary tree algorithm be modified in that case?
4. The framework with enabled intervals has been applied to describe/design tree algorithms. How would an ALOHA-based protocol be represented in the same framework?
5. Describe how the ideas of an Estimating Binary Tree (EBT) algorithm can be applied to a tree a protocol for Class 1 Generation 1 tags.
6. Using the error model in Figure 8.7, explain why having only errors, where single or idle are interpreted as collisions, does not produce missing tags.

References

[1] Want, R. (2006) An introduction to RFID technology, *IEEE Pervasive Computing*, Jan. p. 25.
[2] Buckley, J. (2006) From RFID to the Internet of Things: Pervasive networked systems, DG Information Society and Media Conference, Brussels.

[3] Finkenzeller, K. (2003) *RFID Handbook: Fundamentals and Applications in Contactless Smart Cards and Identification*. Chichester: John Wiley & Sons. Ltd.
[4] Juels, A. (2006) RFID security and privacy: A research survey, *Journal of Selected Areas in Communication*, vol. 24, no. 2, pp. 381–394.
[5] Engels, D. W. and Sarma, S. E. (2002) The reader collision problem, in *Proc. IEEE International Conference on Systems, Man and Cybernetics*.
[6] Waldrop, J. Engels, D. W. and Sarma, S. E. (2003) Colorwave: A MAC for RFID reader networks, in *Proc. IEEE Wireless Communications and Networking Conference (WCNC03)*.
[7] Dong-Her, S. Po-Ling, S. David, Y. and Shi-Ming, H. (2006) Taxonomy and survey of RFID anti-collision protocols, *Computer Communications*, **29**(11): 2150–2166.
[8] Hayes, J. F. (1978) An adaptive technique for local distribution, *IEEE Trans. Commun.*, **26**: 1178–1186.
[9] Capetanakis, J. I. (1979) Tree algorithms for packet broadcast channels, *IEEE Trans. Inform. Theory*, **IT-25**: 505–515.
[10] Tsybakov, B. S. and Mikhailov, V. A. (1978) Free synchronous packet access in broadcast channel with feedback, *Probl. Peredach. Inform.*, **14** (4): 32–59.
[11] Hush D. R. and Wood, C. (1998) Analysis of tree algorithms for RFID arbitration, in *Proc. IEEE International Symposium on Information Theory*, Boston, USA.
[12] Micic, A. Nayac, A. Simplot-Ryl, D. and Stojmenović, I. (2005) A hybrid randomized protocol for RFID tag identification, in *Proc. 1st IEEE Int. Workshop on Next Generation Wireless Networks (WoNGeN 05)*.
[13] Bonuccelli, M. A. Lonetti, F. and Martelli, F. (2006) Tree slotted Aloha: a new protocol for tag identification in RFID networks, in *Proceedings of the 2006 International Symposium on World of Wireless, Mobile and Multimedia Networks (WOWMOM)*.
[14] Chiang, K. W. Hua, C. and Yum, T.-S. P. (2006) Prefix-randomized query-tree protocol for RFID systems, in *Proc. IEEE Int. Conf. on Communications (ICC)*.
[15] Myung, J. Lee, W. and Srivastava, J. (2006) Adaptive binary splitting for efficient RFID tag anti-collision, *IEEE Commun. Lett.*, **10** (3): 144–146
[16] Choi, J. H. Lee, D. and Lee, H. (2006) Bi-slotted tree based anti-collision protocols for fast tag identification in RFID systems, *IEEE Commun. Lett.*, **10** (12): 861–863
[17] Ryu, J. Lee, H. Seok, Y. Kwon, T. and Choi, Y. (2007) A hybrid query tree protocol for tag collision arbitration in RFID systems, in *Proc. IEEE Int. Conf. on Communications (ICC)*.
[18] Popovski, P. Fitzek, F. and Prasad, R. (2007) A class of algorithms for collision resolution with multiplicity estimation, *Algorithmica*, **49** (4): 286–317.
[19] Popovski, P. (2008) Tree protocols for RFID tags with generalized arbitration spaces, in *Proc. 10th Symp. on Spread Spectrum Techniques and Applications (ISSSTA)*.
[20] Cidon I. and Sidi, M. (1988) Conflict multiplicity estimation and batch resolution algorithms, *IEEE Trans. Inform. Theory*, **IT-34**: 101–110.
[21] Wireless LAN Medium Access Control (MAC) and Physical Layer (PHY) Specification, IEEE Std. 802.11, Aug. (1999).
[22] Berstekas D. and Gallager, R. (1992) *Data Networks*, 2nd edn. New Jersey: Prentice-Hall.
[23] Massey, J. L. (1981) *Collision-Resolution Algorithms and Random-Access Communications, ser. CISM Courses and Lectures*. Berlin: Springer-Verlag, no. 265, pp. 73–137.
[24] Greenberg P. F. A. G. and Ladner, R. E. (1987) Estimating the multiplicities of conflict to speed their resolution in multiple access channels, *J. ACM*, **34** (2): 289–325.
[25] Gallager, R. G. Conflict resolution in random access broadcast networks, in *Proc. AFOSR Workshop Commun. Theory Appl.*, Provincetown, MA, Sep. 1978 pp. 74–76.
[26] Dobkin, D. M. (2007) *The RF in RFID*. Oxford: Newnes.
[27] *ISO/IEC 18000: Information Technology Automatic Identification and Data Capture Techniques - Radio Frequency Identification for Item Management Air Interface*, International Organization for Standardization. Std., 2003.
[28] Jacobsen, R. Fyhn, K. Popovski, P. and Larsen, T. (2009) Reliable identification of RFID tags using multiple independent reader sessions, in *Proc. IEEE RFID 2009*.
[29] Puterman, M. L. (1994) *Markov Decision Processes: Discrete Stochastic Dynamic Programming*. New York: John Wiley & Sons, Ltd.

9

A Comparison of TTF and RTF UHF RFID Protocols

Alwyn Hoffman, Johann Holm, and Henri-Jean Marais

Northwest University, Potchefstroom, South Africa

9.1 Introduction

The design of a communications protocol can be viewed as the challenge to find an optimal compromise between the required functionality of intended applications, current state-of-the-art capabilities of electronic technology, and constraints implied by regulatory issues, moderated by the creativity of the designer. Long-read range passive RFID represents a very interesting variation on this theme, as the capabilities of technology and the availability of bandwidth must be stretched to the ultimate in order to allow the candidate protocols to provide acceptable solutions to the envisaged applications, with RFID tag technology cost as an important constraint.

Several texts have appeared in recent years describing the design of UHF RFID systems in general [5, 7]. Much has also been reported in the literature regarding proposed schemes to address the anti-collision problem that is commonly experienced with long-read range passive RFID systems ([8–11] and many others not cited in this text). This chapter will focus only on aspects related to the choice of a suitable protocol. While a brief overview will be provided of RFID protocols in general, the main focus is on the challenges to find a suitable protocol for long-read range passive UHF RFID. The motivation for focusing on UHF RFID is that both short-range magnetic induction-based passive RFID and active UHF RFID have long been established and are thus currently not facing the same challenges as passive UHF. As the working of existing ISO standardized protocols are covered elsewhere in this volume, the focus will furthermore be on candidate protocols that are not yet part of the ISO18000-6 standard for passive UHF RFID, and how they

RFID Systems: Research Trends and Challenges Edited by Miodrag Bolić, David Simplot-Ryl, and Ivan Stojmenović

compare with the most popular standardized protocol (ISO18000-6C or EPC Class 1 Generation 2).

The chapter starts with an overview of the requirements for RFID protocols, to provide a framework within which candidate protocols can be evaluated and compared against the existing standards forming part of ISO18000-6. The different approaches that can be taken in the implementation of an RFID protocols are briefly covered, followed by an overview of the requirements for the effective operation of RFID, with the focus mainly on passive UHF. A detailed description is then provided of the operation of more than one stochastic protocol, including protocols that do not require modulated interrogation signals from the reader, and the potential benefits of such protocols are highlighted. The theoretical and practical performances of such protocols are then compared against the performance achieved by the ISO18000-6 protocols, more specifically the ISO18000-6C or EPC Class 1 Generation 2 protocol. Finally, conclusions are drawn based on the comparison between the different candidate protocols.

9.2 Requirements for RFID Protocols

The requirements to be satisfied by RFID protocols are largely dependent upon the type of functionality and performance intended to be offered by a specific type of RFID. It therefore does not make sense to state one set of requirements for all types of RFID – a more productive approach would be to divide RFID into the major categories and state the requirements for the protocol based on the target applications for each category.

9.2.1 Categories of RFID Technology

For the purpose of this discussion it is sufficient to divide RFID into three primary categories:

- Short-range magnetic induction-based passive RFID, operating in the LF and HF bands.
- Medium to long-range passive RFID operating in the UHF band.
- Long-range active RFID operating in different UHF bands.

Again, the main focus of this chapter is on passive UHF protocols, primarily because this is the technology that has attracted most attention over the past decade. The protocols used by the other two categories of RFID have been deployed in practice for a substantial period of time and seem to be largely stable technologies, with limited recent effort from industry to launch new technologies justifying new standards within those categories.

9.2.1.1 Passive LF and HF

Initial deployments of passive RFID operating in the LF band mostly involved short-range applications that require only one tag to be read at one point in time. Based on this point of departure, a number of standards have been adopted by ISO over the years for the LF band, including ISO 11784, ISO 11785 and ISO 14223-1. These protocols were, in general, not designed to provide any significant anti-collision capability, as the intended applications at the time generally assumed that only one tag would be visible to the reader at one point in time. The low data rates that can be supported by 125–134 kHz band carriers make LF fundamentally unsuitable for implementing fast anti-collision protocols involving substantial collections of tags. This limiting factor, combined with the lack of read range beyond 0.5–0.6 m, largely excludes LF technology from the debate about more effective anti-collision protocols.

The situation is somewhat different for HF (typically around 13.56 MHz). While the read range that can be achieved with HF, using readers that are moderate in terms of both complexity and cost, is similar to that of LF, the much higher carrier frequency (13.56 MHz) does in principle allow the implementation of very fast and efficient anti-collision protocols. This has led to the adoption of three major ISO standards for this band, namely ISO14443, ISO15693 and ISO/IEC 18000 Part 3 Mode 2. The latter is primarily implemented through the product range of [19], and claims very fast anti-collision for short-range applications such as the counting and identification of gambling chips. The combination of this powerful anti-collision protocol with technological implementations that allow tags to be read when stacked directly on top of each other has resulted in some very successful commercial deployments, the most prominent case in point being the gambling industry.

The issue to be considered regarding the relevance of anti-collision protocols developed for HF beyond short-range applications, is whether this technology does have any significant scope for applications that require medium to long read ranges (typically supply chain applications such as bulk stock control). HF has been selected by prominent global business entities, such as Benetton, for pilot implementation in the garments industry, as well as by a number of global pharmaceutical companies for pilots that focus on product authentication. Neither of these initiatives has yet resulted in major commercial deployments, partly due to privacy issues (the Benetton garments tagging project was cancelled several years ago) and partly due to cost pressures in the case of pharmaceuticals [12]. Vendors such as Magellan have made claims of achieving read ranges that allow the implementation of portals 8 feet wide, but it would appear that such implementations are very costly (as effective screening has to be constructed to screen off potential sources of noise). The limiting factor in this case has nothing to do with the protocol as such, but results from the limited strength of the energizing magnetic field that can be set up by HF readers and that comply with existing spectrum regulations, combined with the fact that the HF band suffers from a very wide variety of unintentional radiators. These electro-magnetic culprits cause high noise levels that compete with relatively weak tag signals at the reader's receiver. These unintentional radiators include, for example, most kinds of induction motors, PC power suppliers, noise radiated by power distribution lines, in addition to the noise generated by the energizing signals from the reader itself. In this respect, HF is in a much worse position compared to UHF, and therefore cannot be realistically expected to be a serious future contender for RFID applications requiring read ranges beyond 1 meter.

An interesting technology to take note of in the LF/HF band is the so-called dual frequency (DF) technology developed by IPICO Inc., with EM-Microelectronics acting as manufacturer of transponder chips to support this technology [13]. Dual frequency technology is based on the known technique of using different frequencies for the energizing of tags and for tag-to-reader communications (e.g. [21]), In the case of this specific dual frequency implementation (that will be referred to as DF for the purpose of this discussion), the energizing signal is generated at 125 kHz (or a frequency close to 125 kHz), while the tags respond at a frequency in the HF band (in this case, 6.8 MHz). The carrier frequency for tag communications, which is generated in the transponder chip as a harmonic of the energizing signal, is comparable to the carrier frequency used by 13.56 MHz systems, and thus has the ability to achieve similar anti-collision capabilities and signal-to-noise performance. The fact that the energizing signal is located at 125 kHz, however, allows the powering up of tags at distances roughly double the ranges that can be achieved using standard HF systems. As DF readers do not directly contribute to the noise levels visible to their own receivers, DF systems will in principle suffer somewhat less from the effects of noise, allowing them to practically achieve the longer read ranges allowed by their longer energized ranges (compared to HF). Given the fact that the operation of DF systems is based in total on magnetic induction, it is for all practical purposes not affected by the presence of liquids or humidity (unlike UHF systems), and thus creates opportunities for this technology to be deployed in supply chain applications that require the covering of standard warehouse doors. HF systems that are required to read through liquids, or highly humid media, do not perform as well as DF systems that will be able to energize and read at longer ranges, while UHF systems are affected even more adversely by such conditions. A prominent example is the fresh produce industry, where product is typically transported in plastic bins that are stacked in multiple layers. This application represents a very challenging scenario for both HF and UHF, while it falls relatively easily within the capabilities of DF.

The air-interface protocol currently used by the above DF implementation is not part of existing ISO standards. It is based on the same IP-X tag-talk-first approach as used by IPICO Inc. in its range of UHF products [13]. This protocol will be discussed in detail in the sections that focus on alternative UHF protocols, and will therefore not be covered in any depth in the section. It should be mentioned that the transponder bit rate for DF tags is 128 kbaud, compared to a bit rate of 256 kbaud as implemented in the latest version of UHF IP-X products. The reason for this is the fact that the carrier frequency used for DF products is 6.8 MHz, which limits the maximum baud rate that can be supported. The results displayed in the section on UHF IP-X should therefore be translated by a factor of two when the anti-collision capabilities of the DF implementation of the IP-X protocol are evaluated, for example, when the number of tags read per second or the time required to detect a specific proportion of all tags is determined.

9.2.1.2 Passive UHF

Passive RFID is viewed as the primary candidate technology to fulfill the aim of ubiquitous tracking of items, thereby realizing the dream of "the Internet of things" [14]. Compared to magnetic induction-based versions of passive RFID, UHF comes much closer to fulfilling the functional requirements that will make the widespread use of RFID a reality. UHF achieves much longer read ranges compared to LF or HF and the physical construction of UHF transponders also allows the production of low cost tags. Recent efforts show that the ideal of the so-called "5 cent tag" is already very close to reality [15, 16, 17].

This does not, however, mean that the effort to produce UHF RFID technology (that fulfills these requirements in practice) is without challenges. A great deal of resources has been spent over the past 10 years to develop chip, transponder and reader technologies to the level where both cost and functionality expectations can be met, hence enabling the widespread deployment of a variety of RFID-based tracking and identification systems. Among these efforts is counted the process to drive RFID standards to a point where at least one widely supported standard is now in place, namely ISO18000-6C, also referred to as EPC Gen 2. While it has been demonstrated that this standard comes close to satisfying most requirements for UHF RFID, there is still by no means complete consensus regarding the suitability of this standard to successfully support the deployment of important applications in all regulatory jurisdictions.

The primary debate in defining a global standard for passive UHF RFID has revolved around issues of the most suitable approach to find a compromise between RFID system performance and regulatory restrictions. These regulatory restrictions are defined in terms of the allowed output power and amount of frequency spectrum allocated to passive UHF RFID in some very prominent jurisdictions, including the European Union and China. Earlier versions of the EPC protocol were defined based mainly on the regulatory situation in the USA, where UHF RFID can be used in 52 bands, each of 500 kHz bandwidth, between 902 and 928 MHz. The situation in the EU is very different, allowing only 10 channels of 200 kHz each. The successful deployment of systems within this regulatory environment hence presents a far greater challenge, specifically in terms of restricting reader modulation spectra from causing interference in adjacent channels, as well as allocating channels to readers in such a way that large numbers of readers operating in close proximity do not interfere with each other. These issues will be discussed in more detail in Section 9.3.

9.2.1.3 Active RFID

The technical challenges posed by the air-interface protocols, used by active RFID systems, are very different from those applicable to passive RFID. First, active transponders that have their own power sources, are characterized by much more complex functionality compared to passive transponders, allowing the use of protocols that also require more processing power (e.g. CDMA) than what is viable for passive RFID. Second, active readers do not need to transmit as powerful signals as required for passive readers, since the reader signals are only used for communication, while passive reader signals must in the first place provide power to transponders. Given the much longer read range of active RFID systems, transponders will normally have more time to respond to reader signals compared to passive transponders, making the time required to successfully identify transponders less critical. More emphasis is placed on techniques to conserve transponder power in order to maximize battery lifetimes, which is not an issue with passive RFID. Given the relatively low power levels emitted by both active readers and transponders, the technical challenges to provide active RFID systems with sufficient bandwidth to support the deployment of large systems is quite different from those applicable to passive UHF RFID. While the long read ranges of active RFID systems do lead to the problem of cross-reads between different readers (i.e. two readers picking up the same transponder), the average amplitude of reader and transponder signals are of the same order of magnitude. The presence of another active RFID reader in the same frequency band and transmitting at the same time will therefore not prevent an active reader from successfully detecting the desired transponder signals. This is in contrast to passive UHF RFID, where time and frequency overlap of reader and transponder signals may cause severe interference problems, due to the much smaller amplitudes of passive transponder signals.

Active RFID protocols are mainly covered by ISO standard ISO 18000-7, while several application-specific standards also exist (e.g. ISO 18185 for electronic shipping container seals based on active RFID transponders). For the reasons discussed above, the detailed working of these protocols will not be covered in this chapter.

9.2.2 Requirements for Passive UHF RFID

As recent efforts to arrive at a global standard for RFID have mostly focused on passive UHF, the remainder of this chapter will be devoted to this topic. Before the different approaches to arrive at an efficient UHF protocol are investigated, it will be useful to study the requirements to be satisfied by passive UHF RFID, as derived from some of the most promising application domains. In this respect two extreme scenarios will be considered:

- *Supply chain management*, where the major challenges to RFID involve the identification of large numbers of tagged items, moving at low to moderate speeds, and including potentially large numbers of readers operating in close proximity.
- *Transportation*, where the challenges revolve around the reading of tags moving at high speeds relative to readers, in most cases involving only a small number of tags visible to a reader at one time, and with small to moderately sized collections of readers operating in one location.

9.2.2.1 Supply Chain Applications

For supply chain applications, the primary performance criterium that is directly impacted by the air-interface protocol can be viewed as the overall throughput achieved by the total RFID system operating in one location. The definition of "one location" can be stated as an area of such geographic extent that readers can potentially impact on the performance of other readers operating within that same area. In the process of calculating effective throughput, it will be necessary to estimate the total number of tags that can be identified successfully by a number of independently operating sets of readers, all within a specified time period. A typical example will be readers deployed in different warehouses that could be located around a large port, assuming that it will not be possible to time-synchronize the operation of sets of readers operating on different premises. A typical reader will be monitoring a warehouse door and could be observing relatively large collections of tags (up to 1,000 tags, but more typically 50 to 100 tags per batch) that are moving past readers within a few seconds and at random time intervals.

The challenge presented to the air-interface protocol in this case can be summarized as follows:

- The anti-collision capabilities must be sufficiently powerful to allow the maximum number of tags that may be encountered in one group to be identified in the available period of time (worst case up to 1,000 tags to be read within 1–2 seconds).
- It must be possible for large numbers of readers to operate independently of each other, in order to allow the warehouse operations to continue without disruptions while tags

are being identified. Up to 200 readers could be found per warehouse, and up to 50 warehouses located in an area small enough for different readers located at the extremes of this area to potentially interfere with each other if no precaution is taken, that is, up to approximately 1,000 potentially interfering readers.

It should be noted that the above criteria are not independent of each other: increasing the rate at which readers can identify tags implies an increase in the required data rates for reader–tag communications, which implies more bandwidth used by each reader, thus leaving fewer communication channels for the remaining readers, given that the overall regulated bandwidth will be fixed. The challenge is therefore to find a compromise between an acceptable rate of identifying tags in large tag populations, and maximizing the number of readers that can operate independently at one site. Other factors, such as the directionality of reader antennas and the degree to which reader signals can be isolated from adjacent channels, as well as the average duty cycle of a reader, will all impact upon the effective throughput of the aggregated system.

9.2.2.2 Transport Applications

In the case of transportation systems, the primary challenge is the ability to read tags moving at high speeds, implying a short period of time that any tag will be visible to a particular reader. Road safety regulations typically require readers to be mounted between 5 and 6 meters above the road surface. The radiation patterns of patch antennas – that are normally used with long-range passive UHF readers – result in reader fields with an effective width of around 4 m at the typical height at which tags will move through the beam. Currently, RFID will mostly be used for law enforcement within this application domain, and it is essential for the RFID system to be able to identify tags that are moving at speeds that are well in excess of legal speed limits. A typical requirement will be to detect tags at speeds of up to at least 220 km/h (more or less 140 mph). In addition to reading a tag's unique ID, it may be required to also extract additional information from such tags at speed. This is done in order to allow a law enforcement officer to verify the legal status of a vehicle without first performing an online verification on a central database. A typical need could therefore be to read at least 256 bits of information from any tag at maximum speed. Given the physical nature of dense traffic environments, it can be assumed that less than 10 tags will be visible to a reader at any point in time. However, when tagged cargo is verified at high speed, the number of tags in a reader's field may increase significantly – this is currently not a viable application of UHF RFID when regulatory limitations are taken into consideration.

The nature of typical transportation read stations will imply the presence of only a moderate number of readers in one location, as different independently operating read stations will typically be located several hundred meters if not several kilometers apart. At the same time it will, however, be possible for a highway read station to be located within close proximity to several warehouse management systems, for example, in the case of a highway running past a large port. It will therefore be necessary to determine the minimum required spatial separation between different read stations to prevent reader-to-reader interference. Once again, the need to read tags at high speeds will imply the use of high communication data rates, which will increase the bandwidth requirements per reader, thus leaving fewer bands for the operation of other read stations. The optimal solution will once again have to be a compromise between these conflicting requirements.

The question may be asked why the above challenges to the deployment of passive RFID in transport applications are not avoided altogether by rather reverting to the use of active or even semi-passive technologies for transport applications. The answer to this lies mainly in the cost issue: many transport applications can only deliver maximum benefits if all vehicles in a specific jurisdictions are obliged to carry the same type of RFID transponder. A good example in this respect is the use of RFID as part of vehicle licensing. As vehicle licenses are frequently renewed, there is much cost pressure on the choice of technology, and passive RFID obviously has a significant cost benefit over both active and semi-active RFID. It is therefore logically expected, in spite of technical challenges, that passive RFID will become the technology of choice for mass deployments of automated identification in the transport market.

9.3 Different Approaches Used in UHF Protocols

The different air-interface protocols that have been developed over the years for UHF RFID, are mostly derivatives of protocols that were developed for use in general computer networking applications. The adaptations that are required to arrive at an efficient RFID protocol are necessitated by those factors that characterize passive RFID reader–tag communications:

- Tags must be assumed to be simple devices incapable of performing the type of data processing required by most computer networking protocols.
- As readers have to power tags in addition to receiving and transmitting communications messages, reader signals are in general orders of magnitude stronger than tag signals. The protocol therefore has to ensure that reader and tag communications are separated in either the time domain or the frequency domain, or both, to allow readers to successfully detect tag signals against the background of potentially interfering signals from other readers operating in the same geographic location.
- Tags will communicate only with readers, not with other tags, so instead of peer-to-peer type of communications, the nature of RFID protocols will tend to be master–slave, with one reader (the master) communicating with many tags (the slaves).
- The nature of communication messages are in general not ad hoc communications, as in the case of computers on a network exchanging information with one another as and when required, but rather short bursts of communications from tags to reader to inform the reader about their presence. This normally happens in a short time period as tagged items move past a reader, with more complex two-way exchange of information mostly only happening when tags are initially programmed by a reader, in which case the communication will mostly be one-on-one at close range.

From the above discussion, it is clear that the air-interface protocol must primarily satisfy two important criteria:

1. When a large number of tags moves past a reader in a short period of time, the anti-collision capabilities of the protocol must allow all tags to successfully communicate

their presence to the reader at least once. Alternatively the read scenario may involve a small number of tags but moving at high speeds, thus leaving the tags very little time to communicate with the reader.

2. When a number of readers are deployed at the same location (i.e. sufficiently close to each other so that signals from other readers can potentially interfere with the tag signals that a particular reader needs to detect), the protocol must allow all readers to successfully communicate with the tags that they are supposed to detect, even in the presence of interfering signals from other readers communicating with their own sets of tags.

In general (at least in the case of reader-talk-first protocols), these two primary criteria tend to pose conflicting requirements on the system as a whole: the successful detection of a large number of tags in a short period of time will benefit from a reader that sends out very fast communication signals to tags; such high speed reader communications, however, directly result in a wider reader transmission spectrum, increasing the possibility of interference between different readers, which is in conflict with the second criterion of limiting potential reader interference. The paragraphs below discuss different approaches that have been taken to arrive at a satisfactory solution to the above problem.

9.3.1 Deterministic versus Stochastic

Most communication protocols can be categorized as being either deterministic or stochastic. Deterministic protocols follow an exact and repeatable algorithm, which can be guaranteed to result in the successful transmission of a message within a predetermined time interval, as long as the physical communication link is working. Stochastic protocols are statistical in nature, and there is a non-zero chance that any particular effort to transmit a message may not be successful or may be delayed until a later stage, with the exact time delay also being stochastic in nature.

While the deterministic approach may appear to be the preferred one, in practice, superior performance is often obtained using the stochastic approach when all practical limitations and restrictions are taken into consideration. This can be seen from the definition of protocols that have successfully passed a rigid standards process. For this reason, commonly used protocols in the computer networking domain are mostly stochastic in nature (e.g. the Ethernet protocol), as the stochastic nature provides it with the flexibility to optimally utilize available resources (in this case, available bandwidth) without prior information about the number of network participants or the nature of the messages that will be transmitted. The drawback of deterministic protocols, in general, is the fact that a rigid procedure must be followed that will also cater for worst-case scenarios, even though such scenarios will seldom appear. As a result, deterministic protocols have not survived the same rigid standards processes that stochastic protocols have passed. Good performance (in the case of RFID, the fast identification of tags by readers) can thus only be guaranteed through brute force, that is, high speeds of communications. Some early stage passive UHF protocols that followed the brute force approach, like the EPC Class 0 protocol as implemented in the products of Matrics during the early years of UHF RFID, required high reader data rates. These high rates resulted in readers that transmitted high power modulated signals outside of the regulated band of the carrier, and hence in systems that were not compliant with regulations that restrict reader communications to specific regulated bands. More recent protocols therefore did not persist with this approach.

Stochastic protocols, on the other hand, follow a more flexible approach that makes realistic assumptions in respect of the likely scenarios to be encountered in practice. These protocols are designed to effectively adapt to each practical scenario, either by changing the rate of tag communications (in the case of tag-talk-first protocols) or by adapting the queries sent out by the reader (in the case of reader-talk-first protocols). In this way, close to optimal performance can be achieved for a wide range of tag identification scenarios without having to rely on overly fast reader communications, as is the case with, for example, the dense reader mode of the EPC Class 1 Generation 2 or ISO18000-6C protocol. As an example of the benefits offered by stochastic protocols, the stochastic approach is more bandwidth efficient, as the only modulation signals generated by this protocol come from the RFID tags and not the reader (the reader emits a fixed carrier frequency used to power up tags). The signals backscattered from tags would normally fall below the regulated power levels. This approach offers the potential of providing good performance with limited available spectrum (as is the case in several important regulatory jurisdictions). More will be said about the I-PX protocol in the next section.

9.3.2 RTF versus TTF

The distinction between reader-talks-first (or RTF) and tag-talks-first (or TTF) protocols is even more fundamental than the distinction between deterministic and stochastic protocols, as discussed in the previous section. The arguments on which RTF protocols are based, can be summarized as follows:

- The objective to detect large populations of tags within acceptable time periods (e.g. when reading items leaving a warehouse) is only achievable if the reader is allowed to interrogate the tags, using in most cases a combination of deterministic and stochastic principles, rather than leave it to the tags to decide when to communicate with the reader (which implies the use of some stochastic algorithm to prevent tag collisions). In a sense, a "communication session" is opened between tags and readers until all tag information has been read.
- It is, at least in principle, possible to deploy large RFID systems that involve large numbers of readers that operate in relatively close proximity, and to restrict the interference between such readers within all important regulatory jurisdictions. This can be done when each reader interrogates its own set of tags independently from other sets of readers and tags.
- TTF protocols suffer from too many limitations in terms of lack of flexibility as readers do not have the option to control the way in which tag detections occur but have to rely on the behavior of tags with very limited built-in intelligence. This results in systems that are not sufficiently intelligent to support fast universal adoption.

TTF protocols, on the other hand, depart from the following set of arguments:

- The bandwidth available to UHF RFID readers in some regulatory jurisdictions is limited compared to the number of readers that must operate independently and within

"hearing range" of each other. This effectively makes it impossible to deploy large numbers of readers in close proximity without either running out of bandwidth or severely restricting the duty cycle of individual readers (e.g. by using some kind of listen-before-talk measure). The solution to this dilemma is to not require readers to interrogate tags in the normal course of operation and to leave it to the tags to make their presence known to readers through some stochastic algorithm that is programmed into the tags.
- The time that a tag will be visible to a reader may in some cases be very short (e.g. in the case of tagging vehicle on a highway), and the field through which a tag moves is often very non-uniform [18] due to the inherent nature of UHF multipath propagation. Faster effective identification of tags will be possible through a stochastic TTF protocol compared to an RTF protocol that operates within regulatory limitations, as RTF protocols tend to be less efficient in respect of the time constraints (more specifically when also faced with limited available bandwidth).
- The implementation of TTF protocols, from a protocol point of view, is less complicated compared to RTF protocols, which results in less complex and potentially cheaper transponder chips, and therefore more affordable systems overall.

Both of these two sets of competing arguments contain at least some element of truth, and it is easy to demonstrate that both approaches can offer superior performance for specific applications. It is also interesting to note that the weight of some of the arguments supporting the two opposing points of view is closely related to the current state-of-the-art of integrated micro-electronic and RF technology. As technology has progressed over the past decade, the disadvantages posed by the additional complexity of RTF systems have become less of a challenge, and in some cases even a moot point. The verdict is, however, still out regarding the all-important question: "Which of these two approaches will eventually prove to be good enough to support the deployment of successful large-scale RFID systems (by satisfying the two primary criteria of the marketplace: low cost and reliable operation)?"

9.4 Description of Stochastic TTF Protocols

TTF anti-collision protocols must rely on the stochastic behavior of transponders to avoid collisions in transponder communications, as the reader does not directly control when transponders will transmit their IDs. Such stochastic protocols theoretically suffer from the disadvantage that there is no upper limit to the period of time that it may take for a tag to be detected, as each stochastic effort by the tag to communicate may end in a collision (although this is unlikely), without either reader or tag being aware of this situation. Figure 9.1 describes the behavior of a typical free-running TTF protocol using stochastic tag delay times.

Several alternative variations on the standard TTF free-running protocol have been patented to relieve the above problem, and some have been practically implemented over the past 10 years. Two of these will be described in more detail, namely Supertag [2], which represents a collection of alternative approaches to TTF operation, and IP-X [3], which is actually a special case of the Supertag protocol.

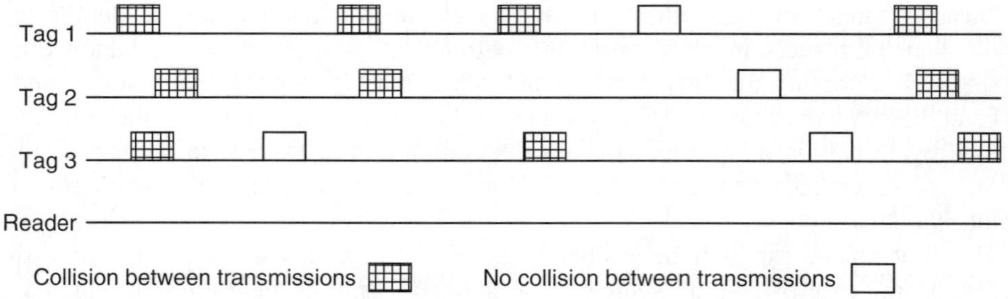

Figure 9.1 Tag transmissions operating under a free-running TTF protocol [3].

9.4.1 Supertag

The earliest implementation of a TTF-type UHF RFID protocol was the so-called Supertag protocol as patented by Atkins, Marais and Smit in [2], which is an example of an unslotted Aloha protocol. The Supertag patent provides a number of alternative and complementary approaches to achieve optimal performance by using the underlying approach that tags are not initially prompted by an interrogator, but start transmitting their IDs in pseudo-random fashion as soon as they are energized. Supertag allows for both free-running versions of the protocol (i.e. tags behave totally independently from reader instructions) or versions where the reader may instruct tags to change their behavior once tag communications have been detected. The common denominator between the different approaches that are proposed in the Supertag patent is the fact that the tags are prompted to start communicating by the energizing signal of a reader, which can be a purely continuous wave (excluding any modulation), and that the tags then transmit their IDs in pseudo-random fashion. Each tag is programmed with an initial maximum waiting interval between consecutive transmissions of its own ID. Upon wake-up a tag will generate a pseudo-random number and use this number to calculate a pseudo-random waiting interval, which will be a fraction of the above maximum waiting interval. After waiting for this pseudo-random period, the tag ID is transmitted and the process of generating another pseudo-random interval is then repeated successively. This system hence relies on the fact that pseudo-random time delays are generated on each tag, that allows all tags to eventually be identified by the reader based on the time diversity achieved through the stochastic nature of tag transmissions. It is accepted that, depending on the number of tags present and the duty cycle of each tag transmission, a specific proportion of tag communications will end up in collisions. The options allowed for in the Supertag protocol are differentiated by (1) the way in which the tags respond once they have started transmitting their IDs, as well as by (2) the way in which the reader is allowed to control the behavior of tags once at least some of them have been successfully identified.

The first possibility involves the reader retransmitting the tag IDs that have been successfully detected. As soon as a tag receives its own ID from the reader, it realizes that it has been detected and stops further transmissions of its own ID until it shuts down (due to loss of energy) and wakes up again (typically upon leaving and then re-entering a reader beam or sometimes moving through a null in the electro-magnetic field). The fast version of Supertag involves the reader sending out a mute signal to all transponders once it has started to detect a tag transmission. The transmitting tag will continue its transmission while all other tags will suspend their waiting period countdowns for 128 clock cycles to allow the current tag transmission to be detected successfully before any other tag will start transmitting.

Another option, patented as part of the Supertag protocol, is the fact that each tag may have to alter its maximum waiting time until the next ID transmission takes place to allow the adaptation of the protocol to specific circumstances in order to optimize performance. This includes the possibility of dynamically altering the waiting time upon reception of an instruction from a reader. If a reader detects that there are initially too many tag collisions (which will be based on the number of failed tag detections due to tag signal overlap in time), it may send out a signal to tags requesting them to increase the maximum waiting period in order to reduce congestion, as depicted in Figure 9.2 [2]. In this example, initial signals from tags 1, 2 and 3 clash and therefore cannot be identified. The reader, sensing that the protocol is over-congested, sends out a "slow down" instruction, resulting in tags doubling the maximum waiting period which increases the chance of successful detection.

The switch-off variation of the Supertag protocol allows tags to be switched off by the reader once the respective tag has been successfully detected. Fast switch-off involves a combination of the fast protocol described above and the switch-off approach. While both of these contribute to a significant reduction in collisions and average faster detection times, it also increases the error rate. The primary problem is the issue of when to allow a switched-off tag to start transmitting again [1]. If the switched off state is terminated at power-on reset, the tag may start transmitting again after moving through a null in the reader field, in the process defeating the objective of reduction in unnecessary tag transmissions. An alternative approach is for tags to be woken up by a reader signal, but due to the unreliability of reader-to-tag communications (following from to the normally non-uniform nature of the reader field), there is no guarantee that a switched-off tag will

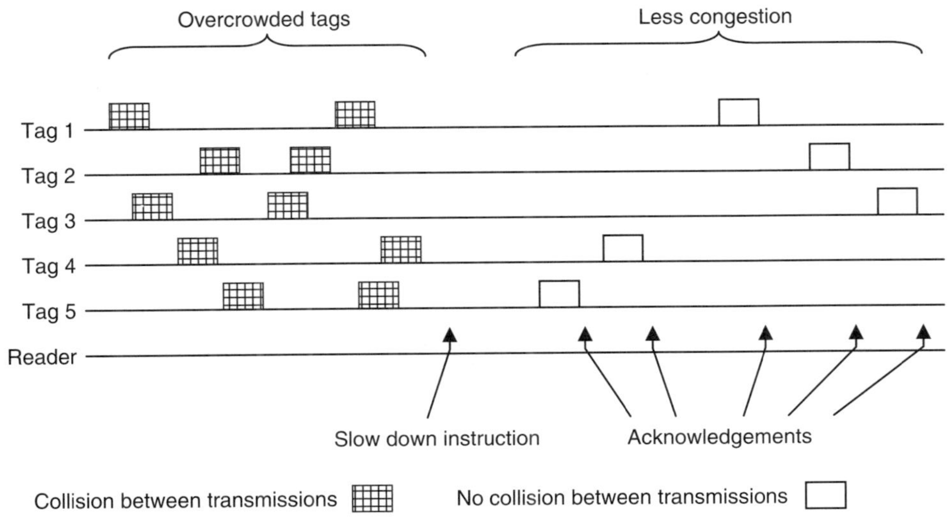

Figure 9.2 Description of dynamic changing of the waiting period under the Supertag protocol [1].

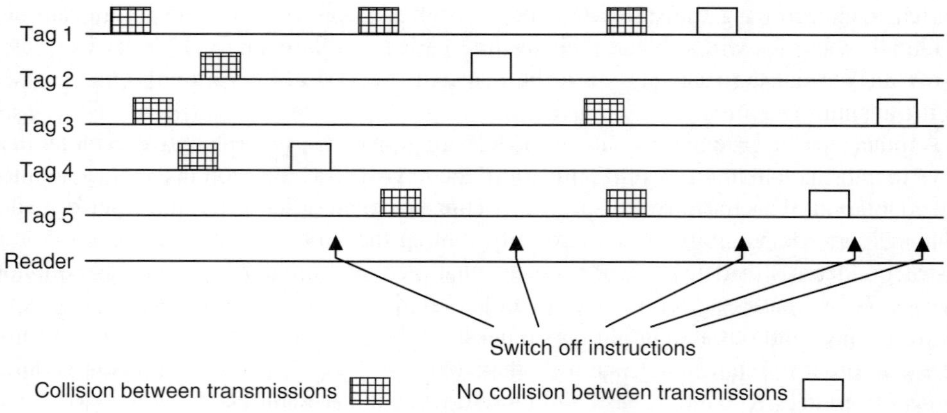

Figure 9.3 Switch-off Supertag protocol sequence [1].

in fact wake up when instructed by a reader. This will result in an unacceptable proportion of tags not being detected at some read station subsequent to being switched off at an earlier read station. Should frequent wake-up calls by the reader be employed to avoid this situation, the TTF protocol will effectively start behaving like an RTF protocol, losing the benefits of operating in TTF mode while not being as effective in other respects as true RTF protocols. Figure 9.3 describes the switch-off mode for a TTF protocol.

Yet another variation provided for by the Supertag patent is for tags to automatically change the waiting time, for example, by doubling the initial maximum allowed waiting time. This may involve the tags incrementing or decrementing the waiting time successively in a number of stages. One option would be to initially transmit with a shorter maximum waiting time, and to subsequently increase the maximum waiting time by repeatedly doubling the period, based on the assumption that at least one of the initial ID transmissions has been successfully detected by the reader. This will lead to a very simple implementation of the TTF protocol, as the reader is not required to instruct transponder to change their behavior, thus retaining all of the inherent benefits of a TTF protocol. The benefits associated with this option are discussed in more detail in the next section.

9.4.2 IP-X

The IP-X protocol, as implemented in a series of transponder chips produced by EM-Microelectronics, can be viewed as a special case of the Supertag protocol as it also implements a free-running unslotted Aloha protocol. IP-X transponders wake up in so-called free-running TTF mode (actually TTO or tags-talk-only, as no communication from the reader is required to read tags), and will start to transmit their IDs at a pseudo-random rate without waiting for any signal from a reader. The pseudo-random delays between subsequent tag ID transmissions are achieved as follows: The transponder ICs are programmed with a maximum delay period. Upon wake-up, the transponder will generate a pseudo-random number between 0 and 1 that is used to determine a pseudo-random fraction of the maximum delay period, and will then transmit its ID after waiting for this pseudo-random time period. In order to initially speed up the rate at which readers will detect such tags, a so-called acceleration version of the Supertag protocol is used. This involves the transponder initially transmitting with a maximum delay period that is an 8th of the eventual maximum delay time for which the chip was programmed. After every second transmission the maximum delay period is doubled until the programmed maximum delay period is reached, after which transponders keep on transmitting at this maximum delay interval. The benefit of this approach is that transponders get the opportunity of being detected very soon after waking up (when the tag is transmitting in accelerated mode), while tags that have already transmitted their IDs slow down in order to reduce the chance of collisions with tags that woke up more recently. The non-uniform nature of the reader field, combined with the fact that spatial diversity will ensure that all tags do not wake up at the same time, assists the IP-X protocol in avoiding collisions during the start-up phase when tags are transmitting in accelerated mode.

This very simplistic anti-collision protocol (which is a special case of one of the simpler options provided for in the Supertag protocol, avoiding instructions from the reader to change the rate of tags transmissions) provides surprisingly good performance when applied either to fast-moving tags or to tag populations of up to 200 tags at a time. In the series of chips developed by IPICO Inc. in cooperation with EM-Microelectronics that support this protocol, provision has been made for four different maximum delay intervals (measured in terms of the number of bits that can fit into the maximum interval): 1 kbits, 4 kbits, 16 kbits and 64 kbits (Table 9.1). This allows transponders to be implemented with different transmission characteristics suited to different applications. The fast variation (1 kbit maximum delay), combined with the acceleration protocol, allows the detection of tags traveling at speeds of theoretically up to 600 km/h. The other extreme (64 kbits maximum delay) can accommodate large numbers of tags given the small tag duty cycle – theoretically up to 800 tags can be allowed before the protocol will saturate, with tags travelling at 5 m/s [22, 23].

Another benefit offered by the TTF approach is the fact that user programmable memory fields, in addition to the tag ID, can be read without any reader communications. Assuming that the tag memory has already been programmed (this would typically happen at the point where tags are issued and associated with specific tagged objects), the tag can be configured to transmit not only its ID in TTO mode, but also additional user-defined memory fields. While this will increase the tag duty cycle – requiring either a reduction in the number of tags or an increase in the maximum delay time – it still offers the benefit

Table 9.1 Different implementations of the IP-X protocol.

Protocol version	Baud rate	Maximum interval
V1	64 k	4 k
V2	64 k	16 k
V3	256 k	4 k
V4	256 k	16 k
V5	256 k	64 k

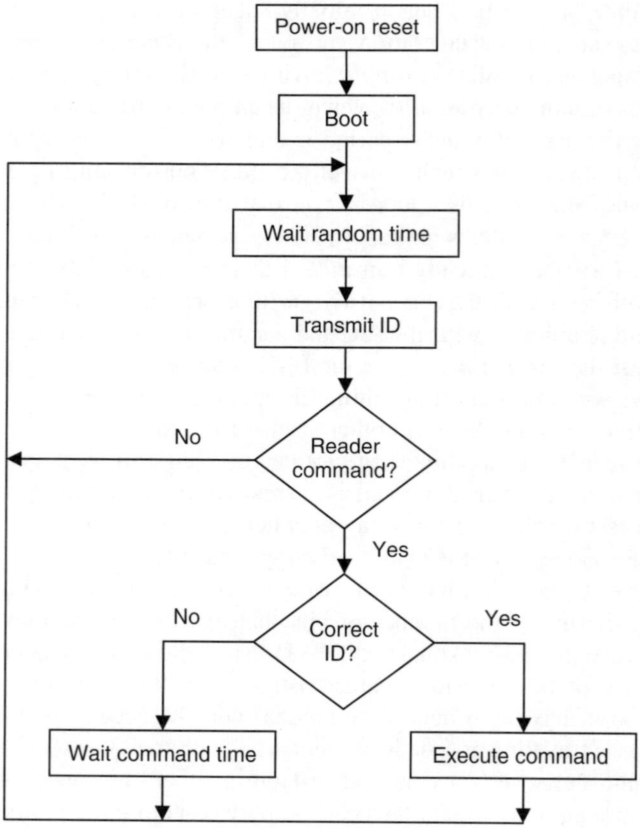

Figure 9.4 iP-X R/W protocol flow diagram. Reproduced from Van Eeden, H.L., "Passive UHF RFID systems," Ph.D. thesis, Northwest University, Potchefstroom, South Africa, 2004.

of operating readers in a mode where only energizing signals have to be transmitted. The fundamental benefit of using very little spectrum for reader operation is therefore retained.

Alternatively, the reader has the option, after successfully detecting a specific tag, to interrupt the anti-collision sequence of all tags through a mute command, and then to communicate with the specific singulated tag by using the last 48 bits of the tag ID as identifier [1], as displayed in Figure 9.4. While this approach provides the user with more flexibility in terms of for example, the specific area of tag memory to address, it also results in the loss of the benefits of communicating in TTO mode as the reader will have to transmit instructions and, as a result, will occupy similar frequency spectrum as is the case for RTF protocols.

9.4.3 TOTAL

The TOTAL protocol ("tag only talks after listening") was proposed to ISO as an addition to the existing set of ISO standards for UHF RFID (i.e. ISO18000-6A, B and C). TOTAL involves the option of adding a TTF protocol, based on the IP-X protocol, to ISO18000-6 compliant readers in addition to one or more of the other ISO18000-6 protocols. The specific motivation for adding the TOTAL option is to make provision for readers that do not generate any reader modulation, and that therefore utilize very little spectrum as such readers only transmit a continuous wave energizing signal when listening for tags.

As TOTAL has to be compliant with the operation of other protocols (such as EPC Gen 2) that, at this stage, are already part of the ISO18000-6 standard, it is not possible to simply operate in the same way as the standard IP-X protocol. The reason for this is the fact that an IP-X tag that is present within a population of RTF tags, and that is energized by an RTF reader, will transmit its ID in stochastic fashion while the RTF reader tries to interrogate the RTF tags. This will result in significant disruption of the operation of the RTF protocol. Practical tests have proven the degree of such disruption to be unacceptable. It was therefore necessary to add an additional aspect to the behavior of standard TTF tags: when energized, the TTF tags must first listen for interrogations from an RTF reader. If such reader communication is detected, the TTF tag must either respond in the required RTF fashion (if it was designed to be a dual-protocol tag) or else it must keep quiet for a prescribed period of time before it can again listen for reader communications. If no reader communication is detected, it can commence to transmit its ID in the same fashion as it would have for a standard TTF tag.

The critical aspect of the TOTAL protocol is obviously the required waiting time for the tag while it is verifying whether an RTF reader is present. Should this time period be too long, the TOTAL protocol will lose one of the primary benefits of a TTF protocol, that is, the ability for a tag to be detected within a very short time period after the tag has been energized. This is a requirement to allow the tag to be read reliably when traveling at very high speeds. If, however, this waiting period is too short, the risk is that an RTF interrogation process may in fact have been ongoing while the TOTAL tag was listening, resulting in disruption of the RTF process as described above.

The main features of the TOTAL proposal that are being considered by the ISO Committee at the present time, can be summarized as follows:

- It is based on the IP-X anti-collision protocol with a LBT (listen-before-talk) feature: tags listen for RTF modulation and remain mute if such modulation is detected in order to minimize interference with RTF applications.
- The minimum listening time is 5 ms after power-up, minimum 25 ms mute time if reader modulation is detected, and a maximum tag ID duty cycle of 2%.
- There is no forward (reader-to-tag) link specification: any of 18000-6 A, B, C or proprietary forward links may be used.
- The return (tag-to-reader) link must be either PPE (pulse position encoding) or Miller encoding at 256 kbit/s.
- Structured or unstructured data is allowed.

If the IP-X options described in the previous section are studied against the background of the above constraints for TOTAL, the following observations can be made regarding the expected performance of a TOTAL tag compared to an IP-X tag:

- The fast version of IP-X (V3 with 4k maximum delay time and 256 kb/s bit rate) has a maximum standard delay time of between 15 and 16 ms, but an initial maximum delay time (when implementing the acceleration part of the protocol) of just below 2 ms.

The prescribed listening time of 5 ms of the TOTAL protocol will therefore make this protocol significantly slower compared to standard IP-X in terms of the ability to detect fast-moving tags.

- The duty cycle of standard IP-X tags programmed according to the fast V3 mode is just more than 2% when responding in standard (non-accelerated) fashion, which is already marginally in excess of the duty cycles that are allowed by TOTAL. The acceleration portion of IP-X will increase this duty cycle by a factor of 8 for the initial two tag transmissions. It is therefore likely that the acceleration portion of IP-X will be deemed to be illegal under TOTAL for tags programmed to run in the 4k mode (as is the case for the V3 version).

The conclusion at this early stage must therefore be that the proposed TOTAL protocol will not be able to achieve quite the same performance for fast-moving tags as the IP-X protocol. At the same time it must be mentioned that the V3 version of IP-X can theoretically detect tags – up to 6 tag reads maximum – at speeds of around 600 km/h (375 mph). A legal implementation of TOTAL, with a total transmission length of 96 bits (including preamble and CRC), baud rate of 256 kb/s, that has an initial listening time of 5 ms and then a maximum delay time of 16 ms to satisfy the 2% duty cycle limit (compared to the initial 2 ms maximum delay of IP-X V3 in accelerated mode) will still be able to read tags at speeds in excess of 200 km/h (125 mph for a maximum of 3 tag reads). If a theoretical limit of maximum 6 tag reads is deemed to be the criterion for reliable detection in potentially high density traffic, the maximum speed for a TOTAL tag will be just above 100 km/h (around 63 mph). This will further deteriorate with the addition of further data pages, over and above the tag ID.

9.4.4 Comparison between Different TTF Protocols

Figure 9.5 displays the typical average reading time distribution of a free-running TTF protocol (i.e. no interruption from the reader to modify the behavior of tags). The long "tail" of the distribution illustrates that the longest possible time to detect a tag is potentially much longer than the average time, which is the primary drawback of this version of TTF protocols.

In order to improve the maximum reading time, different approaches are possible. The Supertag switch-off and fast switch-off protocols will instruct tags that have been detected to stop transmitting, hence reducing the length of the "tail" of the distribution by eliminating unnecessary tag transmissions (see Figures 9.6 and 9.7).

The effect of saturation of TTF protocols with an increasing number of tags is displayed in Figure 9.8. It is clear that for the 64 kb/s version of a free-running protocol, saturation takes place after about 50 tags, while the switch-off protocols do not show significant saturation. In the case of a 256 kb/s implementation of the same protocol, saturation will take place at around 200 tags as the faster bit rate will reduce the tag duty cycle by a factor of 4.

The drawback of the switch-off protocols is, however, the increase in tag error rate that is encountered, resulting from tags being switched off without successful detection (Figure 9.9). As the optimal switch-off period to ensure reliable detection is difficult to implement, this variation of TTF protocols is not suitable for applications requiring read

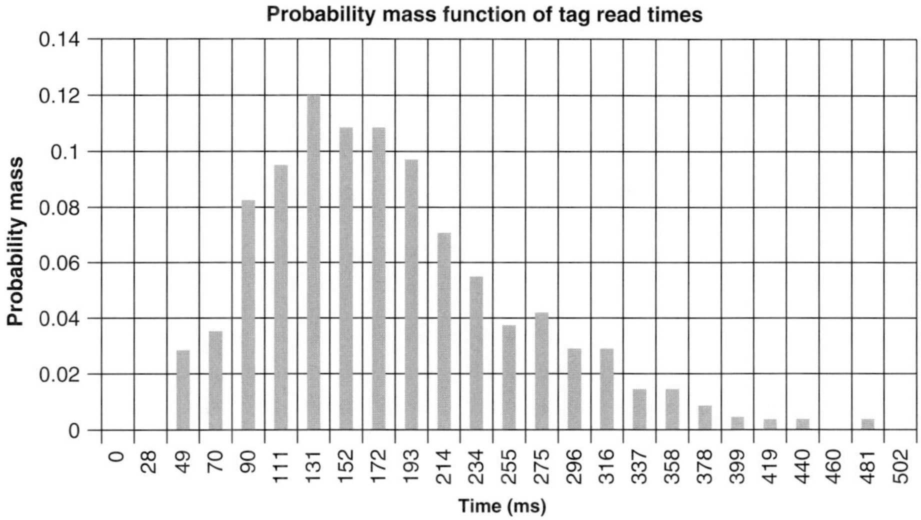

Figure 9.5 Average reading time distribution for the Supertag Free-running protocol (10 tags, 64 k baud rate, 64 k interval) generated using a protocol simulator, with 1,000 samples, resulting in average read time of 172.7 ms, longest read time of 480.9 ms and standard deviation of 70.2 ms. Courtesy of [1]. Reproduced from Van Eeden, H. L., "Passive UHF RFID systems," Ph.D. thesis, Northwest University, Potchefstroom, South Africa, 2004.

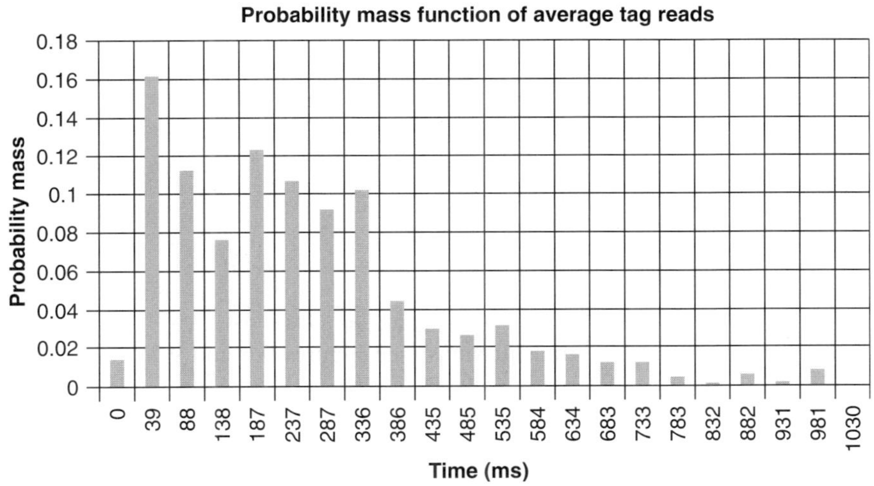

Figure 9.6 Average reading time distribution for the Supertag Switch-off protocol (10 tags, 64 k baud rate, 64 k interval) generated using a protocol simulator, with 1,000 samples, resulting in average read time of 107.0 ms, longest read time of 1030.4 ms and standard deviation of 52.6 ms. Courtesy of [1]. Reproduced from Van Eeden, H. L., "Passive UHF RFID systems," Ph.D. thesis, Northwest University, Potchefstroom, South Africa, 2004.

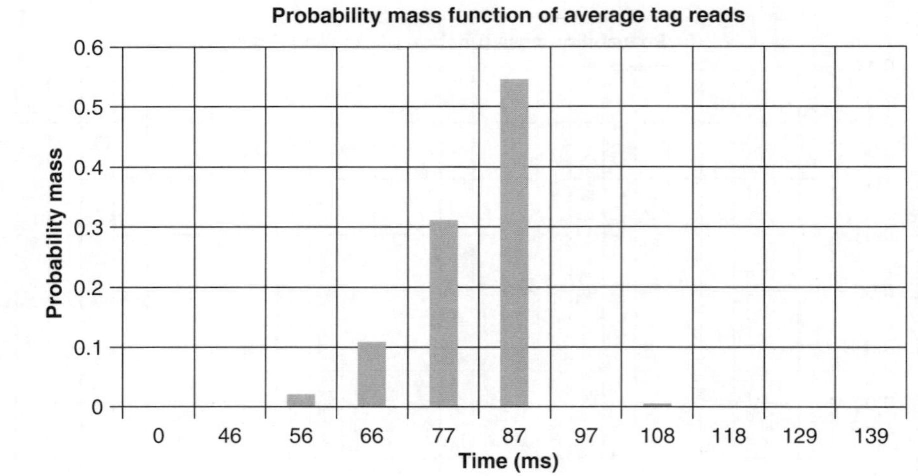

Figure 9.7 Average reading time distribution for the Supertag Fast Switch-off protocol (10 tags, 64 k baud rate, 64 k interval) generated using a protocol simulator, with 1,000 samples, resulting in average read time of 72.6 ms, longest read time of 138.9 ms and standard deviation of 8.1 ms. Courtesy of [1]. Reproduced from Van Eeden, H. L., "Passive UHF RFID systems," Ph.D. thesis, Northwest University, Potchefstroom, South Africa, 2004.

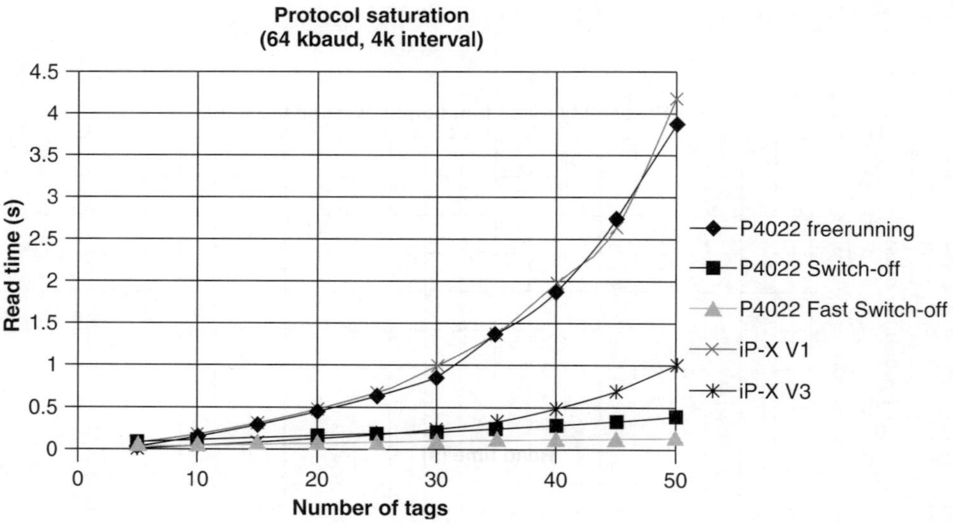

Figure 9.8 Protocol saturation in TTF protocols [22]. © EM Microelectronic-Marin SA.

reliabilities of close to 100%. A better alternative would be to use faster implementations of the IP-X protocol.

Figure 9.10 displays average reading times for IP-X implementations differing with respect to baud rate and maximum delay interval as previously described in Table 9.1. It can be seen that the V3 version, designed for very fast moving tag applications, has a

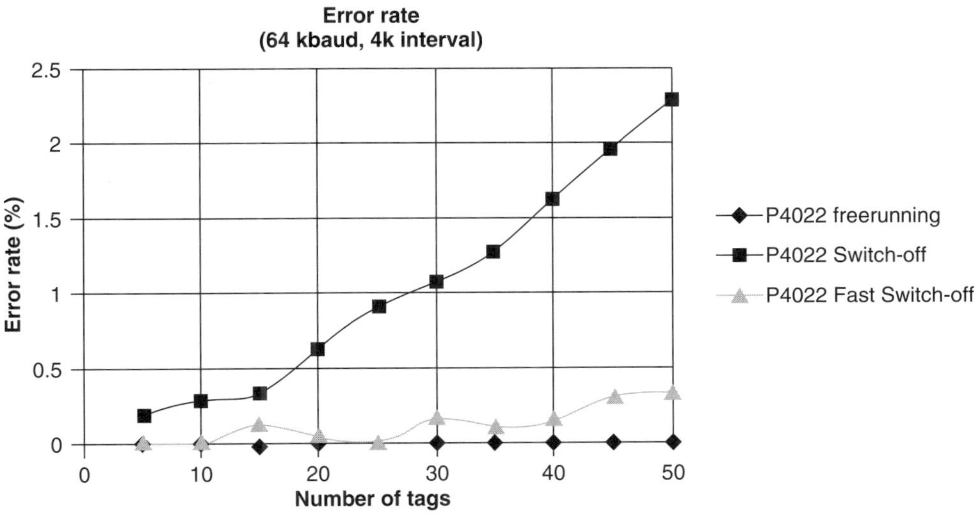

Figure 9.9 Error rates for TTF protocols [22]. © EM Microelectronic-Marin SA.

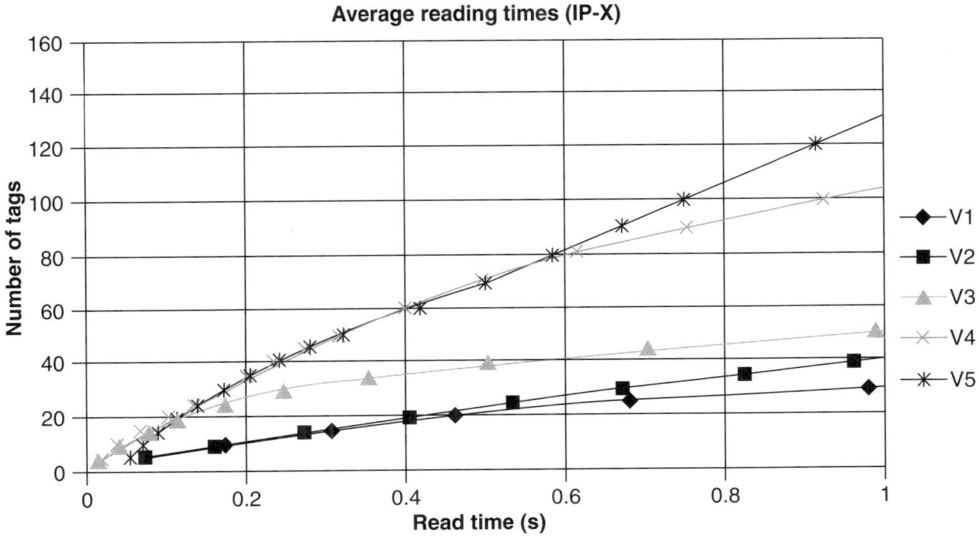

Figure 9.10 Average reading times for different implementations of the iP-X protocol [23]. © EM Microelectronic-Marin SA.

very low reading time for low tag populations but quickly saturates as the number of tags increase beyond about 20. The V4 and V5 versions, designed for large tag populations, only start displaying saturation for much higher numbers of tags. Using these versions of IP-X tags, more than 100 tags can on average be detected in less than one second.

Figure 9.11 displays the maximum read rate using different implementations of the IP-X protocol. For small tag populations, more than 300 tags can be read per second

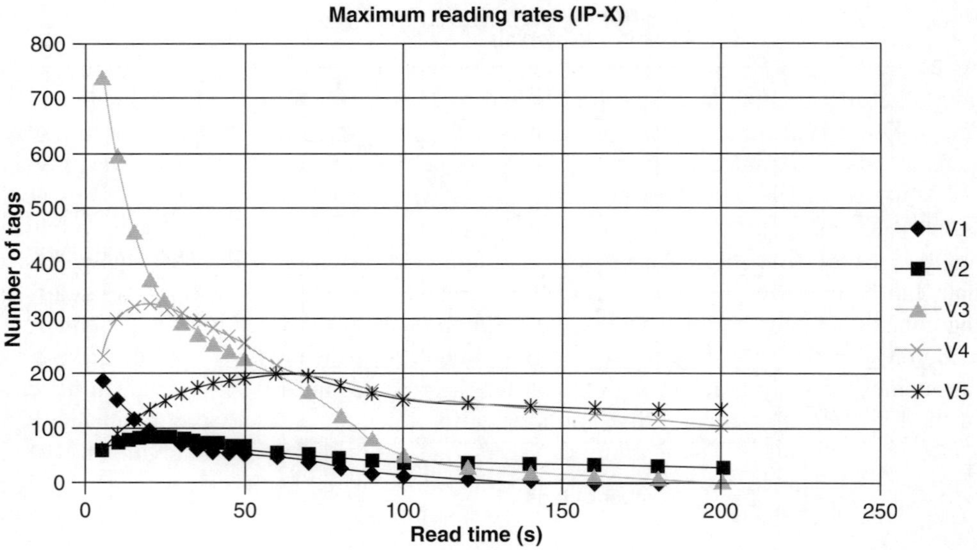

Figure 9.11 Maximum reading rates for the iP-X protocols [23]. © EM Microelectronic-Marin SA.

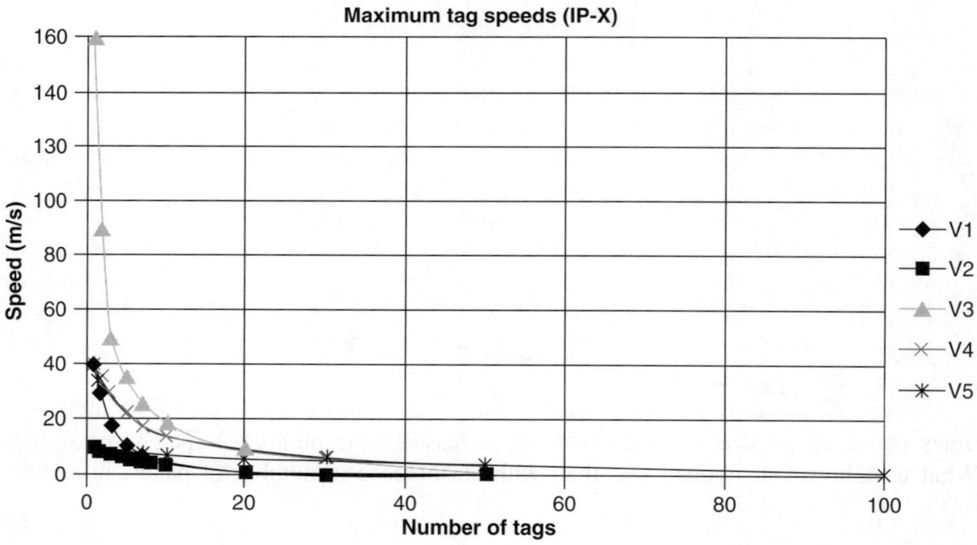

Figure 9.12 Maximum tag speeds for the iP-X protocol [23]. © EM Microelectronic-Marin SA.

using the V3 and V4 versions. Similarly, Figure 9.12 displays the maximum tag speed for different IP-X implementations for different tag populations. If this is converted to km/h, it can be seen that tag speeds in excess of 200 km/h can be handled if only a small number of tags are visible to the reader.

9.4.5 TTF Performance with Additional Data Pages

A feature of the TTF approach is the fact that, in addition to the tag ID, any number of additional memory pages can be transmitted by tags in TTO mode (that is, without any interrogation from the reader). This can be achieved by pre-configuring tags to always transmit a predetermined number of memory pages in addition to the tag ID. This provides TTF protocols with an additional benefit over RTF protocols, as the latter requires additional tag interrogations, over and above the normal operation of the protocol, to extract additional data from tags. An obvious question is: "How much additional user-defined data can be extracted from tags in TTO mode, and to what extent will the reliability of tag communications be impacted by the addition of memory pages to be transmitted?"

Before the experiment involving different numbers of memory pages is described, the tag memory fields provided for by the different protocols must be explained. In the case of the EPC Gen 2 protocol, provision is made for a compulsory user programmable EPC code, for a compulsory tag identifier or TID [20] as well as for optional additional general purpose user-defined memory. While the EPC code will identify the specific type of product to which the tag will be attached, the TID identifies the tag manufacturer, and will typically be hard-coded at the factory to ensure the integrity of tag identities. In the case of the IP-X protocol, only a unique factory programmed tag identifier is compulsory, while additional user programmable memory can be used, for example, to store an EPC code.

A practical experiment was performed using IP-X protocol tags, configured in a mode that allows the transmission of a selectable number of data pages in TTO mode, in addition to the tag ID [4]. In this case, no specific provision was made in the hardware implementation of the protocol on the transponder chips to inform the reader regarding the number of additional data pages that could be expected. This increases the possibility that a reader that has successfully received the tag ID plus some data pages from a particular tag will confuse further data pages with data pages received from a different tag that has also started transmitting. For each experiment, the tags were configured to each transmit its tag ID plus a specific number of additional pages, with repetition of the experiment for different numbers of additional pages, this number ranging from 0 to 5. Three tags at a time, configured in this way, were exposed to a reader for 60 seconds, and the number of times that each page was successfully received was recorded. Figure 9.13 displays the results that were achieved (repeat rate indicates the number of times the respective data page was received).

As could be expected, there is a more or less proportional decline in the number of times that each page is received with an increase in the number of pages transmitted. What was, however, noticed was that, with an increase in number of pages, in addition to the tag ID, a marked reduction in the reliability of detection of pages 4 and 5 was observed. This is evident from Figure 9.14, where all pages were detected with more than 98% reliability compared with the tag ID for up to 3 additional pages, whereas this figure dropped below 70% for the 5th of 5 additional pages. This is likely to be a result of the fact that the additional data pages contain no information linking them to a particular tag ID – it is up to the reader to make this association based on the time differences between consecutive pages that are detected after the initial detection of a tag ID. As more than one tag is present, the probability of another tag starting with a transmission while the first tag is still transmitting some of its data will obviously increase with the number of additional pages transmitted. The longer the time lag between the tag ID and

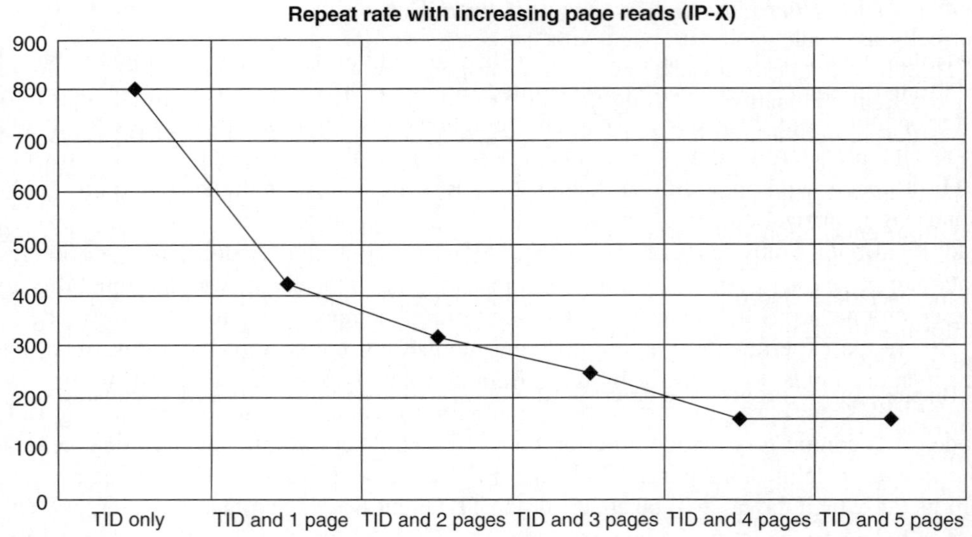

Figure 9.13 Average repeat rate of tag backscatter transmissions with an increase in the number of user pages. Courtesy of Naudé and Marais [4]. © 2008 Northwest University.

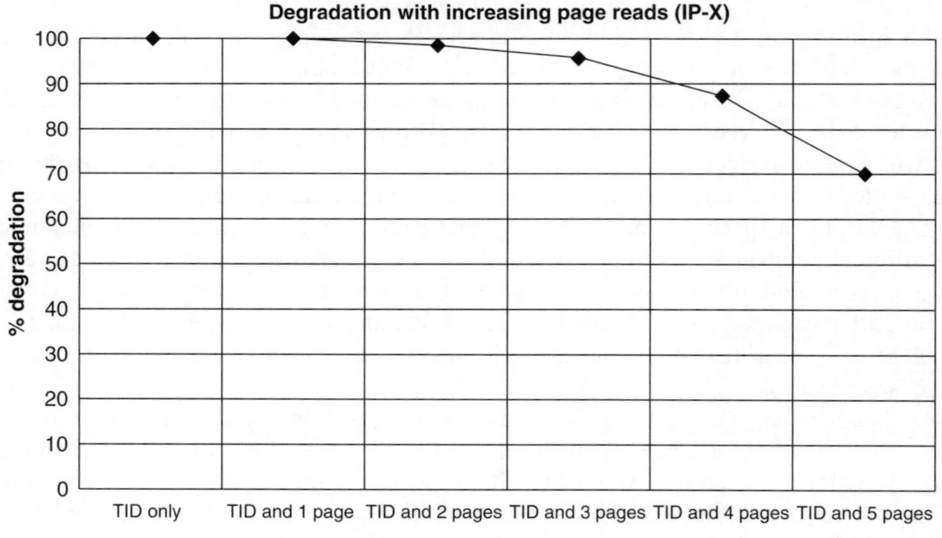

Figure 9.14 Degradation of read performance with an increase in the number of user data pages. Courtesy of Naudé and Marais [4]. © 2008 Northwest University.

an associated data page, the higher the likelihood that the reader will not associate the additional page with the correct tag ID due to data from other tags being received in the meantime. It would therefore seem that, for collections of 3 tags at a time, 3 additional pages is the maximum number that should be used in practical applications if a read reliability of more than 98% is required. With an increase in size of tag population, this number will decrease even further unless a more elegant and robust approach is used to correctly associate each data page with the correct tag ID.

The obvious question is: "How many additional data pages will normally be required for practical applications?" While the tag ID will suffice for many applications, there are cases that require a high level of security where it will be beneficial to extract encrypted codes from tags at high speed (e.g. in the detection of tags that form part of vehicle license discs, where reads must take place at speed to identify illegal vehicles). Given the length of encryption keys of adequate strength, the number of bits to be extracted per tag may be as high as 256, which in this case would require 4 pages if no CRCs are included in the user data, and 6 pages if CRCs are provided for. Such applications may therefore suffer from marginal read performance as the read rates reflected in Figures 9.13 and 9.14 are expected to significantly deteriorate in dense traffic environments.

Because of the progressive deterioration observed in the reliability of tag transmissions with the increase in the number of additional data pages, the proposed TOTAL standard provides for two mechanisms aimed at improving the reliability of reading additional data pages, namely, (1) CRC bits to be included in each data page (to allow the reader to verify if a page has been successfully received); and (2) for so-called link bits that inform the reader about the number of data pages that should be expected in addition to the tag ID. The availability of CRC and link bits will enable the reader to at least be aware if it should be searching for additional data pages in subsequent free-running transmissions from the tag, should it not have successfully received all data pages, possibly due to collisions with transmissions from other tags.

9.5 Comparison between ISO18000-6C and TTF Protocols

The ISO18000-6C protocol (better known as EPC Class 1 Generation 2) is currently the UHF RFID protocol that has achieved the widest adoption in the global RFID community. The purpose for which this protocol was developed, was to provide a universal standard that would be suitable for all "open loop" applications of UHF RFID, that is, applications where systems used by different end-users and supplied by different vendors must be fully interoperable. The two most prominent application domains for UHF RFID are supply chain management and transportation or traffic management. Both of these tend in general to be "open loop" in nature, with potentially large numbers of independent stakeholders that could benefit from accessing the same RFID data, hence justifying a global standard.

9.5.1 Areas of Comparison

ISO18000-6C, like all other RTF protocols, potentially suffers from a number of deficiencies, both in terms of performance as well as in terms of the ability to deploy large systems while complying with frequency spectrum regulations. Three prominent areas of performance can be mentioned where RTF protocols can potentially suffer from deficiencies when compared with TTF protocols:

1. **The reliability of reading tags in large populations, when moving through very non-uniform RF fields:** The potentially inferior performance of RTF protocols under

such conditions is caused by the much more complex process flow required for successful execution of an RTF protocol compared to the much simpler TTF protocols. This can be appreciated by comparing the state diagram for a typical free-running TTF protocol (Figure 9.15) with the state diagram for the ISO18000-6C protocol in Figure 9.16. The much more complicated RTF state diagram implies that the transponder chip must be able to remember in which of a number of states it last was when it has to respond to the next reader command. While this will not be an issue in relatively uniform RF fields, it does become an issue in non-uniform fields where the tag not only can temporarily lose its communication link with the reader when moving through a weak spot in the field, but can even switch off and wake up again without knowing what its state last was in terms of the execution of the protocol. This can result in tags responding in the wrong way or not responding at all, with the implication that tags may pass undetected even though they were visible to the reader for a significant period of time. In comparison, a TTF tag only has to be seen once for a very short period of time, during which period it can transmit its ID in one brief communication packet, without having to wait for the completion of a fairly complex interrogation process where potentially hundreds of tags must respond in coordinated fashion for the detection process to be completed successfully. While earlier versions of RTF protocols definitely suffered from this phenomenon, the question is to which extent the more sophisticated EPC Gen 2 protocol managed to address this weakness of earlier versions. The other question is to what extent improvements in technology have made UHF RFID protocols more immune against the fluctuation nature of RF

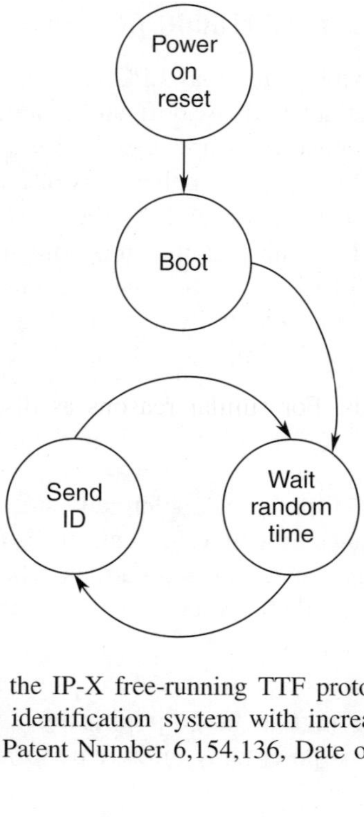

Figure 9.15 State diagram for the IP-X free-running TTF protocol [1]. Reproduced from Van Eeden, H.L., "Free running RF identification system with increasing average inter-transmission intervals," United States Patent, Patent Number 6,154,136, Date of patent: November 28, 2000.

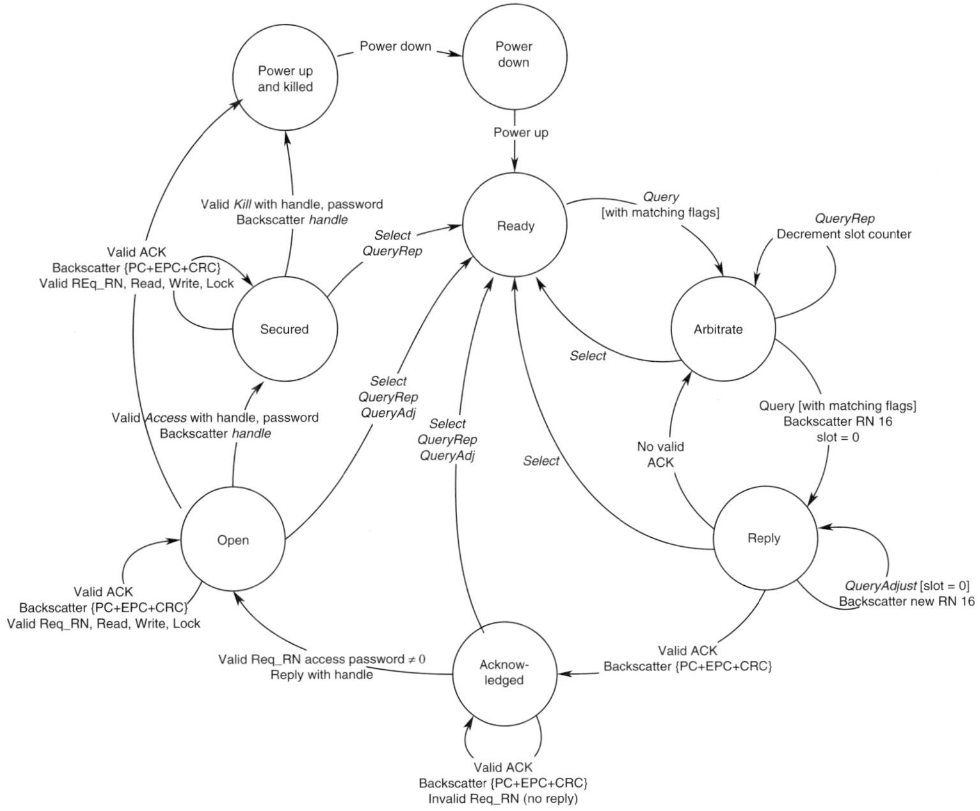

Figure 9.16 State diagram for EPC as adapted from the ISO18000-6C standard [20]. Reproduced from EPC™ Radio Frequency Identity Protocols, Class-1 Generation-2 UHF RFID, Protocol for Communications at 860–960 MHz, Version 1.0.9, January 2005.

fields – more specifically the significant reduction in RF field strength that more recent tag versions need to wake up and stay awake compared to earlier versions.

2. **The ability to reliably detect tags moving at high speeds:** The second area of potential weakness for RTF protocols compared to TTF is the ability to reliably detect tags moving at high speeds. For similar reasons as discussed above, RTF protocols tend to take a longer period of time to complete an entire interrogation cycle compared to TTF protocols, where tags respond in their own time in quick bursts with low duty cycle. The comparison that must be done in this case is to measure the time required for an RTF reader to successfully search through a tag population of specific size, against the time required for a similar number of TTF tags to be successfully detected against the background of possible tag collisions. While both supply chain applications (involving large numbers of tags moving at relatively low speeds) and traffic applications (involving small numbers of tags moving at high speeds) can place severe constraints on RFID protocols, the scenario for which a comparison is easiest is the case of a small number of tags moving at high speeds, which allows the reader

limited time in which to successfully detect tags. In the case of traffic law enforcement, the emphasis would be on detecting vehicles moving at speeds high above the legal speed limit and to do so in dense traffic environments.

3. **The spectrum requirements to support the respective protocols:** The third area for comparison, and possibly the most fundamental limitation on the ability to successfully deploy large RFID applications, is the spectrum requirements to support the respective protocols. As explained in Section 9.4, an inherent benefit of free-running TTF protocols is their ability to detect tags without requiring interrogations from the reader, resulting in very limited spectrum requirements. This comes at a cost, though, since the RF sections of the readers have to be more robust and sufficiently accurate in terms of frequency in order to co-exist. This attribute will allow the operation of potentially large numbers of readers in the same channel and at the same location. Doing the same with RTF readers can potentially result in overlap between the spectral content of reader communications and the spectral content of unrelated tag communications (i.e. tags communicating to other readers in the immediate vicinity), which can be relieved by physically synchronizing RTF readers, which may not always be practical. If the regulatory limitations on RFID channels do not allow the complete spectral separation of tag and reader communications, the implication is that only one reader will be able to operate in a channel and within an area of such size that other reader signals will be of the same order of magnitude or larger than tag signals. The physical size of such an area can be determined by calculating the reduction in secondary reader power density at the primary reader with increase in distance between them (using the $1/r^2$ rule), and comparing this power level with the typical power received from a tag at the primary reader. With no specific measures to limit the size of unwanted reader signals in the tag communication band (i.e. if the reader communication spectral band fully overlaps with the tag communication band), the required separation to prevent interference can be as large as 7 km [1]. Such a situation will make the deployment of large RFID systems totally impractical, against the background that some regulatory jurisdictions, for example, ETSI, only provide for as few as 10 channels. For this reason, the ISO18000-6C standard provides for different types of spectral masking applicable to reader signals, ensuring that unwanted reader radiation will fall below the level of tag signals for much smaller distances between different readers.

9.5.2 The Impact of Progress on Technology

During the early years of UHF RFID the factors discussed in the previous section provided TTF protocols with an undeniable advantage over RTF protocols, even taking into account potential issues around the undetermined maximum time period to detect a TTF tag. An early implementation of the IP-X protocol was the first to be applied in large applications in the European regulatory jurisdiction, based on the fact that the limited spectrum availability did not impact on the operation of this protocol. During this same period (2002–2005), the early version EPC protocols could not effectively operate outside of FCC regulatory areas, as the nature of their reader communications made these protocols illegal to operate in Europe and in other jurisdictions with similar spectrum regulations.

A number of important technological advances have, however, taken place in the past 10 years, resulting in a drastic change in the UHF RFID landscape. A number of these developments and their implications for the comparative situation of RTF versus TTF will be discussed below.

9.5.2.1 Improvements in Silicon Technology

The more simplistic logic of TTF protocols compared to RTF protocols results in TTF transponder chips requiring a factor of 3 to 4 times fewer transistors to implement the protocol (typically 3,000–4,000 transistors for a TTF chip compared to approximately 12,000 chips used in an EPC Gen 2 chip). This results in smaller size, lower energy consumption and lower cost. This was also the primary reason for early commercial successes for TTF protocols in Europe. TTF tags implemented in 0.5 micron processes were approximately 0.25 mm^2 in size compared to RTF chips with areas of around 1 mm^2. The 0.5 mm × 0.5 mm TTF chip sizes were ideal for flip-chip manufacturing operations, as this was about the smallest size that inlay production lines could handle. They required less power to operate, resulting in more reliable operation in the presence of weak and fluctuating RF power, as they switched on in weaker fields (requiring around −1 dBm in power to operate, about 3 dB less than their RTF counterparts).

Over the past five years three significant technological advances have changed this scenario:

- Silicon processes moved from 0.5 micron to 0.35 micron and then on to 0.18 micron process technology. The immediate implication for RFID chips was a reduction in size almost by a factor of 10, resulting in even more complex new RTF chips (e.g. EPC Gen 2 chips) shrinking in size to the inlay manufacturing limit of around 0.5 mm × 0.5 mm. TTF chips did not benefit to the same degree as they were already approaching optimal size in the older 0.5 micron technology. The cost benefit of TTF was therefore reduced.
- Voltage levels for transponder chips to switch on dropped from around 2 V to less than 1 V, resulting in tag power-up levels to drop to −12 dBm and in some cases even less. This development largely eliminated the unsatisfactory behavior of RTF protocols in the presence of fluctuating RF fields, as the tendency of tags to switch off was greatly reduced in the type of RF fields practically experienced by tags.
- Transponder antenna technology became more sophisticated, with the old model linear dipole type antenna being replaced by much more sophisticated designs that further improved the performance of tags in non-ideal environments. This includes more robustness with respect to environmental change, wider bandwidths than what dipoles antennas can achieve, and a reduction in size as opposed to dipoles. This improved performance resulted in more reliable operation of the complex RTF protocols. Once again, the much simpler TTF protocols did not benefit to the same degree from these developments, as they already performed fairly reliably (within their own set of limitations) with older technology.

9.5.2.2 Improvements in DSP Technology

As is evident from the above discussions, RTF protocols will suffer severely from reader interference if a scheme is not devised to effectively separate reader communication signals from tag signals. As discussed before, such separation can take place either in the time or in the frequency domain. Without such separation, the problem can be as bad as requiring physical separation of unpractical proportions between co-channel readers, making the deployment of large systems impractical. As early versions of the EPC Gen 2 protocol did not provide for proper spectral separation between reader and tag signals for co-channel readers, efforts in Europe were aimed at deploying systems based on time separation using LBT (listen-before- talk) measures. These, however, met with limited success as it implied that readers could only communicate with very low duty cycles, less than what is required by either supply chain or traffic management systems. These low duty cycles result from the fact that most of the time an LBT reader, that would like to interrogate tags, will have to keep quiet due to activity detected from other readers in all available LBT channels.

It is therefore clear that large deployments of RTF-based protocols are only possible if sufficient spectral separation can be achieved between reader and tag signals – a form of frequency multiplexing. The effective separation of reader and tag communications requires high performance filters that can only be implemented using digital signal processing techniques. These filters must typically satisfy strict spectral requirements, which implies not only a sharp cut-off but also very low ripple in stop bands by smoothing the shape of pulses as required by the respective modulation technique. The real-time implementation of DSP techniques for such applications and for sampling rates of typically 5 Msamples/sec is beyond the capabilities of low-end DSP processors, and requires implementation of the DSP algorithms on FPGA chips of sufficient capacity. While such FPGA components would have added significantly to the cost of a reader 5 to 10 years ago, is does not currently provide any significant obstacles in terms of implementing high performance readers at acceptable cost. It was therefore practically possible for the ISO18000-6C standard to incorporate spectrum masks that allow separation of 60 dB between reader and tag communications, even for 200 kHz channels as required by ETSI regulations. Readers that are compliant with the so-called "EPC dense interrogator mode" can therefore operate within ETSI regulations and still achieve 60 dB suppression of unwanted reader signals in the tag communication band. This drastically reduces the required physical separation between readers to prevent reader interference by as much as a factor of 1000, making practical reader co-existence a real possibility.

9.5.2.3 More Intelligent Utilization of the Available Spectrum

The first attempts at creating standards for UHF RFID did not meet much success, partly because those efforts tried to achieve too much with a single uniform standard. The drastic differences between UHF RFID spectrum allocation in different regions alone were reason enough why this was not an achievable objective. The more thorough approach followed in the compilation of the ISO18000-6C protocol addressed this issue by allowing for several different options in the same standard, with the optimal choice for a specific jurisdiction depending on the spectrum that is available as well as on the type of RFID deployments that are expected to occur:

- Single interrogator mode is based on the assumption that there are a sufficient number of independent channels available so that a reader is unlikely to encounter other readers in adjacent channels. It therefore does not impose any requirements on a reader, in addition to those already imposed by the regulatory authority.

- Multiple-interrogator mode is designed for those cases where the number of simultaneous collocated readers is modest compared to the number of available channels, and imposes some constraints on the transmitted spectrum, sufficient to minimize interference in adjacent or second-adjacent channels [5].
- Dense-interrogator mode is designed to allow successful tag reading even when every channel is occupied by a reader.

Another innovation in ISO18000-6C to limit reader interference is the use of Miller-modulated subcarrier (MMS) encoding. Miller encoding has the effect of displacing the tag spectrum on a subcarrier away from the carrier. The benefit of this approach is to increase the spectral separation between reader and tag signals, in the process reducing the possibility of interference between tag signals and unwanted reader signals. In the ETSI regulatory environment a scheme has already been proposed whereby readers may only occupy every third reader channel in order to limit adjacent channel interference. The implication is that the associated tag communications will effectively take place in channels where no reader communications are allowed (as the MMS technique will effectively displace the tag spectrum from the reader carrier into the adjacent channels). Any number of readers can therefore operate in the designated reader channels, as their signals will have no spectral overlap with tag spectra occurring in the adjacent channels. The problems of reader interference and of too few available channels are effectively eliminated in the process. Shown in Figure 9.17 is a 25 us TARI spectrum and ETSI spectral mask at 30 dBm output. The spectral mask was scaled for 30 dBm reference power at the output connector of the reader vs 33 dBm ERP that is normally used. All ETSI power measurements, both spurious and intentional, are specified as ERP and the power at the connector of the reader would be 3 dB lower than the normally radiated power. Thus, 3 dB must be subtracted from the spectral limits (e.g. −36 dBm becomes −39 dBm). The continuous wave is shown as a reference.

9.5.3 A Comparison between RTF and TTF for Fast Moving Tags

In the domain of ISO standards, RTF-based protocols have until now been preferred over TTF-based protocols within the global RFID community. The primary reason for this is most likely the fact that RTF-based protocols provide more flexibility in handling a variety of scenarios for reading tags. The fact that the reader can instruct tags how to respond allows the RFID system to moderate tag behavior in a way which suits the requirements of any specific read scenario. In contrast, TTF-based protocols require the system to be designed around a more rigid set of rules, as tags must be configured prior to entering the system to respond in a specific way, while readers can only observe the presence of tags, without being allowed to respond to potentially dynamic system requirements.

As described in earlier sections, the TTF approach does offer specific benefits to compensate for a somewhat less flexible system response. The primary benefits will be summarized again for the sake of completeness:

1. More robust behavior in the presence of fluctuating RF field strengths.
2. Faster tag response times, allowing tags to be read at higher speeds.
3. Less spectrum required per reader, allowing a larger number of readers to be deployed within the same set of regulatory constraints.

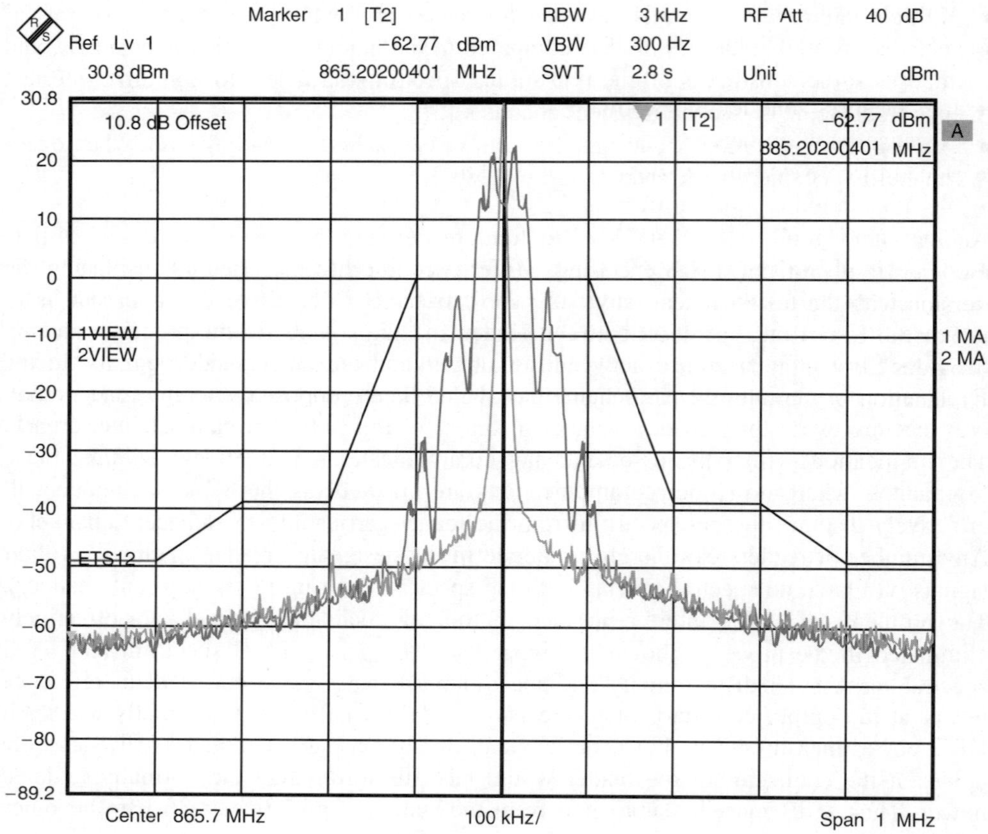

Figure 9.17 Spectrum of a Miller-encoded EPC reader with a spectral mask. Courtesy of Impinj, Inc.

In order to compare the performance of RTF and TTF protocols against at least some of the above claimed benefits for TTF protocols, a simulation was conducted using the CISC RFID protocol simulator. The simulation was aimed at determining the maximum speed at which a tag can be reliably detected using either an RTF or a TTF protocol. As EPC Gen 2 or ISO18000-6C is currently the RTF standard enjoying most support, this was the choice of RTF protocol included in the simulations. To represent current contenders for TTF protocols, IP-X V3 (non-standards based) and the proposed TOTAL protocol (potentially part of the ISO18000-6 standard) were included.

9.5.3.1 Set-Up and Procedure for Simulations

A conveyor belt set-up was used to simulate tags moving at specific speeds. The simulations were done so that only a single tag is present within the interrogation zone at a specific time. Multipath effects were not taken into account, thus the assumption was made that the tag entering the interrogation zone will always be energized.

The physical read scenario applicable to both the IP-X and the EPC Gen 2 protocols was set up using default settings with the following changes:

- Interrogation zone length = 4 m.
- Distance between tags = 4 m.
- Speed of conveyor belt varies between 60 km/h to 250 km/h.
- Tag population comprised of 4 tags.

For the IP-X protocol, the V3 2kbaud version was used for the evaluation, that is, the version with the fastest repeat rate that is most suitable for high speed applications.

For the EPC Gen 2 protocol, some additional settings were required as this protocol provides for several parameters to be adjusted based on the expected read scenario. A short explanation of the relevant parameters is provided before the selected values are given:

- Q: a parameter that the reader uses to regulate the probability of a tag response. Q is an integer in the range 0–15; the probability of a tag responding in the next slot equals 2^{-Q} [20].
- C: a value that is added to the floating point version of Q in order to adjust the probability of a tag response [20].
- Data link rates: these are the bit rates of communication for the downward link (reader-to-tag) and the upward link (tag-to-reader) in kbits/s.

Parameter values used for this evaluation were:

- Data link rates evaluated: 128/640, 40/160, 30/40, 32/64 and 32/53.
- Starting Q values evaluated: 0, 3, 6, 9, 12 and 15.
- C values evaluated: 0.1 to 0.9 in intervals of 0.1 (where applicable).

The EPC parameters described in the set-up section above were evaluated for each combination of speed of the conveyor belt, algorithm, data link rate, starting Q, and C. From these simulations, the EPC combination that performed the best at each movement speed was chosen for comparison with IP-X.

Figure 9.18 displays the number of singulations for EPC Gen 2 tags at different tag speeds and for different link rates. It is clear that the 128/640 link rate (128 kb/s for reader downlink, 640 kb/s for tag uplink) provides more than adequate performance for speeds of up to 250 km/h. The 40/160 link rate also provides acceptable performance for speeds up to 160 km/h, but starts failing as from 200 km/h. The slower link rates provide acceptable performance up to about 120 km/h, but are down to only one tag singulation at a speed of 250 km/h, which will result in unreliable performance in practical scenarios.

Next an experiment was performed using a population of 4 tags and using different link rates. The result in Figure 9.19 shows that only the 128/640 link rate was able to detect all 4 tags for speeds in excess of 200 km/h.

Figure 9.20 compares the most successful of the EPC Gen 2 options (128/640 link rate) with a pure IP-X implementation of the TOTAL concept (to be contrasted with a TOTAL implementation that is fully in line with the current ISO proposal). It is clear that, while both protocols successfully detected all tags for speeds of up to 250 km/h, the IP-X

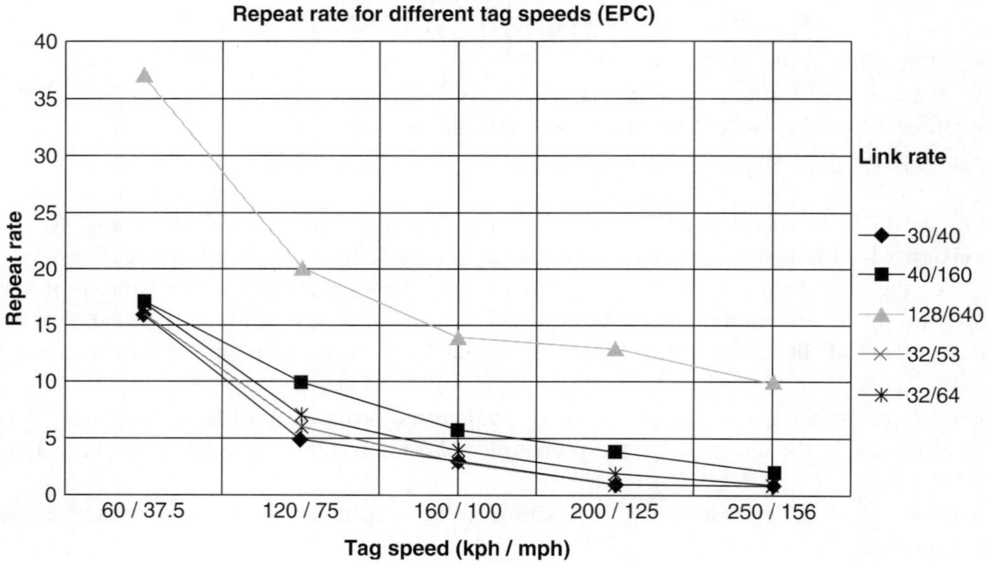

Figure 9.18 Total number of EPC tag singulations for different link rates.

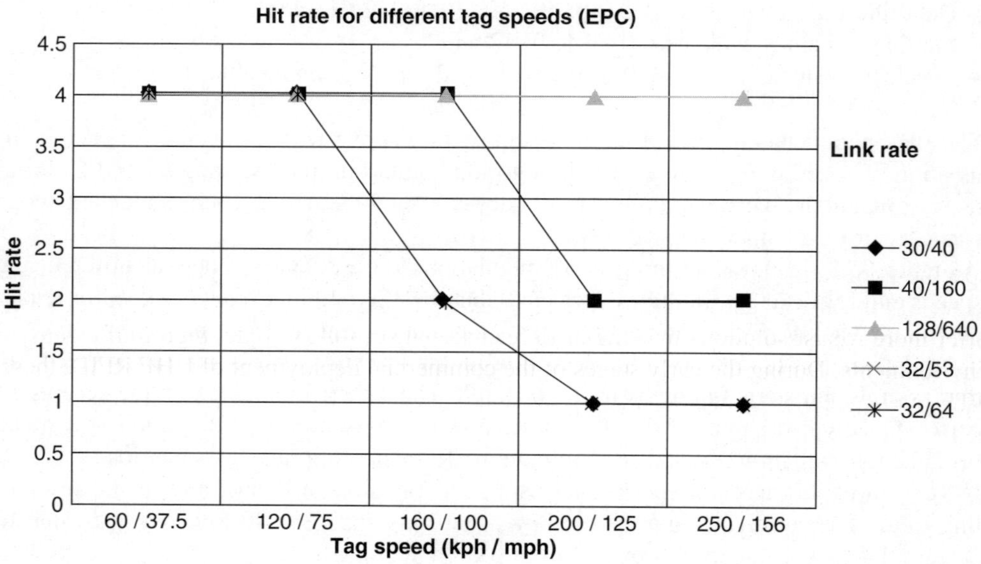

Figure 9.19 Number of tags detected from population of 4 using EPC Gen 2 with different link rates.

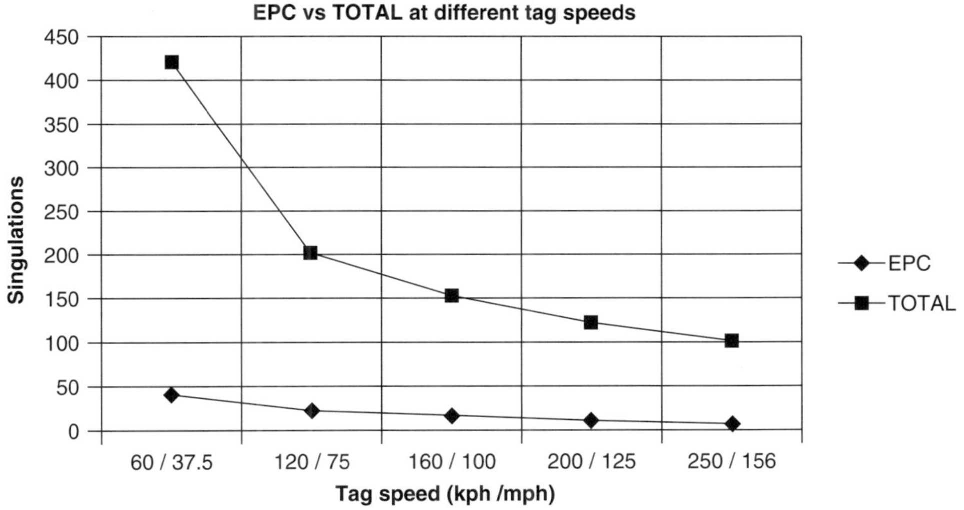

Figure 9.20 Comparison of EPC and IP-X in terms of total number of tag singulations.

protocol outperforms all EPC Gen 2 options, achieving about 10 times the number of tag singulations compared to EPC Gen 2. This will result in more robust performance in dense traffic environments where the practical number of tag singulations tends to decrease significantly compared to the theoretical figure. A fully legal TOTAL implementation is, however, expected to perform similarly to EPC Gen 2 using a 128/640 link rate, as it will have to exclude the IP-X acceleration protocol.

9.6 Conclusion

This chapter provided an overview of TTF and RTF protocols and performed a comparison of these protocols both in philosophical and empirical terms. The above comparison shows that protocols other than those currently included in the ISO18000-6 standard for UHF RFID do have some interesting properties that justify more than passing attention. For specific applications, most significantly in the case of small populations of fast moving tags, TTF protocols can outperform RTF protocols in some respects, and may prove to offer more robust solutions in difficult to read scenarios, for example, high traffic density environments. During the early stages of the commercial deployment of UHF RFID, these TTF implementations indeed had a clear edge over the RTF offerings that were available at the time, both in terms of performance as well as cost and robustness in practical settings. The inability of the standards community to find an acceptable solution for the deployment of UHF RFID in regulatory jurisdictions similar to ETSI, at some stage created a situation where TTF alternatives came close to becoming a *de facto* standard for such jurisdictions.

Over the past three to five years, however, it would seem that a combination of technological advances as well as more effective utilization of available frequency spectrum by proposed standards, have turned the tables and have put RTF protocols firmly back in a dominant position. While the inherent benefits of TTF-based protocols, most notably simplicity of implementation, very fast response time by tags and very little spectrum requirement for readers, still stand firm, it appears as if RTF protocols are now performing sufficiently well in these respects as a more flexible approach to a wide variety of applications. The proposed TOTAL addition to ISO18000-6C represents a last-ditch effort by a TTF protocol to become part of the global standard, although the philosophy of TTF was adapted in the process. The restrictions placed on TOTAL will, however, remove some of the benefits that IP-X, as strongest TTF contender, still offers over the existing ISO18000-6 protocols, and may result in TOTAL finding limited practical adoption. It is therefore likely that TTF standards will be mostly restricted to niche applications where the specific functional advantages offered by such protocols can outweigh the disadvantages associated with non-standardized or proprietary technologies.

Problems

1. Discuss the applicability of passive LF (low frequency), passive HF (high frequency), passive UHF as well as active UHF RFID technology to solve the following automated identification problems:
   - Reading bolus tags that are swallowed by livestock and that remains in the stomach of the animal for its entire lifetime.
   - Bulk reading of tags embedded into the caps of bottles filled with pharmaceutical products, typically containing liquids and being stacked multiple deep and multiple high.
   - Identifying tagged vehicles moving through lanes at a multiple-lane customs gate, requiring the verification of the specific lane that each vehicle moved through.
   - Identifying tagged shipping containers at a depot, requiring the implementation of an electronic stock-take of containers at the press of a button.
2. Identify the different limiting factors that determine the maximum read range of a passive RFID transponder, and describe the circumstances that will cause each of these factors to become the dominant determinant of read range.
3. Discuss the different criteria that a passive UHF RFID air-interface protocol should comply with, and describe the differences between these criteria for transportation (vehicle tagging) versus goods supply chain (item tagging) applications.
4. Consider an application of passive UHF RFID in the field of open-lane tolling of vehicles traveling on a highway. In ideal circumstances it has been determined that a reader can read tags at a distance of 10 m. The maximum speed of travel at which tags must be read is 220 km/h. Assume that the reader antennas are mounted 6 m from the road surface and are facing directly downward with a viewing angle of 60°. Determine the minimum period of time that a reader will have to identify a transponder on a passing vehicle. Discuss the implication of these results on the requirements for the air-interface protocol.
5. Consider an application of passive UHF RFID in the field of item level tagging of goods in a warehouse. Trolleys are stacked with up to 24 bins, each containing 4 manufactured items. The trolleys are moving through dock doors that are 3 m in width, and the UHF RFID readers installed at the dock doors must accurately detect

all tagged items on the trolleys. If the readers employ the IP-X V4 air-interface protocol and the reader beams can energize tags over a minimum width of 1m in the doorway, calculate the maximum speed at which trolleys may be allowed to move through the dock doors to ensure that all items are read.

6. Discuss the benefits and limitations of the Supertag Switch-off protocol, compared to the IP-X free-running protocol, in terms of the saturation levels of the protocol as well as the time required to identify all tags in a population.
7. Discuss the inherent benefits and drawbacks of RTF versus TTF protocols, taking into account regulatory requirements in different jurisdictions, interference between different readers, maximum number of tags in a population that can be handled, as well as maximum allowed speed of travel of tags.
8. Describe the motivation for the use of TOTAL (tag only talks after listening) variation in the use of the TTF free-running protocol, in applications that require the mixing of RTF and TTF tags in the same population. Discuss how an optimal waiting time for the TOTAL protocol can be arrived at. Also discuss the implications of this waiting time on the expected performance level of the system employing the TOTAL TTF protocol, compared to a TTF protocol that does not have to employ the TOTAL approach.
9. A UHF RFID application requires the generation of user-defined information to be written to and read from transponders. The designer of the system must make recommendations regarding the required read scenarios for the writing of user-defined data to transponders, the reading back of only the transponder ID, and the reading back of the transponder ID plus user defined data. Discuss the different factors that will have to be considered in arriving at a reliable system design, and compare these considerations for a system using an RTF versus a system using a TTF air-interface protocol.
10. Discuss the impact that progress in electronic technology has had on the performance that can be achieved by UHF RFID systems employing different types of air-interface protocols, and compare the impact of these factors on RTF and TTF based systems.

References

[1] Van Eeden, H.L. (2004) Passive UHF RFID systems, Ph.D. thesis, Northwest University, Potchefstroom, South Africa.

[2] Atkins, R.C., Marais, M.A. and Smit, H. van Z. (1999) Identification System, International application published under the Patent Cooperation Treaty, International Publication Number WO 99/26081, International Publication Date, 27 May.

[3] Van Eeden, H.L. (2000) Free running RF identification system with increasing average inter transmission intervals, United States Patent, Patent Number 6,154,136, date of patent: November 28, 2000.

[4] Naudé, C.C. and Marais, H.J. (2008) Degradation in read performance with an increase in the number of user data pages, internal research report, Faculty of Engineering, Northwest University.

[5] Dobkin, D. (2008) *The RF in RFID: Passive UHF RFID in Practice*. Oxford: Elsevier Inc.

[6] Naudé, C.C. (2008) Comparison of EPCglobal and iPX protocols with respect to movement speed of tags and different data link rates using the CISC RFID ASD Simulator, internal research report, Faculty of Engineering, Northwest University.

[7] Curty, J-P., Declercq, M., Dehollain, C., and Joehl, N. (2007) Design and optimization of UHF RFID systems, in *Springer Science & Business Media*. New York: Springer.

[8] Carbunar, B., Ramanathan, M., Koyuturk, K.M., Jagannathan, S. and Grama, A. (2009) Efficient tag detection in RFID systems, *Journal of Parallel and Distributed Computing*, **69**: 180–196.
[9] Shih, D., Sun, P., Yen, D. and Huang, S. (2006) Taxonomy and survey of RFID anti-collision protocols, *Computer Communications*, **29**: 2150–2166.
[10] Lee, H. and Kim, J. (2006) QT-CBP: a new RFID tag anti-collision algorithm using collision bit positioning, *Lecture Notes in Computer Science*, **4097**: 591–600.
[11] Chung, H.B., Mo, H., Kim, N. and Pyo, C. (2007) An advanced RFID system to avoid collision of RFID reader, using channel holder and dual sensitivities, *Microwave and Optical Technology Letters*, **49**(11): 2643–2647.
[12] Available at: http://www.rfida.com/weblog/labels/standard.htm: RFID Barriers to Pharma Adoption: Cost Benefits Standards (accessed April 19, 2007).
[13] Available at: http://www.ipico.com/index.cfm?id=5894.
[14] The Internet of Things. Available at: http://www.rfid-weblog.com/50226711/the_internet_of_things.php.
[15] More Than 5 Cents, January 12, 2007. Available at: http://www.rfid-asia.info/2007/01/more-than-5-cents.htm.
[16] Yan, Lu, and Zhang, Yan (eds.) (2008) *The Internet of Things: From RFID to the Next Generation Pervasive Networks*. New York: Auerbach Publications, Taylor & Francis Group.
[17] Available at: http://www.scdigest.com/ASSETS/ON_TARGET/09-01-27-2.PHP?cid=2201&ctype=content, "RFID News: The five-cent tag is here, the five cent tag is here! Well, almost", January 27, 2009.
[18] Marais, H.J. (2009) Characterisation of a UHF fading channel for an RFID-based EVI system, Master's thesis, Northwest University.
[19] Available at: http://www.magellan-rfid.com/pjm-technology/standards.
[20] EPC™ Radio Frequency Identity Protocols, Class-1 Generation-2 UHF RFID, Protocol for Communications at 860–960MHz, Version 1.0.9, January 2005.
[21] Available at: http://www.alibaba.com/showroom/Rfid_Dual_Frequency.html.
[22] Available at: http://www.emmicroelectronic.com/products, EM-Microelectronic – Marin SA data sheets for EM4022 integrated circuit.
[23] Available at: http://www.emmicroelectronic.com/products, EM-Microelectronic – Marin SA data sheets for EM4222, EM4122 and EM4123 integrated circuits.

Part Three

Reader Infrastructure Networking

10

Integrating RFID Readers in Enterprise IT[1]

Christian Floerkemeier and Sanjay Sarma

Massachusetts Institute of Technology

With the increasing adoption of RFID, certain deployment challenges are coming into focus. The management and control of a large number of RFID readers are becoming an issue from a network administration perspective. This includes monitoring of the health of RFID readers and maintaining a consistent configuration across all RFID readers. The shared nature of the wireless medium may require coordination among the RFID readers to minimize interference and ensure compliance with local radio regulations. Readers also do not operate independently but are sometimes triggered by external sensors that need to be configured. RFID readers also create significant data volumes. The data captured by the RFID readers needs to be filtered and aggregated due to the presence of redundant and unwanted data. Most importantly, the captured RFID data has to be interpreted in an application context to make the data capture meaningful in the first place.

There are today a number of different (partly proprietary) solutions to address these challenges. This chapter presents a taxonomy of the corresponding system services and system architectures. We also discuss how the system architectures relate to emerging standards and how these standards could possibly be enhanced to support future RFID deployments. We aim to provide a comprehensive understanding of common industry practices and thus facilitate the deployment of RFID systems.

The chapter is organized as follows. Section 10.1 discusses related work. Section 10.2 presents a number of different services and components that are commonly found in RFID deployments. In Section 10.3, we discuss the services offered by today's RFID

[1] This chapter is based on "An Overview of RFID System Interfaces and Reader Protocols," by C. Floerkemeier and S. Sarma which appeared in IEEE International Conference on RFID, 2008. © 2008 IEEE.

RFID Systems: Research Trends and Challenges Edited by Miodrag Bolić, David Simplot-Ryl, and Ivan Stojmenović

reader products. Section 10.4 presents different system architectures that we identified. In Section 10.5, we discuss EPCglobal specifications that aim to standardize system interfaces. Before we conclude in Section 10.8, Section 10.6 discusses the adoption of today's EPCglobal reader protocol standards and Section 10.7 presents potential enhancements to existing specifications.

10.1 Related Work

There are a number of previous publications that discuss requirements for an RFID infrastructure and propose system architectures [1, 2, 3, 4]. The work presented in this chapter differs from previous work in that we do not propose a particular system architecture but compare different system architectures. We also focus in particular on emerging standards in our analysis.

Within the EPCglobal community, technology vendors and end users that deploy RFID technology have been working jointly on the development of specifications that standardize the interfaces between RFID tags, readers and enterprise IT systems. The *Architecture Framework* published by EPCglobal [5] provides a comprehensive overview of the EPCglobal standards. It shows how the different interface standards are related and outlines the principles that have guided the design of the standards. The *EPCglobal Architecture Framework* does not dictate a particular system architecture, but leaves this to implementers who can choose the system architecture that is most appropriate for their deployments. The analysis presented in this chapter is complementary to the *EPCglobal Architecture Framework* because it provides insights into which system architectures are actually deployed. The *Architecture Framework* document distinguishes a number of different roles, such as *RFID Reader*, *Filtering & Collection* and *Capture Application*, each of which implements a set of different services. Rather than classifying the services as roles which are implemented on a particular device as in the *EPCglobal Architecture Framework*, we categorize them as base, configuration, monitoring and data processing services.

In WLAN access point deployments, the management, monitoring, and control of a large number of wireless access points also represent a challenge. In the WLAN domain, there are a number of different approaches to administer WLAN access points as discussed in [6]. This chapter shows that today's RFID system architectures have some similarities with the WLAN access point architectures. However, the chapter also shows that RFID systems offer a number of services that do not have equivalents in the WLAN domain, such as application-dependent data processing. In WLAN, the need for central control stems from wireless node mobility and network security, neither of which apply in the same way to RFID. We also leverage some of the concepts introduced in the WLAN domain, such as the distinction of different service sets.

10.2 RFID System Services

In this section, we discuss services provided by different components within an RFID system. This includes data and device management, control, and "over-the-air" services. In the remainder of the chapter, we will show how these services are provided by different

Table 10.1 Base service set (BSS).

Service	Description
Transponder Singulation	Collects the identification numbers (ID) of (selected) transponders in range
Transponder ID Programming	Writes identification numbers to transponders
Transponder Memory Access	Reads from and writes to the general purpose memory on a transponder
Transponder Deactivation	Disables the transponder for privacy reasons

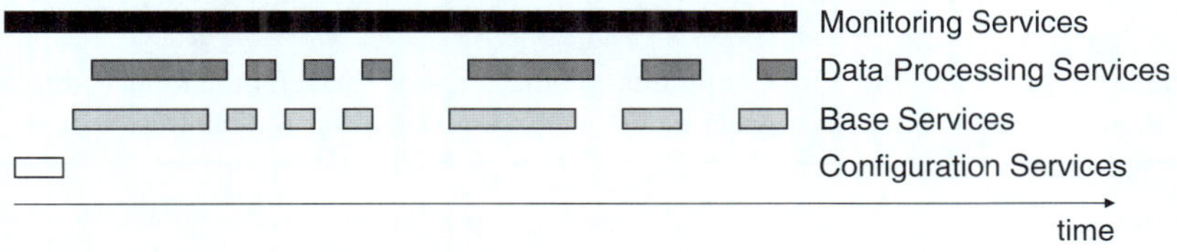

Figure 10.1 Timeline of RFID operation featuring base, configuration, monitoring and data processing services.

entities in different RFID system architectures. Our analysis of the RFID system services begins with the base service set (BSS) specified in the air interface communication standards. These air interface standards that specify the "over-the-air" interface between one or more transponders and a single reader [7] define the services listed in Table 10.1. Base services are in practice triggered in a number of different ways. These include external sensors, external applications, timers on the readers, the detection of certain tags, or humans who press a button on a handheld.

The base service set (BSS) is supported by a number of additional services that configure, control and monitor the system components that carry out these services and process the data captured. We distinguish different service sets according to different phases of the RFID operations (Figure 10.1): Configuration services are executed before any base services are executed, monitoring services are running while base services are executed and data processing services are executed once the base services have returned captured data. In case of memory access and deactivation, the processing of the captured tag IDS can trigger subsequent base services such as tag memory access.

Prior to any data capture, the networking and radio module embedded in a reader needs to be configured appropriately. The configuration service set (CSS) is responsible for setting network parameters, such as IP addresses, but also RF parameters such as transmitter power and frequency channel and air interface protocol-specific parameters, such as timing and coding parameters (Table 10.2). The configuration phase might also comprise specifying which base reader services such as tag identification or tag memory access are executed upon the appropriate triggers. Readers can also be configured to only use certain antennas and select a particular tag population over the air interface so that unwanted data does not need to be filtered out post-capture (Figure 10.2).

Table 10.2 Configuration Service Set (CSS).

Service	Description
Network Interface Configuration	Discovers and sets reader networking parameters and identity, for example, the IP address
Firmware Management	Distributes and manages firmware version on readers
Antenna, Tag Population & Memory Selection	Specifies reader antennas and tag population to be inventoried. In case of tag memory access, specifies memory fields to be accessed
Base Service Set Scheduling	Sets how different BSS services, such as tag inventory, access, and deactivation, are triggered and stopped
RF Transmitter Configuration	Sets transmission channel, hop sequence, transmitter power for readers
Air Interface Protocol-Specific Configuration	Configures timing, coding and modulation parameter of a specific air interface protocol on the readers

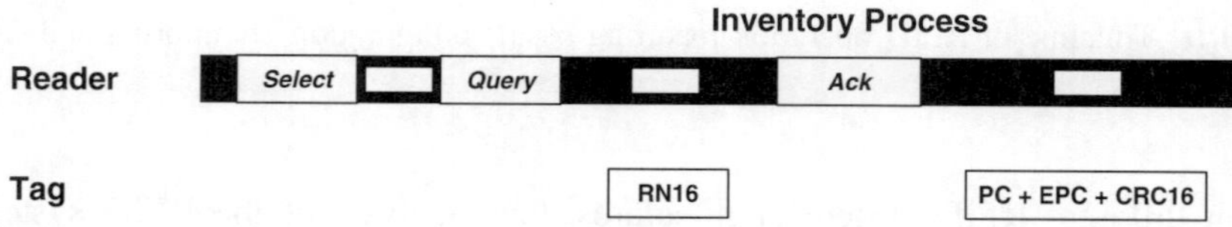

Figure 10.2 "Select" command defined in the EPCglobal UHF Class 1 Generation 2 Protocol (ISO 18000-6C) [7]. A particular tag population is selected with the "Select" command before the inventory process is initiated with the "query" command. Only tags matching the mask of the "Select" command, reply with a random number (RN16) and with their unique ID (EPC).

Parameters characterizing the network interface or regulatory region are configured once and are unlikely to change during the operation of the RFID reader. However, there are also some configuration parameters that will be changed frequently depending on the type of application. Examples include transmitter power adaptation for distance estimates, transmission channel changes for interference avoidance and tag population selection to avoid reading a "parked RFID tag" continuously. The result is that the frequency at which the reader configuration needs to be modified and the resulting coupling between base and configuration services are use case dependent. On the one hand, there are those applications that are unlikely to require changes to the initial configuration of the reader (Figure 10.1). After the initial configuration, the reader executes base services and asynchronously notifies the data consumers about tag data captured. Such applications allow for the separation of the reader control and data processing. On the other hand, some applications require tight coupling between base service execution and configuration services with frequent updates to the configuration (Figure 10.3).

The data processing services (DPSS) include services that clean the data captured by filtering out tag IDs of no interest to applications and compute aggregates over the tag data captured (Table 10.3). This includes aggregates in the time domain, where entry and exit

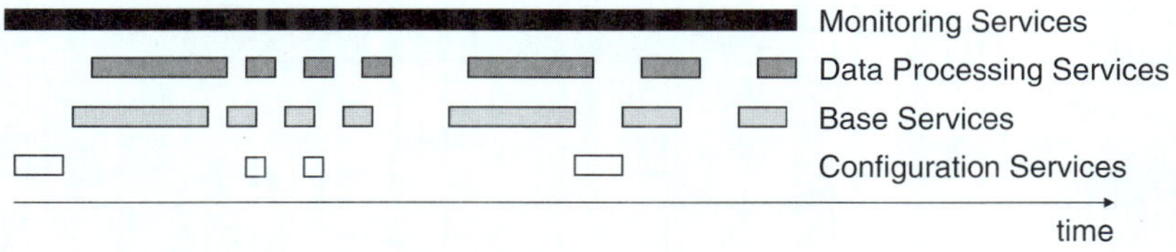

Figure 10.3 Timeline of RFID operation featuring base, configuration, monitoring and data processing services with frequent changes to the reader configuration.

Table 10.3 Data Processing Service Set (DPSS).

Service	Description
Filtering	Removes unwanted tag identifiers from the set of tag identifiers captured, for example, based on the product type or manufacturer encoded in the identifier
Aggregation	Computes aggregates in the time domain (entry/exit events) and the space domain (across reader antennas and readers) and generates the corresponding "super"-events
Identifier Translation	Translates between different representation of the identifier such as from the raw tag object identifier in hexadecimal format to the EPC in URN notation
Persistent Storage	Stores RFID data captured for future application requests
Reliable Messaging	Allows RFID data to be delivered reliably in the presence of software component, system and network failures
Location/Movement Estimation	Detects false positive reads of far-away tags that are outside the "typical" read range and estimates the direction of movement
Application Logic Execution	Interprets the RFID data captured in an application context and generates the corresponding application events, for example, whether a shipment is complete

of tags in the read range are determined, and aggregates in the space domain, where the data captured across multiple reader antennas or even readers is computed (Figure 10.4). Since there is frequently some uncertainty about the true location and movement of an RFID transponder relative to the reader, there are additional services that eliminate so-called "false positive" reads and that possibly even estimate the movement of RFID transponders. Other services include tag data translation and persistent storage.

Messages exchanged between different services in an RFID system travel over a number of communication nodes. This means that some messages may be lost in transit. Additionally, it is possible that either the recipient's or the sender's system will fail while a message is in transit, leaving the overall RFID system in a state of confusion as to whether a given message carrying tag data has been processed or not. Reliable messaging protocols provide guaranteed end-to-end delivery of RFID data. Reliable messaging refers to the ability to deliver a message once and only once to its intended receiver, to deliver messages in order, and to inform the sender and receiver about the failure to deliver a message.

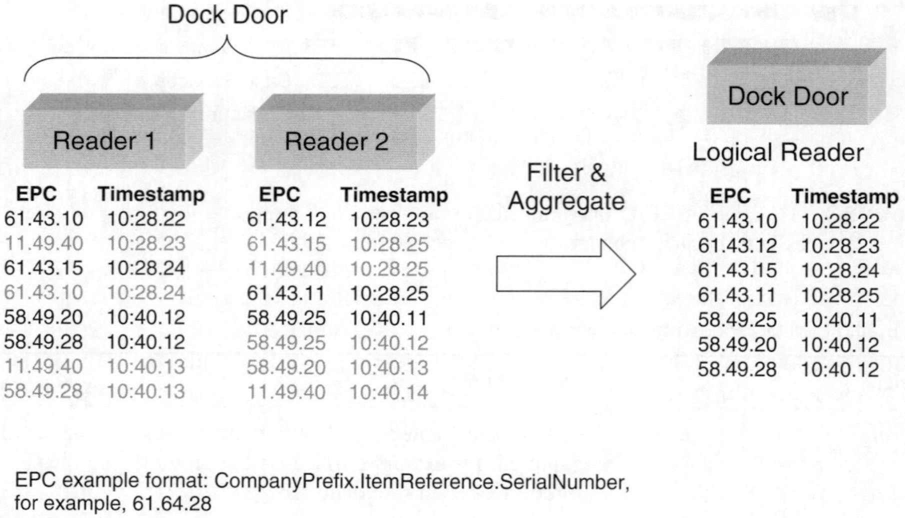

Figure 10.4 Filtering and aggregation of RFID data: data from two dock door readers are combined, duplicate EPCs are eliminated, EPC 11.49.40 is filtered out, and quantities of product categories are calculated. Eliminated "reads" are shown in grey.

Table 10.4 Monitoring Service Set (MSS).

Service	Description
Network Connection Monitoring	Checks network connection between different RFID system components
RF Environment Monitoring	Checks RF noise and interference levels
Reader Monitoring	Checks that the reader is up and running and executing BSS as configured

The DPSS also includes services that interpret the RFID data captured in an application context to generate the corresponding application events. For a supply chain application, this might include matching the detected tag identifiers against a list of identifiers in an electronic advance shipping notice. The result of the data interpretation can be the generation of a business event such as "Shipment complete" to an enterprise resource planning system or a immediate feedback to a local staff via a display. Application logic is typically executed after other DPSS services such as filtering, aggregation and tag identifier translation have preprocessed the captured RFID data.

While filtering, aggregation and tag data translation functionality typically use predefined operators, data interpretation often relies on custom-developed application logic that processes the incoming RFID data. This results from the significantly broader scope of this service when compared to aggregation or filtering. However, there have been a number of (commercial) efforts to define standard workflows for typical RFID applications such as dock door receiving. While standardized workflows can reduce the amount of custom application software development required for each deployment, variations from the standardized workflow in real-world processes typically still result in customization and additional software development.

It is worthwhile mentioning that the "filtering" can also be performed by limiting data capture to a subset of reader and reader antennas in the first place and by selecting a specific tag population (Table 10.2 and Figure 10.2). The filtering is then effectively carried out over the air interface by configuring the reader appropriately. In ISO 18000-6C, this is achieved by executing one or more "Select" commands before an inventory round is initiated with a "Query" command. Even aggregation in the time domain can be performed over the air interface using advanced features in air interface protocols. In ISO 18000-6C, there are a number of different inventory flags that allow the user to minimize redundant identifications of the same tag. In many applications, RFID tags are still identified multiple times while they are in the read range due to multipath effect. As a result entry/exit aggregates need to be computed in software. Since tags might miss a "Select" command and only receive the subsequent "Query" command, "filtering" is typically also performed in software.

Monitoring services (MSS) observe the health of the reader, the RF environment and the network connection to the reader (Table 10.4). This is carried out via heartbeat messages exchanged between reader and monitor to detect network failures and via the monitoring of antenna status, memory overflows and reboot alarms. The monitoring services also receive information about the current RF noise floor and possible RF interference.

10.3 Reader Capabilities

Before we present different RFID system architectures that provide the services listed in the previous sections, we discuss the different RFID reader categories available on the market today. This is important because system architectures depend heavily on the type of readers deployed.

A typical reader is comprised of a radio module, a general purpose computing module, a network interface, and general input/output pins. The general purpose computing unit can be a low-end microcontroller or an embedded processor with significant computing resources. The general input/output pins are used to interface with local sensors and actuators. All readers provide the base service set mentioned earlier. The majority of RFID readers currently available provide some limited data processing services (DPSS), such as pre-defined filters and aggregates over the RFID data captured. This includes the post-capture elimination of redundant reads and the accumulation of tag reporting across antennas [8]. Readers also compute entry/exit aggregates [10]. Some RFID reader products also provide limited persistent storage space so no data is lost during a communication failure with the backend IT systems. RFID readers with significant computing resources execute application code on the reader platform [8, 9, 11]. The result is that the RFID reader can independently control all local interaction via with sensors and displays. The reader only transmits the application dependent high-level events that result from the data processing. The reader verifies, for example, a shipment against an advance shipping notice and sends a "shipment complete" event to the enterprise resource planning system. Some reader vendors only allow preferred partners to execute application code on the RFID reader.

Most RFID readers allow users to configure RF transmitter settings, network interface, and antenna and tag selection parameters (Table 10.2). Base service scheduling without network access, such as the immediate writing to memory upon seeing a particular tag ID, is typically only found on RFID readers that allow users to run custom application code on the reader or those that support the EPCglobal Low Level Reader Protocol.

10.4 RFID System Architecture Taxonomy

In this section, we present different RFID system architectures that are currently deployed throughout the industry. The analysis is based on interviews with companies which have installed RFID systems. Based on our analysis, we distinguish two different architecture types: An autonomous and a centralized architecture (Figures 10.5 and 10.6). In practice, there are also a number of hybrid architectures that feature elements of both architecture types. The deployment diagrams that illustrate these architecture types have a number of different dimensions (Figure 10.5). Each box in the deployment diagram represents a separate device. Each device has a number of services which are grouped into base, configuration, monitoring, and data processing services. The deployment diagrams also show the communication link type (LAN/WAN) between different devices and group devices which are hosted within the same facility such as a store or distribution center.

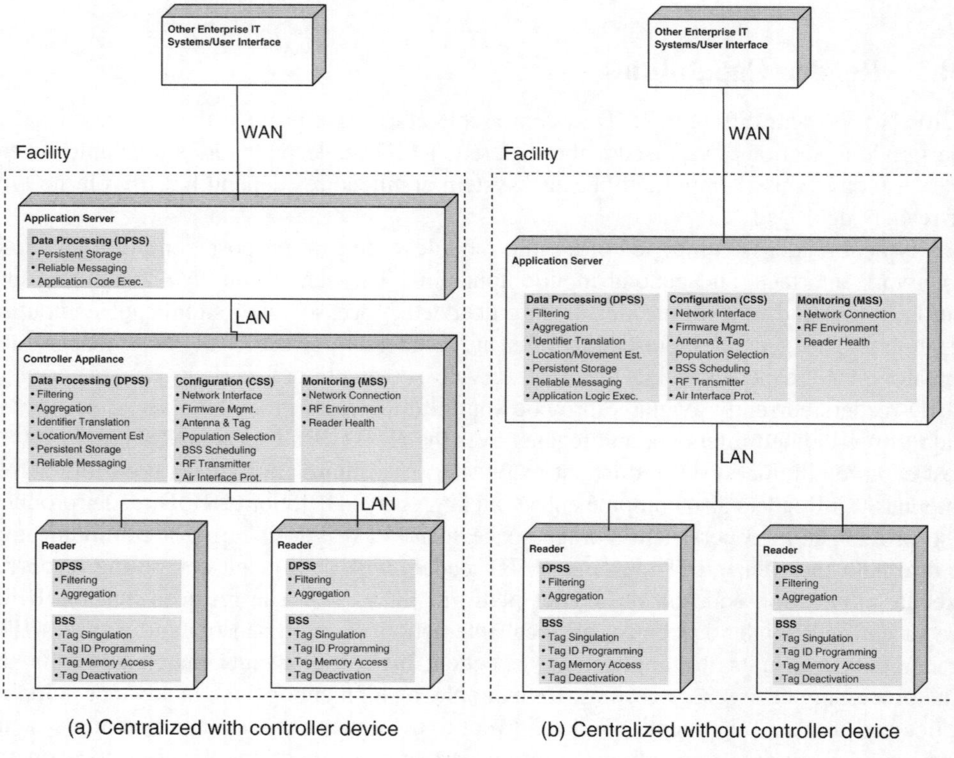

Figure 10.5 Centralized Architecture (with and without controller).

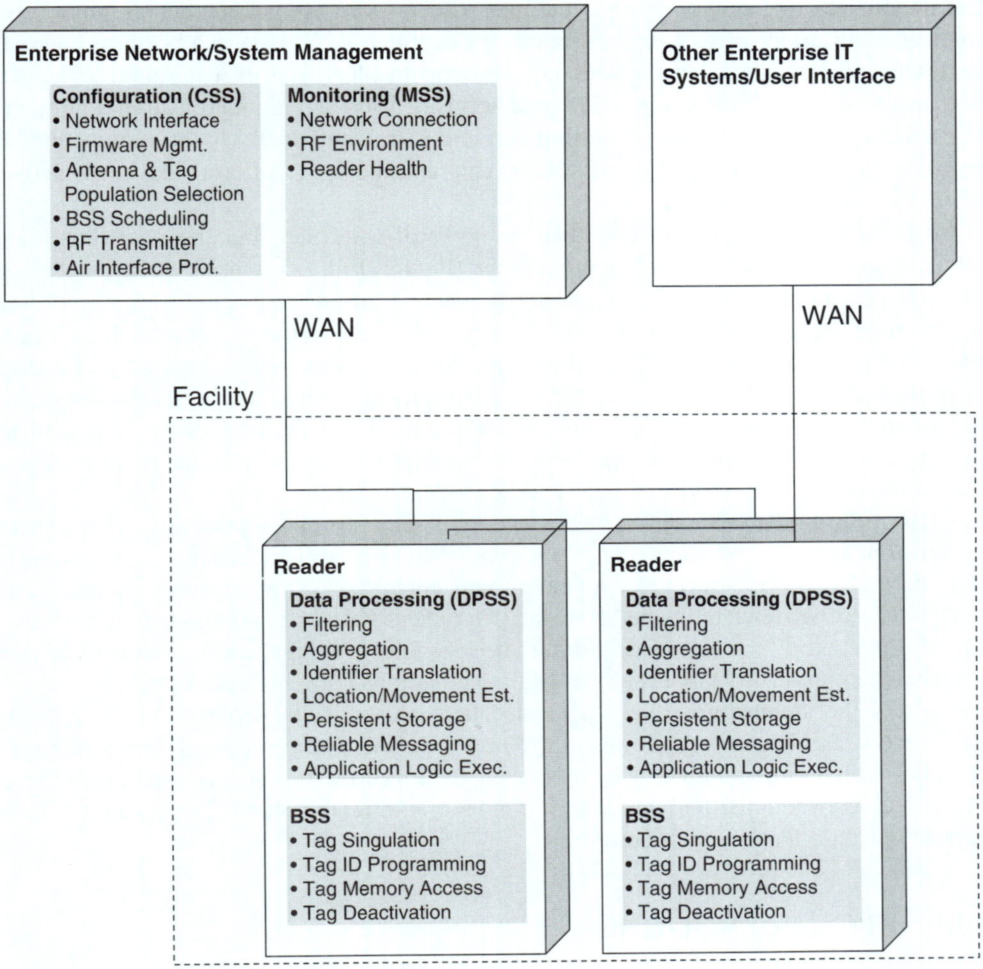

Figure 10.6 Autonomous Architecture.

Figure 10.5 shows an architecture with a dedicated controlling device at each facility, where one or more RFID readers are deployed. We call this the centralized architecture because a central device provides CSS, MSS and the majority of the DPSS services. Existing enterprise IT monitoring systems do not monitor reader devices directly. Figure 10.5 presents two variations of this architecture type. Figure 10.5(a) features a separate application server[2] and a controller. The controller provides the CSS and MSS services and application-agnostic DPSS services such as persistent storage of tag reads, identifier

[2] The term "Application Server" refers here to a server that provides a runtime environment for the application. While the term "application server" is often associated with enterprise-class servers that implement the Java Enterprise Edition or Microsoft.NET framework, application servers hosted on-site in RFID systems often feature only a more lightweight execution environment. In an RFID context, these servers are often also referred to as "Edge Servers."

translation and aggregation across multiple readers. The application server hosts customized application software that processes the captured RFID data in a business context. In Figure 10.5(b), these data processing services are deployed on the same device. In both cases, the RFID readers provide base services, such as tag identification and memory access, and limited data processing services. The latter includes the elimination of redundant reads by computing entry/exit events and the aggregation of tag reads across different antennas.

In the autonomous architecture, there is no local controller, but the readers operate autonomously once configured appropriately (Figure 10.6). Extensive data processing takes place on the RFID readers themselves, where a custom application code processes the captured tag data. The readers send locally computed business events such as "shipment complete" to the enterprise information systems. Before the captured RFID data is processed in a business context, the data is typically filtered and aggregated and tag identifiers are translated. To deal with network and system failures, there is a need to provide reliable messaging and persistent storage services. The readers are monitored and also configured via enterprise system and network management tools.

Each of the two architecture types has its own strengths and weaknesses with respect to performance, ease of maintenance and cost. The most suitable system architecture is thus dependent on the specific application and enterprise IT organization. Installations with hundreds of readers in the same facility typically favor a centralized architecture. The deployment of isolated readers in remote locations or in applications with significant local interactions with staff benefit from the autonomous architecture. In practice, many hybrids of the architectures types presented here are typically deployed. This includes autonomous architectures for readers in remote locations and centralized architectures in facilities with a large number of RFID readers deployed by the same organization. There are also cases where the RFID readers are monitored remotely, but the data processing is carried out on a local server/controller.

10.5 EPCglobal Standards

The *EPCglobal Architecture Framework* defines a number of roles: *RFID reader*, *filtering&collection*, *reader management* and *EPCIS capture applications* [5]. In the "centralized with controller" architecture we presented above, each of the roles maps to one individual device in the system architecture except for the controller that implements both the filtering&collection and reader management role (Figure 10.7(a)). In the variation of the centralized architecture without a separate controller device (Figure 10.7(b)), the filtering&collection, reader management and EPCIS capture application role are combined on the same device. In the autonomous architecture, the RFID reader, filtering&collection and EPCIS capture applications roles of the EPCglobal Architecture Framework are implemented on the RFID reader (Figure 10.7(c)). Each of the different roles in the EPCglobal architecture framework is associated with one or more software specifications that standardize interfaces. In the following subsections, we discuss each of these specifications in the context of the service taxonomy presented earlier.

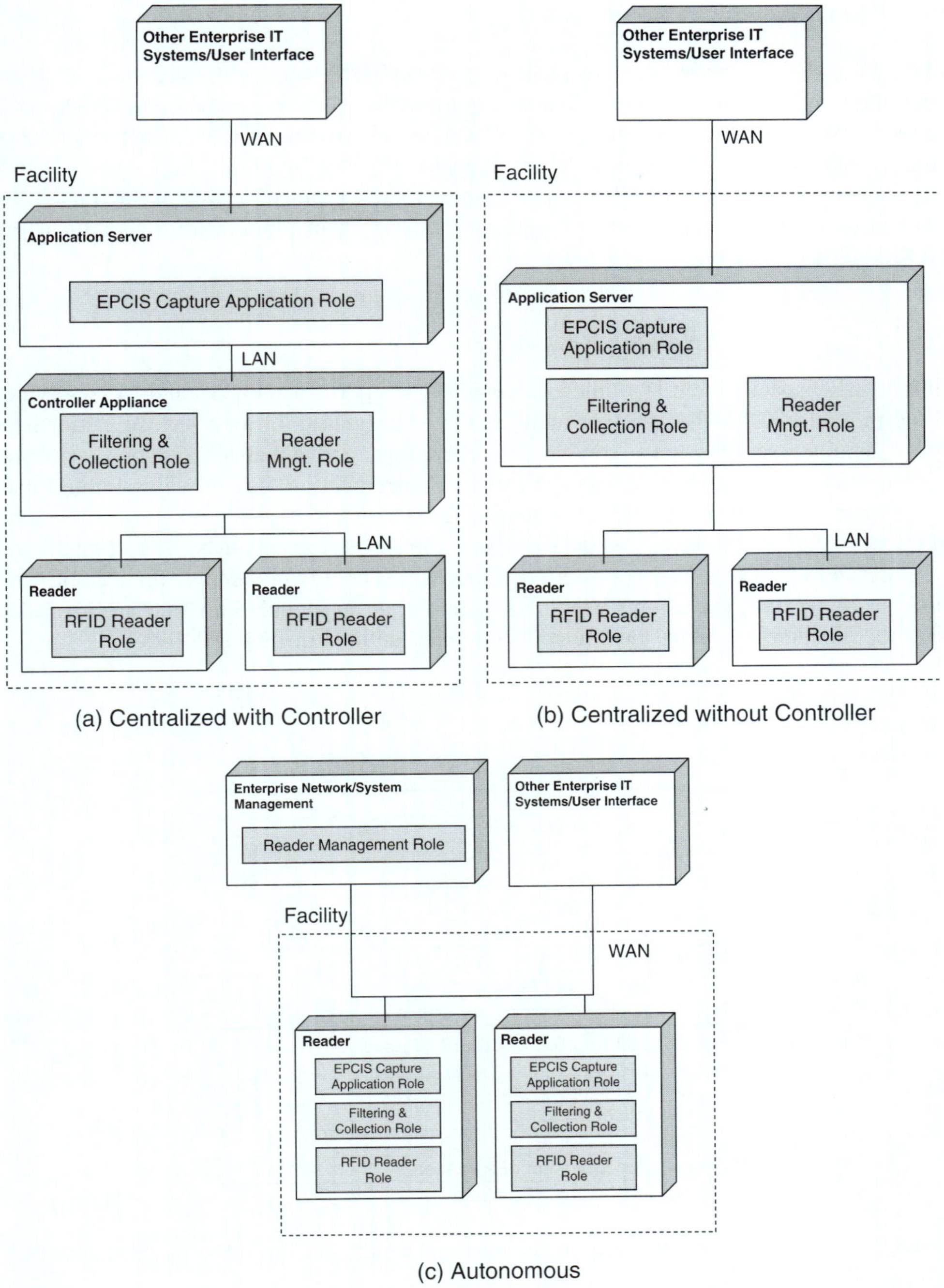

Figure 10.7 Centralized and Autonomous Architectures with associated EPCglobal Architecture Framework Roles [5].

10.5.1 Discovery, Configuration and Initialization (DCI) and Reader Management (RM)

The EPCglobal DCI specification [12] supports the CSS services of network interface configuration and firmware management (Figure 10.8). The EPCglobal RM protocol [13] supports MSS services[3] (Figure 10.8). RM does not reveal any air interface protocol specific statistics, but reports noise levels, transmitter power and failed tag memory access and deactivation operations. RM supports the SNMP protocol and thus facilitates the integration of RFID readers into existing enterprise system monitoring tools which rely on the SNMP protocol.

10.5.2 Low Level Reader Protocol (LLRP)

In the LLRP specification [14], there is extensive support for the CSS and MSS services listed in Table 10.2 and 10.4. This includes the configuration of air interface parameters, RF transmitter settings, base service set scheduling, and antenna, tag population and tag memory selection (Figure 10.8). The LLRP specification also provides limited data processing capabilities. The *Accumulation of TagReportData* in the LLRP specification allows for the collection of tag data across multiple reader antennas. The computation of entry and exit event aggregates is not supported. There is no support for the software filtering of tag IDs, but clients can specify the target tag populations that are identified over the air interface using the "Select" command defined in ISO 18000-6C.

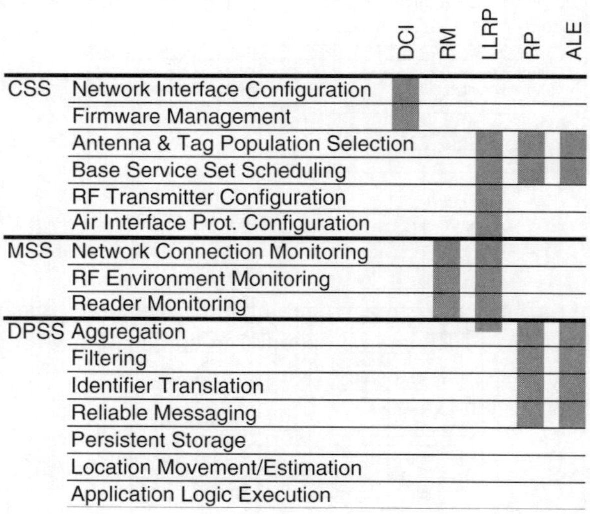

Figure 10.8 High-level overview of the services supported by the EPCglobal specifications DCI, RM, LLRP, RP and ALE.

[3] In addition to the support for monitoring services, the EPCglobal RM Specification also allows hosts to selectively switch off reader antennas via the *ReadPoint.setAdminStatus* method. Strictly speaking, the latter represents a configuration service according to our service taxonomy.

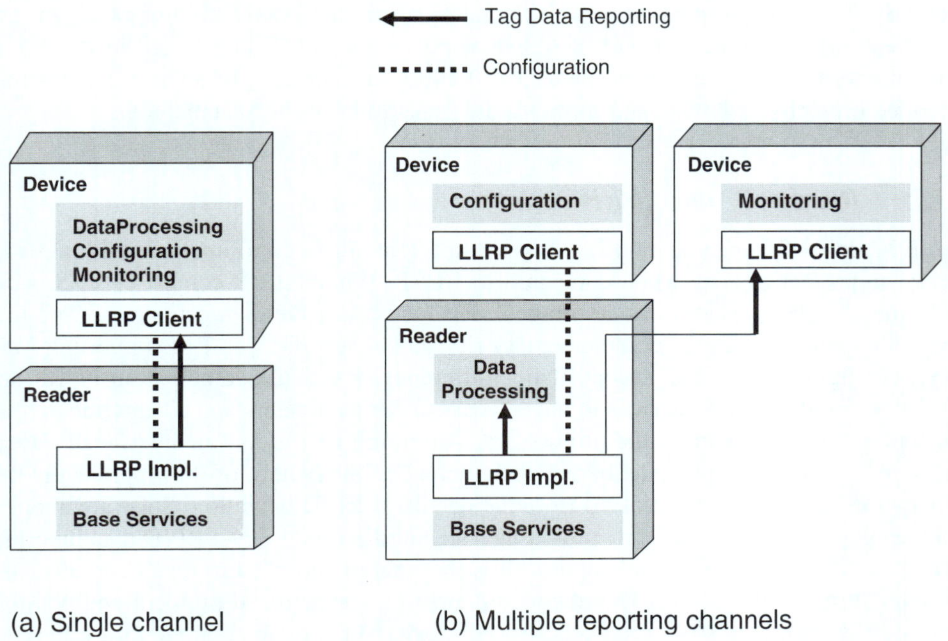

Figure 10.9 Simultaneous notifications to multiple data consumer are not supported by today's LLRP specification.

LLRP is well suited to standardize the reader interface in the centralized architectures. LLRP allows controllers and software deployed on local application servers to schedule base services and to optimize performance by adjusting the RF transmitter settings and air interface parameters. The LLRP specification envisions only a single connection to a client at a time (Figure 10.9(a)). This allows for one device configuring the reader and subsequently for another device to connect to the reader to receive and process the *TagReportData* notifications. It does not permit sending the *TagReportData* to multiple data consumers (cf. Figure 10.9(b)). Thus, LLRP by itself cannot be used to support a deployment scenario where applications on the reader process the data captured and remote system monitoring services receive notifications to monitor the health of the reader and connection. However, LLRP can be operated in conjunction with RM. This provides SNMP access to the reader status and statistics on successful and failed base service operations for remote system monitors.

LLRP can also generate significant network traffic since there is no way to restrict *TagReportData* messages to tags entering and exiting the range of the reader when low notification latencies are desired. The latter is not an issue for local area connections, but can be problematic for wide area connections. The binary communication protocol allows for efficient implementations on RFID readers, but requires additional high-level tools for application developers.

LLRP features an extension mechanism that allows reader vendors to define their own proprietary extensions that expose custom feature sets. In our opinion, this is essential since RFID reader vendors will only consider substituting their proprietary reader protocols for a standardized protocol if the latter allows for extensions supporting proprietary features which are not part of the standard feature set. While LLRP in its current version only features support for the ISO 18000-6C air interface, the support for other air interface protocols is envisioned in the future generation of the specification.

10.5.3 Reader Protocol (RP)

The EPCglobal RP specification [15] takes a different approach to interface with a reader in two important aspects when compared to LLRP. While LLRP comprises CSS, MSS and limited DPSS services in one protocol, RP focuses on DPSS functionality (and limited CSS services) with the complementary EPCglobal specification RM supporting MSS services (Figure 10.8). The second major difference is that the combination of RP and RM does not allow CSS services to have access to RF parameters and air interface protocol settings of the reader[4] – the protocols are often referred to as "air interface protocol unaware." This represents a deliberate design choice. The benefit of this approach is that application developers are shielded from the details of RFID operation. Software application development is facilitated by providing a standardized high-level RFID data reporting interface. Another benefit of this approach is that there is no need to reach consensus on the configuration and control features to support in a standardized protocol and possibly across different air interface protocols. An HF reader vendor can continue to use his own proprietary reader protocol to configure its readers. The tag data reporting takes place via a "high-level" reader protocol and is identical to the tag data reporting of UHF Gen2 or ISO 18000-6B readers which also use the EPCglobal RP to report captured data.

The drawback of this design choice is that the reader cannot be controlled and configured to the extent possible with today's proprietary reader protocols or with the recently released LLRP. It is not possible to select a particular air interface protocol or an interference avoidance feature such as dense reader mode remotely. It is also not possible to select a particular frequency channel to minimize reader collisions. All of these represent standard features offered by today's vendor-specific reader protocols. Since access to these features is necessary at least during the initial deployment, RP requires the use of another vendor-specific protocol to configure the reader appropriately (Figure 10.10). In some applications, the configuration via another "air interface aware" (proprietary) protocol is only required at the time of deployment with no changes during operation. In other applications, RF interference or changes in tag populations require frequent changes to the reader configuration to maintain optimum performance. This results in frequent use of the additional "air interface aware" (proprietary) protocols.

RP supports DPSS services that are not included in the LLRP specification. RP allows for the computation of entry and exit events, different tag ID representations and tag report notifications to one or more data consumers. RP also provides an XML message binding that facilitates software application development. LLRP relies on a binary protocol that requires fewer resources on RFID readers. RP supports reliable messaging for the asynchronous notification channels.

[4] RP only allows for the CSS services of antenna and tag population selection and limited base service scheduling. Via RP, clients can schedule inventory ID operations, but memory access operations cannot be scheduled and need to be triggered remotely.

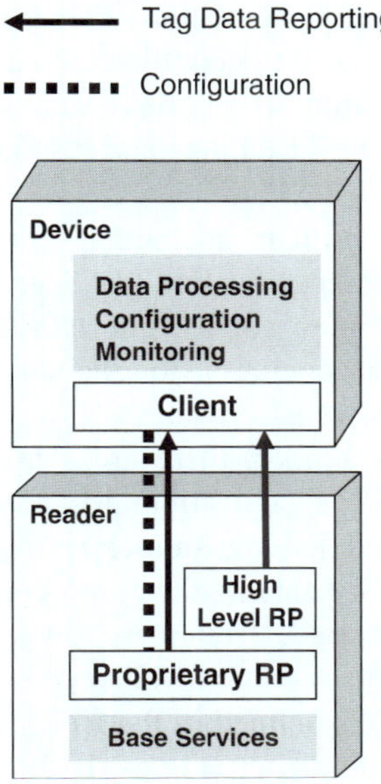

Figure 10.10 High-level reader protocols such as EPCglobal RP and ALE require an additional (proprietary) protocol to configure the reader.

Nearly all of the DPSS services in RP are optional in the specification. This design choice seems to be a direct result of the heterogeneity of the reader landscape. Reader devices with significant computing resources were envisioned to provide the majority of the optional features. Low-end reader devices would only support the mandatory features and possibly a small number of the optional features. However, this design choice also makes application development with RP a challenge. Different reader types will likely support a different combination of optional features.

10.5.4 Application Level Event (ALE)

The EPCglobal ALE 1.0/1.1 [16, 17] specifications provide a standardized interface to application-agnostic DPSS services. ALE also supports the selection of readers, reader antennas and tag populations, which represents a CSS service in our taxonomy. ALE 1.0 comprises a feature set similar to the EPCglobal RP specification: filtering, aggregation, tag identifier translation, buffering of messages and notifications to multiple consumers, but the actual implementation is different. ALE relies on a web service (SOAP) transport protocol and provides a convenient "subscribe" mechanism where clients can register their standing queries/notifications with a single command. In RP, this requires a sequence of individual messages. ALE 1.0 does not support tag user memory.

ALE 1.1 represents a major advance over ALE 1.0. ALE 1.1 supports reading and writing to user memory on tags. ALE 1.1 also provides new interfaces for defining tag memory fields, for reader/reader antenna to location mapping and for access control.

In LLRP and ALE 1.1, the primary interaction between client and implementation for both tag identification and memory access are similar. For tag identification, the

ALE/LLRP client provides the implementation with a specification (*ECSpec* in ALE/ *ROSpec* in LLRP) that defines boundary conditions, channels to be used and the desired content and structure of the asynchronous reports (Figure 10.11). The implementation executes this specification, captures the RFID data and responds by returning the information in the reports as requested (Figure 10.12).

For the tag memory access, the primary interaction sequence is also similar for ALE 1.1 and LLRP. The client transmits a specification (*ECSpec for reading/CCSpec for writing* in ALE/*AccessSpec* in LLRP) (Figure 10.13). The ALE/LLRP implementations respond by carrying out the memory access operations on the tags and return reports that describe which memory access operations were performed.

While the basic interaction is similar, there are a number of conceptual differences between the LLRP and ALE 1.1 specification. LLRP allows the client to optionally specify air interface protocol and RF transmitter settings in *ROSpec* and *AccessSpec* (Figure 10.11). The only mandatory air interface protocol parameter is the air protocol to be used. The ALE specification abstracts from these low level settings and does not expose them. The ALE 1.1 specification uses "high-level" representations for tag identifiers, such as the URI representations of EPCs defined in the EPCglobal Tag Data Standard [18]. In LLRP, EPCs are represented as bit arrays (Figure 10.12). ALE 1.1 includes the "logical reader API" which decouples the identity of reader devices and antennas from the names of the channels used in ALE subscriptions and reports. This permits the replacement of an RFID reader or a change of networking parameters without the need to update the application software. As mentioned earlier, ALE also provides additional data processing services, such as count and entry/exit aggregates, which are not available in LLRP (Figure 10.12).

For tag user memory, ALE 1.1 provides predefined memory field names for elements specified in the EPCglobal Tag Data Standard [18] and support for ISO 15962, both of which facilitate the programming of memory access operations. In LLRP, tag user memory needs to be addressed using memory banks, pointers and length. Masks need to be specified as bit arrays (Figure 10.13). An ALE implementation can also service multiple data consumers simultaneously and disseminate captured tag data to them. LLRP is a network protocol that supports only a single established connection at a time. It does not support multiple clients with different "subscriptions."

The main use case of the EPCglobal ALE specification is on controllers and application servers in the centralized architectures as an application agnostic interface (Figure 10.14), where ALE exposes all those DPSS services that are common across different RFID applications such as filtering, aggregation and tag identifier translation. The ALE interface also allows clients to select readers, reader antennas and tag populations and define tag memory operations. ALE does not provide access to RF transmitter and air protocol settings. At the interface to software and hardware controllers, the control and data processing can often be separated and it is appropriate to abstract from the RFID operational settings. There is no need to expose air protocol settings because the controller implementation is responsible for RFID performance optimization.

ALE can also be deployed on an RFID reader as a "high-level" reader protocol with the similar benefits and weaknesses as mentioned in the subsection on RP. ALE provides a high-level starting point that is well suited for writing application logic, freeing the developer from the kind of low-level programming that would be necessary to code directly to LLRP. The application developer benefits in particular from the high-level tag

```
<ale:ECSpec>
  <boundarySpec>
    <startTrigger>http://example.com/trigger1</startTrigger>
    <repeatPeriod unit="MS">20000</repeatPeriod>
    <stopTrigger>http://example.com/trigger2</stopTrigger>
    <duration unit="MS">3000</duration>
  </boundarySpec>
```

Start/Stop Conditions

Both LLRP and ALE 1.1 provide a number of different start and stop trigger mechanisms.

```
  <logicalReaders>
      <logicalReader>dock_1</logicalReader>
  </logicalReaders>
```

Channels

In ALE 1.1, multiple physical readers and reader antennas can be mapped to a single logical reader.

```
  <reportSpecs>
      <reportSpec reportName="ENTRY_TAGS">
          <reportSet set="ADDITIONS"/>
      </reportSpec>
      <reportSpec reportName="EXIT_TAGS">
          <reportSet set="DELETIONS"/>
          <groupSpec>
              <pattern>urn:epc:pat:sgtin-96:X.X.X.*</pattern>
          </groupSpec>
          <output includeCount="true"/>
      </reportSpec>
  </reportSpecs>
</ale:ECSpec>
```

Report Formatting

Both LLRP and ALE 1.1 allow the subscriber to define the content of the asynchronous tag reports. ALE supports a number of different aggregates: entry/exits, count and grouping. ALE 1.1 is air protocol unaware and does not expose air protocol or RF settings.

(a) ALE1.1

```
<llrp:ADD_ROSPEC Version="1" MessageID="4">
<ROSpec>
  ...
  <ROBoundarySpec>
    <ROSpecStartTrigger>
      <ROSpecStartTriggerType>Periodic</ROSpecStartTrigger...>
        ......
    </ROSpecStartTrigger>
    ........
  </ROBoundarySpec>
```

Start/Stop Conditions

Both LLRP and ALE 1.1 provide a number of different start and stop trigger mechanisms.

```
  <AISpec>
     <AntennaIDs>1 2</AntennaIDs>
     ...
```

Channels

In LLRP, the AISpec defines which antennas are used

```
     <InventoryParameterSpec>
        <ProtocolID>EPCGlobalClass1Gen2</ProtocolID>
     </InventoryParameterSpec>
     <AntennaConfiguration>
        <AntennaID>1</AntennaID>
        <RFTransmitter>
           <TransmitPower>200</TransmitPower>
        </RFTransmitter>
     </AntennaConfiguration>
  </AISpec>
```

Air Interface and RF Configuration

LLRP allows users to specify air interface parameters and RF transmitter settings such as the transmission power.

```
  <ROReportSpec >
     <ROReportTrigger>Upon_N_Tags_Or_End_Of_ROSpec
   </ROReportTrigger>
   .....
     <TagReportContentSelector>
       <EnableLastSeenTimestamp>true</EnableLastSeenTimestamp>
       ....
</ROSpec>
</llrp:ADD_ROSPEC>
```

Report Formatting

Both LLRP and ALE 1.1 allow the subscriber to define the content of the asynchronous tag reports.

(b) LLRP

Figure 10.11 Specification of tag identification in ALE 1.1 and LLRP. The LLRP *ADD_ROSPEC* message is represented in the LLRP LTK XML format (www.llrp.org).

```
<ale:ECReports>
<reports>
    <report name="ENTRY_TAGS">
    <group>
    <groupList>
    <member>
    <epc>urn:epc:id:sgtin:0614141.112345.3
    </epc>
    </member>
    <member>
    <epc>urn:epc:id:sgtin:0614141.112345.4
    </epc>
    </member>
    </groupList>
    </group>
    </report>
    <report name="EXIT_TAGS">
    <group
    <groupList>
    <member>
    <epc>urn:epc:id:sgtin:0614141.112345.5
    </epc>
    </member>
    </groupList>
    </group>
</report>
</ale:ECReports>
```

```
<llrp:RO_ACCESS_REPORT MessageID='7'>
<TagReportData>
  <EPC_96><EPC>300833B2DDD9CAFEBABE8041
  </EPC></EPC_96>
  <PeakRSSI><PeakRSSI>-45
   </PeakRSSI></PeakRSSI>
  <FirstSeenTimestampUTC>
    <Microseconds>1173287084425922
    </Microseconds>
  </FirstSeenTimestampUTC>
  <TagSeenCount><TagCount>696
    </TagCount></TagSeenCount>
  <C1G2_PC><PC_Bits>12288
    </PC_Bits></C1G2_PC>
  <C1G2_CRC><CRC>39494</CRC></C1G2_CRC>
</TagReportData>
<TagReportData>
  <EPC_96><EPC>300833B2DDD9CAFEBABE8025
   </EPC></EPC_96>
  <PeakRSSI><PeakRSSI>-47
   </PeakRSSI></PeakRSSI>
  <FirstSeenTimestampUTC>
    <Microseconds>1173287084425977
     </Microseconds>
  </FirstSeenTimestampUTC>
  <TagSeenCount><TagCount>307
   </TagCount></TagSeenCount>
  <C1G2_PC><PC_Bits>12288</PC_Bits></C1G2_PC>
  <C1G2_CRC><CRC>46692</CRC></C1G2_CRC>
</TagReportData>
</llrp:RO_ACCESS_REPORT>
```

Asynchronous Tag Data Reporting

In LLRP, the EPC is encoded in binary/hexadecimal notation. ALE 1.1 supports the different EPC formats defined in the EPCglobal Tag Data Standard.

ALE 1.1 supports entry/exit aggregates, aggregation across readers and reader antennas and the grouping of EPCs. In LLRP, reports can only be accumulated across reader antennas.

LLRP provides a number of physical layer parameter associated with the tag identification, for example, received signal strength.

Figure 10.12 Tag data reports in ALE 1.1 and LLRP. The LLRP *RO_ACCESS_REPORT* message is represented in the LLRP LTK XML format.

```
<ale:ECSpec>
  <boundarySpec>... </boundarySpec>

  <logicalReaders>... </logicalReaders>

  <reportSpecs>
     <reportSpecname="report1">
        <reportSetset="CURRENT"/>
        <filterSpec>...
           <filter>
<includeExclude>INCLUDE</includeExclude>
             <fieldspec>
                 <fieldname>afi</fieldname>
             </fieldspec>
<patList><pat>xC1</pat></patList>
          </filter>...
        </filterSpec>
        <output> ...
           <fieldList>
              <field><fieldspec>
<name>@3.urn:oid:1.0.15961.12.3</name>
              </fieldspec></field>
        </output>
    </reportSpec>
</reportSpecs>
</ale:ECSpec>
```

```
<llrp:ADD_ACCESSSPEC Version="1" MessageID="5">
<AccessSpec>
    <AccessSpecID>1</AccessSpecID>
    <AntennaID>1 2</AntennaID>
    <ProtocolID>EPCGlobalClass1Gen2
     </ProtocolID>
    <CurrentState>Active</CurrentState>
    <ROSpecID>1</ROSpecID>
    ...
    <AccessCommand>
      <C1G2TagSpec>
        <C1G2TargetTag>
          <MB>1</MB>
          <Match>1</Match>
          <Pointer>24</Pointer>
          <TagMask>1111 1111</TagMask>
          <TagData>1100 0001</TagData>
        </C1G2TargetTag>
      </C1G2TagSpec>
      <C1G2Read>
        <OpSpecID>1</OpSpecID>
        <AccessPassword>0</AccessPassword>
        <MB>3</MB>
        <WordPointer>0</WordPointer>
        <WordCount>1024</WordCount>
      </C1G2Read>
    </AccessCommand>
</AccessSpec>
</llrp:ADD_ACCESSSPEC>
```

Start/Stop Conditions and Channels

In LLRP, access operations inherit the start/stop conditions from the corresponding ROSpec.

Tag Memory Access

The example shows a read operation on a selected number of tags.

ALE 1.1 uses pre-defined and user-defined memory field names and supports ISO 15692 encoded strings. In LLRP, the patterns need to be defined in binary notation and memory locations are addressed with pointers. Bit arrays returned need to parsed with custom code.

Figure 10.13 Tag memory access (reading) in ALE 1.1 and LLRP. The LLRP *ADD_ACCESSSPEC* message is represented in the LLRP LTK XML format.

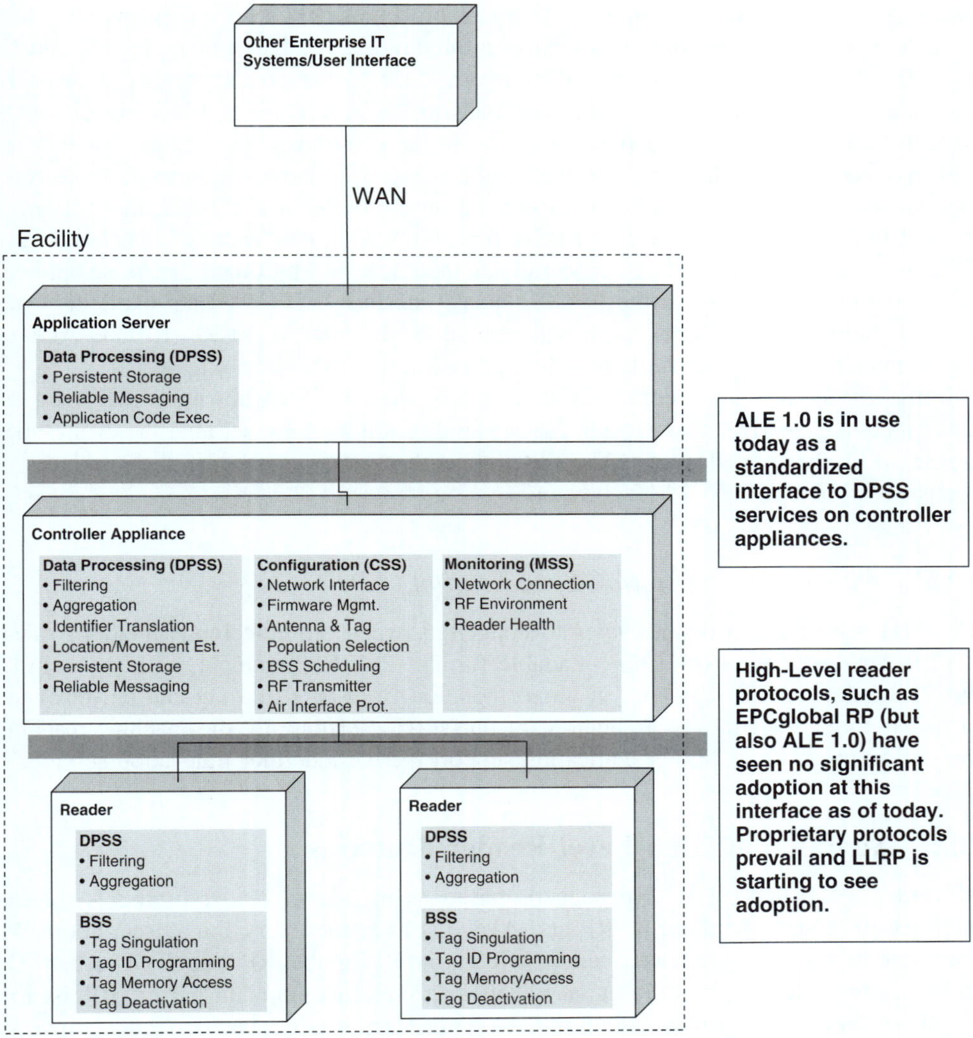

Figure 10.14 Adoption of EPCglobal specifications ALE, RP and LLRP in the centralized architecture.

identifier and user memory representation in ALE 1.1. In LLRP, these are represented in binary notation. The drawback of ALE as a high-level reader protocol is the need for an additional (proprietary) protocol that configures the reader, since the configuration of the RF transmitter and air protocol settings is at least required during the initial deployment (Figure 10.10).

10.5.5 EPC Information Service (EPCIS)

EPCglobal's EPCIS specification [19] deals not just with raw RFID observations, but defines events that link the raw observations reported by LLRP, ALE, or RP with meaning relative to specific steps in business processes. EPCIS-level events stored in an EPCIS repository are thus a result of combining RFID data captured with knowledge about the significance of an "RFID read." Due to the diverse nature of business processes, the actual data processing that leads to the generation of EPCIS-level events typically relies on a custom application code. In the centralized architecture, such "capture applications" are deployed on application servers. In the autonomous architecture, the corresponding application code is executed on the reader itself. The reader transmits the EPCIS-level events to a remote EPCIS repository using the "capture" interface of the EPCIS specification. The EPCIS repository is then accessed by other enterprise applications or trading partners via the EPCIS "query interface." EPCIS deals explicitly with historical data stored in a persistent data repository. In contrast, LLRP, RP and ALE of are oriented exclusively towards real-time processing of EPC data.

In the service taxonomy defined in this chapter, the EPCIS specification thus supports data processing services that execute custom application logic by defining the event types generated by these applications. The EPCIS specification also provides the interfaces to persistent storage for EPCIS-level events.

10.5.6 Tag Data Translation Specification (TDT)

The TDT specification defines rules to convert between different representations of an EPC in an unambiguous machine-readable format [20]. For example, with the help of the TDT rules an EPC captured in binary notation from an RFID tag can be translated to its pure-identity URI representation. In our service taxonomy, the machine-readable translation rules of the TDT specification support the tag identifier translation service.

10.6 Adoption of High-Level Reader Protocols

RP and also ALE 1.0/1.1 have been proposed as high-level reader protocols for the interface to readers. In this role RP and ALE 1.0 have seen very limited adoption. We are aware of only a single reader vendor offering an EPCglobal RP compliant reader [21] and a single reader vendor offering an EPCglobal ALE 1.0 compliant reader [10]. Nearly all RFID system deployments today rely on proprietary vendor-specific reader protocols (Figure 10.14). In our opinion, the lack of adoption of "high-level" reader protocols is due to a number of reasons:

- *Focus on UHF Gen2 by reader vendors:* The development of these high-level reader protocol standards coincided with the ratification of the air interface protocol Gen 2. This led reader vendors to focus on the development of new UHF Gen 2 reader products. The result was that the development of the specifications took place with very limited participation from reader vendors.
- *Limited computing resources on first-generation UHF readers:* High-level reader protocols require significant computing resources which were not always available on first-generation UHF readers. The demand for significant computing resources results from the use of XML messaging, asynchronous notifications to multiple data consumers and buffering, aggregation and filtering on the reader.

- *Uncertainty about the "right" protocol to support:* The availability of two specifications with nearly identical features, RP and ALE 1.0, but very different nomenclature and implementations led to some confusion about the "right" protocol to support. The ongoing development of yet another reader protocol, LLRP, added to the uncertainty.
- *Limited functionality in high level reader protocols:* ALE 1.0 did not support tag user memory, which was added in the recently ratified ALE 1.1.
- *Limited demand by end users and RFID "middleware" companies:* Neither end users nor RFID "middleware" companies requested reader vendors to support any of the two high-level reader protocols. In interviews with a small set of end-users we conducted, no company representative indicated that neither RP nor ALE 1.0 addressed their needs as reader protocols. The end user companies we interviewed disliked the idea of high-level reader protocols and requested access to RF transmitter and air protocol settings. End user companies mentioned that such high-level interfaces are more appropriate as an interface to software and hardware controllers in the centralized architecture.
- *High-level reader protocols are not a replacement for existing proprietary reader protocols:* Since reader vendors need to continue offering their proprietary protocols to configure and monitor their readers in addition to the high-level reader protocols for reporting aggregated and filtered RFID data, the support of a high-level reader protocol meant the de facto support of two reader protocols (Figure 10.15).

Major RFID reader vendors we contacted during this investigation mentioned that they have no intention of supporting RP or ALE on their readers in the short term. Instead, there seems to be a trend to support the LLRP protocol instead [22, 8]. Similarly, end user companies mentioned that they would not request the support of RP or ALE by reader vendors in the near future.

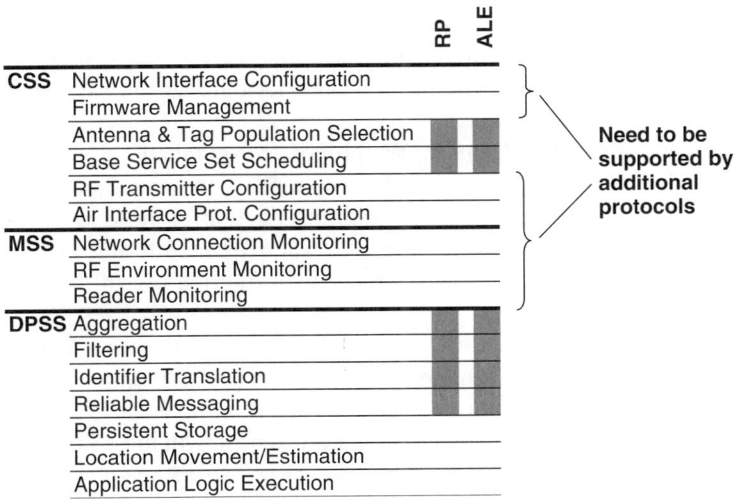

Figure 10.15 High-level reader protocols such as ALE or RP require additional protocols for configuration and monitoring purposes.

It remains to be seen whether the recently ratified specification ALE 1.1 will see more adoption on reader devices than RP and ALE 1.0 in the past. ALE 1.1 addresses the shortcomings of ALE 1.0 and provides a convenient high-level interface for tag user memory operations. Computing resources on today's readers also increased. The above analysis, however, indicates that high-level reader protocols will be adopted as interfaces to reader devices only if they are augmented by an additional protocol that supports fine-grained configuration services – effectively dismissing the idea of a 'high-level only' reader protocol as the sole interface to a reader device. ALE 1.1 will thus also require an additional (proprietary) protocol to configure and monitor the reader (Figure 10.15). As noted above, high-level interfaces such as ALE do have an important role to play at a higher level in the software stack.

10.7 Potential Future Standardization Activities

There are a number of potential future standardization activities. This includes enhancements of existing standards but also the standardization of new interfaces. In our opinion, one should consider enhancing the LLRP specification with a few additional optional features:

- *Entry/Exit aggregates:* In our previous description of LLRP, we already proposed to add optional support for entry/exit aggregates. We believe that this would reduce the network traffic which is important for remotely deployed readers.
- *Filtering:* LLRP currently only allows selecting a tag population over the air interface. Optional filtering in software would help to eliminate those tag reads that resulted from RFID tags missing the Gen 2 *Select* command.
- *XML representation of the LLRP tag data reports:* LLRP uses a binary messaging protocol. To facilitate RFID application development, one could consider standardizing an XML interface for the LLRP tag data reports. Figure 10.12 shows such a high-level representation that was developed in the *LLRP Toolkit Project*.[5] Having an XML interface for the asynchronous notification channel only would not require an XML parser on the reader.
- *Multiple tag data reporting channels:* We propose to optionally allow for more than a single connection at a time. This would allow readers to send captured data to more than one event consumer (Figure 10.9). The introduction of optional multiple tag data reporting channels would allow for the separation of monitoring and data plane in applications where this is appropriate without requiring the use of RM.

In those cases, where RM and LLRP are implemented on the same reader device, we also propose to specify a common approach to deal with the "monitored objects" in RM that are specific to RP such as "Sources" and "NotificationChannels." Otherwise, different implementations of RM only (without RP) are likely to deal with these "monitored objects" in different ways. One might also consider making RM "air protocol aware" by exposing some of the RF transmitter settings and air protocol parameters via the SMNP interface of the RM specification.

[5] www.llrp.org.

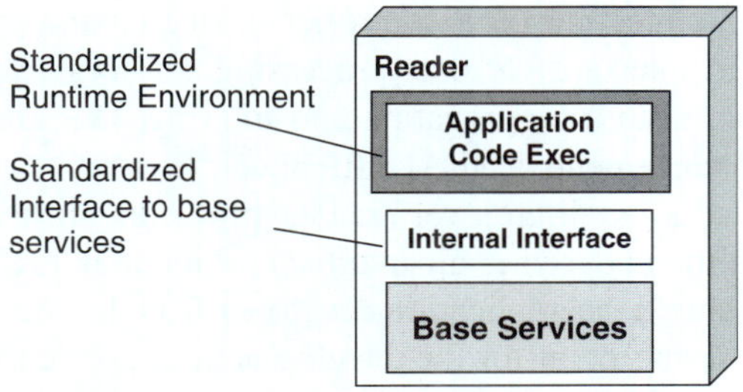

Figure 10.16 RFID reader with standard runtime environment for application code and internal interface to reader module providing the base services such as tag inventory and memory access.

Future standardization efforts might also want to consider specifying the runtime environment on RFID readers to support autonomous architectures (Figure 10.16). There are already many RFID reader products on the market that allow users to execute application codes on the reader. Agreement on a common virtual machine on these devices and a mechanism to upload applications would greatly facilitate the application development. The development of custom application logic on the reader would also benefit from a standardized internal interface to receive captured tag data and to configure the reader base services. This could be realized with an LLRP implementation that supports multiple reporting channels (Figure 10.9). Alternatively, the ALE 1.1 interface with its high-level representation of tag identifiers and user memory is also a suitable candidate. It would need to be augmented by an additional interface to configure and monitor the reader module.

10.8 Conclusion

RFID technology is increasingly often used to identify and locate objects. As businesses start depending on this information technology infrastructure, the management and control of these RFID systems are becoming an important issue. In this chapter, we categorize the services that typically comprise an RFID system into base, configuration, monitoring and data processing services. We use this service taxonomy to distinguish different system architectures deployed in the industry today. While we do not believe that the list of services we provided is necessarily exhaustive, we believe that they provide an adequate framework to analyze RFID deployments.

In our analysis, we identify two main types of system architectures. We distinguish a centralized architecture with a local hardware or software controller managing the RFID readers which provides configuration, monitoring, and data processing services and an architecture where the reader operates independently, has significant data processing capabilities itself and is configured and monitored remotely. There also exist a number of hybrid architectures that combine elements of the centralized and autonomous architecture type.

The EPCglobal community has developed a set of specifications that aim to standardize the interfaces to the configuration, monitoring and data processing services. In our opinion, the combination of DCI and LLRP specification is well suited to standardize the communication interface between RFID readers and local hardware and software controllers in centralized architectures. The EPCglobal ALE specification is today already in use in the centralized architecture on a controller device or an application server where it provides a convenient interface for data processing services across RFID readers that are application agnostic, such as aggregation and filtering. Our analysis also discusses a number of potential enhancements to the LLRP specification.

This chapter also discusses "high-level" reader protocols such as EPCglobal RP and ALE that abstract from the underlying air interface protocol and RF details. These "high-level" reader protocols shield the application developer from low-level details of RFID and provide a convenient starting point for the development of application logic. As of today, neither EPCglobal RP nor ALE 1.0 have seen significant adoption by reader vendors. We argue that the main reasons for this lack of adoption are the need to supplement them with an additional protocol for configuration and monitoring purposes and the lack of demand from end users and RFID middleware vendors. It remains to be seen whether the recently ratified ALE 1.1 with its enhanced feature set will lead to increased demand and adoption as a high-level reader protocol on reader devices. As noted above, high-level interfaces such as ALE1.1 play an important role at a higher level in the software stack.

Future standardization efforts might want to consider specifying the runtime environment on RFID readers to support autonomous architectures. Today an application code to be executed directly on a reader device needs to be customized for each reader platform. A common runtime environment and upload mechanism and a standardized interface to the reader module would greatly facilitate application development in such an autonomous architecture. Similar to today's mobile phones, future RFID readers might thus feature a standardized virtual machine supporting an RFID reader device profile.

Problems

1. Discuss the strengths and weaknesses of the centralized and the autonomous RFID system architecture. Describe scenarios in which the centralized architecture is more appropriate and those where the autonomous RFID system architecture is more appropriate.
2. Explain the difference between the EPCglobal Low-level Reader Protocol and the Application Level Events Protocol.

NB. The solutions will not be provided on the book's website.

References

[1] Bornhövd, C., Lin, T., Haller, S. and Schaper, J. (2004) Integrating automatic data acquisition with business processes - experiences with SAP's Auto-ID Infrastructure. *In Proceedings of the 30th International Conference on Very Large Data Bases (VLDB)*, pp. 1182–1188. VLDB Endowment, Toronto, Canada.

[2] Floerkemeier, C., Roduner, C. and Lampe, M. (2008) RFID Application Development with the Accada Middleware Platform, *IEEE Systems Journal: Special Issue RFID*.

[3] Sarma, S. (2004) Integrating RFID. *ACM Queue* **2**(7): 50–57.

[4] Wang, F. and Liu, P. (2005) Temporal management of RFID data, in *Proceedings of the 31st International Conference on Very Large Data Bases (VLDB)*, pp. 1128–1139. VLDB Endowment.

[5] Architecture Review Committee (2007) The EPCglobal Architecture Framework. Available at: www.epcglobalinc.org.

[6] Yang, L., Zerfos, P. and Sadot, E. (2005) Architecture Taxonomy for Control and Provisioning of Wireless Access Points RFC4118. Network Working Group.
[7] EPCglobal (2005a) Class 1 Generation 2 UHF Air Interface Protocol Standard Version 1.0.9. Available at: www.epcglobalinc.org.
[8] Impinj (2009) Speedway Reader User Guide 3.2. Available at: www.impinj.com.
[9] Intermec (2009) IF61 Enterprise Reader Product Profile. Available at: www.intermec.com.
[10] Alien (2007) Alien API Reference Guide.
[11] Thingmagic (2008) MercuryOS Advance User Guide. Available at: www.thingmagic.com.
[12] EPCglobal (2009) Discovery, Configuration and Initialization (DCI). Available at: www.epcglobalinc.org.
[13] EPCglobal (2007c) Reader Management 1.0.1. Available at: www.epcglobalinc.org.
[14] EPCglobal (2007b) Low Level Reader Protocol (LLRP), Version 1.0.1. Available at: www.epcglobalinc.org.
[15] EPCglobal (2006c) Reader Protocol Standard, Version 1.1. Available at: www.epcglobalinc.org.
[16] EPCglobal (2005b) The Application Level Events (ALE) Specification, Version 1.0. Available at: www.epcglobalinc.org.
[17] EPCglobal (2008) The Application Level Events (ALE) Specification, Version 1.1. Available at: www.epcglobalinc.org.
[18] EPCglobal (2006b) EPCglobal Tag Data Standards Version 1.3.1. Available at: www.epcglobalinc.org.
[19] EPCglobal (2007a) EPC Information Services (EPCIS) Version 1.0. Available at: www.epcglobalinc.org.
[20] EPCglobal (2006a) EPC Tag Data Translation (TDT) Standard. Available at: www.epcglobalinc.org.
[21] Elektrobit (2007) EB RFID Reader URP-1000 ETSI. Available at: www.elektrobit.com.
[22] Impinj (2007) LLRP - Reader Control Simplified. Available at: www.impinj.com.

11

Reducing Interference in RFID Reader Networks

Sung Won Kim and Gyanendra Prasad Joshi

Yeungnam University, Republic of Korea

11.1 Introduction

RFID is popular in various application areas as a key technology for object tracking. Because of its various advantages, wireless mobile RFID readers are being deployed and stationary readers are becoming more functional. RFID does not require line of sight and can process items at a greater speed than legacy bar code systems. Its diverse applications and numerous advantages are expanding the scope of RFID. For example, RFID is now being used in supply chain management, inventory control systems, manufacturing processes, security systems, warehousing, healthcare facilities, transportation, etc.

All of these diverse applications necessitate the deployment of a large number of readers in small geographical areas. There are some application areas where the readers have to work in close proximity to each other, depending on the space and situation. Furthermore, sometimes mobile readers may inadvertently collect in the same interrogation region. The interrogation region of a reader is the area within which it can read tags.

RFID systems have come to handle a wide range of products and services, including some very important and sensitive sectors (e.g. medical, security, etc.). Consequently, a more sophisticated RFID system is needed, which can provide a high tag read rate with the correct reading of multiple tags in the readers' interrogation region and ensure sufficient quality of service (QoS).

Basically, an RFID system contains three primary components: tags, readers and a data processing subsystem. Figure 11.1 shows an RFID system that consists of a reader, a tag and a back-end system. Tags, which are also called transponders, consist of an integrated circuit with a unique identification information and an antenna to receive/transceive RF

RFID Systems: Research Trends and Challenges Edited by Miodrag Bolić, David Simplot-Ryl, and Ivan Stojmenović

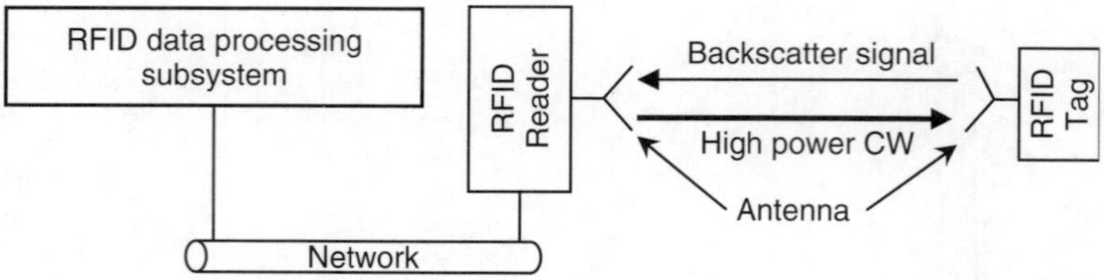

Figure 11.1 Basic RFID operation.

waves. Tags can be active, semi-passive or passive depending on their functionality. Active and semi-passive tags are onboard battery-powered and have a moderate size antenna. They can be read from a great distance, but the onboard battery increases their cost and size and limits their lifetime. Active tags can initiate transmission to readers. Many high frequency tags are passive.

Most ultra high frequency (UHF) tags are passive and operate without an onboard battery. The readers transmit a higher power continuous wave (CW) to energize the passive tag, which contains a very small antenna. The higher power CW is just enough to obtain the information from the tags by means of the signals backscattered to the reader over a very small distance as compared to active and semi-passive tags. The tag receives the energy and transmits the stored data by backscattering communication with the reader. The information from the tags is modulated onto the reader carrier signal and reflected back to the reader in the backscatter communication. Passive tags only respond to the reader; they cannot initiate any communication. Most commodities are equipped with low functionality passive tags, because they are inexpensive, small in size and as thin as a banknote.

11.2 Interference Problem in RFID Reader Networks

When there are a number of RFID readers which are deployed in or enter the same interrogation region, they form a dense RFID System. The formation of a dense RFID System is more prone to occur in the case of mobile RFID systems. This dense RFID system introduces a number of new challenges. In a dense RFID system, the lack of information exchange among the different RFID readers causes signal interference. Generally, one reader cannot know the status of the other readers, because they do not exchange their information with each other. Passive tags, which lack frequency tuning circuitry, cannot distinguish between readers and also cause reader interference. In this way, they reduce the efficiency and reliability of the RFID system, resulting in misreading, unsuccessful reading and an increase in the tag interrogation time. It is very important in any tracking system to achieve perfect accuracy, but such interference can prevent the readers from achieving the desired read rate with accuracy.

There are two types of reader interference: (1) frequency interference, which occurs when two or more readers communicate on the same frequency at the same time. This type of interference is called reader–reader collisions; and (2) tag interference, which arises when two or more readers attempt to communicate with a particular RFID tag at the same time regardless of the differences in frequency. This type of interference is

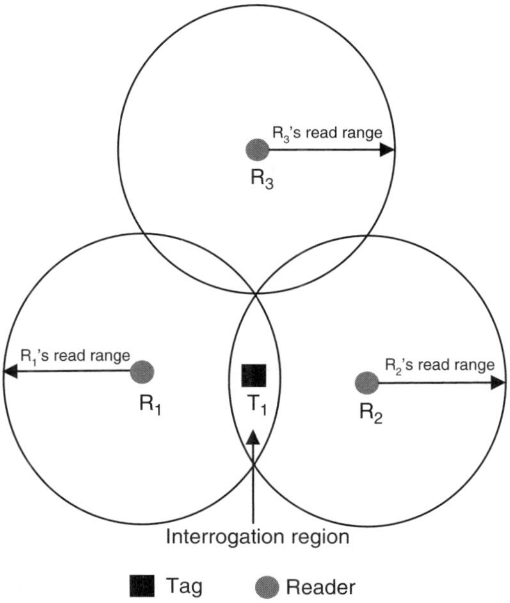

Figure 11.2 Reader-to-tag collision.

called reader–tag collisions. Both types of reader interference caused by the operation of an RFID reader are referred to as reader collisions [1]. In this chapter, we refer to the reader interference and reader collision problems interchangeably. Figures 11.2 and 11.3 illustrate the reader collision problems in detail. In Figure 11.2., tag T_1 is in the same interrogation region as the two readers, R_1 and R_2. When both readers R_1 and R_2 try to read the tag T_1 at the same time, tag T_1 cannot respond to either of the queries.

In Figure 11.3(a), tag T_1 is in the interference region of reader R_2. The in-band backscattering signal from tag T_1 to reader R_1 can be distorted by the interrogation signal from reader R_2. This kind of interference can arise when there is an unwanted transmission from a nearby reader, even when the read ranges of the two readers do not overlap. Similarly, T_1 is in the interference range of R_2 and the read range of R_1. When reader R_1 wants to read T_1 and R_2 wants to read a tag within R_2's read range at the same time and at the same frequency, their signals may interfere with each other and, thus, R_1 cannot read tag T_1. In Figure 11.3(b), reader R_1 and reader R_2's interference ranges overlap and cannot communicate at the same time and at the same frequency. These reader collisions persist in RFID systems [10–17, 38].

The reader collision problem is related to the frequency assignment problem [2–8]. The problem of allocating frequencies over time to RFID readers has been well studied and presented in [8]. Let the RFID reader network be an undirected multigraph G(V, E), where set of vertices V is the set of RFID readers and set of edges E is the set of frequency interference between readers. The frequency assignment problem is equivalent to the simple graph coloring problem that is a well-known NP problem [9]. In addition to the reader

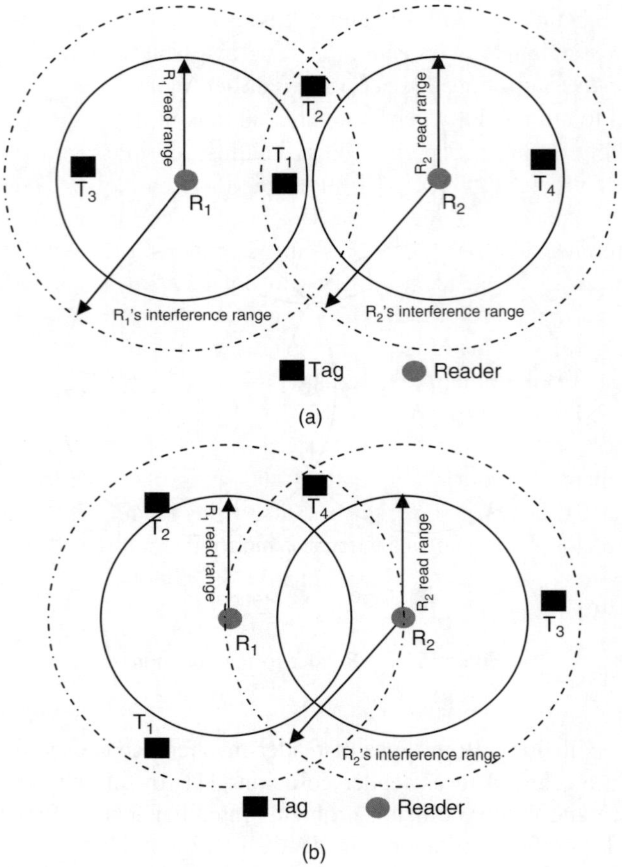

Figure 11.3 Two kinds of reader–reader collision.

anti-collision algorithm, many other algorithms have been proposed to solve the frequency assignment problem. Some examples of these algorithms are neural networks, simulated annealing, genetic algorithms, localized approximation algorithms, reader synchronization algorithm, and distributed power control algorithm [32–36].

11.3 Access Mechanism, Regulations, Standards and Algorithms

Many multiple access schemes such as Time Division Multiple Access (TDMA), Frequency Division Multiple Access (FDMA), Code Division Multiple Access (CDMA) and Carrier Sense Multiple Access (CSMA) have been proposed to solve the collision problem. However, these existing schemes cannot be used directly in RFID systems because of the following problems.

In FDMA, several transmission channels using various carrier frequencies are allocated to various readers and tags. As passive tags cannot choose a particular frequency, they cannot select a particular reader to establish a communication link.

In the TDMA scheme, the RFID reader and tags are allocated at different time slots in order to avoid simultaneous transmission. The available channel capacity is divided for the readers in the same interference range chronologically in such a way that each reader is assigned to a time slot wherein it can read tags in its range, while the rest of the readers remain silent during that period, thus avoiding any interference or collision. This scheme is similar to the graph theory of allocating different colors, where the frequency represents the color, such that none of the readers can pick the same color. This reduces co-channel interference. Graph coloring is an assignment of colors to the vertices of a graph such that any two adjacent vertices do not have the same color. The smallest number of colors needed to color a graph G is called its chromatic number, $\chi(G)$, of the graph. Determining the chromatic number of an arbitrary graph G is called simple graph coloring problems. Computing the chromatic number is a well-known combinatorial problem which is NP-hard [39].

Assigning time slots to readers is equivalent to the graph coloring problem. Determining the pairs of interfering readers and assigning TDMA slots is relatively easier in the fixed RFID readers' scenario. But, in the mobile scenario, non-interfering readers may move closer and start interfering which will require reshuffling of time slots in a dynamic topology. Having such dynamically distributed time slots will reduce the read rate of the RFID system.

CDMA uses spread spectrum modulation techniques based on pseudo-random codes. It spreads the data over the entire spectrum. While CDMA would be suitable in many ways, it adds quite a lot of complexity and would be too computationally intensive for RFID tags. CDMA requires extra circuitry, which is not cost effective for low cost, practical RFID tags.

CSMA avoids collisions by detecting whether the channel is busy or idle and waiting in the contention mode to access the channel. However, carrier sensing is not very effective in RFID systems, because of the well-known hidden terminal problem. The traditional collision avoidance techniques such as RTS (request to send) and CTS (clear to send) cannot be applied directly to RFID systems, because when one reader sends an RTS to the tags in its reading range, multiple tags respond with the CTS message. Another collision avoidance mechanism would be required to avoid collisions among the RFID tags, thus making the system more complicated. Moreover, the carrier sensing mechanism shuts down lots of sub-bands even if such bands were functional in another system.

11.3.1 Regulations

Although frequency selection in RFID depends upon the application, there are certain regulations which need to be applied when it comes to selecting the frequencies. These regulations are necessary to avoid interference between different radio systems. To provide worldwide interoperability, a particular set of frequencies are allocated, which are called ISM (Industrial, Scientific, and Medical) frequencies. The three most common ISM frequencies, which are available in most countries and are most commonly adopted in RFID systems, are 135 kHz, 13.56 MHz, and 2.45 GHz [18] for the low, intermediate, and high frequency bands, respectively.

11.3.1.1 ETSI 302 208

Listen-Before-Talk (LBT) is a multiple access scheme that works on the principle of CSMA. LBT senses the carrier of channels before transmission. If a reader finds that the channel is idle, it occupies the channel by the contention with other readers. LBT was standardized as ETSI EN 302 208 [19] according to European regulations. ETSI EN 302 208 allocates the frequency band of 865–868 MHz that is divided into 15 sub-bands, each spanning a total of 200 kHz. However, if a reader is operating at the maximum total radiated power, which is 2 W ERP (Effective Radiated Power) or equivalently 3.2 W EIRP (Effective Isotropic Radiated Power), only 10 sub-bands are available, while the remaining 5 are used as guard bands in which only a low ERP is allowed. In this standard, all readers must listen to the ongoing transmission in the channel before accessing it.

The listen time comprises a fixed period of 5 ms plus a random time of 0 to 5 ms in 11 steps. If the sub-band is free, the random time is set to 0 ms [19]. To perform communication efficiently, the channel is occupied for up to 4 sec, after which it must release that sub-band for at least 100 ms. Since the time delay and collision probability are high in this standard, it is inefficient for dense RFID systems. Due to the rapid development in RFID industry, multiple readers can transmit simultaneously on the same channel. Therefore, in the current version of European standard, EN 302 208-1 Ver. 1.3.1 (Electromagnetic compatibility and radio spectrum matters (ERM); radio frequency identification equipment operating in the band from 865–868 MHz with power levels up to 2 W), the use of LBT is optional [20].

11.3.1.2 FCC

In the Title 47 Part 15.247 of the FCC (Federal Communication Commission) regulations, a given spectrum is divided into different frequency bands and uses the frequency hopping spread spectrum (FHSS) [21]. This is one of the most efficient ways to avoid the effect of interference and to avoid causing interference to users of a shared spectrum. The transmitted energy is distributed, thereby reducing the likelihood of interference arising with other systems. Likewise, all readers are expected to randomly alternate between these bands in order to reduce the probability of collisions. The receiver frequency varies continually in order to avoid the effects of other users blocking the reader's receiver. FHSS significantly reduces the frequency interference, but cannot overcome the tag interference issues, as it hops between the different channels used by the readers. Nevertheless, there are different regulations in different countries and there is no single global public regulation for RFID.

11.3.2 Standards

11.3.2.1 EPC Class 1 Generation 2 UHF RFID Protocol

The EPC Radio-frequency Identification Protocol, a Class 1 Generation 2 UHF RFID Protocol [22], is an open and global standard protocol developed by EPCglobal [23] for RFID systems operating in the 860–960 MHz frequency range. EPCglobal is a non-profit organization formed as a joint venture between GS1 (formerly EAN International) and GS1 US (formerly Uniform Code Council, Inc.). This Class 1 Generation 2 UHF RFID protocol describes the spectrum management of RFID operations in a dense reader environment. Frequency hopping is suggested for efficient frequency utilization. In a dense reader environment, interrogator transmissions operate in even-numbered channels and tag backscatters are located in odd-numbered channels.

This protocol separates the reader transmission and tag transmission into separate frequency channels, so the reader-to-tag collisions should never happen. However, in a dense reader environment, when two readers use two separate frequencies to communicate with the tag, the tag will not be able to tune to a particular frequency and, hence, collisions can occur at the tag. Thus, in this standard, the reader-to-tag collision problem remains unsolved.

11.3.2.2 ISO 18000 Standards for RFID

The International Standards Organization (ISO) defines the air-interface communication between RFID readers and tags. Unlike EPCglobal, which addresses only the UHF specification, ISO has defined a standard range of frequencies to be used. The ISO 18000 standards [24] for RFID are summarized in Table 11.1.

11.3.3 Reader Anti-Collision Algorithms

Many reader anti-collision algorithms have been developed and published in the literature to reduce interference in RFID reader networks. Some of the existing anti-collision algorithms and their functionalities are discussed below.

11.3.3.1 Classification of Reader Anti-Collision Algorithms

Each reader anti-collision algorithm proposed in the literature has its own unique properties and functionalities. Some operate by means of scheduling, some in a distributed way, and some work on the principle of a notification mechanism of broadcasting control packets. Figure 11.4 shows the classification of the existing approaches and examples of some of the well-known protocols.

Table 11.1 The ISO 18000 standards for RFID.

ISO 18000 standard	Frequencies	Spectrum
ISO/IEC 18000-2:2004	Below 135 kHz	Low frequency
ISO/IEC 18000-3:2008	At 13.56 MHz	High frequency
ISO/IEC 18000-4:2008	At 2.45 GHz	Microwave
ISO/IEC 18000-6:2004	At 860 MHz to 960 MHz	UHF
ISO/IEC 18000-7:2008	433 MHz	UHF

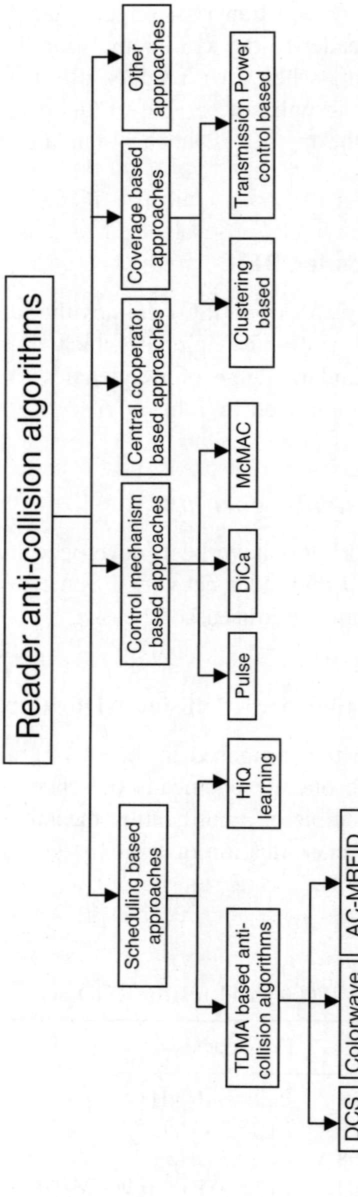

Figure 11.4 Classification of existing RFID reader anti-collision algorithms.

11.3.3.2 Scheduling-Based Approaches

The available system resources such as the frequencies and time are allocated among the readers to prevent them from transmitting simultaneously. This kind of approach can reduce the possibility of reader collisions effectively. The following algorithms fall in this category.

TDMA-Based Anti-Collision Algorithms

In the TDMA-based anti-collision algorithm, time slots are divided into reader-to-tag communication period and reader-to-reader communication period. Figure 11.5 shows the time slot structure of TDMA-based anti-collision algorithm. By the reader-to-reader communication in the reader-to-reader communication period, each reader decides its time slot for transmission. The reader-to-tag period is used for reading the tags of a reader.

The following are the distributed TDMA-based RFID reader anti-collision algorithms. The goal of these algorithms is to color a reader network so that each reader node has the smallest possible number of adjacent nodes with the same color. A color is a periodic reservation for collision-free transmission of data.

DCS

Distributed color selection (DCS) [11, 12] allows easy reservation of time slots. A reader with a queued request for transmission transmits only in its color (time slot). If the transmission collides with another reader, the transmission request is discarded and the reader chooses a new color and reserves it. If any of neighboring readers has the same color, it has to choose another color, thereby clearing the time slot for the reader the next time the reader needs to transmit. This switch and reservation action is referred to as a "kick." In DCS, the maximum number of colors is fixed. Each reader keeps track of what color it believes the current time slot to be. Figure 11.6 shows subroutines to manage collision, color reservation, and kick resolution.

For example, in Figure 11.3(b), two readers R_1 and R_2 are neighbors. Assume that reader R_1 has a queued request for transmission. First of all, it waits for its color time slot and starts transmission. If the transmission experiences collision, reader R_1 selects one of the colors from the maxcolor (DCS requires 8–10 average colors to achieve 98% successful transmissions at 100% transmission [12]). Reader R_1 broadcasts kick packet with newly selected color. Also, if reader R_2 receives the same color as its current color, it randomly changes to different color within maxcolor. DCS is a "greedy" algorithm – a node's chances of colliding immediately after experiencing a collision are minimized at the expense of its neighbors.

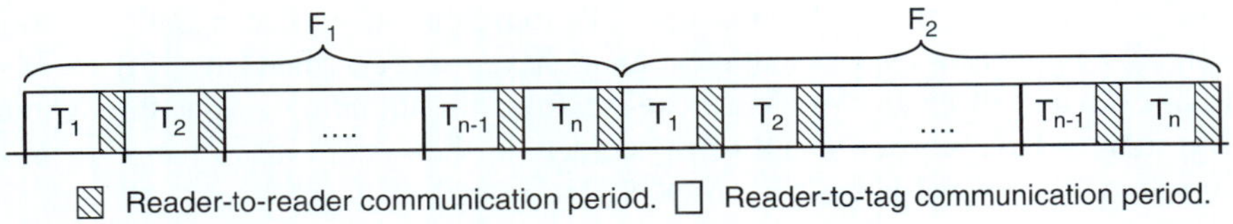

Figure 11.5 Frame structure of TDMA-based anti-collision algorithm.

```
DCS Subroutine

  • If experienced collision
      current color == random(maxcolors)
            broadcast kick with new color
  • If kick packet received with current color
      select different color randomly within maxcolors
```

Figure 11.6 DCS pseudocode.

Variable-maximum Distributed Color Selection (VDCS or Colorwave)
The maximum colors variable is fixed and does not change throughout the algorithm's execution in DCS. A single input for color does not provide flexibility to RFID reader networks with variable transmission probability. Thus, a mechanism for dynamically changing the maximum number of colors at a reader is needed. In [11, 12], Colorwave was suggested as an advanced version of DCS. In Colorwave, each reader monitors the percentage of successful transmissions. Five inputs to the algorithm determine when a reader changes its local value of maximum number of colors (maxcolors):

1. **UpSafe**: the safe percentage at which to increase maxcolors.
2. **UpTrig**: the trigger percentage at which to increase maxcolors if a neighboring reader is also switching to a maxcolors higher than that of this reader.
3. **DnSafe, DnTrig**: analogues of UpSafe, UpTrig, except that they are used for decreasing maxcolors.
4. **MinTimeInColor**: the minimum number of time slots before the Colorwave algorithm will change the value of maxcolors again after initialization or changing maxcolors.

Each reader chooses a random color from 0 to maxcolors to transmit. If it collides, it selects a new time slot and sends a kick (small control packet) to all of its neighbors to indicate the selection of a new time slot. If any neighbor has the same color, it chooses a new color and sends a kick, and so on. In DCS, as described above, each reader keeps track of what color it believes the current time slot to be. However, all the participating readers monitor the percentage of successful transmissions in Colorwave.

When a reader executing Colorwave reaches a Safe percentage to change its own value for maxcolors, it will send out a kick to all of its neighboring readers. If the phenomenon that is causing it to exceed a Safe percentage is local to that reader, the other readers will not have passed their own Trig percentages and will not respond. However, if the phenomenon causing the collision value to exceed a Safe threshold is widespread, neighboring readers will most likely have exceeded their own Trig thresholds, and a kick wave will ensure. As kicks spread from the initiating reader throughout the entire system, a large portion or all of the readers in the system may change their value of maxcolors. Colorwave is built upon DCS algorithm.

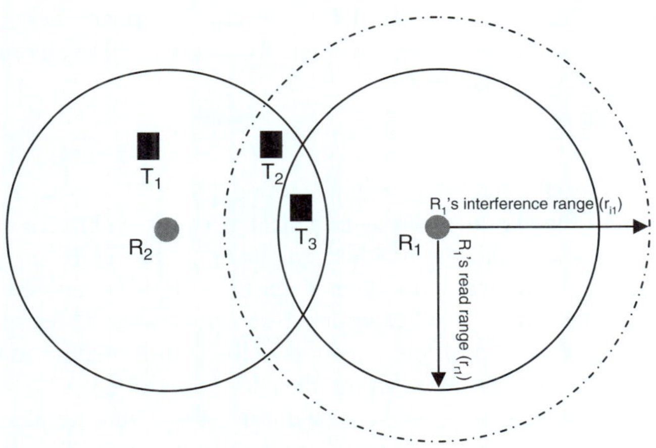

Figure 11.7 An illustration of the AC-MRFID protocol.

Anti-Collision Algorithm for Mobile RFID (AC-MRFID)

Similar to Colorwave, AC-MRFID [25] is also a distributed TDMA-based algorithm, where each reader chooses a random time slot to transmit. Colorwave works fine in the case of fixed or barely moving RFID readers. However, in the case of Mobile RFID readers, reader collisions may occur frequently when they form a dense RFID system and, hence, the value of maxcolors becomes unnecessarily high. An improved version of Colorwave, AC-MRFID, controls the number of time slots allocated to a reader in a frame by setting its value according to the number of readers, which are located in the interference range of the reader.

As shown in Figure 11.7, r_{i1} is the interference range and r_{r1} is the read range of reader R_1. If reader R_1 attempts to read tag T_3 and collides with reader R_2, reader R_1 calculates the number of time slots in a frame by Equation (11.1) and changes the value of max_timeslots. In Equation (11.1), α is the number of readers in the read range.

$$\max\_timeslots = \alpha + \left\lfloor \alpha \times \frac{(r_i^2 - r_r^2)}{r_r^2} \right\rfloor + 1 \qquad (11.1)$$

At the same time, because reader R_2 also experiences a collision, it also changes its number of timeslots in a frame to max_timeslots by calculating Equation (11.1). After changing their numbers of time slots, readers R_1 and R_2 randomly select their time slot (current_timeslot) for reading tags in the next frame and let the other readers, which are located in their interference ranges, know their selection of current_timeslot by broadcasting the information. If a reader, except for the readers that are broadcasting (in this case, readers R_1 and R_2), receives information including a new current_timeslot, it randomly selects its new current_timeslot within the limit of its max_timeslot.

DCS, Colorwave and AC-MRFID are simple, and distributed protocols. However, it requires the system to establish and maintain information over the network, which is time- and energy-consuming. In mobile RFID systems, overhead cost due to time slot reselections increases continuously and significantly, because these algorithms need tight time synchronization among the readers.

HiQ-Learning

HiQ-learning [10] based on reinforcement learning [26, 27] was used to develop an online learning algorithm to use in a hierarchical network architecture. HiQ provides a solution to the reader collision problem by learning the collision patterns of the readers and assigning frequencies to them over time. It is composed of three basic hierarchical layers: the reader, the reader-level server and the Q-learning server. The reader, which is in the lowest position, transmits collision information to the reader-level server. Each individual reader-layer server then assigns resources to its readers. The readers communicate when they are assigned time slots and frequencies. They are aware of the frequency and time slots that are allocated to them. They detect collisions by communicating with the other readers in the overlapped interrogation zone. If two readers communicate using the same time slot and frequency, they will experience both tag and frequency interference. Each reader stores the number of collisions it has experienced. This information is known to the reader-level server, as the reader directly communicates with it. Q-servers allocate the resources to the server below them, that is, to the reader-level servers. Regardless of the number of Q-servers, there is always a root Q-server that has global knowledge of the frequency and time and is able to allocate them. Figure 11.8 shows the hierarchical structure of Q-learning.

In Figure 11.8, readers R_1, R_2, R_3 and R_4 are within the interrogation range of reader R_5. Whenever reader R_5 has to read the tags, it sends the request for frequency and time. It starts reading after getting frequency and time. At the same time it also pings other readers in the same interrogation zone. Reader R_1 is responsible for collision detection processing. After getting the ping, readers R_1, R_2, R_3 and R_4 send back the response with their current state, that is, whether they are reading or not and which frequency and time slot they are using.

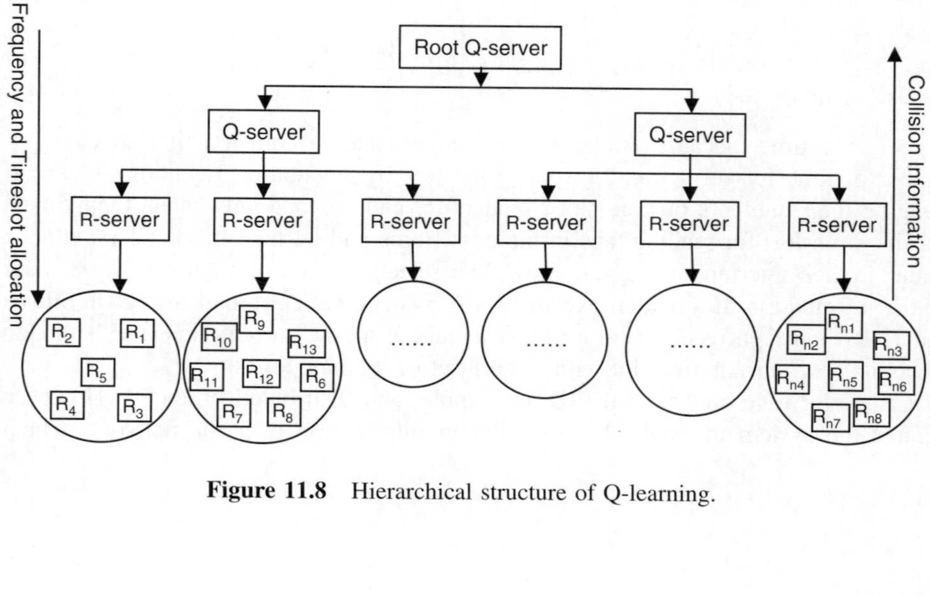

Figure 11.8 Hierarchical structure of Q-learning.

The main drawback with this hierarchical approach is that additional management of the overall hierarchy is required for even a slight change in the lower layer. It is not favorable to highly mobile environments, because in such an environment the management overhead increases exponentially. Another overhead in this protocol is time synchronization, which requires the use of time slots and that all of the readers be synchronized.

11.3.3.3 Control Mechanism-Based Approaches

This approach mitigates the problem of collisions between readers by transmitting notification control packets such as beacon signals. After receiving a beacon signal, the interfering readers interrupt their ongoing communication and wait for the next cycle. This approach efficiently solves the problem of reader-to-reader collisions. However, the actual communication takes place between the reader and tag. This type of protocol does not address the reader-to-tag collision problem, which lowers the RFID performance. The following algorithms fall into this category.

Pulse Protocol

Pulse [13, 14] is a CSMA-based notification protocol that attempts to solve the reader collision problem using two separate channels in the RFID system. One channel, called the data channel, is used for reader-to-tag communication and the other is the control channel that is used for reader-to-reader communication. Broadcasting messages in the control channel does not affect the ongoing communication in the data channel. This protocol mitigates the reader collision problem by continuously transmitting beacon signals through the control channel, while the reader is communicating with the tag through the data channel.

Each reader goes into the waiting state and waits for a DIFS (DCF Interframe Space) time. If it does not receive any beacon signal, the reader concludes that there is no other reader reading the tag. In this case, it enters the contention phase. If the reader receives the beacon signal at this stage, then it waits for a DIFS time in the next cycle until a randomized back-off time is over. If the reader does not receive any beacon signal, it starts reading the tag. While it is reading the tag, it sends a beacon signal in the control channel. The beacon range is equal to the interference range. After reading the tag, it goes back to the waiting state.

Pulse mitigates the reader collision problem significantly, but the hidden terminal problem and exposed terminal problem still exist. Since a beacon does not have any destination address in its structure, it is just a broadcast message on a control channel. When the validity time of a beacon has elapsed, the reader concludes that there is no other reader in the neighborhood which is reading the tag. A beacon is just a means of solving the collision problem. Therefore, a beacon may collide with another beacon from another reader. Furthermore, whenever two channels are used, a transceiver may be required for each channel. A large amount of energy is consumed during carrier sensing, receiving the beacons, and overhearing the beacons.

DiCa (Distributed Tag Access with Collision Avoidance)

DiCa (Distributed Tag Access with Collision Avoidance) [15] is a distributed and energy-efficient collision avoidance algorithm. Similar to the Pulse protocol, it also has a data channel and a control channel. Each reader contends for use of the data channel through the control channel. The winner reads the tags through the data channel while the other readers wait until the channel is idle. The following packets are exchanged for the purpose of collision avoidance:

- *BRD_WHO*: Packet used to identify whether a reader reading tags exists in the same network or not.
- *BUSY*: To indicate that the reader is now reading tags.
- *BRD_END*: Packet used to indicate that the channel is idle after the tags have been read.

A reader that has to read tags (say, R_1) broadcasts the BRD_WHO message to the readers in the same interrogation area to discover the presence of other readers reading tags. After getting the BRD_WHO message, the reader in the same interrogation area (say, R_2) sends back a BUSY message if it is reading tags. Now, R_1 has to wait until it gets BRD_END from reader R_2. If R_1 does not receive a BUSY message after sending a BRD_WHO message for a certain period of time, it assumes that the data channel is free and it starts reading tags. Each reader sends a BRD_END message after completion of reading process. After receiving a BRD_END message, readers wait for random back-off time before sending a BRD_WHO message.

Unlike Pulse protocol, DiCa takes into account the hidden and exposed terminal problems by adjusting the control channel range to twice the radius from the first reader. This channel adjustment in DiCa reduces the energy consumption, so it is more suitable for energy-constrained mobile RFID systems. DiCa consumes less energy than Pulse, CSMA, and ALOHA.

DiCa requires sufficient time to exchange the contention message. It has no fixed tag data size and with a small data size there may not be sufficient time to exchange contention messages, which increases the collision probability.

Multi-Channel MAC Protocol (MCMAC)

The Multi-Channel MAC protocol (MCMAC) [16] is a contention-based MAC protocol for RFID systems. In MCMAC, there are N-1 non-overlapping data channels with the same bandwidth and a control channel. Similar to the Pulse protocol, the control channel is a sub-band of the RFID spectrum and is only used for reader-to-reader communication. Readers can communicate simultaneously through the data and control channels.

As shown in Figure 11.9 (MCMAC's working principle), MCMAC works in a similar fashion to the conventional LBT. MCMAC broadcasts a control message after it wins contention in a control channel and thereby gains access to the data channel. The control message informs other neighboring readers within the interrogation zone that this particular channel is occupied for a certain time. After receiving a control message from a neighboring reader, the other readers do not use that channel for a certain period of time and try to gain access to another channel.

This approach mitigates the reader-to-reader problem significantly. However, the reader-to-tag problem still exists with this approach. Since passive RFID tags are unable to discriminate between two data channels, multiple data channels cannot be used directly in a passive tag environment.

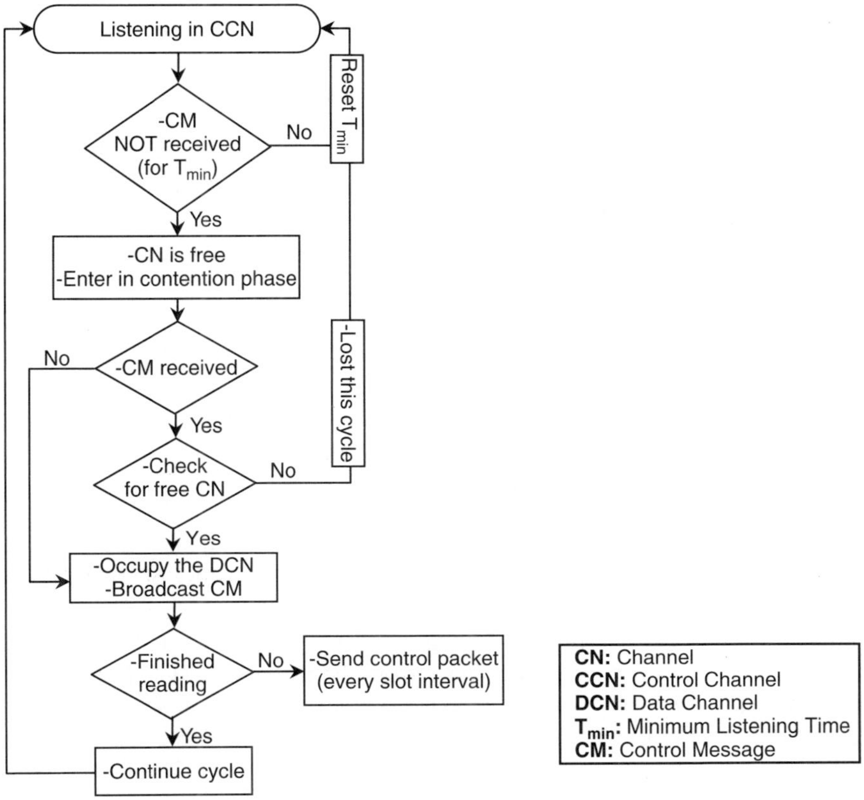

Figure 11.9 MCMAC working principle.

11.3.3.4 Coverage-Based Approaches

Adaptive transmission range-based RFID anti-collision protocols and cluster-based RFID anti-collision protocols come under this approach. In the cluster-based approach, the coverage ranges of the clusters are dynamically adjusted. A cluster head is elected to communicate with the server in an ad hoc network fashion. In adaptive transmission range-based RFID anti-collision protocols, the read ranges of the readers are dynamically adapted to reduce the overlapped areas between adjacent readers. This approach usually needs a central node to calculate the distance between each pair of readers and adjust their reading ranges, which increases the complexity and cost of the system [17]. This approach is nevertheless energy-efficient.

An example of this approach is the array-based reader anti-collision scheme (ARCS) [28]. This scheme prevents collisions by grouping the readers and reducing the read cycle time. In ARCS, a reader agent controls every reader, that is, it turns the reader on or off according to the ARCS procedure. The reader agent is an interface that connects the physical reader device and the middleware. At first, ARCS turns on all the readers sequentially and synchronizes them. Each reader scans the tags and registers them in the registration table. Then it makes a network topology array of the

RFID system that marks the tags that present in the interrogation regions of two or more readers. After analyzing and marking the relationship from the registration table, it creates a scheduling table for all of the readers. ARCS prevents the occurrence of collisions with the initial request signal, rather than responding to collisions after the event. This approach needs very strict time synchronization.

Another unique approach to the coverage-based RFID reader anti-collision mechanism is the protocol proposed in [29]. This is a localized clustering coverage protocol that mitigates the reader collision problem in a homogeneous RFID. There is no communication between the readers, so this protocol cannot solve the hidden terminal and exposed terminal problems completely.

11.3.3.5 Central Cooperator-Based Approach

A central cooperator-based solution is proposed in [30]. In this central cooperator (CC)-RFID system, a central device is used to communicate between the tags and the readers. The problem of collisions is dealt with by adopting a multipoint to single point (MP2P) classical collision problem. The reading queries of multiple readers are multiplexed by the central co-operator and the tag information can be stored and shared among adjacent readers. The central co-operator controls the entire working process of the RFID system.

As shown in Figure 11.10, CC is a power device. Therefore, it can charge the tags in its vicinity to extend the reading range of the readers. The CC device has six modules: (1) CC Receiver Module1: to receive queries from the readers; (2) CC Relay Module: to multiplex and relay the readers' queries; (3) CC Sender Module1: to send out the readers' queries to the tags; (4) CC Receiver Module2: to receive the feedback from the tags; (5) CC Storage Module: to store the data from the tags; and (6) CC Sender Module2: to send the tags' data back to the readers in a sub-carrier.

As shown in Figure 11.11, two kinds of working schemes are proposed for two different RFID application environments: (1) CC-RFID scheme one; and (2) CC-RFID scheme two. CC-RFID scheme one is proposed for highly mobile tag environment. Multiple readers compete for the access to the channel and send their reading requests to CC. Then CC demodulates the request information and multiplexes them into a single reader's query. CC sends the multiplexed query to read tags and the tags respond with their identification information. After receiving the identification information, CC selects the tags' response information by examining the original requests from readers and sends it to each reader separately.

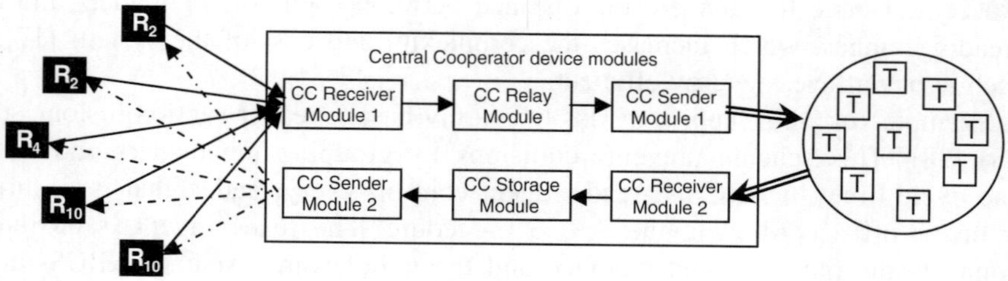

Figure 11.10 CC-RFID system architecture.

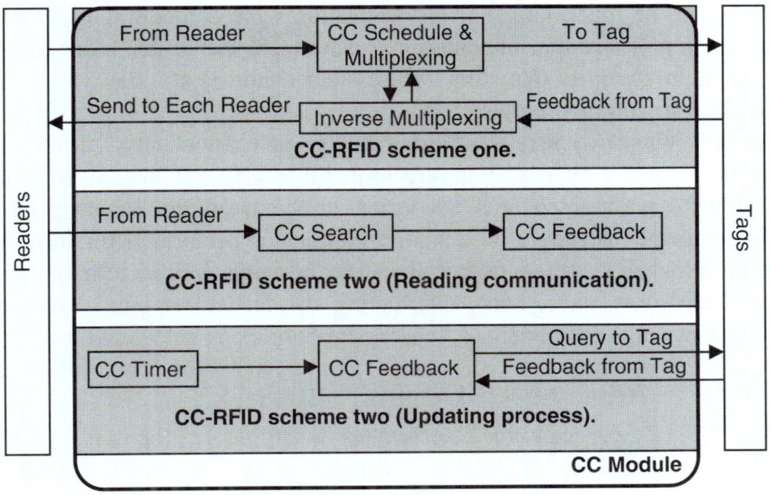

Figure 11.11 Two working schemes of CC-RFID.

CC-RFID scheme two has two independent communication processes: (1) the reading communication between readers and tags; and (2) the updating process between readers and tags (Figure 11.11). In the first communication process, multiple readers compete to send their reading requests to CC. The CC receives readers' requests and looks through the CC Storage Module for the corresponding tags' ID requested by readers. After collecting all the required information, CC sends them back to readers and the reading process is over. In the second communication process, a renewal timer starts to update the tags information stored in CC. As the updated tags' information is necessary, CC sends updating query to read tags and tags respond by sending their latest information to CC. The CC-RFID scheme two is appropriate for static or less mobile tag environments.

This approach has additional overhead associated with the use of an extra device, namely the central cooperator. It might have a scalability problem, because the central device is mounted in a specific place and has a fixed cooperation capacity.

11.3.3.6 Some Other Approaches

The Adaptive Channel Hopping Algorithm (ACHA) [31] is a preliminary approach to prevent RFID collisions. It is multichannel protocol. This algorithm combines the LBT algorithm, a random back-off mechanism with a specific hopping method. In the dense RFID system, where density of readers per channel (number of readers/number of channels) is more than one, simply to select a channel randomly is not efficient, because the reader might perform a lot of channel hopping in order to find the idle channel when the channel utilization is high. If all the channels are busy, readers may keep on hopping to busy channels repeatedly. Furthermore, the reader may check the same channel several times.

A reader waits in the current channel until it becomes idle and it tries to occupy the channel when the current reader finishes the transmission. If the probability that some of the other channels are idle is high, just waiting in that channel is not efficient. The waiting time could be much longer if the reader loses the contention again with another reader after waiting. Therefore, unconditional waiting of the reader is inefficient. On the other hand, if the probability that some of the other channels are idle is high, hopping to find an idle channel can be a better choice.

So, in ACHA, whenever a reader finds the current channel busy, the reader takes a decision on the basis of channel hopping probability (HP). It may wait in the current channel or hop to another channel according to the hopping sequence. The hopping sequence is generated whenever the channel hopping is performed for the first time by calculating the channel hopping probability. The hopping sequence helps the reader to check the channel only once, hence preventing the reader hopping the same channel repeatedly (i.e. looping). The channel hopping probability is calculated as:

$$HP(i,n) = 1 - CLP(i,n) \cdot U(i,n) \tag{11.2}$$

$$CLP(i,n) = (1-\alpha) \cdot CLP(i,n-1) + \alpha \cdot CL(i,n) \tag{11.3}$$

$$Where,\ CL = \begin{Bmatrix} 1, \text{unsuccessful contention} \\ 0, \text{successful contention} \end{Bmatrix}$$

And,

$$U(i,n) = (1-\alpha) \cdot U(i,n-1) + \alpha \cdot U_t(i,n) \tag{11.4}$$

where i is the reader's ID and n is the number of calculations. CLP is contention loss probability, and U_t is the channel utilization for a certain period of time. U is the channel utilization, that is, how long the channel is occupied. Equation (11.3) uses the exponential averaging method that applies weighting factor to the recent value.

After selecting a particular channel, it senses the channel by performing LBT. This approach improves performance and reduces waiting time significantly. This approach works effectively when there are many sub-bands, so that readers can hop from one channel to another, whereas in the case of RFID, the number of sub-bands allocated in the UHF standard is very limited, except in the US.

11.4 Comparison

As discussed in Section 11.3, the existing reader anti-collision protocols used in RFID systems have many fundamental differences. Some of them rely on centralized control for communication, while others function as distributed algorithms with fixed or dynamic channel assignment.

Table 11.2 shows a comparison between eight promising protocols for resolving the reader collision problem based on their nature and operation. As the reader collision problem is similar to the frequency assignment problem in mobile communication systems, most reader anti-collision algorithms allocate frequencies over time to a set of readers to mitigate the collision problem. To eliminate the collision problem in a dense RFID system, the readers should work cooperatively with the tags.

As shown in Table 11.2, the major overheads in the RFID anti-collision protocols are as follows. Colorwave and AC-MRFID require tight time synchronization. HiQ-learning has high management overhead in a mobile environment. In a highly mobile environment,

Table 11.2 Comparison of reader anti-collision algorithms.

Algorithm	Function used	Carrier sensing	Major overhead	Distributed control	Fixed Channel Assignment (FCA)	Dynamic Channel Assignment (DCA)	Tag side consideration for collision avoidance
Colorwave	Color number	×	Time synchronization	√	√	√	×
AC-MRFID	Color number	×	Time synchronization	√	√	√	×
HiQ Learning	Cost function	×	Management overhead	×	×	√	×
Pulse	Beacon frame	√	Energy consumption and additional channel resource	√	√	×	×
DiCa	Energy aware	√	Additional channel resource and time	√	√	×	×
MCMAC	LBT	√	Additional channel resource and computational overhead	√	√	√	×
CC-RFID	MP2P	×	Special hardware	×	×	×	√
ACHA	Probability based channel hopping	√	Multiple channel resources	√	√	√	×

the management overhead increases exponentially with increasing the numbers of readers. The pulse protocol requires more energy for contention and the probability of collisions occurring in the control channel is also high. DiCa spends a lot of time in control channel negotiation. MCMAC and ACHA require multichannel capability and a large amount of computation for channel hopping. Finally, CC-RFID needs special hardware and a database called the central cooperator.

Table 11.3, taken from [37], shows a comparison between the different reader anti-collision protocols in terms of their channel assignment. All of the protocols in Table 11.3 are multichannel protocols. MCMAC and ACHA use multiple data channels for data communication, whereas Pulse and DiCa use only one data channel. However, all of them need one dedicated control channel for control signal transmission. Pulse and DiCa seem to be promising protocols except that a certain amount of time is wasted. They have similar approaches, however, DiCa offers some improvements over Pulse.

Figure 11.12, taken from [37], shows a comparison between the different protocols in terms of the network throughput. Some of the representative protocols discussed above (Colorwave, 1-persistent CSMA, HiQ, Pulse, DiCa, McMAC, ACHA and CC-RFID) are compared. The number of readers varies between 2 to 20 and the readers are deployed in different random topologies. The result shown is for a total of 50 simulations per protocol. Tags are also varied from 100 to 400 and are placed in a grid for the simulations in 10×10 meter area. In case of ACHA, the total number of channels are 4, LBT time is 5 ms, waiting time after transmission is 0.1 sec, and trigger distribution of the reader is Poisson distribution. Read range and interference range of the readers are 1.6 meter and 5.4 meter respectively. In Figure 11.12, except for the protocol using a special device (i.e.

Table 11.3 Comparison in terms of channel assignment.

Criterion → algorithm ↓	Multi channel	Multi-data channel	Dedicated control channel requirements	Indispensable initiative	Optimization technique
MCMAC	Yes	Yes	Yes	LBT in multi channel	Multiple data channel assignment
ACHA	Yes	Yes	Yes	LBT & channel hopping	Multiple data channel hopping
Pulse	Yes	No	Yes	Beaconing in control channel	Control signaling
DiCa	Yes	No	Yes	Handshaking	Improved control channel range

Reproduced with permission of © 2008 IETE from *IETE Technical Review*, vol. 25, no. 5, Sept.–Oct. 2010.

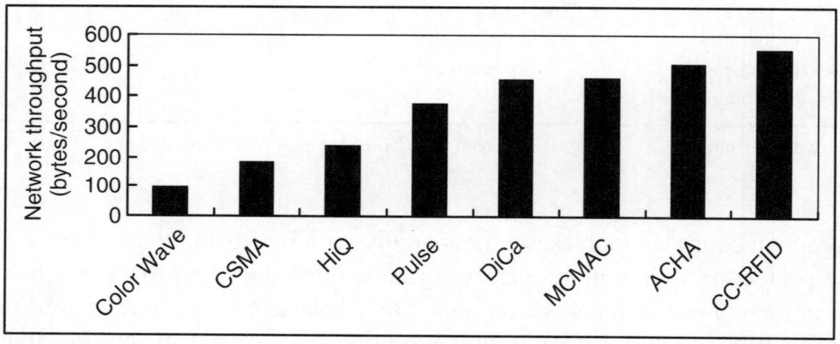

Figure 11.12 Throughput comparison. Reproduced with permission of © 2008 IETE from *IETE Technical Review,* vol. 25, no. 5, Sept.–Oct. 2010.

CC-RFID), DiCa gives the highest throughput of the single data channel protocols. CC-RFID shows the best throughput performance of all of the protocols. However, CC-RFID needs a special device called a central cooperator, as described above. Among the multichannel protocols, ACHA has the highest throughput. Since, it is not fair to compare single data channel protocols and multiple data channel protocols in terms of their throughput, in overall, DiCa shows the better performances.

11.5 Conclusion

In this chapter, the reader collision problem in RFID systems has been discussed. The existing solutions to the reader collision problem are surveyed and the current regulations and standards are outlined. The current solutions are also classified and their characteristics, functions, working principles, and limitations are listed. The problems that are not addressed by the current standards are pointed out. It is found that although many schemes have been proposed as novel solutions to the reader collision problem, still some more

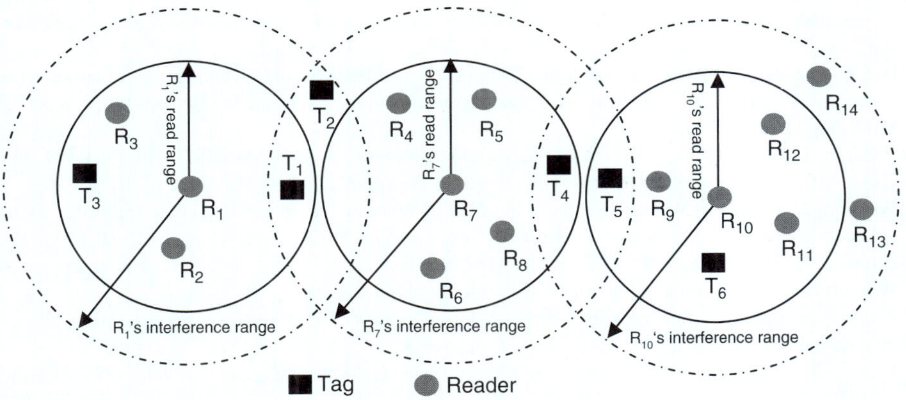

Figure 11.13 Deployment of UHF RFID readers and tags.

improvements need to be made. The use of inexpensive passive tags is one of the reasons why RFID has become popular. Nevertheless, due to their limited functions, the collision problem still exists and needs to be completely solved. Of the above-mentioned protocols, AC-MRFID, DiCa and ACHA seem to be the most promising protocols in terms of the throughput.

Problems

1. What is the main difference between DCS, Colorwave and AC-MRFID?
2. Consider Figure 11.13, which shows UHF RFID readers and tags. Assume that all the readers are homogenous having a read range of 2 meter and an interference range of 3 meter. For AC-MRFID, calculate the maximum time slot for reader 1, reader 7 and reader 10.

References

[1] Engels, D.W. and Sarma, S.E. (2002) The reader collision problem, in *Proceedings of IEEE International Conference on Systems, Man and Cybernetics*, Oct. 6–9, Hammamet, Tunisia.

[2] Metzger, B.H. (1970) Spectrum management technique, paper presented at 38th National ORSA meeting, Detroit, MI.

[3] Zoellner, J.A. (1973) Frequency assignment games and strategies, *IEEE Trans. Electromagnetic Compatibility*, EMC, **15**(4): 191–196.

[4] Hale, W.K. (1980) Frequency assignment: theory and applications, in *Proceedings of the IEEE. 1980*, **68**: 1497–1514.

[5] Yacoub, M.D. (1993) *Foundations of Mobile Radio Engineering*. Boca Raton, FL: CRC Press.

[6] Katzela, I. and Naghshineh, M. (1996) Channel assignment schemes for cellular mobile telecommunication systems, *IEEE Personal Communications*, **3**(3): 10–31.

[7] Malesinska, E. (1997) Graph-theoretical models for frequency assignment problems, Ph.D. thesis, Technischen Universität Berlin.

[8] Engels, D.W. *The Reader Collision Problem*. White Paper, Auto-ID Center MIT. Available at: http://www.autoidlabs.org/uploads/media/MIT-AUTOID-WH-007.pdf.

[9] Zhou, S., Luo, Z., Wong, E., and Tan, C.J. (2007) Interconnected RFID reader collision model and its application in reader anti-collision, in *IEEE International Conference on RFID 2007*, March 26–28, pp. 212–219.

[10] Ho, J., Engels, D.W, and Sarma, S.E. (2006) HiQ: A hierarchical Q-learning algorithm to solve the reader collision problem, in International Symposium on Applications and the Internet Workshops; SAINT Workshops 2006, Jan. 23–27.

[11] Waldrop, J., Engels, D.W., and Sarma, S.E. (2003) Colorwave: A MAC for RFID reader network, in *IEEE Wireless Communications and Networking Conference, WCNC 2003*, March 20, **3**: 1701–1704.

[12] Waldrop, J., Engels, D.W., and Sarma, S.E. (2003) Colorwave: an anticollision algorithm for the reader collision problem, in *IEEE International Conference on Communications*, May 11–15, **2**: 1206–1210.

[13] Birari, S.M. and Iyer, S. (2005) PULSE: A MAC protocol for RFID networks, in 1st International Workshop on RFID and Ubiquitous Sensor Networks, Nagasaki, Japan.

[14] Birari, S.M. (2005) Mitigating the reader collision problem in RFID networks with mobile readers, M.S. dissertation, Indian Institute of Technology Bombay.

[15] Hwang, K.I., Kim, K.T., and Eom, D.S. (2006) DiCa: Distributed tag access with collision-avoidance among mobile RFID readers, in EUC Workshops 2006, LNCS.

[16] Hongyue, D., Shengli, L., and Hailong, Z. (2007) A multi-channel MAC protocol for RFID reader networks, in *WiCom 2007*, Sept. 21–25, pp. 2093–2096.

[17] Acharya, S., Han, N., Shon, S.H., and Kim, J.M. (2006) Elimination of the reader on reader collision problem by transmitting a beacon like signal through a control channel in RFID system, in *Proceedings of IEEK Fall Conference 2006*, Nov., pp. 299–302.

[18] Finkenzeller, K. (2003) *RFID Handbook*, 2nd edn. Chichester: John Wiley & Sons Ltd, pp. 195–219.

[19] ETSI EN 302 208-1 v1.1.1, September, 2004. Available at: http://www.etsi.org.

[20] ETSI EN 302 208-1 v1.3.1, July, 2009. Available at: http://www.etsi.org.

[21] FCC Code of Federal Regulations, Title 47, Vol. 1, Part 15, Sections 245–249.47CFR15. Oct 1, 2000.

[22] EPCglobal. EPCTM radio-frequency identity protocols class-1 Generation-2 UHF RFID protocol for communications at 860 MHz–960 MHz version 1.1.0. EPCglobal Standard Specification, 2007.

[23] Electronic Product Code. Available at: http://www.epcglobalinc.org.

[24] International Organization for Standardization. Available at: http://www.iso.org.

[25] Shin, K.C., Park, S.B., and Jo, G.S. (2009) Enhanced TDMA based anti-collision algorithm with a dynamic frame size adjustment strategy for mobile RFID readers, *Sensors*, **9**(2): 845–858.

[26] Haykin, S. and Nie, J. (1999) A dynamic channel assignment policy through Q-learning, *IEEE Transactions on Neural Networks*, **10**(6): 1443–1455.

[27] Sutton, R.S. and Barto, A.G. (1998) *Reinforcement Learning: An Introduction*. New York: A Bradford Book, pp. 51–52.

[28] Chen, N.K., Chen, J.L., and Lee, C.C. (2009) Array-based reader anti-collision scheme for highly efficient RFID network applications, *Wireless Communication and Mobile Computing*, **9**(7): 976–987.

[29] Kim, J., Kim, S., Kim, D., Lee, W., and Kim, E. (2005) Low-energy localized clustering: An adaptive cluster radius configuration scheme for topology control in Wireless Sensor Networks, in *Proceedings of VTC'2005*, Stockholm, Sweden, May–June, **4**: 2546–2550.

[30] Wang, D., Wang, J., and Zhao, Y. (2006) A novel solution to the reader collision problem in RFID system, in *International Conference on Wireless Communications, Networking and Mobile Computing. WiCOM 2006*, Sept. 22–24, pp. 1–4.

[31] Ok, C.Y., Quan, C.H., Mo, H.S., Choi, K.Y., and Lee, C.W. (2008) Adaptive channel hopping algorithm for reader anti-collision in RFID systems, in *10th International Conference on Advanced Communication Technology, ICACT 2008*, **1**: 90–94.

[32] Kunz, D. (1991) Channel assignment for cellular radio using neural networks, *IEEE Trans. Vehicular Technology*, **40**(1): 188–193.

[33] Duque-Anton, M., Kunz, D., and Ruber, B. (1993) Channel assignment for cellular radio using simulated annealing, *IEEE Transactions on Vehicular Technology*, **42**(1): 14–21.

[34] Lai, W.K., and Coghill, G.G. (1996) Channel assignment for a homogenous cellular network with genetic algorithms, *IEEE Transactions on Vehicular Technology*, **45**: 91–96.

[35] Carbunar, B., Ramanathan, M.K., Koyuturk, M., Hoffmann.C., and Grama. A. (2005) Redundant reader elimination in RFID systems, in *Proceedings of the 2nd Annual IEEE Conference on SECON*, September, pp. 176–184.

[36] Leong, K.S., Ng, M.L., Grasso, A.R., and Cole, P.H. (2006) Synchronization of RFID readers for dense RFID reader environments, in *Proceedings of 2006 Symposium on Applications and the Internet Workshops*, pp. 48–51.
[37] Joshi, G.P. and Kim, S.W. (2008) Survey, nomenclature and comparison of reader anti-collision protocols in RFID, *IETE Technical Review*, **25**(5): 285–292.
[38] Cha, K., Zawodniok, M., Ramachandran, A., Sarangapani, J., and Saygin, C. (2006) Interference mitigation and read rate improvement in RFID-based network-centric environments, *Sensor Review*, **26**(4): 318–325.
[39] Garey, M.R. and Johnson, D.S. (1979) *Computers and Intractability: A Guide to the Theory of NP Completeness*. New York: W.H. Freeman.

12

Optimal Tag Coverage and Tag Report Elimination

Bogdan Carbunar[1], Murali Krishna Ramanathan[2], Mehmet Koyuturk[3], Suresh Jagannathan[4], and Ananth Grama[5]

[1]*Applied Research and Technology Center, Motorola, Illinois*

[2]*Coverity, San Francisco*

[3]*Case Western Reserve University, Cleveland*

[4,5]*Purdue University, Indiana*

12.1 Introduction

RFID technologies have evolved significantly in the past few years. Fundamentally, RFID systems consist of two major components: tags and readers. The tags are small circuits equipped with an antenna, able to store small amounts of data and easily mountable on objects. The readers are more sophisticated devices, able to access the data stored on tags in their vicinity, process data locally and forward it to other devices for further actions. The factors motivating the use of RFID technology when compared to existing solutions, such as barcode scanning, include their speed of detection, ease of use and accuracy even when the tags are not within the line-of-sight of readers. Regarding accuracy, for instance, apparel vendor Gap documented an increase of accuracy from 85% to 99.9% when switching to RFID technology [1]. By accuracy we mean the number of tags that can be correctly identified per time unit.

Of paramount importance to the wide adaptability of RFID systems is the ability to reduce the cost and size of components while improving the reliability of read operations.

RFID Systems: Research Trends and Challenges Edited by Miodrag Bolić, David Simplot-Ryl, and Ivan Stojmenović

For instance, by reducing the cost of tags to a few cents, companies like Wal-Mart anticipate savings of billions of dollars from tagging pallets and individual products. Similarly, the International Air Transport Association (IATA) projects annual industry savings of $760 million from RFID luggage tags [2].

Notably, several solutions for wireless readers are now supported, including quarter-sized wireless readers [3] compatible with Crossbow [4] sensors, Bluetooth-enabled readers [5], SD card reader/writers [6] attachable to phones as well as reader-enabled phones [7]. Such tetherless technologies allow dynamically deployed heterogeneous readers to interconnect in an ad hoc fashion. For instance, Figure 12.1 illustrates a reader network consisting of reader-enabled Mica motes, Wi-Fi or Zigbee enabled readers and cellphones equipped both with readers and Wi-Fi, Zigbee and Bluetooth network interfaces. Such devices can communicate information collected from tags over wireless connections established in an ad hoc manner. This information, locally processed by intermediate routers, can be relayed to a collection point (henceforth called a sink). The sink can use other network interfaces (e.g. cellular) to relay collected information to a central data storage and processing facility.

The scenarios mentioned above imply dense tag and reader deployments in environments where the tag population can be constantly changing. A warehouse or a luggage sorting facility in an airport is likely to contain large numbers of tags in small areas. Tags can frequently join, leave or move within the area covered by the reader network. Under such scenarios, static deployments of readers may not be desirable. This may be due not only to high initial deployment overheads but also to the inability of static reader deployments to adapt to sudden tag population changes.

Consequently, while convenient, a reader network solution needs to address several complex challenges. First, as discussed above, it needs to support dense tag and reader deployments and be adaptive to topological changes. Second, since the lifetime of

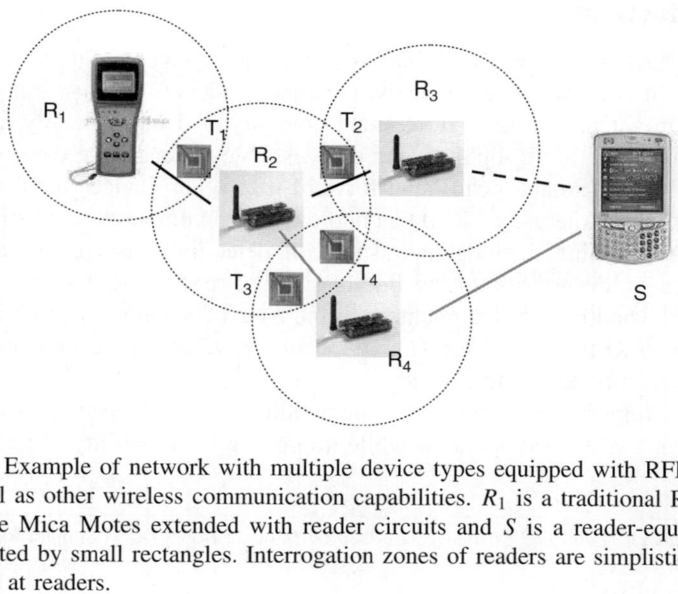

Figure 12.1 Example of network with multiple device types equipped with RFID reading capabilities as well as other wireless communication capabilities. R_1 is a traditional RFID reader, R_2, R_3 and R_4 are Mica Motes extended with reader circuits and S is a reader-equipped cellphone. Tags are denoted by small rectangles. Interrogation zones of readers are simplistically denoted by disks centered at readers.

tetherless readers depends on adequate battery management algorithms, an efficient solution needs to carefully consider the option of alternately switching off the RFID interfaces of reader nodes whose presence is not necessary. This option offers the opportunity to reduce both the number of tag reading operations and the number of wireless transmissions required to forward this information to the sink node. However, the third essential requirement of reader networks is to accurately detect the deployed tags. This requirement forces the periodic re-activation of nodes whose RFID interfaces were turned off, to allow the system to react to novel tag deployments as well as existing reader failures. This chapter focuses on two fundamental, closely related problems that arise from these considerations.

1. **Optimal Tag Coverage**: This problem consists of deciding which reader should be responsible for identifying each tag in the system, with a view to optimizing the number of readers that need to be active at a given time. This problem stems from the observation that a reader that covers only tags covered by other readers as well is *redundant*. Consequently, the number of readers that need to use their RFID interfaces can be reduced by identifying a minimal subset of readers in which no reader is redundant. The optimal tag coverage problem is illustrated in Figure 12.1. Observe that, in this figure, readers R_1, R_2, R_3 and R_4 are all redundant, covering tags that are also covered by other readers. The optimum coverage is provided by reader R_2, which covers all the tags on itself.
2. **Optimal Tag Reporting**: This problem consists of determining the optimal way to report the tags detected in a system deployment. For instance, consider the example in Figure 12.1. If all the readers have a direct communication channel with the sink and all decide to report the tags detected, 8 tags will be reported. However, only 4 tags exist, the remaining tag reports are duplicates. In this instance, the optimal tag reporting problem is equivalent to the optimal tag coverage problem. However, this is not always the case, since the solution to the optimal tag coverage problem may include readers with intersecting interrogation zones.

   More precisely, we define the optimal tag reporting problem in a reader network as one of finding mapping between tags and readers, such that for each tag, the number of hops traversed by the tag report to reach the sink node is minimized. Consider again the network in Figure 12.1. Tags identified by R_1 are forwarded by R_2, tags of R_2 are forwarded by R_4, and R_3 and R_4 communicate directly with the sink S. Even if redundant tag reports are eliminated by intermediate readers, tag T_2 will be reported by both R_3 and R_2. Note that, declaring R_2 as the only reader that should report tags does not provide an optimal solution to the optimal tag reporting problem for this problem instance. This is because, tag T_2 should be reported by R_3, which provides a shorter route (1 hop) to the sink node than R_2 (2-hops).

   Note that our work focuses on passive RFID systems with mobile readers. Our solutions can be also used for statically deployed reader networks and active tags. In fact, relaxing the constraints considered in this chapter may simplify the solutions.

The chapter is organized as follows. In Section 12.2, we describe in detail the direct reporting and ad hoc reader network models. Subsequently, we present algorithms for solving tag and reader collision problems in Section 12.3 and Section 12.4 respectively.

In Section 12.5 we compare the tag coverage and reporting problems in RFID systems with traditional coverage problems in wireless sensor networks. Then, in Section 12.6, we describe our reader and tag deployment assumptions. In Section 12.7 we formalize the tag coverage and reporting problems and show that the optimal tag coverage problem is NP-hard. In Section 12.8 we describe a greedy, centralized heuristic for these problems and study its properties. In Section 12.9 we study distributed algorithms for solving tag coverage and reporting problems in both the direct reporting and the ad hoc reader networks models. In Section 12.10 we show how readers can implement a duty-cycle for their RFID interface, allowing them to both save battery power and be adaptive to topological changes. Since the solutions proposed assume the existence of writable tags, in Section 12.11 we describe a simple optimization procedure that reduces the number of write operations required to solve the coverage problems. We also discuss its performance by comparing it to the previously described algorithms. In Section 12.12, we present related work and provide pointers for further reading. We conclude our discussion in Section 12.13. Problems can be found in at the end of the chapter.

12.2 Overview of RFID Systems

RFID tags, with limited memory resources, store a unique identifier and information relating to the objects with which they are associated (attached). They also contain an antenna, used for communication with readers. There are two kinds of tags: passive and active. Passive tags do not need batteries, but are powered by RF energy from the reader. In addition to the memory chip and antenna, active tags are equipped also with a battery and a transmitter, allowing them to initiate transmissions. Most current RFID applications use mostly passive tags, thus, in the following we only focus on passive tags.

Signals from RFID readers activate compatible tags within their interrogation zone. The interrogation zone is defined to be the area around a reader where tags can receive the reader's signal, process it, and send back a response that can be correctly decoded by the reader. Figure 12.2 shows examples of tag and reader deployments and reader interrogation zones. The information decoded by a reader is passed to host computing systems where it is further processed, according to the application. In addition to locating, activating, and receiving transmissions from RFID tags, RFID reader-writers also have the ability to send data back to read/write-capable tags in order to append or replace data.

Readers may themselves be fixed or portable. Fixed readers are usually attached to antennae that are designed to detect the tags within a specified area. These units typically collect data from products traveling through loading dock doors, conveyor belts, gates and doorways, or even cars on tollways. In contrast, portable wireless readers can be moved to detect remote tags, in areas where wiring or antenna placement could be difficult. Readers with various wireless communication capabilities exist in today's market. SkyeTech's SkyeRead M1 [3] reader is compatible with Mica Motes [4] and major cellular phone manufacturers offer models with embedded RFID readers [7].

Since the interrogation zones of short-range readers may not overlap significantly, in the following we focus mainly on medium and long-range RFID systems.

We consider two models of wireless reader network deployments, each with a different set of networking capabilities. The first model, illustrated in Figure 12.2(a), consists of direct reporting reader networks. All the RFID readers are equipped with an additional

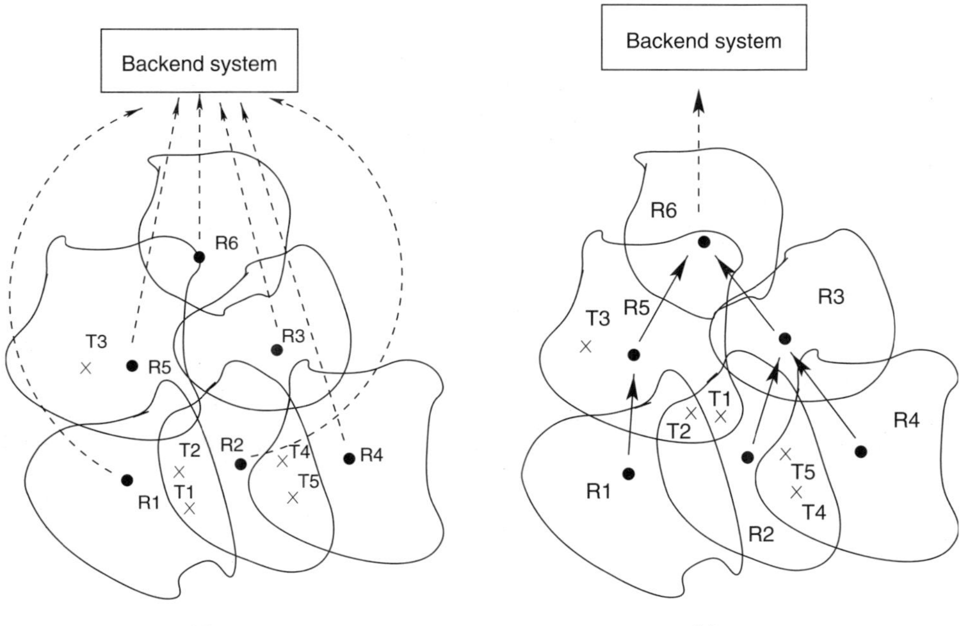

Figure 12.2 Reader network examples. Readers are denoted by black dots and their RF interrogation zones by irregular shapes. Dotted arrows represent cellular links and plain arrows represent reader-to-reader wireless links. Tags are denoted by small crosses. (a) Direct reporting reader network. Wireless readers may only communicate with the host system. (b) Ad hoc reader network. The same reader and tag placement but wireless readers may now communicate with each other and with the host system. The sink node R_6 needs to communicate with the host system. All other readers route the identified tags toward R_6. Figure redrawn from "Efficient Tag Detection in RFID Systems" B. Carbunar *et al.*, Journal of Parallel and Distributed Computing, vol. 69(2), 180–196pp, 2009. © 2009 Elsevier.

network interface (e.g. cellular interface). Such readers may locally detect tags and then report them to the host system using the cellular data link. In this model, the readers are assumed to be unable to communicate with each other, except through the host system.

The size, portability, and price of wireless readers also enable the deployment of a second model, called the ad hoc reader network model. In this model, depicted in Figure 12.2(b), reader nodes equipped with WLAN, Bluetooth, or 802.15.4 can locally communicate and establish ad hoc networks. Readers not only detect tags but also need to relay tag reports of remote readers toward the host system.

The solutions we discuss in this chapter can be applied both to static and mobile reader networks. In fact, assuming a static reader deployment, where location and interference information may be available, may simplify our solutions.

Reader networks are currently supported by a number of vendors. PGS Electronics [9] provides a complete solution for wireless RFID integration and Reva Systems [10] recently proposed a protocol called Simple Lightweight RFID Reader Protocol (SLRRP). SLRRP defines how readers convey configuration, control, status, and tag information between

RFID reader network device managers and other readers in an IP-based network, either wired or wireless.

12.3 Tree Walking: An Algorithm for Detecting Tags in the Presence of Collisions

We begin by briefly describing two well-known collision problems that occur in distributed RFID system deployments. In this section we study the tag collision problem and in the following section the reader collision problem. Solving these problems is not only a prerequisite for addressing the tag coverage and reporting problems studied later, but some of the randomization techniques they introduce are used as building blocks by our solutions.

To understand the tag collision problem, consider that readers have the ability to accurately read (and report) data stored at multiple tags placed within their interrogation zone. Since readers broadcast their queries, multiple tags may end up responding simultaneously, leading to interference and inaccurate decoding of responses.

Several techniques have been proposed to solve the tag collision problem [11–13]. A popular solution, known as the tree walking algorithm (TWA) [11] is based on recursive traversal of the binary name tree of tag identifiers.

A reader initially sends a broadcast query containing the "0" string. All tags in its interrogation zone whose identifier starts with a "0" bit reply. If a reply is received, or a tag collision is detected, the reader recurses on both subtrees of "0," rooted at "00" and "01." However, if no reply is received, the reader concludes the absence of "0"-prefixed tags in its interrogation zone and subsequently sends a "1" query. For a reader, the complexity of TWA is proportional to the number of tags present in its interrogation zone and to the length of the binary representation of tag identifiers. While TWA was designed to accurately read tag identifiers, it can also be used to read arbitrary data stored on tags, as long as the data's bit length is the same across all tags.

12.4 Reader Collision Avoidance

The tag collision problem occurs when one reader simultaneously interrogates multiple tags. In large-scale environments, tags may be covered by more than one reader. Such tags may be incorrectly detected or even escape detection if readers with overlapping interrogation zones simultaneously query them (possibly during the course of TWA). This scenario, known as the reader collision problem [14], is due to reader signal interference occurring at tags. Avoiding reader collisions is an essential requirement for error-free operation of an RFID system.

We now briefly describe a randomized, distributed, and localized solution to the reader collision problem. The algorithm, named RCA [15, 8] (Reader Collision Avoidance), is presented in the context of TWA [11] (see Section 12.3). Similar to TWA, a reader running RCA sends a broadcast query containing a certain prefix expected to match the identifiers of tags in its interrogation zone. However, unlike TWA where the lack of an answer is considered to denote absence of matching tags, RCA backs off for a random number of time frames and repeats the query. The purpose of the random back-off and query repetition is to ensure with high probability (w.h.p.) that the choice of a time frame is not picked by another reader, thus alleviating the impact of reader collisions.

A reader divides time into distinct epochs and each epoch is further divided into multiple distinct time frames. The duration of a time frame is set to exceed the time a reader query needs to reach a reader's entire interrogation zone, plus the time a tag needs to process a query (one string comparison) and the time a tag answer needs to get back to the reader. During each epoch, a reader picks a frame uniformly at random and sends its query in that frame. This ensures that the queries of any two, potentially interfering readers will collide only if both readers choose the same time frame during the current epoch. If no tag answer is received, the reader repeats the query in a randomly chosen time frame of the next epoch. Since queries are repeated in the randomly selected slots, the probability of repeated reader collisions quickly declines with increasing number of epochs.

Let the number of readers in the system be ψ. Then, it can be shown that if the number of time frames per epoch is equal to ψ and a query is not answered $o(\log \psi)$ times, then w.h.p., there are no tags matching the query in the interrogation zone of the reader. If, however, an answer is received, either as a clear tag response or by detecting a tag collision, in the next epoch the reader recursively moves to the next query, as in the TWA algorithm.

12.4.1 Implementation

Algorithm 1 presents the pseudocode for RCA using an Orca-like [16] syntax. Orca is a parallel programming language for distributed systems that provides elegant constructs for expressing reactive behavior, such as *guards*. Operations consist of guards with syntax

guard `expression` **do** `statementSeq` **od**,

where `expression` is a boolean expression and `statementSeq` is a sequence of statements. The operation containing guards blocks until one or more guards are true. Then one of the satisfied guards is randomly chosen and its statements are executed atomically.

The operation of a tag is shown in Algorithm 1, lines 1–10. A tag replies only to queries containing strings whose prefixes match its own identifier (lines 5–9). `inQ.first` is used to denote the packet currently received by the tag. The operation of a reader is shown in Algorithm 1, lines 11–31. Time is divided into epochs, with each epoch containing a fixed number, n, of time frames. The duration of a time frame is equal to the time necessary for a query to propagate from a reader to a tag. For each prefix queried, the reader waits for a maximum of `e` epochs (line 18) and in each epoch sends exactly one broadcast message containing the prefix. During each epoch, the broadcast message is sent in a randomly chosen time frame (lines 19–22).

The lack of a reply may denote either the absence of a tag matching the queried prefix in the interrogation zone, or the occurrence of reader collisions at such tags. If less than `e` queries with the current prefix have been sent, the reader waits until the beginning of the next epoch to repeat the above process (lines 27–29). If no reply or collision is detected after `e` rounds, the reader ignores the subtree rooted at the queried prefix. However, the receipt of an individual reply or the detection of a tag collision stops this process. The reader can then safely recurse on the two children of the employed prefix (lines 23–26).

Algorithm 1 Pseudocode for RCA. Reader and tag operation for avoiding reader collisions. `getRandom`(v_1,v_2) returns a random integer value between v_1 and v_2 and `bCast(packet)` is used to broadcast `packet`. The timeout T_{out} value should be equal to the frame size. We have experimented with several e and n values. We have observed that for dense reader and tag deployments RCA performs well even for values of $e = n = \log \psi$. That is, for 500 readers e and n can be as small as 9.

```
Object implementation RFIDTag;
Tid: integer; #tag identifier
inQ: queue; #queue of incoming packets
Operation run()
   guard inQ.first.type = query do
      if prefixMatch(inQ.first.tid,Tid) then
         bCast(new packet(TAG));
      fi
   od
end

Object implementation RFIDReader;
count,e: integer; #epochs per bit read
frame,n: integer; #time frames in each epoch
T,Tout: integer; #time out value
inQ: queue; #queue of incoming packets
Operation treeWalk(prefix:integer)
   count:=0;
   while count + + < e do
      frame:=getRandom(0,n);
      sleep(frame);
      T=getTime();
      bCast(new packet(query,prefix));
      guard inQ.first.type = TAG_COL || TAG do
         treeWalk(prefix+"0");
         treeWalk(prefix+"1");
         exit;
      od
      guard getTime()-T ≥ Tout do
         sleep(n-frame-1);
      od
   od
end
```

12.5 Coverage Redundancy in RFID Systems: Comparison with Sensor Networks

The problem of coverage of a set of entities has been studied in a variety of contexts. Of particular interest to the problems studied in this chapter are coverage problems in wireless sensor networks (WSNs). In this section we briefly compare the setting of RFID coverage problems to the assumptions made by most coverage problems in WSNs.

In wireless sensor networks, the coverage of a sensor is defined to be a compact circular area around the location of the sensor. Formally,

Definition 12.5.1 *The* coverage of a sensor *s with planar coordinates* `(x,y)` *and sensing range r is a disk with center* `s(x,y)` *and radius r. We call this disk the* coverage or sensing disk, *and call its border the* coverage or sensing circumcircle, *denoted by* $\mathcal{C}$`(s)`. *We say that a point* `p` *is covered by a sensor s if* `dist(s,p)` $\leq$ `r`.

Coverage in sensor networks is illustrated in Figure 12.3(a), where the coverage areas of sensors are denoted by disks. Then, the coverage of a WSN is the union of coverage disks of all the sensors in the network. Formally,

Definition 12.5.2 *The* coverage of a network *is the area* $\mathcal{A}$ *with the property that for any point* `p` $\in \mathcal{A}$, *there exists at least one sensor* `s` *in the network such that* `p` *is covered by the coverage disk of* `s`.

The definition of a redundant sensor follows naturally:

Definition 12.5.3 *A sensor is said to be redundant if its sensing area is* completely *covered by other sensors.*

In Figure 12.3(a), the central sensor, whose coverage area is completely subsumed by the coverage areas of other sensors, is redundant. The most important coverage problem studied in WSNs attempts to identify the minimum set of sensors that are needed in order to cover a certain area, for example, the entire area covered by the network. Coupled with a mechanism for scheduling the duty cycle of sensor nodes, this approach allows the network deployer to turn off the redundant nodes, save their power and extend the

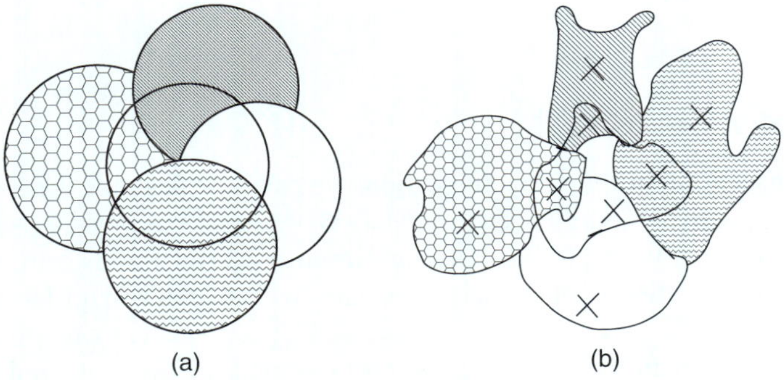

Figure 12.3 (a) Example WSN coverage. The sensing (coverage) areas of sensors are shown as disks centered at the sensors. The central sensor is defined to be redundant since its sensing disk is completely subsumed by the other sensors. (b) Example RFID coverage. The interrogation zones of readers are shown as regions of irregular shapes, around the locations of the readers. Tags are denoted by x marks. The coverage of a reader is defined as a discrete set of elements, the tags in its interrogation zone. The central reader, whose interrogation zone is not completely covered by the interrogation zones of the other readers, is still defined as redundant. This is because the tags covered by its interrogation zone are also covered by other readers.

lifetime of the network.

Consider now the coverage problems defined in the context of RFID readers, using the example deployment shown in Figure 12.3(b). Even though the interrogation zone of the central reader is not completely covered by the interrogation zones of the other readers, the reader is still redundant. This is because we define the coverage of readers in terms of discrete sets of points, which are the tags in their interrogation zones. This leads then to the following major differences between the RFID and WSN coverage problems.

- *Shape of sensing areas*: Most coverage problems in sensor networks assume that the coverage areas of individual sensors are contiguous disks. In contrast, in RFID systems the coverage of readers is defined in terms of the tags covered.
- *Location information*: Location of individual sensors, either absolute or relative, is extremely important in determining the coverage and coverage redundancy in wireless sensor networks. This is due to defining the coverage in terms of contiguous disks. Detecting whether a sensor is redundant requires knowing the positions of sensing disks of other sensors. In RFID systems, where the coverage of readers consists of discrete sets of tags, determining a reader's redundancy requires only knowing whether the tags are also covered by other readers. As we will see later, this can be achieved without using reader location information.
- *Communication capabilities*: Determining the coverage of a sensor network requires sensors to be able to exchange information. Sensors use collected information in order to infer the coverage areas of other sensors. Inter-reader communication capabilities are not required in RFID systems in order to determine coverage redundancies. As we will see later in this chapter, in RFID systems, even though the readers may not be able to communicate with each other, they can communicate with tags and use the tags to effectively communicate with other readers for the purpose of determining each other's coverage.

12.6 Network Model

Based on the discussion of the previous section, we can now describe the assumptions we make about tags, readers and their deployment. We assume the presence of passive tags only, as opposed to active tags (the latter are more powerful but also more expensive). Passive tags use the energy of signals received from readers in order to respond to their queries.

Tags have limited memory, which can be on the order of a few hundred bytes. Part of a tag's memory is read-only, used to store a unique identifier, and part of it is writable. Furthermore, we assume that tags are capable of performing comparison and prefix matching operations. Prefix matching is a requirement of the tree walking algorithm (see Section 12.3) and the circuits performing comparison are similar in terms of their functionality and complexity.

We make no assumptions on the number of readers and tags, or on the underlying reader deployment or the distribution of tags. We do not assume the presence of a centralized entity capable of collecting information about the topology of the reader network or controlling the behavior of individual readers. Thus, our algorithms do not rely on inter-reader communication capabilities.

Finally, we assume that any reader is able to detect tag collisions, occurring when multiple tags respond to one of its queries. In fact, this is a basic assumption necessary for any algorithm that resolves tag collisions and is a requirement for the tree walking algorithm.

12.7 Optimal Tag Coverage and Tag Reporting

12.7.1 Problem Definition

We now provide a formal framework for coverage problems in RFID systems. First, let s denote a binary string and $|s|$ denote the bit length of s. Similarly, let $|\mathcal{A}|$ denote the cardinality of set $\mathcal{A}$. We denote an RFID system as $S = (\mathcal{R}, \mathcal{T}, C, \beta)$, where $\mathcal{R}$ represents the set of readers and $\mathcal{T}$ represents the set of tags in the system. We denote the number of readers in the system by $\psi = |\mathcal{R}|$ and the number of tags by $\gamma = |\mathcal{T}|$. $C : \mathcal{R} \to \mathcal{T}^*$ is the coverage function, such that for any $R \in \mathcal{R}$, $C(R) \subseteq \mathcal{T}$ denotes the set of tags that are within the interrogation zone of R. For completeness, we also assume that for any tag $T \in \mathcal{T}$, there exists at least one reader $R \in \mathcal{R}$ such that $T \in C(R)$. Note that we do not make any assumption on the interrogation zone radii of readers.

Finally, β denotes the bit length of tag identifiers. We denote the set of all binary strings of length at most β with B^{β}. We use the notation $s \trianglerighteq s\prime$ to denote that binary string s is a prefix of binary string s'. We discretize time as the set $\mathcal{U}$ of successive time frames, where the length of a time frame is sufficient for a query to propagate from a reader to any tag within its interrogation zone. Using this notation, we define a redundant reader formally as follows.

Definition 12.7.1 Redundant Reader. *Given $R \in \mathcal{R}$, if for all $T \in C(R)$, there exists $R' \in \mathcal{R}$ such that $T \in C(R')$, then R is said to be* redundant.

The tag coverage problem corresponds to the problem of finding a set of readers that cover all the tags in the system, such that no reader is redundant. Specifically,

Definition 12.7.2 Optimal Tag Coverage Problem. *Given an RFID system $S = (\mathcal{R}, \mathcal{T}, C, \beta)$, determine a set $M \subseteq \mathcal{R}$ of readers satisfying the following properties.*

(i) For any tag $T \in \mathcal{T}$ there exists a reader $R \in M$ such that $T \in C(R)$.
(ii) There exists no redundant reader $R \in M$, that is, the set $M - \{R\}$ no longer satisfies property (i).
(iii) M is of minimum cardinality among all sets that satisfy properties (i) and (ii).

Example 12.7.1 To understand Definitions 12.7.1 and 12.7.2, consider the tag and reader deployment of Figure 12.1. All readers are redundant since each reader covers tags that are also covered by other readers. However, the set $M = \{R_2\}$ is the reader set providing optimal coverage.

A related problem consists of avoiding redundant tag reports in an RFID system. It would be desirable to have each tag reported only once, effectively decreasing the number and duration of reader wireless transmissions. We now formally define this problem. Let

$p : \mathcal{T} \rightarrow \mathcal{R}^*$ be the function that maps each tag into the set of readers that cover it. That is, for any tag T, $p(T) = \{R | T \in C(R)\}$. Let $m : \mathcal{R} \rightarrow \mathbb{R}$ be a cost function associated with readers. For example, in an ad hoc reader network, $m(R)$ can be the number of intermediate routers on the shortest path from R to the sink node. Then, we define the tag reporting problem as follows.

Definition 12.7.3 Tag Reporting Problem. *Given an RFID system $S = (\mathcal{R}, \mathcal{T}, C, \beta)$, with cost function m associated with readers, determine a function $r : \mathcal{T} \rightarrow \mathcal{R}$ such that for all $T \in \mathcal{T}$ we have that $T \in C(r(T))$ and $r(T) = \{\arg\min R_i \in p(T)\}$.*

Informally, a solution to the tag reporting problem maps each covered tag to exactly one reader among those that cover it. That reader is the one that is associated with the minimum cost. Clearly, such a mapping is facilitated by global knowledge on the topology of tag distribution and reader deployment. However, in the absence of inter-reader communication, a solution to this problem is not straightforward. Observe that the optimal tag coverage problem may be considered a variation of this problem, in which the range of function r is minimized.

Example 12.7.2 Consider the network in Figure 12.2(b). If $m(R)$ is the number of hops between R and the sink node, then R_5 should report tags T_1, T_2 and T_3. This is because $m(R_5) = 2$ whereas $m(R_1) = m(R_2) = 3$.

12.7.2 Problem Complexity

In the following we show that the optimal tag coverage problem is NP-hard even in a setting where global information about the distribution of tags and deployment of readers is available. Although this problem is closely related to the minimum set cover problem, a reduction from minimum set cover is not straightforward since the coverage sets of readers are constrained by the geometry of the two-dimensional Euclidean space. In the following, we show that in spite of this constraint, the problem is NP-hard, by reduction from the geometric disk cover problem. We first provide the following lemma.

Lemma 12.7.4 *Given a set of n points, $p_1, p_2, \ldots, p_n$, placed inside a circle of radius ρ, there exists a subset of 3 of the n points, p_i, p_j, p_k, such that all the n points are placed inside $c(o_{ijk}, \rho)$. Here, o_{ijk} is the mass center of p_i, p_j, p_k and $c(x, \rho)$ denotes the circle centered at x with radius ρ.*

Proof. We provide a constructive proof. If all the points are covered by a circle of radius ρ, then a circle of radius ρ going through 2 of the points and covering all the other points exists. If the circle has a third of the n points on its perimeter, then we have completed the proof. Otherwise, shrink the circle until its perimeter touches a third point. The resulting circle has radius less than or equal to ρ, is the circumcircle of three of the n points, and covers all the other points.

We can now prove the following important result.

Theorem 12.7.5 *The optimal tag coverage problem is NP-hard.*

Proof. By reduction from the geometric disk cover (DC) problem, known to be NP-hard [17]. The input to the DC problem consists of a set of m points and a value ρ. The output is the minimum number of disks of radius ρ that cover all the points. We use the following polynomial time reduction from DC to the optimal tag coverage problem. Add a disk of radius ρ centered at each point in the input set of DC. Then, for all combinations of 3 points in the input set, add a disk of radius ρ, centered at the mass center of the 3 points. Let Σ denote the set of all disks created. The points input to DC form the input tags, and the disks in Σ form the input reader interrogation zones for the tag coverage problem. The reduction has $O(m^3)$ complexity. It is clear that the disks in Σ cover all the input points. Moreover, as a direct consequence of Lemma 12.7.4, the disks that form the solution to the DC problem are contained in Σ. Hence, the optimal solution to DC is also contained in Σ. Now consider the optimal solution for the tag coverage problem. If it does not correspond to an optimal solution for DC, then it should be possible to replace more than one of the disks in the solution to the tag coverage problem with a single disk of radius less than or equal to ρ. If so, the solution to tag coverage would not be optimal, since the 3-point disk derived from such a disk according to the construction of Lemma 12.7.4 is also in the input for the tag coverage problem.

Algorithm 2 Centralized heuristic (GREEDY). The *run* operation greedily selects readers with the highest covered tag count first. At the end of the operation, all unselected readers are redundant. The *mapped* structure stores a mapping between each tag in the system to one of the readers covering it. That reader will be responsible for reporting the tag.

```
Object implementation CentralHost;
R: array[integer] of integer; #list of RFID readers
mapped: array[integer] of integer; #reader reporting each tag
done: boolean; #true when all tags are mapped
Operation run()
  for i = 1 to mapped.size do mapped[i] := -1; od
  done := false;
  while (R.size > 0 && done = false) do
      #sort readers by decreasing covered tag count
      sort(R);
      candidate := R[1];
      R.removeFirst();
      candidate_tags := candidate.getCoveredTags();
      for each t in candidate_tags do
         mapped[t] := candidate;
         for i = 1 to R.size do
            if R[i].covers(t) then R[i].remove(t); fi
         od
      od
      done := true;
      for i = 1 to mapped.size do
         if mapped[i] = -1 then done := false; fi
    od
  od
```

12.8 Redundant Reader Elimination Algorithms: A Centralized Heuristic

In the following we study several redundant reader elimination algorithms for the optimal tag coverage problem. The algorithms work by identifying redundant readers and provide a mapping between each tag to a single reader. The mapping specifies which reader needs to cover and report each tag. Readers that are not present in any such mapping are redundant and do not need to report any tag.

In Section 12.7.2 we have shown the optimal tag coverage problem to be NP-hard even in a centralized setting, where a single host has knowledge about all readers, including all the tags each reader covers. Before studying distributed solutions for the tag coverage problem, we briefly focus on a greedy heuristic in the centralized setting. Our greedy approach is based on the observation that by trying to keep active the readers covering more tags, it is expected that more readers will be detected to be redundant.

Algorithm 2 shows the pseudo-code for a central host implementation of the greedy heuristic using an Orca-like [16] syntax. The central host has knowledge of all the readers (line 2), stored as an array of unique reader identifiers. The central host also maintains a mapping between each tag in the system, to one of the readers (line 3). The *mapped* structure is indexed by tag identifiers, assumed for the sake of simplicity to be contiguous integers ranging from 1 to the total number of tags.

The *run* operation (lines 5–21) implements the greedy approach. Initially, each entry in the *mapped* structure stores -1 for each tag (line 6), signifying that the tag is not covered (mapped) by any reader. The operation proceeds iteratively, by sorting first all the readers in the system in decreasing order of the number of tags they are currently covering (line 9). The reader with the highest tag cover count is then selected and removed from the list of readers (lines 10–11). The tags covered by this *candidate* reader (line 12) are mapped (in the *mapped* structure) to the *candidate* reader (line 14). All the tags mapped to the candidate reader are then removed from the coverage sets of all the remaining readers in the system (lines 15–17). The operation then scans the *mapped* structure in search of unmapped tags (lines 18–21). If no tags are left unmapped, the operation stops (line 8). Otherwise, it iterates on the remaining readers, until it runs out of readers or all tags are mapped (line 8). At the end of the operation, all the readers left in the R structure are redundant. The ones selected are a solution and need to be active in order to report the tags mapping to them.

Example 12.8.1 Consider the system shown in Figure 12.4, consisting of 4 readers and 10 tags. Initially, reader R_1 covers the most tags, 6. GREEDY selects R_1 to be active, then removes 4 of the tags that are covered by R_1 from the coverage set of R_2. While initially R_2 was covering 5 tags, at this point it only covers 1. Since R_4 covers more tags, 3, it is selected next by GREEDY. Then, tag T is also removed from R_2's coverage, since it was mapped to R_4. Since R_2 now covers no tag, reader R_3, covering one tag, is next selected by GREEDY. Following this step, GREEDY runs out of unmapped tags, exits the *while* loop and stops. Since reader R_2 was not selected, it is declared to be redundant.

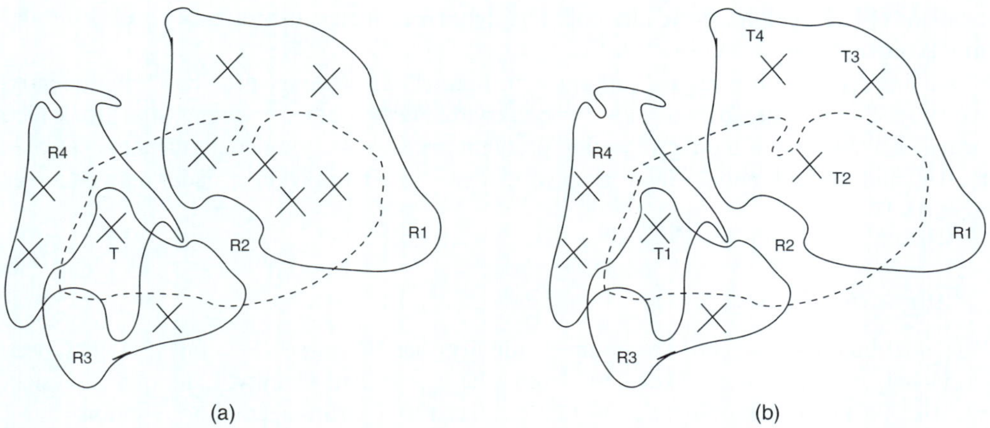

Figure 12.4 (a) RFID system where GREEDY provides an optimal tag coverage solution, by detecting reader R_2 to be redundant. (b) RFID system where RRE provides an optimal tag coverage solution.

12.8.1 Analysis

We can prove that GREEDY satisfies the first condition from Definition 12.7.2. That is, we need to show that for any tag T in the system, GREEDY assigns a reader. In our RFID system definition from Section 12.7.1 we assume that any tag is covered by at least one reader. Then, the only time when GREEDY may exit the *while* loop (line 8) is when it provides a fully assigned *mapped* structure. Let T be a tag, covered by the readers from the set $p(T)$. The first time GREEDY selects one of the readers from $p(T)$ as a candidate reader, T will be mapped to that reader. If GREEDY never selects a reader from $p(T)$, tag T will remain uncovered and the tag count of the readers from $p(T)$ will be at least one. Then, when GREEDY runs out of readers with tag count larger than one, it will need to select at least one reader from $p(T)$.

Note, however, that GREEDY does not satisfy the second condition of Definition 12.7.2. This should certainly be true, otherwise GREEDY would be a solution to a NP-hard problem. For a counter-example, consider the system in Figure 12.3(b). The central reader, covering 4 tags is selected first by GREEDY. This is followed then by each of the remaining readers, each covering an additional tag not covered by anyone else. That is, GREEDY selects all the readers to be active. However, it is easy to see that the central reader is redundant and its presence is not needed to cover and report the tags.

12.9 RRE: A Distributed Solution

GREEDY assumes global knowledge of the RFID system. Since this information may not always be available, we study a distributed and localized alternative. The solution, called RRE, assumes no centralized server and does not require direct communication between readers. RRE is deterministic in nature, but it relies on randomization to avoid reader collisions. The randomized reader collision avoidance mechanism used by RRE is similar to the one used by RCA (see Section 12.4).

We first present RRE as an algorithm for the optimal tag coverage problem. Subsequently, we show that the idea behind RRE can be generalized to obtain a variant, RRE-HC, which provides a solution to the tag reporting problem based on the hop count distance between readers and the sink node. This variant of RRE can be used in ad hoc reader networks (see Section 12.2).

12.9.1 RRE

The distributed RRE algorithm assumes initially that RCA (see Section 12.4) has been previously executed by all readers to identify the tags in their vicinity. Later in this section, we discuss a simple modification to RRE to relax this synchronization assumption.

In RRE, each tag is marked with a value representing the highest number of tags covered by any reader in the tag's vicinity (that is, any reader whose interrogation zone contains the tag). This is done by each reader writing its tag count on its tags and by requiring tags to overwrite locally stored values only with higher ones. The reader that issues the highest count for a tag, *holds* the tag. Therefore, rather than a centralized mechanism that chooses the reader with maximum tag coverage, each tag chooses the locally optimal reader in its vicinity. A reader holding none of its covered tags is declared redundant. Furthermore, a reader reports only the tags that it holds.

RRE consists of two steps. In the first step, each reader attempts to write its tag count (number of covered tags) on each of its covered tags. A tag only stores the highest value seen, along with the identity of the corresponding reader. For this, each reader issues a write command containing its unique reader identifier and its tag count. Similar to RCA, the write operation is performed during $O(\log \psi)$ consecutive epochs, once per epoch, where ψ is the total number of readers. During each epoch, the time frame for sending the write request is randomly chosen. As mentioned in Section 12.4, this process ensures w.h.p. that at least one write command issued by each reader is correctly received by all its covered tags. Thus, after $O(\log \psi)$ epochs, each tag stores the largest number of tags covered by a reader situated in its vicinity, along with the identity of that reader, called the holder of the tag.

In the second step, a reader queries each of the tags in its interrogation zone and reads the identity of the tag's holder. Each read query issued by a reader for each of its tags is similarly repeated during random time frames for $O(\log \psi)$ consecutive time epochs to avoid reader collisions occurring at queried tags. At the end of this step, a reader that holds at least one tag is responsible for monitoring the tag and will have to remain active. However, a reader holding no tag can safely turn off its RFID interface. This is because all the tags covered by that reader are already covered by other readers that will stay active.

12.9.1.1 Implementation

Algorithm 3 shows the pseudocode for RRE-TC using an Orca-like [16] syntax. The functionality of a writable tag is shown in operation `run` of `WritableRFIDTag` (lines 4–13). When a writable tag receives a `write` command containing the identifier of the reader issuing the command and its tag count, it saves the values only if the tag count is larger than the value currently stored. When the command received is a `read`, the tag returns a packet containing its identifier followed by the reader's identifier and count value stored locally.

The detection of redundant readers is exhibited in operation `isRedundant` of `RFIDReader` (lines 18–39). First, a reader selects a random time frame during `e` consecutive epochs, and sends a broadcast `write` command containing its identifier and tag count (lines 19–24). Subsequently, it queries each of its covered tags, using a `read` command, for `e` consecutive time epochs in order to find the tag's holder (lines 25–37). Note that after sending a `read` command, at the chosen time frame, the reader waits either to receive a reply from the queried tag or for the epoch to end (lines 31–35).

Example 12.9.1 We illustrate the operation of RRE using the example systems shown in Figure 12.4. For the system shown in Figure 12.4(b), RRE finds an optimal tag coverage solution. Since from all the readers that cover tag T_1, R_4 has the highest tag count, R_4 will hold T_1. Similarly, tag T_2 will be mapped to reader R_1, since R_1 covers 3 tags but R_2 covers only 2. This effectively makes R_2 redundant, since both its tags will be reported by other readers. However, RRE does not find an optimal tag coverage solution for the system shown in Figure 12.4(a). This is because even though R_2 is redundant, in RRE, tag T will be mapped to R_2, since R_2 covers 5 tags, more than any other reader covering tag T.

Algorithm 3 Pseudocode for RRE-TC. Reader and writable tag operation for providing a solution to the tag coverage and optimal tag reporting problem.

```
1.  Object implementation WritableRFIDTag;
2.   Rid : integer; #identifier of locking reader
3.   count = 0: integer; #count of highest bidder
4.   Operation run()
5.     guard inQ.first.type = write do
6.       if inQ.first.c > count then
7.         Rid := inQ.first.rid;
8.         count := inQ.first.c;
9.       fi;
10.    guard inQ.first.type = read do
11.      bCast(new packet(Tid,Rid,count));
12.    od
13. end

14. Object implementation RFIDReader;
15. Rid: integer; #reader identifier
16. tags: array[integer] of integer; #covered tags
17. redundant = true: boolean; #is reader redundant?
18. Operation isRedundant(prefix:integer)
      #first step of RRE
19.    while count++ < e do
20.      frame:=getRandom(0,n);
21.      sleep(frame);
22.      bCast(new packet(write, Rid,tags.size));
23.      sleep(n-frame-1);
24.    od
```

```
    #second step of RRE
    for i in 1..tags.size do
        while count++ < e do
            T=getTime();
            frame:=getRandom(0,n);
            sleep(frame);
            bCast(new packet(read,tags[i]));
            guard inQ.first.tid = tags[i] do
                if inQ.rid == R_id then
                  redundant := false;
                od
            guard getTime() - T > n do od
        od
    od
    if redundant = true do turnOff(); fi
end
```

12.9.2 *RRE-HC*

We now briefly describe a variant of the distributed redundant reader elimination algorithm, where the tag mapping decision is made based on the distance in hops between readers and the sink node. The variant, named RRE-HC, assumes that each reader maintains its hop count distance to the sink node. This could be achieved by each node maintaining a parent node and each node, starting with the sink and advertising its hop count distance to the sink to each of its neighbors.

The operation of RRE-HC is very similar to the operation of RRE. It also runs in two steps. In the first step, a reader issues a write command to each tag in its interrogation zone. The command contains the reader's identifier and its hop count distance to the sink node. A tag will only overwrite the stored hop count with a strictly smaller value. In the second step, the reader interrogates all its tags to determine the reader identifier that each of them stores. The reader is redundant only if none of its tags stores its identifier.

Example 12.9.2 We illustrate the outcome of RRE-HC using the reader network from Figure 12.2(b). If reader R_1, with a hop count of 3, is the first to issue a *write* command, it will hold the tag T_2. If later, R_2 issues a similar write command, its identifier and hop count (also 3), will not overwrite the values stored on T_2. However, when later, R_5, with a hop count of 2, issues a write command, its identifier and hop count will overwrite the values stored on T_2. Then, T_2 will be reported by R_5 and will cross only 2 hops, instead of the 3 it would have crossed if R_1 or R_2 would have held it. Similarly, tag T_1 will also be mapped to and reported by R_5 and not by R_2.

12.9.3 *Analysis*

We provide a proof of the accuracy of RRE and RRE-HC, and analyze their performance. The following theorem establishes the accuracy of the RRE variants in the absence of reader failures or arrival of new tags, assuming that readers are synchronized. We elaborate on handling system update and synchronization issues in the next section.

Theorem 12.9.1 *When $m(R)$ is defined to be the hop count distance between R and the sink node,* RRE-HC *is an optimal solution to the tag reporting problem for ad hoc reader networks.*

Proof. We need to show that each covered tag T is mapped to exactly one reader and that reader is the one with the smallest hop count among all the readers covering T. It is straightforward to see that the first condition holds. For the second condition, if the *write* command from each reader from $p(T)$ reaches T, T will only store the smallest hop count. With high probability, each *write* command will reach tag T. This is because a *write* command is issued in a random chosen slot during e consecutive time frames.

Lemma 12.9.2 Complexity of RRE. *The total number of epochs in both* RRE *variants,* T_{RRE} *is* $O(\gamma e \beta \log \psi)$.

Proof. The first step of RRE, where each reader sends a write command to all its tags, takes $e \log \psi$ epochs. The second step, where readers send queries to each of their tags, takes $\gamma e \beta \log \psi$ epochs. Thus, $T_{RRE} = O(\gamma e \beta \log \psi)$.

Lemma 12.9.3 Write Complexity of RRE. *The total number of tag write operations performed by a single run of either variant of* RRE *in an RFID system is* $O(\psi \gamma)$.

Proof. In the worst case, each tag T is written to once by each of the readers in the set $p(T)$ of readers covering it. This is because from the set $p(T)$, the reader holding the least number of tags may be the first to issue a write command, followed by second ranked and so on. Thus, in the worst case, each tag may be written to $O(\psi)$ times, which proves the theorem.

In the following we study two extensions to the distributed RRE heuristic. For simplicity, we focus only on RRE, since the results can similarly be applied to RRE-HC.

12.9.4 Dependency on RCA

We have assumed until now that before running RRE, each reader has already executed RCA, detecting all the tags in its interrogation zone. This assumption ensures that on completion of the first step of RRE, tags placed in the vicinity of at least two readers store the highest number of tags covered by the readers. However, this may not always be the case and the following example shows how this can affect the correctness of RRE.

Example 12.9.3 Consider the RFID system from Figure 12.4(b). The count value stored at the completion of RRE on tag T_2 should be 3, from reader R_1. However, if we assume that initially, readers are not aware of the identity of adjacent tags and RCA needs to be executed just before RRE, the following scenario may occur: since R_2 only covers two tags, whereas R_1 covers three, R_2 will complete RCA and also the first step of RRE before R_1. Then, R_2, upon identifying itself to be the holder of T_2, will also decide to stay active, despite being redundant.

12.9.4.1 Extension of RRE

In order to solve this problem, we require readers to maintain a list of tags held, and to passively listen for tag responses to queries initiated by other readers. The duration of this phase is upper bounded by the time taken by a reader to get an estimate of the maximum number of tags a reader can cover. The listening phase proceeds as follows: assume that a reader R overhears a tag response to a query initiated by another reader and the query content is R_x, T_y, c (see Algorithm 3 line 11). This query indicates that the holder of tag T_y is R_x with a tag count c. Then, if c is larger than its own tag count and $R_x \neq R$, reader R removes tag T_y from its list of held tags. When the list is empty, the reader becomes redundant and can be safely turned off.

Example 12.9.4 Using the example in Figure 12.4(b), assume R_2 has T_1 and T_2 in its list of held tags on completion of its first step of RRE. During R_1's execution of the first step of RRE, R_1 chooses at least one time frame during e epochs to write when no other reader is transmitting. When this occurs, R_2 overhears the reply of T_2 and removes the tag from its coverage list. This is because T_2's reply will contain a count higher than its own and a reader identifier different from its own. The situation is identical for tag T_1 that will eventually be covered by R_4.

Note that this solution may not be applicable to all RFID systems (e.g. UHF).

12.10 Adapting to Topological Changes

As mentioned before, readers detected to be redundant by RRE can switch off their RFID interface. However, in a real deployment, tags and readers may fail and new components may be randomly deployed. Scenarios where new tags are deployed or existing readers and tags move are particularly important. Thus, in order to adapt to topological changes, readers need to periodically execute RRE.

Our solution consists of providing a duty-cycle for the RFID interface of readers. The RFID interfaces of readers are periodically activated, allowing them to run RRE and discover new tag deployments. Between active intervals, the RFID interface is placed into sleep mode, effectively saving battery power.

In the following, we consider each duty cycle to consist of an active interval followed by k sleep intervals. k is a parameter, denoting the ratio of sleep to active intervals. We assume that all active and sleep intervals have the same length and we only consider the case of a single active interval per cycle. Figure 12.5 shows an example duty-cycle with a 1:1 ratio of sleep to active intervals. Unfortunately, a simple duty-cycle for the RFID interface does not handle reader failures or relocations well. This is illustrated in the following example.

Example 12.10.1 Consider the system deployment of Figure 12.4(b). Assume that after each reader executes RRE once, active reader R_1, fails. Before failing, R_1 was holding tags T_2, T_3 and T_4, with a count of 3. After R_1 fails and when readers R_2, R_3 and R_4 execute RRE again, tags T_2, T_3 and T_4 will still have a count value of 3. Then, reader R_2 will again discover, this time inaccurately, that it is redundant and will not report

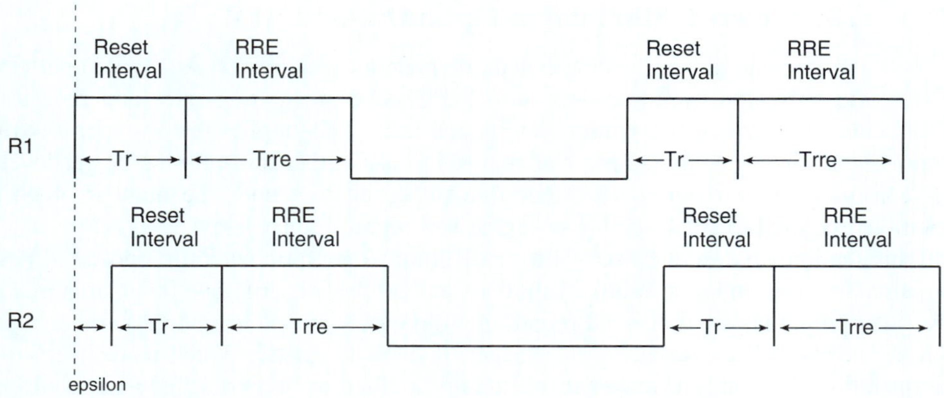

Figure 12.5 Example reader duty-cycle with a 1:1 ratio of sleep to active intervals. Each active interval consists of a first, shorter sub-interval for resetting all tags to 0, followed by a longer sub-interval dedicated to running RRE. The duty-cycles of two readers shown are off by at most ϵ, due to clock drifts.

tag T_2. Then, T_2 will not be reported, even though it is still covered by a functioning reader, R_2.

12.10.1 Tag Count Resetting

We propose a solution to this problem, which assumes a coarse clock synchronization between readers. Let ϵ be the synchronization error we decide to support. We assume all readers start their active intervals at the same time (within the ϵ error bound). We divide each active interval into two sub-intervals. During the first sub-interval, called the tag reset interval, each reader contacts each of its covered tags and resets their counters to 0. If a tag is covered by multiple readers, it may be reset multiple times. Let T_r denote the length of the tag resetting interval. A reader will attempt to reset its tags only in the first $T_r - \epsilon$ section of its tag resetting interval. During the last ϵ section of the sub-interval, a reader will not issue *write* commands. During the second sub-interval, called the RRE *interval*, each reader executes RRE. The length of the RRE interval, T_{rre} needs to be at least $O(\gamma \log \beta \log \psi)$, to allow RRE to complete.

For any reader, at the beginning of each of its RRE sub-intervals, all its covered tags have their counters set to 0. It is straightforward to see that if no two reader clocks drift by more than ϵ, the silent period between the tag reset and the RRE sub-intervals ensures that no resetting command is issued after a reader has written its tag count on any of its covered tags. A tag can then be held and reported by different readers after different active intervals. Then, even if a reader fails, the subset of its tags covered by other readers will still be reported.

Note that a reader does not have to always report all its covered tags. Instead, for slowly changing topologies, a reader can implement the following strategy. The reader reports all covered tags after its first active interval but subsequently reports only changes in coverage. A change consists in the detection of a new tag or the disappearance of a previously covered tag.

12.11 The Layered Elimination Optimization (LEO)

As shown in Lemma 12.9.3, the distributed RRE heuristic requires $O(\psi\gamma)$ write operation for a single execution by all readers in an RFID system. Writing rates and ranges of writable tags are smaller than reading ranges and rates. This implies that in order to write to tags in their interrogation zones, readers need to use higher transmission power levels, thus consuming more energy. It is therefore important to reduce the number of write operations required to solve the tag coverage and reporting problems.

In this section we describe LEO, a Layered Elimination Optimization proposed in [18]. LEO attempts to extend RRE in order to reduce the number of write operations required to solve the optimal tag coverage and reporting problems. The idea behind LEO is that while each tag needs to be reported by one reader, it does not matter which reader performs this operation. Instead, the reader that ends up holding and reporting any tag could be the first reader that attempts to write to that tag. Then, LEO functions in the following manner. Each tag stores a single value, the identifier of the reader holding it. The initial value for that identifier is NULL for any tag. Before a reader writes its identifier on any of its covered tags, it first reads the reader identifier stored on that tag. The reader then proceeds to write its identifier on the tag, only if the reported reader identifier is NULL.

12.11.1 Implementation

Algorithm 4 shows the pseudocode for LEO using Orca-like [16] syntax. The pseudocode simplifies the presentation of read and write operations performed by a reader, hiding the details of choosing a random time frame for transmission, during *e* consecutive epochs. Specifically, the *read* operation used in the pseudocode specifies the tag identifier and returns the value stored on the tag. Similarly, the *write* operation specifies the destination tag and the value to be written. The differences from RRE are the following. When a tag receives a *write* command, it records the specified reader identifier only if the stored reader identifier is NULL (lines 3–6). When a tag receives a *read* command, it returns only its identifier and the reader identifier locally stored (lines 7–9). The operation of an RFID reader differs from RRE in the following fashion. For each tag covered, the reader first reads the tags and only attempts to write its identifier if the identifier stored on the tag is NULL (lines 13–17). Afterward, the reader decides to be redundant only if its identifier is not stored on any of its covered tags (lines 18–22).

Algorithm 4 Pseudocode for LEO. Reader and tag operation for reducing the number of operations performed to provide a solution for the tag coverage and reporting problems.

```
1. Object implementation WritableRFIDTag;
2.  Operation run()
3.      guard inQ.first.type = write do
4.          if R_id = NULL then
5.              R_id := inQ.first.rid;
6.          fi;
7.      guard inQ.first.type = read do
8.          bCast(new packet(T_id,R_id));
9.      od
10. end
```

```
Object implementation RFIDReader;
Operation isRedundant(prefix:integer)
    for i in 1..tags.size do
        if read(tags[i]) = NULL then;
            write(tags[i],Rid);
        fi
    od
    for i in 1..tags.size do
        if read(tags[i]) = Rid then;
            redundant := false;
        fi
    od
    if redundant = true do turnOff(); fi
end
```

12.11.2 Analysis

Similar to GREEDY and RRE, LEO provides a non-optimal solution to the tag coverage problem. However, note that LEO significantly reduces the number of write operations when compared to RRE. Specifically, the number of write operations is γ, that is, each tag is written only once. While this improvement is desirable, we note that LEO introduces several problems when compared to RRE.

First, by greedily assigning tags to the readers holding most tags, RRE may require fewer nodes to stay active. This is because readers-assigned tags need to be active. By giving preference to readers assigned a higher number of tags, RRE is likely to require fewer readers to stay active than a random assignment algorithm as the one proposed by LEO. Second, since in LEO a tag can only be written once, the algorithm becomes vulnerable when readers fail. If a reader fails, the tags it was assigned will not be overwritten by other readers. Such tags will be subsequently ignored. Third, the test-and-write operation proposed in LEO introduces race conditions. If two readers detect a tag that is un-assigned at roughly the same time, both will attempt to mark the tag by issuing write commands. Even though a single write command will succeed, both readers will believe that they are holding the tag. This can lead to both discovering fewer redundant readers and to generating duplicate tag reports.

As a final note, the advantage of LEO lies in the fact that each tag is written to only once, instead of multiple times as it is likely to happen in RRE. Note that a similar effect could be achieved in RRE if readers that have detected more tags choose earlier slots for transmission when writing. This approach can effectively prevent readers with a lower tag count from issuing write commands before heavier readers.

12.12 Related Work

12.12.1 Coverage Problems in WSNs

Coverage in sensor networks is a well-covered research topic. Tian and Georganas [19] were the first to propose an algorithm for detecting sensors whose coverage area is completely covered by other sensors. In their solution, a sensor turns itself off only when each sector of its coverage disk is covered by another sensor. Zhang and Hou [20] turned off "redundant" sensors by using bitmaps indicating the coverage redundancy of small zones belonging to the coverage disks of sensors.

For the interested reader we also recommend the papers of Slijepcevic and Potkonjak [21], Ye *et al.* [22] and Gallais *et al.* [23], which study various aspects and approaches addressing the coverage problem in sensor networks. For further reading on the approach briefly covered in Section 12.5, we direct the reader to the work of Carbunar *et al.* [24].

12.12.2 Collisions in RFID Systems

The reader-collision problem in RFID systems, which is briefly studied in Section 12.4, was first documented in [14]. The solution proposed, of allocating different frequencies to interfering readers, is centralized. A simple decentralized version, where readers listen for collisions and use randomized back-off when detecting one, is discussed. For further reading on this topic, including various centralized and decentralized, time and frequency division approaches for solving this problem, we direct the reader to works by Waldrop *et al.* [25], Ho *et al.* [26] Zhou *et al.* [27], Deolalikar *et al.* [28] and Birari and Iyer [29]. For more challenges in managing RFID data we recommend the work of Derakhshan *et al.* [30].

Moreover, Medium Access Control protocols for wired and wireless networks share several details with the reader collision avoidance algorithm. For the interested reader we recommend reading the ALOHA [31], MACA [32] and CSMA/CD [33] MAC protocols.

12.13 Conclusion

In this chapter we have studied two important problems in wireless RFID systems. The first problem consists of finding the minimal set of readers that cover all the tags in the system. This problem relates to extending the lifetime of the reader network by detecting and temporarily disabling the RFID interfaces of redundant readers. The second problem consists of determining the optimal way to report tags. We defined coverage in terms of discrete sets of points, tags, and proved that the optimization version of the first problem is NP-hard. For both problems we studied first a centralized heuristic, then distributed and localized algorithms. The distributed algorithms are based on a randomized querying technique, that ensures, w.h.p., the accurate receipt of reader queries by tags. Since writing rates and ranges are smaller than reading rates and ranges, we have also discussed a layered elimination optimization technique and studied its properties.

Future research direction may consists of: (i) investigating in detail the energy cost of the read and write operations invoked by RRE; (ii) determining the difference in range for read and write operations; and (iii) developing algorithms that further reduce the number of tag (read/write) operations.

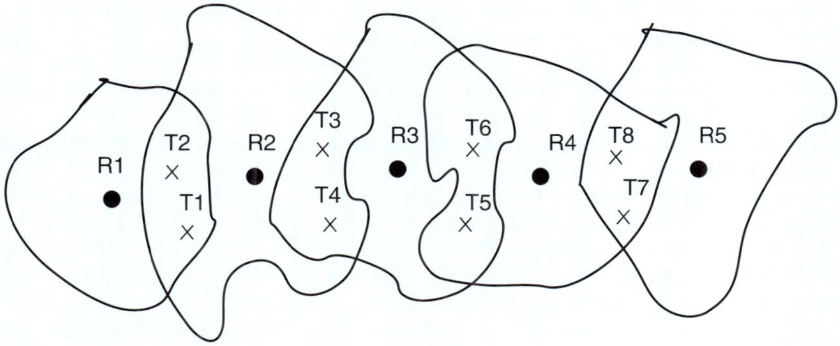

Figure 12.6 Difficulty of consistently breaking ties. The optimal solution keeps only R_2 and R_4 active. However, in a scenario where R_2, R_3, and R_4, each covering 4 tags, hold a different set of tags, all of them will have to be active.

Problems

1. How and why does RCA differ from TWA? How does RCA avoid collisions and ensure that reader transmissions are received by tags?
2. What is the difference between the optimal tag coverage and the tag reporting problems?
3. Show that RRE does not provide an optimal tag coverage solution for the RFID system shown in Figure 12.4(a).
4. Use the pseudocode of Algorithm 3 to implement the RRE-HC algorithm, assuming that each reader has a local variable storing its distance in hops to the sink host. Then modify the Algorithm 2 to provide a centralized, greedy version of the RRE-HC algorithm.
5. Show that in a network where r nodes can provide full coverage (cover all the tags) a worst case run of RRE can require $O(r)$ nodes to be active, where the constant hidden by the big-oh notation is larger than 1. Hint: use the example system from Figure 12.6 to find the optimum tag coverage solution and determine the worst case RRE result.

NB The solutions are provided on the book's website.

References

[1] Bednarz, A. (2002) Wireless technology reshapes retailers. *Network World*, 12 August.

[2] Zhang, T., Ouyang, Y., and He, Y. (2008) Traceable air baggage handling system based on RFID tags in the airport, *JTAER*, **3**(1): 106–115.

[3] SkyeTek. (2004) Available at: http://www.skyetek.com/Portals/0/Documents/Products/SkyeModule_M1Mini_DataSheet.pdf, January.

[4] Crossbow Technology Inc. RFID and Asset Tracking. Available at: http://www.xbow.com/.

[5] Baracoda An efficient way to add RFID reader/encoder to Bluetooth PDA and mobile phones. Available at: http://www.baracoda.com/baracoda/products/p_21.html.

[6] Wireless Dynamics Wireless Dynamics Inc. announces the mini-SDiD. Available at: http://www.wdi.ca.

[7] *Mobile Magazine*. Nokia 5140 RFID Reader. Available at: http://www.mobilemag.com/content/100/104/C2607.

[8] Carbunar, B., Ramanathan, M.K., Koyutürk, M., Jagannathan, S., and Grama, A. (2009) Efficient tag detection in RFID systems, *J. Parallel Distrib. Comput.*, **69**(2): 180–196.

[9] PGS Electronics Wireless RFID Systems. Available at: http://www.pgselectronics.com/PGSRFID.htm.

[10] O'Connor, M.C. *RFID Journal*. Reva Announces RFID Network Design. Available at: http://www.rfidjournal.com/article/articleview/1638/1/1/.

[11] Sarma, S.E., Weis, S.A., and Engels, D.W. (2003) RFID systems and security and privacy implications, in *CHES '02*. Berlin: Springer-Verlag, pp. 454–469.

[12] Micic, A., Nayak, A., Simplot-Ryl, D., and Stojmenović, I. (2005) A hybrid randomized protocol for RFID tag identification, in *Proceedings of the IEEE International Workshop on Next Generation Wireless Networks (WoNGeN)*.

[13] Pupunwiwat, P., and Stantic, B. (2009) Unified q-ary tree for RFID tag anti-collision resolution, in *ADC*.

[14] Engels, D.W., and Sarma, S.E. (2002) The reader collision problem, in *IEEE International Conference on Systems, Man and Cybernetics*.

[15] Carbunar, B., Ramanathan, M.K., Koyuturk, M., Hoffmann, C., and Grama, A. (2005) Redundant reader elimination in RFID systems, in *2nd Annual IEEE Communications Society Conference on Sensor and Ad Hoc Communications and Networks (SECON'05)*.

[16] Bal, H.E., Bhoedjang, R., Hofman, R., Jacobs, C., Langendoen, K., Ruhl, T., and Kaashoek, M.F. (1998) Performance evaluation of the Orca shared-object system. *ACM Trans. Comput. Syst.*, **16**(1): 1–40.

[17] Fowler, R., Paterson, M., and Tanimoto, S. (1981) Optimal packing and covering in the plane are NP complete. *Information Processing Letters*, **12**(3): 133–137.

[18] Hsu, C.-H., Chen, Y.-M., and Yang, C.-T. (2007) A layered optimization approach for redundant reader elimination in wireless RFID networks, in *APSCC '07: Proceedings of the the 2nd IEEE Asia-Pacific Service Computing Conference*.

[19] Tian, D., and Georganas, N.D. (2002) A coverage-preserving node scheduling scheme for large wireless sensor networks, in *Proceedings of the 1st ACM WSNA*, ACM Press, pp. 32–41.

[20] Zhang, H., and Hou, J. (2004) Maintaining coverage and connectivity in large sensor networks, in *International Workshop on Theoretical and Algorithmic Aspects of Sensor, Ad hoc Wireless and Peer-to-Peer Networks*.

[21] Slijepcevic, S., and Potkonjak, M. (2001) Power efficient organization of wireless sensor networks, in *IEEE ICC*.

[22] Ye, F., Zhong, G., Lu, S., and Zhang, L. (2003) PEAS: a robust energy conserving protocol for long-lived sensor networks. In *23rd IEEE ICDCS*.

[23] Antoine Gallais, Jean Carle, David Simplot-Ryl, and Ivan Stojmenović (2008) Localized sensor area coverage with low communication overhead, *IEEE Trans. Mob. Comput.*, **7**(5).

[24] Carbunar, B., Grama, A., Vitek, J., and Carbunar, O. (2006) Redundancy and coverage detection in sensor networks, *ACM Trans. Sen. Netw.*, **2**(1).

[25] Waldrop, J., Engels, D.W., and Sarma, S.E. (2003) Colorwave: a MAC for RFID reader networks, in *Wireless Communications and Networking (WCNC)*.

[26] Ho, J., Engels, D.W., and Sarma, S.E. (2006) HIQ: a hierarchical q-learning algorithm to solve the reader collision problem, in *Proceedings of the Applications and the Internet Workshops*.

[27] Zhou, Z., Gupta, H., Das, S.R., and Zhu, X. (2007) Slotted scheduled tag access in multi-reader RFID systems, in *Proceedings of ICNP*.

[28] Deolalikar, V., Mesarina, M., Recker, J., and Pradhan, S. (2006) Perturbative time and frequency allocations for RFID reader networks, in *Proceedings of Emerging Directions in Embedded and Ubiquitous Computing*.

[29] Birari, S.M., and Iyer, S. (2005) Pulse: A MAC protocol for RFID networks, in *EUC Workshops*.

[30] Derakhshan, R., Orlowska, M.E., and Li, X. (2007) RFID data management: Challenges and opportunities, in *Proceedings of IEEE RFID*.

[31] Abramson, N. (1970) The ALOHA system: Another alternative for computer communications, in *AFIPS Conf. Proc., Fall Joint Computer Conf*.
[32] Karn, P. (1990) MACA a new channel access method for packet radio, in *ARRL/CRRL Amateur Radio 9th Computer Networking Conf*.
[33] Carrier sense multiple access with collision detection (CSMA/CD) access method and physical layer Specifications, Institute of Electrical and Electronics Engineers, 1996.

13

Delay/Disruption-Tolerant Mobile RFID Networks: Challenges and Opportunities

Hongyi Wu and Zhipeng Yang

The Center for Advanced Computer Studies, University of Louisiana at Lafayette

13.1 Motivation

Wireless sensor networks have been widely employed in wildlife research, where sensors are attached to animals of interest for pervasive information gathering [1]. While the effectiveness and efficiency of such sensor networks have been demonstrated in multiple research projects, they are limited in applications for large animals only. Recent studies (to be discussed next) have revealed that battery-powered sensors have fundamental limitations that hinder their use in tracking small wildlife.

Radio Frequency IDentification (RFID) has received growing attention lately, as a result of reduced tag costs and sizes as well as expanded reader communication ranges. This chapter presents the latest research that deals with a *Featherlight Information Network with Delay-Endurable RFID Support (FINDERS)*, composed of passive RFID tags which are ultra light, durable, and flexible, without power supply for long-lasting applications [2]. It investigates the use of RFID gear for wireless network construction, aiming to find events of interest and gather aggregate information. Like modern smart sensors [1], RFID tags and readers together may constitute a distributed sensor system for pervasive information gathering, subject to different data acquisition, delivery, and storage approaches. Unlike typical active sensors that rely on battery power supply and indispensably require sturdy casing for their protection, however, a passive RFID tag is particularly suitable for environments with strict weight constraints, for example, in the studies of small wildlife,

RFID Systems: Research Trends and Challenges Edited by Miodrag Bolić, David Simplot-Ryl, and Ivan Stojmenović

where the weight of the sensor must be under 5% of the weight of the animal to avoid hindering the wildlife movement or the welfare of the animals themselves. This requires very small sensors to be employed, which is in sharp contrast to other projects that target large animals including zebras [3], whales [4], and deers [5], and human beings [6]. Earlier investigation has revealed that most of these small animals cannot carry any active devices (such as GPS receivers or the smallest Crossbow sensors) with battery and casing, since overweight sensors often lead to high mortality rates of the animals being studied. Similar weight constraint also applies in many other fields of scientific studies and industrial applications. While efforts have been made to develop miniature sensors in laboratories, the lowest weight of any active sensor that aims to achieve a given communication range and lifetime is bounded inevitably by its battery and casing, where the latter must be heavy-duty for the protection of the power source and powered electronic circuits under harsh environments. This becomes a key hurdle that limits the applicability of active sensors in various applications with strict weight constraints, not to mention extra hassle involved in ensuring adequate battery power.

The proposed FINDERS system intends to address such fundamental challenges in wireless sensor networks by leveraging RFID and modern computer network technologies to establish a featherlight information network for pervasive data gathering under strict weight constraint.

13.2 Overview of FINDERS

In this section, we outline the overall architecture of FINDERS, including its constituent components and basic communication schemes. The details of its network protocols are omitted, as it is still an open issue to design the most appropriate protocols for FINDERS.

FINDERS consists of two types of nodes, RFID readers and tags, as illustrated in Figure 13.1. The readers are deployed at strategic locations according to specific applications. For example, in wildlife and biological studies, the readers can be set up at natural "choke points" where the animals have to move past or through, because of significant

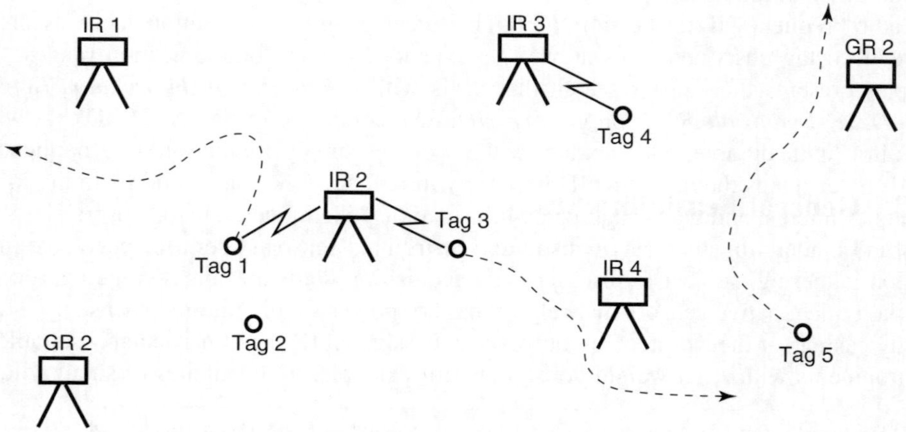

Figure 13.1 An overview of FINDERS.

movement barriers otherwise. Since FINDERS is often deployed under harsh and complex field environments, it is not practically viable for most readers to establish reliable connections to communicate with each other or to access the backbone networks. As a result, they become isolated readers, or IRs (see IRs 1-4 in Figure 13.1). Power supply is challenging for the IRs too. Most of them rely on solar panels and high capacity batteries, which can be recharged/replaced but will post a limit on the daily duty circle of the readers. Only a few readers at convenient locations are equipped with a reliable power source and network connections, and they are dubbed gateway readers (GRs; see GRs 1 and 2 in Figure 13.1). GRs serve as the gateways to deliver data from FINDERS to the destination (e.g. a data server). The readers have relatively large storage space compared with tags, and each of them maintains a data queue. While all readers are fixed, the tags are attached to moving targets and thus become mobile (see Tags 1-8 in Figure 13.1).

Many off-the-shelf passive tags that are light and durable and have sufficient reading/writing ranges can be employed in FINDERS. For example, we have adopted the Alien passive RFID system for this research, which consists of ALR-9900 reader and ALN-9540 Squiggle^{TM} (passive) tags. The ALR-9900 reader is a powerful device that possesses 64 MB RAM and 64 MB flash memory and supports up to 4 antennae and 50 communication channels. Being very thin and light, an ALN-9540 tag measures $8.15 \times 94.8 \times 0.05$ mm and weights less than one gram. Its memory can hold up to 20 bytes of data. With such small storage space, the tag maintains a simple buffer that can be read and written in one operation, without queuing. The tag is well engineered and packaged, exhibiting excellent survival skills under such harsh environments as underwater, underground tunnels, and extreme temperatures ranging from $-25°$C to $+50°$C. Our lab and field experiments have revealed that the Alien system can achieve a reading/writing distance of some 20 ft.

Since the readers are static and most of them are isolated from each other, the communications in FINDERS rely on the mobility of the tags to establish time-varying opportunistic links with nearby readers, forming a delay-tolerant network (DTN; [7]) for data delivery. For example, in the scenario shown in Figure 13.1, Tags 1 and 3 are within the reading/writing range of IR 2. Thus IR 2 may read the data from Tag 1 and write them into Tag 3. When Tag 3 passes by IR 4, it unloads its data to the latter, which in turn writes the data into Tag 5 when it comes into the writing range. Tag 5's trajectory passes through GR 2. The data can thus be delivered to the destination via this gateway.

In general, FINDERS may serve to find events of interest and to gather aggregate system information through sporadic wireless communications between its constituent passive tags and readers.

13.3 General Feasibility Study

While the featherlight weight of passive RFID tags provides unique opportunities to support various applications with strict weight constraints, the capacity of FINDERS is unsurprisingly low compared with many other data networks, due to its extremely limited network resources. To understand its capacity and accordingly the applications that it can support, we first carried out a feasibility study of the FINDERS system based on a trivial communication scheme. This feasibility study aims to gain insights into the basics of FINDERS and reveal its general capacity, without consideration of specific protocols and network environments.

To this end, we need to investigate why a data delivery effort may fail and where the data packet may be dropped. First, since the GRs connect to the conventional networks, they usually have sufficient communication bandwidth (for our target applications) and can safely deliver the received data to end users. The tags keep data packets in their nonvolatile memory. As long as they are not falsely overwritten, the data will not be lost (if we ignore any physical failures). On the other hand, each IR maintains a data queue, which may overflow when packet arrival is too high. Therefore, we focus on the queuing behavior of the IRs in our feasibility study.

1. **Queuing Model:** We consider a general queuing model (i.e. a $G/G/1$ queue) for each IR. While it is difficult to derive solutions for the G/G/1 queue without a detailed knowledge of packet arrival and service (which depend heavily on the data transmission protocol), we intend to obtain an estimation of the network capacity (i.e. the amount of data that can be handled by the system) based on the condition of queue stability. More specifically, the arrival rate must be lower than the service rate in order to keep the queue stable. This study only requires to discover the average data service rate, to be discussed next.
2. **Average Data Service Rate:** A data packet is served (i.e. sent out from the IR's queue) when the IR meets a suitable tag. Thus, it is critical to understand how often the meeting between the IR and a suitable tag can happen. Consider a FINDERS with K IRs, J GRs, and T tags, where each tag has a capacity of m, that is, the tag can store up to m data packets. For simplicity, we assume that a tag can receive data (i.e. becomes a suitable tag) only if it is empty (while more sophisticated schemes for data transmission will be developed in the future). Moreover, we consider that Tag i moves by following a randomly chosen path. For analytic tractability, the time to travel through such a path is assumed to be a random variable under exponential distribution with the rate of λ_i. This assumption, thought not always true under all network environments, greatly reduces the analytic complexity. Note that our feasibility study does not intend to provide an accurate analysis, but instead to gain a general observation of the capacity of FINDERS. Besides, we assume Tag i has a long-term statistical probability of γ_{ik}^I to meet IR k and the probability of γ_{ij}^G to meet GR j. It is easy to verify that the time intervals to meet IR k and GR j are exponentially distributed with the rate of λ_{ik}^I and λ_{ij}^G, respectively, where $\lambda_{ik}^I = \lambda_i \gamma_{ik}^I$ and $\lambda_{ij}^G = \lambda_i \gamma_{ij}^G$. In addition, let $\lambda_i^I = \sum_{k=1}^K \lambda_{ik}^I$, which is the total rate for Tag i going to any IRs, and similarly $\lambda_i^G = \sum_{j=1}^J \lambda_{ij}^G$.

   Based on above assumptions, a Markovian model (see Figure 13.2) is established to analyze the meeting events between Tag i and the readers, and accordingly the data service rate of the IR's queue. The model consists of $K+2$ states. State S_k $(1 \leq k \leq K)$ is the state that IR k can transmit m packets to Tag i (i.e. the tag is empty). State S_N is for the state where Tag i meets any IRs, but it is not empty and thus cannot receive data packets. State S_G indicates that Tag i meets a GR and delivers the data packets it carries. Once the packets are delivered to the GR, they are removed from the tag, which thus becomes empty and ready to receive new data packets.

   Let's denote P_k, P_N, and P_G the steady state probabilities of States S_k, S_N, and S_G, respective. Based on the Markovian model, we can derive the following state

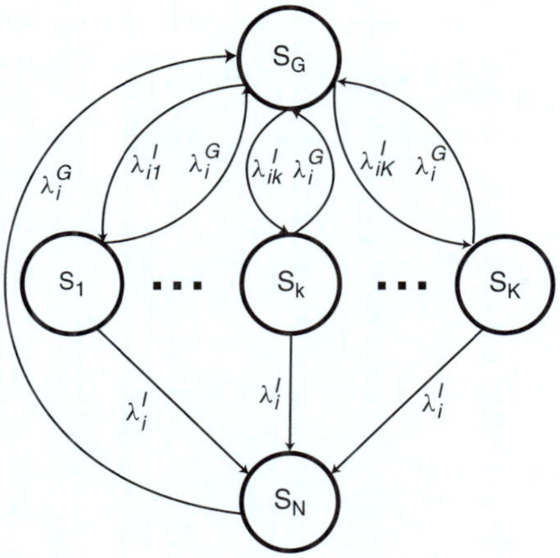

Figure 13.2 Markovian model for Tag i.

equations:

$$\begin{cases} P_k(\lambda_i^G + \lambda_i^I) = \lambda_{ik}^I P_G, \forall 1 \leq k \leq K, \\ P_N \lambda_i^G = \sum_{k=1}^K \lambda_i^I, \\ P_G \lambda_i^I = P_N \lambda_i^G + \sum_{k=1}^K (\lambda_{ik}^I P_k), \\ \sum_{k=1}^K P_k + P_N + P_G = 1. \end{cases} \tag{13.1}$$

To solve them, we have $P_G = \frac{\lambda_i^G}{\lambda_i^I + \lambda_i^G}$, $P_k = \frac{\lambda_{ik}^I \lambda_i^G}{(\lambda_i^G + \lambda_i^I)^2}$, and $P_N = \frac{\lambda_i^{I2}}{(\lambda_i^G + \lambda_i^I)^2}$. To facilitate the understanding of FINDERS' general capacity, we further simplify the above results by letting $\gamma_{ik}^I = \gamma_{ij}^G = \gamma$ (i.e. Tag i has equal probability of meeting all readers), and accordingly $\lambda_{ik}^I = \lambda_{ij}^G = \lambda$. Then we arrive at $P_k = \frac{J}{(K+J)^2}$, which is the steady state probability that Tag i becomes empty and arrives at an IR. With an overall rate of $(K + J)\lambda$ for Tag i to visit any readers, it has a rate of $(K + J)\lambda P_k$ to meet IR k when it is empty. Assume all tags are independent, the aggregated arrival rate of empty tags at an IR is $T(K + J)\lambda P_k$. Since each tag can carry m packets, we have the service rate of an IR's queue, $\mu = \frac{J}{J+K}\lambda m T$.

3. **Estimated Capacity:** To ensure a stable queue at the IR, the service rate must be greater than the arrival rate. Therefore, the entire FINDERS with K IRs can accept no more than $\frac{JK}{J+K}\lambda m T$ packets, or

$$A \leq \frac{JK}{J+K}\lambda m T, \tag{13.2}$$

which can serve as an estimation of the network capacity. It shows that, under typical scenarios, FINDERS can easily achieve a data rate of multiple kbps or higher

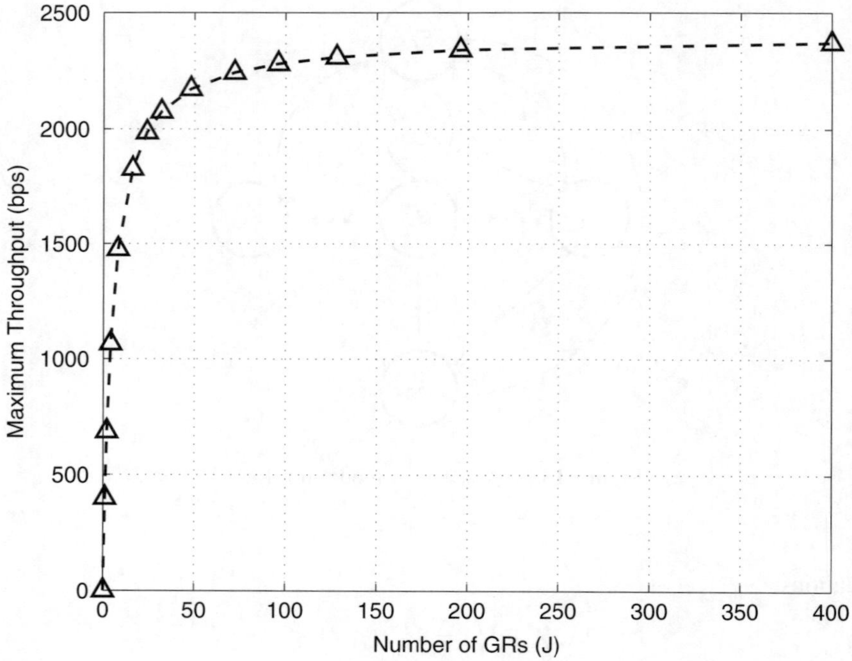

Figure 13.3 Throughput vs. J.

(see Figures 13.3 & 13.4 where by default a tag can hold 96 bits, $K = 5$, $m = 1$, $T = 100$, $\lambda = 0.05$, and $J = 2$), which is sufficient to support a wide range of applications, such as wildlife tracking and lightweight monitoring. The capacity increases linearly with m, λ, and T, and nonlinearly with J and K.

In addition to the overall network capacity, we also have several interesting findings on the data generation in individual events. In general, new data are fed into the IR's queue only when it meets a tag. Without loss of generality, let's assume α packets arrive at the IR when such meeting event occurs. Thus, we have $\alpha\lambda T \leq \mu$, or $\alpha \leq \frac{Jm}{J+K}$. In other words, when a tag meets an IR, no more than $\frac{Jm}{J+K}$ packets can be generated and fed into the IR's queue. This result provides a guideline for traffic control in FINDERS and leads to two interesting observations. First, α is not related to the number of tags. This is because the tags contribute to both data arrival and service of individual IR. Second, increasing the number of IRs, K, results in a lower α. Though this is a little anti-intuitive (as the IRs help data relaying), it can be explained below. The IRs directly contribute to the network traffic load (when they meet tags), but do not directly consume the data (which can be successfully delivered only when they arrive at the GRs but not IRs). Thus, more IRs decrease the maximum allowed α. However, a larger K does help improve the overall network capacity, because they relay and buffer data packets, thus increasing the opportunity to deliver data to GRs. Additionally, a sufficiently large K is usually required by the applications, in order to achieve the needed coverage and granularity in data acquisition.

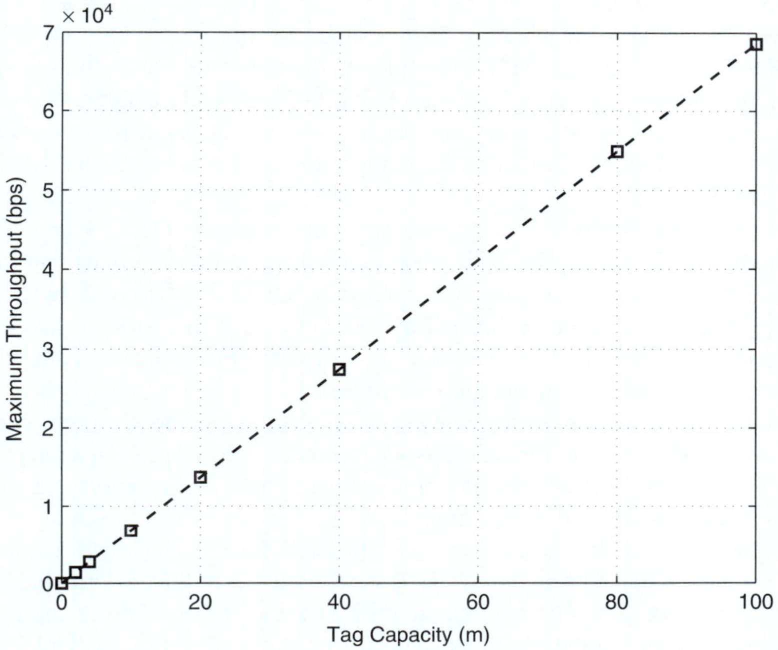

Figure 13.4 Throughput vs. m.

13.4 Unique Challenges and Tactics

The above analysis based on a trivial communication scheme has shown that FINDERS, though with extremely limited resource, can provide sufficient capacity and effectively support our target applications. However, to design the most appropriate protocols for FINDERS still remains an open problem.

FINDERS is a very unique wireless information network that distinguishes itself from conventional communication networks, sensor networks, and delay-tolerant networks by the following characteristics:

- *Nodal heterogeneity:* FINDERS consists of two very different types of nodes. A reader is a static and powerful device, with large storage, high computing power, and long-lasting (but still limited) battery power. On the other hand, the passive tag can be mobile and has extremely limited resource.
- *Asymmetric communication:* The communication in FINDERS can be established between a tag and a reader only, but not tags to tags or readers to readers. The communication must be initiated by a reader, in contrast to symmetric transmissions in most conventional networks. In addition, reading and writing ranges are usually different, according to our experiments.
- *Nodal mobility:* All readers are static, while the passive tags can be attached to moving objects with various mobility patterns, leading to a dynamic network topology. This is similar to that of a mobile ad hoc network but with very frequent partitions.

- *Intermittent connectivity:* The connectivity of FINDERS is very low and intermittent, forming a sparse network where a tag is connected to a reader only occasionally. In fact, FINDERS is a special DTN with unique communication and storage constraints.
- *Intermittent computation:* Besides connectivity, the computation at the tag is also intermittent. It is available only when the tag is powered up by a nearby reader. Thus, such continuous functions required by many protocols as counters and timers cannot be implemented here.
- *Critical network resource:* The fundamental philosophy in DTN is to trade storage for sporadic connectivity. The DTN node is assumed to have sufficient storage space, which can hold a large volume of data to alleviate the needs of immediate transmission. This, however, is no longer valid in FINDERS, because the buffer space of the tags (the main vehicle for data transportation) is so limited that it may become the critical network resource and communication bottleneck.
- *Delay tolerability:* Data delivery delay in FINDERS is potentially high, due to loose connectivity and extremely limited network resources. However, such delay, though not desirable, is usually tolerable by the applications which aim at pervasive information gathering from a statistical perspective.
- *Fault tolerability:* Redundancy may exist in FINDERS during data acquisition and delivery. Thus, a data packet may be lost without degrading information gathering performance. At the same time, there are critical needs in data filtering and aggregation at IRs and GRs, due to the likely existence of multiple copies of the same data.

The above characteristics make the development of FINDERS a very unique, interesting, and challenging problem, calling for effective solutions to overhaul the data acquisition and delivery schemes in such a featherlight information network with extremely limited resources. While various sensor networks [3, 1, 4, 6] have been investigated and deployed, data collection, storage and communication solutions devised therein are not directly applicable to FINDERS. Few earlier studies have ever been conducted on information networks composed of RFID gear with sporadic communication between moving tags and nearby readers. In particular, two sets of design issues specific to FINDERS must be addressed:

1. **Effective and Reliable Data Delivery:** FINDERS faces unprecedented challenges in communication and networking. First, the sporadically available wireless links render it impossible to form a well-connected mesh network for end-to-end communications, which are the basis of mainstream sensor network technologies [1]. Moreover, the unique asymmetric communication paradigm, the intermittent computation capability and the extremely small buffer size of a passive tag overthrow the fundamental principle of DTN (where ample buffering is employed under the intermittent connectivity to alleviate the needs of immediate transmission), resulting in inefficiency if the existing DTN protocols are employed in FINDERS. To this end, a series of interesting problems related to data delivery need to be investigated:
   - *Routing metrics:* Routing metrics are usually adopted by the routing protocol as an indicator of the available resource of a given path or link. For example, path length and end-to-end delay are popularly used in conventional networks. These metrics, however, do not reflect the unique network resource of FINDERS, and thus may

lead to poor network performance or even failure if used for routing. New routing metrics must be explored, possibly based on meeting probabilities between tags and readers and available memory of the tags. Maintaining and updating such metrics must be simple and consume insignificant storage space.

- *Duplication control:* In FINDERS and many other DTN networks, replication is necessary during data delivery for achieving a given success ratio. However, replication increases the overhead, and worse yet, excessive replication may even degrade the delivery ratio due to frequent buffer overflow. Duplication control thus becomes a key issue to be tackled in FINDERS. It reaches a system-wide optimization problem. While the data delivery schemes in FINDERS are (and have to be) much simpler compared with the protocols for other wireless or wired networks, such optimization is non-trivial and its performance largely determines the effectiveness and feasibility of FINDERS.
- *Queue management:* Each reader maintains a data queue. Given the limited communication resource (due to sporadic connectivity and extremely limited storage for tags) and the relatively high loading factor (due to newly acquired and duplicated data), it is an overriding design issue in FINDERS to differentiate the packets in the queue by a simple and efficient parameter which signifies their importance and determines which packet to transmit if a communication opportunity becomes available or which packet to drop if the queue is full. Novel approaches need to be explored. For example, the prioritization parameter may be based on the probability that at least one copy of given information can be delivered to the GRs. Note that the goal of such queue management is to facilitate efficient communication in the network with redundancy, in contrast to the QoS-aware algorithms which aim to prioritize raw information.
- *Protocol design:* Routing metrics, data duplication, and queue management are not independent. For instance, data duplication will clearly affect queue management, and at the same time, itself is also influenced by routing. Built on these three basic components, the data delivery protocol needs to be carefully devised by synergizing the interaction among them, in order to achieve efficient network resource utilization.
- *Protocol optimization:* Regardless of the protocol yet to be developed, several key constraints and enhancements must be investigated and explored to optimize network performance, including traffic engineering, power optimization, clustering, erasure coding and network coding, and the employment of heterogeneous types of tags with different levels of resources. Each of these issues becomes an interesting problem in the unique context of a FINDERS system.

2. **Efficient Data Acquisition, Processing, and Storage:** Besides the above networking issues, data acquisition, processing, and storage in FINDERS must be carefully handled to achieve high efficiency. The related research problems are outlined below:
   - *Data acquisition:* Various types of application-specific data may be generated in FINDERS. Some of them can be readily acquired. For example, a passive sensing unit coupled with an RFID tag may generate a reading when the tag is powered up, and the output can be written into the tag's memory accordingly. However, the acquisition of data is not always straightforward. For example, the location information is important to many applications for tracking moving objects. While the RFID reader may be equipped with a GPS receiver and thus provide coarse

location information about a tag when the tag moves into its reading range, this solution results in errors of some 20 ft or higher. To achieve a finer degree of localization is nontrivial.

- *Data compression/aggregation:* The data delivery in FINDERS relies on the transportation of tags. Given the extremely limited memory of each tag, the data format must be carefully designed by considering the tradeoff between data accuracy and storage efficiency. As typical data compression algorithms do not work well here due to the small data size, novel schemes will be sought for compact data storage. For example, effective models will be explored to represent raw data, by storing only a few characteristic parameters of a given object trajectory.
- *Data reading/writing:* With potentially high mobility of the targets being studied and the short radio range in RFID, the communications between tags and readers should be completed as fast as possible. However, this problem becomes challenging due to the considerable delay in tag arbitration, reading, and writing, especially under high tag density and harsh communication environments. A unique niche in FINDERS is that the tags are not equally important. During tag arbitration, it is desirable to minimize the collision probability of those tags which carry important (fresh) data or are good vehicles for data transportation. After arbitration, the tags should be designated to appropriate states and given priority for successful communications (especially for those with short sojourn time with the reader), avoiding unnecessary overhead due to repeated reads/writes. This can be very useful to expedite the arbitration of tags.

13.5 Related Work

FINDERS is related to two emerging technologies: RFID and DTN. RFID has achieved increasingly widespread adoption recently, with several interesting research problems being explored in the past few years. Tag arbitration has been extensively studied, yielding solutions fall into two categories, slotted ALOHA [8, 9, 10, 11], and tree splitting algorithms [12, 13, 10, 14]. The problem of counting RFID tags by using statistic approaches was studied in [15, 16]. Meanwhile, Tan *et al.* [17] and Sheng *et al.* [18] aim to discover missing tags and tag popularity, respectively. A load balancing scheme is introduced in Dong *et al.* [19] to distribute the communication load evenly among readers. In addition, several algorithms have been developed to locate the tags, given that object tracking and localization is one of the most important applications of the RFID system. For example, Hahnel *et al.* [20] propose to deploy reference tags inside a building and equip the robot with an RFID reader, which scans nearby reference tags to obtain location information. Yamano *et al.* [21] study the same problem by employing the support vector machine scheme to reduce errors. A location sensing prototype system, called LANDARC (LocAtioN iDentification based on dynaMic Active Rfid Calibration), is presented in Linel *et al.* [22], where a k-nearest neighbor approach is taken to estimate the location of the target tag. In Wang *et al.* [23], we have proposed two RFID-based 3-D positioning schemes, namely, the active scheme and the passive scheme, to localize the reader and the tag, respectively, in a three-dimensional space. All of the RFID localization schemes developed so far require reference tags, and thus are difficult to be applied in FINDERS, because deploying reference tags with known locations in the harsh environment is very costly, if not impossible.

DTN is a sporadically connected network with frequent partitions [7, 24]. Originally developed for deep space communication in high-delay environments, DTN technology has been recently introduced into wireless sensor networks [25, 26, 27, 28, 29, 30, 4, 31, 6, 32, 33, 34, 35] and mobile ad hoc networks [36, 37, 38, 39, 40, 41, 42, 43, 44, 45]. An in-depth investigation into existing delay-tolerant data delivery protocols can be found in [6, 46]. Almost all of these studies assume sufficient storage space at the DTN nodes, which can hold a large volume of data to alleviate the needs of immediate transmission. But this fundamental philosophy of paying storage for sporadic connectivity is no longer valid in FINDERS, because the available memory could be extremely limited at the tags. Several DTN-based network architectures most relevant to FINDERS are briefed below. Infostation is proposed as a complement to the cellular network, to provide high speed wireless access in isolated coverage areas [38]. It aims to support information distribution, instead of data acquisition and delivery. Data MULE [30] is proposed to collect data in sparse sensor networks, where a mobile entity (called a data mule) travels in the network, receives data from its nearby sensors, and delivers them to sinks. The data mule has sufficient storage space and battery power, and its mobility is controlled in a way to reach efficient data transportation. Similarly, Message Ferrying [45] intends to exploit non-random nodal mobility to help deliver data in partitioned mobile ad hoc networks. Those earlier DTN systems are in sharp contrast to FINDERS, where a tag moves at will (dictated by the object to which it is tagged) and usually possesses intermittent computing capability and only some tens of bytes of memory for user data storage.

13.6 Conclusion

We have introduced a Featherlight Information Network with Delay-Endurable RFID Support (FINDERS), composed of passive RFID tags which are ultra light, durable, and flexible, without power supply for long-lasting applications under strict weight constraints and harsh environments. It expands the use of RFID gear for wireless network construction, aiming to find events of interest and gather aggregate information. We have conducted analysis based on a trivial communication scheme to show that FINDERS, though with extremely limited resource, can provide sufficient capacity and effectively support our target applications.

At the same time, it remains an open issue to design most appropriate protocols for FINDERS, which faces unprecedented challenges in communication and networking, due to its sporadic wireless links, unique asymmetric communication paradigm, intermittent computation capability, and extremely small memory of tags. Several issues in protocol design including routing metrics, queue management, and duplication control must be carefully addressed. Furthermore, testbed implementation and experiments need to be carried out to investigate practical problems due to unreliable and asymmetric RFID communications and develop corresponding solutions.

Problems

1. Please develop a simple simulation program for the network described in Section 13.3. Use the same parameters to produce simulation results and compare with the analytic

results given in Figures 13.3 & 13.4. Identify the factors that lead to the difference between simulation and analytic results.

2. Vary K, m, T, λ, and J to observe their impacts on network performance and explain why you have such observations.
3. Propose five techniques that can improve the capacity of the delay/disruption-tolerant mobile RFID networks.
4. Experiment the techniques proposed by you and/or discussed in the class in your simulation program. Explain why or why not they are effective.

NB The solutions are not provided on the book's website.

References

[1] Akyildiz, I., Su, W., and Sankarasubramaniam, Y. (2002) A survey on sensor networks, *IEEE Comm. Magazine,* **40**(8): 102–114.

[2] Yang, Z., and Wu, H. (2009) Featherlight information network with delay-endurable RFID support (FINDERS), in *Proc. of 6th Annual IEEE Communications Society Conference on Sensor, Mesh and Ad Hoc Communications and Networks (SECON)*, pp. 1–9.

[3] n.d.a. Available at: http://www.princeton.edu/ mrm/zebranet.html.

[4] Small, T., and Haas, Z.J. (2003) The shared wireless infostation model – a new ad hoc networking paradigm (or where there is a whale, there is a way), in *Proc. of ACM International Symposium on Mobile Ad Hoc Networking and Computing (MOBIHOC)*, pp. 233–244.

[5] n.d.b. Available at: http://www.wu.ece.ufl.edu/projects/DeerNet/DeerNet.html.

[6] Wang, Y. and Wu, H. (2006) DFT-MSN: The delay fault tolerant mobile sensor network for pervasive information gathering, in *Proc. of IEEE Conference on Computer Communications (INFOCOM)*, pp. 1–12.

[7] Cerf, V., Burleigh, S., Hooke, A., Torgerson, L., Durst, R., Scott, K., Fall, K., and Weiss, H. (2004) Delay tolerant network architecture. Available at: www.draft-irtf-dtnrg-arch-02.txt.

[8] Ali, K., Hassanein, H., and Taha, AEM. (2007) RFID anti-collision protocol for dense passive tag environments, in *Proc. of IEEE Conference on Local Computer Networks (LCN)*, pp. 819–824.

[9] Jacomet, M., Ehrsam, A., and Gehrig, U. (1999) Contactless identification device with anti-collision algorithm, in *Proc. of International Technical Conference on Circuits, Systems, Computers and Communications (CCNC)*.

[10] Myung, J. and Lee, W. (2006) Adaptive splitting protocols for RFID tag collision arbitration, in *Proc. of ACM International Symposium on Mobile Ad hoc Networking and Computing (MOBIHOC)*, pp. 202–213.

[11] Vogt, H. (2002) Efficient object identification with passive RFID tags, in *Proc. of International Conference on Pervasive Computing*.

[12] Hush, D.R., and Wood, C. (1998) Analysis of tree algorithms for RFID arbitration, in *Proc. of IEEE International Symposium on Information Theory*, p. 107.

[13] Law, C., Lee, K., and Siu, K.Y. (2000) Efficient memoryless protocol for tag identification, in *Proc. of 4th International Workshop on Discrete Algorithms and Methods for Mobile Computing and Communications*, pp. 75–84.

[14] Namboodiri, V., and Gao, L. (2007) Energy-aware tag anti-collision protocols for RFID systems, in *Proc. of IEEE International Conference on Pervasive Computing and Communications (PERCOM)*, pp. 23–36.

[15] Kodialam, M., and Nandagopal, T. (2006) Fast and reliable estimation schemes in RFID systems, in *Proc. of ACM International Symposium on Mobile Ad hoc Networking and Computing (MOBIHOC)*, pp. 322–333.

[16] Qian, C., Ngan, H.L., and Liu, Y. (2008) Cardinality estimation for large-scale RFID systems, in *Proc. of IEEE International Conference on Pervasive Computing and Communications (PERCOM)*.

[17] Tan, C.C., Sheng, B., and Li, Q. (2008) How to monitor for missing RFID tags, in *Proc. of International Conference on Distributed Computing Systems (ICDCS)*.

[18] Sheng, B., Tan, C.C., Li, Q., and Mao, W. (2008) Finding popular categories for RFID tags, in *Proc. of ACM International Symposium on Mobile Ad Hoc Networking and Computing (Mobihoc)*.
[19] Dong, Q., Shukla, A., Shrivastava, V., Agrawal, D., and Banerjee, S. (2007) Load balancing in largescale RFID systems, in *Proc. of IEEE Conference on Computer Communications (INFOCOM)*, pp. 2281–2285.
[20] Hahnel, D., Burgard, W., Fox, D., Fishkin, K., and Philipose, M. (2004) Mapping and localization with RFID technology, in *Proc. of IEEE International Conference on Robotics and Automation*, pp. 1015–1020.
[21] Yamano, K., Tanaka, K., Hirayama, M., Kondo, E., Kimuro, Y., and Matsumoto, M. (2004) Self-localization of mobile robots with RFID system by using support vector machine, in *Proc. of IEEE/RSJ International Conference on Intelligent Robots and Systems*, pp. 3756–3761.
[22] Linel, M. Ni, Yunhao Liu, Y.C.L., and Patil, A.P. (2003) Landmarc: Indoor location sensing using active RFID, in *Proc. of IEEE Conference on Pervasive Computing and Communications*, pp. 407–415.
[23] Wang, C., Wu, H., and Tzeng, N.F. (2007) RFID-based 3-D positioning schemes, in *Proc. of IEEE Conference on Computer Communications (INFOCOM)*, pp. 1235–1243.
[24] Fall, K. (2003) A delay-tolerant network architecture for challenged internets, in *Proc. of ACM SIGCOMM Conference*, pp. 27–34.
[25] n.d.c. Available at: http://www.cens.ucla.edu/.
[26] n.d.d. Available at: http://down.dsg.cs.tcd.ie/sendt/.
[27] n.d.e. Available at: http://www.sics.se/cna/dtnsn/index.html.
[28] Ho, M. and Fall, K. (2004) Poster: delay tolerant networking for sensor networks, in *Proc. of IEEE Conference on Sensor and Ad Hoc Communications and Networks*.
[29] Mainwaring, A., Polastre, J., Szewczyk, R., Culler, D., and Anderson, J. (2002) Wireless sensor networks for habitat monitoring, in *Proc. of ACM International Workshop on Wireless Sensor Networks and Applications (WSNA)*, pp. 88–97.
[30] Shah, R.C., Roy, S., Jain, S., and Brunette, W. (2003) Data MULEs: modeling a three-tier architecture for sparse sensor networks, in *Proc. of The First International Workshop on Sensor Network Protocols and Applications*, pp. 30–41.
[31] Small, T., and Haas, Z.J. (2005) Resource and performance tradeoffs in delay-tolerant wireless networks, in *Proc. of ACM SIGCOMM Workshop on Delay Tolerant Networking and Related Topics*, pp. 260–267.
[32] Wang, Y. and Wu, H. (2007) Delay/fault-tolerant mobile sensor network (DFT-MSN): a new paradigm for pervasive information gathering, *IEEE Transactions on Mobile Computing* **6**(9): 1021–1034.
[33] Wang, Y., Lin, F., and Wu, H. (2005) Poster: efficient data transmission in delay fault tolerant mobile sensor networks (DFT-MSN), in *Proc. of IEEE International Conference on Network Protocols (ICNP'05)*.
[34] Wang, Y., Wu, H., Lin, F., and Tzeng, N.F. (2008) Cross-layer protocol design and optimization for delay/fault-tolerant mobile sensor networks, *IEEE Journal on Selected Areas in Communications (JSAC), Special Issue on Delay and Disruption Tolerant Wireless Communication*. A preliminary version was presented in IEEE ICDCS'07.
[35] Wu, H., Wang, Y., Dang, H., and Lin, F. (2007) Analytic, simulation, and empirical evaluation of delay/fault-tolerant mobile sensor networks, *IEEE Transactions on Wireless Communications* **6**(9): 3287–3296.
[36] Burns, B., Brian, O.B., and Levine, N. (2005) MV routing and capacity building in disruption tolerant networks, in *Proc. of IEEE Conference on Computer Communications (INFOCOM)*, pp. 398 – 408.
[37] Ghosh, J., Philip, J., and Qiao, C. (2007) Sociological orbit aware location approximation and routing (SOLAR) in MANET, *Ad Hoc Networks* **5**(2): 189–209.
[38] Goodman, D.J., Borras, J., Mandayam, N., and Yates, R. (1997) Infostations: A new system for data and messaging services, in *Proc. of IEEE Vehicular Technology Conference*, pp. 969–973.
[39] Hui, P., Chaintreau, A., Scott, J., Gass, R., Crowcroft, J., and Diot, C. (2005) Pocket switched networks and human mobility in conference environments, in *Proc. of ACM SIGCOMM Workshop on Delay Tolerant Networking and Related Topics (WDTN)*, pp. 244–251.
[40] LeBrun, J., Chuah, C.N., and Ghosal, D. (2005) Knowledge based opportunistic forwarding in vehicular wireless ad hoc networks, in *Proc. of IEEE Vehicular Technology Conference (VTC) 2005 Spring*, pp. 1–5.
[41] Lindgren, A., Doria, A., and Schelen, O. (2004) Probabilistic routing in intermittently connected networks, in *Proc. of The First International Workshop on Service Assurance with Partial and Intermittent Resources (SAPIR 2004)*.
[42] Musolesi, M., Hailes, S., and Mascolo, C. (2005) Adaptive routing for intermittently connected mobile ad hoc networks, in *Proc. of IEEE 6th International Symposium on a World of Wireless, Mobile and Multimedia Networks (WOWMOM)*, pp. 1–7.

[43] Spyropoulos, T., Psounis, K., and Raghavendra, C.S. (2005) Spray and wait: an efficient routing scheme for intermittently connected mobile networks, in *Proc. of ACM SIGCOMM Workshop on Delay Tolerant Networking and Related Topics*, pp. 252–259.

[44] Wang, Y., Jain, S., Martonosi, M., and Fall, K. (2005) Erasure-coding based routing for opportunistic networks, in *Proc. of ACM SIGCOMM Workshop on Delay Tolerant Networking and Related Topics*, pp. 229–236.

[45] Zhao, W., Ammar, M., and Zegura, E. (2004) A message ferrying approach for data delivery in sparse mobile ad hoc networks, in *Proc. of the 5th ACM International Symposium on Mobile Ad Hoc Networking and Computing (MOBIHOC)*, pp. 187–198.

[46] Wang, Y., Dang, H., and Wu, H. (2007) A survey on analytic studies of delay-tolerant mobile sensor networks, *Wireless Communications and Mobile Computing (WCMC) Special Issue on Disruption Tolerant Networking for Mobile or Sensor Networks*, **7**(10): 1197–1208.

Part Four

Addressing Other Challenges in RFID Systems

14

Improving Read Ranges and Read Rates for Passive RFID Systems

Zhiguang Fan[1,2], Fazhong Shen[1], Jianhua Shen[1], and Lixin Ran[1]

[1]*Zhejiang University, China*

[2]*Tektronix (China) Co., Ltd*

14.1 Introduction

Like other wireless communications or RADAR systems, radio frequency identification (RFID) systems also rely on electromagnetic (EM) waves to work. Similarly, antennas are used to transmit inquiry energy to and sense modulated responses from tags. According to Maxwell's theory, in the space surrounding an antenna, the distribution of the EM field depends on the distance to the antenna, r. The region immediately surrounding the antenna, estimated by $r \leq \lambda/6$, where λ denotes the wavelength of EM field, is called the near-field zone, where reactive fields dominate. Low frequency (LF) and high frequency (HF) RFID systems work in this range with standard frequencies of 125 KHz and 13.56 MHz, respectively. The outer region next to the near-field is called the far-field zone, where radiating fields dominate. Ultra-high frequency (UHF) and microwave RFID systems work in this range, whose typical frequencies are 915 MHz and 2.4 GHz, respectively. The field strength decays linearly with $1/r$ in the far-field zone but exponentially with $1/r^2$ or even $1/r^3$ in the near-field zone, which is the reason why a UHF or microwave RFID system generally has a longer operational distance than a LF or HF RFID system.

From the tag's point of view, RFID systems can be classified into passive and active. In an active RFID system, the tag has its own power supply, and therefore the communication between the reader and the tag behaves much like a traditional wireless system. However,

RFID Systems: Research Trends and Challenges Edited by Miodrag Bolić, David Simplot-Ryl, and Ivan Stojmenović

in a passive RFID system, the tag needs to collect enough energy from the EM fields radiated by the reader to power up itself, and then communicates with the reader by changing its own backscatter cross-section, which makes it a more complex process than in an active RFID system.

In this chapter, we discuss "read range" and "read rate" issues of passive RFID systems working in far-field ranges. In EPC's terminologies [1], the "read range" refers to the distance at which a reader can communicate with a tag. A more complicated definition is

> the maximum distance between a reader and a tag in which the radiation field from the reader is strong enough to power up the tag and the backscatter signal from the tag is strong enough to be detected correctly by the reader.

From this definition, we can simply conclude that the read range of an RFID system is affected not only by the performance of the reader, but also by that of the tag to be detected. Also in EPC's terminologies, the "read rate" refers to the maximum rate at which data can be read from a tag, generally expressed in bits-per-second (bps). However, in this chapter, the "read rate" is redefined as the statistical ratio of the quantity of tags successfully read versus that of all the tags.

The read range and the read rate characterize the performance of RFID systems from different points of view, and are actually closely related to each other. In most cases, the attempt to improve the read range of an RFID system will also improve the read rate since they are both related to the signal-to-noise ratio (SNR) of the backscatter signals received by the reader. Theoretically, for each read of a tag, the readability is determined by the SNR, and the read range is the maximum distance where the SNR starts to decrease below a specified threshold value. However, in a real application environment, it frequently occurs that not every tag inside the read range can be detected, yielding a read rate lower than 100 %. There are many reasons for this, for example, the polarization and gain variation of a tag's antenna with different tag orientations, the non-homogeneous field distribution in closed spaces, the existence of various obstacles, different background materials for tags, interference from other wireless systems, read collisions, and so on. The issue of low read rate is an open technical challenge and has plagued the passive RFID industry since its inception.

This chapter is arranged as follows. First, we focus on in-depth studies on the working principles of passive RFID systems in the far-field range. The signal descriptions and formulations of SNR and read range are theoretically presented. Based on those equations, we further discuss some issues regarding read range and read rate. Apart from some conclusions that can be drawn directly from those equations, other research efforts related to improving read rates and read ranges are also introduced. At the end of this chapter, two examples are presented to illustrate cost-effective designs of long-range RFID readers utilizing off-the-shelf chips that were originally designed for wireless communications.

14.2 Signal Descriptions and Formulations for Passive Backscatter RFID Systems

It is well known that RFID systems began their explosive development only in the last decade. However, RFID actually has a long history of more than half a century [3]. Passive backscatter RFID readers are interrogators for tags, or transponders, based on

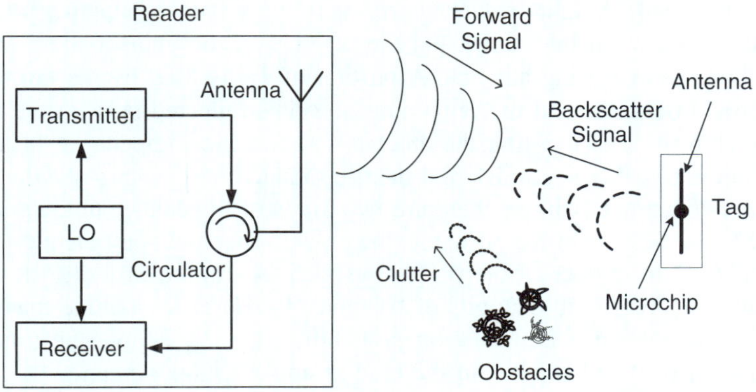

Figure 14.1 Basic architecture for a passive, far-field RFID system. Reproduced courtesy of © 2007 EMW Publishing.

the principles of wireless power transmission [4] and backscatter communication [5–7]. The reader transmits electromagnetic waves into the air and the tag draws energy from radiation fields. The electromagnetic waves are partially backscattered and modulated by the tag, through which the reader remotely retrieves the tag's information. The basic architecture of such a passive RFID system is shown in Figure 14.1, which consists of a reader, a tag and surrounding obstacles.

14.2.1 Signal Descriptions

Generally, the tag modulates the backscatter waves by toggling the termination impedance of the tag's antenna, yielding amplitude shift keying (ASK) or phase shift keying (PSK) modulations [8]. The reader in Figure 14.1 is composed of a local oscillator (LO), a transmitter, a receiver and a receive-transmit duplex antenna plus a circulator as duplexer. A two-antenna solution is also commonly utilized, where the circulator is replaced by two directivity-optimized antennas to isolate the receiver and transmitter channels. However, in practice, the receive-transmit isolation hardly can be ideal. The limited isolation characteristic of the circulator and the impedance mismatch of the transmitter antenna, and the coupling between the transmitter and receiver antennas inevitably result in the leakage of transmitted power into the receiver. The leakage may seriously degrade the performance of the receiver since it is possibly much stronger than the backscatter signal when the tag is far away from the reader.

To begin with, we define a power transmission coefficient α and a power leakage coefficient β to characterize the insertion loss and the isolation of the duplexer, respectively. Then, if P_T denotes the signal power feeding into the transmit antenna, βP_T will be the power leaking into the receiver, and if P_B denotes the backscatter power received by the reader antenna, αP_B will be the power entering the receiver. Then, we define ξ that is equal to α/β and call ξ the receive-transmit isolation coefficient of the reader.

According to the Friis equation, the power received by the tag antenna, denoted as P_R, is given by

$$P_R = \left(\frac{c}{2\omega r}\right)^2 P_T G_T G_R \tag{14.1}$$

where c is the speed of light in free space, ω is the angular frequency of EM wave, r is the distance between the reader and the tag, G_R is the gain of the reader antenna and G_T is the gain of the tag antenna. A portion of P_R is used by the tag to power up its own microchip circuits and the remaining power is reflected back to the tag antenna by the microchip through toggling the antenna's termination impedance in terms of the stored data in the tag, that is, the ID, and then re-radiated into air to make the backscatter modulation. As mentioned above, there are two kinds of impedance modulations, that is, ASK and PSK. In the case of the ASK tag, the power reflection coefficient typically takes a value of "0"/"Γ" for a data bit of "0"/"1," respectively; in the case of PSK tag, it takes the same value of Γ for both data bits of "0" and "1." Here, we assume that data bits of "0" and "1" have the same transmission probability. Let P_A denote the time-averaging absorption power of the tag. Then, in the case of an ASK tag, we can get

$$P_A = \left(1 - \Gamma/2\right) P_R; \tag{14.2a}$$

in the case of a PSK tag, we can get

$$P_A = (1 - \Gamma) P_R. \tag{14.2b}$$

As can be seen in Figure 14.1, the electromagnetic wave signals picked up by the reader antenna consist of two components: one is the backscatter signal from the tag and the other is the clutter that is the sum of the backscatter signals from surrounding obstacles. The backscatter signal from the tag is indeed composed of the modulated backscatter signal due to the tag's impedance modulation, as mentioned above, and the unmodulated backscatter signal due to the induced current on the surface of the tag antenna [9]. Both the backscatter signal and the clutter enter the receiver via the circulator. In addition, the leakage signal between the transmitter and receiver ports of the circulator also enters the receiver. These input signals can be summed up as two components: one is the unmodulated signal $X_U(t)$ that is the sum of the clutter from the obstacles, the structural backscatter signal from the tag antenna and the receiver-transmit leakage signal, which all act as harmful interferences to the receiver; the other is the modulated backscatter signal $X_M(t)$ that conveys the tag's data information. For a passive backscatter RFID reader, $X_U(t)$ is usually much stronger than $X_M(t)$, which makes it barely possible for the reader's receiver to retrieve from the received signals a LO signal $X_S(t)$ being synchronized with $X_M(t)$. In other words, $X_S(t)$ indeed will mainly track the phase of $X_U(t)$, but not that of $X_M(t)$. Thus, a non-coherent demodulation scheme needs to be adopted for the receiver design.

In practice, the leakage component generally dominates in $X_U(t)$. Thus, we can ignore other components and approximate $X_U(t)$ as

$$X_U(t) = A_U \sin(\omega t + \theta_U), \tag{14.3a}$$

where θ_U denotes the signal phase and A_U denotes the signal amplitude that is given by

$$A_U = \sqrt{2\beta R_0 P_T}, \tag{14.3b}$$

where R_0 is the input resistance of the receiver. The modulated backscatter signal $X_M(t)$ can be expressed as

$$X_M(t) = A_M[S(t)] \sin(\omega t + \theta_{M0} + \theta_M[S(t)]), \tag{14.4}$$

where $S(t)$ denotes the tag's binary data sequence of "0"/"1," θ_{M0} denotes the unmodulated part of signal phase, $\theta_M[S(t)]$ denotes the part of signal phase modulated by $S(t)$ and $A_M[S(t)]$ denotes the signal amplitude modulated by $S(t)$. In the case of ASK modulation, $A_M[S(t)]$ takes a value of "0"/"A_M" while the tag transmits a data bit of "0"/"1," respectively, and $\theta_M[S(t)]$ takes a constant value of θ_M. In the case of PSK modulation, $\theta_M[S(t)]$ takes a value of $-\theta_M/\theta_M$ while the tag transmits a data bit of "0"/"1," respectively, and $A_M[S(t)]$ takes a constant value of "A_M." Using the Friis equation again and Equation (14.1), A_M can be given by

$$A_M = \left(\frac{c}{2\omega r}\right)^2 \sqrt{2\Gamma\alpha R_0 P_T G_T G_R}. \tag{14.5}$$

The LO signal $X_S(t)$ of the receiver can be expressed as

$$X_S(t) = A_S \sin(\omega t + \theta_S), \tag{14.6}$$

where θ_S denotes the signal phase and A_S denotes the signal amplitude. In practice, $X_U(t)$, $X_M(t)$, and $X_S(t)$ always have phase and amplitude noises. Those signals can be regarded as stochastic processes and their phases θ_U, θ_{M0}, and θ_S as time-dependent random variables to denote their phase noises that mainly originate from the LO. The amplitude noises are usually much smaller than the phase noises. In order to simplify the analysis in the later section, we will ignore the former's effects and denote these signals' amplitudes as constant quantities.

14.2.2 SNR and Read Range Formulation

Based on the analysis in Section 14.2.1, a block diagram for realizing a passive, far-field RFID reader is given in Figure 14.2. In the RF front-end, the most important component is the dual-channel (i.e. I channel and Q channel) non-coherent quadrature demodulator, which consists of an in-phase power splitter, a $\pi/2$-phase-shifting power splitter and two mixers. The baseband for an RFID system is generally realized with digital circuits, generally having an analog-digital-converter (ADC), a digital signal processor (DSP), a central processing unit (CPU), and so on.

The read range is actually determined by two conditions: (1) the tag could draw enough energy from incident electromagnetic waves to power up itself; and (2) the modulated backscatter signal from the tag should be strong enough so that the SNR of the demodulation output signal in the reader's receiver meets the user-specified value.

For Case (1): it is obvious that increasing the transmitting power P_T of the reader will directly improve the absorption power P_A of the tag. However, in practice, P_T is generally constrained by regional regulations to be no larger than a specified power value P_{EIRP} (EIRP: Effective Isotropic Radiated Power). For example, the P_{EIRP} is 4 watt within UHF ISM band (902–928 MHz) in the United States. Let P_T take the maximum allowed value, that is, $P_T = P_{EIRP}/G_R$. Then, using Equations (14.1), (14.2a) and (14.2b), in the ASK case, we get

$$P_A = (1 - \Gamma/2)\, P_{EIRP} G_T \left(\frac{c}{2\omega r}\right)^2 \tag{14.7a}$$

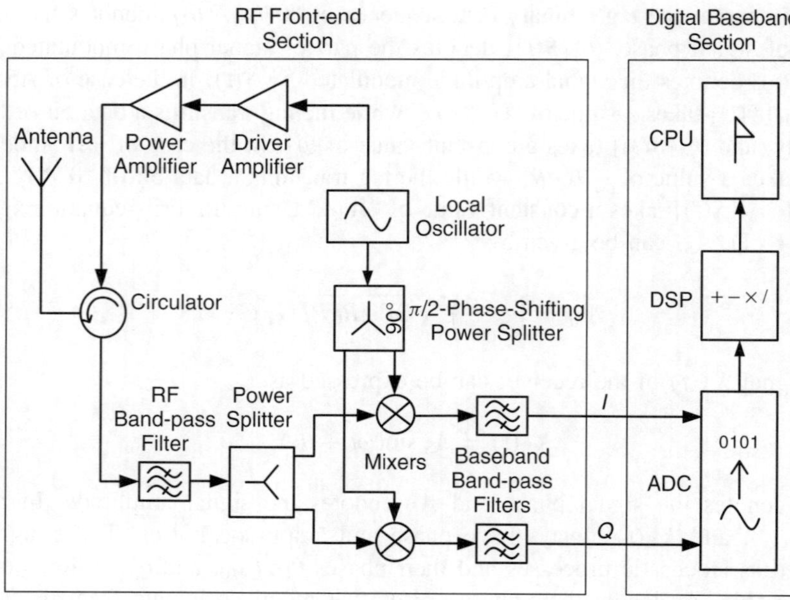

Figure 14.2 Reader architecture. Reproduced courtesy of © 2007 EMW Publishing.

And in the PSK case, we get

$$P_A = (1 - \Gamma) P_{EIRP} G_T \left(\frac{c}{2\omega r}\right)^2. \tag{14.7b}$$

Example 14.2.1 Calculating the power absorbed by an ASK tag.
Consider an ASK tag working at 2.5 GHz, and a reader radiating a P_{EIRP} power of 20 dBm. Assume the tag's antenna has a gain of 3 dBi and $\Gamma = 1$. Utilizing Equation (14.7a), we find that this tag will receive a power of −40.4 dBm when it is 10 m away from the reader's antenna.

In order to power up the tag, P_A must be no less than a specific tag turn-on threshold power P_{TH}, that is, $P_A \geq P_{TH}$, which gives an upper limit of operational distance and is called the tag-determined read range, denoted by r_{TAG}. Then, using Equations (14.7a) and (14.7b), for the ASK tag, we get

$$r_{TAG} = \frac{c}{2\omega}\sqrt{\frac{(1 - \Gamma/2)\, P_{EIRP} G_T}{P_{TH}}} \tag{14.8a}$$

And for the PSK tag, we get

$$r_{TAG} = \frac{c}{2\omega}\sqrt{\frac{(1 - \Gamma) P_{EIRP} G_T}{P_{TH}}}. \tag{14.8b}$$

Example 14.2.2 Calculating the tag-determined read range r_{TAG} for an ASK tag. Consider the same case in Example 14.2.1. If this tag includes a chip having a turn-on threshold power P_{TH} of -15 dBm, we see that the tag will not work because its absorption power P_A, that is, -40.4 dBm, is much lower than P_{TH}. In this case, the tag-determined read range r_{TAG} of such a tag can be calculated through using Equation (14.8a) as below:

$$\begin{aligned} r &= \frac{c}{2\omega}\sqrt{\frac{(1-\Gamma/2)P_T G_T}{P_{TH}}} \\ &= \frac{3\times 10^8}{2\times 2\pi \times 2.5\times 10^9}\sqrt{1-1/2}\times 10^{(20+3+15)/2/10} \\ &= 0.54 \qquad meters \end{aligned}$$

Case (2): assume the baseband band-pass filter utilized in the receiver has a sharp frequency selectivity with low-end and high-end cut-off frequencies of f_L and f_H, respectively. Referring to Figure 14.2 and using Equations (14.3a), (14.4) and (14.6), we can express the demodulation output signals of I and Q channels as

$$X_I(t) = k_D A_S A_U \cos(\theta_U - \theta_S) + k_D A_S A_M[S(t)]\cos\left[\theta_{M0} - \theta_S + \theta_M[S(t)]\right] + n_0(t), \tag{14.9a}$$

$$X_Q(t) = k_D A_S A_U \sin(\theta_U - \theta_S) + k_D A_S A_M[S(t)]\sin\left[\theta_{M0} - \theta_S + \theta_M[S(t)]\right] + n_0(t), \tag{14.9b}$$

respectively, where θ_U, θ_{M0} and θ_S are all time-dependent random variables to denote the phase noise of the LO, $n_0(t)$ denotes the internal thermal noise of the receiver and k_D denotes the transfer coefficient of the receiver that takes into consideration the total loss or gain of the RF band-pass filter, the power splitter, the mixer and the baseband band-pass filter. In practice, $X_U(t)$ is usually much larger than both $X_M(t)$ and $n_0(t)$, which means that the SNR of the demodulation output signal is mainly determined by the phase noise of $X_U(t)$ and then we reasonably can ignore the noise contributions of $n_0(t)$ and $(\theta_{M0} - \theta_S)$ terms in Equations (14.9a) and (14.9b). In addition, since $X_U(t)$ and $X_S(t)$ originate from the same LO, we can assume that their phase noise can be characterized with the same stochastic process except for a time delay. Then, we can rewrite Equations (14.9a) and (14.9b) as

$$\begin{aligned} X_I(t) \approx{} & k_D A_S A_U \cos\left[\omega\Delta t + \theta_P(t+\Delta t) - \theta_P(t)\right] \\ & + k_D A_S A_M[S(t)]\cos\left[\theta_{MOS} + \theta_M[S(t)]\right] \end{aligned} \tag{14.10a}$$

$$\begin{aligned} X_I(t) \approx{} & k_D A_S A_U \sin\left[\omega\Delta t + \theta_P(t+\Delta t) - \theta_P(t)\right] \\ & + k_D A_S A_M[S(t)]\sin\left[\theta_{MOS} + \theta_M[S(t)]\right] \end{aligned} \tag{14.10b}$$

respectively, where $\theta_P(t)$is a stochastic process characterizing the phase noise of the LO, Δt denotes the time delay between $X_U(t)$ and $X_S(t)$, and θ_{MOS} denotes the phase difference between $X_M(t)$ and $X_S(t)$. Then, the SNRs of $X_I(t)$ and $X_Q(t)$ can be given in the frequency domain by

$$SNR_I = \frac{\int_{f_L}^{f_H} \mathbf{S}\{A_M[S(t)]\cos[\theta_{MOS} + \theta_M[S(t)]]\}\,df}{\int_{f_L}^{f_H} \mathbf{S}\{A_U \cos[\omega\Delta t + \theta_P(t+\Delta t) - \theta_P(t)]\}\,df} \tag{14.11a}$$

$$SNR_Q = \frac{\int_{f_L}^{f_H} \mathbf{S}\{A_M[S(t)]\sin[\theta_{MOS} + \theta_M[S(t)]]\}\,df}{\int_{f_L}^{f_H} \mathbf{S}\{A_U \sin[\omega\Delta t + \theta_P(t+\Delta t) - \theta_P(t)]\}\,df} \tag{14.11b}$$

respectively, where $\mathbf{S}\{\}$ denotes the operator of calculating the power density spectrum (PDS) of a stochastic process, that is, the Fourier transform of the auto-correlation function of a stochastic process. It is obvious that the SNRs of both $X_I(t)$ and $X_Q(t)$ strongly depend on the unknown θ_{MOS} and Δt. In other words, there is an inherent uncertainty in terms of the SNR performance of an uncorrelated receiver. Thus, two channels could have different SNRs and the reader's baseband should choose the one having a higher SNR to retrieve the tag's ID information. In the worst case where θ_{MOS} takes a value of $\pi/4$, $\omega\Delta t$ takes the values of $(2n+1)\pi/2$ in Equation (14.11a) or $n\pi$ in Equation (14.11b), where n is an integer number, and the stochastic processes $\theta_P(t)$ and $\theta_P(t+\Delta t)$ are assumed to be uncorrelated to each other, considering that $\theta_P(t)$ and $\theta_P(t+\Delta t)$ are mostly taking values near to zero for a practical LO having a low phase noise, the minimal achievable SNR can be approximately expressed as

$$SNR_{MIN} \approx \frac{\int_{f_L}^{f_H} \mathbf{S}\{A_M[S(t)]\cos[\pi/4 + \theta_M[S(t)]]\}\,df}{2\int_{f_L}^{f_H} \mathbf{S}\{A_U\theta_P(t)\}\,df}, \tag{14.12}$$

which gives a lower boundary of the SNRs of the demodulation output signals. Here, we ignore the correlation effect between $\theta_P(t)$ and $\theta_P(t-\Delta t)$, which indeed would dramatically reduce the demodulation output noises [9]. In order to characterize the correlation effect, we define a phase noise improvement factor ψ, which can be achieved experimentally as shown in the later section. Then, we can rewrite Equation (14.12) as

$$SNR_{MIN} = \frac{\int_{f_L}^{f_H} \mathbf{S}\{A_M[S(t)]\cos[\pi/4 + \theta_M[S(t)]]\}\,df}{\psi\int_{f_L}^{f_H} \mathbf{S}\{A_U\theta_P(t)\}\,df}. \tag{14.13}$$

Now, substituting Equation (14.3b) and (14.5) into Equation (14.13), in the case of ASK tag, we get

$$SNR_{MIN} = \frac{\Gamma\xi G_T^2 G_R^2}{2\psi}\left(\frac{c}{2\omega r}\right)^4 \frac{\int_{f_L}^{f_H} \mathbf{S}\{S(t)\}\,df}{\int_{f_L}^{f_H} \mathbf{S}\{\theta_P(t)\}\,df} \tag{14.14a}$$

in the case of PSK tag, we get

$$SNR_{MIN} = \frac{\Gamma\xi G_T^2 G_R^2}{\psi}\left(\frac{c}{2\omega r}\right)^4 \frac{\int_{f_L}^{f_H} \mathbf{S}\{\cos[\pi/4 + \theta_M[S(t)]]\}\,df}{\int_{f_L}^{f_H} \mathbf{S}\{\theta_P(t)\}\,df} \tag{14.14b}$$

Example 14.2.3 Calculating the SNR$_{MIN}$ for the reader
Consider a passive RFID system composed of an ASK tag and a reader working at 950 MHz. Assume the gains of the tag and reader antennas are 3 dBi and 15 dBi, respectively, and the isolation of the circulator is 20 dB and $\Gamma = 1$. The SNR$_{MIN}$ can be calculated by Equation (14.14a). In the calculation, the integration items $\int_{f_L}^{f_H} \mathbf{S}\{S(t)\}\,df$, $\int_{f_L}^{f_H} \mathbf{S}\{\theta_P(t)\}\,df$ and the improvement factor ψ can be obtained from experimental data. As measured in Section 14.5, when $\int_{f_L}^{f_H} \mathbf{S}\{S(t)\}\,df = -6.8\,\text{dB}$, $\int_{f_L}^{f_H} \mathbf{S}\{\theta_P(t)\}\,df = -39.9\,\text{dB}$ and $\psi = -37.5\,\text{dB}$ and the tag is 10 meters far from the reader, the SNR$_{MIN}$ is

$$SNR_{MIN} = \frac{\Gamma \xi G_T^2 G_R^2}{2\psi}\left(\frac{c}{2\omega r}\right)^4 \frac{\int_{f_L}^{f_H} \mathbf{S}\{S(t)\}\,df}{\int_{f_L}^{f_H} \mathbf{S}\{\theta_P(t)\}\,df}$$

$$= (20 + 3 \times 2 + 15 \times 2 - 6.8 + 39.9 + 37.5)$$

$$+10\lg\left[\left(\frac{3 \times 10^8}{2 \times 2\pi \times 0.95 \times 10^9 \times 10}\right)^4 /2\right] = 19.6\,\text{db}$$

The method for calculating $\int_{f_L}^{f_H} \mathbf{S}\{S(t)\}\,df$, $\int_{f_L}^{f_H} \mathbf{S}\{\theta_P(t)\}\,df$ and ψ from experimental data can be found in Section 14.5.

In Equations (14.14a) and (14.14b), the integral items in the denominators denote the single-sideband in-band phase noise power of the LO (a dimensionless quantity normalized to the carrier power), expressed as P_{PN}, which can readily be obtained by means of numerical integral calculation on the known LO phase noise data or direct measurement with a spectrum analyzer (SPA) device; the integral items in the numerators denote the single-sideband in-band signal power of the tag's binary data sequence (also treated as a dimensionless quantity), expressed as P_{DATA}, which can be derived according to specific data coding schemes such as unipolar/bipolar coding, return-to-zero/non-return-to-zero (RZ/NRZ) coding, Manchester coding, Miller coding, FM0 (bi-phase-space) coding, etc. [8, 10]. Then the read range can be calculated through requiring the SNR_{MIN} be no lower than a user-specified SNR_{USER}, that is, $SNR_{MIN} \geq SNR_{USER}$, which is called the reader-determined read range, denoted by r_{READER}. Then, for the ASK tag, using Equation (14.14a), we get

$$r_{READER} = \frac{c}{2\omega}\left(\frac{\Gamma \xi G_T^2 G_R^2}{2\psi SNR_{USER}} \frac{\int_{f_L}^{f_H} \mathbf{S}\{S(t)\}\,df}{\int_{f_L}^{f_H} \mathbf{S}\{\theta_P(t)\}\,df}\right)^{1/4} \qquad (14.15a)$$

and for the PSK tag, using Equation (14.14b), we get

$$r_{READER} = \frac{c}{2\omega}\left(\frac{\Gamma \xi G_T^2 G_R^2}{\psi SNR_{USER}} \frac{\int_{f_L}^{f_H} \mathbf{S}\{\cos[\pi/4 + \theta_M[S(t)]]\}\,df}{\int_{f_L}^{f_H} \mathbf{S}\{\theta_P(t)\}\,df}\right)^{1/4} \qquad (14.15b)$$

Example 14.2.4 Calculating the reader-determined read range r_{READER}.
Consider the same case as in Example 14.2.3. Assume the reader requires a minimum SNR of 12 dB to correctly decode the tag data. Then, a reader-determined read range of 11.95 meters can be achieved by calculating Equation (14.15a).

Now we have theoretically obtained the tag-determining read range r_{TAG}, the reader-determining read range r_{READER} and the minimal achievable SNR_{MIN} of the demodulation output signal in the receiver for both ASK and PSK cases. In the next sections, based on those equations, we will investigate how to improve the read range and the read rate for a passive RFID system.

14.3 Improving the Read Range of a Passive RFID System

As discussed in last section, the read range of the passive RFID system is actually determined by the relation between r_{TAG} and r_{READER}, which also gives two kinds of RFID systems. One is called a tag-determined passive RFID system, where the tag-determined read range is smaller than the reader-determined one, shown as $r_{TAG} < r_{READER}$. This situation happens when $P_A \geq P_{TH}$ cannot be met earlier than $SNR_{MIN} \geq SNR_{USER}$ while increasing the operational distance. The other is called a reader-determined system, which is just the opposite. So, it is obvious that we need to improve both r_{TAG} and r_{READER} to extend the read range for an RFID system.

First, we begin with a discussion about what can be done on the reader's side. We assume the reader works at a transmitting power P_{EIRP} of 36 dBm and an operating frequency ω of 915 MHz. Then, based on Equation (14.15a) for r_{READER}, in order to improve the read range, some key design considerations can be presented as follows:

- A larger reader antenna gain G_R has many advantages: (1) A larger G_R will efficiently improve the SNR of the demodulation output signal of the reader and consequently the read range. As shown in Figure 14.3 and Figure 14.4, using Example 14.2.3, the SNR_{MIN} increases log-linearly with the G_R. With the increase of the SNR_{MIN}, the read range is also improved. (2) A larger G_R can reduce the interference between multiple readers, especially in dense-multiple-reader RFID systems. A larger G_R generally means a higher antenna directionality, which can help to focus the electromagnetic energy in a specified area and decrease the inferences among readers. (3) For a given P_{EIRP}, a larger G_R requires a lower output power from the power amplifier (PA), which saves the cost of using a high-power RF amplifier and alleviates the leakage power into the receiver. In addition, it should be noted that simply increasing the transmitting power P_T of the reader would not improve the receiver's SNR performance. The reason is that the increase of P_T simultaneously increases both the modulated signal power and the noise power according to Equation (14.13).
- A large receive-transmit isolation coefficient ξ helps to realize a long read range due to its important effect on the SNR of the demodulation output signal, as shown in Figure 14.5. Generally, the isolation coefficients of commercial circulators vary from 15 to 25 dB. For example, a circulator MAFRIN0461, designed for UHF RFID systems by MIA-COM Company, has an isolation coefficient of 22 dB. In the case of using a circulator as duplexer, the receive-transmit isolation coefficient of the reader is

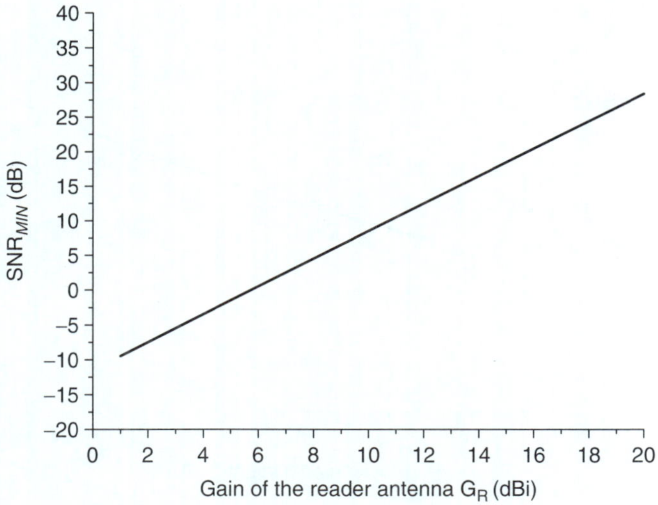

Figure 14.3 The calculated SNR_{MIN} vs. the gain G_R of the reader antenna when an ASK tag is 10 meters away. Note: In the calculation, $G_T = 2.15\,dBi$, $\xi = 20\,dB$, $P_{DATA} = -6.8\,dB$, $P_{PN} = -39.9\,dB$, $\psi = -37.5\,dB$, r = 10 m and $\Gamma = 1$ are assumed.

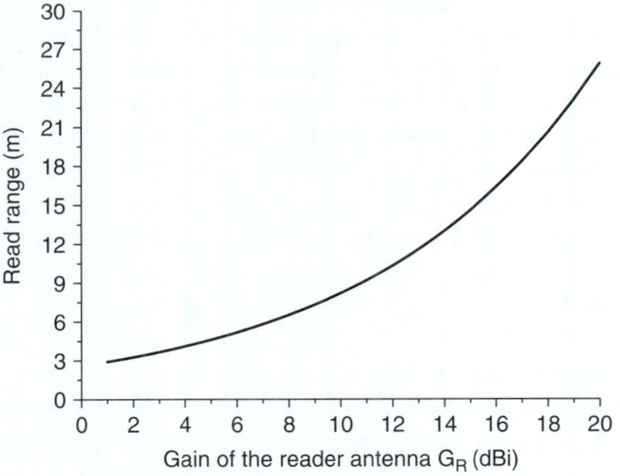

Figure 14.4 The reader-determined read range vs. the reader antenna gain G_R (for ASK tag). Note: In the calculation, $G_T = 2.15\,dBi$, $\xi = 20\,dB$, $P_{DATA} = -6.8\,dB$, $P_{PN} = -39.9\,dB$, $\psi = -37.5\,dB$, $SNR_{USER} = 12\,dB$ and $\Gamma = 1$ are assumed.

mainly limited by the circulator's performance. A circulator having both a low inherent leakage and a low insertion loss are preferred for the reader. In addition, since the mismatch reflection signal from the reader antenna also enters the receiver and consequently deteriorates the circulator's isolation performance, great attention should be paid to the impedance matching between the reader antenna and the circulator. For the two-antenna solution (Bi-static), the isolation between two antennas should also be optimized through antenna design, antenna placement, and antenna orientation.

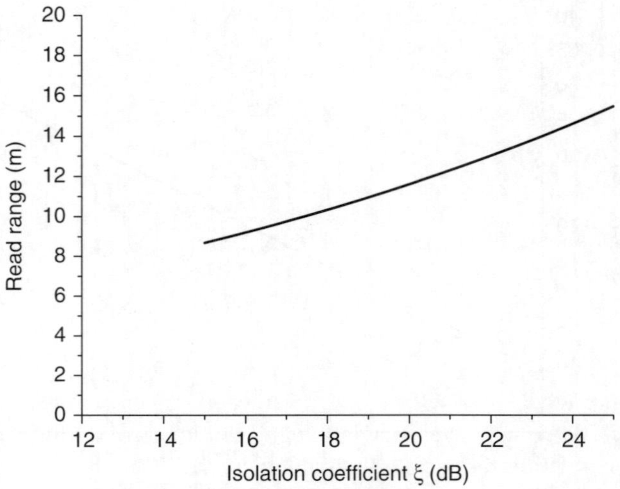

Figure 14.5 Reader-determined read range vs. receiver-transmitter isolation coefficient ξ (for ASK tag). Note: In the calculation, $G_T = 2.15\,\text{dBi}$, $G_R = 13\,\text{dBi}$, $P_{DATA} = -6.8\,\text{dB}$, $P_{PN} = -39.9\,\text{dB}$, $\psi = -37.5\,\text{dB}$, $SNR_{USER} = 12\,\text{dB}$ and $\Gamma = 1$ are assumed.

- A low phase noise LO is strongly recommended. The phase noise is more severe than the internal thermal noise of the receiver, and has a more significant impact on the demodulation output of the reader. Incidentally, considering that the phase noise is concentrated within a narrow band surrounding the carrier frequency and consequently the receiver noise is mainly of low frequency component, the data coding and modulation schemes adopted in a tag should be capable of suppressing the DC and low frequency components in its data signal as much as possible. That would facilitate utilizing band-pass filters in the receiver to suppress harmful noises and simultaneously preserve desired data signals, which consequently can relax the requirement of the phase noise of the LO.

Example 14.3.1 Selecting a proper G_R for an RFID reader
Generally, G_R can be chosen from 2 dBi (for example, a standard dipole antenna) to 20 dBi (for example, an aperture antenna or an antenna array), or even higher. From Figure 14.4, we see that the larger G_R, the longer read range. However, a larger G_R also means a higher reader directivity, that is, less angular coverage for the tags.

On the other hand, for a specified P_T, the larger G_R, the larger P_{EIRP} ($P_{EIRP} = P_T\ G_R$). However, the P_{EIRP} needs to conform to RFID regulations. For example, Part 15 of FCC of USA advises that the P_{EIRP} should not exceed 36 dBm. Therefore, the selection of G_R is also limited.

A high gain antenna generally costs more than a low gain one. As for a tag-determined RFID system, a large G_R probably means a low cost-efficiency. In this case, the read range is equal to the r_{TAG}, but not the r_{READER}. However, increasing G_R and simultaneously keeping P_{EIRP} equal to the power limit set by the regulation actually cannot improve the r_{TAG}. As an example, using the same conditions as those in calculating the curve in Figure 14.4, Equation (14.8a) and $P_{TH} = -15\,\text{dBm}$, we can get the r_{TAG} as 8.3 meters.

Figure 14.4 shows that the r_{READER} is larger than 8.3 meters with $G_R = 12$ dBi. So, it is unnecessary to select a G_R larger than 12 dBi.

The aforementioned comments are focused on the optimization designs of RFID readers. However, in a tag-determined RFID system, small improvements in tag performance will greatly improve the read range. Thus, let's continue our discussions to explore the potential improvements that can be achieved at the tag side. We begin by assuming the tag works with a specified reader having a transmitting power of P_{EIRP} and a specified operating frequency of ω. Then, based on Equation (14.8) for r_{TAG}, in order to improve the read range, we provide the following design considerations:

- A larger tag antenna gain G_T will efficiently increase the tag received power P_R and consequently the read range. Of course, a larger G_T also constrains the tag "visible" to the reader only in a narrower viewing angle. So, it is required to weigh the pros and the cons while adopting higher gain tag antennas in a specific application.
- A smaller tag power reflection coefficient Γ will help to increase the tag absorption power P_A and consequently the tag can get enough energy to work at a longer read range. However, from Equation (14.15), it is obvious that a smaller Γ decreases the performance of r_{READER} at the same time. Thus, in order to get the optimum read range for an RFID system, a compromise needs to be reached between r_{READER} and r_{TAG} through tuning the reflection coefficient Γ of the tag. As seen in Figure 14.6 and Figure 14.7, the power absorbed by the tag decreases with the increase of reflection coefficient, while the SNR performance of the reader is to the contrary. Then the power reflection coefficient can be optimized based on those two figures.
- An appropriate data coding scheme should be chosen for a tag to improve the read range. It is well known that different coding schemes have different power density spectrums. In addition, the receiver baseband filter of the reader always has a limited pass-band in order to suppress DC, low and high frequency noises. So, the value of

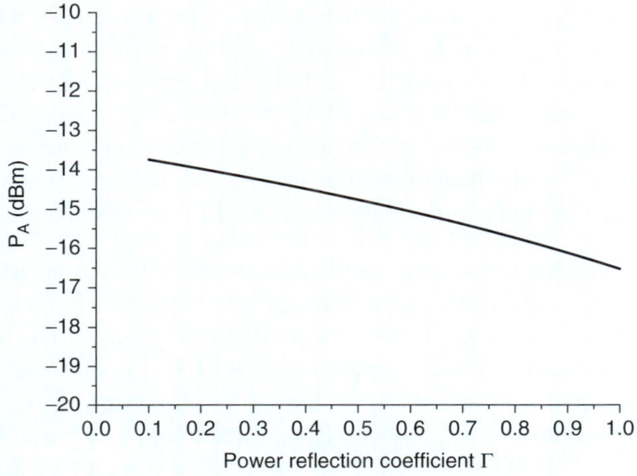

Figure 14.6 The power absorbed by the tag P_A vs. the ASK tag's reflection coefficient Γ. Note: In the calculation, $r = 10$ m, $P_{EIRP} = 36$ dBm, $P_{TH} = -15$ dBm and $G_T = 2.15$ dBi are assumed.

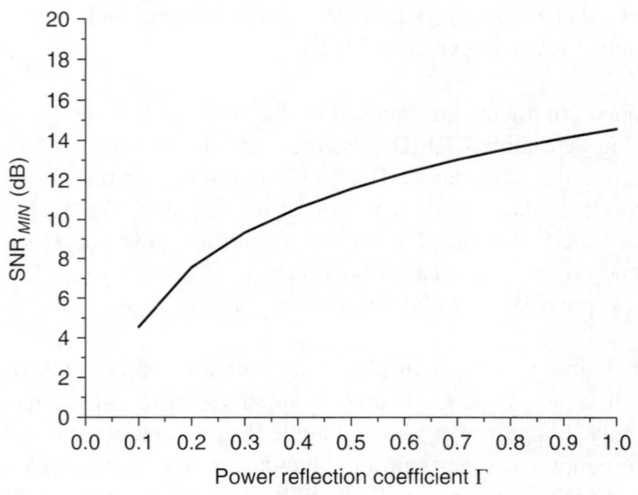

Figure 14.7 The SNR of the reader vs. the ASK tag's reflection coefficient Γ. Note: In the calculation, $G_T = 2.15$ dBi, $G_R = 13$ dBi, $P_{DATA} = -6.8$ dB, $P_{PN} = -39.9$ dB, $\psi = -37.5$ dB, r = 10 m and $\xi = 20$ dB are assumed.

the numerator of Equation (14.13), which denotes the single-sideband in-band signal power of the tag's demodulated data sequence, will vary with the coding scheme, which consequently impacts the SNR_{MIN} of the demodulation output signal and thus the read range. So, it is worthwhile comparing different tag data coding schemes while optimizing an RFID system to extend its read range.

After presenting our considerations for improving the read range, here we also want to include a brief discussion about other RFID researches found in the literature. In the reference [11], through optimizing the quality factors of tag antenna and tag chip, a design trade-off is achieved between the turn-on voltage of tag chip and the backscattered power from the tag, which helps to increase the read range. In reference [12], an approach which reconfigures the tag antenna as a Yagi antenna by adding parasitic elements at appropriate separations can dramatically increase the tag antenna gain and therefore the tag's read range. But it also increases the tag size, which confines its application. In references [13–17], considerable research has been conducted on several tag antenna types, including covered slot antenna, circular patch antenna and meander antenna. In addition, a compact reader antenna with dual polarizations is investigated in the reference [18].

Example 14.3.2 Calculating the power reflection coefficient Γ in an RFID system.
From the above analysis, we learn that a proper power reflection coefficient Γ is important for an RFID system. Assume the tag is placed 10 meters away from the reader and the parameters are the same as those in the calculations of Figures 14.6 and 14.7. Consider: (1) enough power should be absorbed by the tag to power up itself, which means P_A should be larger than the turn-on threshold (−15 dBm here), consequently Γ should be less than 0.58; and (2) the SNR$_{MIN}$ in the receiver should be larger than the SNR$_{USER}$, that is, SNR$_{MIN}$ > 12 dB, which means Γ should be larger than 0.55. Therefore, we should select a Γ between 0.55 and 0.58.

14.4 Improving the Read Rate of a Passive RFID System

The read rate of an RFID system is defined as the statistical ratio of tags successfully read versus the number of all the tags, which describes how well the reader can detect all the tags in the read range. In EPC terminology, "readability," which refers to the ability of a reader to obtain data from a tag, has a similar meaning and can be regarded as the read rate in the single-tag case.

The users of passive RFID systems are very concerned about the read rate. The performance of passive RFID systems is far from satisfactory regardless of the hype and high expectations of early stage RFID systems. The problem lies in non-100 % read rate, which has hampered the widespread use of passive RFID systems. For example, Wal-Mart's RFID plan was reported as a revolution in commercial supply chain. But, according to the report by EPCglobal, during early field trials at Wal-Mart using its top 100 suppliers, only 89 % reading rate was achieved. Therefore, in this section, we will investigate the low read rate issue and present some solutions to improve the read rate of a passive RFID system.

The read range and the read rate of a passive RFID system are closely related. Based on the fundamental principles of wireless communication, the probability of reading a tag successfully, that is, the readability, is determined by the signal noise ratio (SNR) of the backscatter signal received by the reader, and the distance where the threshold SNR occurs corresponds to the maximum read range. It can be reasonably deduced that a reader of longer read range will help to improve its read rate of the tag at a specified distance compared with that of short read range. So, the considerations we propose to improve the read range in previous section are also definitely applicable to improving the read rate.

Many researchers, RFID system suppliers and RFID users have presented their observations about the low read rate issue. Based on that information, it is generally thought that an RFID system can reach a satisfactory read rate only under ideal experimental conditions, but the read rate may substantially deteriorate in real-world applications. Obviously, the difficulty lies in the complex environment surrounding the tags and the readers.

Now let's examine the environmental factors that can affect the RFID performance. Environmental factors include the atmosphere, the electromagnetic environment and the surroundings and obstacles between tags and readers. The atmospheric factors, such as wind, rain and temperature, need to be considered while trying to achieve a higher readability. For example, Ultra High Frequency (UHF) tags are impacted much more by wind and sun than High Frequency (HF) tags [19]. Temperature can affect the performance of chips and then the read rate. Among many environmental factors, the surroundings and obstacles are the biggest troubles, which can absorb and reflect the energy of electromagnetic waves, and thus cause interference or multi-path fading effects.

Obstacles in the environment can exhibit various electromagnetic behaviors depending on their material compositions. Generally, they are electromagnetically transparent, reflecting, and/or absorbing. Metals and water cause the biggest issues. They reflect and/or absorb electromagnetic waves respectively, which means the passive RFID tag has no chance of receiving sufficient operating power. In practical applications, many goods contain metal and/or water, such as canned food, detergents and drinks. Experiments show that the water has a large impact on the performance of tag, especially in the UHF range. It is found that tags perform well while reading through frozen beef, but not when the beef is thawed, illustrating the liquid water problem [20].

When an RFID tag is placed in close proximity to a metallic surface, the metal will detune the tag antenna and lower its performance. Currently, there are several methods to solve this issue. One is to use a foam separation. For example, if tags and the metal surface are separated by a piece of foam of about 8–10 mm thickness, the detuning effect will weaken dramatically. However, one shortcoming of this technique is that it also will increase the thickness of tags. Another method is to use wave-absorbing material as the substrate of tags, but wave-absorbing material is very expensive at present. Alternatively, a better solution is to use the metallic surface as the tag backplane and then make it behave as an integral part of the tag antenna [21].

Local surroundings can create unwanted multipath signals and deteriorate the read rate. The multipath signals alter, modify or disrupt a message when it travels between a reader and a tag, similar to the interference that one experiences with a car radio when driving past an intense RF emission source [20].

The effect of surroundings and obstacles generates the need to study the stack-up orientation of tags. Many tag manufacturers have declared that the tag orientation has little effect on the read range and the readability of their tags. However, there are some disputes about the importance of relative orientation of a tag's antenna relative to the reader's antenna. In reference [22], Huang *et al*. studied the above problem experimentally. They adopted the standard procedure that Wal-Mart mandates its suppliers to follow [22]. The experiment is as following: Twenty cartons of goods are affixed with passive RFID tags and are loaded onto a pallet. The pallet is passed through two reader antennas. The reader was equipped with a middleware (RFID application software) to read the ID information from the tags. The reader antennas form a 3 m-wide channel for pallets to pass through. There are different palletizing methods which cause different orientations of tags with respect to the reader antenna. The authors compared several palletizing methods in such a situation and concluded that the read rate of an RFID system is mainly affected by relative positions of the tags and the reader antennas.

Some researchers have managed to solve the low read rate issue from a novel point of view. They improved the RFID read rate/reliability through a systematic error detection approach. An intelligent middleware solution can be implemented to detect missed tags. But a shortcoming is that it requires some prior information (the number of tagged items that are entering the interrogation zone), which might be unavailable in some cases. So they proposed a new solution where an RFID reader works along with a normal weighing machine. The idea is to compare the gross weight of the tagged items against the gross weight (of the same items) stored in a back-end database. If some tags are not read at all, these weights would vary and hence incorrect readings could be identified [19].

Passive RFID technology has begun to be adopted in supply chain applications. Although it would be more cost-efficient if a 100 % read rate could be achieved through setting up only one read point, it is not a realistic expectation at the current maturity level of RFID technology. One solution to improve the read rate is to add a few more read points in the supply chain. For instance, if an RFID system can achieve a 50 % read rate at an individual point, then according to the basic probability theory, it can reach a 97 % read rate (about 1 in 33 products missed) with 5 read points or a 99.9 % read rate (about 1 in 1000 products missed) with 10 read points in the end of the supply chain [23].

Recently, the array processing technology in radar system has been introduced into RFID systems. In fact, an RFID system also can be regarded as a radar. This technology has the potential to greatly improve RFID performance, including read rate and read range. A typical array processing technology is the beam forming. In some cases, we need the reader antenna to radiate electromagnetic waves in a specific pattern to decrease interferences among tags. The ordinary antenna's radiation pattern is fixed. However, the antenna array system could steer the radiation beam to selectively cover targeted transponders and thus reduce reading errors and collisions among tags. This technology exploits the spatial diversity and the polarization diversity of tags. The directional beam also reduces the effects of multi-path fading [24].

A novel technology will always face plenty of problems at the beginning. For instance, the bar code technology also had its own problems during early implementations. At present, the EPC Generation 2 RFID tag has shown a remarkable improvement in its ability to identify multiple items. It also has achieved higher read rates and read speeds compared to the Generation 1 tag. We think that, through researchers' efforts, it looks hopeful that the low read rate problem for passive RFID will be solved in near future.

14.5 Two Design Examples for RFID System

Based on the aforementioned discussions, we built two RFID readers with off-the-shelf components, one working at the frequency of 915 MHz and the other at 2.45 GHz, and set up the systems with commercial UHF RFID tags. Those two readers are shown in Figure 14.8 and Figure 14.9, respectively.

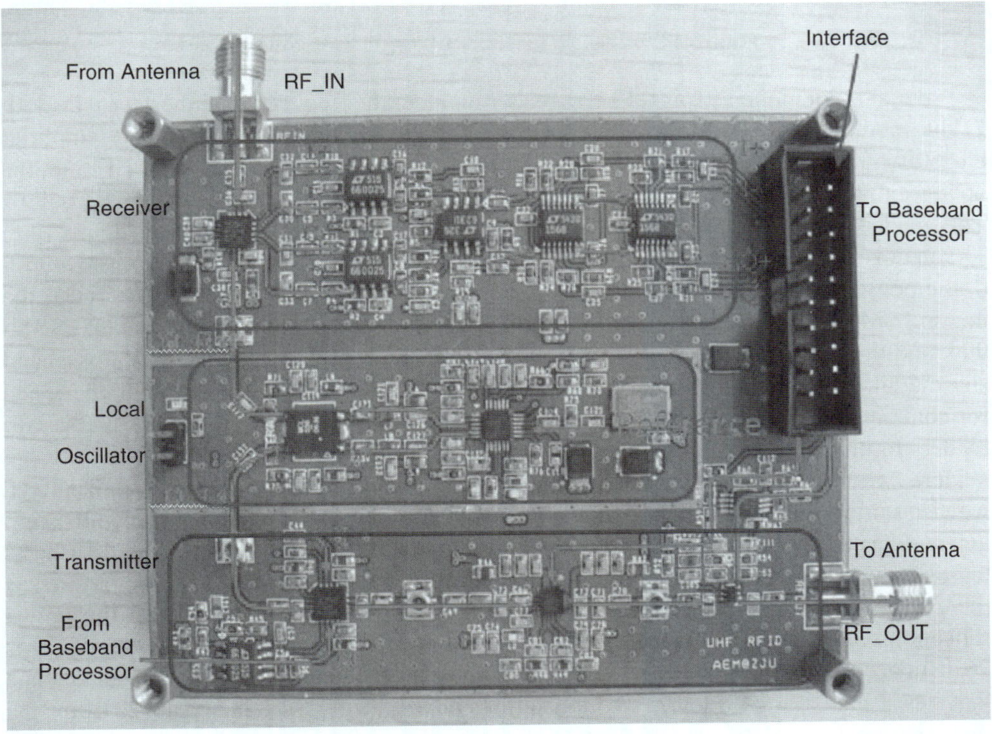

Figure 14.8 The 915 MHz RFID reader prototype.

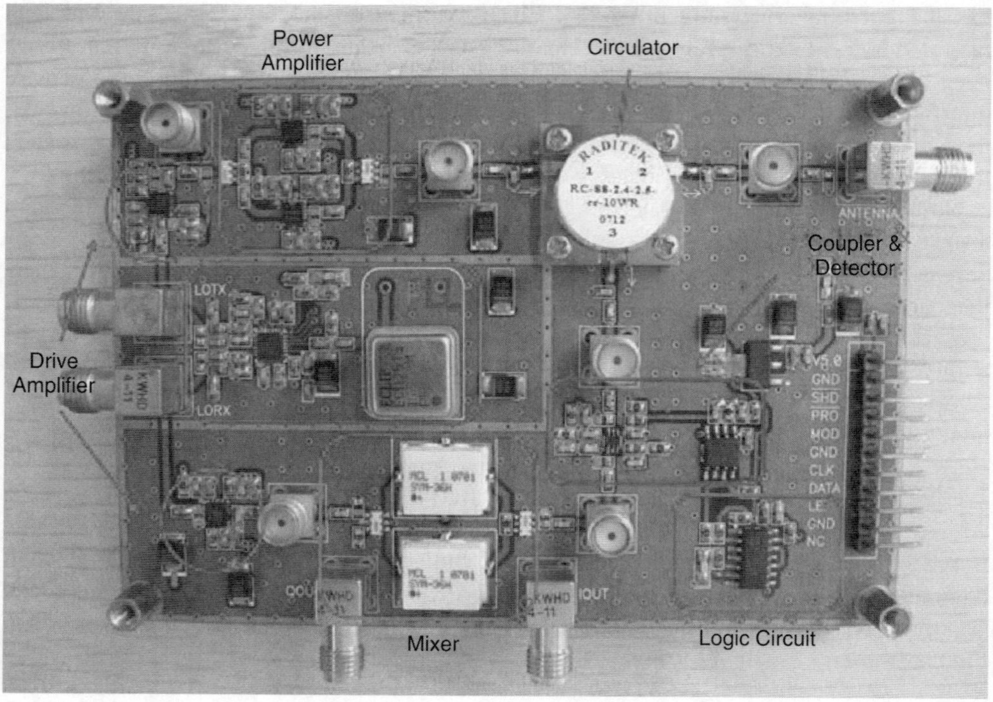

Figure 14.9 The 2.45 GHz RFID reader prototype.

The major components for 915 MHz reader include: the frequency synthesizer AD4360-7, the modulator LT5568, the demodulator LT5575, the first-stage low noise amplifier LT6600-2.5, the second-stage low noise amplifier LT6231, the baseband filter LT1568, the power amplifier SKY65111-348LF, and the directional coupler and detector DD02-999. This reader design also has a circulator sub-module, which is not shown in Figure 14.8. The major components for 2.45 GHz reader include: the frequency synthesizer AD4360-0, the power splitter SCN-2-27, the 90-degree phase-shifting power splitter QCN-27, the mixer SYM-36H, the power amplifier PA2455, the directional coupler and detector DD02-999, and the circulator RC-SS-CC-2.4-2.5-10WR. This reader design also has a baseband sub-module, which is not shown in Figure 14.9. Most of those chips are originally designed for commercial wireless communications, which makes it feasible to design cost-effective readers to promote RFID applications.

Here, we use the 915 MHz reader as an example and present its design details. First, we list some key design parameters as follows: for the tag, its half-wavelength dipole antenna has a gain G_T of 2.15 dBi, its input threshold power level P_{TH} is -15 dBm, its tag data rate is 160 kbps, the unipolar FM0 coding is utilized for the tag's data sequence, the ASK modulation method is utilized for backscatter communication and the power reflection coefficient Γ takes a value of 1; for the reader, the allowed maximum P_{EIRP} is 36 dBm according to the US regulations, it has a vertical-linear-polarization panel antenna that has a gain G_R of 13 dBi, its transmitting power P_T is constrained to 23 dBm, the receive-transmit isolation coefficient ξ of the reader is 20 dB, which is mainly limited by the circulator's performance, its baseband band-pass filter approximately has an ideal rectangular transfer function, whose low-end cutoff frequency is 10 kHz and high-end cutoff frequency is 320 kHz.

Using Equation (14.8a) and based on the parameters given above, we can calculate that the tag-determining maximum operational distance takes a value of 8.3 m and the reader-determining maximum operational distance takes a value of 11.8 m, which shows that this is a tag-determining RFID system.

However, it is not an easy task to evaluate r_{TAG} directly by measuring the time-averaging absorption power P_A of the tag because the tag IC is really very tiny and testing fixtures may have a big impact on the tag antenna. However, for the purpose of only judging whether the system is tag-determining or reader-determining, we just conduct two comparative operational distance experiments as illustrated in Figure 14.10 (a) and Figure 14.10 (b), which utilize the receiver-attenuation and the transmitter-attenuation, respectively.

We observe that the reader cannot detect the tag's data signal earlier in the case of transmitter-attenuation than in the case of receiver-attenuation while the operational distance increases. Thus, we can conclude that this is a tag-determining RFID system.

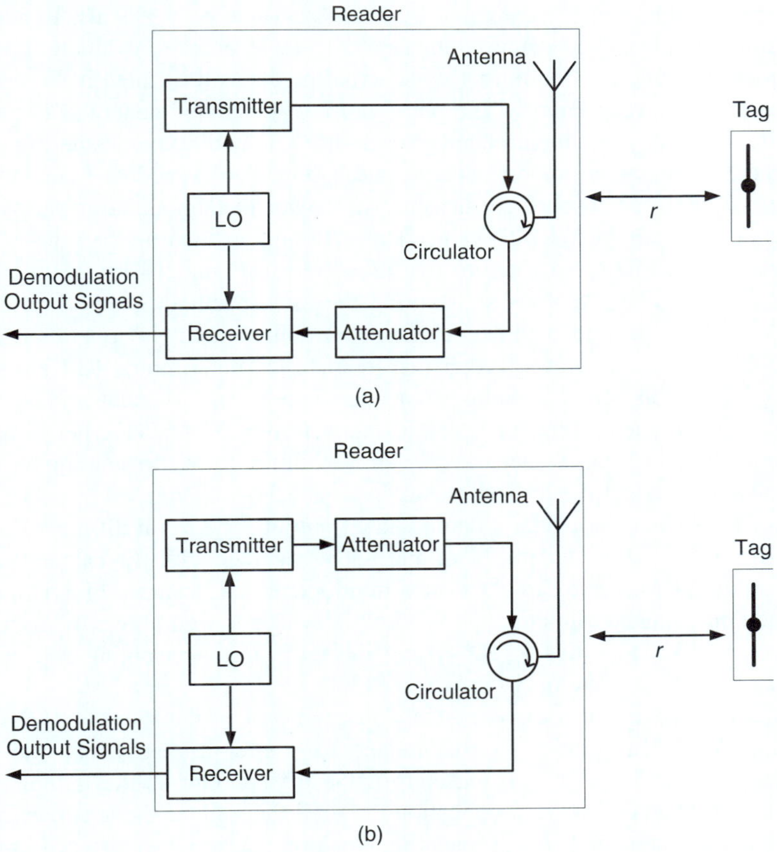

Figure 14.10 Comparative operational distance experiments. (a) Receiver-attenuation case. (b) Transmitter-attenuation case. Reproduced courtesy of © 2007 EMW Publishing.

In addition, using the transmitter-attenuation, we get a maximum operational distance of 4.1 m while the attenuation takes a value of 6 dB. Then, based on Equation (14.7a), we can get an estimated value of 8.2 m for r_{TAG}, which is close to the above calculated value for r_{TAG}.

Before using Equation (14.15a) to calculate the reader-determining maximum operational distance r_{READER}, we need to know the single-sideband in-band signal power P_{DATA} of the tag's binary data sequence and the band-limited phase noise power P_{PN} of the LO. First, we manage to figure out the PDS of P_{DATA}. Since each tag has a unique data sequence of finite-length (64 bits for the current tag), one tag's PDS is generally different from another. Here, instead of directly calculating the PDS of a specific tag, we evaluate it in a statistical-averaging-sense through constructing a random-generated data sequence of infinite length. We assume that this sequence is a wide-sense stationary ergodic stochastic process. Then, we can achieve a good approximate result of its PDS through applying discrete fast Fourier transform (DFFT) to its any segment of finite-length (1024 bits here). Then, we numerically integrate its PSD within the receiver's pass-band of [10 kHz, 320 kHz], which means that P_{DATA} takes a value of -6.8 dB. Second, we utilize an SPA directly to measure the normalized phase noise power of the LO within the receiver's bandwidth, which means that P_{PN} takes a value of -39.9 dB. In addition, we need to know the phase noise improvement factor ψ. In order to evaluate ψ, we set up two comparative demodulation output noise experiments as illustrated in Figure 14.11(a) and Figure 14.11(b), respectively. The phase noise of the signal generator is much lower than the LO's and they are uncorrelated to each other. In addition, considering that a SPA device generally does not support accurate measurement of very-low-frequency signals, we utilize a digital oscilloscope to measure the power of demodulation output noise in terms of root mean squared (RMS) value. Then, through evaluating the ratio of the noise power measured in Figure 14.11(b) to that measured in Figure 14.11(a), we find that ψ takes a value of -37.5 dB. Now, substituting the user-specified SNR_{USER} of 12 dB into Equation (14.15a), we find that the reader-determining maximum operational distance r_{READER} is 11.8 m, which shows further that this is a tag-determining RFID system.

Then, let's put this RFID system in an open indoor environment and test its SNR performance in practice. Here, in order to characterize the reader's performance in a practical operational scenario where there is a lot of clutter from surrounding obstacles, we do not choose to conduct the measurements in an anechoic chamber. Figure 14.12 shows the measured SNRs of the reader's demodulated output for several different distances of 2.0 m, 3.0 m, 4.0 m, 5.0 m, 6.0 m, 7.0 m and 8.4 m together with the calculated SNR_{MIN} using Equation (14.14a). We can get a maximum operational distance of 8.4 m, where the SNR of the demodulated output signal is 17.5 dB, which is well above the user-specified SNR_{USER} of 12 dB. Thus, this RFID system indeed is tag-determining and accordingly the read range takes a value of 8.4 m.

In Figure 14.12, we can see some small offsets between the measured SNR and the calculated SNR_{MIN}. In addition, the calculated r_{TAG} is a little less than the measured read range. There mainly are two reasons for the offsets. One is the indoor multi-path effects of electromagnetic wave propagation [25–27] that reduce the accuracy of the free space Friis electromagnetic wave propagation equation that has been utilized in Equations (14.1) and (14.5). The other is the inherent SNR uncertainty of the uncorrelated demodulation scheme as discussed in Section 14.3. However, these calculated results indeed have

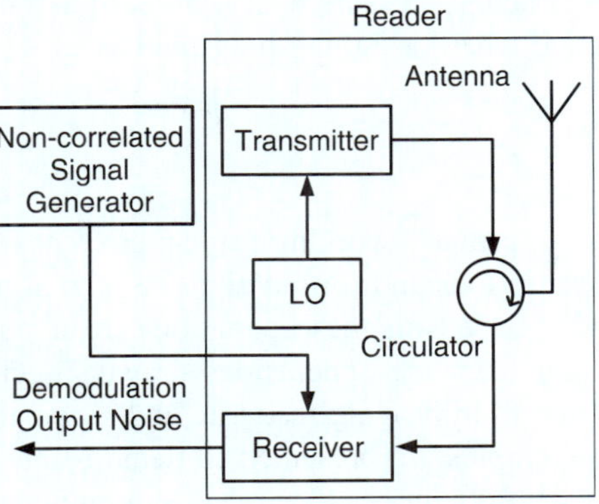

(a)

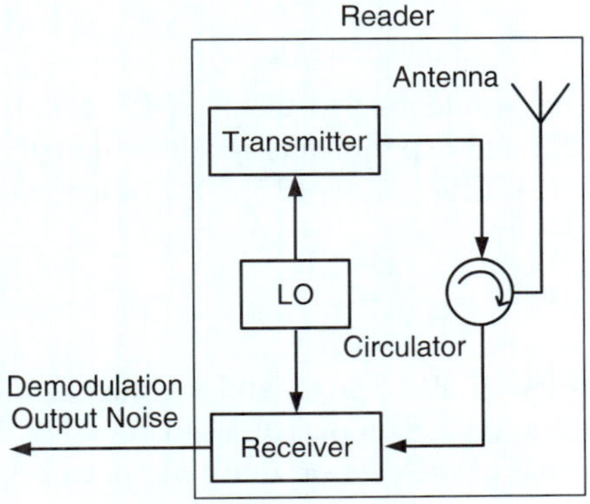

(b)

Figure 14.11 Comparative demodulation output noise experiments. (a) Uncorrelated case. (b) Correlated case. Reproduced courtesy of © 2007 EMW Publishing.

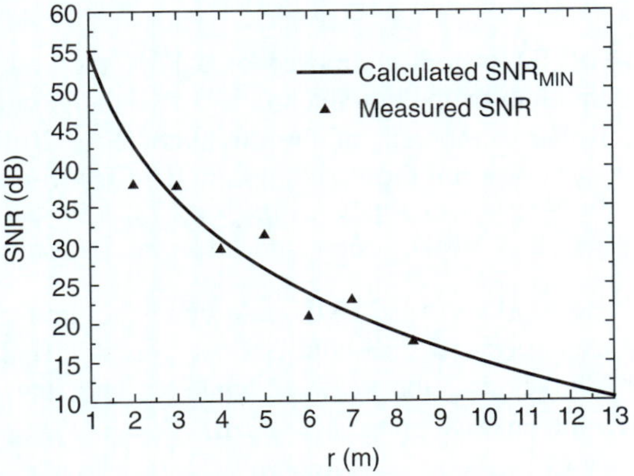

Figure 14.12 Calculated SNR_{MIN} and measured SNR of the reader's demodulator output signal for several different distances r. Reproduced courtesy of © 2007 EMW Publishing.

good approximations to the actual measurements, which illustrate the effectiveness of the methodology presented here when designing long read range readers.

14.6 Conclusion

This chapter investigates the primary working principles of passive RFID systems in far-field range. The signal descriptions and formulations for signal noise ratio and read range are theoretically presented. Based on these equations, read range and read rate issues are discussed further. Apart from the conclusions drawn directly from those equations, other research efforts related to improving read rate and range are also introduced. At the end of this chapter, two examples are presented to demonstrate cost-effective designs of long-range RFID readers with off-the-shelf wireless communication chips.

Acknowledgements

The authors of this chapter would like to thank EMW Publishing for kindly allowing us to reuse some theoretical and experimental results originally published in *Progress in Electromagnetics Research*, PIER71, 109–127, 2007, referenced as [2] in this chapter.

Problems

1. Consider a PSK tag working at 2.5 GHz, and a reader radiating a power of 20 dBm. Assume the tag's antenna has a gain of 3 dBi and $\Gamma = 1/2$. If the r_{TAG} is designed to be 10 m, what is the value of the turn-on threshold power of the tag?
2. Consider a passive RFID system composed of an ASK tag and a reader working at 950 MHz. Assume the gain of the tag antenna is 3 dBi, and the user-specified threshold $\text{SNR}_{USER} = 12\,\text{dBm}$, calculate the gain of the reader antenna if the r_{READER} is 10 m. (Suppose $\xi = 20\,\text{dB}$, $\Gamma = 1$, $P_{DATA} = \int_{f_L}^{f_H} \mathbf{S}\{S(t)\}\,df = -6.8\,dB$, $P_{PN} = \int_{f_L}^{f_H} \mathbf{S}\{\theta_P(t)\}\,df = -39.9\,dB$ and $\psi = 37.5\,\text{dB}$).
3. Consider a passive RFID system composed of a PSK tag and a reader working at 950 MHz. Assume the gain of the tag antenna is 3 dBi, the gain of the reader antenna is 15 dBi and the isolation efficient of the circulator ξ is 20 dB, find the SNR_{MIN} performance of the system when the tag is placed 10 m far from the reader. (Suppose $\Gamma = 1/2$, $P_{DATA} = \int_{f_L}^{f_H} \mathbf{S}\{S(t)\}\,df = -6.8\,dB$, $P_{PN} = \int_{f_L}^{f_H} \mathbf{S}\{\theta_P(t)\}\,df = -39.9\,dB$ and $\psi = 37.5\,\text{dB}$).
4. Consider an ASK tag working at 915 MHz. Assume the gain of the reader antenna is 13 dBi and the user-specified threshold SNR_{USER} is 12 dB. If the tag antenna is a half wavelength dipole, find the r_{READER} while varying the tag antenna's orientation to the reader antenna. (Suppose $\xi = 20\,\text{dB}$, $\Gamma = 1$, $P_{DATA} = \int_{f_L}^{f_H} \mathbf{S}\{S(t)\}\,df = -6.8\,dB$, $P_{PN} = \int_{f_L}^{f_H} \mathbf{S}\{\theta_P(t)\}\,df = -39.9\,dB$ and $\psi = 40.9\,\text{dB}$).
5. In Section 14.5, a 915 MHz passive RFID system is introduced and a group of specific values for its parameters were designed, calculated and measured. Then, the problem

for the readers is as follows: figure out the values of the parameters of a 2.45 GHz RFID system, which has a reader-determined read range $r_{READER} = 15\,\text{m}$ and a tag-determined read range $r_{TAG} = 10\,\text{m}$.

References

[1] RFID Implementation Cookbook. Available at: http://www.epcglobalinc.org/what/ cookbook/.

[2] Fan, Z.G., Qiao, S., Huangfu, J.T., and Ran, L.X. (2007) Signal descriptions and formulations for long range UHF RFID reader, *Progress In Electromagnetics Research*, PIER **71**: 109–127.

[3] Landt, J. (2005) The history of RFID, *IEEE Potentials*, **24**(4): 8–11.

[4] Brown, W.C. (1984) The history of power transmission by radio waves, *IEEE Trans. Microwave Theory Tech*, **32**(9): 1230–1242.

[5] Stockman, H. (1948) Communication by means of reflected power, *Proc. IRE*, Oct. pp. 1196–1204.

[6] Harrington, R.F. (1964) Theory of loaded scatterers, *Proc. Inst. Elect. Eng.*, **111**(4): 617–623.

[7] Koelle, A., Depp, S., and Freyman, R. (1975) Short-range radio-telemetry for electronic identification using modulated backscatter, *Proc. IEEE.*, **63**(8): 1260–1260.

[8] Finkenzeller, K. (2003) *RFID Handbook: Radio-Frequency Identification Fundamentals and Applications*, 2nd edn. New York: John Wiley & Sons, Ltd.

[9] Ruck, G.T., *et al*. (1970) *Radar Cross Section Handbook*, vols. 1–2. New York: Plenum.

[10] Saunders, W.K. (1990) CW and FM radar, in M. Skolnik (ed.) *Radar Handbook*. 2nd edn. New York: McGraw-Hill.

[11] Lee, J.W., Kwon, H., and Lee, B. (2006) Design consideration of UHF RFID tag for increased reading range, in *Proceedings of IEEE MTT-S International Microwave Symposium*, San Francisco, USA, June, pp. 1588–1591.

[12] Cheng, C.H., and Murch, R.D. (2007) Antenna modifications for enhancing RFID tag reading range, in *Proceedings of 2007 IEEE Antennas and Propagation Society International Symposium*, Honolulu, HI, USA, June, pp. 1084–1087.

[13] Rao, K.V.S., Nikitin, P.V., and Lam, S.F. (2005) Antenna design for UHF RFID tags: a review and a practical application, *IEEE Trans. Antennas Propagat.*, **53**(12): 3870–3876.

[14] Foster, P.R., and Burberry, R.A. (1999) Antenna problems in RFID systems, in *IEE Colloquium on RFID Technology* (Ref. No. 1999/123), **3**: 31–35.

[15] Karthaus, U., and Fischer, M. (2003) Fully integrated passive UHF RFID transponder IC with 16.7-μW minimum RF input power, *IEEE Journal of Solid State Circuits*, **38**(10): 1602–1608.

[16] De Vita, G., and Iannaccone, G. (2005) Design criteria for the RF section of UHF and microwave passive RFID transponders, *IEEE Trans. Microwave Theory Tech.*, **53**(9): 2978–2990.

[17] Curty, J.P., *et al*. (2005) Remotely powered addressable UHF RFID integrated system. *IEEE Journal of Solid State Circuits*, **40**(11): 2193–2202.

[18] Zhang, M., Chen, Y., Jiao, Y., and Zhang, F. (2006) Dual circularly polarized antenna of compact structure for RFID application, *Journal of Electromagnetic Waves and Applications*, **20**(14): 1895–1902.

[19] Potdar, V., Hayati, P., and Chang, E. (2007) Improving RFID read rate reliability by a systematic error detection approach, in *Proceedings of the 1st RFID Eurasia Conference*, Istanbul, Turkey, pp. 1–5.

[20] Clarke, R.H., Twede, D., Tazelaar, J.R., and Boyer, K.K. (2006) Radio Frequency Identification (RFID) performance: the effect of tag orientation and package contents, *Packaging Technology and Science*, **9**: 45–54.

[21] Daily, J., and McCann, R. (2007) Improving RFID read rate in metallic tractor-trailer applications, in *Proceedings of IEEE Region 5 Technical Conference*, Fayetteville, AR, USA, pp. 404–408.

[22] Huang, C.T., Lo, L.W., Wang, W.L., and Chen, H.L. (2008) A study for optimizing the reading rate of RFID tagged cartons in palletizing process, in *IEEE International Conference on Industrial Engineering and Engineering Management*, Singapore, pp. 1138–1142.

[23] Cook, C., and Brown, M. (2006) *Practical Performance Expectations for Smart Packaging*, Texas Instruments and RFID4U White Paper, December.

[24] Karmakar, N.C., Roy, S.M., and Ikram, M.S. (2008) Development of smart antenna for RFID reader, in *IEEE International Conference on RFID*, Las Vegas, NV, USA, pp. 65–73.

[25] Kim, D., Ingram, M.A., and Smith, W.W. (2003) Measurements of small-scale fading and path loss for long range RF tags, *IEEE Trans Antennas Propagat*., **51**(8): 1740–1749.
[26] Yarkoni, N., and Blaunstein, N. (2006) Prediction of propagation characteristics in indoor radio communication environments, *Progress In Electromagnetics Research*, PIER **59**: 151–174.
[27] Martinez, D., Las-Heras, F., and Ayestaran, R.G. (2007) Fast methods for evaluating the electric field level in 2D-indoor environments, *Progress In Electromagnetics Research*. 2007; PIER **69**: 247–255.

15

Principles and Techniques of RFID Positioning

Yimin Zhang, Xin Li, and Moeness Amin

Villanova University

15.1 Introduction

Radio frequency identification (RFID) systems, which basically consist of readers and tags, were originally developed for the identification of tagged objects, as the name RFID implies. Recently, precise positioning and tracking of RFID tags or readers have received considerable attention from both academia and industry. For example, finding the position of RFID tags is an important task in various real-time locating systems (RTLS) to locate and track products, assets, and personnel with attached RFID tags in an area covered by the RFID readers, for example, [1, 4, 15, 17, 51, 52, 69, 74, 75, 85]. In other applications, it is desirable for a reader to identify its own position with the assistance of reference tags, for example, [16, 18, 26, 42, 66, 73, 80, 81, 87]. Numerous RFID localization products have been developed for various applications. RFID positioning techniques and applications have been surveyed in [9, 13, 46, 89, 64].

It is pointed out that, depending on applications, the required positioning accuracy may differ. For some applications, such as parcel tracking in a warehouse, accuracy of 1 meter is acceptable and considered sufficient [35], whereas an accuracy of several centimeters is desired to unambiguously identify tagged parcels placed on a conveyer belt [85]. Therefore, it is important to select the appropriate positioning technique that meets the varying positioning accuracy and cost requirements.

This chapter provides a comprehensive introduction of the principles and techniques of RFID positioning. The majority of RFID positioning systems are based on the fusion of multiple pieces of relevant information. Examples of such information include range,

RFID Systems: Research Trends and Challenges Edited by Miodrag Bolić, David Simplot-Ryl, and Ivan Stojmenović

direction-of-arrival (DOA), and the propagation characteristics which are obtained from signal strength, time, and/or phase measurements at a single or multiple antenna positions.

One of the commonly used approaches to locate an RFID tag is trilateration. This approach determines the tag position by incorporating the range information of an RFID tag estimated at multiple reader antenna positions. Range information can be obtained, for example, through the received signal strength (RSS), round-trip time-of-flight (TOF), time-difference-of-arrival (TDOA), and/or phase-difference-of-arrival (PDOA) of the RFID signals. RSS is an easily measured quantity which provides range information based on the fact that a radio frequency (RF) signal emitted or backscattered from a tag attenuates with a law related to the distance which the RF signal traveled. Passive (including semi-passive) RFID systems can use round-trip TOF of a signal transmitted from an RFID reader and backscattered from the RFID tag for the estimation of the round-trip distance between the reader and the tag. In active RFID systems, the estimation of one-way TOF from the tag to the reader is often difficult because it requires precise synchronization between the reader and the tag. A rather practical solution, referred to as TDOA, is to compute the difference of the time-of-arrival (TOA) between multiple reader antennas which receive the same signal transmitted by the RFID tag. Accurate range estimation using time-based techniques, such as TOF and TDOA, may require a wideband signal to be used. Range estimation using PDOA utilizes the different phase delays exhibited by signals with different carrier frequencies when propagating over the same distance between the reader and the tag.

Tag localization can also utilize the triangulation technique based on the DOA information observed at multiple reader antenna positions. Array processing techniques that exploit the phase information of signal arrivals observed at multiple collocated antennas can achieve high DOA estimation accuracy. When high positioning accuracy is not required, the use of directional antennas is a low-cost alternative approach to obtain DOA information. RFID tag positions can also be determined at a single reader position by combining range and DOA information. Other RFID tag locating techniques include proximity and radio map matching. The former locates a tag by finding the closest reader antenna, whereas the latter compares the RSS or other signal signatures of a tag with that of reference tags whose positions are known *a priori*.

It is worth noting that a very challenging problem in RFID positioning lies in the effect of complicated wave propagation due to the presence of various obstacles and reflectors in the environment. Walls, human bodies, furniture and supplies that contain metallic and liquid materials, such as partitions, cabinets, bookshelves, water containers, may cause obstruction and reflections of electromagnetic waves. When a signal transmitted/backscattered from an RFID tag arrives at an RFID reader over a multiplicity of paths, it extends the delay profile and results in fluctuation in the RSS as well as the received signal phase [58, 90]. Similar effects can be observed for downlink propagation from a reader to a tag. Multipath propagation alters both signal strengths and phase. As such, RFID positioning techniques based on RSS and/or signal phase may become inaccurate. The localization performance of the TOF- and TDOA-based techniques may also be compromised due to multipath.

A number of approaches have been developed to improve the RFID positioning performance in a multipath environment. A class of such approaches is based on the comparison of the RSS, tag count/tag detection rate, and/or distribution probability of RSS corresponding to an unknown RFID tag with those measured for multiple known reference tags located in its vicinity [28, 51, 68, 79]. Because tags in the same area and close proximity are likely to be similarly affected by the propagation channel, multipath propagation effect can be mitigated through calibration using the information collected from the reference tags [61]. A well-known example of such approaches is LANDMARC [51]. Another approach is to use ultra-wideband (UWB) signals that allow high-resolution delay profile estimation and thereby achieve effective discrimination of multipath signals. As a result, the range information can be estimated based only on the TOF of the direct path [48, 50, 53]. The use of frequency hopping signals can achieve frequency-domain diversity to combat multipath effects [3].

RFID-based positioning and tracking techniques share similarities with those exploiting other technologies, such as acoustic, ultrasound, vision, infrared, laser, radar, GPS and WLAN (see, for example, [10, 27, 55, 78]). In particular, positioning of active RFID systems bears great similarity to WLAN-based positioning techniques. The concept of a passive RFID system, on the other hand, is analogous to an active radar system for the estimation of tag range and DOA, whereas differences exist as well. An RFID system is usually required to operate in much reduced complexity compared to other systems. Contrasting with radar systems which are typically wideband, non-UWB RFID systems use a much narrower frequency bandwidth. In addition, RFID systems have anti-collision capabilities and, as a result, only one tag needs to be considered at a time. While range-Doppler radars use the Doppler frequencies associated with moving targets to mitigate clutter, many RFID systems do not benefit from Doppler frequencies for clutter mitigation as they are typically stationary or move at a low-speed. Rather, backscattered signals from passive RFID tags are modulated and thus their spectra are shifted from that of the energizing carrier signal, enabling the suppression of strong carrier presence from the reader and clutter reflection.

Non-RFID positioning techniques can be incorporated into an RFID system to enhance its positioning capability, such as coverage extension and positioning accuracy improvement. For example, incorporating GPS and other global navigation satellite system (GNSS) receivers into RFID tags can provide positions in wide outdoor areas. WLAN infrastructure can be used to construct RTLS in indoor environments [20]. Passive RFID and laser range scanner are jointly used to improve the localization accuracy of mobile robots and persons [26]. A laser-activated RFID-based indoor localization system was developed for mobile robots, where a number of laser-activated active tags are placed in the environment as landmarks [87]. A real-time identification and localization system, named LotTrack, uses active RFID and ultrasound technologies to improve tracking visibility for logistics in a wafer fabrication clean-room [74].

The aim of this chapter is to provide a comprehensive introduction of the principles and techniques involved in RFID positioning. We first summarize key principles of information acquisition, and then introduce RFID positioning algorithms and techniques. As depicted in Figure 15.1, an RFID positioning system typically involves two major functional blocks, that is, location sensing and positioning processing. The location sensing block senses the tag location in terms of range and/or DOA using proper location metrics. This function block is discussed in Sections 15.2 and 15.3, respectively, for range and DOA estimation techniques. The objective of the positioning processing block is to find the location of an RFID tag or reader based on the information obtained from the location sensing

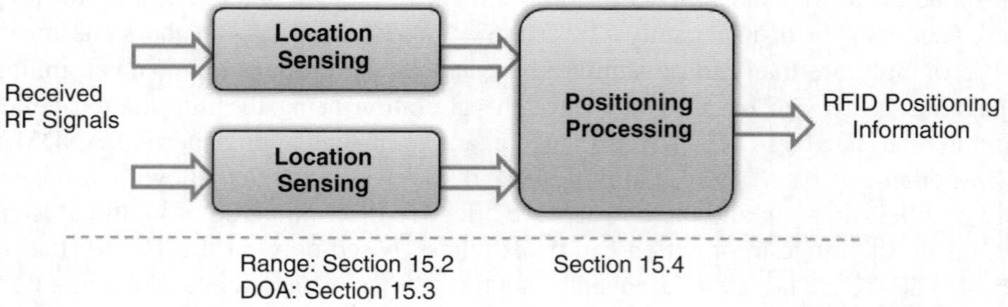

Figure 15.1 Block diagram of an RFID positioning system.

block. The positioning processing algorithms, techniques, and applications are addressed in Section 15.4. In Section 15.5, possible measures for the improvement of positioning accuracy are discussed. The chapter is concluded in Section 15.6.

15.2 Tag Range Estimation Techniques

Many RFID positioning techniques are based on the range information of tags, evaluated from a single or multiple RFID reader antennas. The accuracy of range estimation, therefore, directly affects the performance of the positioning of RFID tags. In this section, range estimation techniques based on RSS, phase, and time measurements are presented.

15.2.1 RSS-Based Techniques

For an active RFID system operating in a free space environment, the signal power received at the reader is expressed as [19]

$$P_{RX,\text{reader}} = P_{TX,\text{tag}} G_{\text{tag}} G_{\text{reader}} \left(\frac{\lambda}{4\pi d} \right)^2 \tag{15.1}$$

where $P_{TX,\text{tag}}$ is the transmit power at the active tag, G_{tag} and G_{reader} are the antenna gain of the tag and the reader, respectively, λ is the wavelength, and d is the range between the tag and reader.

For a passive RFID system, the signal is transmitted from the reader and backscattered at the tag. Thus, a round-trip path loss should be considered. The received signal power becomes

$$P_{RX,\text{reader}} = P_{TX,\text{reader}} \eta G_{\text{tag}}^2 G_{\text{reader}}^2 \left(\frac{\lambda}{4\pi d} \right)^4 \tag{15.2}$$

where $P_{TX,\text{reader}}$ is the transmit power from the reader and η is the power transfer efficiency of the passive tag. The typical value of η is 1/3 or $-5\,\text{dB}$ [19], but this value may change as technology advances.

Equations (15.1) and (15.2) clearly show that the RF signal power decays with d. In the above two equations, which consider free-space propagation, the one-way power attenuation is proportional to d^2, whereas the round-trip power attenuation is proportional to d^4. The actual attenuation rate varies, however, depending on the environment where the RFID system is deployed. In this case, Equation (15.2) is revised as

$$P_{RX,\text{reader}} = P_{TX,\text{reader}} \eta G_{\text{tag}}^2 G_{\text{reader}}^2 \left(\frac{\lambda}{4\pi d} \right)^{2n} \tag{15.3}$$

which shows that the signal strength is inversely proportional to d^{2n}, where n is referred to as the path loss exponent. The typical value of n is between 1.6 and 1.8 for line-of-sight (LOS) indoor environments, and between 2 and 6 for outdoor propagation environments [58]. As a result, when the path loss exponent is known or can be estimated for a specific environment, range d can be estimated from RSS measurements. The ratio between the backscattered signal power a reader receives from a passive RFID tag and the power the reader transmits is illustrated in Figure 15.2, where different values of path loss exponent n are considered. The following parameters are used: $f = 915\,\text{MHz}$, $\eta = 1/3$, $G_{\text{tag}} = 1$ (or 0 dB), and $G_{\text{reader}} = 4$ (or 6 dB). In practice, tags are expected to have low directivity, whereas a reader antenna gain of about 6 dB is commonly used because FCC regulations require proportional reduction of the transmit power when the transmit antenna gain exceeds 6 dB.

The use of RSS for range estimation is simple and handy [5, 18, 25, 30, 31]. Many RFID readers make the RSS information available. Moreover, the tag detection rate or tag count patterns also provide information related to the RSS and thus can be used to estimate the range [68, 79]. That is, the number or rate of successful readings of a tag is associated with the RSS information. The tag detection rate and the tag count patterns can be fused with the RSS measurements to increase the localization reliability [70]. Multiple

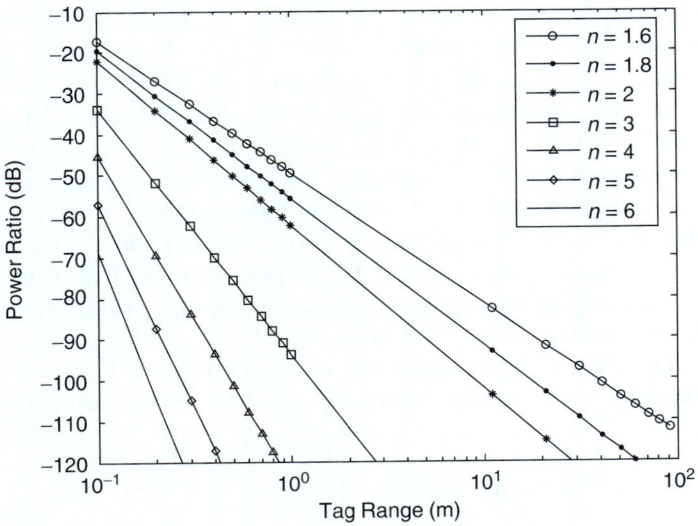

Figure 15.2 Ratio between the received power and the transmitted power of a passive RFID reader for different path loss exponents ($f = 915\,\text{MHz}$, $\eta = 1/3$, $G_{\text{tag}} = 0\,\text{dB}$, and $G_{\text{reader}} = 6\,\text{dB}$).

measurement results can be used to improve the reliability of RSS-based range estimations, particularly in a multipath environment. A convenient way is to use a frequency hopping method in which a reader sends out bursts at specified intervals in a frequency hopping manner and measures response at the corresponding frequency each time [3]. As such, multiple RSS values evaluated at these frequencies become available and the reliability of range estimation is improved by selecting the highest, average, median values, or by using other combining approaches.

The range estimation performance from RSS is, however, not robust, particularly when the RFID system is operated in a complex propagation environment. Obstruction of LOS between the reader and the tag yields additional signal loss, often referred to as shadowing. The effect of shadowing is typically characterized by using the log-normal distribution with variance σ_{sh}^2. In addition, RSS is also sensitive to reflection and scattering from walls, furniture, and various conductive materials in the propagation environment. Such reflection and scattering yield the multipath fading phenomenon. The effect of multipath fading can be described using the Rayleigh distribution or the Ricean distribution. The latter assumes that a dominant direct path is present, whereas the former assumes no dominant path. Various models are available to describe and predict the RSS in different indoor and outdoor propagation environments [58].

For unbiased parameter estimators, the Cramer-Rao lower bound (CRLB) is known to provide the minimum achievable mean square error (MSE) of a set of parameters, given the probability density function (PDF) of random variables involved in the problem [56]. For the range estimation $\hat{d}$, the CRLB of the range estimation error due to log-normal shadowing effect is given, in the form of root mean square error (RMSE), as [24, 57].

$$\mathrm{RMSE}(\hat{d}) \geq \frac{\ln 10}{10} \frac{\sigma_{sh}}{n} d \tag{15.4}$$

It shows that the error increases with the standard deviation σ_{sh} of the shadowing, and decreases with the path loss exponent. In addition, the accuracy of RSS-based range estimation degrades as the range d increases.

Figure 15.3 shows simulated RSS results to illustrate the significance of path obstruction and reflection for a downlink scenario [90]. The reader transmits a 1 W RF signal from its antenna located at $x = 0$ m, $y = 0.5$ m, and $z = 0.5$ m. The signal power received by a tag at various locations is shown in different gray levels in dBm. In a free space, as depicted in Figure 15.3(a), the received signal power is monotonically attenuated as the range increases. When a metallic cabinet, consisting of perfect conductor surfaces, is placed at the side of the link path, as depicted in Figure 15.3(b), the RSS behind the cabinet is significantly reduced due to path obstruction. On the other hand, when the cabinet is in front of the transmitter, as depicted in Figure 15.3(c), the received power oscillates with observation locations. In the latter two cases, the RSS no longer shows a unique, monotonic relationship with the range, making RSS-based range estimation difficult.

15.2.2 Phase-Based Techniques

PDOA-based approaches allow coherent signal processing and, therefore, at least in theory, can improve range estimation performance of passive RFID tags compared to RSS-based techniques [38]. During a time period designated for uplink data transmission,

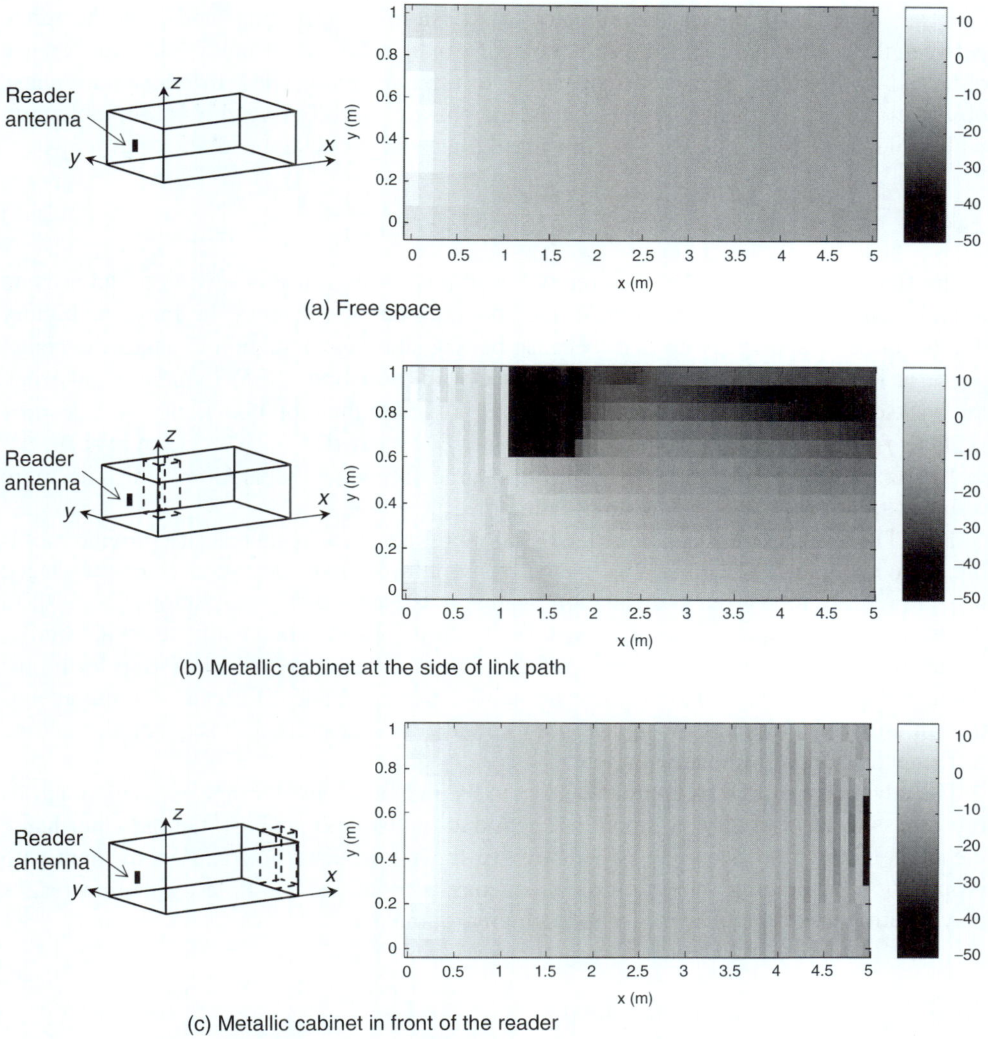

Figure 15.3 Simulated RSS in free space and in the presence of a metallic cabinet.

a reader transmits two continuous-wave (CW) signals which are then backscattered by a tag and received at the reader. The two CW signals propagate over the same distance, but their phase delays are proportional to their respective carrier frequencies. Therefore, a reader can estimate the tag range based on the phase difference observed at the two frequencies. Note that the response time that an IC-based passive tag takes is irrelevant to the phase difference used in PDOA-based range estimation. PDOA-based approaches share the same concept as the dual-frequency radar techniques for range estimation [2, 83].

Consider that an RFID reader transmits two CW signals at frequencies f_1 and f_2. Without considering the modulation performed at the RFID tag and the receiver noise, the phase of the uplink signal at frequency f_i can be expressed as $\phi_i = 4\pi f_i d/c$, where

$i = 1, 2$, $c = 3 \times 10^8$ m/s is the velocity of RF signal propagation, and d is the range between the reader and tag. Therefore, range d can be estimated from the phase difference observed at the return signal corresponding to the two frequencies. In reality, the phase observation is subject to wrapping, that is, the phase at each frequency is observable only within the range $0 \leq \phi_i < 2\pi$. As a result, the tag range is estimated as

$$\hat{d} = \frac{c\Delta\phi}{4\pi(f_2 - f_1)} + \frac{cm}{2(f_2 - f_1)} \tag{15.5}$$

where $0 \leq \Delta\phi = \phi_2 - \phi_1 < 2\pi$ is the wrapped phase difference observation and m is an unknown integer. The second term in the above expression denotes the range ambiguity due to phase wrapping. Note that, because backscattering modulation changes the signal phase at both carrier frequencies in the same way, Equation (15.5) remains valid when the backscattering modulation is applied. The maximum unambiguous range is $d_{\max} = c/[2|f_2 - f_1|]$. For example, when $\Delta f = |f_2 - f_1| = 10\,\text{MHz}$, $d_{\max}$ is 15 m. When $\Delta f = 1\,\text{MHz}$, $d_{\max}$ becomes 150 m. Clearly, a large frequency separation, which is more resistant to noise [45], yields a small value of $d_{\max}$.

The PDOA approaches can be extended to more than two frequencies to provide multiple frequency pairs. The ranges estimated from multiple frequency pairs can be averaged to yield a more robust range estimation against noise and other perturbations [38, 39]. In this case, equal frequency separation is desirable to obtain range estimates with similar variance. Multiple frequencies can also be designed to have unequal separation for robust range estimation over an extended unambiguous tag range [45]. Instead of simultaneous transmission of multiple CW signals, they can also be transmitted in sequential, yielding a frequency hopping implementations [84].

The advantage of PDOA approaches lies in its high accuracy in range estimation and its robustness to the variation of signal strength due to obstructions [18]. On the other hand, the range estimation accuracy is sensitive to the phase distortion caused by multipath propagation. Averaging over multiple frequency pairs or/and multiple estimation results may improve the reliability in a multipath environment.

15.2.3 Time-Based Techniques

Measurement of round-trip TOF in a passive RFID system can be used to estimate the tag range as

$$\hat{d} = c \cdot \frac{\text{TOF}}{2} = \frac{c \cdot (\text{TOP} - T_p)}{2} \tag{15.6}$$

where TOP is the overall round-trip time delay which includes the round-trip propagation time, TOF, as well as the signal processing time consumed at the tag circuitry, denoted as T_p. The measurement of round-trip TOF or TOP only utilizes the clock at the reader and thus does not require clock synchronization between the reader and the tag.

For active RFID tags, on the other hand, measurement of one-way TOA requires that the reader and the tag have precisely synchronized clocks. In many active RFID systems, however, achieving precise synchronization between a reader and a tag is impractical.

Rather, it is often feasible to synchronously process the data received at multiple readers or reader antennas, and thus the TDOA related to different tag-reader antenna paths can be estimated. The TDOA information obtained from a pair of reader antennas corresponding to the same signal transmitted from an active RFID tag yields hyperbola location trajectories with the foci positioned at the two reader antennas. Thus, by utilizing multiple reader antennas to form multiple antenna pairs, the tag position can be determined as the intersection of the respective hyperbolas.

For a conventional narrowband RFID system, immediate application of time-based techniques (e.g. TOA and TDOA) for the localization of RFID tags is often difficult because of the poor time resolution limited by the frequency bandwidth. In addition, time-based techniques may experience additional challenges in the presence of multipath [22, 23, 44, 47, 49, 62].

Nevertheless, time-based range estimation techniques could be promising when sufficient signal bandwidth is available, for example, when the UWB techniques are used [3, 48]. UWB is defined by FCC and ITU-R in terms of a transmission from an antenna for which the emitted signal bandwidth exceeds 500 MHz or 20% of the center frequency. FCC approved license-free use of low-power UWB radio transmission with an enormous bandwidth of 7.5 GHz at the frequency band 3.1–10.6 GHz [21]. Such a wide bandwidth provides an excellent means for wireless positioning due to its high time-domain resolution. Specifically, for a single-path additive white Gaussian noise (AWGN) channel, it was shown that the RMSE of the range estimate $\hat{d}$ derived from the TOA estimation is lower bounded by [24]:

$$\mathrm{RMSE}(\hat{d}) \geq \frac{c}{2\sqrt{2}\pi\sqrt{\mathrm{SNR}}\mathrm{BW}} \tag{15.7}$$

where SNR is the signal-to-noise power ratio and BW is the effective signal bandwidth. Therefore, UWB RFID systems can achieve a high range resolution by utilizing a wide bandwidth, although the SNR is usually low. Note that, unlike RSS-based techniques, the RMSE of the range estimate obtained from time-based approaches is independent of d.

UWB signaling can be carrier-based or impulse-based. Both types of UWB RFID systems can provide good immunity against signal distortion and multipath effects. Carrier-based UWB RFID systems can achieve frequency diversity to alleviate the impact of multipath fading [24]. On the other hand, due to high time-domain resolution, impulse-based UWB RFID systems can resolve multipath components to eliminate the effect of reflection and scattering paths.

15.3 DOA Estimation Techniques

Some RFID tag positioning techniques are based on the DOA information of the RF signal, observed at multiple reader antenna positions. In these techniques, the positioning performance of RFID tags is affected by the accuracy of DOA estimation. DOA estimation is typically achieved using directional antennas, phased arrays and smart antennas. Utilization of directive beams also helps to enhance the read range and to reduce interference as well as multipath effects.

15.3.1 Directional Antenna

A directional antenna can transmit energy to or receive energy from a small angular sector so as to improve the radiation (or reception) efficiency or to mitigate interference. In an RFID system, when a tag enters the area covered by a directional reader antenna, the reader can sense it and thus determine its rough DOA. The accuracy of the DOA depends on the antenna beamwidth. An antenna with a narrower beamwidth yields a higher DOA accuracy.

15.3.2 Phased Array

A phased array is a group of antennas in which the relative phases of the respective signals feeding or weighted to the antennas are varied in such a way that an effective radiation/reception pattern of the array is reinforced in the desired direction, whereas low array sensitivity is exhibited in other directions. Some phased arrays steer, without physical movement, the beams to fixed directions (known as the switched beam technique), whereas some can electronically steer the beams to any directions. For example, the bidirectional electronically steerable phased array (BESPA), developed by RF Controls, steers beams with a phase shift network and firmware control algorithm [59].

15.3.3 Smart Antenna

Smart antennas, also known as adaptive array antennas or adaptive arrays, are antenna arrays with sophisticated signal processing capability. They can be designed to adaptively steer beams and nulls toward arbitrary directions and to provide high-resolution DOA estimations. The number of antennas and the size of the array aperture are key parameters that determine the capability of an adaptive array.

An RF signal generates time delays when it propagates in the space. For a narrowband signal, such time delay can be equivalently considered as a phase delay across multiple antennas. Thus, the measurement of the phase difference between signals received at different array antennas can be used for DOA estimation. Commonly used DOA estimation techniques include maximum likelihood (ML), multiple signal classification (MUSIC), estimation of signal parameters via rotational invariance technique (ESPRIT), minimum variance distortionless response (MVDR), matrix pencil method or one of their derivatives [32, 76]. For a narrowband tag signal arrived from DOA θ observed at an RFID reader equipped with an N-element uniform-linear array (ULA) with inter-element spacing l, the RMSE of the estimated spatial frequency, defined as $\omega = \frac{l}{\lambda}\sin(\theta)$, is lower bounded by the CRLB [67],

$$\mathrm{RMSE}(\hat{\omega}) \geq \sqrt{\frac{6}{KN^3\mathrm{SNR}}} \tag{15.8}$$

where K is the number of available data snapshots. It shows that the accuracy of DOA estimate can be improved by increasing SNR, the number of snapshots, and the number of array antennas.

Note that most of the literature considers far-field DOA estimation problems where the unknown object is located at the far field of the antenna array. When the tag is closely placed around a reader, RFID tag positioning and tracking may involve rather complicated near-field DOA estimation problems, where the phase difference between different array antennas is a function of both DOA and the range [77]. For a two-antenna array, however, it is shown that the phase difference is approximately a function of only the DOA, thus simplifying the DOA estimation problem. Another important issue to be taken into account in near-field DOA estimation is the effect of the radiation field pattern, particularly the phase, of the antennas. Because the tag may be viewed by different array antennas from different angles, the observed phase difference should first be compensated by the phase difference in the antenna patterns before it is used for DOA estimation [77].

In the presence of multiple tags, simultaneous estimation of their DOAs, in principle, requires more antennas than the number of tags. However, in practice, multiple tags are often discriminated in the process of collision avoidance and, as a result, only a single tag should be considered at each time. With the use of multiple antennas, nevertheless, it is possible to design a reader to simultaneously resolve multiple tags without collision, yielding faster tag reading and DOA estimations.

15.4 RFID Positioning Techniques

Using the estimated range and/or DOA information, a tag or reader can be localized using various techniques. In this section, we provide an overview of different RFID positioning techniques.

15.4.1 Trilateration/Multilateration

15.4.1.1 Principles and Algorithms

The trilateration/multilateration method determines the position of a tag or reader using the range information estimated at several spatially separated reference points (reader antennas or reference tags). As previously mentioned, the range can be estimated using RSS-, phase-, or time-based techniques. Specifically, to unambiguously localize a tag in an n-dimensional space, range information from at least $n+1$ reference points is required. However, in some cases, n reference points may suffice if the range ambiguity can be resolved by other means. For example, bilateration that uses only two reference points yields two intersections in a two-dimensional (2-D) plane [74], and the tag position may be uniquely determined if the other intersection is out of the interested area.

Consider a tag positioning problem in a 2-D space as an example. Figure 15.4 shows how the tag position can be estimated using the trilateration method, where the range of the unknown tag to reference points (reader antennas) $p_1(x_1, y_1)$, $p_2(x_2, y_2)$, and $p_3(x_3, y_3)$ are estimated as d_1, d_2 and d_3, respectively. The location of the unknown tag, denoted as (x, y), can be determined by solving the following three equations

$$(x - x_i)^2 + (y - y_i)^2 = d_i^2, \quad i = 1, 2, 3 \tag{15.9}$$

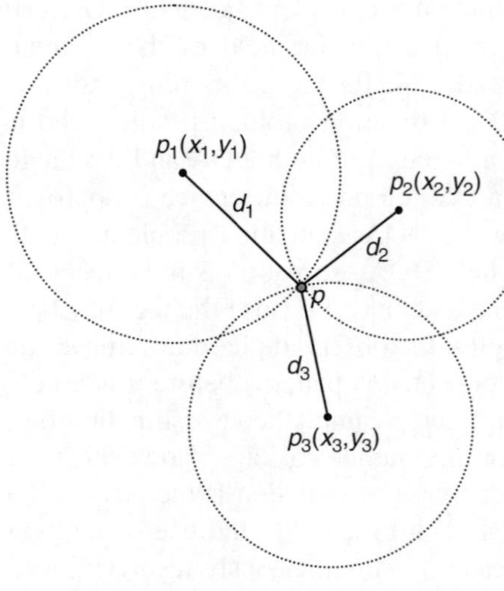

Figure 15.4 Tag positioning using trilateration. All circles are intersected at the tag position.

As a result, the coordinate of the unknown tag is obtained as

$$\begin{cases} x = -\dfrac{1}{2}\dfrac{(d_1^2 - y_1^2 - x_1^2)(y_2 - y_3) + (d_2^2 - y_2^2 - x_2^2)(y_3 - y_1) + (d_3^2 - y_3^2 - x_3^2)(y_1 - y_2)}{x_1(y_2 - y_3) + x_2(y_3 - y_1) + x_3(y_1 - y_2)} \\ y = -\dfrac{1}{2}\dfrac{(d_1^2 - y_1^2 - x_1^2)(x_2 - x_3) + (d_2^2 - y_2^2 - x_2^2)(x_3 - x_1) + (d_3^2 - y_3^2 - x_3^2)(x_1 - x_2)}{y_1(x_2 - x_3) + y_2(x_3 - x_1) + y_3(x_1 - x_2)} \end{cases} \tag{15.10}$$

In reality, the range estimates would have measurement errors, thus yielding an erroneous estimation of the tag position. One way to improve the positioning accuracy in this case is to use the multilateration method as described below, where more than three reference points are utilized.

For $M > 3$ reference points, M equations can be established similar to Equation (15.9). In this case, it becomes an over-determined problem. For convenience, we express them in a vector form as [71]

$$\mathbf{A}\begin{pmatrix} x \\ y \end{pmatrix} = \mathbf{b} \tag{15.11}$$

where $\mathbf{A} = \begin{pmatrix} x_1 - x_2 & y_1 - y_2 \\ \vdots & \vdots \\ x_{M-1} - x_M & y_{M-1} - y_M \end{pmatrix}$ is an $(M-1) \times 2$ matrix, and $\mathbf{b} = \dfrac{1}{2}\begin{pmatrix} d_2^2 - d_1^2 + x_1^2 - x_2^2 + y_1^2 - y_2^2 \\ \vdots \\ d_M^2 - d_{M-1}^2 + x_{M-1}^2 - x_M^2 + y_{M-1}^2 - y_M^2 \end{pmatrix}$ is an $(M-1) \times 1$ vector. The least-square solution of Equation (15.11) is given by

$$\begin{pmatrix} x \\ y \end{pmatrix} = (\mathbf{A}^T\mathbf{A})^{-1}\mathbf{A}^T\mathbf{b} \tag{15.12}$$

where $(\cdot)^T$ denotes the transpose of a matrix or a vector.

15.4.1.2 Applications

SpotON [33, 34] uses RSS to localize long-range active RFID tags in a three-dimensional (3-D) space. Multiple receivers collect RSS measurements and use the trilateration/multilateration method to estimate the tag locations. The Local Position Measurement (LPM) method [69] localizes outdoor active tags based on TDOA measurements using at least four synchronized reader antennas with known positions. Positioning accuracy can be improved by increasing the number of reader antennas through the minimization of the weighted mean square error. A Patient Management and Tracking System (PMTS) using the RSS-based trilateration technique is developed in [41]. It is reported that the average accuracy is less than 1 m in an open-space test.

UWB-based RFID systems have been developed for RTLS [48, 53]. The Sapphire DART system developed by Multispectral Solutions achieves a 200 m read range with an accuracy of about 0.3 m even in a multipath environment [48]. Recently, development of passive and semi-passive UWB-based RFID techniques was also reported. For example, Martec developed a UWB-based carrierless RFID system, named Passpulse [50], for both passive and semi-passive operations.

Passive SAW (Surface Acoustic Wave) ID-tags use TOA measurements to localize a tag [15]. SAW is an electromechanical device constructed of a piezoelectric crystal or ceramic to convert an RF signal to mechanical wave, which has a much smaller wavelength than that of the RF signal and thus is convenient to implement or measure the delay with a miniaturized size. Localization of a SAW tag is achieved by analyzing the round-trip TOF observed at three separate reader antennas through trilateration. Each tag has a fixed code described as its unique impulse response. Thus, a reader interrogates a tag by transmitting the time inverse of the tag-specific impulse response, and the tag then retransmits the correlated signal with a high peak to be easily detected at the reader. A 20-cm position accuracy was reported in [15] for a signal of carrier frequency 2.5 GHz and signal bandwidth of 40 MHz.

RSS-based trilateration/multilateration technology can also be used to locate the position of an RFID reader. Using multiple reference tags which are placed at known positions, a mobile reader can localize itself based on the RSS measurements corresponding to two or more tags [18].

15.4.2 Triangulation

15.4.2.1 Principles and Algorithms

Triangulation is a process to determine the location of a radio transmitter by measuring the DOA of the received signal from two or more known reference points. As we

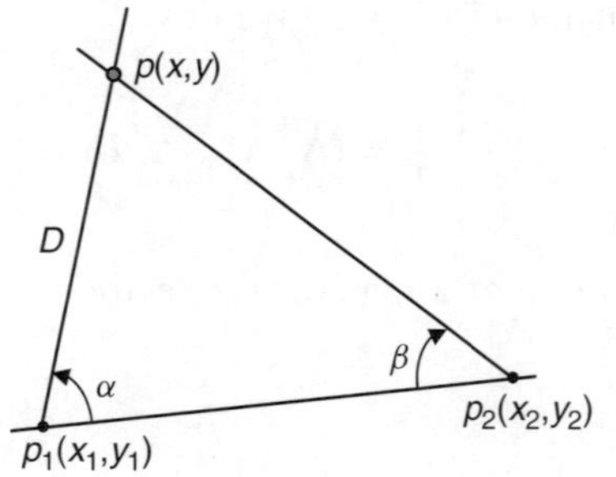

Figure 15.5 Tag positioning using triangulation. The tag location is obtained by intersecting the two lines.

discussed in Section 15.3, DOA information can be obtained using directional antenna, phased array, or smart antennas.

Figure 15.5 illustrates the basic principle of triangulation. In this example, the DOAs of the tag signal measured at two reference points $p_1(x_1, y_1)$ and $p_2(x_2, y_2)$ are respectively α and β with respect to the line determined by the two reference points (observation antennas). The tag can be localized by intersecting the two rays. The coordinate of the tag is thus given by

$$\begin{cases} x = x_1 + D\cos(\alpha + \gamma) \\ y = y_1 + D\sin(\alpha + \gamma) \end{cases} \tag{15.13}$$

where $D = \frac{\sin(\beta)}{\sin(\alpha+\beta)}\sqrt{(x_1 - x_2)^2 + (y_1 - y_2)^2}$ is the distance from the unknown tag to p_1 and $\gamma = \tan^{-1}\left(\frac{y_1 - y_2}{x_1 - x_2}\right)$.

When there are $M > 2$ reference points, $p_i(x_i, y_i), i = 1, \ldots, M$, and the ith reference point has the measured DOA α_i with respect to the x-axis, as shown in Figure 15.6, the unknown tag location can be obtained by intersecting the multiple lines, each passing through an observation position with the slope determined by the respective DOA. Thus, the tag location is estimated by solving the following equations

$$k_i x - y = k_i x_i - y_i, i = 1, \ldots, M \tag{15.14}$$

where $k_i = \tan(\alpha_i)$ refers to the slope of the line passing through $p_i(x_i, y_i)$. Stacking Equation (15.14) for the M observations yields the following over-determined problem

$$\mathbf{C}\begin{pmatrix} x \\ y \end{pmatrix} = \mathbf{d} \tag{15.15}$$

where $\mathbf{C} = \begin{pmatrix} k_1, -1 \\ \vdots \\ k_M, -1 \end{pmatrix}$ is an $M \times 2$ matrix, and $\mathbf{d} = \begin{pmatrix} k_1 x_1 - y_1 \\ \vdots \\ k_M x_M - y_M \end{pmatrix}$ is an $M \times 1$ vector. Note that, for the case of $|k_i| = \infty$, that is, $\alpha_i = 0.5\pi$ or 1.5π, Equation (15.14) becomes $x = x_i$, and the corresponding row of $\mathbf{C}$ and $\mathbf{d}$ should be [1, 0] and $[x_i]$, respectively.

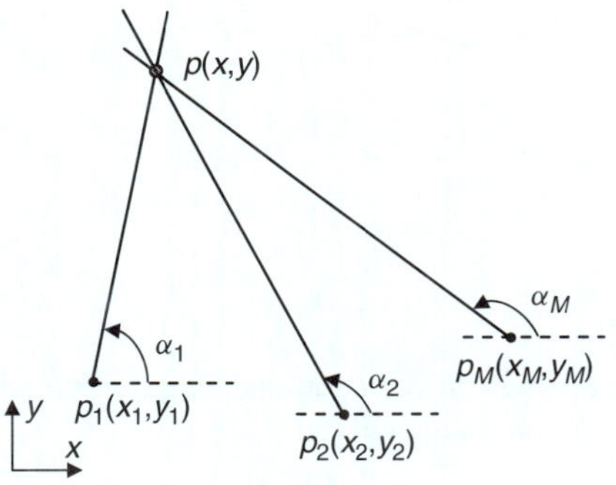

Figure 15.6 Tag positioning using triangulation with more than two reader antennas. The tag location is obtained by intersecting the multiple lines.

Similar to Equation (15.12), the least-square solution of Equation (15.15) is given by

$$\begin{pmatrix} x \\ y \end{pmatrix} = (\mathbf{C}^T\mathbf{C})^{-1}\mathbf{C}^T d \tag{15.16}$$

The localization accuracy can be improved by incorporating more reference points.

15.4.2.2 Applications

The FAST Tag Over-the-Conveyer RFID Tunnel System developed by Accu-Sort uses narrow-beam antennas to locate tags between closely spaced cartons [6]. In [85], DOA information obtained at two separate arrays is used to locate and track a moving tag on a conveyor belt. With the use of two sets of obliquely oriented two-element arrays, tags on the belt can be accurately localized through a triangulation operation. At the expense of higher signal processing complexity, it can be formulated as a near-field DOA estimation problem using a four-element array to yield more accurate positioning estimation [77]. In [4], the triangulation principle is used to locate indoor tags. Each reader uses two directional antennas to identify which side the tag is located. Although the angle estimation based on one reader does not have a high resolution, the use of multiple readers is expected to improve the estimation. In another example, three directional antennas are used to determine the rough area in which the tag is located [37].

15.4.3 Hybrid Direction/Range Methods

15.4.3.1 Principles and Algorithms

When both the DOA and range information of a tag are available at a reference point (reader antennas), the location of the tag can be uniquely determined. The principle is illustrated in Figure 15.7, where the DOA of the backscattering signal from the tag is α with respect to the horizontal axis of the reference coordinate and the estimated range is d. The tag is localized by intersecting the incident ray and range curve and thus one can

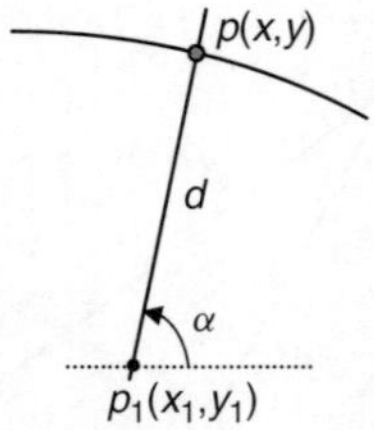

Figure 15.7 Tag positioning based on direction and range information. The tag is localized by intersecting the incident ray and the range curve.

get its coordinate as

$$\begin{cases} x = x_1 + d\cos\alpha, \\ y = y_1 + d\sin\alpha. \end{cases} \tag{15.17}$$

When the range and direction information at multiple reference points $p_i(x_i, y_i)$, $i = 1, \ldots, M$ are available, the tag location can be estimated by solving the following equations

$$\begin{cases} x = x_i + d_i\cos\alpha_i \\ y = y_i + d_i\sin\alpha_i, \quad i = 1, \ldots, M \end{cases} \tag{15.18}$$

where d_i is the range from the tag to the reference point p_i, and α_i is the DOA of the tag measured at reference point p_i. Similar to Equation (15.14), stacking the above equation for the M reference points renders the following over-determined problem

$$\mathbf{E}\begin{pmatrix} x \\ y \end{pmatrix} = \mathbf{f} \tag{15.19}$$

where $\mathbf{E} = \begin{pmatrix} 1 & 0 \\ 0 & 1 \\ \hline 1 & 0 \\ 0 & 1 \\ \hline \vdots & \vdots \\ \hline 1 & 0 \\ 0 & 1 \end{pmatrix}$ is a $2M \times 2$ matrix, and $\mathbf{f} = \begin{pmatrix} x_1 + d_1\cos\alpha_1 \\ y_1 + d_1\sin\alpha_1 \\ \hline x_2 + d_2\cos\alpha_2 \\ y_2 + d_2\sin\alpha_2 \\ \hline \vdots \\ \hline x_M + d_M\cos\alpha_M \\ y_M + d_M\sin\alpha_M \end{pmatrix}$ is a $2M \times 1$ vector.

The least-square solution of Equation (15.19) is given by

$$\begin{pmatrix} x \\ y \end{pmatrix} = (\mathbf{E}^T\mathbf{E})^{-1}\mathbf{E}^T\mathbf{f} \tag{15.20}$$

which results in

$$\begin{cases} x = \frac{1}{M}\sum_{i=1}^{M}(x_i + d_i\cos\alpha_i) \\ y = \frac{1}{M}\sum_{i=1}^{M}(y_i + d_i\sin\alpha_i) \end{cases} \tag{15.21}$$

The localization accuracy can be improved by incorporating more reference points.

15.4.3.2 Applications

Trolley Scan developed an RFID Radar system for identifying and locating passive RFID tags [75]. For 2-D applications, a reader is equipped with three high-gain patch antennas. One of them is used to transmit energizing signal, whereas the other two receive backscattered tag signals to enable range and DOA estimations. For 3-D applications, a third receive antenna is required. In a static situation where tags are relatively stationary, the radar achieves high range and DOA estimation accuracy by taking a long integration time. It is reported that the accuracy of range estimation is less than 0.5 m and DOA accuracy is better than 1 degree, at the maximum coverage range of 100 m.

A scheme for real-time 2-D localization of a SAW tag using a single reader antenna was demonstrated in [1]. In this scheme, the range is estimated based on the TOF measurement, and the DOA is implicitly obtained using the angular rotation of the reader's antenna and performing an operation of complex pattern matching (maximum correlation) between the received tag signal response and the ideal signal response (pattern) saved in the reader. This scheme requires that the tag is normal to that of the LOS between the tag and the reader antenna, and that the tag's response pattern is known.

15.4.4 Radio Map Matching Methods

15.4.4.1 Principles

Radio map matching methods are also known as "scene analysis" approaches. They are composed of two distinctive steps. In the first step, the radio scene information or RF fingerprints in the environment are collected to form a radio map, described as

$$M = \{(\mathbf{m}_1, \mathbf{p}_1), (\mathbf{m}_2, \mathbf{p}_2), \cdots, (\mathbf{m}_{N_R}, \mathbf{p}_{N_R})\} \tag{15.22}$$

where $\mathbf{m}_i$ denotes the measurement vector corresponding to the ith known position $\mathbf{p}_i = [x_i, y_i, z_i]^T$, and N_R is the total number of elements in the radio map. The elements of vector $\mathbf{m}_i$ correspond to the RF fingerprint measurements at multiple reader antennas. The RSS and tag count can be used as the fingerprints [79]. In the second step, unknown tags are localized by matching the measured data corresponding to the unknown tags with an appropriate subset of fingerprints recorded in the radio map. Two major fingerprinting-based matching methods are the k-nearest-neighbor (kNN) and the probabilistic methods. The kNN method [51] uses the fingerprint (say, the RSS measurement) of an unknown tag, recorded as $\mathbf{m}$, to find its k closest matches $(\mathbf{m}_{m_j}, \mathbf{p}_{m_j})$, $j = 1, \cdots, k$, in the radio map according to

$$m_j = \underset{i \in \{1, \cdots N_r\}, i \notin \{m_1, \cdots m_{j-1}\}}{\arg\min} ||\mathbf{m} - \mathbf{m}_i|| \tag{15.23}$$

The estimated location of the unknown tag, $\hat{\mathbf{p}}$, is then obtained as the weighted sum of the positions corresponding to the k nearest neighbors, that is

$$\hat{\mathbf{p}} = \sum_{j=1}^{k} w_j \mathbf{p}_{m_j} \tag{15.24}$$

where w_j is the weighting factor for the jth nearest neighbor with $\sum_{j=1}^{k} w_j = 1$. There are several ways to determine the weighting factors. Specifically, when the uniform weighting scheme is used, that is, all w_j's take the same value of $1/k$, it becomes the arithmetic average of the k nearest neighbors as

$$\hat{\mathbf{p}} = \frac{1}{k}\sum_{j=1}^{k} \mathbf{p}_{m_j} \tag{15.25}$$

The probabilistic methods [14, 26, 37, 52, 60, 70, 73], on the other hand, are to find the location of a tag from multiple possible locations to yield the highest posterior probability. In this case, vector $\mathbf{m}_i$ describes the joint PDF of the measurement or measurement error, observed at multiple reader antennas, corresponding to the ith reference tag at $\mathbf{p}_i = [x_i, y_i, z_i]^T$. In this way, the probabilistic radio map models the distribution of the measurement in different geographical positions. By exploiting the Bayesian rule, location estimation errors can be mitigated for improved tag positioning. Let $\mathbf{o} = [x_o, y_o, z_o]^T$ denote the observation of the position $\mathbf{p} = [x, y, z]^T$. For any given reference position $\mathbf{p}$, the probability distribution of the observation variable $p(\mathbf{o}|\mathbf{p})$, namely, the likelihood function, can be obtained through measurements at multiple readers. By using the Bayesian rule, the posterior probability of the position $\mathbf{p}$ is

$$p(\mathbf{p}|\mathbf{o}) = \frac{p(\mathbf{o}|\mathbf{p})p(\mathbf{p})}{p(\mathbf{o})} \tag{15.26}$$

where $p(\mathbf{p})$ is the prior probability of being at position $\mathbf{p}$ before knowing the value of the observation variable. When a uniform prior distribution is assumed, that is, the distribution of unknown tag does not have any preference for any particular position, the likelihood function completely determines the posterior distribution of the location. Further, the posterior distribution can be used to choose an optimal estimator of the unknown tag. For example, the position of the unknown tag can be estimated as

$$\hat{\mathbf{p}} = \sum_{i=1}^{N_R} p(\mathbf{p}_i|\mathbf{o})\mathbf{p}_i \tag{15.27}$$

Consequently, the posterior probability is maximized and the squared localization error is minimized.

In practice, the RSS measurements often suffer from multipath propagation and shadowing. By taking these factors into account in the pre-stored radio map, their effects on the location estimation can be mitigated. In addition to the RSS, the spatial signatures can also be used into map matching. The primary advantage of map matching methods lies in the corporation of the environment effect, such as NLOS propagation and multipath. However, the radio map should be constructed based on dense reference tags to represent the current environment and should be periodically updated to reflect the environmental dynamics [82].

15.4.4.2 Applications

Radio map matching methods have been widely used for RFID positioning. LANDMARC [51] is a well-known approach that uses several readers and a number of reference tags to locate indoor active RFID tags. The RSS of the reference tags are first recorded at each reader. When an unknown tag is present, its RSS is measured and the Euclidean distance relative to the RSS vector of each reference tag is calculated. The unknown tag is localized using the kNN method that takes a weighted sum of the coordinates of the k reference tags with the smallest Euclidean distance. Empirically, the reference tag with the shortest Euclidean distance takes the highest weight. The advantages of using reference tags are multi-fold. First, by using low-cost reference RFID tags, it maintains a low number of expensive RFID readers. Second, the effect of environment dynamics can be mitigated because the unknown tag and the reference tags are subject to similar propagation characteristics. The radio map of the reference tags can be dynamically updated to maintain the accuracy. Accurate positioning requires proper and dense distribution of reference tags and that the RSS of the reference tags is adequately updated.

Several variants of LANDMARC have been developed to improve the positioning accuracy and/or to reduce the system complexity. For example, the positioning error can be reduced through the removal of dissimilar reference tags [86, 88]. For the localization of stationary tags, the use of mobile readers for the measurement of the RSS of all tags from different locations is proposed [14]. Further, to reduce the effect of RSS fluctuation and measurement noise on the position estimation, Kalman filtering and probabilistic map matching can be utilized [14].

In the Flexible Localization EXplOits Rfid (FLEXOR) scheme, the area of interest is divided into a number of hexagonal cells [72]. Reference tags are placed at the center as well as on the vertices of each cell. Two localization modes are provided: the region mode finds a cell tag nearest to the unknown tag, whereas the coordinate mode determines the coordinate of the unknown tag through the weighted average of the coordinates of three reference tags, one at the center and two on the vertices of the same cell. It is reported that this scheme reduces computation overhead and provides similar accuracy as LANDMARC. A smart Book-LOCating System (BLOCS) was developed to locate tagged books on bookshelves [65]. The single book mode localizes a tagged book by minimizing the RSS-based Euclidean distance between bookshelf tags and the tagged book, whereas the book list mode routinely provides a list of the bookshelves and the misplaced books to help a librarian to localize all misplaced books.

When reference tags can be simultaneously detected by a set of readers, the difference between their actual and RSS-based estimated locations can be used as a correction factor to mitigate the position estimation error of the unknown tag [36]. The use of multi-power level transmission in a LANDMARC-based tag positioning system was proposed in [80]. To locate an unknown tag, the readers start with the lowest power level and gradually increase the transmit power until they receive the response from the unknown tag. Only the reference tags that are activated by the same power level as the unknown tag but not activated by a lower level are selected to estimate the range of the unknown tag. Such range estimations obtained from multiple readers are then used to trilaterate the position of the unknown tag.

As we discussed above, the use of a large number of reference tags can improve the localization accuracy of a LANDMARC system. The VIrtual Reference Elimination (VIRE) scheme develops the concepts of virtual reference tags, instead of placing more physical reference tags, to improve the positioning accuracy without increasing the number of actual reference tags [88]. In this scheme, the entire sensing area is divided into a number of small regions. Each region is centered by a physical reference tag and contains many virtual reference tags whose RSS values corresponding to each reader are determined through interpolation operations.

15.4.5 Proximity

15.4.5.1 Concept

An RFID reader has a limited read range and thus can only reach those tags that are located within a limited coverage area around the reader antennas. Therefore, observing whether a tag is within the reach of a reader antenna yields the proximity (or connectivity) information of the tag [63]. While high accuracy of tag positions in this case requires dense deployment of reader antennas with a small coverage area, this approach is easy to implement. The simplest implementation may be the distributed antenna scheme, where a variety of antennas, regardless of being omnidirectional or directional, are distributed in an area of interest. Each antenna senses tags in its respective coverage area. When a tag is sensed by a reader antenna, the tag location is assumed to be the same as this antenna. When a tag is detected by more than one antenna, it is considered that the tag is close to the antenna with the strongest RSS. Alternatively, the tag position can be localized using a weighted average of coordinates of those antennas. As thus, the positioning accuracy is on the order of the size of the antenna coverage area or smaller. In this case, the position of the unknown tag is estimated as

$$\hat{\mathbf{p}} = \sum_{i=1}^{N_R} w_i \mathbf{p}_i \tag{15.28}$$

where $\mathbf{p}_i = [x_i, y_i, z_i]^T$ denotes the coordinate of the ith reader antenna, N_R is the number of reader antennas. The weighting factors are subject to $\sum_{i=1}^{N_R} w_i = 1$. The ith weighting factor is set to $w_i = 0$ if the ith antenna cannot sense the unknown tag.

15.4.5.2 Applications

Mojix's STAR system consists of a single STAR receiver and multiple transmitters (known as eNodes) via wired connection [54]. A STAR receiver can manage up to 512 eNodes, which are daisy-chained with STAR receiver to cover a large geographical area or mounted in a facility in an orientation to enable the STAR receiver determining a tag's location in three dimensions. The eNodes transmit RF signals that interrogate and power up the tags, which then backscatter the signal to a receiver as far as 600 feet (200 m) away. The STAR system can localize a tag by determining which eNodes can activate the tag.

Robot search and rescue suffers from hostile conditions encountered after a disaster. A robot equipped with an inertial measurement unit can record its own trajectory. By further utilizing RFID readers in the robots and deploying RFID tags in the area, the robots can use the deployed tags as common references so as to minimize the trajectory error and better coordinate their exploration [42]. An embedded navigation system composed of

GPS and active RFID is proposed to localize pedestrians [43]. With the use of Kalman filtering, the GPS is used outdoors to adjust errors in position and direction, whereas the RFID is used for indoor localization. RFID readers are placed on fixed positions so that the pedestrian with an active tag can be localized by detecting the ID signal from the tag.

A 3-D localization scheme is proposed for locating passive or active RFID tags based only on the connectivity information associated with multiple readers [11, 12]. The estimated position of the unknown tag is obtained by simply averaging the corresponding positions of the virtual landmarks contained in the bounded space. Clearly, the accuracy depends closely on the density of the readers and on the size of virtual landmarks.

A localization system using geographical location for wireless sensor and RFID networks is demonstrated in [40]. In this scheme, an RFID reader first obtains its own location information from the messages broadcast by the nearby reference nodes. The reader then reads nearby tags and reports data upward until it arrives at the location server. Thus, the location system can monitor and track RFID tags based on the general purpose geographical location identifier for each region.

An RFID-based human-probe positioning system for urban sensing was proposed in [66]. In this system, RFID tags are deployed as landmarks in an urban area. When a person who carries an RFID reader moves into the area, the reader can be self-localized based on the proximity principle.

15.5 Improving Positioning Accuracy

As we discussed in Section 15.1, an RFID positioning system, in general, consists of the location sensing and the positioning processing. The former obtains necessary information, such as range and DOA, using various resources transmitted and received at single or multiple antennas. Such information is then fed to the positioning processing block for data fusion to yield the location information of the RFID tag or reader of interest. Therefore, positioning accuracy of an RFID system can be improved from both location sensing and positioning processing perspectives.

The selection of appropriate sensing techniques, under the resource limitations and system constraints, is critical in achieving satisfactory range and DOA estimations. For example, RSS-based range estimation techniques are simple but its sensitivity reduces as the range increases, whereas time-based range estimation techniques require a large signal bandwidth, and the accuracy is not sensitive to the range. A high SNR is always beneficial to improve the positioning performance. In addition, improved location sensing can be achieved by collecting more and diversified information in terms of time, frequency, space, and polarization. It is desirable to have over-determined measurement data sets that are more than the minimum requirement:

- *Time diversity:* A simple way to combat noise is to accumulate over a longer time, provided that the reader and tag are stationary [33, 75].
- *Frequency diversity:* The use of a wider frequency bandwidth provides frequency diversity for enhanced robustness against noise as well as multipath fading. In particular, the use of UWB signals enables the achievement of significant frequency diversity or,

equivalently, time resolution in discriminating reflection paths from the direct path [48, 50, 53]. Using frequency diversity at the same antenna, however, is not effective in combating path obstructions.

- *Spatial diversity:* Multiple spatially separated readers or reader antennas can be exploited to improve the RFID positioning in the presence of path obstruction because the probability that the LOS to all the readers is obstructed is very low. Moderately separated antennas (one wavelength or larger inter-element spacing) are effective in combating multipath fading. The use of more collocated antennas in a smart antenna enables higher degree of processing capability to provide more accurate DOA estimation. The antennas at different locations can be physically placed, or can be synthesized by moving antennas in different positions [29].
- *Polarization diversity:* The use of antennas with different polarizations is another effective way to combat multipath fading. Compared to spatial diversity, polarization diversity does not have an inter-element spacing requirement because different polarizations experience different propagation characteristics even when the antennas are closed spaced.

Better positioning processing can be achieved by fusing the collected data in an optimal or suboptimal way, for example:

- *Utilizing as much information as collected at the location sensing block:* Use all observation data to globally optimize a criterion, such as the mean square of positioning error or the likelihood function, and consider positioning methods that exploit the probabilistic characteristics of the measurement data and the historical status [26, 70, 73]. For example, Bayesian algorithm can be used to maximize the posterior probability of the positioning estimation. A Kalman filter is a convenient approach to exploit the dynamics of an RFID tag or reader, particularly when they are in motion [14].
- *Exploiting optimization algorithms to solve over-determined localization problems with noisy observations generally yields performance improvement:* Using convex optimization techniques for target and node localization is an important area in radar and wireless network communities [7, 8]. Weighted averaging operations incorporating the reliability of the measurement data or processed results usually improve the positioning accuracy.

The collection of more location sensing data as well as advanced positioning processing requires more investment in terms of hardware and/or signal processing capability. The decision is to best trade-off between the affordable system complexity and the required system performance.

The presence of various obstacles and reflectors in the environment alters the propagation decay and delay profile and thereby imposes significant challenges to RFID positioning. By using diversified information in terms of time, frequency, space, and polarization, as summarized above, the RFID positioning performance in a multipath environment can be improved. When it is available and feasible, the use of signals with excessive signal bandwidth, such as UWB signals, allows separation and discrimination of multipath signals and thereby enables elimination of the multipath effects. The use of directional antennas can also reduce the significance of multipath effects for performance improvement.

15.6 Conclusion

In this chapter, we have reviewed the principles, algorithms, and techniques of RFID positioning. In most RFID positioning problems, the location information of RFID tags is interested, but some applications involve the localization of RFID readers. An RFID positioning system, in general, consists of the location sensing component and the positioning processing component. The location sensing component provides necessary information, such as range and DOA, which is then fed to the positioning component for data fusion to yield the location information of the RFID tag or reader of interest. Different range and DOA estimation techniques have been introduced. RFID positioning principles and techniques based on trilateration, triangulation, hybrid direction/range, radio map matching, and proximity methods were presented. Potential approaches for improving positioning accuracy were also discussed.

Problems

1. What are the two major functional steps in a typical RFID positioning system? Please briefly describe the function of each step.
2. What are the three important tag range estimation techniques? Briefly describe how to improve the range estimation accuracy for each technique in the presence of multipath propagation.
3. What is the range ambiguity problem in the phase-based range estimation techniques?
4. Why can carrier-based and impulse-based UWB signaling improve positioning accuracy?
5. From the CRLB of the DOA estimation for a narrowband tag signal, what approaches can be used to improve the DOA estimation accuracy?
6. To unambiguously localize a tag in an n-dimension space, what is the required minimum number of reference points from which range information is provided?
7. Which positioning techniques can be used for tag positioning?
8. Consider a two-dimensional positioning scenario in which four reference points are located at $p_1(12.2, 3.8)$, $p_2(1.8, 4.9)$, $p_3(7.5, 2.8)$, and $p_4(6.3, 9.2)$, respectively, and the corresponding tag range estimates are 6.2, 4.5, 2.9, and 4.0. All the units are in meters. Find the position of the tag by using the multilateration method.
9. Consider a positioning scenario in which four reference points are also located at $p_1(12.2, 3.8)$, $p_2(1.8, 4.9)$, $p_3(7.5, 2.8)$, and $p_4(6.3, 9.2)$, respectively, and the corresponding DOAs, relative to the horizontal axis, are 151, 29, 111, and 254 degrees. The units of the reference positions are in meters. Find the position of the tag by using the triangulation method.
10. What is the principle of radio map matching methods for tag positioning? Describe the advantages and disadvantages of this kind of method.
11. What are the two primary approaches to improve positioning accuracy?

References

[1] Arumugam, D., Ambravaneswaran, V., Modi, A., and Engels, D. (2007) 2D localization using SAW-based RFID systems: a single antenna approach, *Int. Journal of Radio Frequency Identification Technology and Applications*, **1**(4).
[2] Ahamd, F,. Amin, M., and Setlur, P. (2006) Through-the-wall target localization using dual-frequency CW radars, *Proc. SPIE*, 6201, April.
[3] Amir, I. and Naim, A. Frequency hopping range estimation with low power consumption, US Patent, 7061428 B1.
[4] Amir, I. Dual antenna base station for improved RFID localization, US Patent, 20090027209 A1.
[5] Assad, M.A. (2007) A real-time laboratory testbed for evaluating localization performance of WiFi RFID technologies [thesis], Worcester Polytechnic Institute.
[6] Accu-Sort Systems, Accu-Sort FAST Tag Over-the-Conveyor (OTC) RFID Tunnel. Available at: http://www.marktecprods.com/PDF/fast_tagotc.pdf.
[7] Cheung, K.W., Ma, W.-K., and So, H.C. (2004) Accurate approximation algorithm for TOA-based maximum likelihood mobile location using semidefinite programming, in *Proc. IEEE Int. Conf. Acoustics, Speech, and Signal Processing (ICASSP)*, May, pp. 145–148.
[8] Biswas, P., Liang, T., Toh, K., Wang, T., and Ye, Y. (2006) Semidefinite programming approaches for sensor network localization with noisy distance measurements, *IEEE Trans. Automation Science and Engineering*, **3**(4): 360–371.
[9] Bhatia, A., Mehta, B., and Gupta, R. (2009) Different localization techniques for real time location sensing using passive RFID. Available at: http://filedb.experts-exchange.com/incoming/2009/02_w09/110353/RFID-DTOA.pdf.
[10] Bahl, P. and Padmanabhan, V.N. (2000) RADAR: an in-building RF-based user location and tracking system, in *Proc. IEEE InfoCom*, Mar, 2, pp. 775–784.
[11] Bouet, M. and Pujolle, G. (2008) 3-D localization schemes of RFID tags with static and mobile readers, in *Proc. Int. Federation for Information Processing* (IFIP).
[12] Bouet, M. and Pujolle, G. (2008) A range-free 3-D localization method for RFID tags based on virtual landmarks, in *Proc. Int. Symp. Personal, Indoor and Mobile Radio Communications* (PIMRC), Sept.
[13] Bouet, M. and Santos, A.L. (2008) RFID tags: Positioning principles and localization techniques, in *Proc. IFIP Wireless Days 2008*, Nov.
[14] Bekkali, A., Sanson, H., and Matsumoto, M. (2007) RFID indoor positioning based on probabilistic RFID map and Kalman filtering, in *Proc. IEEE Conf. Wireless and Mobile Computing, Networking and Communications* (WiMob), Oct., pp. 21–21.
[15] Bechteler, T., and Yenigun, H. (2003) 2-D localization and identification based on SAW ID-tags at 2.5 GHz, *IEEE Trans. Microwave Theory and Techniques*, **51**(5): 1584–1590.
[16] Chon, H., Jun, S., Jung, H., and An, S. (2004) Using RFID for accurate positioning, *Journal of Global Positioning Systems*, **3**(1–2): 32–39.
[17] Contractor, B.V. (2008) Two-dimensional localization of passive UHF RFID tags [thesis], Wright State University.
[18] Duron, M. and Bridgelall, R. Reverse infrastructure location system and method, US Patent, 20080157972 A1.
[19] Dobkin, D.M. (2008) *The RF in RFID*. Burlington, MA: Newnes.
[20] Ekahau, The case for real time location systems – from technology to business reality, White Paper, 2008. Available at: http://www.integratedsolutionsmag.com/index.php?option=com_docman&task=doc_view&gid=98.
[21] Federal Communications Commission Report and Order, FCC 02-48, Feb. 2002. Available at: http://hraunfoss.fcc.gov/edocs_public/attachmatch/FCC-02-48A1.pdf
[22] Guvenc, I., Sahinoglu, Z., and Orlik, P.V. (2006) TOA estimation for IR-UWB systems with different transceiver types, *IEEE Trans. Microwave Theory and Techniques*, **54**(4): 1876–1886.
[23] Ge, F.X., Shen, D. X., Peng, Y.N., and Li, V. O. K. (2007) Super-resolution time delay estimation in multipath environments, *IEEE Trans. Circuits and Systems- I:Regular Papers*, **54**(9): 1977–1986.
[24] Gezici, S., Tian, Z., Giannakis, G., Kobayashi, A., Molisch, H., Poor, H., and Sahinoglu, Z. (2005) Localization via ultra-wideband radios: A look at positioning aspects for future sensor networks, *IEEE Signal Processing Magazine*, **22**(4): 70–84.

[25] Hatami, A., Alavi, B., Pahlavan, K., and Kanaan, M. (2006) A comparative performance evaluation of indoor geolocation technologies, *Interdisciplinary Information Sciences*, **12**: 133–146.

[26] Hahnel, D., Burgard, W., Fox, D., Fishkin, K., and Philipose, M. (2004) Mapping and localization with RFID technology, in *Proc. IEEE Int. Conf. Robotics and Automation* (ICRA), **1:** 1015–1020.

[27] Harter, A., Hopper, A., Steggles, P., Ward, A., and Webster, P. (1999) The anatomy of a context-aware application, in *Proc. ACM/IEEE Int. Conf. Mobile Computing and Networking* (MobiCom), Aug., pp. 59–68.

[28] Huang, X., Janaswamy, R., and Ganz, A. (2006) Scout: outdoor localization using active RFID technology, in *Proc. Int. Conf. Broadband Communications, Networks and Systems* (BROADNETS), Oct., pp. 1–10.

[29] Hinske, S. and Langheinrich, M. (2008) Using a movable RFID antenna to automatically determine the position and orientation of objects on a tabletop, in *Proc. European Conf. Smart Sensing and Context*, **5279**, pp. 14–26.

[30] Hatami, A. and Pahlavan, K. (2005) A comparative performance evaluation of RSS-Based positioning algorithms used in WLAN networks, in *Proc. IEEE WCNC*.

[31] Heidari, M. and Pahlavan, K. (2007) Performance evaluation of WiFi RFID localization technologies, in *RFID Technology and Applications*. Cambridge: Cambridge University Press.

[32] Hudson, J.E. (1981) *Adaptive Array Principles*. London: Peter Peregrinus.

[33] Hightower, J., Vakili, C., Borriello, G., and Want, R. (2001) Design and calibration of the SpotON Ad-Hoc location sensing system, University of Washington CSE Technical Report, Aug. Available at: http://www.cs.washington.edu/homes/jeffro/pubs/hightower2001design/hightower2001design.pdf.

[34] Hightower, J., Want, R., and Borriello, G. (2000) SpotON: An indoor 3D location sensing technology based on RF signal strength, University of Washington CSE Technical Report, Feb. Available at: http://seattle.intel-research.net/people/jhightower/pubs/hightower2000indoor/hightower2000indoor.pdf.

[35] Ingram, S.J., Harmer, D., and Quinlan, M. (2004) Ultra wideband indoor positioning system and their use in emergencies, in *Proc. IEEE Position Location and Navigation Symp*., Monterey, CA, April, pp. 706–715.

[36] Jin, G., Lu, X., and Park, M. (2006) An indoor localization mechanism using active RFID tag, in *Proc. IEEE Int. Conf. Sensor Networks, Ubiquitous, and Trustworthy Computing*, June, pp. 40–43.

[37] Jia, S.. Sheng, J., and Takase, K. (2008) Improvement of performance of localization ID tag using multi-antenna RFID system, in *Proc. Annual Conf. Society of Instrument and Control Engineers* (SICE).

[38] Knox, M. and Bridgelall, R. (2006) Object localization based security using RFID, Int. Patent WO 2006/039119 A1.

[39] Knox, M., Bridgelall, R., Duron, M., Knadle, R., and Bender, J. (2006) Angle of position object location system and method, Int. Patent WO 2006/026518 A2.

[40] Kim, S., Ko, D., and An, S. (2008) Geographical location based RFID tracking system, in *Proc. Int. Symp. World of Wireless, Mobile and Multimedia Networks (WoWMoM)*, June, pp. 1–3.

[41] Kim, D., Kim, J., Kim, S., and Yoo, S. (2008) Design of RFID based the patient management and tracking system in hospital, in *Proc. IEEE Annual Int. Conf. Engineering in Medicine and Biology Society*, Aug., pp. 1459–1461.

[42] Kleiner, A., Prediger, J., and Nebel, B. (2006) RFID technology-based exploration and SLAM for search and rescue, in *Proc. IEEE/RSJ Int. Conf. Intelligent Robots and Systems*, pp. 4054–4059.

[43] Kourogi, M., Sakata, N., Okuma, T., and Kurata, T. (2006) Indoor-outdoor pedestrian navigation with an embedded GPS-RFID-self-contained sensor system, in *Proc. Int. Conf. Artificial Reality and Telexistence (ICAT)*, pp. 1310–1321.

[44] Li, J. and Wu, R. (1998) An efficient algorithm for time delay estimation, *IEEE Trans. Signal Processing*, **46**: 2231–2235.

[45] Li, X. Zhang, Y. and Amin, M.G. (2009) Multifrequency-based range estimation of RFID tags, in *Proc. IEEE Int. Conf. RFID*, April.

[46] Miller, L. (2006) Indoor navigation for first responders: a feasibility study, Technical Report, National Institute of Standards and Technology, Feb.

[47] Moghaddam, P.P., Amindavar, H., and Kirlin, R.L. (2003) A new time-delay estimation in multipath, *IEEE Trans. Signal Processing*, **51**(5): 1129–1142.

[48] Multispectral Solutions, Sapphire DART Ultra Wideband Precision Asset Location System. Available at: http://www.multispectral.com/pdf/Sapphire_Revolution.pdf.

[49] Manickam,T.G., Vaccaro, R.J., and Tufts, D.W. (1994) A least-squares algorithm for multipath time-delay estimation, *IEEE Trans. Signal Processing*, **42**(11): 3229–3233.

[50] Muchkaev, A. and Waverly, G. Carrierless RFID system, US patent 7,385,511 B2.

[51] Ni, L., Liu, Y., Lau, Y., and Patil, A. (2004) LANDMARC: indoor location sensing using active RFID, *ACM Wireless Networks*, **10**(6): 701–710.

[52] Oktem, R., Aydin, E., and Cagiltay, N. (2008) An RFID based location finding and tracking with guidance, in *Proc. Int. Conf. Wireless Communications, Networking and Mobile Computing* (WiCOM), Oct., pp. 1–4.

[53] O'Connor, M.C. (2004) FCC certifies Ubisense's UWB, *RFID Journal*, Dec. Available at: http://www.rfidjournal.com/article/articleview/1285/1/1/.

[54] O'Connor, M.C. (2008) New RFID technology helps Kraft, P&G, Kimberly-Clark go the distance, *RFID Journal*, April. Available at: http://www.rfidjournal.com/article/articleview/4041/.

[55] Priyantha, N., Chakraborty, A., and Balakrishnan, H. (2000) The cricket location-support system, in *Proc. ACM Int. Conf. Mobile Computing and Networking (MobiCom)*, Aug., pp. 32–43.

[56] Poor, H.V. (1994) *An Introduction to Signal Detection and Estimation*. New York: Springer-Verlag.

[57] Qi, Y. (2004) Wireless geolocation in a non-line-of-sight environment, thesis. Princeton University, Dec.

[58] Rappaport, T.S. (2002) *Wireless Communications: Principles and Practice*, 2nd edn. Upper Saddle River, NJ: Prentice Hall.

[59] RF Controls, Bidirectional electronically steerable phased-array antennas for passive UHF RFID systems, *RFID Journal*, White Paper, Jan. 2009. Available at: http://www.rfidjournal.com/whitepapers/download/260.

[60] Roos, T., Myllymaki, P., Tirri, H., Misikangas, P., and Sievanen, J. (2002) A probabilistic approach to WLAN user location estimation, *Int. Journal of Wireless Information Networks*, **9**(3): 155–164.

[61] Sypniewski, J. (2000) The DSP algorithms for locally deployable RF tracking system, in *Proc. Int. Conf. Signal Processing with Applications*, Oct.

[62] Sarac, U., Harmanci, F.K., and Akgul, T. (2008) Experimental analysis of detection and localization of multiple emitters in multipath environments, *IEEE Antennas and Propagation Magazine*, **50**(5): 61–70.

[63] Song, J., Haas, C.T., and Caldas, C.H. (2007) A proximity-based method for locating RFID tagged objects, *Advanced Engineering Informatics*, **21**(4): 367–376.

[64] Sanpechuda, T. and Kovavisaruch, L. (2008) A review of RFID localization: Applications and techniques, in *Proc. Int. Conf. Electrical Engineering/Electronics, Computer, Telecommunications and Information Technology (ECTI-CON)*, 2, pp. 769–772, May.

[65] Sue, K. and Lo, Y. (2007) BLOCS: A smart book-locating system based on RFID in libraries, in *Proc. Int. Conf. Service Systems and Service Management*, June, pp. 1–6.

[66] Suzuki, R., Martins, M., Ishida, Y. Tobe, Y., Konomi, S., and Sezaki, K. (2008) An RFID-based human-probe positioning system, in *Proc. Int. Conf. Networked Sensing Systems (INSS)*, June, pp. 248–258.

[67] Stoica, P. and Nehorai, A. (1989) MUSIC, maximum likelihood, and Cramer-Rao bound, *IEEE Trans. Acoustic, Speech, and Signal Processing*, **37**(5): 720–741.

[68] Schneegans, S., Vorst, P., and Zell, A. (2007) Using RFID snapshots for mobile robot self-localization, in *Proc. 3rd European Conf. Mobile Robots (ECMR)*, Freiburg, Germany, Sept.

[69] Stelzer, A., Pourvoyeur, K., and Fischer, A. (2004) Concept and application of LPM-a novel 3-D local position measurement system, *IEEE Trans. Microwave Theory and Techniques*, **52**(12): 2664–2669.

[70] Subramanian, S.P., Sommer, J., Schmitt, S., and Rosenstiel, W. (2008) RIL: Reliable RFID based indoor localization for pedestrians, in *Proc. Int. Conf. Software, Telecommunications and Computer Networks (SoftCOM)*, Sept., pp. 218–222.

[71] Sayed, A.H., Tarughat, A., and Khajehnouri, N. (2005) Network-based wireless location, *IEEE Signal Processing Magazine*, **22**(4): 24–40.

[72] Sue, K., Tsai, C., and Lin, M. (2006) FLEXOR: A flexible localization scheme based on RFID, in *Proc. Int. Conf. Information Networking (ICOIN)*, Jan., pp. 306–316.

[73] Schneegans, S., Vorst, P., and Zell, A. (2007) Using RFID snapshots for mobile robot self-localization, in *Proc. European Conf. Mobile Robots (ECMR)*, Germany.

[74] Thiesse, F., Fleisch, E., and Dierkes, M. (2006) LotTrack: RFID-based process control in the semiconductor industry, *IEEE Pervasive Computing*, **5**(1): 47–53.

[75] Trolley Scan, RFID-radar. Available at: http://rfid-radar.com/introduc.html.

[76] Van Trees, H.L. (2002) *Optimum Array Processing, Part VI of Detection, Estimation, and Modulation Theory*. New York: John Wiley & Sons. Ltd.

[77] Wang, J., Amin, M., and Zhang, Y. (2006) Signal and array processing techniques for RFID readers, in *Proc. SPIE Symp. Defense and Security*, Orlando, FL, April.

[78] Want, R., Hopper, A., Falcao, V., and Gibbons, J. (1992) The active badge location system, *ACM Trans. Information Systems*, Jan., pp. 91–102.

[79] Wilson, A., Prashanth, D., and Aghajan, H. (2007) Utilizing RFID signaling scheme for localization of stationary objects and speed estimation of mobile objects, in *Proc. IEEE Conf. RFID*, March, pp. 94–99.
[80] Wang, C. Wu, H., and Tzeng, N. (2007) RFID-based 3D positioning schemes, in *Proc. IEEE InfoCom*, pp. 1235–1243.
[81] Xu, B. and Gang, W. (2006) Random sampling algorithm in RFID indoor location system, in *Proc. IEEE Int. Workshop Electronic Design, Test and Applications*, Jan.
[82] Yin, J., Yang, Q., and Ni, L.M. (2008) Learning adaptive temporal radio maps for signal-strength-based location estimation, *IEEE Trans. Mobile Computing*, **7**(7): 869–883.
[83] Zhang, Y., Amin, M.G., and Ahmad, F. (2008) Time-frequency analysis for the localization of multiple moving targets using dual-frequency radars, *IEEE Signal Processing Letters*, **15**: 777–780.
[84] Zhang, Y., Amin, M.G., and Ahmad, F. (2008) Narrowband frequency-hopping radars for the range estimation of moving and vibrating targets, in *SPIE Symposium on Defense and Security Symposium*, Orlando, FL, March.
[85] Zhang, Y. Amin, M.G., and Kaushik, S. (2007) Localization and tracking of passive RFID tags based on direction estimation, *Int. J. Antennas and Propagation*, Dec.
[86] Zhang, T., Chen, Z., Ouyang, Y., Hao, J., and Xiong, Z. (2008) An improved RFID-based locating algorithm by eliminating diversity of active tags for indoor environment, *The Computer Journal Advance Access*, Aug.
[87] Zhou, Y., Liu, W., and Huang, P. (2007) Laser-activated RFID-based indoor localization system for mobile robots, in *Proc. IEEE Int. Conf. Robotics and Automation*, pp. 4600–4605.
[88] Zhao, Y., Liu, Y., and Ni, L. (2007) VIRE: active RFID-based localization using virtual reference elimination, in *Proc. Int. Conf. Parallel Processing* (ICPP).
[89] Zhou, J. and Shi, J. (2008) RFID localization algorithms and applications – a review, *Journal of Intelligent Manufacturing*, DOI:10.1007/s10845-008-0158-5, Aug.
[90] Zhang, Y., Yemelyanov, K., Li, X., and Amin, M. (2009) Effect of metallic objects and liquid supplies on RFID links, in *Proc. IEEE AP-S Symp.*, June.

16

Towards Secure and Privacy-Enhanced RFID Systems

Heiko Knospe[1] and Kerstin Lemke-Rust[2]

[1] *Cologne University of Applied Sciences*

[2] *Bonn-Rhein-Sieg University of Applied Sciences*

16.1 Introduction

RFID is a technology for automated identification over a radio interface and drives many applications in ubiquitous and pervasive computing today. RFID systems are applied in security-sensitive areas such as ticketing, electronic passports, product anti-counterfeiting or entry systems.

In this chapter we follow a generic approach to security and privacy of RFID systems. We also call it a generic (or low-level) approach because it considers generic threats and solutions to an RFID system and does not discuss application-level security protocols.

This chapter is organized as follows. Section 16.2 provides an introduction to the notions of security and privacy. Section 16.3 classifies RFID transponders with respect to chip area and security functionality. In Section 16.4 we evaluate the attacks to security and privacy that are crucial to generic RFID systems. For each attack, recommendations on possible countermeasures are provided. Section 16.5 presents lightweight cryptography and discusses feasible implementations for RFID tags.

16.2 Security and Privacy

Security aims at minimizing risks to the assets of computer systems. Any security design faces the following questions:

1. Which threats apply to the assets of the computer system?

RFID Systems: Research Trends and Challenges Edited by Miodrag Bolić, David Simplot-Ryl, and Ivan Stojmenović

2. Which security objectives should be addressed to counteract the threats?
3. Which security functions should implement the security objectives?
4. To what extent are the implemented security functions effective in counteracting the threats?
5. What are the remaining vulnerabilities and risks to the assets of the computer system?

Therefore, any security problem starts with an analysis of threats. Threats are countered by security objectives. The three most widely used security objectives from [1] are

- *Confidentiality:* To protect and preserve the confidentiality of information means ensuring that it is not disclosed or available to unauthorized entities.
- *Integrity:* To preserve the integrity of information means protecting the accuracy and completeness of information and the methods that are used to process and manage it.
- *Availability:* An asset is available if it is accessible and usable when needed by an authorized entity.

Other security objectives exist, for example, authenticity, non-repudiation, accountability. Also privacy forms a special class of security objectives aimed at maintaining the right of self-determination of individuals. Among them are anonymity, pseudonymity, unlinkability, and unobservability. For a detailed terminology on privacy see, for example, [2]. Note that the choice of security objectives for a specific security problem can be subjectively biased.

Security mechanisms are chosen in order to support security objectives and to counteract the threats. There are three generic approaches to implement security as

- a preventive action;
- a detective action;
- a reaction.

Security mechanisms are not solely a technical problem. They also demand supporting organizational security mechanisms.

The effectiveness of security mechanisms is gradually assessed according to the attack potential of an adversary. Attack potential is the combination of expenditure of money, time, equipment, and insider knowledge.

Almost every solution to a security problem leaves some remaining risks to the assets. The provider of a security system is then responsible for managing and controlling these remaining risks. Possible approaches are to accept the risk, to avoid the risk, to transfer the risk, or to reduce the risk (ISO 27001 and ISO 27002 2008).

16.3 Classification of RFID Systems

RFID transponders and RFID readers that are located in the end-user environment constitute the front-end of an RFID system. Optionally, further components for data and application management are used. If so, the RFID readers are connected to a corresponding back-end system. Two solutions exist in practice:

- either the RFID reader has an online connection to the back-end system (*online RFID system*); or
- the RFID reader is offline but contacted in some regular time intervals (*offline RFID system*).

In an online RFID system, the RFID reader can access the back-end system during transaction processing. Because of this, an online RFID system usually facilitates the detection of fraud, for example, the multiple use of tickets. Further, the download of configuration data and software can easily be managed for an online RFID system. In an offline RFID system RFID readers work autonomously. Transactions are downloaded from time to time by maintenance personnel. Only at an inspection, can new configuration data and software be uploaded to a reader.

We provide a classification of current RFID systems with particular focus on the implemented security functionality. The security of an RFID system is mainly influenced by the transponders and their chip area. Chip area is needed for control logic, data and possibly program memory or cryptographic coprocessors. Security mechanisms therefore cause additional costs. Since the actual chip area depends on the technology, the number of gate equivalents (GEs) is generally used to measure the hardware complexity. One GE corresponds to the chip area required by a two-input NAND gate. Based on the chip and the security functionality, RFID systems can be roughly classified into three categories:

1. **Low-end RFID systems**: In this category, we have low-cost tags with typically less than 5000 GEs. Low-end systems are mainly used for automatic identification. Among these, there are one-bit transponders used for anti-theft in stores and tags with a permanent ID but without any security or privacy functions. Other low-end tags provide some elementary functions: a password-protected KILL command, an access password, read, write and lock memory blocks, checksums (e.g. CRC16) and a pseudo random number generator. The prevalent EPC (Electronic Product Code) Class 1 Generation 2 UHF tags (EPCglobal (2008), ISO 18000-6C) belong to this category.
2. **Mid-range RFID systems**: This category includes read-write transponders of moderate cost with data memory ranging from a few hundred bytes to more than 100 kB non-volatile read-write memory. Mid-range RFID systems are used for various applications such as ticketing, access control, and automotive immobilizers. Usually, these tags implement mutual authentication and enforce an access control for the transponder's memory. Session keys are established between the RFID transponder and the reader and transmitted data is encrypted and integrity protected over the radio interface. The transponders are usually implemented as a "hard-wired" state machine and often proprietary cryptographic functions are in use. MIFARE Classic (see Section 16.4.8) is the most prominent example in this category. MIFARE is compliant with ISO 14443 (1–3) and provides read-writable data blocks that are organized in multiple cryptographically separated sectors, thereby allowing for multi-applications.
3. **High-end RFID systems**: These RFID transponders are compliant with ISO 14443 (1–4), contain microprocessor-based smartcard chips and are equipped with a smartcard operating system. Both the chip and its operating system implement security mechanisms. The security mechanisms are application specific and include advanced

cryptographic functions. In the very high-end, we find dual-interface smartcards with crypto-coprocessors allowing for public key cryptography such as digital signature operation.

Most current RFID standards (including ISO 14443, ISO 15693 and ISO 18000) focus on the physical characteristics, the radio interface and transmission protocols and hardly specify any security functions. The standardization of RFID security services is work in progress, for example, in the forthcoming standard ISO 29167.

16.4 Attacks on RFID Systems and Appropriate Countermeasures

Components that are located in an end-user environment are usually at high risk of attack. This concerns the RFID reader, the RFID transponder and the communication between them. The RFID transponder is usually exposed to a higher risk of attack because it is handed over to all the end-users of an RFID system while RFID reader are usually located in environmentally protected areas.

The assets and the attack objectives depend on the concrete RFID application. From a more generic point of view, threats to RFID systems comprise

- *RFID-enabled theft of assets* stored in the RFID transponder,
- *RFID-enabled theft of identity credentials* stored in the RFID transponder, and
- *RFID enabled vandalism* [3].

Security is usually achieved by a combination of technical and organizational security mechanisms. Technical security mechanisms include hardware and software mechanisms of the components and the use of secure cryptographic protocols. Organizational security mechanisms are realized in the environment of the RFID-system and may include guards, alarm systems, auditing, as well as fraud detection in back-end systems.

This section aims at providing an overview of security attacks and corresponding risks. As with any information system there are four generic attacks on the communication flow between a sender and a receiver (see Figure 16.1): (a) eavesdropping of messages (see Section 16.4.1), (b) interruption of messages (see Section 16.4.2), (c) manipulation of messages (see Section 16.4.3), and (d) generation of messages (see Section 16.4.4). Afterwards, we discuss a specific attack on a message flow that applies to wireless communication channels: (e) relay of messages (see Section 16.4.5).

Attacks can also be directed at the data of an RFID system. Tracking infringes the privacy of human individuals due to the fact that active communication initiated from an adversary's RFID reader cannot be notified (see Section 16.4.6). Cloning means that an attacker produces duplicates of RFID transponders. It is discussed in Section 16.4.7. Cloning can be achieved by cryptanalytical and implementation attacks as discussed in Section 16.4.8 and Section 16.4.9.

There are other highly relevant attack paths that apply to almost any computer security system and also to RFID systems. Among them are:

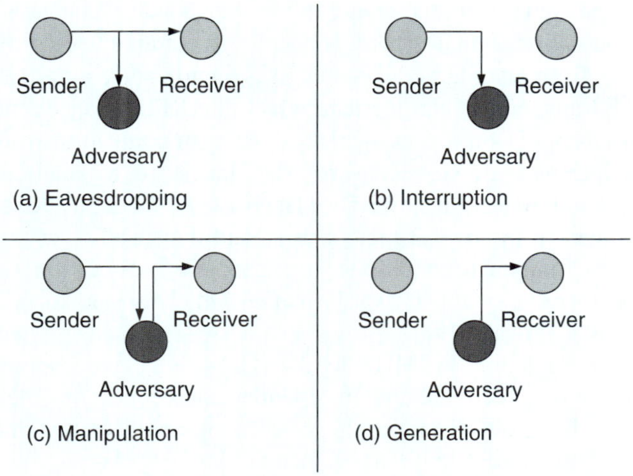

Figure 16.1 Illustration of common attacks to the information flow.

- *social engineering attacks* that spoof human users and administrators in order to thwart system security;
- *network attacks* such as denial-of-service that are conducted via network interfaces on the back-end system and also on the RFID readers in an online RFID system;
- *protocol attacks* that exploit weaknesses of cryptographic protocols used on a communication interface of the RFID system;
- *software implementation attacks* that exploit software vulnerabilities such as buffer overflows and unsecured software updates on the software running in components of an RFID system.

These general attacks on computer systems are beyond the scope of this chapter. For an overview on computer security, [4] is recommended. A valuable survey on RFID security and privacy can also be found in [7].

16.4.1 Eavesdropping of Messages

The communication between the RFID reader and the RFID transponder is made via the radio interface that is subject to interception by an adversary in the vicinity. Such an interception of the data communication is an eavesdropping attack. Note that eavesdropping is a passive attack.

For the RFID-communication two different physical channels are used:

- RFID reader to RFID transponder (*forward channel*): This channel provides the RFID transponder with energy and is used for transferring data from the reader to the transponder.
- RFID transponder to RFID reader (*backward channel*): This channel is used by the RFID transponder to send data from the transponder to the reader.

At 13.56 MHz, the signal from the reader to the transponder is about 80 dB stronger [5] than the load modulation signal which is used for communication on the backward channel. Therefore, from an enlarged distance, it is significantly more difficult to observe data sent from the transponder than the data which the RFID reader sends to it.

However, eavesdropping on the two-channel RF communication from several meters imposes a real threat, for example, recent work by Hancke [6, 8] practically demonstrated that the two-way communication at 13.56 MHz between an RFID reader and an RFID tag can be intercepted from about 3 meters. It is said that results vary on eavesdropping components and environmental conditions. In a concrete setting a faraway eavesdropper may only be able to monitor the forward channel which is said to be possible from a distance up to about 10 meters [9].

In the frequency range of 868 MHz to 2.45 GHz the eavesdropping range can be enlarged with the use of directional antennas. Under good conditions the forward channel is said to be receivable from distances up to a few hundred meters and the backward-channel from distances up to a few tens of meters [5].

In order to prevent eavesdropping attacks it is necessary to apply cryptographic encryption of messages (see Sections 16.5.2 and 16.5.3).

16.4.2 Denial-of-Service

For RFID systems it is essential that the communication channel between RFID transponders and RFID readers remains fully functional. Denial-of-service attacks aim at interrupting the communication between an RFID reader and an RFID transponder. Attacks apply to both the RFID reader and the RFID transponder. A denial-of-service attack may involve the destruction of RFID transponders so that they do no longer respond to reader requests. Destruction of a transponder can be carried out mechanically or chemically to cut off the antenna from the chip, or by a high electromagnetic field strength at the carrier frequency in order to thermally destroy the transponder [5]. Accordingly, an RFID reader can be subject to vandalism.

Another denial-of-service approach is to remove the detachment of RFID tags to valuable goods. Shielding of RFID transponders with aluminium foil has the same effect and may have the advantage that no physical destruction of RFID tags is required.

A jamming transmitter is a third approach to denial-of-service attacks. A blocker tag [9] emulates a high number of transponders in the field of the RFID reader. In order to cope with the presence of multiple transponders in the reader's range an anti-collision mechanism is implemented in an RFID reader. The anti-collision mechanism shrinks the referenced number space until only one tag responds that is afterwards selected for the communication session. A blocker tag, however, emits multiple serial numbers in each referenced number space. Identification of the genuine serial number among emulated serial numbers is thereby prevented. Due to collisions the reader is not able to determine a unique serial number and the reader is stuck with the anti-collision mechanism.

Denial-of-service attacks are a risk to anti-theft protection and access control systems. Prevention and detection of denial-of-service attacks are known to be a difficult problem that cannot be solved with cryptography.

16.4.3 Manipulation of Messages

RFID messages are sent on a radio channel in the near vicinity of an RFID reader and an RFID transponder. Manipulation on this radio channel is in principle possible with a transmitter under the control of the adversary, but may have a high risk resulting in a denial-of-service attack. So far manipulation of messages in transit is not known as a common attack on RFID systems.

In order to detect manipulations of messages it is required that sender and receiver use cryptographic checksums (see Section 16.5.4). Hereby, it is necessary that the receiver of a message checks that the cryptographic checksum is correct.

16.4.4 Generation of Messages

Generation of messages by an adversary is a common attack to RFID systems. Attack objectives include the impersonation of holders of a genuine RFID transponder. Note that an RFID system is based upon machine-to-machine authentication. An adversary that has gained possession of a genuine RFID transponder can impersonate the genuine owner of the RFID transponder. Most prominent is the *replay attack* that retransmits a sniffed original RFID message at a later time and possibly in another context.

In order to detect unauthorized generation of messages it is required that sender and receiver use cryptographic message authentication codes and message integrity. Hereby, it is necessary that the receiver of a message checks that the message authentication code is correct and the message integrity is fulfilled. Unauthorized transfer of ownership of transponders requires additional measures such as human user authentication based on knowledge or biometrics.

16.4.5 Relay of Messages

Authentication with RFID transponders usually assumes that the transponder's owner is in the near vicinity of an RFID reader if it communicates with the RFID transponder. A relay attack aims to fool RFID systems acting on this common assumption.

A relay attack does not manipulate the communication data, but it establishes an additional radio link to enable the RFID communication between a distant RFID reader and a transponder (see Figure 16.2). Thereby, the adversary enables communication between a genuine RFID reader and a genuine transponder that are not positioned at a nearby communication distance. For the relay attack it is required that the adversary locates an emulating reader nearby the genuine transponder and that the adversary has an emulating transponder in the communication range of a genuine reader. By establishing a relay, an adversary makes use of a remote transponder, without the owner's awareness of the remote transponder. Evidence on a practical relay attack can be found, for example, in [10, 6]. A relay attack is also known as *mafia fraud* and *chess grandmaster problem* in literature and applies to all wireless communication systems.

Relay of messages can be prevented similarly to non-destructive denial-of-service attacks in Section 16.4.2: the owner of an RFID transponder can shield the RFID transponder with aluminium foil or make use of a jamming transmitter.

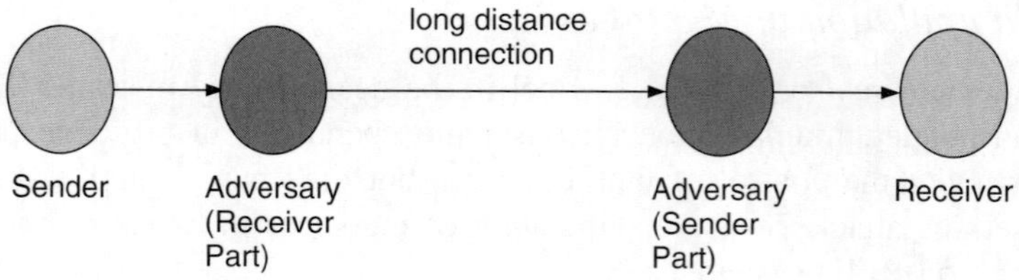

Figure 16.2 A relay attack. An adversary relays messages between distant sender and receiver.

Note that detection of relay attacks cannot solely be achieved using cryptographic schemes. Time-of-flight measurements of signals become crucial. Electromagnetic waves propagate with the speed of light c, which is approximately $c = 3 \cdot 10^8$ m/s. The spatial extension Δr of an electromagnetic field after a time Δt is given as $\Delta r = c \cdot \Delta t$. A location which is 3 meters away from the origin is reached after 10 nanoseconds. Implementing restrictive time-outs on the reader are a low-cost variant for preventing relay attacks.

Brands and Chaum [11] proposed distance bounding protocols. The basic idea is as follows: A series of rapid bit exchanges becomes part of a cryptographic protocol. After all bits are exchanged, the protocol completes by authenticating that in fact all bits have been correctly received. A distance-bounding protocol dedicated to RFID systems is proposed by Hancke and Kuhn [12]. The suitability of distance bounding strongly depends on the frequency of the bit exchange. In the low frequency range of 100 kHz, the time of one clock cycle corresponds to covering a distance of 3000 m at the speed of light so that distance bounding is of very limited practical use. In the high frequency range, distance bounding becomes reasonable in theory. However, to our best knowledge, distance bounding has not yet been implemented in any practical RFID system.

16.4.6 Tracking and Hotlisting

If RFID transponders are associated with human individuals, RFID systems may be in conflict with privacy requirements or laws. This is because any RFID transponder can be activated by a reader without notifying the human individual. As response to the activation, standard RFID transponders reply with a Unique ID (UID). A UID is not only specific to a specific product class, but also to each unique item of the specific product class. Thereby, each UID can be potentially linked to a human individual. In a pervasive environment that is equipped with RFID readers, *tracking* of transactions and movements of humans individuals is a matter of fact. *Hotlisting* is another threat to privacy in this context. Here, the adversary aims to be alarmed if one transponder of a targeted group of transponders shows up in an RFID reader of the adversary.

The technical possibilities create strong concerns about privacy and anonymity in RFID systems. Particular conflicts with the right of self-determination occur if human individuals are obligated to wear RFID tags, for example, at work, in hospital, or while traveling. Note that these concerns apply to the overall RFID system and especially to the databases held in the back-end system.

Preventing the communication between RFID transponder and possibly unauthorized readers is one approach to this tracking and hotlisting problem. These privacy solutions are in fact identical to the denial-of-service attacks in Section 16.4.2: An owner of an RFID transponder can destroy the RFID transponder, shield the RFID transponder with aluminium foil, or make use of a jamming transmitter. For example, the U.S. electronic passport has a metal shielding in its cover.

A low-level solution for privacy enhancement is a randomized UID that is randomly generated by the transponder itself so that the UID cannot be used as an unique identifier anymore. This solution requires a suitable random number generator (see Section 16.5.1), for example, the random number should not solely depend on a program counter. Such a solution is, for example, already incorporated into German e-passports. Authentication and identification mechanisms need then to be incorporated in cryptographically secured protocols.

From the user perspective, four different approaches are summarized by Spiekermann and Evdokimov [13]:

1. **Killing Scheme:** Users physically or logically deactivate RFID tags. The software based kill function is included in some EPC tags.
2. **On-Tag Scheme:** Users trust the privacy management of the RFID system. An RFID tag ensures that access to tag's data is restricted to genuine RFID readers by the use of security functions authentication and encryption.
3. **Agent Scheme:** Users delegate privacy management to their own privacy agent, for example, a smart phone that enforces a user-defined access control list.
4. **User Scheme:** Users authorize each individual access request. Therefore, an RFID tag requires a successful user authentication, for example, with a password, before it allows an RFID reader to access its data.

16.4.7 Cloning of Transponders

Cloning is a high risk to almost all RFID service providers. Cloning means that an adversary produces emulators of a genuine RFID transponder that behave identically and hence cannot be distinguished from the original transponder.

Cloning is easy if there is no cryptographic protection and an adversary can eavesdrop on static message exchanges. A clone is then a simple emulator of an RFID transponder that responds to the requests of an RFID reader with the same static response as the genuine transponder.

In order to defend against fraud, first of all cryptographic mechanisms are needed, especially in an offline RFID system that only has limited fraud detection capabilities. However, even if an RFID system is cryptographically secured, the secret cryptographic keys can potentially also be compromised either by a cryptanalytic attack (see Section 16.4.8) or an implementation attack (see Section 16.4.9). In an online RFID system organizational countermeasures can be implemented, for example, *blacklisting* of transponders is frequently used to block specific transponders that are suspected of fraud.

16.4.8 Cryptanalytic Attacks

Cryptanalytic attacks exploit weaknesses in a cryptographic algorithm or protocol. Cryptanalytic attacks are purely mathematical attacks, given exchanged messages they aim to find the associated secret keys used. Cryptanalytic attacks are possible because of

- weaknesses in the cipher design; or
- insufficient key space.

Weaknesses in the cipher design are assumed to happen if a proprietary cipher design has not been extensively analyzed. It is good practice for a new cipher design to undergoe comprehensive cryptanalysis that should be conducted by many independent cryptologists. A brute force attack completely searches the entire key space until the correct key is found. Current expectation is that about 80 key bits can be cracked with highly sophisticated key search engines by intelligence agencies today [14]. As a consequence, key sizes below 80 key bits can be seen as susceptible to brute force attacks.

Famous cryptanalytic attacks have been conducted on proprietary cryptographic algorithms used in widely employed RFID systems in the last years. Cryptanalytic analysis can be performed with the RFID reader, the RFID transponder or both. Some proprietary algorithms have been known to be susceptible because the key space is obviously chosen too small and the security of these cryptographic algorithms solely depends on the secrecy of the cryptographic algorithm. Because of ongoing reverse-engineering efforts by universities as well as unknown hackers it is not a good choice to base the security on secrecy of proprietary algorithms. Three examples are presented in more detail below.

Example 16.4.1 The Digital Signature Transponder (DST) by Texas Instruments. In 2005, a group of researchers from the United States [15] succeeded in completely reverse-engineering the design of the proprietary stream cipher used in the DST transponder produced by Texas Instruments. Their work started from a rough sketch of the cipher that was presented by Texas Instruments. Because the key-length of that cipher is only 40 bit, the cipher is highly susceptible to brute force once the ciphering algorithm is disclosed. The authors demonstrate that any 40-bit key can be determined given two responses of the transponder to two arbitrary challenges in less than one hour. For this brute force attack, the authors used sixteen FPGAs. The DST transponder is widely used as an ignition key for electronic car immobilizers (e.g. in Ford models) and as payment transponder (e.g. in the SpeedPass™ system).

Example 16.4.2 The KeeLoq Algorithm by Microchip. In 2008, a group of researchers from Belgium and Israel [16] revealed the most efficient key recovery attack on the block cipher KeeLoq. KeeLoq is a proprietary cipher with a 64-bit key length. The attack exploits weaknesses in the KeeLoq design. In total, the attack requires 2^{16} chosen plaintexts and about three days' computation time on 64 CPU cores. The chosen plaintexts can be acquired in about one hour using a transponder running the "Identify Friend or Foe" protocol, that is, a challenge-response protocol between the RFID reader and the RFID transponder. Even worse, knowledge of one transponder key is sufficient to compromise the corresponding master key in some key derivation schemes that are used in practice. The KeeLoq algorithm is widely used in keyless entry systems, car locks, and alarm systems.

Example 16.4.3 The CRYPTO1 Algorithm by NXP Semiconductors. The CRYPTO1 algorithm is a proprietary stream cipher with 48-bit key length. First details were revealed by an implementation attack conducted by researchers from the United States and

Germany [17]. However, these findings did not lead to a complete disclosure of the cipher. Follow-up cryptanalytical work done by researchers in the Netherlands revealed the overall cipher design [18]. Furthermore, several vulnerabilities in the cipher design were found. The most serious attack recovers the secret key from just two authentication attempts with a genuine reader in less than a second without pre-computation [18]. Currently, the most efficient attack on a genuine card requires about 300 queries (or approximately 10 seconds) to find the first key [19]. For many years the CRYPTO1 algorithm has been shipped with MIFARE products. Most MIFARE products are used in public transportation and facility management. According to NXP, more than one billion MIFARE cards have been sold, covering about 85% of the contactless smart card market [18].

These examples show that only established cryptographic algorithms should be used for commercial RFID systems.

16.4.9 Physical Implementation Attacks

Once security designs are implemented in hardware or software, the resulting implementation becomes a physical object that is susceptible to physical attacks. Physical implementation attacks include a battery of attacks on the internal construction of RFID components. For a more detailed description, see [20].

In an RFID system, physical attacks can be directed at each system component. In the end-user environment, this concerns the RFID reader and the RFID transponder. As the RFID reader often stores master keys in its memory, it may be an attractive target of attacks as a successful implementation attack may compromise the security of the entire RFID system.

Implementation attacks are less general than other attacks, for example, cryptanalytical attacks in Section 16.4.8 in consideration of the impact's results. A successful cryptanalytic attack applies to all products that are secured with the broken cryptographic primitive. Results on an implementation attack remain limited as they are specific to the targeted implementation. Different implementations have usually a different susceptibility on the same implementation attack. Nevertheless, some classes of implementation attacks exist that are a threat to almost all implementations.

An implementation attack is called *active* if the adversary actively interferes with the cryptographic implementation, for example, tampers with the implementation under extreme physical conditions. If active interactions with the cryptographic implementation are completely absent and the cryptographic module is operated in its intended environment, the attack is said to be *passive*.

An implementation attack is said to be *invasive* if it breaches the internals of an implementation, for example, penetrates an integrated circuit with physical means. The term *semi-invasive* is also known for some easy preparation steps that do not breach the internals of a cryptographic implementation but some outer coverings, for example, removing the package of a chip. Otherwise, an attack is *non-invasive*.

Passive physical attacks exploit the inherent physical leakage of the cryptographic module. Most prominent is *side channel analysis*. Side channel analysis reveals critical information by measuring the timing, the electromagnetic emanation, or the power consumption of a cryptographic operation.

Active implementation attacks include:

- non-invasive tampering attacks due to anomalous environmental conditions;
- semi-invasive and invasive tampering attacks due to light and electromagnetic radiation;
- invasive reverse-engineering attacks in order to disclose the internal construction of a cryptographic module for further vulnerability analysis;
- invasive penetration attacks in order to probe internal communication lines or test pads of the security component, for example, microprobing on bus lines of a chip;
- invasive penetration attacks in order to destroy internal units, for example, a random number generator;
- invasive modification attacks in order to modify internal units of the security components, for example, security relevant internal memories.

Implementation attacks are often carried out in laboratories and the damage to RFID components is usually accepted by the adversaries. Cloning constitutes the highest risk as result of successful implementation attacks. The countermeasures introduced in Section 16.4.7 apply.

16.4.9.1 Reverse-Engineering

Reverse-engineering is a generic attack on the design of any chip, for example, used in RFID systems. Reverse-engineering efforts are high and only affordable by highly motivated attackers. Usually, these efforts exceed the monetary value stored in an RFID chip by far. Because of that, the objective of reverse-engineering is the disclosure of proprietary designs. Note that knowledge about such designs can be used to produce clones in a high volume. However, reverse-engineering remains a significant threat to RFID reader if master keys of an RFID system are stored within. Unfortunately, there is not much known about chip reverse-engineering in the scientific literature.

Example 16.4.4 Reverse Engineering of the CRYPTO1 Algorithm by NXP Semiconductors. As already introduced in Example 16.4.3 the first details of the proprietary cipher CRYPTO1 were revealed by an implementation attack conducted by researchers from the United States and Germany [17]. For this analysis, the first step is to dissolve the plastic embeddings of an RFID chip. This can be done, for example, with acetone or fuming nitric acid and is the easy step. The chip on the Mifare Classic is found to have a total area of roughly one square millimeter. One quarter of the chip area is used for 1K of flash memory, another quarter is used for the radio front-end and connectivity. The rest of the chip area includes the digital logic including the implementation of the proprietary cipher. Once the blank chip is unpacked, Nohl *et al.* [17] make use of mechanical polishing steps to remove the single layers of the chip, one after the other. This approach is new and replaces chemical etching, for example, with hydrofluoric acid. The MIFARE chip was found to include six layers. The researchers took photos using a standard microscope of each layer and revealed about 70 different types of gates. Following the assumption that the cryptographic circuit contains a 48-bit register with some XOR gates it could be identified. Nohl *et al.* [17] used template matching of the photos in order to find other instances of the same gate in the circuit. Another problem was to identify the metal connections between gates in order to disclose the wiring of the circuit. Also this problem could be mainly solved using automated tools.

Countermeasures against reverse-engineering are based on obfuscation of the chip design. This may be achieved by use of special logic styles and a full custom layout. Interconnections may be made more complicated so that they are not easy to analyze by visual inspection. However, automated attacks as described by Nohl *et al.* [17] may be still applicable.

A generic but until now mostly theoretical countermeasure to cloning is a Physical Unclonable Function (PUF) that is originally due to Pappu [21]. A PUF is a physical object that is inherently unclonable since it contains many uncontrollable parameters during production. When a stimulus is applied to the object, it reacts with a measurable response. Further, this structure changes its behavior if it is tampered with. Several directions of PUF designs are studied in the moment. Among them, the most interesting to chip security is a PUF coating invented in Tuyls *et al.* [22].

16.4.9.2 Side Channel Analysis

Side channel analysis exploits the inherent physical leakage of a cryptographic device. This research area was initiated in 1996 by Paul Kocher who invented timing analysis on cryptographic algorithms [23]. The publication on differential power analysis (DPA) [24] has, however, attracted much more attention, especially from the growing smart card industry. DPA exploits the fact that any microchip in standard CMOS logic produces a data-dependent leakage signal, for example, switching of a gate requires power while holding the same state consumes almost no power. The key idea of DPA is to correlate the externally measured power consumption with some predictable internal state of the cryptographic implementation, for example, with one bit that depends on known data and a subkey hypothesis. As processing of the device is deterministic, one can expect that after a sufficient number of measurements the correlation signal becomes significant for the correct subkey hypothesis. DPA is a divide-and-conquer attack that is applied on subkeys of, for example, six bits in DES and eight bits in AES. Some repetitions of DPA are needed for the overall disclosure of a secret key. Side channel analysis is a signal detection problem. In practice, a certain number of measurements is needed to eliminate the noise sufficiently and to reveal the side channel leakage.

Example 16.4.5 Side Channel Analysis of KeeLoq Transponders. In 2008, researchers from Germany and Iran succeeded in breaking commercially available implementations of KeeLoq that are used, for example, for garage doors and car locks [25]. KeeLoq was found to be implemented in a hardware module of Microchip that is embedded in the RFID transponder (transmitter) and in software on a read-protected PIC microcontroller that is embedded in the RFID reader (receiver). The researchers adapted the differential power analysis attack to KeeLoq. For the attack on the transmitter, about ten measurements suffice. The receiver contains the manufacturer key, that is, a master key. To attack the KeeLoq implementation on the receiver, about one thousand measurements were needed to recover the manufacturer key [25]. Kasper *et al.* [26] presented an improved simple power analysis attack that suffices with only one single measurement on the receiver. Once the manufacturer key is known, one to two eavesdropped communication sessions between a transmitter and receiver are sufficient to compromise the corresponding transmitter key by deriving it from the master key. Different kinds of fraud scenarios such as cloning and denial-of-service result from compromised keys.

Since 1999, side channel analysis has been an established subject matter in cryptanalysis. A valuable overview on different variants of side channel analysis attacks and countermeasures can be found in [27].

Roughly stated, countermeasures are divided into two main classes:

- *Reducing the signal-to-noise level* can be achieved by additional noise sources on the chip, unstable internal clocking, random processor interrupts, internal capacitances, randomization of execution sequence, or random delays. Most promising are special dynamic and differential logic styles in order to equilibrate data dependent leakage already on the gate level, however, at the cost of roughly quadrupling the chip area and power consumption and a full custom layout. These countermeasures do not prevent differential power analysis, but strengthen the security level of the implementation because the adversary needs much more efforts to extract the side channel leakage.
- *Preventing any predictable internal state in the cryptographic implementation* can be achieved by using randomization techniques such as masked logic gates, masked data representation on busses and memories, or at the algorithmic level that is called *masking* if applied at symmetric ciphers and *blinding* if used at asymmetric primitives. However, higher-order side channel analysis remains a possible threat to implementations.

In practice, a combination of both strategies is recommended.

16.4.9.3 Fault Analysis

A key recovery attack based on erroneous cryptograms was invented by Boneh *et al.* [28], but remains purely theoretical for many years. Such kinds of attacks are known as *Fault Analysis* on cryptographic implementations. Practical evidence on the power of the attack was firstly demonstrated by a research group of Infineon [29] who used spikes on the power supply of smart cards for fault induction. At the same time, Skorobogatov and Anderson [30] proposed flash light to induce faults in RAM memory. Today, laser set-ups are used for fault induction in practice [31].

The first application to RFID tags in the HF (High Frequency) and UHF (Ultra High Frequency) band was recently done by Hutter *et al.* [32]. Therein, the write operation to the persistent memory EEPROM was studied in the presence of faults. For fault induction, the authors temporarily applied antenna tearing, electromagnetic interferences, and optical fault induction. For antenna tearing, the RFID chip was separated from its antenna in order to temporarily interconnect the antenna pins of the chip. Electromagnetic interferences were achieved by discharging a high-voltage generator. The following effects are observed as a result of faults:

- The transponder confirms the correct completion of write operation, but in fact the write operation was not carried out.

- The transponder confirms the correct completion of write operation, but in fact a faulty value was written to the EEPROM.

A valuable survey of hardware and software countermeasures can be found in [31]. The generic approach is redundancy, either in hardware by doubling critical circuits and checking their results or by double computation and comparison. Auxiliary countermeasures for the RFID chip include light detectors, supply voltage detectors, frequency detectors, active shields, and execution randomization. To detect errors in memory, checksums can be implemented. Extensive tampering with fault inductions can be inhibited by implementing a disable instruction in the RFID chip that is automatically activated when a predefined number of fault inductions has been detected.

16.5 Lightweight Cryptography for RFID

The feasibility of security and privacy mechanisms for low-end systems is of particular research interest. The EPC Class-1 Gen-2 standard [33] provides the base point. It is generally assumed that the current mechanisms in the EPC standard are (too) weak: the UID (here: Electronic Product Code) can be eavesdropped and is only protected with the checksum CRC16. Furthermore, the 32-bit access password can be computed by an eavesdropper with simple XOR operations. Additional functions providing privacy protection, authentication, encryption, and message integrity are desirable.

Secure and privacy protecting RFID systems require security mechanisms and protocols and a large number of proposals have been made during the last few years. The complete system (including distributed tags, readers and back-end components) has to be considered but it is generally assumed that at least some fundamental security mechanisms and cryptographic primitives are needed on the tag. It is a major challenge to design RFID chips with cryptographic functions considering their various restrictions [34]:

- Low cost Integrated Circuits (ICs) with a small chip area so that hardware efficiency for the cryptographic logic and the memory (registers) are important.
- Low peak and average power consumption.
- Restricted operation performance on the tag, low cryptographic throughput, latency requirements.
- Slow data rates over the radio interface.

On the other hand, the cryptographic operations handle only small amounts of data and require only a moderate level of security in terms of resistance to cryptographic and implementation attacks. Generally speaking, cryptography is the study of mathematical techniques related to aspects of information security such as confidentiality, data integrity, entity authentication, and data origin authentication [35]. Lightweight cryptography considers resource-optimized algorithms for constraint devices like RFID transponders. This section deals with the various cryptographic primitives which are used for RFID security protocols. The focus is on state-of-the-art algorithms which facilitate the development of secure and privacy-enabled RFID systems. We focus on medium and low cost tags and take the various implementation and operation restrictions into account.

The exact conditions depend on the IC and RFID technology, cost or chip area restrictions and intended applications. For typical low cost RFID systems produced on a 0.35 μm CMOS process, no more than 5000 GEs and an average power of 22.5 μW (HF) respectively 4 μW (UHF) are available for cryptographic functionality [36]. Typical RFID systems communicate with data rates between 5 and 106 kbps (HF) respectively 320 kbps (UHF) and require a related cryptographic throughput, at least for encryption and message authentication. With a typical clock rate of 100 kHz, around 1000 clock cycles or 10 milliseconds for a single cryptographic operation (e.g. a 128-bit AES encryption) would provide a throughput of only 12.8 kbps. The strict latency requirements (time to answer requests) require special handling, for example, a separation of requests and responses [37].

A security strength of 80 bits is generally accepted for RFID devices, that is, successful brute force attacks would require on average 2^{79} operations. It should be noted that additional countermeasures against implementation and side-channel attacks are necessary (see Section 16.4.9).

16.5.1 Random Number Generators

Random numbers are often required for privacy-enhanced identification and secure authentication protocols. A Random Number Generator (RNG) or, equivalently, a Random Bit Generator (RBG) is a device or procedure which produces a series of numbers or bits which are statistically independent and identically distributed. A potential bias in the output sequence can be removed with de-skewing techniques but true randomness is hard to realize and also difficult to prove. Hardware Random Number Generators use physical sources and there are propositions for an oscillator-based RNG in RFID tags [38].

A Pseudo Random Number Generator (PRNG) or Deterministic Random Bit Generator (DRBG) is an algorithm which takes a seed value as input and generates an output sequence that appears random. It is cryptographically secure if there exists no polynomial-time algorithm which predicts the next output bit from the previous bits with a probability significantly greater than $\frac{1}{2}$. This also implies that the output cannot be distinguished from a true random sequence. There are secure pseudo random bit generators, for example, the Blum-Blum-Shub generator which performs modular squaring operations and relies on the complexity of the factorization problem. In practice, fast generators without a strong security proof are used. They are subjected to batteries of statistical tests and a PRNG is rejected if the output is revealed as non-random.

The EPCglobal Class-1 Gen-2 standard specifies that tags should implement a 16-bit PRNG with the following properties [33]:

- The probability that any 16-bit pseudo random value shows up as the next output is between $0.8 \cdot 2^{-16}$ and $1.25 \cdot 2^{-16}$.
- Among 10000 tags, the probability that any two or more tags generate the same sequence of 16-bit numbers is less than 10^{-3}. The probability of random collisions (birthday paradox) then implies that the internal state must consist of at least 36 bits, which is a weak requirement (more than 100 bits are common).
- The probability of predicting the next pseudo-random number from the previous outputs is less than $2.5 \cdot 10^{-4}$ (the probability of a random hit is $1.5 \cdot 10^{-5}$).

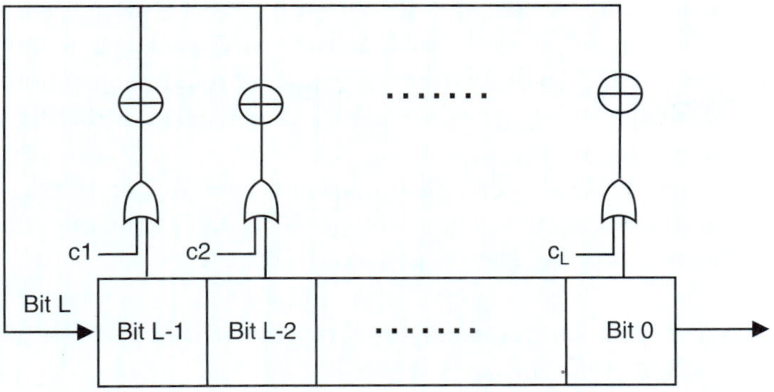

Figure 16.3 Operation of a Linear Feedback Shift Register of length L.

The PRNG of common commercial tags like MIFARE Classic has not been published but was recently reverse-engineered, see Example 16.4.3, thereby breaking security by obscurity. The PRNG was verified to be a 16-bit LFSR with a constant initial value after power-up of the tag [18]. Since random numbers used in the MIFARE authentication protocol are 32-bit long, the first half determines the second half.

A popular and hardware-efficient type of PRNG are Linear Feedback Shift Registers (LFSRs). An LFSR of length L uses L register bits $[x_{L-1} \ldots x_0]$. The register state is updated (each time the LFSR is clocked) by an XOR operation involving certain tap bits of the register, followed by a right shift (see Figure 16.3):

$$x_L = c_1 x_{L-1} \oplus c_2 x_{L-2} \ldots \oplus c_L x_0 \qquad [x_L \ldots x_1] \to [x_{L-1} \ldots x_0]$$

An LFSR can produce an output sequence of maximal period $2^L - 1$, but $2L$ consecutive output (feedback) bits suffice to set up a system of linear equations (modulo 2) and to compute the tap positions (i.e. the feedback coefficients c_i). Then the following output bits can be predicted. Plain LFSRs are hence inadequate for cryptographic purposes but their security is often augmented by adding non-linear functions. Different methods exist to destroy the linear properties of LFSRs [35]:

- Applying a non-linear filter function to the extracted register bits.
- Combining the output of several LFSRs by a non-linear function.
- Controlling the clocking of one (or more) LFSRs by one (or more) LFSRs (irregular clocking).

General Feedback Shift Registers are also employed as building blocks for stream ciphers (see Section 16.5.3). It is a challenging task to analyze the security of such generators. In the past, successful attacks have been launched (e.g. against the clock-controlled GSM A5 generators).

As an example of a dedicated lightweight PRNG satisfying the EPCglobal Class-1 Gen-2 requirements, Peris-Lopez *et al.* [39] developed the 32-bit generator *LAMED* which can be implemented with around 1600 gates and requires less than 200 clock cycles for each output, that is, less than 2 ms at a 100 kHz clocking frequency. LAMED is seeded with an initialization vector and a secret key (32 bits respectively) which can be set during manufacturing of the tag. Three additional registers are used and recursive output is generated by a series of XOR, addition (mod 2^{32}) and bit rotation operations. This PRNG passed the standard batteries of statistical tests (including Diehard and NIST), but a further security analysis is necessary.

Random Bit Generators can also be based on other cryptographic primitives, in particular hash functions and block ciphers. The American NIST has published recommendations for DRBG and specified four different generators [40]:

1. Hashing a counter with an approved hash function (SHA-1, SHA-256, SHA-386, SHA-512, see Section 16.5.4).
2. Iterated HMAC computations using an approved hash function as in 1.
3. Encrypting a counter with an approved block cipher (AES or Three-key 3DES, see Section 16.5.2).
4. Multiplication of a point on an elliptic curve and extracting coordinate bits (see Section 16.5.5).

Only generator 4. is based on a hard number theoretic problem (Elliptic Curve Discrete Logarithm Problem). The other three generators rely on statistical properties of hash functions and block ciphers. All DRBG need a source of entropy input to seed the mechanism and reseeding is necessary after extracting a certain number of bits. As we will see below, only generator 3. is currently feasible for very restricted hardware, although this may change in the future.

16.5.2 Block Ciphers

Block ciphers belong to the most fundamental cryptographic primitives. A block cipher E_k of length n over the binary alphabet is a bijective map $E_k : \{0, 1\}^n \to \{0, 1\}^n$ indexed by a key k. Data of arbitrary length is then enciphered by padding, decomposing the plaintext into blocks of n bits and applying the function E_k. An operation mode can add dependencies on previous ciphertext blocks, initialization vectors or counters. The common Cipher-Block Chaining (CBC) mode produces the ciphertext blocks $c_1, c_2, \ldots$ as follows:

$$c_i = E_k(p_i \oplus c_{i-1})$$

Here $p_1, p_2, \ldots$ denote the plaintext blocks and $c_0 = IV$ is a fixed initialization vector.

Following general design principles of Claude Shannon, a good cipher should provide confusion and diffusion. Confusion refers to a complicated (in particular non-linear) relationship between the ciphertext and the key for a given plaintext. It is usually achieved by substitution operations (S-Boxes). Diffusion refers to the dependency between plaintext and ciphertext bits. Flipping a single bit of a plaintext block should change all ciphertext bits with a probability of $\frac{1}{2}$ (Strict Avalanche Criterion). It can be realized with linear or affine mixing transformations.

There exists a large variety of block ciphers. In practice, the former and the current American FIPS/NIST standards, DES (Data Encryption Standard) and AES (Advanced Encryption Standard) are of particular relevance.

DES is a Feistel cipher with 16 rounds which transforms a plaintext block $p = (L_0, R_0)$ into pairs (L_i, R_i) of $2 \cdot 32 = 64$ bits:

$$L_i = R_{i-1} \qquad R_i = L_{i-1} \oplus f_{K_i}(R_{i-1}) \qquad i = 1, 2, \ldots, 16$$

In each of round of a Feistel cipher, one half of the state is transformed using an inner Feistel function and an XOR operation. $f_{K_i} : \{0, 1\}^{32} \rightarrow \{0, 1\}^{32}$ denotes the inner Feistel function of DES, which consists of a linear extension operation, followed by XOR-ing the key K_i, then applying the DES S-Boxes and finally permutating the bit positions. K_i is a 48-bit round key which is derived from the 56-bit DES key. The ciphertext (R_{16}, L_{16}) then results from the last Feistel round.

DES can be efficiently implemented with around 2300 gates [41] but offers only limited security (less than 56 bits and successful attacks within several days using special hardware). There are strengthened variants of DES with up to three keys k_1, k_2, k_3 which provide more than 80 bits of security:

- Three-Key Triple DES: $3DES_{k_1,k_2,k_3}(p) = E_{k_1}(E_{k_2}^{-1}(E_{k_3}(p)))$.
- Two-Key Triple DES: $3DES_{k_1,k_2}(p) = E_{k_1}(E_{k_2}^{-1}(E_{k_1}(p)))$.
- DESX: $DESX_{k_1,k_2,k_3}(p) = k_3 \oplus E_{k_1}(k_2 \oplus p)$.

The original DES S-Boxes require a significant amount of memory and hence substantial chip area. An optimized variant DESL was developed [41], replacing the eight original DES S-Boxes with a single new S-Box. The cipher DESXL, which combines DESX and DESL, had been implemented in serialized version for constrained environments with only 2168 gates [34]. It requires 144 clock cycles and has acceptable current and power consumption at 100 kHz on a 0.18 μm process. The performance is competitive compared to AES (see below), so that DES or its variants could be employed for RFID applications with lower security requirements.

AES (Rijndael) has been designed for efficient implementations and a high level of security. For RFID systems, we focus on the version with a minimal key length of 128 bits. AES is a substitution-permutation network and uses a 128-bit state, which is organized as a 4×4 matrix of bytes. A byte corresponds to an element in the field $GF(2^8)$. This finite field consists of 256 elements and is defined as the residue classes of the polynomials over the binary field $GF(2)$ modulo a fixed irreducible polynomial of degree 8. The state is initialized with the plaintext block and gives after ten transformation rounds the ciphertext block. Each round (except the initial and the last) consists of the following operations (see Figure 16.4):

- SubBytes (the S-Box of AES): multiplicative inversion in the field $GF(2^8)$, followed by an affine transformation.
- ShiftRows: a rotation of the state rows.
- MixColumns: a $GF(2^8)$-linear transformation of the state columns.
- AddRoundKey: XOR-ing the state with the round key, which is derived from the AES key.

AES can be implemented with 3503 gates and consumes 3.0 μA at 100 kHz and a voltage of 1.5 V on a 0.35 μm CMOS process technology. It requires 1044 clock cycles (10 ms) to encrypt one block [36]. Although a theoretic attack (XSL) has been published exploiting the algebraic simplicity of AES, the cipher still provides 128 bits of security. The cipher represents the current industry standard and AES as well as DES and Triple-DES are currently used for moderate-cost (HF-)tags.

We also present two specifically lightweight block ciphers, the Tiny Encryption Algorithm and PRESENT. They provide better performance than AES but are less recognized in the industry and their security analysis has been less exhaustive.

The Tiny Encryption Algorithm (TEA) and also its revised version (XTEA) is a 64-bit cipher with a 128-bit key. The ciphers have a Feistel structure with a suggested number of 64 rounds. The left- and the right-hand side of the state are transformed with the following hardware-efficient operations:

- XOR;
- bit-rotation;
- addition modulo 2^{32}.

XTEA has an improved key-scheduling and the cipher has safely more than 80 bits of security despite a known differential attack for a reduced number of rounds. XTEA was implemented with 2636 gates and requires 705 clock cycles for a 64-bit block and consumes 3.86 μA at 100 kHz [36].

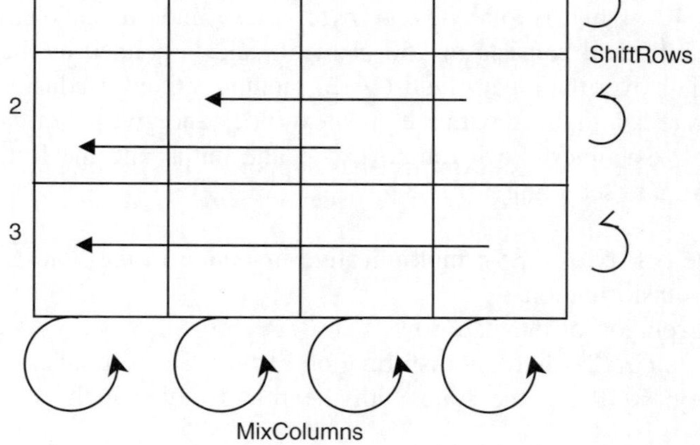

Figure 16.4 Schematic description of AES round steps on the 4 × 4 state matrix of bytes. SubBytes and AddRoundKey are applied to each matrix entry individually.

PRESENT-80 is a 64-bit cipher with an 80-bit key and was developed for extremely constrained environments [42]. The 64-bit state is initialized with the plaintext and then updated in 31 rounds with the following steps:

1. addRoundKey: XOR-ing the state with the round key, which is derived from the PRESENT key.
2. sBoxLayer: applying a 4-bit S-Box to each of the 16 segments.
3. pLayer: a bit-permutation.

The cipher requires 1570 gates, 5 μW power on a 0.18 μm logic process and only 32 clock cycles for the encryption of a 64-bit plaintext block, giving a rather high throughput of 200 kbps at 100 kHz. A further security analysis of this interesting cipher deserves closer attention.

16.5.3 Stream Ciphers

Stream ciphers provide an alternative to block ciphers for symmetric encryption. They are sometimes considered as particularly efficient although this may not generally hold true. Instead of directly enciphering the plaintext, stream ciphers generate a *keystream* of bits which depends on the key, an initialization vector and potentially the plaintext. The ciphertext is then computed by XORing the plaintext and the cipher stream. Enciphering and deciphering are hence identical operations. Many stream ciphers which are currently in use are synchronous, that is, the keystream is generated independently from the plaintext. It is well known that stream ciphers can also be derived from block ciphers by employing certain operation modes (Cipher Feedback, Output Feedback and Counter Mode). Here the cipher stream is generated by encrypting initialization vectors or counters and recursively updating these parameters. Currently, AES in Counter Mode provides the baseline for native stream ciphers.

There exists a variety of stream ciphers but no general standard. The American NIST currently does not recommend a single stream cipher! The RC4 cipher has been in widespread use for many years but is inappropriate for constrained environments since the state contains a permutation of $\{0, 1, \ldots, 255\}$ which requires more than 12,000 gates.

Instead, Feedback Shift Registers are often used when ciphers for constraint hardware have to be constructed. But the proprietary combination of LFSRs can lead to insecure ciphers. The ECRYPT *eSTREAM* project has supported the development and cryptanalysis of new stream ciphers. The hardware-oriented *Profile 2* aims at "stream ciphers for hardware applications with restricted resources such as limited storage, gate count, or power consumption." A key length of 80 bits and an IV of either 32 or 64 bits must be supported. The security criteria require that:

- Any key-recovery attack should be as difficult as exhaustive search.
- The cipher stream should not be distinguishable from a random stream.

A stream cipher with these properties would offer full 80-bit security and also yield a cryptographically secure pseudo-random bit generator. But a strong security proof can hardly be achieved and trust in these ciphers has to be established by thorough cryptanalysis and statistical testing. At the end of a three-year process (2005–2008), three hardware-oriented ciphers remained in the projects' portfolio [43]: Grain v1, MICKEY v2 and Trivium.

The Trivium cipher is considered as leading the field. It combines a simple design (requiring 3090 gates), good performance (176 cycles to encrypt a 128-bit block) and low power consumption (0.68 μA at 100 kHz and a voltage of 1.5 V on a 0.35 μm CMOS process) [36]. Trivium is based on three LFSRs of length 93, 84 and 111, respectively. Two bits are extracted from each register and XORed in order to generate one output bit (Figure 16.5). The state of each register is then updated by an XOR operation of the following bits (see Figure 16.6 for the update of the second register):

- one specific bit of that register;
- two bits of another register;
- the result of an AND operation of the third- and second-last-bit of another register.

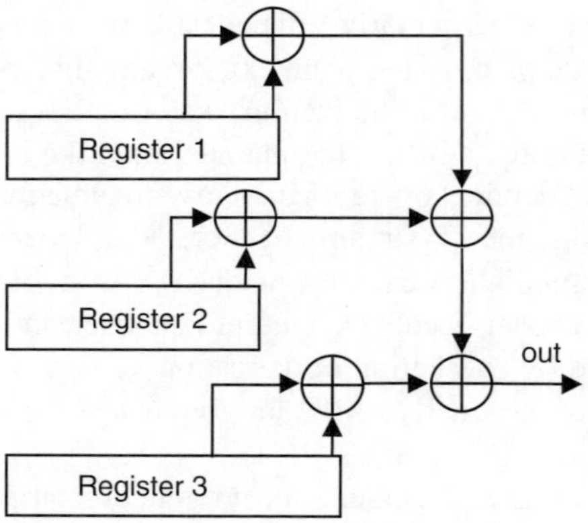

Figure 16.5 Trivium: Bit extraction.

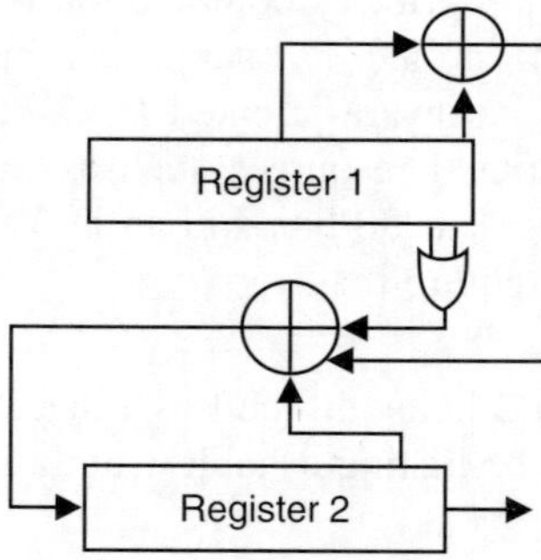

Figure 16.6 Trivium: Update of the second register.

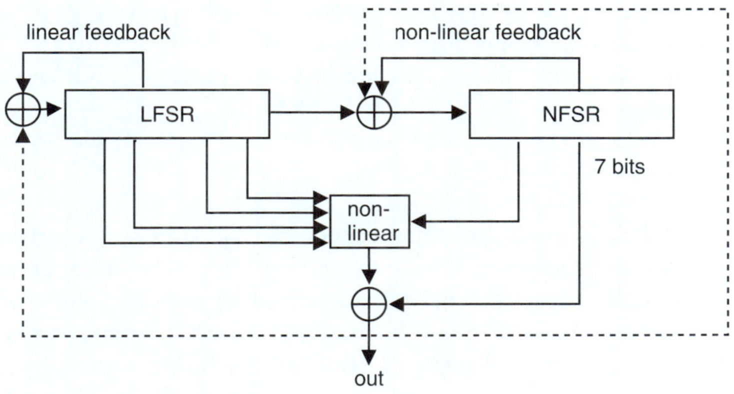

Figure 16.7 Operation of the Grain stream cipher.

Trivium is initialized by loading the 80-bit key and the 80-bit IV into the registers and setting the remaining bits to zero except the last three bits which are set to 1. Then the update operations is applied $4(93 + 84 + 111) = 1152$ times without using the output. The relatively long initialization phase requires 1603 cycles.

Grain has similar hardware properties as Trivium. The cipher was implemented on a 0.35 μm CMOS process technology and requires 3360 gates, 0.8 μA at 100 kHz, 1.5 V and 104 cycles for the encryption of 128 bit [36]. Grain combines a linear (LFSR) and a non-linear feedback shift register (NFSR), see Figure 16.7. The NFSR takes the LFSR output as additional input for its update. Four bits from the LFSR and one bit from the NFSR are extracted and feed a non-linear function which performs XOR and AND operations. The output bit is finally generated by XORing the result with seven bits from the non-linear register. During the initialization phase, the output bit serves as additional input to both registers.

The third cipher in the current eSTREAM portfolio is MICKEY v2 which shows a performance lying behind Trivium and Grain. MICKEY is a based on two feedback shift registers R and S with variable clocking. R acts as a standard LFSR, if a specific control bit is zero. Otherwise, the operation of R is equivalent to clocking the register a certain, high number of times. S is a non-linear feedback shift register and the feedback is influenced by a second control bit. The control bits are computed by an XOR operation of two bits extracted from both registers.

Further research on stream ciphers has to be conducted before new industry standards can be announced.

16.5.4 Hash Functions

Hash functions compute a digital fingerprint and play an important role in connection with message authentication, data integrity, and digital signatures. They are also used in many RFID security and privacy protocol proposals. A cryptographic hash function h maps input data x of variable length to a binary string $h(x)$ of fixed length (e.g. 160 bits) and should satisfy the following requirements [35]:

1. *Preimage resistant* or *One-way*, that is, for essentially all y it is computationally infeasible to find any preimage x such that $h(x) = y$.
2. *Second Preimage resistant*, that is, given x it is computationally infeasible to find a second preimage $x' \neq x$ such that $h(x) = h(x')$.
3. *(Strong) Collision resistant*, that is, it is computationally infeasible to find any x, x' with $x \neq x'$ and $h(x) = h(x')$.

Of course, 3. implies 2. Hash functions cannot be injective and therefore collisions certainly exist, but it is required that it should be very hard to find concrete collisions so that any input data x is almost uniquely characterized by its hash value $h(x)$. It should be noted that random collisions of a n-bit hash values occur after approximately $2^{n/2}$ values (birthday paradox). For example, a 160-bit hash function provides a maximum security of 80 bits.

Most common hash functions follow the Merkle-Damgard construction. They are based on a compression function

$$f : \{0, 1\}^k \times \{0, 1\}^n \longrightarrow \{0, 1\}^n$$

for some k and n (e.g. $k = 512$, $n = 160$). A message m is padded and decomposed into blocks m_i of bit-length k. The compression function produces a sequence of intermediate hash values h_i:

$$h_i = f(m_i, h_{i-1})$$

h_0 is a constant initialization vector, and the output of the last block gives the hash value $h(m)$.

After MD5 has been broken (meaningful collisions with controllable content), the most popular hash function is currently SHA-1 (160 bits). But since it was shown that SHA-1 collisions can be found faster than brute-force, the American NIST encourages the use of the SHA-2 family SHA-224, SHA-256, SHA-386 and SHA-512 with longer bit-lengths.

The SHA-1 compression function updates the internal state of five 32-bit words (= 160 bits) during 80 rounds which involve the operations XOR, AND, OR, rotation and addition modulo 2^{32}. In each round, a part of the message block is inserted. The SHA-2 compression functions have a similar structure but a larger internal state.

A low-power implementation of SHA-256 on a 0.35 μm CMOS process technology requires 10868 gates and consumes a current of 5.86 μA at 100 kHz with a supply voltage of 1.5 V [36]. A hash calculation on a 512-bit block takes 1128 clock cycles (11 ms). The results for SHA-1 are only slightly better (8120 gates) so that the SHA-functions currently do not satisfy the strict resource constraints of low-cost RFID tags.

Further research is necessary regarding secure and efficient hash functions. The American NIST is currently conducting a competition to develop and analyze new hash algorithms and a new standard (SHA-3) is planned for the year 2012.

16.5.5 Public-Key Cryptography

Public-key methods are an integral part of modern cryptography. Public-key cryptography supports non-repudiation, integrity protection, authentication, confidentiality and key exchange without previously established symmetric secret keys. It is based on asymmetric key pairs: a private key (e.g. for signature creation) and a public key (e.g. for signature verification). Public-key methods are generally more complex than symmetric-key operations. Often, the authenticity of public keys is assured with signed certificates that are issued by a trusted certification authority.

Over the last decade, the RSA encryption and signature algorithm have been widely used. RSA is based on the integer factorization problem: given an integer $n = pq$ of at least 1024 (better 2048) bit-length, it is currently computationally infeasible to compute the prime factors p and q. For RSA key generation, one selects large prime numbers p and q as well as exponents e and d with $ed \equiv 1 \bmod (p-1)(q-1)$. Then (e, n) forms the public and (d, p, q) the private part of an RSA key. The exponentiations $x^e \bmod n$ and $x^d \bmod n$ are inverse operations on the integers modulo n. They are used for encryption and signatures:

- $x^e \bmod n$ to encrypt a plaintext x or to verify the signature value x.
- $x^d \bmod n$ to decrypt a ciphertext x or to sign a message with hash value x.

Because of the memory and computing resource requirements for modular exponentiations, it is not feasible to implement RSA for a low-cost RFID tag. This also applies to other public-key algorithms which require modular exponentiations of large integers like ElGamal and DSA (Digital Signature Algorithm). These primitives can be implemented for microprocessor-based contactless smart cards with a cryptographic coprocessor.

Recent work investigated the feasibility of Elliptic Curve Cryptography (ECC) for RFID. ECC is based on the multiplication of points on an elliptic curve over a finite field, instead of exponentiation in the multiplicative group of a finite field or ring. An elliptic curve E over K is a non-singular cubic curve described by a Weierstrass equation

$$E: \; y^2 + a_1xy + a_3y = x^3 + a_2x^2 + a_4x + a_6 \qquad a_i \in K$$

For ECC, the elliptic curve is defined over a finite field of type $K = GF(p)$ for a prime p or $K = GF(2^n)$ for some integer n. It is recommended that n resp. the bit-length of p should be above 160. The points of an elliptic curve (i.e. all $(x, y) \in K^2$ satisfying the Weierstrass equation plus one extra point O at infinity) form an abelian group. Since E is defined by a cubic equation, a line through the points $P, Q \in E(K)$ intersects the curve at a third point R and we set $P + Q + R = O$, that is, $P + Q = -R$. Also, the tangent through a point $T \in E(K)$ intersects at another point $(-2T)$ on the curve. This is illustrated in Figure 16.8 over the real numbers, but the principle holds also true over other (e.g. finite) fields.

The main idea behind ECC is that points on an elliptic curve can be efficiently multiplied, but for given points P and $Q = aP$ of sufficient size it is very hard to compute the factor a (Elliptic Curve Discrete Logarithm Problem). The domain parameters of an elliptic curve cryptosystem consist of a finite field K as above, an elliptic curve (i.e. the coefficients of the Weierstrass equation), a base point $G \in E(K)$, the order n of the point G and the cofactor $\frac{\#E(K)}{n}$. These parameters have to be carefully chosen in order to obtain a cryptographically strong elliptic curve group so that the discrete logarithm problem is computationally intractable. Recommended elliptic curve domain parameters for various bit-lengths are published by the standardization bodies (e.g. ANSI, FIPS, IEEE).

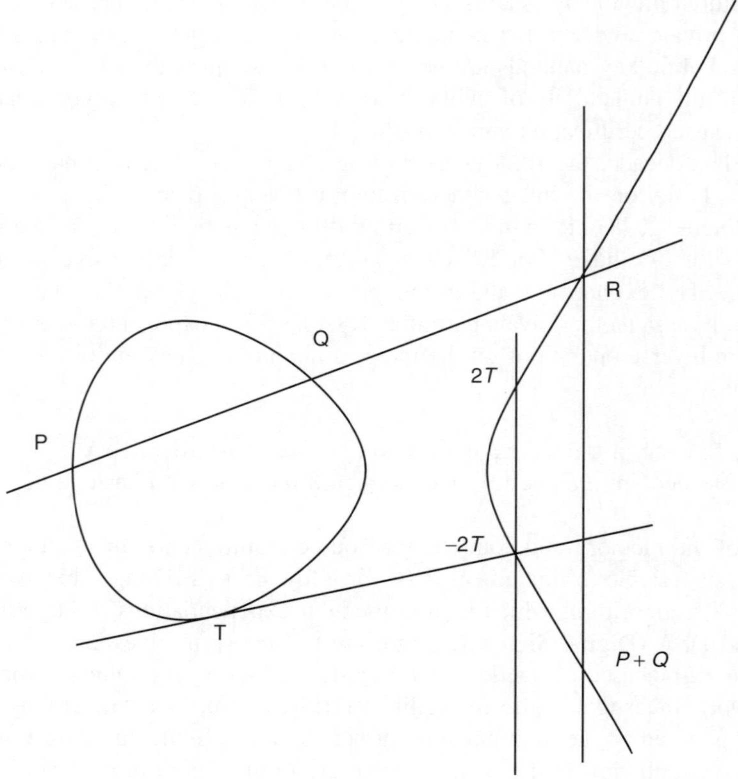

Figure 16.8 The elliptic curve $E : y^2 + y = x^3 - x$ over the real numbers; illustration of point addition $P + Q$ and multiplication $2 \cdot T$.

ECC is in particular used for digital signatures (ECDSA), but also for encryption, key exchange and for other security protocols. Since the point multiplications are relatively complex in terms of computational logic, process steps and memory, the resource requirements are well above those of standard symmetric algorithms. On the other hand, ECC provides a similar security level as other established asymmetric mechanisms (e.g. RSA) with significantly shorter key lengths. An implementation of ECC with 192 bits would require 23,600 gates and point multiplication needs more than 500,000 cycles and 18.85 μA at 100 kHz on a 0.35 μm CMOS process [35]. The high number of cycles would then rather require a clock frequency in the megahertz range. A lightweight 163-bit ECC implementation was developed [44] and a tag with elliptic curve cryptography (compliant with ISO 15693 and ISO 18000-3) is available in industry where the ECC functions are mainly used for anti-counterfeiting. ECC over the smaller field $GF(2^{131})$ has also been studied, which has a security level comparable to 858-bit RSA. An efficient implementation has been achieved by Batina *et al.* [45] requiring 8104 gates. On a 0.13 μm CMOS technology and with an operating frequency of 500 kHz, one point multiplication needs 106 ms and less than 30 μW. ECC is hence feasible for medium-cost RFID tags, active tags and wireless sensors and provides asymmetric cryptography at a reasonable level of security.

16.6 Conclusion

This chapter investigated generic and low-level aspects of RFID security and privacy. The basic notions of security threats, objectives and mechanisms were given and different categories of RFID systems were introduced. Security threats, attacks against RFID systems and possible countermeasures were discussed. The generic threats are eavesdropping, denial-of-service, manipulation, and generation of messages. Then specific attacks were presented such as relaying, cloning, and cryptanalytic attacks against widespread RFID systems. Furthermore, this chapter revisited attacks on the physical implementation of security mechanisms.

Lightweight cryptography forms the basis for future security-enhanced RFID systems. The standard cryptographic primitives including random number generation, block and stream ciphers, hash functions and public key cryptography were recapitulated and the feasibility of realizations for RFID tags was discussed. Recent research shows that lightweight implementations of block ciphers, stream ciphers and random number generators are feasible even for very constrained RFID transponder hardware. Even the standard block ciphers DES and AES are considered possible for low-cost transponders. The hardware demands of hash functions and also of elliptic curve cryptography are slightly higher and currently above the limitations of low-cost tags, but they are feasible for mid-range RFID systems. Further research is necessary to develop new algorithms, to analyze the proposals and to develop efficient and secure implementation.

Problems

1. Consider a replacement of a paper-based ticketing system with an RFID system for a public transport system. What are benefits and drawbacks for the service provider? What are benefits and drawbacks from a customer perspective?
2. What is the primary threat for an RFID-based payment scheme in a canteen that uses:
   (i) RFID tags without any cryptographic function (low-end RFID system)?
   (ii) RFID tags with general-purpose cryptographic functions (mid-range RFID system)?

   For both cases, design appropriate security functions.
3. Summarize advantages and disadvantages for the privacy approaches "Killing Scheme," "On-Tag Scheme," "Agent Scheme," and "User Scheme." For this task, consider the achieved privacy solution, its user-friendliness, after-sales applications, and the required costs of implementing this functionality.
4. Which clock frequency of the RFID reader is at minimum required in order to restrict the distance between a transponder and an RFID reader to 15 metres in a distance bounding protocol? Consider that the time for signal processing is negligible and that the reader precisely controls the timing of the transponder.
5. Should countermeasures against physical implementation attacks be required for RFID readers?
6. Discuss why MIFARE Classic systems are still in wide use though its underlying cryptographic cipher CRYPTO1 is easy to break.

7. How many years would a brute-force attack take on a cryptographic algorithm with a security strength of 80 bits, if special code breaking hardware were to perform 50 billion crypto operations per second?
8. The following successive output bits of a Linear Feedback Shift Register (LFSR) of length 5 are given: 1, 0, 0, 0, 1, 1, 1, 1, 1, 0. Compute the feedback coefficients $c_1, \ldots, c_5$ and the subsequent output bits. What is the period of this LFSR?
9. Consider a linear congruential generator defined by $x_{n+1} \equiv ax_n + b \bmod m$ where a and b are secret parameters. The generator is initialized with a seed integer value x_0. Is this a cryptographically secure pseudo-random number generator?
10. Assume that the keystream of a stream cipher recurs after some low period. What would be the consequence for the security of this cipher?
11. What are the non-linear operations of the Trivium stream cipher?
12. Many RFID tags are able to compute checksums (e.g. CRC16), defined by the remainder of a division over the polynomials of the binary field. Can such a *linear* function satisfy the requirements of a cryptographic hash function?
13. Why is it reasonable to use Elliptic Curve Cryptography for anti-counterfeiting protection in RFID systems?

NB Solutions are provided on the book's website.

References

[1] ISO 27001 and ISO 27002 (2008) Plain English information security management definitions. Available at: http://www.praxiom.com/iso-27001-definitions.htm.

[2] Pfitzmann, A. and Hansen, M. Anonymity, unlinkability, undetectability, unobservability, pseudonymity, and identity management – A consolidated proposal for terminology. Available at: http://dud.inf.tu-dresden.de/Anon_Terminology.shtml.

[3] Rieback, M.R. (2008) Security and privacy of radio frequency identification, PhD thesis, Vrije Universiteit Amsterdam.

[4] Gollmann, D. (2006) *Computer Security*. Chichester: John Wiley & Sons, Ltd.

[5] Finkenzeller, K. (2003) *RFID-Handbook*. Chichester: Wiley & Sons Ltd.

[6] Hancke, G.P. (2006) Practical attacks on proximity identification systems (short paper), in *IEEE Symposium on Security and Privacy 2006*. Available at: http://www.cl.cam.ac.uk/~gh275/SPPractical.pdf.

[7] Juels, A. (2005) RFID security and privacy: a research survey. Available at: http://www.rsa.com/rsalabs/staff/bios/ajuels/publications/pdfs/rfid_survey_28_09_05.pdf.

[8] Hancke, G.P. (2008) Eavesdropping attacks on high-frequency RFID tokens, RFIDSec 2008. Available at: http://www.rfidblog.org.uk/Hancke-RFIDsec08-Eavesdropping.pdf.

[9] Juels, A., Rivest, R.L., and Szydlo, M. (2003) The blocker tag: selective blocking of RFID tags for consumer privacy, in *Proceedings of the 10th ACM Conference on Computer and Communications Security*, ACM, pp. 103–111.

[10] Hancke, G.P. (2005) A Practical Relay Attack on ISO 14443 Proximity cards, technical report. Available at: http://www.rfidblog.org.uk/hancke-rfidrelay.pdf.

[11] Brands, S. and Chaum, D. (1993) Distance-bounding protocols, *Advances in Cryptology – Eurocrypt 1993*, LNCS 765. Berlin: Springer, pp. 344–359.

[12] Hancke, G.P., and Kuhn, M.G. (2005) An RFID distance bounding protocol, *Proceedings of IEEE/CreateNet SecureComm*, pp. 67–73. 2005. Available at: http://www.rfidblog.org.uk/RFIDdistancebound-Securecomm2005.pdf.

[13] Spiekermann, S. and Evdokimov, S. (2009) Privacy enhancing technologies for RFID – A critical investigation of state of the art research, *IEEE Privacy and Security*, **7**(2): 56–62.

[14] ECRYPT Yearly Report on Algorithms and Keysizes (2008) Revision 1.1m, D.SPA.28. Available at: http://www.ecrypt.eu.org/ecrypt1/documents/D.SPA.28-1.1.pdf.

[15] Bono, S.C., Green, M., Rubin, A.D., Stubblefield, A., and Szydlo, M. (2005) Security analysis of a cryptographically-enabled RFID device, in *14th USENIX Security Symposium*, 1–16. Available at: http://usenix.org/events/sec05/tech/bono/bono.pdf.

[16] Indesteege, S., Keller, N., Dunkelman, O., Biham, E., and Preneel, B. (2008) A practical attack on KeeLoq, in *EUROCRYPT 2008*, LNCS 4965. Available at: http://www.cosic.esat.kuleuven.be/publications/article-1045.pdf.

[17] Nohl, K., Evans, D. Starbug., and Plötz, H. (2008) Reverse-engineering a cryptographic RFID tag, in *17th USENIX Security Symposium*, pp. 185–193. Available at: http://www.usenix.org/events/sec08/tech/full_papers/nohl/nohl.pdf.

[18] Garcia, F.D., de Koning Gans, G., Muijrers, R., van Rossum, P., Verdult, R Wichers., Schreur, R., and Jacobs, B. (2008) Dismantling MIFARE Classic, in *ESORICS 2008*, LNCS 5283. Available at: http://www.sos.cs.ru.nl/applications/rfid/2008-esorics.pdf.

[19] Courtois, N.T. (2009) The dark side of security by obscurity and cloning MiFare Classic rail and building passes anywhere, anytime, in SECRYPT 2009-International conference on Security and Cryptography, to be published in LNCS series, Springer.

[20] Lemke-Rust, K. (2007) Models and algorithms for physical cryptanalysis, PhD thesis, Ruhr-Universität Bochum.

[21] Pappu, R. (2001) Physical one-way functions, PhD thesis, Massachusetts Institute of Technology.

[22] Tuyls, P., Schrijen, G.J., Skoric, B., van Geloven, J., Verhaegh, N., and Wolters, R. (2006) Read-proof hardware from protective coatings, in *Cryptographic Hardware and Embedded Systems – CHES 2006*, LNCS 4249. Berlin: Springer, pp. 369–383.

[23] Kocher, P.C. (1996) Timing attacks on implementations of Diffie–Hellman, RSA, DSS, and other systems, in *Advances in Cryptology – CRYPTO '96*, LNCS 1109. Berlin: Springer, pp. 104–113.

[24] Kocher, P.C., Jaffe, J., and Jun, B. (1999) Differential power analysis, in *Advances in Cryptology – CRYPTO '99*, LNCS 1666. Berlin: Springer, pp. 388–397.

[25] Eisenbarth, T., Kasper, T., Moradi, A., Paar, C., Salmasizadeh, M., and Manzuri Shalmani, M.T. (2008) On the power of power analysis in the real world: a complete break of the KeeLoq Code hopping scheme, in *Advances in Cryptology – CRYPTO 2008*, LNCS 5157. Berlin: Springer, pp. 203–220.

[26] Kasper, M., Kasper, T., Moradi, A., Paar, C. (2009) Breaking KeeLoq in a flash: on extracting keys at lightning speed, in *Progress in Cryptology – AFRICACRYPT 2009*, LNCS 5580. Berlin: Springer, pp. 403–420.

[27] Mangard, S., Oswald, E., and Popp, T. (2007) *Power Analysis Attacks*. Berlin: Springer.

[28] Boneh, D., and DeMillo, R.A., and Lipton, R.J. (1997) On the importance of checking cryptographic protocols for faults (extended abstract), in *Advances in Cryptology – EUROCRYPT '97*, LNCS 1233. Berlin: Springer, pp. 37–51.

[29] Aumüller, C., Bier, P., Fischer, W., Hofreiter, P., and Seifert, J.P. (2003) Fault attacks on RSA with CRT 2002: concrete results and practical countermeasures, in *Cryptographic Hardware and Embedded Systems – CHES 2002*, LNCS 2523. Berlin: Springer, pp. 260–275.

[30] Skorobogatov, S.P. and Anderson, R.J. (2002) Optical fault induction attacks, in *Cryptographic Hardware and Embedded Systems – CHES 2002*, LNCS 2523. Berlin: Springer, pp. 2–12.

[31] Bar-El, H., Choukri, H., Naccache, D., Tunstall, M., and Whelan, C. (2004) The sorcerer's apprenctice's guide to fault attacks, in Cryptology ePrint Archive, Report 2004/100. Available at: http://eprint.iacr.org/2004/100.

[32] Hutter, M., Schmidt, J.M., and Plos, T. (2008) RFID and its vulnerability to faults, in *Cryptographic Hardware and Embedded Systems – CHES 2008*, LNCS 5154. Berlin: Springer, pp. 363–373.

[33] EPC Radio-Frequency Identity Protocols, Class-1 Generation-2 UHF RFID (2008) Protocol for Communications at 860 MHz–960 MHz, Version 1.2.0. EPCglobal Inc. Available at: http://www.epcglobalinc.org/standards/uhfc1g2/uhfc1g2_1_2_0-standard-20080511.pdf

[34] Paar, C., Porschmann, A., and Robshaw, M.J.B. (2008) New designs in lightweight symmetric encryption, in *RFID Security: Techniques, Protocols and System-on-Chip-Design* (eds. Kitsos, P. and Zhang, Y.). Berlin: Springer, pp. 349–371.

[35] Menezes, A.J., Oorschot, P.C., and Vanstone, S.A. (2001) *Handbook of Applied Cryptography*, fifth edn. Boca Raton, FL: CRC Press. Available at: http://www.cacr.math.uwaterloo.ca/hac.

[36] Feldhofer, M. and Wolkerstorfer, J. (2008) Hardware implementations of symmetric algorithms for RFID security, in *RFID Security: Techniques, Protocols and System-on-Chip-Design* (eds. Kitsos, P. and Zhang, Y). Berlin: Springer, pp. 373–415.

[37] Feldhofer, M., Dominikus, S., and Wolkerstorfer, J. (2004) Strong authentication for RFID systems using the AES algorithm, in *CHES 2004*, LNCS 3156. Berlin: Springer, pp. 357–370.

[38] Che, W., Deng, H., Tan, X., and Wang, J. (2008) A random number generator for application in RFID tags, in *Networked RFID Systems and Lightweight Cryptography* (eds. Cole, P.H., and Ranasinghe, D.C). Berlin: Springer, pp. 278–287.

[39] Peris-Lopez, P., Hernandez-Castro, J.C., Estevez-Tapiador, J.M., and Ribagorda, A. (2009) LAMED – A PRNG for EPC Class-1 Generation-2 RFID specification. *Computer Standards & Interfaces*, **31**(1): 88–97.

[40] Barker, E. and Kelsey, J. Recommendation for random number generation using deterministic random bit generators (revised) (2007). NIST Special Publication 800-90. Available at: http://csrc.nist.gov/publications/nistpubs/800-90/SP800-90revised_March2007.pdf

[41] Leander, G., Paar, C., Poschmann, A., and Schramm, K. (2007) New lightweight DES variants, in *Proceedings of Fast Software Encryption 2007 – FSE 2007* LNCS 4593. Berlin: Springer, pp. 196–210.

[42] Bogdanov, A., Knudsen, L.R., Leander, G., Paar, C., Poschmann, A., Robshaw, M.J.B., Seurin, Y., Vikkelsoe, C. (2007) PRESENT: An ultra-lightweight block cipher, in *Cryptographic Hardware and Embedded Systems – CHES 2007* LNCS 4727/2007. Berlin: Springer, pp. 450–466.

[43] Babbage, S., De Cannière, C., Canteaut, A., Cid, C., Gilbert, H., Johansson, T., Parker, M., Preneel, B., Rijmen, V., and Robshaw, M. (2008) The eSTREAM Portfolio (rev. 1). eSTREAM, ECRYPT Stream Cipher Project. Available at: http://www.ecrypt.eu.org/stream/portfolio.pdf and http://www.ecrypt.eu.org/stream/portfolio_revision1.pdf

[44] Braun, M., Hess, E., and Meyer, B. (2008) Using elliptic curves on RFID tags, *International Journal of Computer Science and Network Security*, **8**(2): 1–9.

[45] Batina, L., Mentens, N., Sakiyama, K., Preneel, B., and Verbauwhede, I. (2006) Low-cost elliptic curve cryptography for wireless sensor networks, in *Security and Privacy in Ad-Hoc and Sensor Networks – ESAS 2006*, LNCS 4357. Berlin: Springer, pp. 6–17.

17

Cryptographic Approaches for Improving Security and Privacy Issues of RFID Systems

Miyako Ohkubo[1], Koutarou Suzuki[2], and Shingo Kinoshita[2]

[1]*National Institute of Information and Communication Technology*

[2]*NTT Information Sharing Platform Laboratories*

Radio-frequency identification (RFID) is being considered for, or is already being used in many identity management applications. RFID systems have been applied not only in logistics and public transport but also in electronic passports and other identification documents. Moreover, many organizations are exploring the feasibility of tracking items from production to consumption using RFID technology. Another possible application of RFID technology is as a tool for tracking people or animals, for example, tracking cognitively impaired elderly people, children on their way to schools, or a pet. All these applications open up possibilities for tracking people and inventorying their possessions. Therefore, there are privacy concerns including leakage of data about belongings without users being aware of it and ID tracing to monitor an owner's activities. In addition, these applications could open up possibilities for corporate espionage. This has given rise to a myriad of privacy issues. Cryptographic technologies can be useful in solving these privacy issues, while retaining the usability and convenience of the RFID system.

In this chapter we survey state-of-the-art research related to the security and privacy of RFID systems. Applications and systems based on RFID technology should satisfy the requirements for preserving privacy and providing usability. We classify the security

RFID Systems: Research Trends and Challenges Edited by Miodrag Bolić, David Simplot-Ryl, and Ivan Stojmenović

and privacy issues of RFID systems based on privacy issues, authenticity, restriction and delegation of traceability, forward security and other requirements. Then, we survey cryptographic protocols addressing solutions to these issues.

17.1 Introduction

Communication styles have been improving and changing on a daily basis, and in our lifetime the use of wireless communication has expanded dramatically. One technology that utilizes wireless communication is radio-frequency identification (RFID), which is an automatic identification technology. RFID is expected to provide a new communication style for identifying items. Simultaneous scanning of many tags, or identification of a single tag's ID through wireless communications are two examples of this new style of communication style. An RFID system generally consists of wireless tags, which are attached to objects, and wireless readers. (In fact, a reader identifies a tag's output by sending the output to a back-end system and receiving the tag's ID from that system. In this chapter, we assume that the communication between the reader and back-end system is secure, so the RFID system consists of just the tag and reader.)

A tag consists of an integrated circuit (IC) chip and an antenna. The tag sends information that identifies itself (ID code) to a reader via a wireless channel. The reader receives the information sent by the tag via the wireless channel and uses it to determine the identity of the tag. The wireless communication between the tag and the reader uses a low-frequency (LF) band from 124 to 135 kHz, a high frequency (HF) band around 13.56 MHz, or an ultrahigh frequency (UHF) band from 860 to 960 MHz. The frequency band is chosen according to the specifications of the tags and readers. (EPCglobal [1] has established specifications for the ID code named the Electronic Product Code (EPC).) The entire RFID system including tags, readers, Object Name Services (ONS), and EPC Information Services (EPCIS) is called the EPCglobal network.

RFID has enormous, though not yet fully realized, potential to be applied in many different situations. For example, it has been used in supply-chain management instead of bar codes. However, for successful deployment some barriers need to be overcome, including the privacy issue and unit cost, which must be kept low. The privacy issue arises from the basic functions of RFID tags. Each tag can be identified easily through wireless communication and each tag has a unique ID. Therefore, there are risks that consumers or objects could be identified and tracked over wide areas for long periods of time. These privacy issues of RFID tags appear in various scenarios, and can be very serious depending on the application. These issues are becoming a widespread public concern [2–4]. To address these concerns Karygiannis *et al.* [5] presented guidelines for RFID systems that take security and privacy into consideration. In recent years, many investigations of RFID privacy issues have been reported. Additionally, some papers discuss the possibility of using RFID as a tool for breaking privacy. For example, Rotter Daskala, and Compano [6] introduced the challenges of implanting RFID tags in humans or pets. RFID implants have several advantages, but they raise various issues including a lack of privacy. Concerns related to this technology such as the above-mentioned case are becoming serious. Therefore, solving the privacy issue for RFID technology is an essential precondition to the successful expansion of RFID applications available on the market.

Besides the privacy problem, the cost limitation of the tag unit is a serious problem. It means that each tag is likely to have poor computational power, that is, the electric power that each tag can use for computation is too small for complex calculations. This limitation would seem to make it difficult to use ordinary public-key cryptographic primitives, which have been constructed assuming that significant computational resources are available. These circumstances are the motivation behind efforts to solve security and privacy problems by using lightweight cryptographic primitives that require low computational costs. Indeed, various schemes using cryptographic techniques have been proposed.

In this chapter, we introduce studies related to the security and privacy of RFID, and showcase related state-of-the-art research. This chapter also discusses security issues for RFID, including threats, low-level cryptanalytic attacks, and lightweight cryptography primitives. In this chapter, we focus on security threats, required security properties, and cryptographic protocols from the viewpoint of cryptology. To preserve privacy, and also provide usability and convenience for RFID applications, security and privacy requirements are needed. Several surveys of the literature have been done in the past [7–9]. Juels [7] presented a survey of studies on RFID security and privacy. Avoine [8] provided a list of many papers that discuss RFID tag security and privacy. Rotter [9] discussed the security of the RFID tag system from the viewpoint of threats. We classify the security and privacy issues of RFID systems and show the required properties from the following viewpoints: privacy, authenticity, restriction and delegation of traceability, and forward security. We also show cryptographic protocols for providing the required properties.

17.2 Threats against the RFID System

In this section, we describe various threats against RFID systems. Attacks against RFID systems can be classified according to the target (tag, wireless channel, or reader and back-end) and the method (passive reading, active reading, rewriting, cloning, destruction/Denial of Service (DoS), scanning/tracking, or side-channel attack) as listed in Table 17.1. The details of these attacks are described as follows:

Table 17.1 Classification of attacks against RFID system according to target and method of attack.

Method of attack	Tag	Wireless channel	Reader and back-end
Passive reading	–	Eavesdropping	–
Active reading	Tag reading	Replay, Relay, Modification	Reader reading
Rewriting	Tag rewriting, Virus/Malware	–	Reader rewriting, Virus/Malware
Cloning	Tag cloning	–	Reader cloning
Destruction/DoS	Wireless destruction/DoS	Jamming	Wireless destruction/DoS
Scanning/Tracking	Tag scanning, Tag tracking	–	–
Side-channel	Wireless side-channel analysis	–	Wireless side-channel analysis

17.2.1 Passive Reading Attack

A passive reading attack means that an attacker passively reads the message transmitted on a wireless channel, in an attempt to obtain secret information. In a passive reading attack, an attacker passively, that is, without manipulating the communicated messages, tries to read information from a wireless channel between a tag and reader.

- *Eavesdropping*: Eavesdropping is a passive reading attack against a wireless channel between a tag and reader. Since the communication between the tag and reader is done wirelessly, an attacker can easily eavesdrop on it, which is difficult to detect. Information obtained by eavesdropping might enable an attacker to break the identification, authentication, or privacy of the RFID system. Communication between a tag and reader can be protected by encrypting the data before transmission. There are many schemes for secure communication between the tag and reader (See Section 17.4.)

17.2.2 Active Reading Attack

An active reading attack happens when an attacker actively modifies messages sent to a tag, transmitted on a wireless channel or sent to a reader, in an attempt to obtain secret information or to attack the authentication mechanism. In an active reading attack, an attacker actively, that is, by manipulating the communicated messages, tries to wirelessly read information from a tag, wireless channel, or reader.

- *Replay Attack*: A replay attack is an active attack against a wireless channel. An attacker records the output of a tag and replays it to a reader. The replay attack might enable an attacker to break the authentication of the RFID system. Replay attacks can be prevented using challenge-and-response authentication. (See Section 17.5.)
- *Relay Attack*: A relay attack is an active attack against a wireless channel. An attacker relays the communication between a tag and a remote reader, that is, illicitly reads a tag's output using a fake reader, transports the output to another location, and sends the output to the remote reader. The relay attack might enable an attacker to break the authentication of the RFID system. Relay attacks can be prevented using the distance bounding protocol proposed by Brands and Chaum [10] and Clulow *et al.* [11] or physically shielding the tag to prevent illicit reading.
- *Message Modification Attack*: A modification attack is an active attack against a wireless channel. An attacker modifies the communication between a tag and reader by intercepting the communication. An attacker can even block the communication between a tag and reader. The modification attack might enable an attacker to break the identification, authentication, or privacy of the RFID system. Modification attacks can be prevented using an integrity checking mechanism of the communicated data.
- *Tag/Reader Reading Attack*: A tag reading attack is an active attack against the tag. An attacker illicitly reads a tag's output using a fake reader. Since the communication between tag and reader is done wirelessly, an attacker can easily read the tag's output using a fake reader. Information obtained by illicitly reading the tag might enable an attacker to break the identification, authentication, and privacy of the RFID system. Tag reading attacks can be prevented by requiring the tag to authenticate the reader or using physical shielding. We can also consider a reader reading attack, where an attacker illicitly reads a reader's output using a fake tag.

17.2.3 Rewriting Attack

A rewriting attack happens when an attacker rewrites information stored in a tag or a reader, in an attempt to obtain secret information or to fool the authentication mechanism. In a rewriting attack, an attacker tries to wirelessly rewrite information in a tag or a reader.

- *Tag/Reader Rewriting Attack*: A tag rewriting attack is an active attack against a tag. An attacker illicitly rewrites a tag's memory using a fake reader. Since communication between tag and reader is done wirelessly, an attacker can easily perform the attack. The tag rewriting attack might enable an attacker to break the identification, authentication, and privacy of the RFID system. Tag rewriting attacks can be prevented by requiring the tag to authenticate the reader or using a tag with a memory lock, for example, an EPC tag [1]. We can also consider a reader rewriting attack, where an attacker illicitly rewrites a reader's memory using a fake tag.
- *Virus/Malware*: The possibility of transmitting a virus/malware via an RFID tag has been shown by Rieback, Crispo, and Tanenbaum [12]. A reader reads the code of a virus/malware from a contaminated tag. The reader executes the code of the virus/malware and rewrites a copy of that code into other tags, which spreads the virus/malware infection.

17.2.4 Cloning Attack

A cloning attack happens when an attacker clones a tag or reader, that is, creates a copy of the tag or reader, in an attempt to attack authentication. In a cloning attack, an attacker tries to create a copy of a tag/reader using information obtained from that tag/reader.

- *Tag/Reader Cloning Attack*: A tag cloning attack is an attack in which an attacker creates a clone, that is, a complete copy, of a tag using the information obtained from that tag. Tag cloning might enable an attacker to break the identification and authentication of the RFID system. Tag cloning attacks can be prevented using reader authentication, tamper-proof tags, or a physically uncloneable function that is proposed by Bolotnyy and Robins [13]. We can also consider a reader cloning attack, where an attacker creates a clone of a reader using the information obtained from that reader.

17.2.5 Destruction/DoS Attack

A destruction/DoS attack happens when an attacker physically destroys a tag or reader, performs a DoS attack against a tag or reader, or performs jamming against the wireless channel in an attempt to obstruct the service of the RFID system.

- *Tag/Reader Destruction/DoS Attack*: In a tag/reader destruction/Denial of Service (DoS) attack, an attacker tries to physically destroy a tag, possibly wirelessly, or executes a DoS attack against the tag and/or reader so that it cannot communicate. Tag destruction or DoS attacks might enable an attacker to stop the operation of the RFID system. Tag/reader destruction or DoS attacks are difficult to prevent.
- *Jamming Attack*: In a jamming attack, an attacker tries to block or jam the wireless communication between a tag and reader, wirelessly or through some other means.

Jamming attacks might affect normal communication between a tag and reader, and enable an attacker to stop operation of the RFID system. Jamming attacks are difficult to prevent.

17.2.6 Scanning/Tracking Attack

A scanning/tracking attack happens when an attacker scans or tracks a tag by detecting messages transmitted by the tag in an attempt to violate the privacy of the tag owner. In a scanning/tracking attack, an attacker tries to scan a tag to obtain information about a tagged item or to track the tagged item or person carrying the tagged item.

- *Tag Scanning Attack*: In a tag scanning attack, an attacker tries to scan a tag to obtain information about the tagged item, for example, medicine name or book title. Tag scanning attacks might enable an attacker to breach the privacy of the person carrying the tagged item. Tag scanning attacks can be prevented using physical shielding or encrypted tag outputs. (See Section 17.6.)
- *Tag Tracking Attack*: In a tag tracking attack, an attacker tries to track a tag (tag tracking) or a person carrying a tag (person tracking). Tag tracking attacks might enable an attacker to breach the location privacy of the person carrying the tagged item. Tag tracking attacks can be prevented using physical shielding or randomized tag outputs. (See Section 17.6.)

17.2.7 Side-Channel Attack

A side-channel attack means that an attacker gathers and analyzes side-channel information, for example, power consumption or response delay, which varies as devices process secret data, in an attempt to obtain information about the secret data.

- *Tag/Reader Wireless Side-channel Analysis*: In a wireless side-channel attack against a tag/reader, an attacker can observe the field strength (power analysis) or the timing (timing analysis) of wireless waves from the tag/reader. A wireless power analysis of tags has been reported by Oren and Shamir [14]. Side-channel attacks might enable an attacker to obtain secret information stored in a tag/reader. This type of attack can be prevented using a side-channel-proof implementation of the tag/reader.

17.2.8 Attack against Overall System Security

An attacker can attack a reader and back-end system as if they were a normal computer system, that is, non-RFID specific attacks, since the RFID system is also a kind of computer system. Such an attack might enable an attacker to breach the overall system security. This type of attack is studied by Fabian, Günther and Spiekermann [15], where the attack is used on the Object Name Service (ONS) of the EPC global network.

17.3 Required Properties

In this section, the required properties for secure RFID systems are described. The main properties of RFID are outlined below, and shown in Figure 17.1 that is, identification,

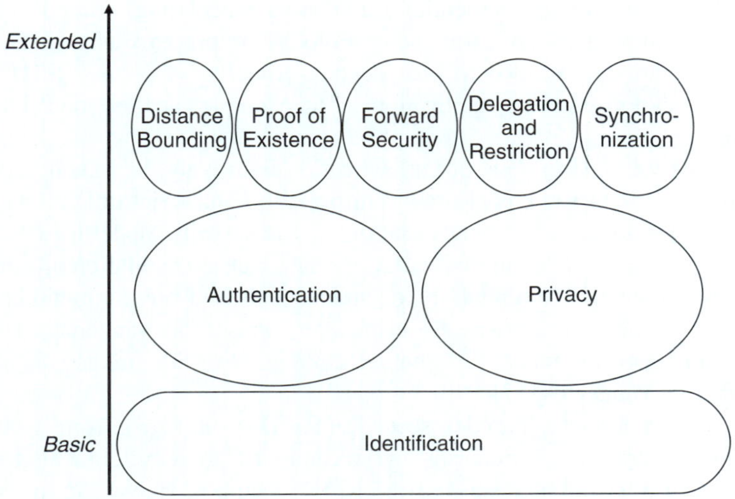

Figure 17.1 Classification of required properties of a RFID system.

authentication, indistinguishability as basic properties, and forward security, delegation and restriction, proof of existence, distance bounding, and synchronization as additional properties.

17.3.1 Identification

The main function of an RFID reader is to identify a unique ID given to each RFID tag. This property is called identification. Intuitively, the identification property means that each RFID tag's ID can be identified from the tag's output. Basically, a tag is assigned a unique ID before shipping, and the ID recorded in the ROM of the tag is difficult to overwrite and assign a different ID.

The property of identification is useful in many situations. For example, many items can be identified by the attached RFID tag, and so RFID is expected to contribute to the improvement of logistics management and inventory systems. Moreover there are possible contributions to applications such as extended services for consumers. Identification can be provided which contains no secret information. Instead only unique data, that is a unique ID or information related to the unique ID, is required. Therefore no cryptographic techniques are needed to provide identification. We should note that identification itself may result in the leaking of secret information. One example is the tag tracking attack.

17.3.2 Authentication

Anyone, whether an honest or malicious person, can obtain the output of tags and readers. However, a reader (respectively, a tag), that receives an output as a tag's output (respectively, a reader's output), cannot determine if the received data is one from a valid tag or not, unless there is a property for ensuring its correctness. To identify and ensure the validity of a tag's output (and/or a reader's output), the communication between tag and reader should incorporate authenticity, that is, a reader (tag) should accept a tag's (reader's) output only if it can ensure its validity. This property is called authenticity, and secret information is needed to provide this property. A secure RFID tag scheme requires authentication in some applications. For example, access control for entrance into buildings.

Below, we show the definitions of authentication intuitively, and Damgård and Østergaard [16] described authenticity. Detailed information and strict definitions were given in Vaudenay [17]. Vaudenay presents variations in the security model for tag authenticity with privacy, that is, classification of variations in attack goals and environments. Eight classes are shown for the combinations of attack goals and environments and the relationships between the eight classes are also given. Paise and Vaudenay [18] presented variations in the security model for mutual authenticity with privacy using a similar strategy to that of Vaudenay [17].

Generally, the properties required to secure an RFID system arise from the threat model which contains several attack scenarios and accounts for the environments the system is found in. We can treat the attacker and the RFID system as players in an attack game; that is, the adversary wins the game if he or she successfully achieves the attack goal in the given environment. Authenticity also can be treated as a game. Below, we describe the authenticity game for a tag. In this case, the prover is a tag and the verifier is a reader. The authenticity game for a reader can be described in a corresponding manner. There are n tags $T_1, ..., T_n$ and a reader R. Tag T_i has one piece secret information s_i and communicates with reader R. On the other hand, reader R has n pieces of secret information, $s_1, ..., s_n$ and communicates with all the tags. Reader R receives the output of tag T_i and outputs either *reject* or *accept tag* T_i. As the attack environment, the adversary is allowed to (i) communicate with reader R, (ii) communicate with tag T_i, and (iii) corrupt tag T_i and obtain its secret information s_i. Note that by corrupting the tag, the adversary may obtain all data, including secret information s_i, recorded in the tag's memory. These three behaviors are allowed to be performed multiple times and in any order. The attack goal of the authenticity game is for the adversary to communicate with reader R and make it output "*accept tag* T_i" under the constraints that tag T_i is not corrupted and is not currently communicating with the reader. In other words, trivial attacks are not allowed, and the authentication must be broken. An RFID scheme can be said to exhibit authenticity if and only if it is computationally infeasible for the adversary to win and thus to achieve the attack goal of the authenticity game in the attack environment.

17.3.3 Privacy

Identification is the most basic property of RFID since the purpose of deploying RFID is to identify a tag, and in turn, the object or person to which the RFID tag is attached. However, because a tag has the following two basic functions:

- each tag can be easily identified through wireless communication;
- each tag has a unique ID enabling it to be identified;

anyone might be able to identify a specific object or person, that is, ID leakage, and track it over a wide area, that is, ID tracking.

Privacy issues arise from these basic functions. There are privacy-sensitive applications of RFID in which the tag is required to provide privacy protection. However, to define privacy is difficult because it depends on what the user requires, how the tag system is used, and how the user feels about the leakage of his/her private information.

Complete protection of privacy is difficult. Preventing the leakage of privacy information during communication and in the physical layer is important, as well as leakage of privacy information in the application layer as introduced by Avoine and Oechslin [19]. From another point of view, it should be noted that leakages occur from the back-end database in which private information is recorded as well as leakages by eavesdropping tag's/reader's output. Careful consideration of privacy is essential for the success of the RFID market because these issues are becoming matters of wide public concern [2–4]. This has prompted research on technical solutions and methods for protecting privacy.

17.3.4 Indistinguishability

The important privacy issues are ID leaking and ID tracing of RFID tags, as described above. To avoid these issues, a tag's output needs to have a property in which it is impossible for an eavesdropper to distinguish between two tags by observing their outputs. This security requirement is called indistinguishability, which is needed to protect the privacy of the tag, that is, the user's privacy.

Avoine [20] presented requirements for the indistinguishability of RFID tags. Moreover, Juels and Weis [21] presented indistinguishability requirements, focusing on correlated secret keys that tags carry.

Just like authenticity, indistinguishability can be treated as a game. The game of indistinguishability can be described as follows: In the following scenario, the distinguisher is a reader. There are n tags $T_1, \ldots, T_n$ and a reader R, which is a distinguisher. Tag T_i obtains its own secret data s_i and communicates with reader R. Reader R has n secret pieces of data, $s_1, \ldots, s_n$ and communicates with all of the tags. Reader R receives the output of tag T_i and outputs either *reject* or *accept tag* T_i. As the attack environment, the adversary is allowed to (i) communicate with reader R, (ii) communicate with tag T_i, and (iii) corrupt tag T_i and obtain its secret data s_i. These behaviors are allowed to be performed multiple times and in any order. The attack goal of the indistinguishability game, given two uncorrupted tags T_i and T_j, is to distinguish these tags without information of the indices i or j, while allowing the reader to communicate with all the tags, including T_i and T_j. An RFID scheme can be said to exhibit indistinguishability if and only if it is computationally infeasible for the adversary to win, that is, to achieve the attack goal of the indistinguishability game in the attack environment.

17.3.5 Forward Security

As an extension of the requirements for indistinguishability and/or authenticity, there is a property called forward security. We should consider what may happen after a tagged item is thrown away or stolen because an attacker might be able to tamper with the tag in the item and obtain the secret information recorded in it. The cost limitation of RFID has made it difficult to implement tamper-resistant properties in RFID tags. Therefore, for applications that require indistinguishability and/or authenticity, we should be aware that secret information registered in a RFID tag might be stolen and that the security and privacy of the tag might be compromised.

There is forward security of indistinguishability and forward security of tag authenticity. Forward security of indistinguishability is defined as follows: If the secret information used for preserving indistinguishability is leaked through tampering, then the tag's past output can be traced using the secret information. An attacker who obtains the secret key of a tag can transform (decrypt) the tag's output in an eavesdropped database by using the tag's secret key and then identify which transaction in eavesdropped data is an output of the tag. Ohkubo, Suzuki, and Kinoshita [22] have pointed out this issue and presented the concept of forward security for RFID tag privacy. Even if the secret information in a tag becomes known to an attacker, the tag's past output should remain inaccessible. In the indistinguishability game described in subsection 17.3.4, we assumed that an adversary is allowed to corrupt tags T_i and T_j only after communicating with them.

Ohkubo, Suzuki, and Kinoshita [22] proposed a scheme that satisfies *forward security* for indistinguishability and tag authentication using the hash chain technique.

17.3.6 Delegation and Restriction

The properties called delegation and restriction come from security demands for applications in which tags are reused and transferred from their original owner to new ones. In such applications, the original owner can delegate the right of tracing a tag or tagged object to a new owner, and then the original owner of a tag or tagged object should not be able to trace the tag after it has been transferred to the new owner.

Molnar, Soppera, and Wagner [23] discussed the issue of ownership transfer and the requirements for restricting the ability to trace a tag within a certain period of time to prevent tag tracing by the previous owner. They also proposed a scheme that satisfies these delegation and restriction requirements. Ohkubo and Suzuki [24] proposed a scheme that can satisfy the combined security requirements of both restricted delegation and forward security. This scheme is based on a hash-chain and binary tree. It prevents a previous owner from breaking the tag's privacy (i.e. the privacy of the tag's current owner) and ensures that no one can trace the tag, even if they know the secret information in the tag.

17.3.7 Proof of Existence

A major application of RFID is supply-chain management, in which each product is tagged and traced by readers along the distribution route. For such applications, a security requirement arises, that is, the need to guarantee the existence of a particular tag in a specific location, at a specific time, with other particular tags.

For instance, Juels [25] proposed the yoking-proof technique, which proves that the outputs of two tags are received by the same reader simultaneously. The reader that receives data from the two tags concludes that the two tags exist together (i.e. in the same place or at the same point). This technique can be used for applications for example, pharmaceutical distribution, where medicine and its description should be distributed together.

17.3.8 Distance Bounding

A relay attack, which is described in subsection 17.2.2, is difficult to prevent using any authentication protocols. To raise the security level against a relay attack, we must limit the accepted distance between any tag and reader. This required property is called distance bounding.

Distance bounding is made by limiting the round trip time of the exchanges between a tag and a reader. There are two types of protocols that provide this property. The first protocol type consists of random one-bit message exchanges and a signature. It has two phases: fast and slow. In the fast phase, the reader and tag exchange random one-bit messages and the reader measures the round trip time. After n rounds, where n is a security parameter, the reader requests a signature from a tag in the slow phase. Schemes presented by Brands and Chaum [10], Munilla, Ortiz, and Peinado [26], Tu and Piramuthu [27], Kim *et al.* [28], and Munilla and Peinado [29] are classified as this type. The false-acceptance rate, which means the rate at which wrong (corrupted) data is accepted, is lower than the second type of protocol.

The second type of protocol also consists of fast and slow phases. In the slow phase, the reader and tag exchange random nonces. Next, both compute two secret registers from the nonces and secret keys. Then, in the fast phase, the reader sends a random bit, and the tag replies with an answer by using the two secret registers. Schemes presented by Hancke and Kuhn [30], and Reid *et al.* [31] are classified as this type. The false-acceptance rate is higher than the first type, but a signature is not required. Avoine and Tchamkerten [32] proposed a low complexity protocol that takes advantage of both types.

17.3.9 Synchronization

The property called synchronization arises from stateful construction. For instance, consider a stateful authentication scheme, where common information (e.g. secret keys) are updated both in the tag and reader. In this case, an attacker can cause authentication failure by delaying and disturbing the communication between a tag and a reader and by choosing a time when only the tag (respectively, reader) updates the secret information stored in its memory, while the reader (respectively, tag) does not but still has the previous version. In such a case, authentication is unsuccessful because the secret information, which should be common to both the tag and the reader, differs.

Synchronization means that once desynchronization occurs, the reader and tag can retain synchronization.

Canard and Coisel [33] proposed a scheme using a hash chain that satisfies this property. Burmester, de Medeiros, and Motta [34] also presented a (mutual) authentication scheme that satisfies this property by preparing back-up data.

17.4 Cryptographic Protocols for Identification with Privacy

In this section, we describe identification protocols that preserve tag privacy (i.e. user's privacy). An RFID tag provides identification as an inherent property. However, depending on the aim of the RFID tag system, some additional properties are required to achieve this purpose. Cryptographic approaches can contribute to efforts to provide these required properties. They include various methods, with the representative ones introduced below.

The rewriting technique is useful if the aim of using the RFID tag system is only identification and the requirements dictate that privacy should be considered. This approach requires a tag to have a rewritable memory (RAM). When the reader receives data as output from the tag, the data is updated and then the updated data is written to the memory in the tag in order to protect privacy.

Inoue and Yasuura [35] presented a rewriting scheme for maintaining user privacy that does not use heavy cryptography. A user can write a random value to the tag's memory when he or she purchases a tagged item. The tag responds to a reader by disclosing this random value so the user receives a tag that cannot leak useful information about the tagged item. However, a table of the random values and corresponding tagged items must be maintained.

Juels and Pappu [36] presented an encrypted ID scheme using the ElGamal encryption and re-encryption technique. ElGamal encryption is a public key encryption algorithm with secret key x and public keys $(g, y(= g^x))$ and is calculated in modulo p, where g and p is a generator and a prime number, respectively, and the ciphertext for message m is $E(m, r) = (g^r, my^r)$, where r is a random value. In this scheme, the tag's output can be randomized by multiplying $(g^r \times g^s, my^r \times y^s)$ using only public key (g, y). This technique is called re-encryption. The ciphertext of ID $E(ID, r)$, encrypted using the ElGamal encryption scheme is stored in the tag. When a tag is interrogated by a reader, it outputs the ciphertext. The reader updates the ciphertext by randomizing it using the re-encryption technique. The readers are not required to have a secret key and can determine the tag's ID by requesting it from the back-end server using the tag's ciphertext. Only entities, that is, a back-end server that has a secret key x, can identify a tag's ID by decrypting the tag's ciphertext output using the secret key x.

Ishikawa *et al.* [37] presented an encrypted ID scheme that works with any encryption scheme. In this scheme, the encrypted ID, that is, $E(ID, r)$, is stored in the tag. The encryption scheme E is a probabilistic encryption algorithm, and a random value r is chosen by the reader. The reader receives the encrypted ID, $E(ID, r)$ and decrypts it to obtain the tag's ID by using a secret key. Then, the reader chooses $r'(\neq r)$ randomly, encrypts the tag's output with the random value r', and rewrites the updated encrypted ID $E(ID, r')$ in the tag. The encrypted ID cannot be linked to the updated one because the new encrypted ID, that is, $E(ID, r')$, looks completely different. Thus, tag tracing can be prevented. Ishikawa *et al.* present another scheme that uses the ElGamal re-encryption technique. However, in that scheme a single key should be used for a group within which tags are indistinguishable, because if multiple keys are used, no one can determine which key corresponds to a particular ciphertext.

Golle *et al.* [38] proposed an encrypted ID scheme that uses the universal ElGamal re-encryption scheme. In that scheme, both public key (g, y) and encrypted ID $E(ID, r)$, encrypted using the ElGamal encryption scheme, are stored in the tag. The reader can update them by randomizing not only the encrypted ID but also the public key. Moreover, multiple public keys can be used. Therefore, the reader does not need to hold any public keys and can obtain them from the tag's output. However, the ciphertext stored in a tag can be subverted if the attacker were to replace the tag's public key with his/her own public key. After that, the attacker would be able to decrypt the tag's output using his/her own secret key to trace the tag.

Ateniese, Camenisch, and de Medeiros [39] proposed an encrypted ID scheme with an unsubvertible encryption technique. Their proposed scheme requires a Certificate Authority (CA). The tag stores a certificate received from the CA, which is a CA's signature on the public key, and the encrypted ID. The reader checks the validity of the CA's certificate and updates all the information stored in the tag by randomizing the certificate, public key, and encrypted ID. Therefore, the stored ciphertext cannot be subverted because an attacker cannot forge the certificate from the CA and cannot replace the public key stored in the tag with his/her own public key.

17.5 Cryptographic Protocols for Authentication without Privacy

In this section, we describe authentication protocols. If applications need only the authenticity property and not privacy, then a simple lightweight scheme can be used for the RFID tag system.

Hopper and Blum [40] proposed an authentication scheme called the HB (Hopper and Blum) protocol. This scheme is intended to provide only authentication. This protocol is based on the learning parity with noise (LPN) problem and uses only exclusive OR (XOR) calculations, which require only two operations. This authentication scheme is described in full below: In the initialization phase, reader R holds secret $x \in \{0, 1\}^k$, where k is a security parameter, while tag T holds secret (x, η), and v is defined as a biased random bit and recorded in both reader R and tag T, where $Prob[v = 1] = \eta$ and $0 < \eta < \frac{1}{2}$. In the authentication phase, tag T and reader R repeat the following steps: (1) reader R selects random $a \in \{0, 1\}^k$ and sends a to tag T as a challenge; (2) tag T receives a and calculates $z = (a \cdot x) \oplus v$, where the inner-product of the vectors a and x is described as $a \cdot x$, and sends z to reader R as a response; and (3) reader R receives z and checks whether $a \cdot x = z$. If it does, the response is accepted. However, the HB protocol is only secure against passive eavesdroppers; it cannot maintain security against an active attacker who can query the tag because there is a very high probability that he/she can learn the error-free values of $a \cdot x$ by repeating the same challenge a times.

Juels and Weis [41] proposed a scheme called the HB+ protocol, which is secure against an active attacker [41]. The proposed scheme is based on an extension of the HB protocol which similarly uses only XOR calculations and requires only three operations. A detailed description is given below. In the initialization phase, reader R holds secret (x, y), tag T holds secret (x, y, η), and v is defined as a biased random bit and recorded in both reader R and tag T, where $Prob[v = 1] = \eta$ and $0 < \eta < \frac{1}{2}$. In the authentication phase, tag T and reader R repeat the following steps: (1) tag T selects random $b \in \{0, 1\}^k$ and sends b to reader R as a blinding factor; (2) reader R receives b and selects random $a \in \{0, 1\}^k$ and sends a to tag T as a challenge; (3) tag T receives a and calculates $z = (a \cdot x) \oplus (b \cdot y) \oplus v$ and sends z to reader R as a response; (4) reader R receives z and accepts it if $(a \cdot x) \oplus (b \cdot y) = z$. Tag T can effectively prevent an active attacker from extracting x or y with a non-random challenge by choosing its own random blinding factor b. However, a more precise analysis of attacks against the HB+ protocol has recently been shown.

Gilbert, Robshaw, and Sibert [42] discussed a man-in-the-middle attack against the HB+ protocol. In this attack, an attacker can obtain secret x by performing the following procedure. First, the attacker selects a constant $\delta \in \{0, 1\}^k$ and perturbs challenge a by sending $a \oplus \delta$ as a in all rounds of the HB+ protocol, pretending to be a reader R. If the authentication process is successful, the result is that $\delta \cdot x = 0$ with overwhelming probability. Here, the attacker can obtain partial information about secret x. Then, by repeating the procedure with different values of δ, the attacker can obtain the entire secret x. Once x has been obtained, the attacker can immediately impersonate the tag by setting the blinding factor as $b = 0$. Furthermore, another effect of the disclosure of x is that the privacy of tag T is compromised. The execution of this attack can be prevented by setting an alarm to go off when the number of failed authentication attempts exceeds a certain threshold. Katz and Shin [43] conducted a strict analysis of the HB+ protocol, and they concluded that the protocol is secure under concurrent and parallel composition as long as $0 < \eta < \frac{1}{4}$. Moreover, Katz and Smith [44] conducted an analysis for the case where $\frac{1}{4} \leq \eta < \frac{1}{2}$.

Gilbert, Robshaw, and Seurin [45] proposed modified HB+ protocols called HB$^{\#}$ and RANDOM-HB$^{\#}$. These protocols are constructed to be secure against the man-in-the-middle attack described above. The RANDOM-HB$^{\#}$ scheme is constructed by generalizing HB+. HB$^{\#}$ is a modification of RANDOM-HB$^{\#}$ for reducing storage costs. The security of the RANDOM-HB$^{\#}$ protocol was proven not only in the detection-based model (DET-model) but also in the GRS man-in-the-middle model (GRS-MIM-model), named after Gilbert, Robshaw, and Sibert. The DET-model is an adversarial model used in current proofs of security for HB+ and its variants. The GRS-MIM-model is an adversarial model that assumes stronger adversary than that in the DET-model and allows an active adversary to manipulate messages from the reader.

However, Ouafi, Overbeck, and Vaudenay [46] showed that HB$^{\#}$ and RANDOM-HB$^{\#}$ are also vulnerable to a general man-in-the-middle attack. In [46], the attack is also applied to various HB-like protocols. Though [46] presents the lower bound on the parameter set at which the attack can be prevented, such a setting is not acceptable in realistic RFID implementations.

17.6 Cryptographic Protocols for Privacy and Other Requirements

This section describes cryptographic protocols using hash functions, symmetric encryption, public-key encryption, and other means for protecting privacy and other required properties.

17.6.1 Approaches with Hash Functions

For privacy protection in an RFID system, various schemes using hash function and/or hash chain construction have been proposed. Early on, the main purpose for using a hash function and/or hash chain construction was to protect a tag's private information. Later, extended properties were also proposed using hash function and/or hash chain construction.

Syamsuddin *et al.* [47] conducted a survey of various hash-chain-based authentication schemes. In Syamsuddin's paper, various schemes proposed in recent years and the characteristics of each scheme are introduced. In early research on RFID security and privacy, Sarma, Weis, and Engels [48] presented the threats and security benefits of RFID systems. They proposed a privacy protection scheme called the hash-lock scheme, which requires only a hash function, and so it can be implemented at low cost. To avoid illegal ID scanning and ID leakage, the validity of reader R is checked using the following procedure: Reader R has a unique key k for each tag. Each tag T holds a hash value $h(= H(k))$, called a *meta-ID*, where H is the hash function. Tag T receives a request for ID access and sends *meta-ID* h to reader R. Reader R searches for key k, which is related to the *meta-ID* h, and sends key k to tag T. The tag receives key k, calculates the hash function using the received key k, and checks the relation of $h = H(k)$ with *meta-ID* h held in the tag. Tag T replies with its own ID to reader R, only if the checked relation holds. The scheme provides a protection scheme against ID leakage at low cost, however, it still cannot prevent an attacker from tracing a target tag, since the *meta-ID* is fixed and the attacker can trace the tag using the *meta-ID*. The *meta-ID* should be changed for each response to prevent this threat. The threat of traceability can be a problem in practical use.

As an extension of the hash-lock scheme, Weis *et al.* [49] proposed a randomized hash scheme. The randomized hash scheme requires a tag to have key k, a hash function, and a random number generator. The procedure of the scheme is as follows: A tag calculates the hash function H using a key k and random value r generated by the random generator, that is, $c = H(k|r)$, and sends reader R the hash value c and the random value r. Reader R maintains a database for key k, which is unique for each tag, and the tag's ID. The reader receives hash value c and random value r and checks the relation $c = H(k|r)$ for all keys k. If there is a key k that satisfies $c = H(k|r)$, then the tag's ID related to key k is determined from the database. The output from each tag T is changed with each response. Therefore, the randomized hash scheme can prevent tracing. However, if tag T is exposed, this scheme allows the location history of tag T to be traced. Therefore the randomized hash scheme cannot satisfy forward security. Moreover, the cost required for the random generator is significant.

17.6.2 Approaches for Forward Security with Hash Chain

Ohkubo, Suzuki, and Kinoshita [22] presented a new security concept for RFID security, called forward security, and proposed a scheme using a hash chain that provides forward security for privacy. The provided property is that an attacker cannot trace a tag back to past events in which the tag was involved, even if the attacker acquires the secret data stored in the tag. For the construction to achieve forward security, Ohkubo, Suzuki, and Kinoshita use the hash chain technique to renew the tag's secret information. A summary of the construction in the i-th transaction is as follows, where H and G are hash functions: Tag T (1) sends output $a_i = G(s_i)$ to reader R. (2) renews its secret information $s_{i+1} = H(s_i)$ as determined from its previous secret information s_i, (3) then, erases that previous secret information s_i The reader maintains the database, in which initial secret information s_0 and the tag ID are included, and the reader receives the tag's output and checks the relation $a_i = G(H^i(s_0))$ of the hash chain, for all initial secret information s_0 of tags as candidates, and then finds the ID related to the tag's output from the database. The required computation of the tag is efficient since the tag calculates only hash operations that can be constructed with small gates. Thus, the proposed scheme is practical, with privacy still preserved even in the face of tampering. However, the reader has to search for all indices i and all tags as candidates, so the calculations required for the reader increase according to the number of tags.

17.6.3 Approaches with Binary Tree

Molnar, Soppera, and Wagner [23] proposed a privacy protection scheme using tree structures, keys which are efficient and are allowed to delegate the ability to identify tags. The keys are generated using a hash tree, and the root value of the hash tree, which is stored in the tag. During each session, the tag generates its own output by calculating the hash tree from the root value. The calculation can be done efficiently through the tree structure. The ability to identify a tag is restricted to be within a certain period (number of sessions) and can be delegated to another person by providing the root value of the sub tree which corresponds to the period (i.e. number of sessions). The ownership of the tag can be transferred to another person without violation of privacy by the delegation property of the scheme. The new owner of the tag makes the tag update its own output a sufficient number of times after ownership is transferred. The previous owner can no longer identify the tag after the new owner uses up the sessions as mentioned above, since the identification ability of the previous owner is restricted to be within a certain number of communication attempts. As another approach, Lu *et al.* [50] proposed a tree-based authentication scheme called SPA(Strong and lightweight RFID Private Authentication protocol). This scheme provides dynamic key-updating, and thus forward security is satisfied.

17.6.4 Approaches with Block Ciphers

Avoine *et al.* [51] presented a symmetric key based authentication scheme. The summary of the scheme is as follows: The set of all tags is divided into groups of equal size. All tags in the same group have a common secret key. Since there are several tags in a group, reader R cannot identify a specific tag. Each tag also has a unique key, which is only shared between the tag and the reader. In a communication session, reader R sends a challenge to tag T. The tag receives the challenge, and computes two ciphertexts. One is the reader's challenge concatenated with a nonce determined by tag T and the tag's identifier, which is encrypted with the group key. The other is the challenge concatenated with the nonce encrypted with the tag's unique key. This second ciphertext is needed to prevent the tag from impersonating other tags in the same group. Moreover, Avoine *et al.* [51] compared the efficiency with a previous tree-based scheme.

17.6.5 Approaches with Lightweight Methods

Juels [52] proposed a significantly lightweight scheme. The proposed scheme requires only XOR calculations and therefore can be implemented at a low cost. In this scheme, reader R and tag T share a common list of random keys, and they confirm that the opponent party (i.e. the reader for tag, respectively, the tag for the reader) has the common list through several interactions. If the check in all of these interactions passes, the tag sends its ID. Large calculations are not required, and the tag needs only to perform the XOR operation. Several interactions between the tag and reader are required and the common list should be completely overwritten as needed to ensure security.

As an extended work of Juels [52], Castelluccia and Soos [53] proposed an identification scheme, which preserves the tag's private information, called the Probabilistic Identification Protocol [53]. The key is used as a bit-vector and is constructed as a combination of the constructions of the schemes of Juels [52] and Molnar *et al.* [23]. This identification scheme is stronger against a passive attacker than the original scheme.

17.6.6 Approaches with Public-Key Methods

Martinez *et al.* [54] presented a scheme based on an elliptic curve. The proposed scheme requires a zero-knowledge proof which is a proof of knowledge without leaking any information related to that knowledge. The heaviest operation that a tag is required is the identity verification of reader R and the generation of the next secret. Tag T renews the secret key so that forward security is ensured. Cui *et al.* [55] presented a tag authentication scheme based on a fast asymmetric encryption scheme, proposed by Niederreiter, which has a Knapsack-type construction. The scheme requires no exhaustive search or synchronization.

17.6.7 Approaches for Proof of Existences

This ability can be provided using the yoking-proof protocol [25], which proves that the outputs of two tags are received by the same reader simultaneously. The reader that receives the two tag's outputs can make sure that the two tags are together (i.e. in the same place or at the same point). The yoking-proof protocol [25] is vulnerable to replay attack. To prevent a replay attack against the yoking-proof protocol, Saito and Sakurai [56] proposed a modified yoking-proof protocol using a timestamp. Moreover, Piramuthu [57] proposed another modified yoking-proof protocol where a verifier should attend the protocol instead of using a timestamp. Afterwards, Cho *et al.* [58] proposed a stronger protocol called the enhanced yoking-proof protocol. They also presented an enhanced yoking-proof protocol for multiple tags. This technique can be used for applications that require this ability, such as pharmaceutical distribution, where medicine and its description should be distributed together.

17.6.8 Mutual Authentication

Generally, there are two types of authentication. One is the tag's authentication by the reader. The other is the reader's authentication by the tag. Most of the above schemes can be used as tag authentication. When both tag and reader authentication are required, it is called mutual authentication.

Recently, various mutual authentication schemes have been presented with privacy protection. However, some of them lack strict security analyses. Ouafi and Phan [59] introduced various schemes and showed the results of their analyses. Additionally, Yousuf and Potdar [60] analyzed mutual authentication schemes and showed the result of their analyses.

Paise and Vaudenay [18] presented a security model for mutual authentication. They present a classified model, show the relationships between some of them including impossible results, and mention case studies of recently proposed protocols.

Burmester, de Medeiros and Motta [34] presented a (mutual) authentication scheme that provides synchronization. In this scheme, reader R can renew the common secret key shared between it and tag T, and tag T also can renew its secret key by synchronizing with the secret key in reader R.

17.6.9 Approaches without Cryptography

A Faraday cage is used to block the wireless communication between a tag and a reader by covering the tag with a metal film. Illegal wireless tag scanning can be prevented using this method to protect privacy. However, it is not convenient for a person to cover all tags he/she carries. What is worse is that he/she may not be aware of some of the tags that he/she carries and may not cover them.

Another physical method for protecting privacy is a jamming signal. A jamming signal can be used to block the wireless communication between a tag and a reader, However, the jamming signal can harmfully affect other wireless devices.

EPC tags [1] are equipped with a kill command to protect user's privacy. This command can be used to permanently deactivate a tag. A PIN should be sent to the tag with this command to prevent illegal use. Only if the PIN is correct is this command executed. Once a kill command is executed, the tag will no longer respond to a reader. And so privacy can be perfectly protected after the kill command is executed. While it is not convenient that all tags carried by a person should be killed, he or she may not be aware of some of the tags. The greatest disadvantage of this method is that the tag is permanently deactivated and cannot be used again.

Juels, Rivest, and Szydlo [61] proposed a blocker tag. A blocker tag is a special tag to block the access of readers around the tag, and it interferes with the anti-collision protocol between a tag and reader in the RFID system. The blocker tag acts like a DoS attack against the reader to block its access to the tag.

17.7 Implementation

In this section, we describe implementation techniques that require only small gates and low power for computation.

Tuyls and Batina [62] presented an implementation of an elliptic curve computation that does not require as much computational power as previous ones. Their results show the possibility of using a public-key-based protocol for an RFID tag scheme. As an example, they show that the Schnorr identification scheme can be implemented based on an elliptic curve cryptography for the RFID tag authentication system. Kumar and Paar [63] constructed an optimized elliptic curve cryptography (ECC) processor for standards-compliant binary field curves. This construction requires only 10–18 kilogates in a 0.35-μm complementary metal oxide semiconductor (CMOS) process. Such studies increase the possibility of RFID tags using ECC, even though ECC is difficult to be implemented at low cost currently. Batina *et al.* [64] showed an implementation of the Okamoto identification scheme based on an elliptic curve. Oren and Feldhofer [65] reported in the implementation of an efficient public-key-based identification scheme called Weizmann-IAIK Public-key for RFID(WIPR). The 1214-bit WIPR requires 7505 gates, a mean current consumption of 10.88 μA, and 300 bits of random access memory (RAM).

McLoone and Robshaw [66] reported on an implementation of a public-key-based authentication protocol for RFID tag authentication using another approach. This protocol is based on an elliptic curve version of the GPS (Girault-Poupard-Stern) identification scheme. GPS is a public-key-based Schnorr-like identification scheme and requires only lightweight operations. Another proposed protocol [66] requires pre-computation. The calculation load required in the online phase is sufficiently light that a tag can perform the calculation. In summary, the protocol computes the first message in the pre-computation phase and stores the calculated value in the tag as a coupon. The following online phase is efficient, and the tag does not require elliptic scalar multiplication. However, each tag must have memory for recording the coupon. As extended work, Girault, Juniot, and Robshaw [67] proposed a prototype implementation of GPS in a field programmable gate array (FPGA). This implementation requires 2600 gates for the cryptographic components. As extended work, Hofferek and Wolkerstorfer [68] proposed a more efficient implementation, which requires only 800 gates equivalents and 560 bytes of storage as a result of relaxed latency requirements. The performance is almost the same as that of the original one, but security against DoS attacks cannot be provided.

Feldhofer, Dominikus and Wolkerstorfer [69] reported on an implementation of a symmetric-key-based scheme. They implemented a lightweight version of a symmetric key encryption scheme, the Advanced Encryption Standard (AES), which can be embedded in even low-performance RFID tags. Their implementation requires 3,595 gates and consumes 8.15 μA at a frequency of 100 kHz, and the number of clock cycles for 128-bit AES encryption is 1,000. They used an 8-bit architecture instead of the usual 32-bit architecture for the AES algorithm. This allows them to reduce the required number of S-boxes from four to one. This results in some good properties, that is, silicon resources can be conserved and power consumption is lower than that required for usual 32-bit operations. A new construction was proposed as a lightweight block cipher based on the data encryption standard (DES) by Poschmann *et al.* [70] that is called DES lightweight extension (DESL), which requires only light calculation and has a compact construction. The idea behind this lightweight implementation is to replace the original eight DES S-boxes with a single S-box and repeat the calculation eight times. As a result, this implementation could be achieved with 50 % smaller chips, 85 % fewer clock cycles, and 90 % less energy compared with the best AES implementation for RFID applications reported by Feldhofer, Dominikus, and Wolkerstorfer [69]. As a specific example, DESL requires 144 clock cycles to encrypt a 64-bit block plaintext. The average power consumption for one encryption at 100 kHz is 0.89 μA and that at 500 kHz is 4.45 μA. The throughput reaches 5.55 KB/s at 100 kHz and 27.78 KB/s at 500 kHz.

Vaudenay [71] proposed a new construction for a lightweight public-key encryption scheme for RFID tag authentication. The scheme requires only a small computation load so a tag can perform all the calculations. Moreover, Vaudenay [71] showed new security definitions and their relationships. This public-key-based encryption scheme is based on the problem of finding a sparse polynomial. The cost for encrypting a block message is only the computation of a linear feedback shift register (LFSR) and the generation of a biased random string; therefore, this scheme can be implemented with a small gate size and its characteristics are suitable for RFID tags.

Shamir [72] proposed a new type of property called SQUASH (short for square hash). SQUASH is suited to challenge-response type authentication protocols and it is proven to be as secure as the Rabin public-key encryption scheme. O'Neill (nee McLoone) [73] presented an architecture for the SHA-1 hash function. This method can be implemented in 130-nm CMOS, and it requires only 5,527 gates and consumes only 2.32 μW of power.

17.8 Real Systems and Attacks

In this section we describe examples of RFID systems already in use, such as e-passport, MiFare Card, KeeLoq, and EPC. We also describe attacks on these systems.

17.8.1 e-Passport

E-passports have already started using RFID for identification and/or authentication. E-passports are based on guidelines described in the International Civil Aviation Organization (ICAO). The cryptographic technique used in e-passports is as follows: First, ICAO [74, 75] described a Basic Access Control for the purpose of preventing skimming and eavesdropping attacks.

Juels, Molnar, and Wagner [76] pointed out some threats related to Basic Access Control. They pointed out that the entropy of the key is too small and it is difficult to revoke a reader's access once the e-passport has been read. Juels, Molnar, and Wagner also introduce measures to strengthen the protection scheme, for example, using Faraday cages, setting a larger secret for basic access control, and using a private collision avoidance protocol.

Bundesamt für Sicherheit in der Informationstechnik (BSI) [77] introduced an Extended Access Control, which uses a certificate of a trusted party called a Document Verifier. However, weaknesses were found by Hoepman *et al.* [78]. One weakness is that an RFID tag cannot check whether a certificate has expired. Another is that the hierarchy of the certificate has to be quite shallow.

Kosta *et al.* [79] discussed key security and privacy issues, and also presented the required technical measures for protecting privacy. Recently, there have been studies on the security and privacy of e-passports, and the results of analyses have been reported. For example, Blundo *et al.* [80] pointed out a weakness in Extended Access Control. They showed the possibility of transferability and reset attacks, and they presented a new notion for security against transferable as well as resettable attacks and a secure scheme that provides this property.

Avoine, Kalach, and Quisquater [81] also pointed out the vulnerability of the access control mechanism. The weakness arises from the key derivation mechanism, so they evaluated the basic access key entropy. Moreover, as an example, they also demonstrated their analysis on Belgian e-passports.

17.8.2 MiFare Card

A MiFare Card is basically a memory card with access control. Threats against cards called MiFare Classic are described, and methods of strengthening the security of MiFare Classic are also presented.

Teepe [82] described state restoration and cloning attacks against MiFare cards and a method for protecting against these attacks was also proposed. The authors stated that the key points for protecting RFID tag security against those threats are the key infrastructure, card-state signing, and card-state verification.

Garcia *et al.* [83] discussed an attack based on reverse engineering. In this attack, the secret key in a tag is extracted from one or two authentication events with a valid reader. Moreover, an attacker, who can obtain the tag's secret key by eavesdropping on the communication between the tag and reader, can decrypt the eavesdropped data. This would enable the attacker to clone the card or restore the attacked card to a previous state.

de Koning Gans, Hoepman, and Garcia [84] pointed out the weakness of using a pseudo-random generator, which enables recovery of the keystream generated using the stream cipher CRYPTO1. This weakness makes the stream cipher malleable.

17.8.3 KeeLoq

KeeLoq is an authentication protocol used as a remote keyless entry system for car doors and garages. In the first paper to point out its weakness, Bogdanov [85] showed two types of attacks: one uses the slide and guess-and-determine techniques and has a computation complexity of $2^{50.6}$ for encryption; the other uses cycle structure analysis in addition to the aforementioned two techniques and has a computational complexity of 2^{37} for encryption.

Indesteege *et al.* [86] discussed a stronger attack using the slide and man-in-the-middle techniques. It needs only 2^{16} known plaintext-ciphertext pairs in the best case, and its computational complexity is $2^{44.5}$.

Eisenbarth *et al.* [87] discussed an attack using side-channel attack techniques with software attack techniques against KeeLoq. By using this attack the secret key can be cloned without physically accessing the device. They also discuss a DoS attack.

17.8.4 Approach to Strengthen EPC

One type of EPC tag is the Class-1 Gen-2 tag. Recently, EPC tags of this type have been incorporated in United States passports and Washington State enhanced drivers licenses (WA EDLs). As a case study of applications of EPC tag, Koscher *et al.* [88] analyzed passport cards and EDLs then showed the results of vulnerability analyses, countermeasures, and recommendations.

Juels, Pappu, and Parno [89] discussed the importance of secure key management and proposed a procedure for distributing keys securely. Their scheme can be applied to actual EPC-based RFID tag systems. Intuitively, two schemes for secret sharing in unidirectional channels were proposed: one for secret-sharing across space and the other for secret-sharing across time. The former can be used as a tool for preserving privacy in RFID tag systems, for example, RFID-enabled supply chains. As an example of the latter, they proposed a family of sliding-window information secret-sharing (SWISS) schemes as an approach for helping authenticate tags and readers.

17.9 Conclusion

In this chapter we have surveyed studies pertaining to the security and privacy of RFID tags, especially ones based on cryptographic techniques. We presented various threats as well as definitions of RFID security and privacy from various viewpoints. Then, we reviewed current studies on cryptographic protocols for the privacy, authenticity, and implementation of cryptographic primitives and protocols.

Technical approaches for security requirements could raise the costs for constructing such tags. These high cost tags are less likely to become commonly used, and so a lightweight cryptographic technique can improve the security and privacy of RFID systems while lowering costs.

On the other hand, RFID systems have not only privacy requirements but also other various requirements. In some cases, a trade-off between privacy requirements and these other requirements such as low-cost and usability may be necessary. It is thus important to balance privacy with those other requirements.

Problems

1. List kinds of threats, then explain how the threats occur.
2. List kinds of required properties of RFID systems, then describe these properties.
3. Explain why forward security, which is one of the required properties, can be provided by using hash-chain schemes.
4. Present the strong and weak points of hash-chain type construction.
5. Show some examples in which cryptographic tools (algorithms) are used practically.
6. Raise some cases of RFID application in which privacy would be an important issue.

NB Solutions are provided on the book's website.

References

[1] EPCglobal. EPCglobal web site; Available at: http://www.epcglobalinc.org.

[2] CASPIAN (Consumers Against Supermarket Privacy Invasion and Numbering); (2002). Available at: http://www.nocards.org.

[3] Associated Press (2003) Benetton undecided on use of "smart tags"; 8 April.

[4] CNET. (2003) Wal-Mart cancels "smart shelf" trial; 9 July. Available at: http://www.cnet.com.

[5] Karygiannis, T., Eydt, B., Barber, G., Bunn, L., and Phillips, T. (2007) *Guidelines for Securing Radio Frequency Identification (RFID) Systems*.

[6] Rotter, P., Daskala, B., and Compano, R. (2008) RFID implants: Opportunities and challenges for identifying people. *IEEE Technology and Society Magazine*, **27**(2): 24–32.

[7] Juels, A. (2005) RFID security and privacy: A research survey, manuscript.

[8] Avoine, G. (2006) Bibliography on security and privacy in RFID systems. Available Online.

[9] Rotter, P. (2008) A framework for assessing RFID system security and privacy risks, *IEEE Pervasive Computing*, **7**(2): 70–77.

[10] Brands, S. and Chaum, D. (1993) Distance-bounding protocols (extended abstract), in *EUROCRYPT*, pp. 344–359.

[11] Clulow, J., Hancke, G.P., Kuhn, M.G., and Moore, T. (2006) So near and yet so far: distance-bounding attacks in wireless networks, in *ESAS*, pp. 83–97.

[12] Rieback, M.R., Crispo, B., and Tanenbaum, A.S. (2006) Is your cat infected with a computer virus? In *PerCom*, pp. 169–179.
[13] Bolotnyy, L. and Robins, G. (2007) Physically unclonable function-based security and privacy in RFID systems, in *PerCom*, pp. 211–220.
[14] Oren, Y. and Shamir, A. Power analysis of RFID tags; 2006. panel discussion in RSA Conference 2006.
[15] Fabian, B., Günther, O., and Spiekermann, S. (2005) Security analysis of the object name service for RFID, in *International Workshop on Security, Privacy and Trust in Pervasive and Ubiquitous Computing – SecPerU'05*.
[16] Damgård, I. and Østergaard, M. (2006) RFID security: Tradeoffs between security and efficiency, *Cryptology ePrint Archive*, Report 2006/234.
[17] Vaudenay, S. (2007) On privacy models for RFID, in *ASIACRYPT*, pp. 68–87.
[18] Paise, R.I. and Vaudenay, S. (2008) Mutual authentication in RFID: security and privacy. In *ASIACCS*, pp. 292–299.
[19] Avoine, G., and Oechslin, P. (2005) RFID traceability: a multilayer problem, in *Financial Cryptography*, pp. 125–140.
[20] Avoine, G. (2005) *Adversary Model for Radio Frequency Identification*. Swiss Federal Institute of Technology (EPFL), Security and Cryptography Laboratory (LASEC); LASEC-REPORT-2005-001.
[21] Juels, A. and Weis, S.A. (2007) Defining strong privacy for RFID, in *PerCom Workshops*, pp. 342–347.
[22] Ohkubo, M., Suzuki, K., and Kinoshita, S. (2003) Cryptographic approach to "privacy-friendly" tags, *RFID Privacy Workshop*.
[23] Molnar, D., Soppera, A., and Wagner, D. (2005) A scalable, delegatable pseudonym protocol enabling ownership transfer of RFID tags, in *Selected Areas in Cryptography*, pp. 276–290.
[24] Ohkubo, M., Suzuki, K. (2006) Forward secure RFID privacy protection scheme with restricted traceability, in *ACNS – Industrial Track Proceedings*, pp. 1–16.
[25] Juels, A. (2004) "Yoking-proofs" for RFID tags, in *PerCom Workshops*, pp. 138–143.
[26] Munilla, J., Ortiz, A., and Peinado, A. (2006) Distance bounding protocols with void-challenges for RFID, in *Workshop on RFID Security – RFIDSec'06*.
[27] Tu, Y.J. and Piramuthu, S. (2007) RFID Distance bounding protocols, in *First International EURASIP Workshop on RFID Technology*.
[28] Kim, C.H., Avoine, G., Koeune, F., Standaert, F.X., and Pereira, O. (2008) The Swiss-knife RFID distance bounding protocol, in *ICISC*, pp. 98–115.
[29] Munilla, J. and Peinado, A. (2008) Distance bounding protocols for RFID enhanced by using void-challenges and analysis in noisy channels, *Wireless Communications and Mobile Computing*, **8**(9): 1227–1232.
[30] Hancke, G. and Kuhn, M. (2005) An RFID distance bounding protocol, in *Conference on Security and Privacy for Emerging Areas in Communication Networks – SecureComm 2005*, pp. 67–73.
[31] Reid, J., Nieto, J.M.G., Tang, T., and Senadji, B. (2007) Detecting relay attacks with timing-based protocols, in *ASIACCS*, pp. 204–213.
[32] Avoine, G. and Tchamkerten, A. (2009) An efficient distance bounding RFID authentication protocol: balancing false-acceptance rate and memory requirement, in *Information Security Conference – ISC'09*.
[33] Canard, S. and Coisel, I. (2008) Data synchronization in privacy-preserving RFID authentication schemes, in *Workshop on RFID Security – RFIDSec'08*.
[34] Burmester, M., de Medeiros, B., and Motta, R. (2008) Robust, anonymous RFID authentication with constant key-lookup, in *ASIACCS*, pp. 283–291.
[35] Inoue, S., Yasuura, H. (2003) RFID Privacy Using User-controllable Uniqueness. RFID Privacy Workshop.
[36] Juels, A., Pappu, R. (2003) Squealing Euros: Privacy Protection in RFID-Enabled Banknotes. In *Financial Cryptography*; pp. 103–121.
[37] Ishikawa, T., Yumoto, Y., Kurata, M., Endo, M., Kinoshita, S., Hoshino, F. *et al.* (2003) Applying Auto-ID to the Japanese publication business, *Auto-ID Center*, KEI-AUTOID-WH-004.
[38] Golle, P., Jakobsson, M., Juels, A., and Syverson, P.F. (2004) Universal re-encryption for Mixnets, in *CT-RSA*, pp. 163–178.
[39] Ateniese, G., Camenisch, J., and de Medeiros, B. (2005) Untraceable RFID tags via insubvertible encryption, in *ACM Conference on Computer and Communications Security*, pp. 92–101.
[40] Hopper, N.J. and Blum, M. (2001) Secure human identification protocols, in *ASIACRYPT*, pp. 52–66.
[41] Juels, A. and Weis, S.A. (2005) Authenticating pervasive devices with human protocols, in *CRYPTO*, pp. 293–308.

[42] Gilbert, H., Robshaw, M., and Sibert, H. (2005) An active attack against HB^+ – A provably secure lightweight authentication protocol, in *IEE Electronic Letters*, **41**(21): 1169–1170.
[43] Katz, J. and Shin, J.S. (2006) Parallel and concurrent security of the HB and HB^+ protocols, in *EUROCRYPT*, pp. 73–87.
[44] Katz, J. and Smith, A. (2006) Analyzing the HB and HB+ protocols in the "large error" case; *Cryptology ePrint Archive*, Report 2006/326.
[45] Gilbert, H., Robshaw, M.J.B., and Seurin, Y. (2008) $HB^{\#}$: Increasing the security and efficiency of HB^+, in *EUROCRYPT*, pp. 361–378.
[46] Ouafi, K., Overbeck, R., and Vaudenay, S. (2008) On the security of HB# against a man-in-the-middle attack, in *ASIACRYPT*, pp. 108–124.
[47] Syamsuddin, I., Dillon, T., Chang, E., and Han, S. (2008) A survey of RFID authentication protocols based on hash-chain method, *Convergence Information Technology, International Conference on*. **2**: 559–564.
[48] Sarma, S.E., Weis, S.A., and Engels, D.W. (2002) RFID systems and security and privacy implications, in *CHES*, pp. 454–469.
[49] Weis, S.A., Sarma, S.E., Rivest, R.L., and Engels, D.W. (2003) Security and privacy aspects of low-cost radio frequency identification systems, in *SPC*, pp. 201–212.
[50] Lu, L., Han, J., Hu, L., Liu, Y., and Ni, L.M. (2007) Dynamic key-updating: privacy-preserving authentication for RFID systems, in *PerCom*, pp. 13–22.
[51] Avoine, G., Buttyán, L., Holczer, T., and Vajda, I. (2007) Group-based private authentication, in *WOWMOM*, pp. 1–6.
[52] Juels, A. (2004) Minimalist cryptography for low-cost RFID tags, in *SCN*, pp. 149–164.
[53] Castelluccia, C. and Soos, M. (2007) Secret shuffing: a novel approach to RFID private identification, in *Workshop on RFID Security – RFIDSec'07*, pp. 169–180.
[54] Martinez, S., Valls, M., Roig, C., Gine, F., and Miret, J. (2007) An elliptic curve and zero knowledge based forward secure RFID protocol, *Workshop on RFID Security – RFIDSec'07*.
[55] Cui, Y., Kobara, K., Matsuura, K., and Imai, H. (2008) Lightweight privacy-preserving authentication protocols secure against active attack in an asymmetric way, *IEICE Transactions*, **91-D**(5): 1457–1465.
[56] Saito, J. and Sakurai, K. (2005) Grouping proof for RFID tags, in *AINA*, pp. 621–624.
[57] Piramuthu, S. (2006) On existence proofs for multiple RFID tags, in *Pervasive Services 2006*, pp. 371–320.
[58] Cho, J.S., Yeo, S.S., Hwang, S., Rhee, S.Y., and Kim, S.K. (2008) Enhanced yoking proof protocols for RFID tags and tag groups, in *AINA Workshops*, pp. 1591–1596.
[59] Ouafi, K. and Phan, R.C.W. (2008) Traceable privacy of recent provably-secure RFID protocols, in *ACNS*, pp. 479–489.
[60] Yousuf, Y. and Potdar, V. (2008) A survey of RFID authentication protocols, in *AINA Workshops*, pp. 1346–1350.
[61] Juels, A., Rivest, R.L., and Szydlo, M. (2003) The blocker tag: selective blocking of RFID tags for consumer privacy, in *ACM Conference on Computer and Communications Security*, pp. 103–111.
[62] Tuyls, P. and Batina, L. (2006) RFID-Tags for anti-counterfeiting, in *CT-RSA*, pp. 115–131.
[63] Kumar, S. and Paar, C. (2006) Are standards compliant elliptic curve cryptosystems feasible on RFID? Printed handout of Workshop on RFID Security – RFIDSec'06.
[64] Batina, L., Guajardo, J., Kerins, T., Mentens, N., Tuyls, P., and Verbauwhede, I. (2007) Public-key cryptography for RFID-tags, in *PerCom Workshops*, pp. 217–222.
[65] Oren, Y. and Feldhofer, M. (2008) WIPR – a public key implementation on two grains of sand, in *Workshop on RFID Security – RFIDSec'08*.
[66] McLoone, M. and Robshaw, M.J.B. (2007) Public key cryptography and RFID tags, in *CT-RSA*, pp. 372–384.
[67] Girault, M., Juniot, L., and Robshaw, M. (2007) The feasibility of on-the-tag public key cryptography, workshop on RFID Security – RFIDSec'07.
[68] Hofferek, G. and Wolkerstorfer, J. (2008) Coupon recalculation for the GPS authentication scheme, in *CARDIS*, pp. 162–175.
[69] Feldhofer, M., Dominikus, S., and Wolkerstorfer, J. (2004) Strong authentication for RFID systems using the AES algorithm, in *CHES*, pp. 357–370.
[70] Poschmann, A., Leander, G., Schramm, K., and Paar, C. (2006) A family of light-weight block ciphers based on DES suited for RFID applications. Printed handout of Workshop on RFID Security – RFIDSec'06.

[71] Vaudenay, S. (2006) RFID privacy based on public-key cryptography, in *ICISC*, pp. 1–6.
[72] Shamir, A. (2008) SQUASH – A new MAC with provable security properties for highly constrained devices such as RFID tags, in *FSE*, pp. 144–157.
[73] O'Neill (nee McLoone), M. (2008) Low-cost SHA-1 hash function architecture for RFID tags, Workshop on RFID Security – RFIDSec'08.
[74] ICAO (2004) Development of a logical data structure – LDS for optional capacity expansion technologies, revision 1.7. ICAO; Technical report.
[75] ICAO (2004) PKI for machine readable travel documents offering ICC read-only access version – 1.1. ICAO; Technical report.
[76] Juels, A., Molnar, D., and Wagner, D. (2005) Security and privacy issues in e-passports, in *SecureComm*, pp. 74–88.
[77] BSI (2006) Advanced security mechanisms for machine readable travel documents – extended access control (eac). BSI; Technical report.
[78] Hoepman, J.H., Hubbers, E., Jacobs, B., Oostdijk, M., and Schreur, R.W. Crossing borders: Security and privacy issues of the European e-passport. CoRR. 2008;abs/0801.3930.
[79] Kosta, E., Meints, M., Hansen, M., and Gasson, M. (2007) An analysis of security and privacy issues relating to RFID enabled epassports, in *SEC*, pp. 467–472.
[80] Blundo, C., Persiano, G., Sadeghi, A.R., and Visconti, I. (2008) Resettable and non-transferable chip authentication for epassports. Workshop on RFID Security – RFIDSec'08.
[81] Avoine, G., Kalach, K., and Quisquater, J.J. (2008) epassport: Securing international contacts with contactless chips, in *Financial Cryptography*, pp. 141–155.
[82] Teepe, W. (2008) Making the best of Mifare Classic, manuscript.
[83] Garcia, F.D., de Koning Gans, G., Muijrers, R., van Rossum, P., Verdult, R., Schreur, R.W., and Jacobs, B. (2008) Dismantling MIFARE Classic, in *ESORICS*, pp. 97–114.
[84] de Koning Gans, G., Hoepman, J.H., and Garcia, F.D. (2008) A practical attack on the MIFARE Classic, in *CARDIS*, pp. 267–282.
[85] Bogdanov, A. (2007) Attacks on the KeeLoq block cipher and authentication systems, Workshop on RFID Security – RFIDSec'07.
[86] Indesteege, S., Keller, N., Dunkelman, O., Biham, E., and Preneel, B. (2008) A practical attack on KeeLoq, in *EUROCRYPT*, pp. 1–18.
[87] Eisenbarth, T., Kasper, T., Moradi, A., Paar, C., Salmasizadeh, M., and Shalmani, M.T.M. (2008) On the power of power analysis in the real world: A complete break of the KeeLoqCode hopping scheme, in *CRYPTO*, pp. 203–220.
[88] Koscher, K., Juels, A., Kohno, T., and Brajkovic, V. (2008) EPC RFID tags in security applications: passport cards, enhanced drivers licenses, and beyond. RSA Laboratories; Manuscript.
[89] Juels, A., Pappu, R., and Parno, B. (2008) Unidirectional key distribution across time and space with applications to RFID security, in *USENIX Security Symposium*, pp. 75–90.

18

Novel RFID Technologies: Energy Harvesting for Self-Powered Autonomous RFID Systems

Raj Bridelall[1] and Abhiman Hande[2]

[1]*Axcess International, Inc.*

[2]*Texas Micropower, Inc.*

18.1 Introduction

Radio Frequency (RF) tags find applications in several areas including pallet/container tracking for supply chain management, identification (ID) badges for personnel access control, vehicle tracking for parking lot access, product condition monitoring for manufacturing quality control, and many more. RFID circuits coupled with transducers are also used for wireless sensing applications that involve monitoring the health of structures such as bridges, roadways, pipelines, and buildings for basic maintenance and more critical defects such as corrosion and fractures. These RFID-based technologies have been deployed to track and monitor the condition of materiel in the military supply chain such as subsistence, ordnance, and replacement parts as they are transported by airplanes, ships, trailers, tanks, and other modes of transportation. Other applications that combine RFID with transducers for wireless sensing include vehicular systems for monitoring wheel hub-odometers, tire pressure and cargo tamper sensors to improve safety, reliability, and reduce fleet maintenance costs.

As these types of applications proliferate, it will become necessary to deploy RFID-based sensors in hard-to-reach places. Once thousands of such RF devices are deployed for any given application, replacing batteries will become an impractical task. Therefore, self-sufficient devices that can operate for an indefinite period of time will be required.

RFID Systems: Research Trends and Challenges Edited by Miodrag Bolić, David Simplot-Ryl, and Ivan Stojmenović

Section 18.2 examines novel low power RFID-based sensor architectures that are designed for energy harvesting (EH) power supplies. Architectures such as Dual-Active RFID incorporate both near-field and far-field physical layers to wake up on demand and transmit information only when entering or passing through predefined control zones. Chapter 2 in this volume provides an in-depth description of this hybrid near-field and far-field operation enabling both proximity communications using magnetic fields, and long-distance communications using far-field propagation. *Micro-Wireless*™ extends the Dual-Active architecture by combining additional physical layer (PHY) types for seamless roaming between multiple wireless infrastructures. This multi-protocol and multi-frequency RFID capability has become essential since existing infrastructures contain one or more type of interrogators in different strategic locations to enable their practical use. For example, proximity door and gate controllers use near-field readers at those locations for access control, while far-field receivers are strategically hidden from view to capture any asset-tamper alerts or personnel panic alerts. UHF backscatter interrogators for passive and semi-passive RFID are generally deployed at the choke-points of a logistical process, for example, dock doors through which inventory and assets must transition. Therefore, combined near-field and far-field interrogators are not practical, due to typically inconsistent requirements for co-location. Even if co-location is possible, replacement is not sufficiently cost effective, for example, when considering that old interrogators must be ripped out and replaced with potentially more expensive devices throughout the entire infrastructure.

Micro-Wireless solutions leverage highly integrated system-on-a-chip (SoC) implementations and software defined radio (SDR) techniques to minimize cost and size. SoC devices facilitate added functionality, larger non-volatile read/write memory, and flexible interfaces for the analog and digital transducers needed for a complete wireless sensor. SDR techniques provide the ability for adaptable air interfaces and real-time functional adaptations. However, Dual-Active and Micro-Wireless architectures rely on stored on-board energy to enable greater range, higher throughput, and more robust performance in electromagnetically unfriendly environments. Therefore, various forms of EH technology are utilized to supplement or fully replace batteries.

Section 18.3 introduces EH approaches suitable for low-power RFID with adaptations that also support wireless sensor network modalities. Traditional RFID technologies communicate directly with networked interrogators while traditional wireless sensor network architectures require that nodes communicate directly with each other, forming a pathway for data transport leading to a final destination which is typically a network connected gateway. Micro-Wireless architectures include such adaptations, and their communications protocols must, therefore, be context- and energy aware. That is, each Micro-Wireless node must be capable of repeatedly recovering from power failures and to seamlessly re-establish connectivity to an available network, as well as support anytime interrogation from an RFID device such as a hand-held. Some forms of Micro-Wireless devices will support RFID interrogation in a passive mode by harvesting RF energy from the interrogator. Since this chapter focuses on novel RFID technologies, other forms of EH transducers such as solar, vibration, and thermoelectric will be explored instead. The energy source will vary with application. For example, experiments find that the level of vibration energy produced from various types of moving vehicles and structures such as bridges are adequate for vibration-based energy harvesting tags and RFID based wireless sensors [1].

Other forms of EH, such as solar and thermoelectric are more suited for applications such as unattended ground sensors where vibrations are absent but a temperature differential between the air and in the ground is available. Technology providers are also developing hybrid approaches that optimally combine various forms of EH transducers to meet these new challenges in self-sufficient wireless sensors. Section 18.4 will cover these and other future trends to construct EH devices from thin film micro-electro-mechanical systems (MEMS) to enable high volume and low-cost roll-to-roll manufacturing techniques.

18.2 Novel Low Power Architectures

Common low power RFID architectures utilize a single PHY type with a fixed multi-tag collision arbitration protocol or media access control (MAC). From Chapter 2, it is evident that no single RFID PHY and MAC combination sufficiently addresses the needs of a given application. This places a burden on the RFID interrogator infrastructure to support multiple combinations of RF links, and this can be impractical and very expensive, as previously described. The alternative approach is to employ the flexibility in the RF tag itself. Therefore, emerging tag architectures are combining several PHY and incorporating an ability to adapt the MAC, based on feedback from an interrogator tied into an enterprise application layer. Unfortunately, a brute force combination of several PHY will tend to increase cost, size, and power consumption. This section examines PHY and MAC combinations that yield synergies to sufficiently minimize power consumption for use with low cost and small form-factor EH technologies.

18.2.1 Dual-Active Standards

Some commercial products utilize separate physical layers for the signal transmission and signal reception. In particular, tags that comply with the ISO18185 and ISO24730 standard receive data from excitation signals near 125 kHz (LF) and transmit at either 433.92 MHz or 2.45 GHz. This approach benefits from the zone control accuracy and dense media signaling robustness of a near-field LF link as well as the long-distance, high data rate characteristics of a far-field link. In addition, the Dual-Active architecture inherently supports both the near-field proximity and the far-field computational positioning methods of RTLS described in Chapter 2.

The Dual-Active architecture marks a rather significant departure from the homogeneous PHY and MAC architectures of traditional RFID tags. Nevertheless, the relevant ISO standards still limit both the near-field and the far-field links to a single PHY and MAC that does not adapt when moving between different infrastructures. Furthermore, these standards limit the near-field link to a receive-only mode. The Dual-Active architecture, however, more naturally facilitates low-power operational modes where the tag wakes up from a sleep mode to transmit information to an interrogator only when it enters pre-determined near-field zones of interest. These types of tags will transmit only when an event occurs, such as when entering a room, leaving a building, or when a transducer parameter exceeds some predetermined value. This mode of operation obviates the need for periodic transmissions at the high duty cycles required for real-time information exchange, henceforth, significantly reducing power consumption.

Unlike passive UHF RFID that detunes near dense media such as wet dirt, an LF near-field link facilitates installations where activation loop antennas can be buried in the ground or concealed within foliage. This ability to work in dense RF media is particularly important for deployments in mines and construction sites. For example, the near-field backscatter mode of a Micro-Wireless tag can be utilized for access control at the mine entrance, and UHF receivers placed periodically along the tunnel can pick up health and condition monitoring signals, including signal strength information to determine relative position within the tunnels. In addition, miners can push a button on the tag to enable a high periodicity UHF transmission signaling a "panic" alert to initiate an emergency response process. Multiple near-field loops can also provide speed and travel direction by analyzing time-stamps as tags move between consecutive near-field zones.

18.2.2 Micro-Wireless RFID

The term *Micro-Wireless RFID* refers to a tag architecture that supports some combination of all four basic RF PHY introduced in Chapter 2. These are near-field, far-field, signal emission, and signal reflection. In addition to implying small size, "micro" relates to their use of a programmable microcomputer element for real-time PHY switching and MAC adaptation as the tag moves between operational modalities and infrastructure [2]. This combination of selectable PHY and adaptable MAC introduces seamless infrastructure roaming capabilities for RFID tags. A seamless roaming feature also addresses concerns about the fragmentation of RFID installations, and the proliferation of standards becoming a barrier to mass adoption.

A brute force combination of PHY and MAC can bloat a design and increase cost, size, and power consumption. Therefore, a Micro-Wireless design focuses on combinations that yield both improved performance and functional synergies for the class of applications targeted. The Micro-Wireless Dot™ tags produced by Axcess International, Inc. are examples of commercial implementations featuring the Micro-Wireless architecture [3]. The SoC solution is based on the architecture illustrated in Figure 18.1. As illustrated, the PHY supported includes a narrow-band far-field transceiver, a wide-band UHF far-field backscatter transceiver that can support a semi-passive mode of operation, and an LF near-field load-modulation transceiver. As with Dual-Active architectures, the

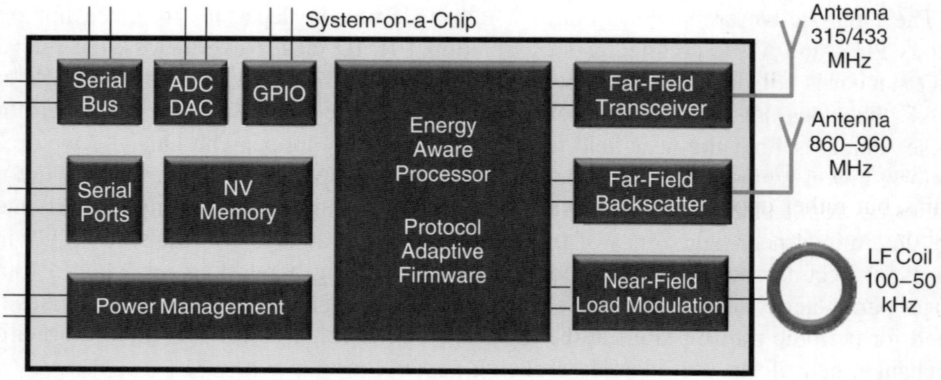

Figure 18.1 Micro-Wireless Dot™. Courtesy, Axcess International.

Micro-Wireless architecture achieves those identical synergies realized from incorporating wake-up power management through zone-based RTLS. The near-field LF modulation simultaneously populates all tags present in that LF zone with its activation identification (AID). Tags are capable of differentiating valid AID modulations from noise while maintaining a low-power sleep state. Once any Micro-Wireless Dot validates the AID, it enters a fully operational state in preparation to transmit its AID and unique identification (UID). Depending on the infrastructure, Micro-Wireless Dot tags transmit AID, UID, and other data by adapting to the appropriate near-field or far-field link.

The near-field AID may optionally initiate changes in the tag's behavioral modes if needed. For example, tags can enter a UHF *beaconing* mode to transmit an identification signal at longer distances than near-field, and at some prescribed duty-cycle if someone enters a mining tunnel. The tag can then cease UHF transmissions to save power after the person exits the tunnel. Multiple far-field receivers strategically placed inside the tunnel optionally capture the tag's signal for course position determination using techniques such as triangulation or trilateration within the tunnel. Attached transducers and actuators such as panic buttons can also initiate far-field UHF transmissions at anytime. Authorized personnel may also use the built in LF near-field load modulation capability to open gates or doors to secure facilities.

18.2.3 Semi-Active

Low-power active and semi-passive tags are well suited for battery substitution with EH power supplies. However, the natural energy sources required may not be necessarily available throughout the life of the tagging or sensing application. The EH device's volumetric conversion efficiency (W/m^3) and the energy source intensity determine the amount of energy that can be harvested. A state-of-the-art vibration EH device, for example, can deliver $100\,\mu W/cm^3$ of power from a vibrating source producing $0.1\,g_{peak}$ between five and 30 Hz [4]. The g force measurement ($1\,g = 9.8\,m/s^2$) is a measure of the excitation (or acceleration) of the structure. A low-power active tag consumes energy at a rate proportional to its transmission periodicity. Therefore, the EH device must be appropriately sized to store sufficient energy in a reservoir when it is available in order to match the application's average energy demand over the device's operating life. The lowest power RFID architectures will require the smallest reservoirs for the same transmission periodicity, and consequently result in smaller overall implementations. Active RFID implementations that can still meet their application demands with a renewable energy source characterized by constraints in its energy availability can be categorized as Semi-Active RFID. The subtle difference between Active and Semi-Active architectures is that the latter is optimized for use with renewable energy sources which are not constant or finite, but rather opportunistic and essentially infinite.

For example, Table 18.1 shows an example of a one cubic centimeter vibration harvester that is capable of supplying 100 micro-Watts of power at low frequencies at about 0.1 g excitation. A semi-active RFID tag that consumes 4.8 milli-amperes at three volts for 40 milli-seconds when operational, and ten micro-amperes in sleep state, will consume a total of 2.7 watt-sec of energy for 200 transmissions per day. The EH device can replenish this 2.7 watt-sec of energy consumed after 7.5 hours of continuous vibrations. If only five hours of sustained vibrations were available per day, then the EH

Table 18.1 RFID/Sensor lifetime calculation based on vibration energy available.

Axcess tag powered by battery	3V
Power available from vibrations	100 μW
Current available from vibrations (i_{source})	33.33 μA
Transmission of ID to receiver takes	40 ms
Current draw during transmission (i_{active})	4.8 mA
Energy necessary/transmission	0.000576 Watt-sec.
Number of transmissions/day	200
Transmission time/day (t_2)	8 sec.
Energy necessary for transmissions	0.12 Watt-sec.
Idle state current draw (i_{sleep})	10 μA
Idle state time/day (t_2)	86392 sec.
Energy necessary for idle conditions/day	2.59 Watt-sec.
Total energy reqd (idle + transmission)/day	2.71 Watt-sec.
Time required to gather this energy from vibrations/day (t_3)	7.52 hrs

device must produce 150 micro-watts of power. Therefore, the same state-of-the-art EH device must be about one half cubic centimeter larger if operations cannot pause during long periods of energy absence. Alternatively, for a given energy harvester, the tag energy requirement can be scaled down by scaling down the duty cycle. Since the average energy availability is constrained by the application conditions, RFID architectures with higher power requirements and/or transmission rates would need larger harvesters with larger reservoirs for the same vibration amplitudes, frequency, and periods of operation indicated above.

18.3 Energy Harvesting Optimized for RFID

The key building blocks of a semi-active sensor are the EH transducer, the power management, and the energy reservoir. Figure 18.2 shows a typical architecture. Maximum power transfer efficiency is realized when the physical transducer characteristics are tuned to those of the natural energy sources. For example, vibration transducers should resonate around the source's spectral response, and the optical filters of a light energy transducer should match those of the source spectra. The power management module must consistently adjust to provide matching impedance characteristics between the source, the reservoir, and the load. The power electronics must have high efficiency otherwise their overhead can result in lack of output power to the reservoir in spite of maximizing power harvester from the EH transducer. The storage device should provide sufficiently large volumetric energy density, fast charge transfer times, and be able to withstand the number of RFID charge-discharge cycles over the expected lifetime. For example, if a capacitor is used as the storage device in the example in Table 18.1, its size will depend upon the required capacitance (C) in Farads. The capacitance, in turn, is obtained by the maximum allowable voltage drop (ΔV) that can be sustained without affecting power delivered to the load. For a maximum ΔV of 0.5 V per day of operation, C can be obtained using the

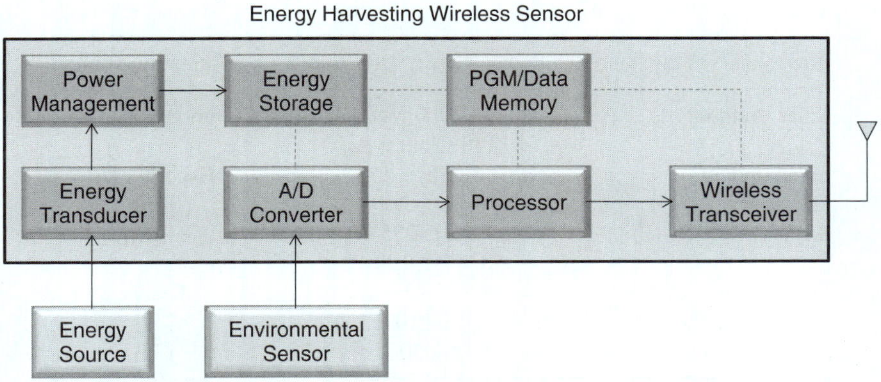

Figure 18.2 Block diagram of an EH sensor. Reprinted with kind permission from © 2008 Springer Science and Business Media.

load parameters as shown below:

$$\Delta V = \frac{1}{C}\left(i_{sleep} \times t_1 + i_{active} \times t_2\right) \quad (18.1)$$

Substituting values in Table 18.1 in Equation 18.1, it can be seen that C must be at least 1.8 Farad. A similar analysis can be obtained by using source parameters as indicated below:

$$\Delta V = \frac{1}{C}\left(i_{source} \times t_3\right) \quad (18.2)$$

For a lower ΔV requirement, C will have a higher value and therefore will have a larger size. Similarly, for lower load power requirements, C will have a lower value and therefore a smaller size.

The load, which is the RFID tag and associated transducers, may demand high power levels for relatively short periods, and just a trickle of energy during sleep modes. The best power management sub-systems continuously adapt to support the tag's energy demand profile across a broad spectrum of applications.

Semi-Active RFID implementations can harvest vibration energy from piezoelectric transducers, light energy from solar cells, and heat flow energy from thermoelectric transducers. These can be either standalone or used in concert with other traditional power sources and storage devices. Table 18.2 compares the power generation potential of some of the typical EH modalities which include ambient radiation [5], temperature gradients [6], light [7], and vibrations [8–11]. Among these, solar EH through photovoltaic conversion, and vibration EH through piezoelectric elements provide relatively higher power densities. The energy harvested from any one source is of the order of a few hundred microwatts using a practical transducer. Therefore, technologists seek to combine multiple sources in order to boost the harvesting capability. However, this requires efficient power management (PM) circuit design and possibly a single PM solution to minimize size and cost.

Table 18.2 Power densities of EH technologies.

Energy harvesting source	Power density ($\mu W/cm^3$)	Information source
Solar (outdoors)	15,000 Direct sun 150 Cloudy day	Commonly available
Solar (indoors)	6 Office desk	Experiments
Vibrations	100–200	[10]
Acoustic noise	0.003 @75 dB 0.96 @ 100 dB	Theory
Daily temp. variation	10	Theory
Temp. gradient	15@10°C	[35]
Piezo shoe inserts	330	[36]

18.3.1 Solar Cells

Figure 18.3 shows a typical EH solution for harvesting solar energy where a DC-DC converter maximizes the energy transfer. Figure 18.4 shows the V-I characteristics of a single 25-mm × 60-mm silicon photovoltaic (PV) panel under different lighting conditions. These curves show the relationship between two key parameters, the open circuit voltage (V_{OC}) and the short circuit current (I_{SC}), that influence energy conversion. Each parameter form the x- and y- intercepts of the V-I curve, respectively as shown in Figure 18.4. The curves demonstrate that a solar panel behaves as a voltage limited current source and that the short circuit current $\mathbf{I_{SC}}$ is proportional to the light intensity.

A power management scheme is required to regulate the power from what is essentially a current source transducer, and deliver it to storage devices such as rechargeable batteries or ultracapacitors. Efficient power management circuits utilize maximum power point tracking (MPPT) techniques to harvest the most energy possible from a given light intensity [12]. For example, Figure 18.4 indicates that the selected panel produces a maximum power of about 200 micro-watts at two volts when the light intensity is about 300 lux. Analog circuits, such as the DC-DC buck-boost converter shown in Figure 18.5, track the optimum voltage V_{MPPT} for different light intensities more efficiently as compared to circuits using higher power consumption microcontrollers [12].

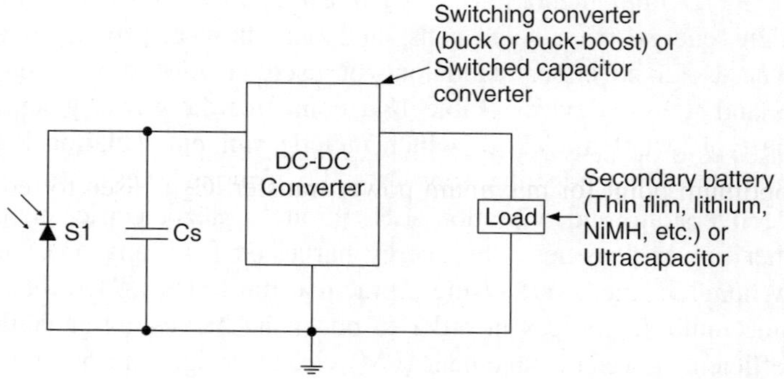

Figure 18.3 EH system for solar cells.

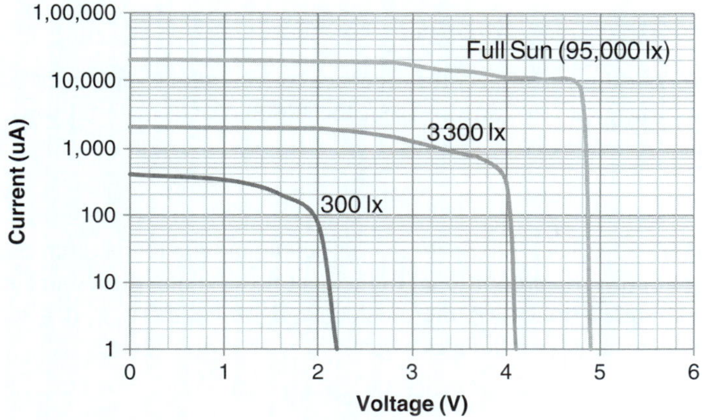

Figure 18.4 Solar panel V-I characteristics.

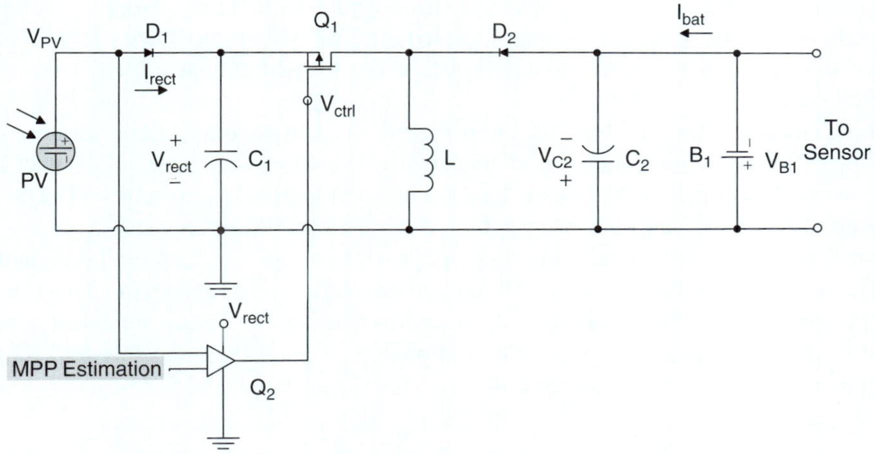

Figure 18.5 DC-DC buck-boost converter for solar EH.

The DC-DC buck-boost converter monitors the voltage across the input capacitor, C_1, until it increases above V_{MPPT}. At this point the MOSFET switch, Q_1, is turned on to route energy to the storage device, B_1. The converter turns off when the input capacitor voltage decreases to drop below V_{MPPT} thus maintaining the average voltage at V_{MPPT}, which is the optimum point for maximum power transfer. C_2 is used to reduce the ripple current across B_1, and therefore preserves its lifetime. Since V_{OC} scales linearly with light intensity and V_{MPPT} scales linearly with V_{OC}, then measuring V_{OC} leads to a simple estimation of V_{MPPT}. Such circuits have been shown to have efficiencies of up to 80% [12]. The designers must complete this characterization for each solar cell prior to its implementation in a control circuit. It is, however, possible to use a fixed reference voltage for the MPP estimation input at comparator Q_2 given known lighting conditions, such as indoor applications.

18.3.2 Thermoelectric Transducers

Thermoelectric transducers utilize the Seebeck effect which produces an electrostatic potential when charge carriers diffuse from the hot end of a material towards the colder end. The voltage produced is,

$$V = \alpha \Delta T \tag{18.3}$$

The parameters α and ΔT are the Seebeck coefficient and the temperature gradient respectively. The output power is proportional to the heat flow, which drives an electrical current across the temperature gradient. A thermoelectric figure of merit, **z**, determines the point of maximum power transfer efficiency where,

$$z = \frac{\alpha^2}{\rho}\frac{1}{\kappa} \tag{18.4}$$

The parameters $\boldsymbol{\rho}$ is the electrical resistivity, and $\boldsymbol{\kappa}$ is the thermal conductivity. Since **z** varies with the temperature, **T**, a more convenient dimensionless figure of merit **zT** is used. The goal is to maximize **zT** to enhance thermodynamic efficiency. Doping crystalline-order semiconductor materials increase the carrier mobility that maximizes the term α^2/ρ. Historically, reducing thermal conductivity below the alloy limit to achieve **zT** above unity has been challenging.

Recent advances have enabled thin-film thermo-electric generator development. However, the efficiencies are much lower due to high electrical and thermal contact resistance losses. Nevertheless, Micropelt has shrunk thermoelectric generators (TEGs) from 1000 mm^3 to 10 mm^3, and increased output voltage from 100 micro-volts to a few volts while producing few milli-watts output power [13]. Based on CMOS production methods, these devices offer good feasibility of integration with power management and storage devices. However, the methodology still requires tens of °C temperature difference to produce few milli-watts of power [14] as implied in Figure 18.6, and this will limit the deployable application environments.

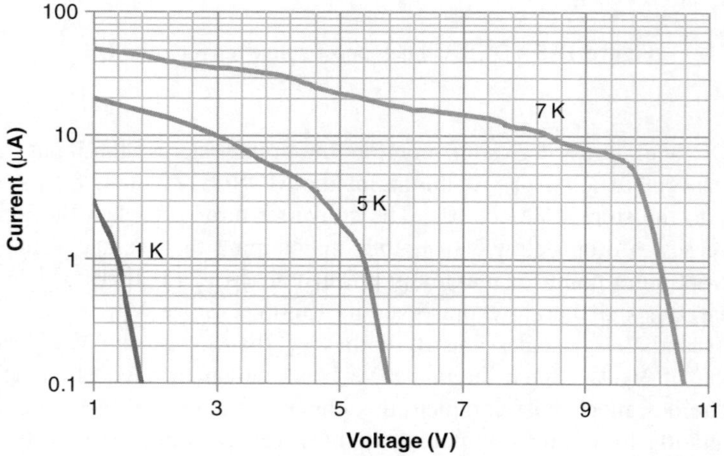

Figure 18.6 A thermoelectric transducer VI characteristics. (1K = 1°C).

18.3.3 Vibration Energy Scavenging Solutions

Previous studies have characterized the performance of the three main types of vibration energy transducers: electro-magnetic, electrostatic, and piezoelectric. In each approach, the vibration changes inductance, capacitance, and strain respectively to produce energy as summarized in Figure 18.7.

18.3.3.1 Electromagnetic

Electro-magnetic generators follow the principles of Faraday's law of electromagnetic induction. Vibration energy produces the movement of a coil through a magnetic field, or a permanent magnet through a coil to induce an electric current in the coil. These devices effectively convert kinetic energy from rotational armatures into electrical energy, but making them smaller tends to reduce efficiency levels. This method can be used to harvest energy using both linear and rotational devices in the micro-watt range, provided the generator size is not constrained. Existing commercial products such as the crank radio and the shaker flashlight works on this principle. Commercial devices for sensors have been shown to produce about one to ten milli-watts of power for industrial health monitoring applications at resonant frequency of 60 or 120 Hz [15–17].

18.3.3.2 Electrostatic

Electrostatic generators can utilize a MEMS structure to modulate the capacitance between two surface areas. This mechanical work amplifies a reference voltage across another

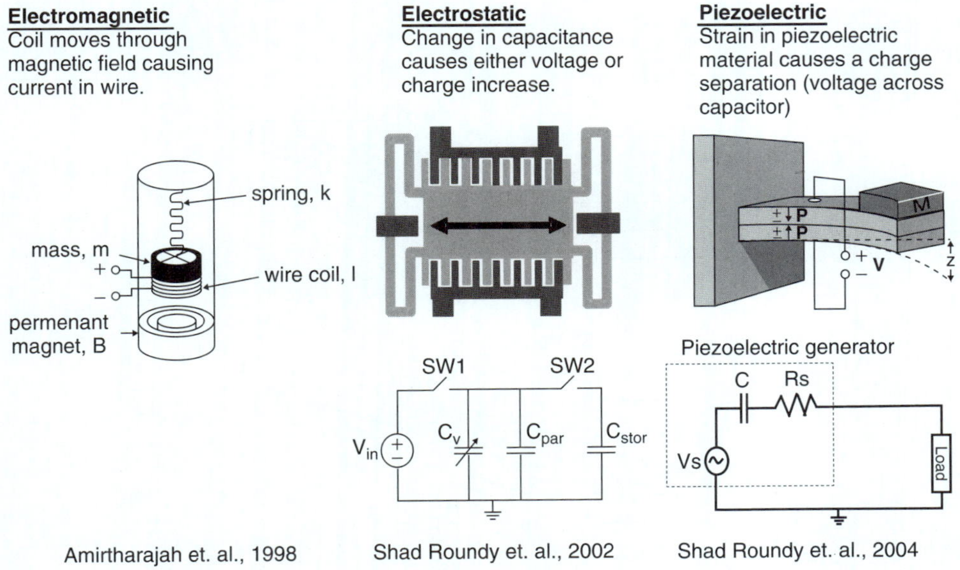

Figure 18.7 Types of vibration EH transducers. Reprinted with permission of © 2003 Elsevier B.V.

fixed-value capacitance. The mechanical work is done against electrostatic forces between the two plates. This initial input voltage must be provided somehow. Although not capable of converting as much power per unit volume as piezoelectric converters, electrostatic generators are attractive for their MEMS implementations.

The University of California at Berkeley has developed three types of MEMS structures and evaluated two types called in-plane overlap and in-plane gap [18]. The study concluded that the former required large spring deflections and high Q-factors to produce maximum power while the latter has an optimum spring deflection and can accommodate lower Q-factors. A high Q-factor is more vibration-frequency selective while a lower Q-factor can accommodate a wider range of vibration frequencies. A significant portion of any initial design involves carefully characterizing the acceleration profile of vibration sources for the intended application. This determines the design parameters and optimization criteria for a vibration transducer that can deliver maximum energy.

18.3.3.3 Piezoelectric

Piezoelectric generators work by straining a piezoelectric material to produce charge separation across the material. For most approaches, the amount of power generated is proportional to the g-force acceleration experienced by an oscillating mass, as well as the oscillating frequency. These converters tend to be more efficient than electrostatic and electromagnetic vibration transducers.

Figure 18.8 shows the two operating modes (d_{33} and d_{31}) in which the piezoelectric material can be used to generate power. In d_{33} mode, both the mechanical stress and output voltage act in the 3 direction. In d_{31} mode, the stress acts in the 1 direction and voltage acts

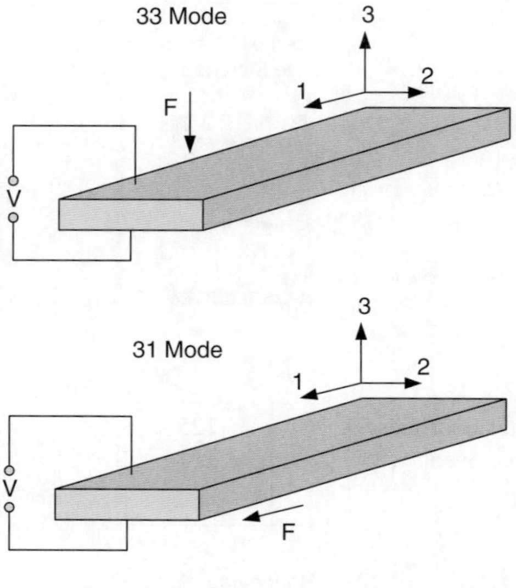

Figure 18.8 Piezoelectric conversion modes.

in the 3 direction. Operation in the d_{31} mode leads to the use of thin bending elements such as bimorphs, in which two separate sheets are bonded together, sometimes with a center shim in between them. The electro-mechanical coupling for d_{31} mode is lower than for d_{33} mode. However, d_{31} systems can produce larger strains with smaller input forces. Also, the resonant frequency is much lower. An immense mass would be required in order to design a piezoelectric converter operating in d_{33} mode with a resonant frequency below 60 Hz. Therefore, the d_{31} mode of operation is more suitable for powering RF tags and sensors.

Experiments show that a practical generator of one cubic centimeter is capable of producing more than 300 micro-watts of power from an oscillating mass that experiences 0.25 g of acceleration at 120 Hertz [19]. The transducer produces the maximum output near its resonance frequency. Therefore, it is essential to first characterize the acceleration spectral profile of the vibration source before optimizing the transducer design. Table 18.3 lists the fundamental frequencies and acceleration levels for common vibration sources in our environment [20]. Low level vibrations occurring on many household appliances and everyday objects in and around buildings appear to have a fundamental mode on the order of 100 Hz [20]. Thus cantilevers for these types of applications should be designed for a resonance frequency near 100 Hz. The cantilever resonant frequency is,

$$\omega_n = \frac{\kappa}{m} = \sqrt{\frac{E}{4m}\frac{h^3 w}{L^3}} \tag{18.5}$$

where, κ is the cantilever spring stiffness, **E** is Young's modulus, **w** is the beam width, **L** is the beam length, **h** is its thickness, and **m** is the tip mass [21]. Therefore, resonant frequency is essentially a function of the cantilever's dimensions and weight. Given a resonant frequency, the maximum output power is approximately,

$$P = \frac{mA^2}{4\omega_n}\frac{\zeta_e}{(\zeta_e + \zeta_m)^2} \tag{18.6}$$

where **m** is the oscillating mass, **A** is the input vibration acceleration amplitude, $\boldsymbol{\xi_e}$ is the electrically induced damping ratio, and $\boldsymbol{\xi_m}$ is the mechanical damping ratio. These equations suggest that heavier transducer devices that resonate at the source's fundamental

Table 18.3 Common vibration sources.

Vibration source	Peak frequency (Hz)	Acceleration (m/s^2)
Car engine compartment	200	12
Base of 3-axis machine	70	10
Door frame (after door closes)	125	3
Small microwave oven	121	2.5
HVAC office vents	60	0.2–1.5
Window on busy road	100	0.7
CD in Notebook PC	75	0.6
2nd story floor in busy office	100	0.2

frequency are capable of producing more energy. Specifically, to maximize power the following aspects must be taken into account:

- Design the cantilever to have a resonant frequency that matches the frequency of the intended application. This can be accomplished by appropriately choosing the cantilever geometry (h, L, w).
- Increase the effective mass of the cantilever. This can be done by either enlarging the cantilever size or by adding a proof mass.

ξ_m is related to the material choice and cantilever size [20, 21]. In addition, the choice of cantilever design over other transducer structures (e.g. diaphragm) decreases the overall stiffness of the structure and improves power harvesting. ξ_e is generally a function of electrical circuit parameters, and to some extent the circuit designer can control its value. Finally, the equations also show that the source acceleration of mechanical vibration is important.

Figure 18.9 shows how the maximum power point varies with excitation amplitude. As with solar cells, the transducer impedance varies with source excitation. Cantilever based transducers are most efficient at their resonant frequency. Disk-shaped piezoelectric transducers with several electrodes enable energy harvesting from multiple mechanical resonances [22]. Matching the impedance (maximum power point) will yield maximum power transfer to the energy storage device. The PM circuit is responsible for adaptively calculating the transducer's output impedance and consequently, adjusting key parameters to maximize charge transfer to the energy storage device.

A typical system for implementing vibration energy harvesting from a piezoelectric element is shown in Figure 18.10. In order to obtain maximum power transfer, the input impedance of the converter Z_{in} should match that of the source Z_o. The transducer

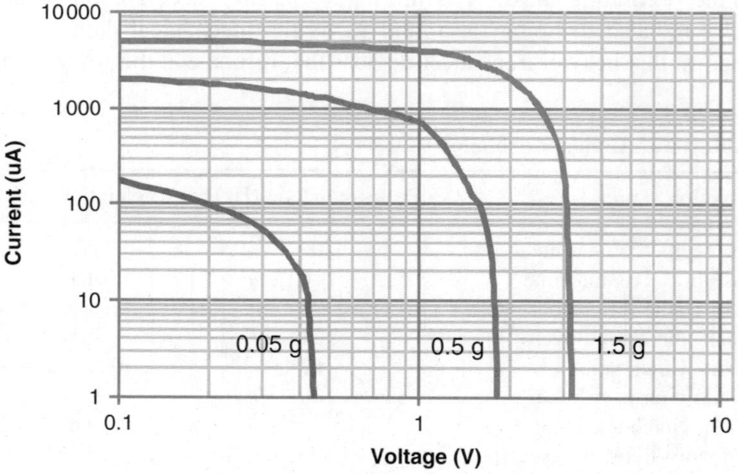

Figure 18.9 Cantilever characteristics at 150 Hz with beam length of 18.9 mm and tip mass of 5 gm.

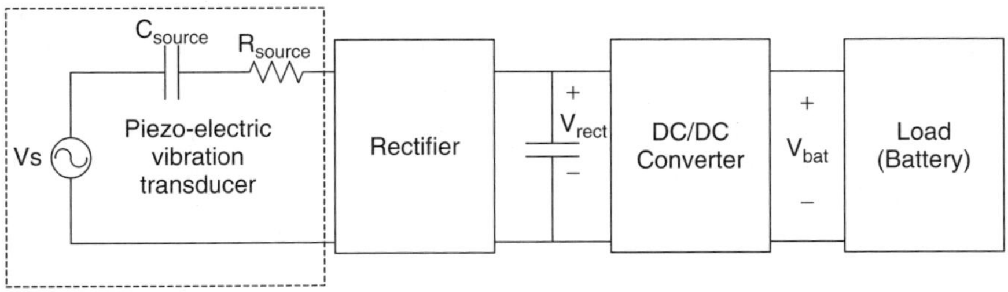

Figure 18.10 Block diagram of a vibration energy harvesting system.

impedance Z_o is,

$$Z_0 = \frac{1}{4 f_{Source} C_{Source}} \tag{18.7}$$

where $\mathbf{f_{Source}}$ is the frequency of a piezoelectric cantilever based vibration transducer, and $\mathbf{C_{Source}}$ is the transducer series capacitance. Therefore, as the frequency decreases, the impedance increases and vice versa. The rectifier output voltage, $\mathbf{V_{rect}}$ determines the amount of power delivered to the converter, and consequently, the load. The power delivered to the converter is,

$$P_0 = \frac{V_{rect}^2}{Z_{in}} \cong V_{Speak}^2 \frac{Z_O}{(Z_O + Z_{in})^2} \tag{18.8}$$

From Figure 18.10 and Equation 18.7, it can be seen that $\mathbf{Z_{in}}$ must be equal to $\mathbf{Z_o}$ to obtain maximum V_{rect} which is approximately half of the peak transducer voltage, V_{Speak}.

Figure 18.11 shows how a switch mode buck-boost converter implements the required impedance matching. Any non-isolated switch mode DC-DC converter such as a buck, boost, buck-boost, or boost-buck topology may be employed as the converter. It is, however, important to fix the converter's operation in the discontinuous conduction mode (DCM) for maximum power harvesting. DCM mode is preferred over continuous conduction mode (CCM), because the former avoids the reverse recovery problem of the

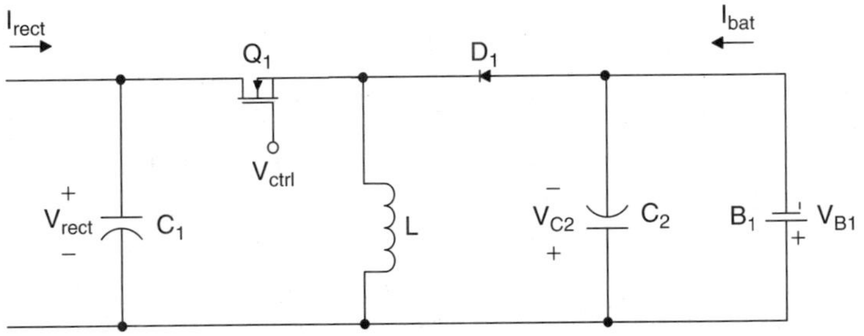

Figure 18.11 DC-DC buck-boost converter with battery load. Reprinted with kind permission from © 2008 Springer Science and Business Media.

diode. Also, for certain topologies such as buck-boost operating in DCM, the average input impedance $\mathbf{Z_{in}}$ does not depend upon the output energy storage device voltage $\mathbf{V_{bat}}$ and this simplifies controller design. Buck converters, however, require a high input to output voltage differential to operate. The optimal converter duty cycle depends on output filter inductance **L**, and the switching frequency $\mathbf{f_{sw}}$.

Under DCM, Z_{in} is given by

$$Z_{in} = \frac{2Lf_{sw}}{D^2} \tag{18.9}$$

where, **L**, $\mathbf{f_{sw}}$, and **D** are the inductance, switching frequency, and the converter duty cycle respectively. This indicates that for a constant switching frequency $\mathbf{f_s}$, inductance **L**, and duty cycle **D**, $\mathbf{Z_{in}}$ is constant and does not depend on the output battery voltage $\mathbf{V_{B1}}$ and current $\mathbf{I_{bat}}$. Similar equations can be derived for buck converters at high excitations. The converter can also operate in CCM with $\mathbf{Z_{in}}$ given by

$$Z_{in} = \left(\frac{1-D}{D}\right)^2 \frac{V_{bat}}{I_{bat}} \tag{18.10}$$

Here, $\mathbf{Z_{in}}$ is not constant and it is necessary to incorporate feedback circuits to measure $\mathbf{V_{B1}}$ and $\mathbf{I_{bat}}$, and adjust **D** to obtain the desired $\mathbf{Z_{in}}$ for appropriate impedance matching. This adaptation results in additional power draw by the pulse width modulation (PWM) control circuits and hence, lower efficiency. However, in DCM, the converter average input impedance $\mathbf{Z_{in}}$ does not depend upon output voltage $\mathbf{V_{bat}}$, and thus minimizes components for the control circuits. The constraint to enable DCM mode of operation is as follows:

$$V_{rect} < \frac{1-D}{D} V_{bat} \tag{18.11}$$

Similar results can be obtained for other topologies. Therefore, **D** can be adjusted to account for source frequency $\mathbf{f_{Source}}$ and transducer capacitance $\mathbf{C_{source}}$ changes. Any change in $\mathbf{f_{Source}}$ and/or $\mathbf{C_{source}}$ will result in a change in $\mathbf{Z_o}$. This means that $\mathbf{Z_{in}}$ must be tuned to match this altered $\mathbf{Z_o}$. For changes in $\mathbf{f_{Source}}$, **D** is controlled by a frequency dependent voltage that needs to be generated. In order to obtain higher power output from piezo cantilevers, synchronized switch harvesting on inductor (SSHI) techniques have also been developed. The technique comprises of a switch and inductor in series with the piezo element and uses LC resonance to enhance energy output from the transducer [23].

18.4 Future Trends in Energy Harvesting

Roadmaps for future implementations suggest integration of EH transducers, circuits and systems, energy storage devices, power management, and Micro-Wireless electronics on flexible thin-film substrates thus realizing smart sensors. Figure 18.12 shows a concept for a vibration-powered smart sensor using MEMS cantilever devices and thin-film piezoelectric deposition [24]. Conceptually, it would also be beneficial to add a layer formed from amorphous solar cells. The availability of low cost, miniature, flexible, integrated EH RFID tags is still a few years away. This is primarily because the energy density of MEMS and nano-based EH transducers are much lower than meso-scale transducers.

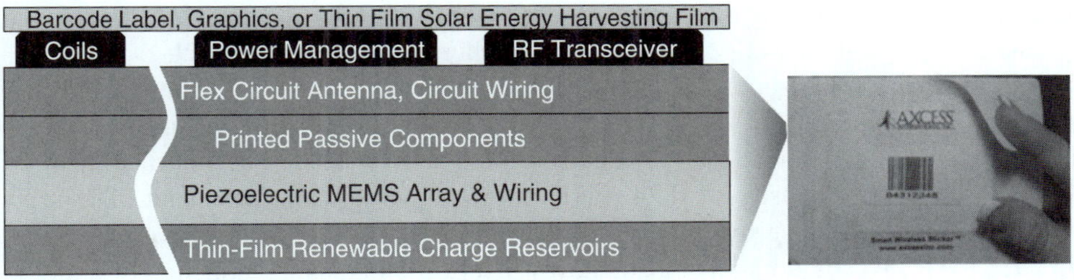

Figure 18.12 Vision for EH-based Micro-Wireless Dot™ sensor.

Further, research and development will be required to verify structural integrity, reliability and performance. Also, packaging and integration with sensors and other electronics are a key challenge. However, meso-scale harvesters can be deployed commercially. These bulk harvesters are relatively larger in volume/size and higher in cost. As discussed in Section 18.3, the choice of harvester depends on the application. For example, vibration EH can be used for deeply embedded sensors in structures such as bridges, roadways, vehicles, aircraft, etc. Similarly, for outdoor applications such as weather monitoring, border surveillance, etc. solar EH can be utilized. The cost of additional EH circuitry is expected to be orders of magnitude lower than the transducer cost. Also, efficient circuits can be employed in integrated circuit (IC) form to obtain high efficiency EH. This section covers details on miniaturization of certain EH transducers and integrated multi-source power management.

Flexible amorphous silicon and other nano-based materials are available to harvest energy from light. Amorphous silicon seems to be the preferred choice currently, but these materials are relatively inefficient under outdoor conditions and are also relatively expensive. However, higher efficiency amorphous panels are available from vendors such as PowerFilm Inc. [25]. Alternatively, companies such as Nanosolar are developing solar cells made from nano-based compounds like CIGS (copper, indium, gallium, selenide) [26]. Other companies such as Konarka have developed nano-enabled polymer photovoltaic materials that are lightweight, flexible and more versatile than traditional solar materials [27].

Thin film solid state energy storage device research and production are also making considerable progress. Infinite Power Solutions (IPS) offer batteries operating with standard 4.2 V_{DC} constant voltage chargers that recharge to 90% of capacity in few minutes [28]. These devices are advertised to operate for thousands of charge–discharge cycles with no memory effect and minimal charge loss over time. For example, the IPS LS101 is rated at 0.7 milli-ampere hour per square inch (110 μm thickness) and the second generation devices (LS201) will be designed for higher capacity. Similarly, Cymbet Corporation has two versions of solid state batteries providing 12, 50, and 85 micro-Ampere hour capacity, and require a 4.1 V_{DC} charging circuit [29]. These devices are surface-mounted chips, and rated for several thousand charge–discharge cycles.

18.4.1 Thin-Film MEMS Piezoelectric Cantilevers

Thin film piezoelectric cantilever design and fabrication are an effective way to reduce transducer size and allow cost-effective manufacturability. For energy from vibrations, piezoelectric MEMS-based transducers seem to be the obvious choice. Although these systems are not commercially available, researchers have designed such cantilevers. A system consisting of a composite micro-cantilever beam with a PZT thin film layer and electrode layer operating in d_{33} mode has been demonstrated [30]. A single piezoelectric micro power generator (PMPG) device could deliver about one micro-watt at 2.36 V_{DC} with an energy density of 0.74 milli-watt-hour per square centimeter. The second generation PMPG could provide 0.173 milli-watts at three V_{DC} with one **g** excitation at 155.5 Hz [31]. Investigators have also shown that at the resonant frequency of about 608 Hertz, a MEMS-based generator prototype can output about 0.89 volts AC peak–peak voltage output with output power of 2.16 mill-watts [32].

The length of the cantilever, the thickness of the individual layers, and the electrode pattern and spacing can all affect the resonant frequency of the beam. A cantilever of the d_{33} configuration has higher electro-mechanical efficiency. Lower resonance frequency cannot be achieved by simply increasing beam width and length as with a bulk cantilever. To achieve the desired resonance frequency, a tip mass will need to be added to the cantilever. On the other hand, the thickness of the individual layers has only a small impact on the resonance frequency as compared with beam dimensions and proof mass adjustments. However, the piezoelectric thin film needs to be relatively thick in order to maximize the value of the piezoelectric constants and hence the cantilever's power output.

The electrode pattern on top of the cantilever is inter-digitated (interlocked, for example, fingers of two hands joined together) in nature, as shown in Figure 18.13. This results in an electric field that is parallel to the surface of the cantilever and the d_{33} mode of operation, which is more efficient than the d_{31} mode of operation. The gaps and number of fingers in the inter-digital pattern were designed to generate sufficient capacitance in order to utilize the pattern for dielectric and piezoelectric property measurement. Figure 18.14 shows the interdigitized electrode pattern on the cantilever stack. It also shows a close-up of a d_{33}

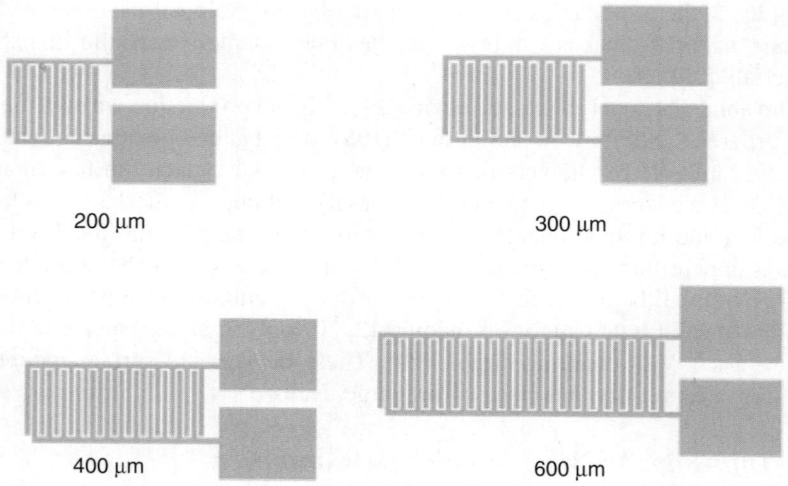

Figure 18.13 Top view of the electrode patterns for cantilevers of different sizes.

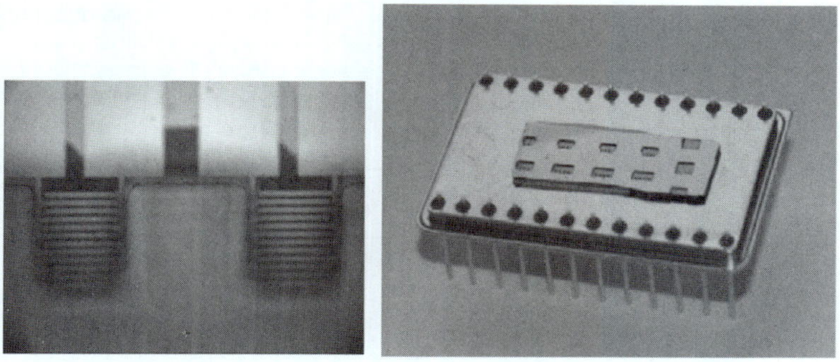

Figure 18.14 MEMS cantilever. Courtesy, Texas Micropower Inc.

Table 18.4 Comparison of power densities of MEMS piezoelectric cantilevers.

Author	Device	Effective are (mm^2)	Power (μW)	Power density (μW $inch^{-2}$)	Power density (μW $inch^{-2}g^{-1}$)	A(g)	f(Hz)
[37]	d33 PZT cantilever	0.0442	1	14596.41	1351.52	10.8	13.7 k
[38]	d31 PZT cantilever	1.92	2.15	722.45	361.22	2.0	462.5
[39]	d31 AIN cantilever	3.8	0.038	6.45	12.90	0.5	204
[40]	d33 PZT cantilever	0.027	1.01	24133.81	2234.61	10.8	13.9 k
[32]	d31 PZT cantilever	0.1992	2.16	6995.72	6995.72	1.0	608
[41]	d31 AIN cantilever	0.552	1.97	2302.48	575.62	4.0	1368
[42]	d31 PZT cantilever	1.845	40	13987.24	7361.70	1.9	-

mode cantilever beam prototype of size 200 × 200 μm developed by the University of Texas at Dallas and Texas Micropower Inc. Table 18.4 compares the power density of various MEMS piezoelectric cantilevers.

18.4.2 Integrated Power Management with Load Balancing

Due to the low power levels, it is desirable to harvest energy from multiple sources to maximize the amount of energy harvested. Most of these systems have high impedance sources resulting in considerable power loss and poor efficiency. Consequently, off-the-shelf power management solutions such as DC-DC converters and rectifiers, when coupled with off-the-shelf vibration transducers such as piezoelectric bimorphs and solar cells, do not yield acceptable results.

Existing system-level solutions that incorporate custom DC-DC converters for impedance matching are tuned to specific transducers and for specific application environments. These power management strategies are not generic and, therefore, cannot be used over a wide spectrum of environmental conditions (e.g. vibration frequencies and light intensity), and transducer types. Moreover, some of these methods use high power consumption digital signal processors (DSPs) for implementing PWM control algorithms.

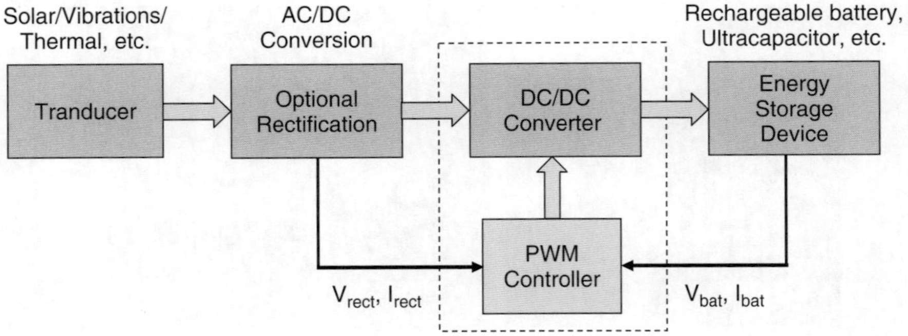

Figure 18.15 Adaptive multi-source power management architecture. Reprinted with kind permission from © 2008 Springer Science and Business Media.

This strategy results in excessive power consumption in the power management solution and, therefore, does not allow feasibility in practical applications.

It is desirable to have an intelligent power management solution that accommodates multiple EH sources, that is adaptive by automatically tuning itself to the source changes, and to facilitate easy interface to different transducer types. The power management solution must consume low power, preferably as an integrated circuit (IC). This approach would lead to high efficiency, low cost, and small form factors.

Figure 18.15 shows a simplified block diagram of the power management (PM) system architecture. It is important to note that no matter what the source energy is, the PM circuit needs to be designed so that its input impedance matches that of the source. The PM circuit is responsible for adaptively calculating the output impedance of the transducer and consequently, adjusting key parameters to maximize charging of the energy storage device. As noted earlier, it is possible to use switch mode DC-DC converters to facilitate such impedance matching and allow implementation of a single PM solution. Preliminary results obtained for EH from vibrations have shown greater than 70% efficiency for low excitation levels of less than $0.3\,g_{peak}$ using discrete circuits [4]. Higher efficiency is expected when the circuits are implemented in IC form. For solar EH, the PM circuits allow operation close to the MPPT point so that maximum power is transferred to the storage device at all times. The emphasis is on using the minimum amount of power to achieve the maximum power transfer and adaptive power management. These circuits can be fabricated as an IC with extremely low power consumption of the order of a few microwatts [33]. Researchers have developed integrated power management systems which employ energy awareness and charge recycling to increase the total energy available, by allowing the system to gather and add together voltages from multiple sources at the same time [34].

Progress made by the above technologies indicates good feasibility towards development of smart sensors. Researchers will focus on developing each layer of the device to accommodate roll-to-roll manufacturing, and then integrate each layer to realize the complete smart sensor. Such devices will reduce the form factor and weight of existing active sensors used in applications such as asset and personnel monitoring, and open the door to new applications such as smart bandages, labels, stamps, and others.

18.5 Conclusion

The design of EH systems for semi-active RFID and Micro-Wireless™ devices is explored in this chapter. These low-power wireless sensor architectures coupled with EH are very important primarily because of the rapid development of RFID and wireless sensor network markets, and second, due to the limited life of batteries. This, in particular, adds cost and limits deployment of these devices. Although the advantages of these devices are enormous in terms of communication range, throughput, and reliability, the battery cost and its replacement can hamper adoption rates. Harvesting energy from natural sources such as vibrations, solar, and thermal gradients will extend battery life and at the bare minimum replenish idle energy consumption which is a dominant factor in battery life reduction. Designs of new RFID devices (e.g. smart labels) and device technology (e.g. communication protocols) are steadily decreasing the energy requirements and therefore, smart power management and EH techniques allow a realistic possibility for extended life. Further, miniaturization of these components allows development of feasible, cost-effective RFID devices that can be utilized for tracking a larger variety of assets, particularly those that are cost- and size-sensitive such as laptops and personal electronic devices. Thinner RFID tags are more desirable for personnel security and safety applications, and these smaller form-factors and thinner profiles will continue to place greater demands to reduce the size and cost of all types of EH techniques.

Problems

1. Consider a semi-active RFID tag operating at 3V that consumes 20 milli-amperes when active and 20 micro-amperes in sleep state. In the active state, the tag is operational for 20 milli-seconds and is programmed in the beacon mode to transmit data once every minute.
   (a) Calculate the amount of energy required by the tag per day.
   (b) Typically, the output power from a vibration harvester scales linearly with the source excitation level. If the excitation level of the source is one g, calculate how many hours of sustained vibrations are required per day for the harvester to replenish energy consumed by the tag. (Assume that the harvester produces 300 micro-Watts per cm^3 per 0.1 g of source excitation.)
   (c) If two hours of sustained vibrations are available per day, calculate power required from the harvester. Estimate volumetric size (in cm^3) for the new harvester.

References

[1] Hande, A., Bridgelall, R., and Bhatia, D. (2008) Energy harvesting for active RF sensors and ID tags, in S. Priya and D.J. Inman (eds.) *Energy Harvesting Technologies*. New York: Springer.
[2] Bridgelall, R. (2008) Introducing a micro-wireless architecture for business activity sensing, in IEEE International Conference on RFID, April 16.
[3] Axcess International Inc. (2007) Dot™ micro-wireless technology for business activity monitoring. Press Release, Nov.
[4] Hande, A. and Bridgelall, R. (2008) High Efficiency Vibrational Energy Harvestor for Active RFID Tags, 2nd Annual nanoPower Forum, June.
[5] Yeatman, E. (2004) Advances in power sources for wireless sensor nodes, in International Workshop Wearable and Implantable Body Sensor Networks, pp. 20–21.

[6] Stevens, J. (1999) Optimized thermal design of small ΔT thermoelectric generators, in 34th Intersociety Energy Conversion Engineering Conference, Society of Automotive Engineers, 01(2564).

[7] Schmidhuber, H. and Hebling, C. (2001) First experiences and measurements with a solar powered personal digital assistant (PDA), in 17th European Photovoltaic Solar Energy Conference, pp. 658–662.

[8] Shearwood, C. and Yates, R. (1997) Development of an electromagnetic micro-generator, *Electronics Letters*, **33**(22): 1883–1884.

[9] Amirtharajah, R. and Chandrakasan, A. (2004) Self-powered signal processing using vibration-based power generation, *IEEE Journal of Solid State Circuits*, **33**(5): 687–694.

[10] Roundy, S., Wright, P., and Rabaey, J. (2003) A study of low level vibrations as a power source for wireless sensor nodes, *Computer Communications*, **26**: 1131–1144.

[11] Ottman, G., Hofmann, H., and Lesieutre, G. (2003) Optimized piezoelectric energy harvesting circuit using step-down converter in discontinuous conduction mode. *IEEE Transactions on Power Electronics*, **18**(2): 696–703.

[12] Brunelli, D. and Benini, L. (2008) An efficient solar energy harvester for wireless sensor nodes, 11th Conference on Design, Automation and Test in Europe, March.

[13] Micropelt GmbH. Available at: http://www.micropelt.com/.

[14] Böttner, H., Nurnus, J., Schubert, A., and Volkert, F. (2007) *New High Density Micro Structured Thermogenerators for Standalone Sensor Systems*. Fraunhofer Institute for Physical Measurement Techniques.

[15] Perpetuum Ltd. Available at: www.perpetuum.co.uk.

[16] Ferro Solutions Inc., VEH-360 electromechanical vibration energy harvester. Available at: http://www.ferrosi.com/files/VEH360_datasheet.pdf.

[17] Lumedyne Technologies Inc. Available at: http://www.lumedynetechnologies.com/Energy%20Harvester.html.

[18] Roundy, S., Wright, P.K., and Pister, K.S.J. (2002) Micro-electrostatic vibration-to-electricity converters, in ASME International Mechanical Engineering Congress & Exposition, Nov.

[19] Roundy, S. and Wright, P.K. (2004) A piezoelectric vibration based generator for wireless electronics, *Smart Materials and Structures*, **13**: 1131–1142.

[20] Roundy, S., Wright, P.K., and Rabaey, J.M. (2004) *Energy Scavenging for Wireless Sensor Networks*. Boston: Kluwer.

[21] Bryzek, J., Roundy, S., Bircumshaw, B., Chung, C., Castellino, K., Stetter, J.R., and Vestel, M. (2006) Advanced IC sensors and microstructures for high-volume applications, *IEEE Circuits and Device Magazine*, 2006: 8–21.

[22] Guilar, N., Amirtharajah, R., and Hurst, P. (2009) A full-wave rectifier with integrated peak selection for multiple electrode piezoelectric energy harvesters, *IEEE Journal of Solid-State Circuits*, **44**(1): 240–246.

[23] Lefeuvre, E., Badel, A., Richard, C., and Guyomar, D. (2005) Piezoelectric energy harvesting device optimization by synchronous electric charge extraction, *Journal of Intelligent Material Systems and Structures*, **16**: 865–876.

[24] Fernandez, E., Baldenegro, L., Alshareef, H., Debray, W., Hande, A., Shah, P., and Gnade, B. (2009) Characterization of ferroelectric Pb(Zr,Ti)O3 thin film deposited on non-conducting surfaces for energy harvesting applications, in 4th Annual Energy Harvesting Workshop, Jan.

[25] PowerFilm Inc. Available at: http://www.powerfilmsolar.com/

[26] Nanosolar Inc. Available at: http://www.nanosolar.com/.

[27] Konarka Technologies Inc. Available at: http://www.konarka.com/.

[28] Infinite Power Solutions. Available at: http://www.infinitepowersolutions.com/.

[29] Cymbet Corporation, EnerChip products. Available at: http://www.cymbet.com/content/products.asp.

[30] Sood, R. (2003) Piezoelectric micro power generator: A MEMS-based energy scavenger, M.S. thesis, Massachusetts Institute of Technology, September.

[31] Xia, R., Farm, C., Choi, W., and Kim, S. (2006) Self-powered wireless sensor system using MEMS piezoelectric micro power generator, in IEEE Sensors Conference, October.

[32] Fang, H., Liu, J., Xu, Z., Dong, L., Wong, L., Chen, D., Cai, B., and Liu, Y. Fabrication and performance of MEMS-based piezoelectric power generator for vibration energy harvesting, *Microelectronics Journal*, **37**(11): 1280–1284.

[33] Hande, A., Shah, P., Fernandez, E., Baldenegro, L., Alshareef, H., and Gnade, B. (2009) Integrated energy harvesting with multisource, adaptive interfaces, 4th Annual Energy Harvesting Workshop, Jan.

[34] Guilar, N., Amirtharajah, R., and Hurst, P. (2009) An energy-aware multiple-input power supply with charge recovery for energy-harvesting applications, in IEEE International Solid-State Circuits Conference (ISSCC), Feb.

[35] Stordeur, M. and Stark, I. (1997) Low power thermoelectric generator - self-sufficient energy supply for micro systems, in *16th International Conference on Thermoelectrics*, pp. 575–577.

[36] Starner, T. (1996) Human-powered wearable computing, *IBM Systems Journal*, **35**(3): 618–629.

[37] Sood, R., Jeon, Y.B., Jeong, J.H., and Kim, S.G. (2004) Piezoelectric micro power generator for energy harvesting, in *Proceedings of the Solid State Sensor and Actuator Workshop*.

[38] Shen, D., Park, J., Ajitsaria, J., Choe, S., Wikle, H., and Kim, D. (2008) The design, fabrication and evaluation of a MEMS PZT cantilever with an integrated si proof mass for vibration energy harvesting, *Journal of Micromechanics and Microengineering*, **18**: 1–7.

[39] Marzencki, M., Charlot, B., Basrour, S., Colin, M., and Valbin, L. (2005) Design and fabrication of piezoelectric micro power generators for autonomous microsystems, in *DTIP7 '05 - Symp. on Design Testing Integration and Packaging of MEMS/MOEMS*, Montreux, Switzerland, pp. 299–302.

[40] Jeon, Y.B., Sood, R., Jeong, J.H., and Kim, S.G. (2005) MEMS power generator with transverse mode thin film PZT, *Sensors Actuators A.*, **122**: 16–22.

[41] Marzencki, M., Ammar, Y., and Basrour, S. (2007) Integrated power harvesting system including a MEMS generator and a power management circuit, in *Proceedings of the International Conference on Solid-State Sensors, Actuators and Microsystems*, Lyon, France, pp. 887–890.

[42] Renaud, M., Sterken, T., Schmitz, A., Fiorini, P., Van Hoof, C., and Puers, R. (2007) Piezoelectric harvesters and MEMS technology: fabrication, modeling and measurements, in *Proceedings of the International Conference on Solid-State Sensors, Actuators and Microsystems*, Lyon, France, pp. 891–894.

19

Simulators and Emulators for Different Abstraction Layers of UHF RFID Systems

Christian Steger[1], Alex Janek[1], Reinhold Weiß[1], Vojtech Derbek[2], Manfred Jantscher[2], Josef Preishuber-Pfluegl[2], and Markus Pistauer[2]

[1]*Institute of Technical Informatics*
Graz University of Technology Austria

[2]*CISC Semiconductor Design+Consulting GmbH*
Klagenfurt Austria

19.1 Introduction

19.1.1 Motivation

The resources of UHF RFID devices are limited in many ways. The data rates and power transmitted by UHF RFID readers are restricted by national regulations. Communication standards define timing and data packets handling. Transponders operating in RF-unfriendly environments are limited in their size and available energy. The performance requirements of individual hardware components are derived from system-level specifications defined by end users. From the system point of view, the development of the hardware components is seen as a bottom level design, while requirements defined by the end user community stand on top of the system design hierarchy. The bottom up design constraints imposed by hardware manufacturers meet during the system integration and technology adoption, with the requirements defined in the process of the top down decomposition.

RFID Systems: Research Trends and Challenges Edited by Miodrag Bolić, David Simplot-Ryl, and Ivan Stojmenović

The design of system components and the integration of UHF RFID into existing manufacturing processes have therefore became a joint activity of both technology providers and end users. A study [7] has shown that implementing RFID systems requires both investments in the tagging and reading systems, as well as significant initial investments in the introduction of new systems to ensure that RFID-stored data can be used. To decrease the investment cost, end users and system integrators look for methods to integrate the UHF RFID technology into its target application without extensive on-site testing efforts. At the same time, tag ASIC and reader manufacturers seek support for hardware and software development in order to verify the low-level functionalities within the scope of entire system.

The extended RFID features of higher class RFID tags (Figure 19.20) such as sensors, high operational distances and computing power find their application mainly in the logistics process (goods tracking and transport conditions surveillance), health care (patient data monitoring) and civil structure monitoring [35, 37].

19.1.2 Goal of the Simulation/Emulation Platform

However, a methodology is missing which would combine the existing simulation and modeling approaches, thus providing a possibility to translate the models automatically onto a hardware platform enabling a real-time verification in the target operating environment.

A methodology has therefore been developed which builds upon models of UHF RFID hardware modules, communication links and operating set-ups. The models support development and integration of readers and tags based on continuous verification. Three main topics are covered: (1) System level optimization; (2) Communication link performance; and (3) novel concepts for an extended UHF RFID transponder architecture equipped with energy harvesting devices. From the perspective of the OSI[1] reference model, the UHF RFID communication link is covered by the physical and data link layers.

The physical layer defines how a reader communicates with a tag and a how a tag communicates with a reader. The features of the physical layer are signaling, modulation and data encoding. The data link layer defines packetizing of logical communications, identification algorithms and collision arbitration.

19.1.2.1 Simulation Cases

Figure 19.1 represents various scenarios under which the model can be simulated. It is possibile to check the conformance of the model with existing hardware parts as well as simulating the influence on application-specific layouts.

Simulation cases 1 (see Section 19.3) and 2 (see Section 19.5) depicts a layout suitable for verification of the UHF RFID system integration for specific applications. One example of the simulation of applications could be a supply chain distribution center where RFID labels are used to identify transport items, for example, by EAN International specified EPC (Electronic Product Code) Label. The tools could be used to optimize the set-up of

[1] Open System Interconnection.

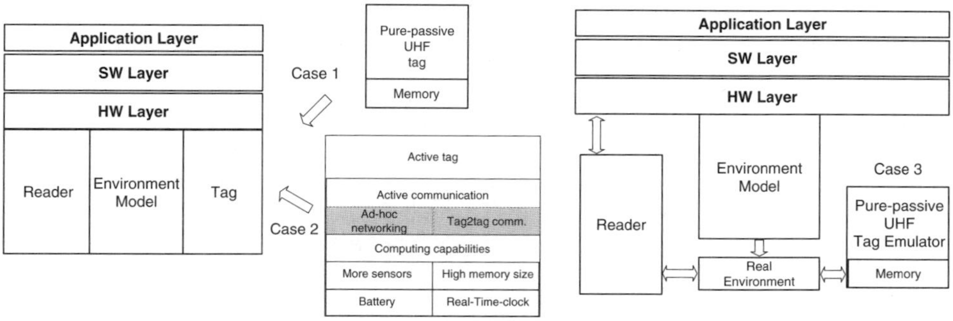

Figure 19.1 Abstract overview of simulation/emulation cases.

readers on conveyer belts and portals without shutting down the distribution center for a long time to evaluate the best placement of the equipment.

In simulation case 3 (see Section 19.4) a scenario is depicted where the communication of a reader with set of emulated tags is influenced by signals from the virtual environment. These signals can represent an environmental noise, signals from other devices or, most often, signals from other virtual tags and readers modeled in the system. A multi-reader arbitration can be checked, for instance, using a layout from the simulation case 2.

19.1.3 Model-Based Design and Verification of UHF RFID Systems

Verification based on prototypes in the target operating environment provided reliable but also costly feedback on design quality. To reduce the design effort in early development stages a more frequently used evaluation method is supported by models. The assumption of UHF RFID specific constraints brings the advantage of simplifying the complex behavior of general wireless systems with minimal information loss. Parts of the model can then be moved to higher abstraction levels to increase the simulation speed. The use of advanced modeling and simulation techniques at the system level achieves more accurate results both for hardware design and for system performance evaluation. Khouri *et al.* proposed a methodology for verification of wireless system designs using HDL behavioral models [50].

19.1.3.1 Real-Time Modeling, Simulation and Verification

Deriving implementations directly from the specification is a desirable step in the design process that significantly decreases both time-to-market and the risk of errors. Several approaches are known and established in the developers community [2, 18]. The challenges are twofold. The first is the development of an accurate model which precisely follows all the real requirements and the second is to obtain a flexible and high performance coupling between the embedded system and the simulation model [8]. Real-time functional design verification of UHF RFID systems is commonly based on emulators [52, 32]. Such emulators are developed on programmable hardware using languages most suitable for the nature of the target platform. The languages used are usually C for microcontrollers or HDL for FPGA.

Execution of RF System Models on Real-Time Platforms

Nowadays, development of prototypes based on models is no longer driven by the target architecture. Developers concentrate on directly solving issues concerning their designs rather than analyzing whether it fits into the constraints of their particular hardware. Target real-time platforms for embedded systems range from digital signal controllers (DSC), digital signal processors (DSP) to multiple core processors and field programmable gate arrays (FPGA). Recent developments in prototyping platforms for mobile phone industry have even provided a combination of DSPs and FPGA with high performing peripherals. Processor speeds achieve a clock rate of several hundreds MHz and input/output data sampling rates range in the order of hundreds of MHz as well. When such a platform also includes an antenna and RF part providing multi-mode broadband/multi-band processing capabilities at the front end with RF sampling, then this system is commonly described as Software Defined Radio (SDR).

Software Defined Radio systems design [40, 16, 49] address the definition of both signal processing functions and reconfiguration management according to the applications. The functional constraints concerned within the SDR design are data rates, quality of service (QoS), and reconfiguration impacts. Energy consumption, for instance, is a technological constraint.

Automatic Generation of Prototypes

Choosing the right platform for the DSP portion of the SDR requires the designer to make decisions such as whether computations are done using integer, fixed-point, or floating-point architecture, what part of the design is implemented in hardware and what part in software, etc.

FPGA soft processors have become an attractive choice for the rapid prototyping [36, 10] and implementation of embedded systems. To gain full leverage from their potential, portions of the design are executed as software programs while other portions are implemented as customized hardware. Commercially available soft processors [4, 66] support various dedicated interfaces and bus protocols, allowing the connection of hardware peripherals to them. Due to their configurability they can be customized for additional instructions and optimized for target applications.

Implementations of prototypes of RF devices are based on models described in C language and optimized to the implementation platform, which is a microcontroller with an RF interface and additional peripherals. A proposal for a novel EPCglobal Class 1 Generation 2 [27] baseband processor power aware architecture is presented in [67].

19.1.4 Higher Class RFID Tags and Energy Harvesting Devices

New RFID applications require high operating ranges and high communication reliability, especially in inhospitable industrial environments [61]. Not only simple identification but also sensing and monitoring capabilities have become real requirements. The convergence between RFID and wireless sensor networks is a new challenge for the future [54]. Such functionality demands a large memory and high computational capabilities, which are necessary to process the measured sensor data. As this data has to be transferred periodically to the requesting RFID interrogator, high data rates and reliable communication links are also required. This has a significant impact on the architecture of RFID transponders, forcing them to become embedded systems of increasing complexity.

GS1 EPCglobal, an industry-driven organization founded in November 2003 and mainly focusing on research related to RFID with the target of developing and establishing the technology, has defined a class structure for RFID systems mainly in the ultra high frequency (UHF) range depending on their functionality [26]. Out of this classification the term *higher class RFID tag* is defined by the functionality offered and the on-board energy supply. Due to the communication distance and performance [62], the UHF frequency range has been chosen for the architecture design presented later and is part of the main focus of research driven by GS1 EPCglobal. Figure 19.2 shows the functionality of RFID tags and the related classification. The border between higher and lower class RFID tags is clearly visible, defined by the offered functionality and the on-board power supply.

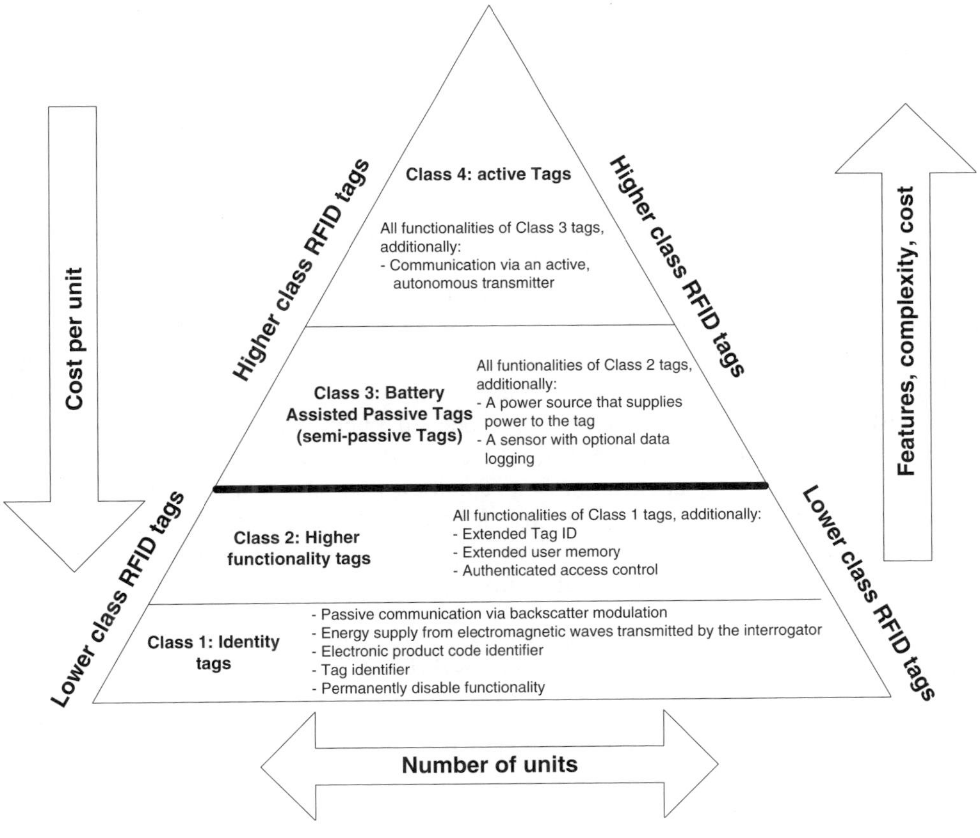

Figure 19.2 Tag Classes: UHF RFID Tags [20].

19.1.4.1 Advantages of Energy Harvesting Devices

Energy harvesting devices address the issue of limited battery source. They provide a small, unstable, non-continuous but nearly infinite energy supply [59]. As the energy is usually non-predictable, a special energy management combined with monitoring of the reliability of the harvesting devices available is absolutely necessary. Additionally, special buffers and storage devices (batteries) are required. The requirements for energy harvesting devices are defined by the energy consumption of the application concerned, by the required size and weight and by the energy sources available (e.g. ambient RF – <1 W/cm, thermoelectric – 60 W/cm, ambient airflow – 1 mW/cm). Energy harvesting is a multidisciplinary area and requires research in different fields [58, 48]. Besides the research on the energy sources it is necessary to take into account the nature of the energy source during the design of the architecture. This includes a flexible and reliable power management as well as large energy containers for storing energy during idle periods of the application and a mechanism for switching between the energy sources depending on the power provided and on their reliability. Additionally, if the focus is on sensor networks, the task management should also be adapted with specific middleware functionality. This is necessary because each node is in a different place with mostly different energy density. Tasks, which require much computing power (and high energy), should be done by sensor nodes with a high amount of energy.

19.1.5 Basics on Conformance, Performance and Interoperability Testing

The testing of standards-based products is crucial in order to ensure efficient and reliable system implementations. Generally, conformance, performance, and interoperability tests are conducted. These three kinds of tests are not unique for RFID systems but are common for communication systems like Bluetooth, GSM, and WLAN [60].

The test environment is an integral part of conformance, performance, and interoperability test specifications. Test environments define the environmental conditions under which tests are conducted. They must reflect the actual purposes of tests. Generally, three groups of test environments can be distinguished [60]: laboratory conditions, application reference conditions, and application field conditions. Conformance, performance, and interoperability test results are very important fundamentals for business confidence [55] and market maturity [57]. Thus, most standards organizations offer certification programs which cover the tests described and officially confirm the test results.

19.1.5.1 Conformance and Performance Testing

> Conformance testing is concerned with the assessment of the extent to which an implementation or system conforms to a specification [30, page 15].

According to the above definition, conformance testing ensures that an implementation meets all the requirements of the underlying standard. Typically, individual parameters such as timings, tolerances, and waveforms, as well as actual operations like multiple-tag reading can be tested. Often, a combination of both leads to the best test results because

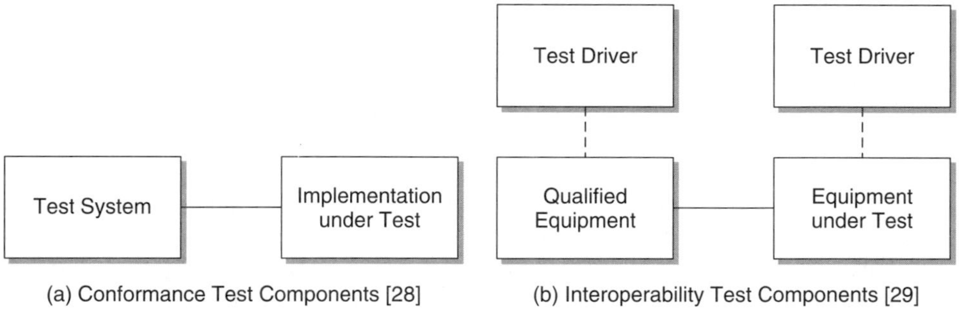

Figure 19.3 Test components.

individual parameters are tested under conditions close to reality [60]. Generally, the results of conformance tests are expressed by fail or pass.

Figure 19.3(a) (modified from [28]) shows a conformance test configuration. The product to be tested is called IUT (Implementation Under Test). The test system includes all the equipment necessary to perform the tests and to record and evaluate the test results. It controls the IUT and monitors its behavior. The test system is connected with the IUT via a standardized interface, which is normally the native communication interface of the IUT. ETS 300 406 describes a standardized methodology for specifying conformance tests [30].

The advantage of conformance testing is that it explicitly allows the requirements of standards to be checked. Clearly defined testing objectives based on single requirements or groups of requirements are transformed into test cases which are tested automatically using the test architecture. Because a conformance test set up includes only one implementation under test, faulty devices are identified directly.

Performance testing deals with the determination of individual parameters which affect system performance as well as with the determination of the overall system performance under defined environmental conditions [60]. In this respect the term system performance has to be defined individually, depending on the characteristics of the system described by a standard. In the area of RFID, system performance could be expressed as an inventory rate (number of tag identifications per second) or inventory reliability (percentage of tags that can be detected in a certain set-up). Individual parameters may be tag sensitivity or the radiation patterns of tags applied to objects. The Bit Error Rate (BER) is an important performance indicator for communication systems in general [17].

19.1.5.2 Interoperability Testing

> Interoperability testing is the activity of proving that end-to-end functionality between (at least) two communicating systems is as defined in the underlying standards [55, page 183].

In other words, interoperability tests deal with testing actual products against other actual products. It may seem that this kind of test is not necessary since two products conforming to one and the same standard are expected to be interoperable. This might be true for a majority of implementations. However, as will be discussed below, there may be issues outside the scope of a standard that can only be covered by interoperability tests.

Figure 19.3(b) (modified from [29]) shows a typical interoperability test configuration. The product to be tested is called EUT (Equipment Under Test). Its functionality is tested against the QE (Qualified Equipment). The tests are conducted and observed by test drivers which operate the QE and the EUT normally. If the EUT does not provide an interface for a test driver, as is the case with RFID tags, only one test driver is utilized. ETSI has also formalized the procedure of interoperability testing [31].

Interoperability tests allow much better consideration of user requirements than conformance tests since interoperability tests are based on interoperable function statements and not just on the requirements dictated by standards [51]. However, this does not mean that interoperability tests ignore the requirements of standards. In fact, conformance tests are typically the basis for subsequent interoperability tests. For example, in the GS1 EPCglobal hardware certification program, devices first have to pass conformance tests before they are admitted for interoperability tests. The drawbacks of interoperability tests are that generally N∗(N-1) tests are necessary in order to test N different products. Further, the reasons for failed tests cannot always be discovered by straightforwardly examining the test results. Additionally, since the test driver(s) in interoperability tests only control and observe the tests, generally no reference system is included in the tests which might cause distrust in the test results [51].

19.1.5.3 Testing Methods Performed in RFID Industry

Generally the testing is divided into two different testing methods. First, the conformance of the EUT (RFID interrogator and RFID tag) to certain predefined standard is tested (e.g. ETSI regulations, ISO standard conformance, EPCglobal standard conformance). Furthermore, this type of test includes performance measurements, for example, reading distance, multi tag handling, probability and speed of identification. These tests are well defined and formalized by ISO (ISO/IEC 18047).

As these tests have been seen to be inadequate for ensuring real application operation (which mostly combines different real product implementations) interoperability testing has been introduced and formalized by ETSI. Interoperability testing means testing different components of the system from different manufacturers against each other. Some example test parameters are identification distance, multi-RFID tag handling, interrogator-to-interrogator communication and the influence of interrogators among each other. The idea is to perform real-life testing with real DUT implementations and to test the hardware against the hardware itself. In other words, each component itself may be implemented so that it works within the defined borders of the standard. That does not, however, mean that it is able to deal with a component manufactured in a different way (and a different standard interpretation). Example: Company X produces RFID tags and interrogators, which are implemented conforming to the ISO/IEC 18000-6 standard. As the implementation is done in one house, the tuning of RFID tags and RFID interrogator is oriented to achieve maximum performance. If the RFID tags of company X, on the other hand, are operated with the RFID interrogator produced by Y – tuned to perform best with the RFID tags of Y, the performance (identification distance and identification rate-tags/second) is low and inadequate for real applications.

The interoperability testing is based on the combination of special pre-qualified hardware implementations on the RFID Tag or RFID Interrogator side (called qualified equipment in Figure 19.3(b)) with the DUT. Special border and limit scenarios are intentionally generated to see the reaction of the DUT. The interoperability testing ensures the interoperability of different real implementations from different manufacturers.

The testing is done in different laboratories specializing in this area. These laboratories are certified by the standards organizations and are equipped with certified testing equipment (e.g. anechoic chamber, automated testing equipment for example CISC MeETS based on National Instruments Test Hardware). An example of such a testing laboratory is the European EPC Competence Center (EECC) operated by METRO Group, Neuss, Germany.

19.2 The Simulation/Emulation Platforms

The verification of designs for UHF RFID deals with two aspects. The requirements need to be fulfilled both at the system level and at the level of individual devices. First, at the system level consistency and optimum performance towards end user requirements are required. Second, at the level of individual devices, designs need to be verified primarily for their conformance to regulations and protocols and optimum internal resource management. The latter is critical especially for UHF RFID tags. Designers of readers leverage from their modular structure. Their designs almost can be modified in run-time by exchanging submodules or modifying firmware (during the development process, not by the user). Tag design procedures are different. The tag ASIC is implemented on a single chip, which is hard-wired to a strip antenna. Tags can only be redesigned in the premanufacturing phase. Therefore special attention was devoted to the real-time verification of tags in the implementation of the methodology proposed.

The modeling and simulation framework [21] of the methodology proposed were implemented for UHF RFID technology with passive transponders and single or two reader environments. Figure 19.4 shows the proposed system level design flow. It starts in the first step with the specification. In the second step it applies the proposed methodology and ends with the analysis and evaluation of the system. The simulation platform and real-time verification platform interact with each other through a bi-directional interface in discrete times during the model-based optimization and verification procedure.

The components of the implemented platforms are shown in Figure 19.5. The simulation platform comprises of libraries and simulation controllers. The libraries are defined

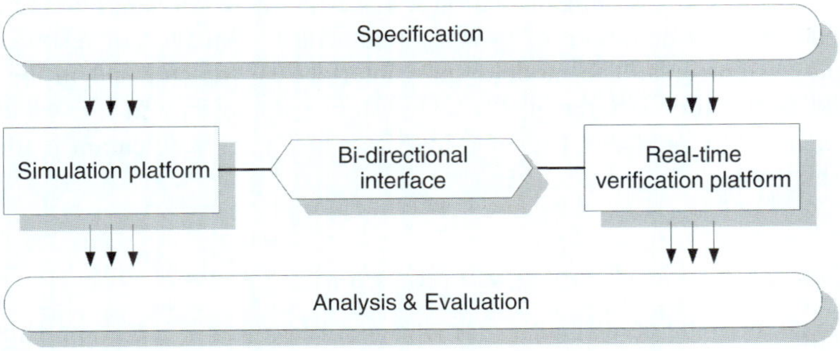

Figure 19.4 System level design flow based on the methodology proposed.

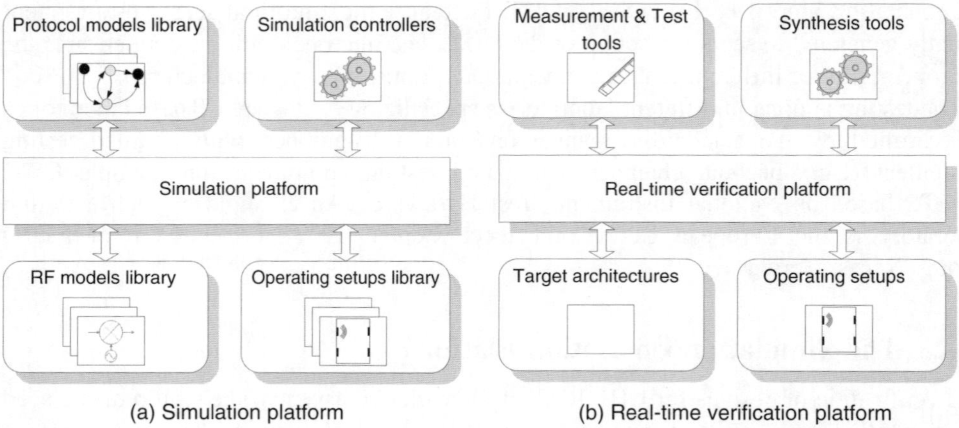

Figure 19.5 Components of implementation of the proposed methodology.

for each layer of the framework. The RF model library is assigned to the RFID hardware layer. It defines models of analog components of the system. The protocol model library is assigned to the RFID software layer. It defines the implementations of reader and tag statemachines for various communication protocols. The operating set-up library is assigned to the RFID application layer and defines the physical set-ups of UHF RFID systems, positioning, antenna switching patterns, tag movement scenarios, etc. The real-time verification platform comprises of target architectures for real-time execution of models and synthesis tools assigned to them. The measurement and test tools were integrated into the flow of the system verification procedure. The platform furthermore embodies physical operating set-ups including physical hardware.

The bidirectional interface between simulation platform and real-time verification platform comprises of two unidirectional interfaces. The model export interface is used to provide a link from the simulation platform which defines the model, a set of test parameters and system configurations in the simulation environment, to the real-time verification platform. The platform implements the model on a Real-time architecture and verifies it in an operating set-up under a certain configuration. The model export interface consists of assignments that are dedicated to the given model. For a selected Real-time architecture, the model is partitioned for implementation in hardware and software that are mapped to the target hardware by definition of system architecture, allocation of resources, parameters and pin assignments. The model export interface is implemented in the form of a resource and parameters assignment list and system parameters and test vector set. The data transfer interface provides a reverse link from real-time verification platform to the simulation platform.

19.2.1 Layers of the Modeling and Verification Framework

A system is an assemblage of entities (or objects), real or abstract, comprising a whole with each and every component (or element) interacting or related to at least one other component (or element) [64]. In these terms, the UHF RFID system is defined by its

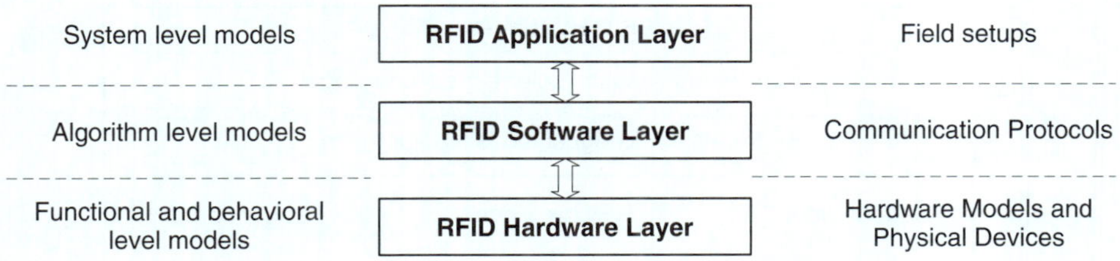

Figure 19.6 Layers of a framework for UHF RFID system modeling.

physical and virtual components and the values of their parameters. This system was abstracted into three layers [22]:

- RFID application layer;
- RFID software layer;
- RFID hardware layer.

The methodology for Real-time verification of the UHF RFID systems was developed based on a modeling framework which reflects the abstraction of the system. The physical components of the UHF RFID system are its readers, tags and their communicating channel. Their models are mapped on the hardware layer of the framework. Its virtual components are the software modules of the physical devices, namely the digital communication protocol implemented in a form of state machines and firmware. These are mapped on the software layer of the framework. The application layer of the framework defines values of parameters of both physical and virtual components that influence the functionality and performance of the system as a whole.

Figure 19.6 shows the assignment of model complexities to individual layers of the proposed framework. The functional and behavioral level models of the continuous parts of the systems are developed on the RFID hardware layer. The RFID software layer defines discrete models at the algorithm level according to the Y-chart for digital systems. System level models of both discrete and continuous models are defined at the RFID application layer.

19.2.1.1 RFID Hardware Layer

The RFID hardware layer is the basic block of the entire system. It defines the electrical properties and physical structure of devices and air interface. The hardware layer models of the UHF RFID system describe physical components with three main independent and exchangeable modules for a reader, a tag, and a wireless communication link. The basic structure of the hardware level components are shown in Figure 19.7.

The model of a reader consists of the following modules:

- antenna;
- demodulator;
- filters;
- modulator;
- digital logic.

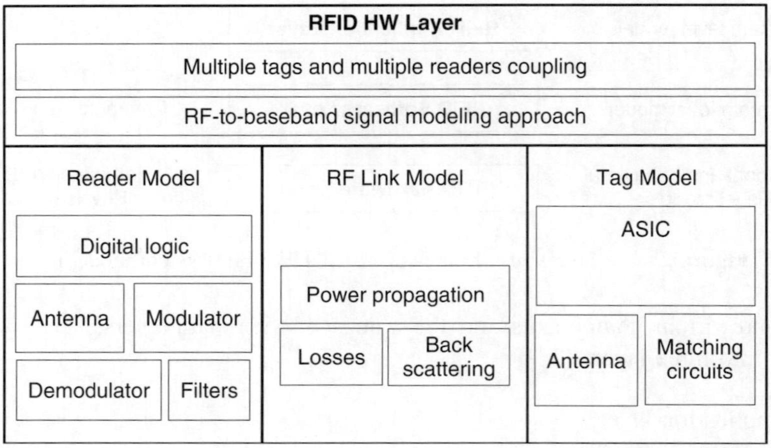

Figure 19.7 Hardware layer structure of the framework for UHF RFID system modeling.

The model of a tag consists of:

- ASIC;
- antenna;
- matching circuitry.

The model of an air interface consists of:

- power propagation;
- field disruptors.

The RFID hardware layer further defines the concepts for the simulation of coupling multiple tag and multiple readers. The RF to baseband signal modeling approach is based on simplifications of the representation of the RF link. The structure of the modules, including the definition of their inputs and outputs, were accommodated to these simplifications to guarantee the performance required for simulations of separate evaluations in time and frequency domains. The trade-off between simulation accuracy and speed was set to fit the objective, that is, a multi-layer simulation of the system. Further details will be discussed in Section 19.3.2.

19.2.1.2 RFID Software Layer

The RFID software layer defines the properties of the digital communication link and data transfers. The data is handled either bitwise or as a packetized information. Models on this abstraction level simulate the data flow and the quality of communication protocols. The model is split into two parts, as shown in Figure 19.8:

- reader software module;
- tag statemachine.

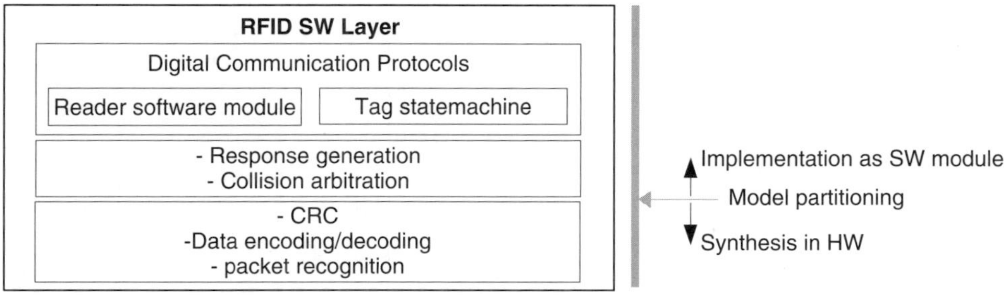

Figure 19.8 Structure of the software layer for UHF RFID system modeling.

The subjects for evaluation at the system level are reader-to-tag and tag-to-reader communications, multi-reader digital data collision arbitration and multi-tag digital data collision arbitration. At the level of an individual device (tag or reader), concepts for data encoding techniques, data error detection algorithms and packet recognition are developed and implemented in the software layer of the UHF RFID system model.

As discussed earlier, one of the requirements is to enable the Real-time simulation of the model and thus implementation of the model in hardware. The partitioning of the model into hardware and software blocks is therefore based on the selected target architecture.

19.2.1.3 RFID Application Layer

The application layer specifies properties of the UHF RFID application set-up. Parameters such as configuration of reader gates, antenna type and position, type of item identification method, etc. are defined here. The parameters of the two underlying layers are set according to the specification of the RFID application layer.

The RFID application layer provides a set of field set-up parameters and configurations. It defines the mechanical arrangements of devices as seen in target operating environments. The layer stimulates the other two layers to the specific conditions of the UHF RFID operating environment, such as for example warehouse or library environments, antenna and read points arrangements.

19.2.2 Implementation Languages

The simulation platform was implemented based on models developed in the Matlab Simulink® environment. The models handle analog and digital UHF RFID system designs. The real-time verification platform was implemented primarily for the verification of UHF RFID system designs of digital subsystems with a limited possibility to verify analog designs due to the simulation speed. Hardware-in-the-Loop (HIL) simulation is used to verify the whole system. The implementation environment furthermore allows for the simulation of heterogeneous models in various languages (Figure 19.9). C++ based models are linked to Simulink® through S-functions. The heterogeneous system is (Simulink®, VHDL) executed by a cosimulation environment. Cosimulation is one of the solutions for verification of heterogeneous systems that has been intensively investigated in recent

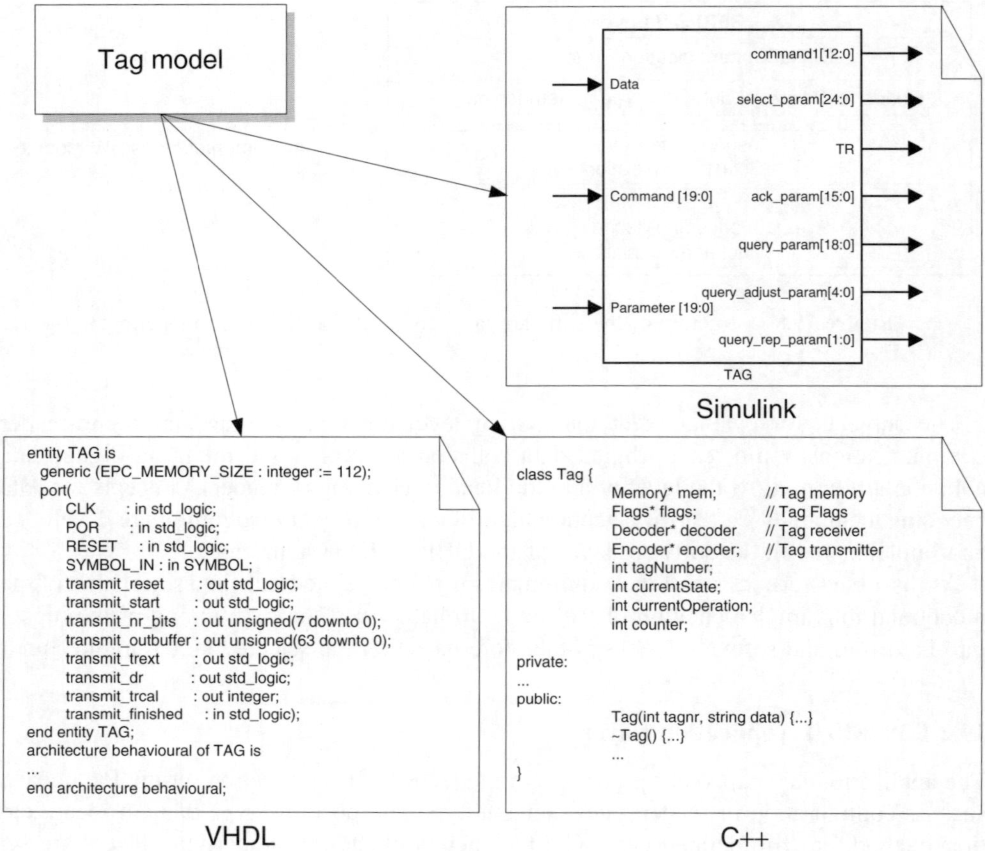

Figure 19.9 Different implementations of a tag ASIC model.

years. In cosimulation, subsystems are simulated using simulators optimised for certain languages. The problem of communication between subsystems has been solved at the simulator level using cosimulation interfaces, which connect involved simulators in one simulation environment. In [47] a design methodology was developed for the verification of heterogeneous embedded systems using an advanced cosimulation technique. Moreover, in order to prove the feasibility of the approach, an application framework has been developed [11], that creates fully automatically a coverification platform. The framework enables a multi-language, model-based, system design and automatic platform generation for distributed cosimulations.

The communication path connects major modules of the system. Figure 19.10 gives the overview of the signal path. The model of the complete system includes time-discrete and time-continuous subsystems. To support multi-layer simulation of the entire system, abstraction of time-continuous models has been limited to the system, functional, and behavioral levels of detail. Time-discrete models are implemented at the algorithm level.

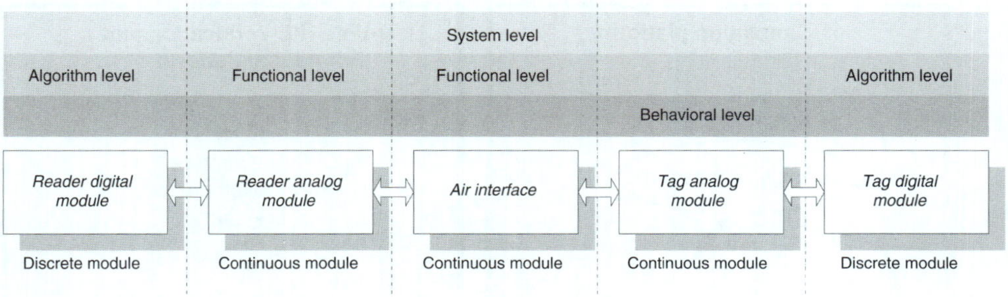

Figure 19.10 Communication path between major modules of the system with indication of abstraction level of corresponding models.

19.3 UHF RFID Simulation Platform

The simulation and modeling of UHF RFID systems (e.g. RFID gate in Figure 19.11) combine the areas of analog RF simulation of transceivers, RF field propagation, digital protocol evaluation, and system performance evaluation. RFID simulation and design space exploration have been done by several other research groups for example [33] (RFIDSim simulation engine), [6] (rapid protyping framework), [44] (RFID tag design flow).

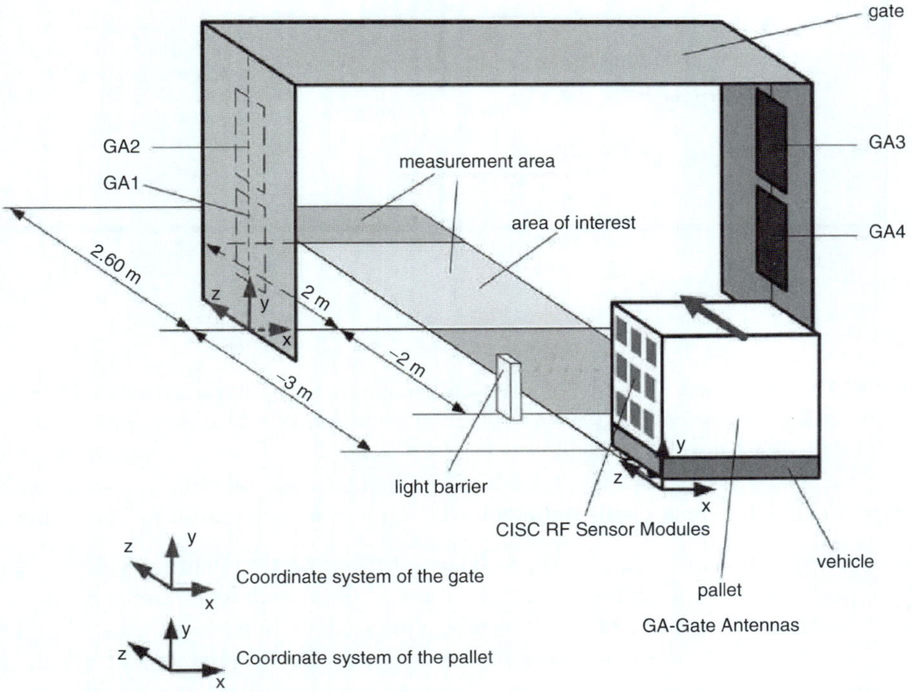

Figure 19.11 Example of an application specific layout.

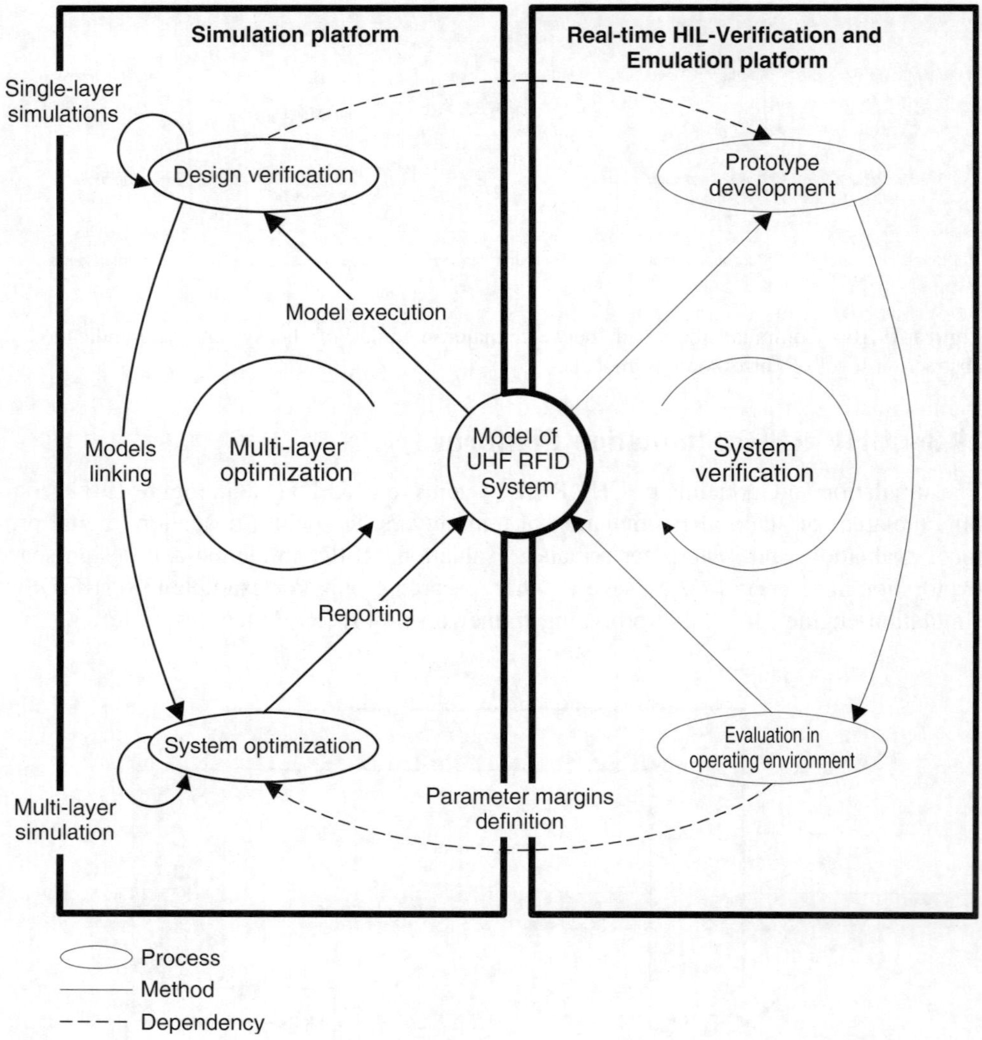

Figure 19.12 Simulation platform for multi-layer optimization (left part, Section 19.3) and a real-time verification platform for system verification (right part described in Section 19.4).

19.3.1 Multi-Layer Optimization

Multi-layer optimization is the kernel procedure of the simulation platform (Figure 19.12). Its processes and methods aim to optimize the design of individual components of the system and of the system as a whole. The procedure starts with a model of the UHF RFID system and it proceeds towards an optimized model in several iterations of simulations. The multi-layer optimization is based on two main processes: (1) design verification, and (2) system optimization.

19.3.1.1 Design Verification

Model Execution
Before design verification can be performed using single layer simulations, the model has to be prepared for execution by assigning model parameters and simulation settings. The selection of simulation platform, type of simulator, type of simulations solver and its settings can be defined individually for different modules of the model.

Single Layer Simulations
Design verifications of individual modules of a single device are based on single-layer simulations of models. Single layer simulations verify designs on the RFID hardware layer, RFID software layer, or RFID application layer separately. For instance, the digital communication protocol is thus validated or verified for its conformance with specifications on the software layer. The verification of analog designs of readers and tags is independent of this.

19.3.1.2 System Optimization

Models Linking
Models of individual subsystems are built on various abstraction levels. Inputs of interacting models are linked together. Dynamic responses of individual subsystems need to be considered when linking subsystems to correctly set the sampling bases of simulators. Subsystems may further be designed using diverse languages in order to preserve the performance of the simulation. Functional verification of such designs is verified using cosimulations. Commercial tools like Link for Modelsim [1] and SyAD [11] are available to take responsibility for the heterogeneous system cosimulations. However, models need to be constructed according to common rules in order to utilize not only the advantage of cosimulation of heterogeneous description languages but also the simulation speed improvement.

Multi-Layer Simulation
Multi-layer simulation takes place to simulate models of RFID hardware, software and application layers of the modeling framework. Models of devices whose designs were individually verified are linked together and simulated as a system. The tool applied for system level simulations is Matlab Simulink®. Block level models are executed by a cosimulation environment [45] implemented within the SyAD® tool. Models defined on various levels of abstraction or in various languages are executed by cosimulation environments [45]. The simulated virtual system is fully reproducible and the set of test patterns is restorable, saved in a test bench table. Thus it is possible to make direct comparisons between simulations with different parameter settings.

Reporting
During the system optimization supported by multi-layer simulation it is determined whether the system is consistent and performing according to specifications. In the first

iteration of the simulation and Real-time verification process the parameter's margins are defined, based on application data sheets. In later iterations, values of parameters measured in operating set ups are fed into the simulation test bench. Finally, the parameter's margins are extracted from the system verification process in the Real-time verification platform. This can first be done after a reconfigurable prototype has been generated from the model. The results of the system optimization process are reported back to the model, which can be updated or its structure can be modified or refined and specifications for individual components of the system can be adjusted or changed.

19.3.2 Modeling and Simulation Techniques

The objective of the methodology is that system components can be tested both separately and in the context of the entire complex system during the development of individual modules. While design verification on individual layers can be done using simulators in time, frequency, or mixed domains, system optimization reduces the choice to simulations in one domain at a time. The limits are imposed by RFID software layer modules, whose performance measures are expected in the time domain. Furthermore, Real-time interaction and the influence of other RFID devices in a specific environment are essential.

In order to fulfill these requirements, models of the UHF RFID system are defined based on mathematical description of their behavior. This description defines outputs as functions of inputs [63]. It describes algorithms that are used to obtain those outputs, but the implementation of the algorithms themselves is not defined by the description. A critical aspect specific to UHF RFID systems is the coexistence of high carrier frequency and low rate data signals. The RF carrier frequency (1 GHz) applicable to the RF subsystem of the tag is 10^3 times faster than the digital circuitry, which only needs to handle the baseband frequencies of up to 640 kHz. This has a major influence on the methodology proposed.

19.3.2.1 RF to Baseband Modeling Approach

Abstracting models from their radio frequency-based representation into baseband representation is the key strategy for achieving short simulation times with reliable description of critical signals. A model of a UHF RFID system that works solely with a baseband signal while preserving information about carrier signal properties on individual components is a part of the framework proposed. Time domain simulation is of primary interest as one of the targets of the methodology is to simulate models on real-time architectures.

As depicted in Figure 19.13(a), an RF signal is propagated through the model as a complex signal of actual envelope carrier frequency and phase values. The carrier is represented by one frequency value and phase, both variable in time.

As radio regulations in most countries require carrier frequencies to have a tolerance of less than 10 ppm, which means $1\,\text{GHz} * 10\,\text{ppm} = 10\,\text{kHz}$ and the baseband frequencies are at least 40 kHz, signals representing frequency and phase variation occupy a lower spectrum band than the envelope signal. Therefore these signals no longer have a determining effect on simulation sampling rates.

Each modeled block has a separate signal path for envelope signal and carrier frequency propagation. Frequency and time domain aspects of the model are handled separately

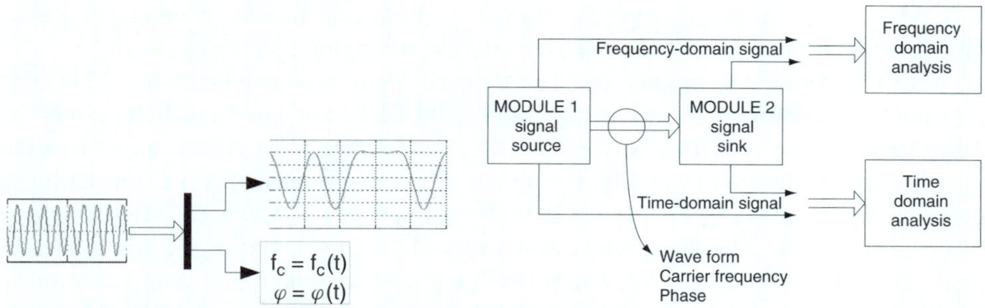

(a) Carrier and base band signals separation in the model -splitting in two components: (i) envelope and (ii) carrier frequency and phase

(b) Separation of frequency and time domain analyses

Figure 19.13 RF to baseband modeling approach.

because no interaction between more carrier waves occurs in the signal path. The separation of frequency and time domain analyses is depicted in Figure 19.13(b). The carrier frequency dependency of individual modules is very strong for most modules of UHF RFID systems. The methodology proposed distinguishes two types of system module:

- Modules with one carrier frequency in the spectrum of the input signal.
- Modules with multiple carrier frequencies in the spectrum of the input signal.

The first type of module processes the envelope signal based on behavioral models considering the presence of a single-frequency carrier signal. The latter type represents signal mixers. Their functionality in the signal flow is to produce a multiplexed output consisting of an envelope and new carrier frequency (instead of multiple frequencies that are physically present at the input of the subsystem). The envelope of this output signal is strictly identical with the envelope of mixed input carrier waves. The resulting output carrier frequency considers the values of input frequencies, maintaining a minimum error when replacing the original spectrum by a composition of envelope and single tone signals. In [24] it is derived that the guaranteed maximum inaccuracy brought into the UHF RFID model is 11.6% resulting in seemingly increased or decreased performance of the tag compared to real values.

19.3.2.2 Models of Individual Modules

UHF RFID systems that work on 860–960 MHz carrier frequency typically have data rates between 40 to 160 kbps in the forward link (reader-to-tag) and depending on given standard and RF regulatory environment up to 640 kbps in the return link (tag-to-reader). Due to the fact that the ratio between data and carrier frequencies is in the order of thousands to tens of thousands, simulating analog subsystem would require proportionally higher sampling rates than digital data simulations. Analog parts including capacitors, inductors and filters are modeled as a general continuous system. If a mathematical model of such circuitry includes feedback loops, simulation may become too slow to be useful for design/testing purposes. Applying the RF to baseband modeling approach leads to significant simplifications (we do not simulate the RF signal).

Passive RFID systems communicate based on carrier amplitude modulation. This allows the building of communication units, especially on RFID tag side, which are simple and require very low power. Both of these are key requirements for cheap mass production and achieving high operating ranges without onboard power supply for the RFID tag. Regarding the simulation, this means that the communication between RFID interrogator and RFID tag can be simplified by considering only the envelope of the communication signal. This allows the movement from the extremely complex and computing power-intensive high frequency carrier signal (868 Mhz) based simulation to a medium frequency (approx. 600 kHz) and thus being able to simulate the sometimes complex and time-consuming communication sequences. This includes the transfer of the 96 bit identification code as well as reading and writing the RFID tags memory (can be 600 bit and more). The simulation is based on the mathematical modeling of the baseband signal but considers in-system-noise generation, free space loss, and antenna angle and gain characteristics. This allows the simulation of the entire communication sequences with several RFID system components such as interrogators and tags in reasonable simulation time.

19.3.2.3 Reader Module

Antenna

An antenna is generally a device of a complex structure and characteristics. At the system level, a behavioral model of UHF RFID reader antenna is characterized by its gain, radiation pattern, resonant frequency, efficiency, bandwidth and polarization, and axial ratio. There is no time dependency of the values of these parameters.

Demodulator

The demodulator of an UHF RFID reader demodulates the low-frequency signal from the received radio frequency wave. The signal received by the reader demodulator is made up of two components: (1) Signal T transmitted by the reader damped by the reader's circulator and coupled back to the receiver and (2) received tag backscattered signal B. The signal R received by the reader (not considering noise for this moment) has the following general form:

$$R \sin(\omega t + \varphi_R) = T \sin(\omega t + \varphi_T) + B \sin(\omega t + \varphi_B), \tag{19.1}$$

where

$$T = \eta T_0$$

$$B = B(d)$$

$$\varphi_B = \varphi_B(d).$$

η is the transmitter-to-receiver coupling factor and T_0 is the amplitude of the transmitted signal. Both amplitudes and phases of its two components T and B are different. The amplitude and phase of the received backscattered signal are further functions of distance

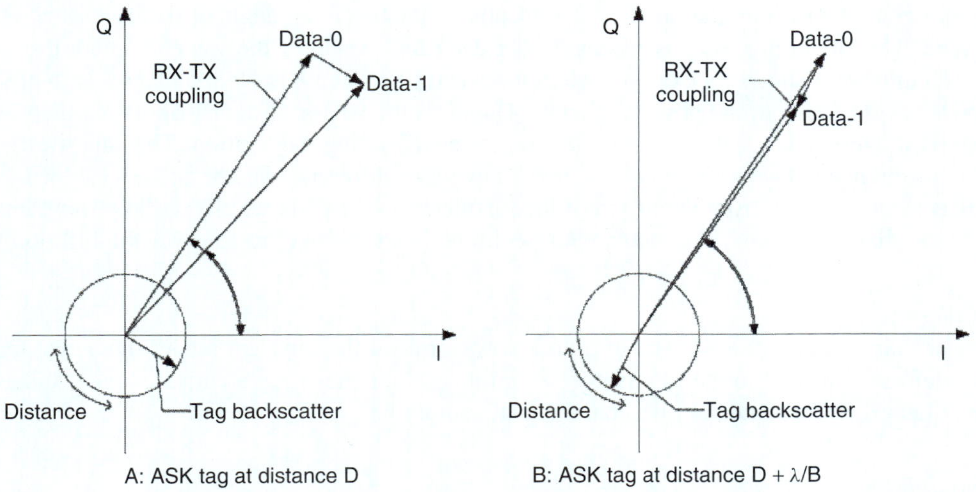

Figure 19.14 Reader demodulator: phase vs. amplitude.

between the reader and the tag. The dependency of the phase φ of the received signal on the reader-to-tag communication distance d is expressed in Equation 19.2.

$$\Delta\varphi = \frac{2\pi}{\lambda}\Delta d \tag{19.2}$$

Although the subject of the simulation is a system with tags that only implement ASK[2] modulation, information received by a real receiver from a backscattering ASK tag is generally present both in amplitude and phase components [25]. This is shown in Figure 19.14. Shifting the tag by the distance of $\frac{\lambda}{8}$ away from/towards to the receiver prolongs the path of the signal by the distance of $\frac{\lambda}{4}$ and thus shifts the phase by $\frac{\pi}{2}$. Phase and amplitude signal paths are therefore included in the model of the return link.[3] Figure 19.14 shows the amplitude representation of the carrier signal being reflected (backscattered) by the RFID tag. The length of the vector itself represents the strength of the carrier signal reflected by the RFID tag. The information (data) which travels to the RFID interrogator can have two main representations (and each level in between). Depending on the distance between RFID interrogator and RFID tag, the information is represented in the phase shift (Figure 19.14(a)) or in the amplitude change (Figure 19.14(b)) of the carrier signal. The phase change is shown as the vector changing its angle compared to the original carrier (Figure 19.14(a)) and the amplitude change as the modification of the length of the vector (with no phase change). The switch between the data coded completely in the phase and the amplitude happens each distance change of Lambda/8. Data-0 represents the low level data representation (binary "0") and Data-1 the high level representation (binary "1"). The data is not transmitted in plain format and the data representation shown describes the levels of the binary coding of the information (based on FM0 coding). The distance is

[2] Amplitude Shift Keying.

[3] reader-to-tag.

represented by the circular arrow as it depends on the receiving angle of the backscattered wave. The situation shown is repeated after each full period of the wave (Lambda).

Figure 19.14 shows an example where the vector between Data-0 and Data-1 is rotated by 90 degrees due to the position change. Data-0 is the vector of the carrier when there is no modulation and Data-1 is the vector of the carrier during modulation. The data itself is the difference between these two vectors shown as a difference vector and also shown as a vector originating from the origin of the IQ diagram. The distance arrow shows how this vector changes through distance and/or position change. It may rotate over the full 360°.

Filters

Filters shape the digital bit stream into a waveform. In the physical reader the generated waveform is passed to the modulator, which then modulates it on a carrier wave. Models of filters therefore maintain baseband signal sampling.

Modulator

The modulator performs modulation of a signal on a carrier in order to transmit it over a transmission line. Data receivers on recently used tags are implemented in principle as envelope detectors. Envelope shaping is the most critical aspect for ASK tags. Information in the signal received by the tag is defined by the variation of the amplitude. The forward link (not considering noise at the moment) is expressed by the following equation:

$$R \sin(\omega t + \varphi_R) = T \sin(\omega t + \varphi_T), \tag{19.3}$$

where

$$T = T(d)$$

$$\varphi_T = \varphi_T(d).$$

The amplitude T is a function of reader-to-tag distance d. The phase φ_T of the received signal is proportional to the distance d. The variation of the phase due to tag movement is limited by the assumption of a maximum speed set to 200 kmph, thus not affecting the behavior of the tag front end. As the ASK tag is not capable of detecting data from a phase change, the forward link model can be simplified for the time domain system performance simulation, not including the phase signal. One of the key performance measures of the forward link is still the frequency spectrum mask. The reader signal has to fit into a predefined spectrum, fulfilling the regulation requirements.

For the purpose of characterizing the tag signal received by reader in I and Q channels, let us first consider a simplified model of the forward link in which the tag signal is represented as a perfect rectangular signal modulated on a carrier with the frequency ωt. Modulated parts (data 1) are then seen as sine wave, with amplitude T1 and phase φ_{T1} and non-modulated parts (data 0) as sine waves with amplitude T2 and phase φ_{T2}. The following will hold for a perfect ASK tag:

$$\mathrm{T1} < \mathrm{T2}$$

$$\varphi_{T1} = \varphi_{T2}$$

As the signal is transmitted using backscattering from the tag to the reader over distance d, the phase of the received signal additionally becomes a function of the distance between the transmitter (tag) and the receiver (reader). The forward link can then be described by expression 19.3. Noise and waveform shaping are accounted in the model later on.

Digital Logic

The digital logic of an UHF RFID reader implements a physical hardware for the reader's software module. Power concepts need to be considered for handheld readers that are supplied by batteries. Power consumption of the digital logic is largely given by the implementation of the communication protocol. The protocol aspects are accounted in the software layer of the framework.

19.3.2.4 Tag Module

Antenna

The high frequency parts of the tag (i.e. the tag antenna and ASIC's RF frontend) were modeled as a behavioral RLC circuit [23]. The antenna model is a behavioral description of the equivalent circuit for the UHF RFID antenna. The behavioral description is further applied to baseband signals that determine corresponding low frequency sampling rates.

ASIC and Matching Circuitry

The ASIC and matching circuitry component values are limited by die size, power consumption, etc. The tag ASIC is built of a modulator, demodulator, voltage controllers, rectifier, voltage multiplier, and digital logic. Modern tags also embed sensors on the chip. Proposed structuring of the model allows for modeling of any potential architecture of the ASIC based on the description of the current through individual blocks as a function of the input voltage. Signals are represented by their envelopes. A model of the RF frontend was replaced by a behavioral description of the baseband circuitry.

19.3.2.5 Air Interface

The model of the air interface is given by a mathematical description of the power propagation effects at the functional level. Field variation in time is given by the variation of the signal transmitted and changes in the arrangement of the operating environment. Variation of the transmitted signal is described using above discussed baseband models. Field variation due to changes in the operating environment arrangements is caused by the movement of physical objects. The variation is limited by assumptions of maximum speeds and therefore has no impact on reader and tag signal sampling rates.

It can be concluded that the maximum sampling rate $f_{s\,\max}$ of signals of the models is limited by the maximum sampling rate $f_{sb\,\max}$ required to represent the fastest baseband signal in the system.

$$f_{s\,\max} = f_{sb\,\max} \tag{19.4}$$

The fastest baseband signal is given by the maximum data rate [27], data encoding technique and maximum frequency of the behavioral model of the UHF RFID system according to Equation 19.5.

$$f_{sb\,\max} = 640\,kbps \cdot 2\frac{char.}{bit} \cdot 400\frac{samples}{char.} = 5120\,kHz \tag{19.5}$$

The maximum sampling rate of signals from the models is the sampling rate of the Matlab/Simulink models. Simulink models are implemented with certain parameters and under certain rules.

The modeling of the air interface has not been a main topic for the modeling; therefore it is modeled more simply. This is also a way of focusing the simulation accuracy on the digital communication, multi tag handling and the behavior of the RFID tag interface itself. The channel model is based on simple propagation loss, which is the main factor limiting the operating range of passive RFID tags. This is because the energy required for their operation is provided by the field strength emitted by the RFID interrogator (Battery-less operation) and so mainly limits the communication range. The power or field strength required to operate an RFID tag is perfectly known – if the field strength corresponds to this defined level at a certain location in the simulation environment, the RFID tag works – otherwise it does not. In the same way, a "fading channel" is simulated (other simulation scenario) – the underlying statistics describing field nulls and holes come from measurements of the field distribution in a real set-up (based on the CISC RFID Field Recorder [14]). This is truly a compromise between accuracy and flexibility but the results have shown good correspondence with real measurements.

19.3.3 Model for the Simulation of the UHF RFID System

The methodology proposed was applied to a simulation model of the UHF RFID system. The model gives an insight into the behavior of the analog and digital structures of the UHF RFID system in terms of providing envelope signals of power transmitted by the reader, power received by the tag, backscattered signal, tag ASIC DC voltage level over time and many others. The system model consists of models of a reader and tags and a model of the communication channel between the reader and each tag. The reader's main modules are the *Modulator*, *Demodulator*, and *Software Module*. Since the communication channel has different parameters for each tag (due to position, orientation, etc.), each tag is assigned a separate model of a communication link. The tags are grouped in the *Tags* module. Simulation of the model provides a time domain analysis of an ASK system. Configurable modules allow exchanging individual blocks to analyze the system performance under various designs, parameter sets and set-up conditions. This includes, for example, the modulation filter type, modulation depth, minimum DC voltage level threshold of the ASIC and size of the tag DC voltage buffer capacitor, etc. Details of the reader and tag models are accessible through the hierarchical structure of the model. Behavioral models of the tag are intended to verify and optimize the performance of the system based on functional and system-level specifications of the design of the physical devices.

19.3.4 Use Case: UHF RFID Systems

The methodology was applied to a model of a UHF RFID system whose kernel was implemented in Matlab Simulink. The test cases presented in this chapter represent various scenarios under which the methodology was evaluated. The target of the RFID hardware layer implementation is an ASK system. The RFID software layer is implemented for EPCglobal Class 1 Generation 2 protocol [27].

19.3.4.1 Evaluation of Time-discrete and Mixed-signal Simulation Methods

The reader software module, tag state machine and digital circuits implement the communication protocol. The channel between these digital modules consists of the analog circuitry and RF frontend of the reader and the tag and, further, of an air interface. The tag's digital modules are capable of receiving, processing and transmitting data if they are provided with DC voltage sufficient for their operation. In that case the tag is said to be in operating mode. The DC voltage is dependent on the immediate power density in the tag's location and, furthermore, on the structure and energy dissipation of the analog and DC circuitry and other factors. The DC voltage is desired to be within certain limits. The lower voltage limit, called the voltage power-on-reset (V-POR) threshold, is critical for the tag's operability. If the V-POR threshold is exceeded, the tag is put in operating mode and stays there as long as the DC voltage level is above the V-POR threshold. In the case of underrun, the tag is reset. The V-POR threshold correlates with the power-on-reset (POR) threshold. The POR threshold determines the minimum RF power required by the tag in order to be operable. By applying standard techniques such as voltage buffering and multiplying at the rectifier [9], short drop-offs of the RF power (e.g. due to modulation) are compensated by the tag analog circuitry to prevent them from resetting the DC modules. The reader analog, RF and digital modules are powered from the AC power. Therefore their power aspects are not considered.

It is important to also consider environmental aspects in the simulation of the system performance. Figure 19.15(a) shows how system parameters, set-up arrangements and the path of the tags' movement are processed for time-continuous simulation of mixed-signal system model and time-discrete simulation of digital system models. Each tag is provided with a set of patterns representing time-varying magnitudes of power available to the tag. The power patterns consider known set-ups, antenna radiation patterns, power damping due to surrounding materials and other RF effects. A tag is provided with a separate pattern for each radiator. A reader commonly has four antennas that are time-multiplexed. Therefore RF effects due to interaction between multiple transmitters are not an issue in a single reader set up. For time-discrete simulations, power patterns are further processed into POR patterns based on defined POR thresholds.

In the time-discrete model the data link is represented bitwise by signals of two discrete levels: high and low. For time-discrete simulation no noise is applied either on the communication channel or on the analog and RF parts of the reader and tags. Furthermore no modulation and signal filtering is applied, meaning digital data transmitted by the reader digital module is received by the tag digital module if the tag is in operating mode.

The time-discrete simulation is mainly to simulate the digital behavior of the RFID system – focusing on multi-RFID tag handling (anti-collision). This is a direct connection between the RFID Interrogator and multiple RFID tags, incorporating pre-simulated times of their activation. The simulation is done as follows. First, tags are placed in a virtual environment, usually on a pallet with different products. Then the simulation environment is set up, usually a reading gate construction equipped with one RFID

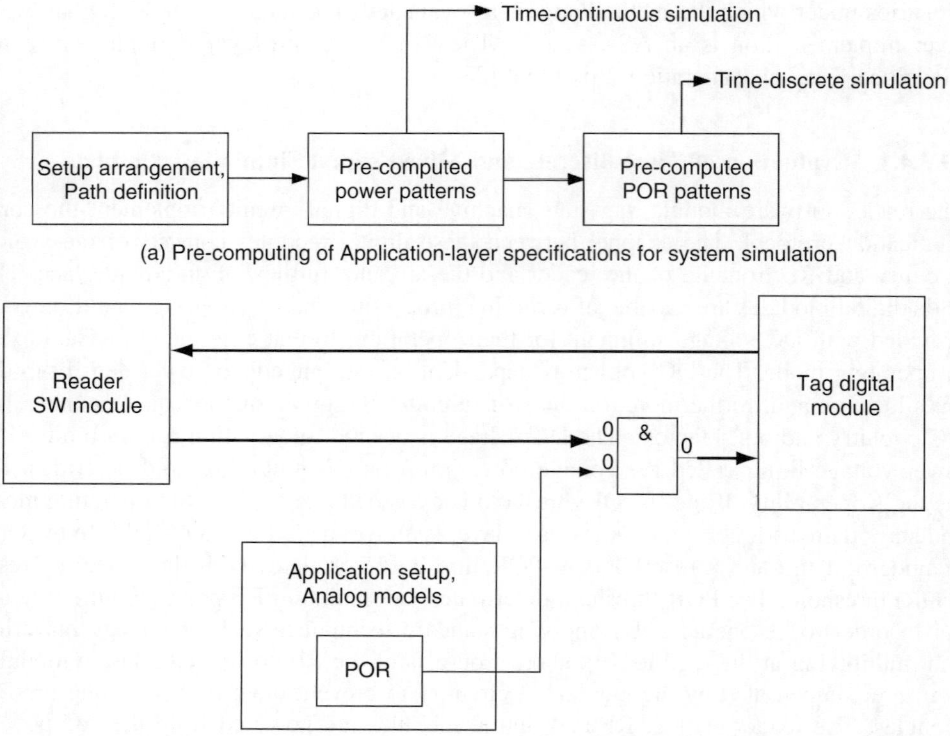

(a) Pre-computing of Application-layer specifications for system simulation

(b) Gated data and clock signals represent tag inputs considering power-off times in time-discrete models

Figure 19.15 Pre-computing and considering power off times.

interrogator multiplexing four antennas. When the simulation starts, the field strength (power) theoretically available to the RFID tag is pre-simulated (precomputed) to have the information at which point of the environment an RFID tag would work for each point in the simulation environment (three-dimensional). Afterwards the pallet is moved virtually through the gate. If a RFID tag reaches a region where the precomputed field strength would be high enough to operate the RFID tag, the tag is activated and starts communicating directly with the RFID interrogator. The activation is stochastic, so it fits very well with real-life identification scenarios. As usually more RFID tags are in the same area close to each other, they are activated at the same time and the anti-collision mechanism becomes necessary.

The time-continuous simulation deals with the analog representation of the signal meaning the envelope considering noise and the analog RFID tag components. This is usually done to find the analog environmental parameters and test different RFID tag implementations. In the simulation usually just a few tags are combined with one single RFID interrogator.

Environmental aspects are considered in the simulation based on pre-computed POR patterns. Power drops caused by modulation are omitted from the POR pattern, assuming

Table 19.1 Simulation and system performance measures of time-discrete and time-continuous models.

	Time-discrete	Time-continuous
Simulation time	1 hr 40 min	55 hrs 35
Number of tags in simulation	10	10
First tag detected	575 ms	512 ms
Last tag detected	1.31 s	2.30 s
Total number of detections	10	10
Number of different tags detected	10	10
Anticollision rate	13.69 Tags/s	5.60 Tags/s

that tags are adapted from an analog design by the DC control circuitry. Power drop-offs due to any other reason causes an immediate tag reset. In the implementation, tag input data and clock signals are gated using the POR signal according to Figure 19.15(b).

19.3.4.2 Evaluation of Simulation Results

To evaluate the time-continuous and time-discrete models, a model of a population of ten tags was equally distributed in a cuboid of the dimensions of a loaded Euro Pallet[4] and passed through a model of a standard portal reader set-up consisting of four antennas and one reader. To assure objective results the simulation was performed on a free space model. The parameter settings for the time-continuous and time-discrete models were identical. Table 19.1 shows the results. Compared to the time-continuous simulation of the mixed-signal model, the time-discrete simulation of the communication between the reader and tags provides significantly faster simulation. As the tags were arriving into the interrogation zone[5] the first of them were detected in both models approximately 500ms after the beginning of the simulation. The last tag was detected after 1.3 s in the time-discrete model whereas the detection time of the last tag in the time-continuous model was 2.3 s.

Unlike the simulation of protocol performance in digital models with respect to the field distribution, mixed signal simulation provides a more comprehensive and accurate result when considering unequal signal strengths, bit-timing errors due to noise in signal, applied methods for modulation, demodulation and data detection based on models of analog and RF components, etc. Slightly delayed detection of the first tag in the time-discrete simulation observed in Table 19.1 is one of the inaccuracy trade-offs compared to the time-continuous model. Multiple tags arrived in the reader field at approximately the same time which caused a concurrent response to a reader command and finally resulted in a collision of responses in the time-discrete model. As the tags could not be detected simultaneously, a collision arbitration process took place and the first tag could be singulated first at 575 ms.

The performance metrics in this case are the time required for the simulation. This is because reasonable simulation time is one key parameter when building a system model.

[4] A Euro Pallet is a block-style pallet measuring 1200 mm × 800 mm built to European Pallet Association (EPA) specifications. Only pallets built with an EPA branded logo are considered authentic EURO pallets.

[5] The interrogation zone is the operating field of the reader and its antennas respectively.

It does not make sense to make a time-continuous (analog) simulation with 100 RFID tags and 2 RFID readers when the computing of the simulation results takes more than 300 hours. Therefore the simulation time was one of the key requirements in the development of the RFID system model. Table 19.1 mainly shows the difference in the complexity comparing the required simulation time (and so computing power) of the time-continuous (analog) simulation and the discrete (digital) simulation. The two seconds (2 s) are the virtual (simulated) time required to identify the 10 RFID Tags. After 2 seconds all the tags have been virtually identified – which corresponds to the identification rate given. The 55 hours and 100 minutes are the real time required to simulate this virtual scenario (1.3 and 2.3 seconds).

Comparing the result with the time-continuous simulation on a mixed-signal model, the first tag was already detected at time 512 ms. The simulated situation was identical. Tags arriving into the reader field sent their responses concurrently, but one tag was significantly closer to the reader antenna. The weak responses were filtered out by the reader demodulator and a tag was detected with no collision, due to the analog representation of the backscatter signals. This is because of the adaptive matching of data detection level and noise threshold on the reader demodulator modeled in the mixed-signal model. The distances between the different RFID tags in the application are usually very low (a few centimeters) and so multiple response can be received and decoded by the RFID interrogator at the same time (interrogator sensitivity is approximately -110 dBm).

The anticollision rate describes the mean amount of RFID tags per second which could have been identified within the scenario. This is a very important parameter as it defines how many tags could be identified within a certain amount of time. Usually the real identification scenario contains numerous RFID tags placed on a pallet and they all have to be identified while the pallet moves through a gate. The area of identification usually is approximately 6 meters. From this it follows that assuming a slow motion of 0.5 m/s, all the RFID tags on the pallet must have been identified within 12 seconds. For our example with the anticollision rate of 13 tags/second, 154 tags could theoretically be on the pallet. The real number would be much lower. The time of the simulation is a key parameter to allow flexible changing of parameters and several consecutive simulation steps. If the complexity of the models is increased, the simulation time increases exponentially. The most important topic for this simulation was to simulate the behavior of multi-tags and anti-collision combined with the possibility of tuning parameters and simulating the related effects. Using more complex models would have caused very long simulation times (1 week and more) and made the iteration of several simulations combined with tuning of parameters very difficult. The protocol is implemented to fully conform with the most important HF and UHF standards for passive RFID: ISO/IEC 18000 Part 3 and Part 6 and EPCglobal Class 1 Generation 2 HF and UHF.

19.3.5 RFID Application and System Design Kit+Library®

The CISC RFID Application and System Design (ASD) Kit+Library® [12] is a simulation software solution for the design and evaluation of RFID systems and applications. It targets the initial development phase. It helps to determine how to arrange and configure a workable RFID system. It is possible to reduce the time-to-market cycle significantly with the ASD Kit+Library®.

19.4 Real-Time HIL-Verification and Emulation Platform

The simulation platform was focused on a multi-layer system optimization by simulating models. Simulations in the time domain were executed on a virtual time base. Every component of the system was required to be assigned to a model in order to execute the simulation. Exact models of operating set-ups are impossible to develop. The reasons are twofold. First, detailed specification of set-ups are not always available. This also applies to physical devices that are used in the set-ups. Second, the complexity of the system would require enormous computing power to execute simulations with the required high granularity of details. Another option for testing designs within the true image of real systems and to verify models to the satisfaction of Real-time constraints is to execute them directly in the operating environment. The Real-time verification platform (Figure 19.12 right part) enables a cosimulation of models with physical devices. The model of a subsystem synthesized on a hardware simulation platform is configured and set up for a verification in a target operating environment. The verification is performed in Real-time with real-world stimuli (Figure 19.16). The test bed consists of commercially available UHF RFID devices, reference devices and measurement equipment.

Hardware parts (e.g. a reader module) connected to the simulation model place high demands on the speed of the simulator. Coupling models specified at various levels

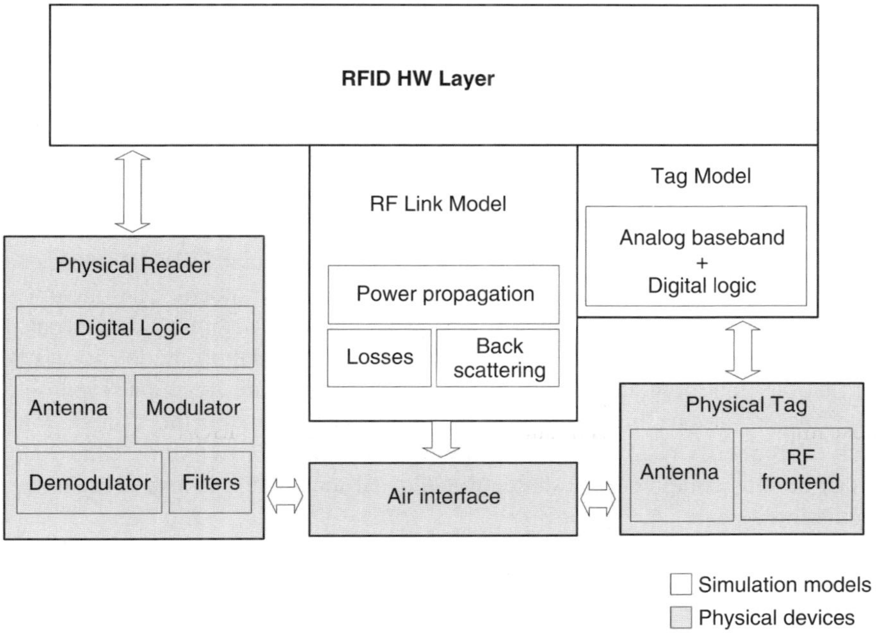

Figure 19.16 Cosimulation of the model with physical devices.

of abstraction with hardware parts requires a Real-time simulation environment. This is solved by mapping the models onto a Real-time simulation hardware that includes a combination of a processor, a set of ASICs, and reconfigurable hardware blocks to execute the model. Implementation is realized through digital signal processors (DSP) and field programmable gate arrays (FPGA).

Those models of system components that have passed the initial design verification and optimization process in the simulation platform are then transferred to the Real-time verification platform. System verification is the central topic of the Real-time verification platform. The objective of the method is to verify model-based designs in the target operating environment. The model must be executed in Real-time as it is interfaced directly to the system.

19.4.1 Timing Analysis

The requirements for the target system hosting the simulation are mainly defined by the minimum sampling time of the system model. In [19] it was determined that three different time constraints have to be fulfilled:

1. Received data has to be sampled at the input of the real-time simulation platform quickly enough to ensure that no signal transitions are missed.
2. The sampled pattern must be evaluated and the response packet must be generated within a defined time frame to ensure that no subsequent packets are missed.
3. Computed data must be transferred from the processing unit to the output of the real-time simulation platform within a defined time frame.

The shortest of these time constraints determines the required step size for the model. The sampling frequency is then:

$$f_{sampling} \geq 2 \cdot f_{\max} \tag{19.6}$$

19.4.1.1 Performance Requirements and Design Constraints

In the case of EPCglobal Class 1 Generation 2 [27] compliant UHF RFID devices, the communication link timing is determined by the respective protocol. The maximum time allowed from the end of the transmitters data packet until the beginning of the transponder response defines the computation time allowed for the simulation model. It was derived in [65] that the minimum and maximum time constraint for delivering a response can be computed based on data link rate f_{Link} duration of protocol specific symbol RT_{cal} and variation of frequency (FT) allowed. For the response time T_1 two time values, $T_{1_{min}}$ and $T_{1_{max}}$, can be thus computed that define the allowed maximal response delay targeted for the simulation device.

$$\begin{aligned} \max(RT_{cal}, 10\frac{1}{f_{LinkMax}})(1 - FT) + 2 \cdot 10^{-6} \leq T_1 \\ \leq \max(RT_{cal}, 10\frac{1}{f_{LinkMin}})(1 + FT) + 2 \cdot 10^{-6} \end{aligned} \tag{19.7}$$

$$T_{1_{min}} = \max(6.25 \cdot 10^{-6}, 10 \cdot \frac{1}{640 \cdot 10^3})(1 - 0.22) - 2 \cdot 10^{-6} = 14.19 \cdot 10^{-6}\, s \quad (19.8)$$

$$T_{1_{max}} = \max(25 \cdot 10^{-6}, 10 \cdot \frac{1}{40 \cdot 10^3})(1 + 0.04) + 2 \cdot 10^{-6} = 262 \cdot 10^{-6}\, s \quad (19.9)$$

The response to an incoming packet has to be available at the output of the Real-time simulation hardware within $T_{1_{min}} = 14.19\,\mu s$ and $T_{1_{max}} = 262\,\mu s$ for the maximum and minimum data link rates respectively. The time is measured from the last rising edge of the last bit of the received packet to the first rising edge of the expected response. The tag's receiver module has to collect serial data transmitted by the reader and to detect the preamble, command sequence and parameters. After calculation and evaluation of the cyclic redundancy check (CRC), the tag state machine can process this command and create the response. The tag has to fulfill the task within the time T1, which varies from 14,19 μs to 262 μs depending on the rate of received and transmitted data. Then the transmitter can start to send the response to the reader. The transmission of the response has to start in the *Response Time*. After the transmission of the response command of the first tag, the software can start with the receiver of the second tag. At the sequential call of tasks on one processor there is no more time to receive the command and to calculate and transmit the response signal for another tag.

Figure 19.17 shows a time flow of the partially parallel design. The receiver and the transmitter are realized in hardware. Thus the receiver and transmitter of every tag can

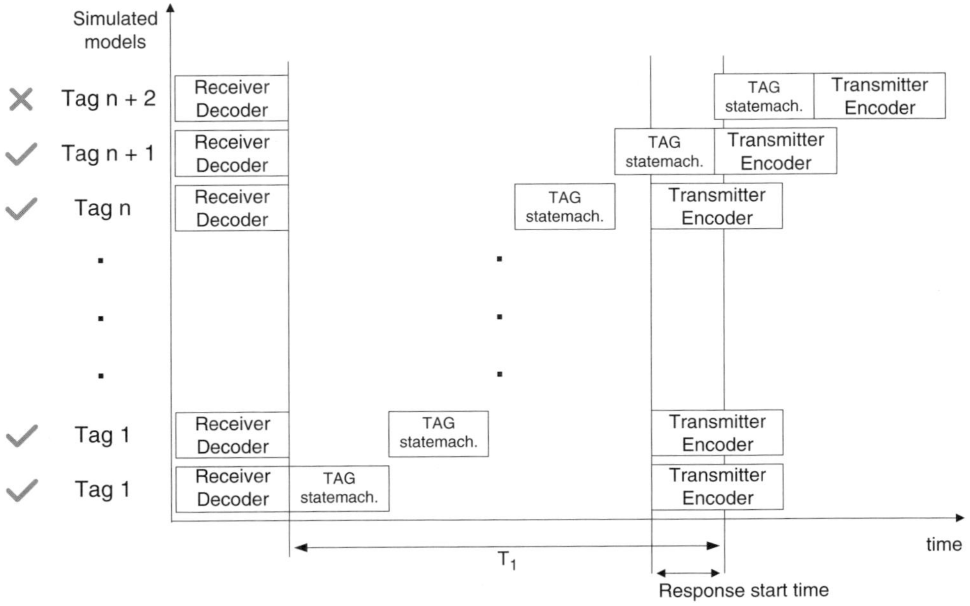

Figure 19.17 Parallel simulation of tags.

be processed in parallel. Only the non-time critical tag statemachine runs sequentially for every tag. This implies that a limited number of tags can be processed until the *Response Time* is reached.

19.4.2 Use Case: Multi UHF Tag Emulator

Applying multiple tags in the reader field provides vital information for the evaluation of the quality of the collision arbitration. Time-critical functions of the protocol have to be parallelized to simulate multiple RFID tags on one hardware simulator. We propose a hardware software codesign architecture for model-based development of UHF RFID prototypes based on a FPGA platform to solve the problem. The major characteristics are:

- The design supports high level languages.
- Prototype implementation using HW-SW codesign flow is based on automatically synthesizable models.
- One model is used both for software simulation and for real-time verification in hardware.
- Model partitioning is optimized for UHF RFID specifics.

This design work flow is based on commercially available Quartus II [5] software and Cyclone II [3] hardware tool chain by Altera®. The platform can emulate multiple Matlab Simulink® models of RFID tags on a single FPGA board.

19.4.2.1 Model Mapping

A hardware/software (HW/SW) partitioning is proposed which realizes parts of the model on the Nios II soft core processor [4]. The design implements time critical RFID functions as hardware units. These run parallel for every simulated RFID tag. The HW/SW codesign of the proposed RFID tag model is implemented in Matlab Simulink® which is a standard tool for DSP design. The VHDL code of these parts is generated automatically. The software parts are implemented on the Altera Nios II soft core processor. The benefit of the Nios II processor is the communication between the hardware and the software. Namely, NIOS II is synthetized on the same FPGA chip and like the hardware parts, the complete communication is on chip. The communication between custom logic blocks and the NIOS II processor is done over a bus, all implemented on a single FPGA chip.

19.4.2.2 Hardware and Software Part

Figure 19.18 shows the system design for multiple tags. The time critical IP blocks, are called TAG1, TAG2 and TAGn. They consist of a receiver, a transmitter and a CRC unit. The non-time critical classes like the TAG Statemachine, Random, and the Flags class run on the Nios II processor. A C++ code of the RFID tag model is used for the software classes. The software part, which is realized on the Nios II processor (Figure 19.18), is based on the *TAG Statemachine*. This statemachine processes all the emulated tags. If multiple tags receive commands at the same time the tag state machine processes these commands sequentially. The *Memory* block emulates the memory of an RFID tag. This

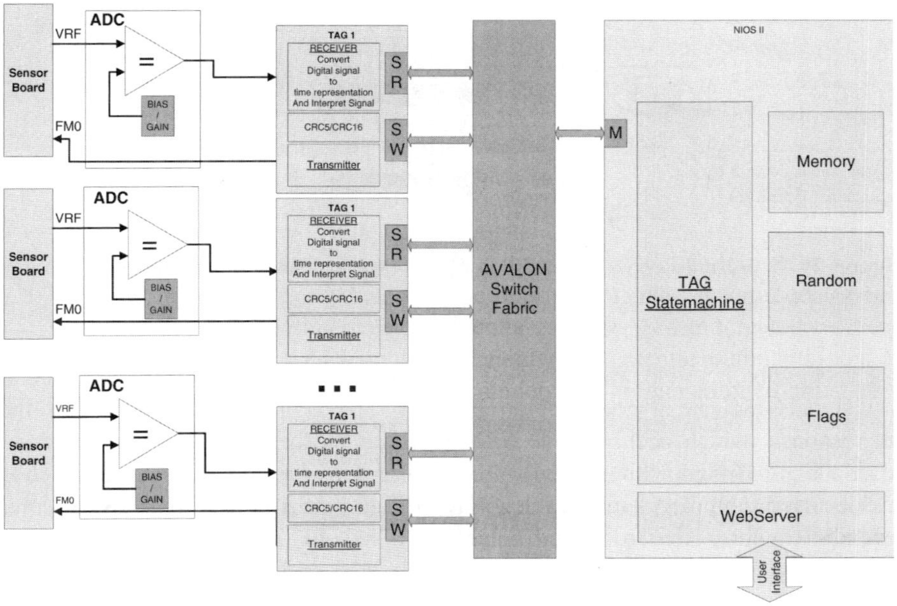

Figure 19.18 Complete system of the tag emulator.

memory stores data like the EPC, PC and CRC number. The memory can also store data sent by the reader. The *Random* class is a random number generator. Every tag needs a 16-bit random number to respond to a new *inventory round*.[6] This 16-bit random number is used for the anticollision algorithm. The website [26] explains this algorithm in detail. The *Flags* class stores different flags which can be set by the reader commands or the state machine.

The *receiver* and *transmitter* blocks in this software module encapsulate the HW/SW communication from the rest of the software model. Every IP tag block also has an SR (Avalon Slave Read) and an SW (Avalon Slave Write) block. These two blocks make it possible to connect a custom logic, like the TAG IP, to the Avalon bus. The Avalon switch fabric is an on chip bus system. It enables the connection of a custom logic hardware and other IPs with the Nios II processor. An RF sensor module [13] is used to retrieve data from the RF air interface, which is attached to a signal optimizing circuit converting analog baseband signal to a digital stream. The RF sensor module is based on an analog front-end of a tag providing the signals for data detection and RF field strength level measurements and supporting the back link via a modulation. The antenna is a dipole antenna, trimmed for the European frequency band at approximately 868 MHz.

19.4.2.3 Experimental Results

Figure 19.19 shows an oscilloscope screenshot of a correct reader-tag communication with one emulated tag. The timing matches with the simulation.

[6] An inventory round is initiated by Select and Query commands send by the reader.

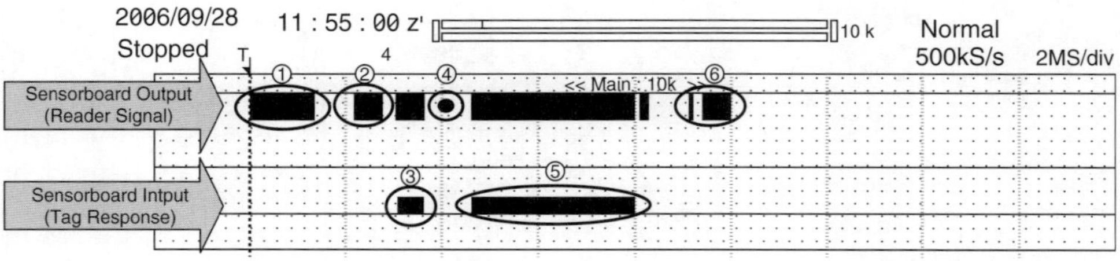

Figure 19.19 Oscilloscope screen shot of a correct reader - tag emulator communication. Reproduced by permission of © IEEE.

The signals 1 and 2 are the select and query command sent by the reader. These two signals indicate a new inventory round. The signal marked with 3 is the RN16 tag response, which occurs at the time 3ms afer the beginning of the inventory round. After the RN16 signal the reader answers directly with an ACK signal as no collision has occurred. Number 4 marks this ACK in Figure 19.18. The ACK signal contains the sent RN16 value. After receiving a correct ACK signal, the tag sends the EPC, marked with the number 5, back to the reader. The reader receives this EPC and starts a new communication. Number 6 marks a new query command. The tag has been already identified and it does not answer again.

In more detail: The RN16 signal is sent by the RFID tag 3 ms after the start of the overall inventory round (total simulated time). This means the tag starts transmitting the first time after 3 ms. This 3 ms includes the SELECT and QUERY commands (requiring together approximately 2.5 ms to transmit) transmitted by the RFID interrogator. T1 describes something completely different; the maximum time between the RFID interrogator finishing the transmission of the command and the RFID tag starting to answer (Figure 19.19, time difference between the end of (2) and the beginning of (3)). It can be seen as the maximum time the RFID tag can require to react to interrogator commands. Let us assume that there are more RFID tags in the range of the RFID interrogator, and that with the ending of Figure 19.19 just one of them has been handled (identified). The RFID interrogator continues transmitting at least one Query command to be sure that no other unidentified RFID tag (being part of the inventory round defined by the parameters of the SELECT command) is in its range. An RFID tag, which has been identified successfully should not reply to a query/query rep/query adjust command until a new inventory round is started (new SELECT command).

19.4.3 RFID Tag Emulator®

The CISC RFID Tag Emulator® emulates up to 16 EPCglobal Class 1 Generation 2 tags in a real environment. The main module is a powerful multi-core processing unit based on a highly flexible FPGA with high-speed analog inputs and outputs capturing reader data and generating tag responses propagated by the CISC RF Sensor Modules. Up to 16 CISC RF Sensor Modules can be connected. The CISC RF Sensor Modules replace real RFID tags to test readers and infrastructure. The emulation report shows how the RFID set up performs and how it should be designed to gain the best response rate from the

RFID tags with the existing readers. The following applications can be carried out with the CISC RFID Tag Emulator®: (1) RFID reader testing with the CISC RFID 50 Ohm Sensor Modules and RF Sensor Modules, (2) test, analysis, and verification of RFID system installations (3) recording of RFID sequence in real-time, and (4) evaluation of RFID application or system via stress testing.

19.5 Higher Class Tag Architecture Based on Energy Harvesting

Using energy from the environment and converting it to electrical energy allows the combination of a high lifetime and high computational performance with reasonable battery size and costs [42]. On the other hand, the strongly fluctuating and unreliable energy supply requires a new design of the power subsystem of the tag [43]. The higher class UHF RFID tag is equipped with energy harvesting devices; state-of-the-art (active) UHF RFID tags are mostly only equipped with a primary battery. Higher class tags including energy provided from the environment allow a significant extension of the tag's lifetime (as shown in [43], 7 years instead of 4 years).

Based on the above considerations three main constraints for the design of an architecture of a higher class UHF RFID tag can be identified: power awareness, communication performance, and optimization for the incorporation of energy harvesting devices. The extended functionality and high communication performance required cause high power consumption, which requires the use of large batteries to guarantee a reasonable lifetime. Economical aspects require reducing the size of the battery and the whole system demands the power-aware design of the architecture. As shown in the next section and in [38], a strict power-aware design allowing a year-long lifetime requires a significant reduction in the communication performance. As mentioned before, the solution proposed is to use environmental energy as shown in [42], which requires a complete restructuring of the tag's power subsystem (see [43]).

The following architecture [20] can be defined (Figure 19.20) which includes the considerations mentioned and those influences that come into play by energy harvesting devices. Figure 19.20 shows a functional block diagram including all the blocks to satisfy the described requirements of a future tag architecture.

19.5.1 Proposed Mapping of Functional Blocks to Tag ASIC Architecture

Figure 19.21 also shows the proposed hardware partitioning of the functional blocks. Most of the blocks may be combined into a mixed signal ASIC. The external components are the ultracapacitors, the primary battery, the energy source and the antenna, as well as external sensors. The interface to the power subsystem consists of the power net, analog measurement lines and only one switch-control line. The antenna interface requires just one HF capable line. The interface connecting the sensors may be more complex, depending on which type of sensor is connected; Plug&Play capable sensors require a bus connection like I^2C bus, simpler types need only one or two analog feedback lines.

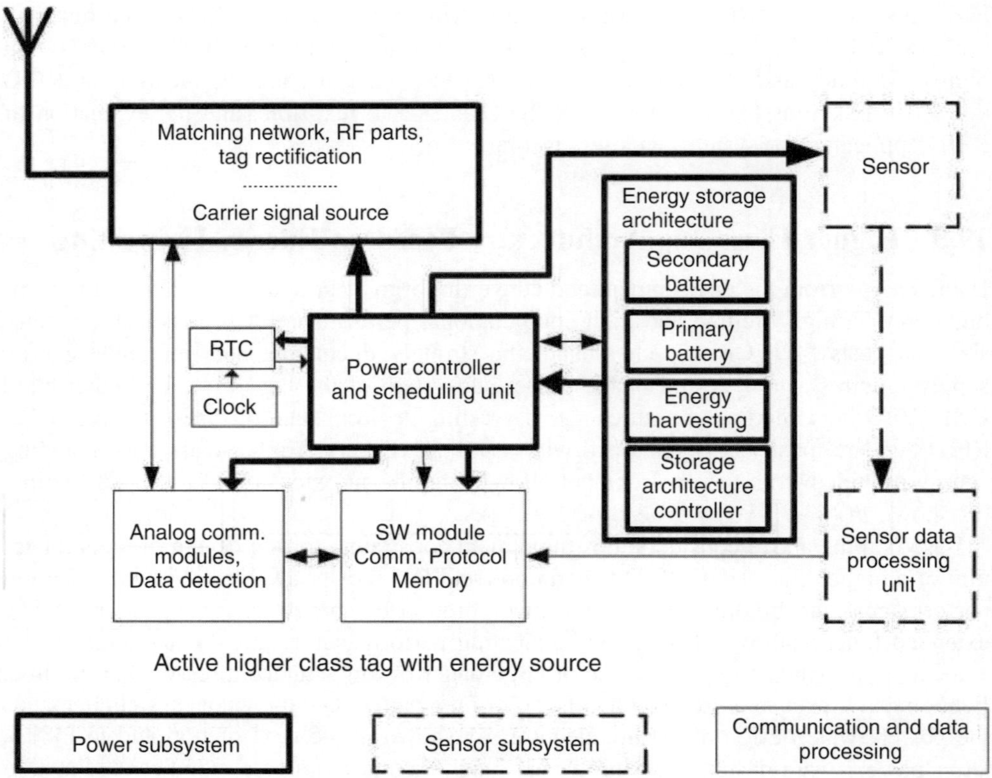

Figure 19.20 Structure of a higher class (active) UHF RFID tag. Reproduced from "Simulation Based Verification of Energy Storage Architectures for Higher Class Tags Supported by Energy Harvesting Devices." C. Steger *et al*. *Microprocessors and Microsystems* 32 (2008): 330–339, © 2008 Elsevier B. V. All rights reserved.

19.5.2 Cosimulation for Functional Verification: The Partitioning of the UHF RFID System Simulation Model

As mentioned previously, a key requirement for a higher class UHF RFID system is the lifetime, and thus the power consumption of the tag. It is essential for the development process of a tag architecture to be able to accurately estimate the power consumption of the various functional blocks and to verify their functionality. This requires detailed and accurate simulation models of some blocks of the tag. On the other hand, one leading factor for the energy consumption is the communication with the reader. Therefore it is also important to consider reader-triggered communication processes. However, for the development of the tag architecture, which is the most critical part in the higher class UHF RFID system, let us look at the reader in detail; a modeling approach on a higher abstraction level can be used. So the system model proposed includes the reader as a behavioral high level model and parts of the tag in a hardware level description.

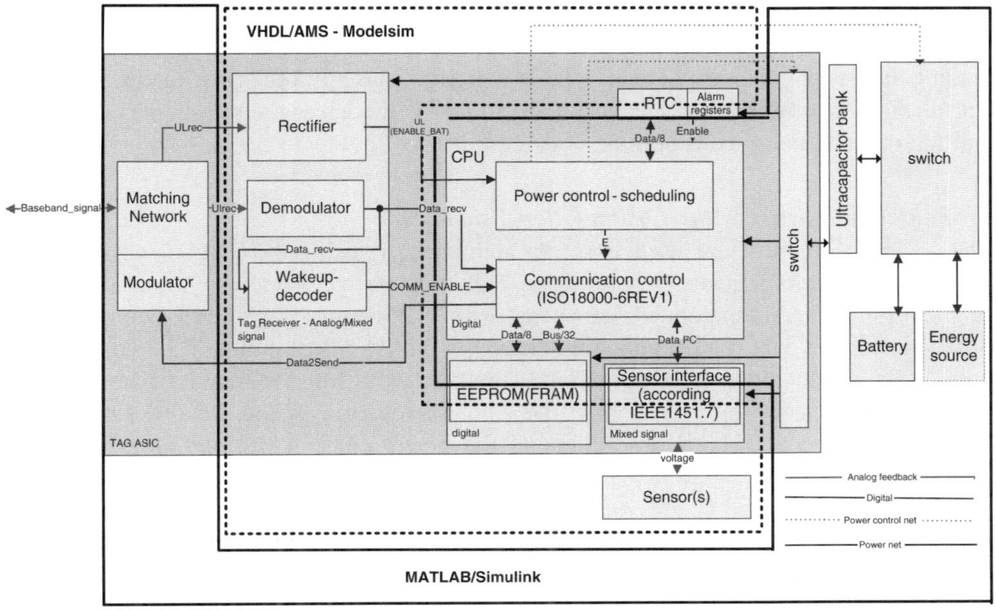

Figure 19.21 Mapping of the functional blocks on the tag hardware units. Reproduced from "Functional Verification of Future Higher Class UHF," 2008 IEEE International Conference on RFID, The Venetian, Las Vegas, Nevada, USA, April 16–17. 2008, © 2008 IEEE.

19.5.2.1 Description of Partitioned Simulation Blocks as Shown in Figure 19.21

MATLAB/Simulink™ Part - System Level Model of the Communication and the UHF Environment

Based on the method description in Section 19.3, MATLAB/Simulink™ has been chosen as the simulation environment for the system level simulation of the communication between reader and tag. The model includes the full digital CPU block of the tag including the Air-Interface communication protocol statemachines according ISO/IEC 18000-6REV1 [39], the power control and scheduling block, the digital memory and the sensor metadata description as well as a simple RF model of the UHF environment. The model of the battery, the ultracapacitors and the power switches are included additionally as well as the model of the energy source as described in [43]. The power switches are controlled by the power control and scheduling block.

VHDL/AMS Part - Low Level Hardware Based Modeling of the Power Consumption Defining Parts of the Tag

As shown in [56] and [15] VHDL/AMS is well suited for modeling analog and digital components of the hardware level (including RF components). Therefore it has been chosen to describe key modules of the tag. The main focus of the modeling process is on the simulation of the power consumption of various hardware blocks depending on actions triggered by the on-tag running applications and the communication process. The key blocks are the wake-up decoder and the real-time clock, as both run permanently. The energy required by both the demodulator (as it is active most of the time) and for reading and storing data in the non-volatile memory is also an important factor. Thus one cell of the non-volatile (NV) memory (EEPROM) is modeled, the behavior is simulated and the overall energy consumption is determined.

Cosimulation - SyAD®, the Cosimulation Framework

The cosimulation structure itself is set up automatically using the SyAD® cosimulation framework. It automatically generates the required interfaces and the connecting environment [46]. The simulators are connected with TCP/IP links together for the exchange of the simulation data. The simulation is based on MATLAB/Simulink™ 7.3 cosimulated with VHDL/AMS modules compiled and simulated with MentorGraphicsAdvanceMS™ V4.5. A cosimulation approach allow the combination of a system level description of the reader with a low level description of the power consumption defining functional blocks of the tag.

19.5.2.2 Simulation Model Implementation

The cosimulation approach described here is based on a simple cosimulation structure; the communication system modeled in MATLAB/Simulink™ is cosimulated with one main analog component of the tag: the demodulator. The key requirement of the demodulation circuit is to convert the baseband signal into the corresponding digital representation as it comes out of the matching network. The analog baseband signal is generated by the MALAB/Simulink™ baseband model as an outcome of the reader commands sent through the communication channel.

MATLAB/Simulink™ Part of the Cosimulation Structure

The baseband signal is an outcome of the reader communication sequence, which is sent through the communications channel to the upper layer model. In the other direction, the converted digital data stream coming from the VHDL/AMS demodulator model is the input for the MATLAB/Simulink™ block and evaluated by the digital receiver implementation.

VHDL/AMS Part of the Cosimulation Structure

The demodulator was implemented in VHDL/AMS. The demodulator defines the tag sensitivity and is frequently active. Therefore it is important for the determination of the communication performance as well as for the energy estimation that it is modeled in detail. The comparator sets the reference level according to the mean value of the received baseband signal and converts it into a digital data stream.

19.5.2.3 Results

Figure 19.22 depicts the simulated communication sequence (tag inventory according to [39]). It shows the baseband signal containing the reader commands as produced by the matching network (Figure 19.22, upper scope line) and the field strength produced which

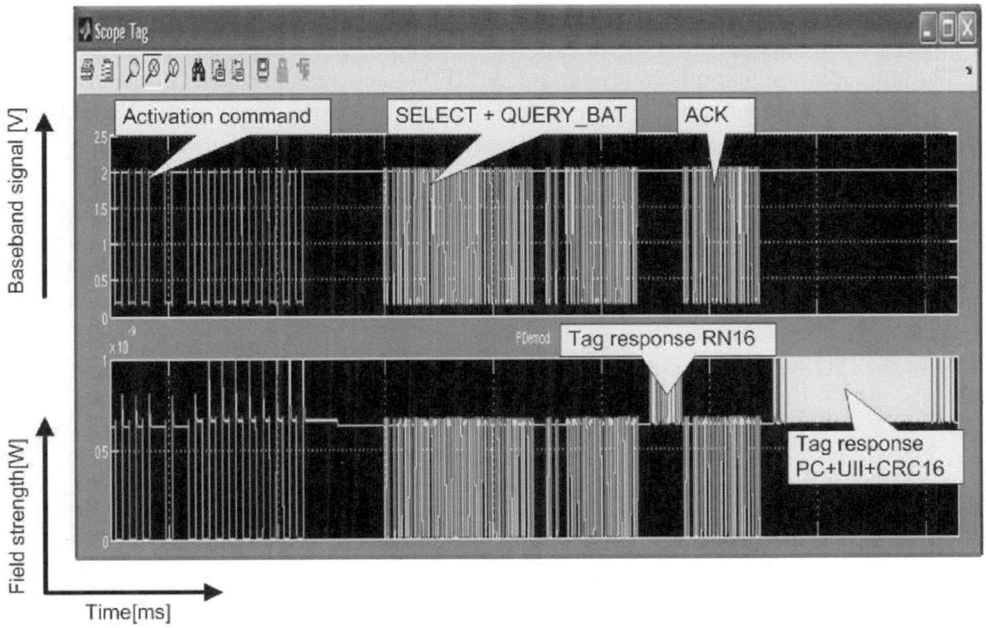

Figure 19.22 Full baseband tag inventory sequence according to [39]. Reproduced by permission of © IEEE.

contains both reader commands and related tag responses (Figure 19.22, lower scope line). The tag uses the cosimulated demodulator implementation for the demodulation of the reader commands in the baseband signal.

19.5.3 Two-Level Simulation Method for Verification and Improvements Evaluation

One of the major issues of this was the verification and evaluation of the improvements achieved. As the ongoing optimization process required high system design flexibility, a simulation-based evaluation was chosen. The analysis of the system's behavior and the functionality-related power consumption demonstrated the need for a highly accurate, activity-related, short-term power consumption simulation. On the other hand, one of the main factors influencing the operational lifetime is the standby power. Its impact can just be evaluated in a long-term, application-related simulation. As the approach of merging both highly accurate short-term and application-related long-term simulation into one simulation model did not deliver the desired results, we decided to split the two simulation levels. The simulation set-up of the implemented simulation model targeting the modeling of a higher class UHF RFID system is shown in Figure 19.23(a). The short-term, activity-related simulation model implementation was done at the hardware level. It is based on a cosimulation of hardware-description language modeled analog modules of the higher class RFID tag with a higher level system model including digital statemachines and the connecting RF environment [41]. This hardware level model generates power consumption profiles related to certain RFID interrogator triggered (commands) and autonomously performed activities (periodical sensor readings, etc.). These power profiles are stored in a power profile database.

The simulation scenario is controlled by the RFID interrogator, which is based on a predefined command input script. It transmits commands at a certain time stamp to the higher class RFID tags in the simulation environment. The connection of the RFID interrogator with the RFID tags is made with the propagation model. The RFID interrogator and tag are associated with a certain coordinate tuple in three dimensions, which define their positions. The propagation model considers environmental disturbances and distances by using a simple path loss calculation. Furthermore, obstacles and a linear movement (two coordinate positions for one higher class RFID tag with two different timestamps) can be modeled. The simulation model shown in Figure 19.23(b) depicts the model implementation for the ISO/IEC 18000-7 active higher class UHF RFID communication protocol [38].

19.5.4 Use Case Logistics: A Container Transport

The architecture for the active higher class RFID tag shown in Figure 19.20 has been modeled in the hardware level simulation for short-term communication and activity

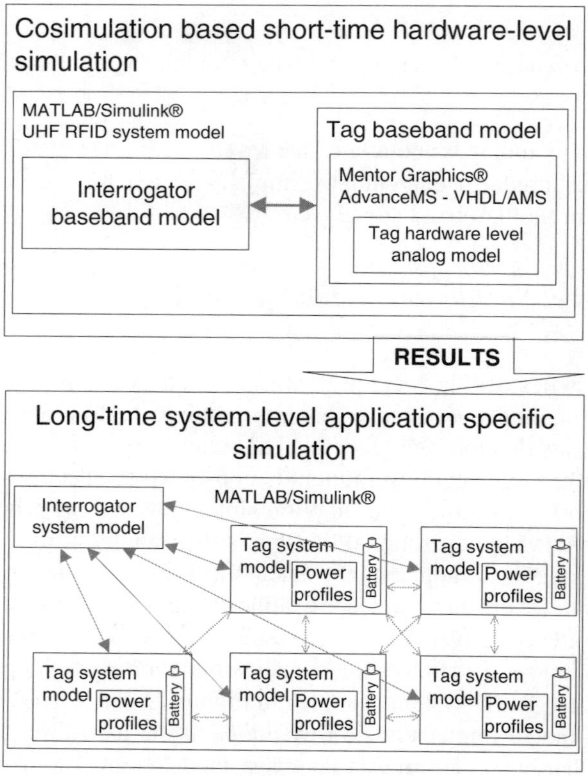

(a) Two level simulation setup

Figure 19.23 Lifetime simulation of higher class tags.

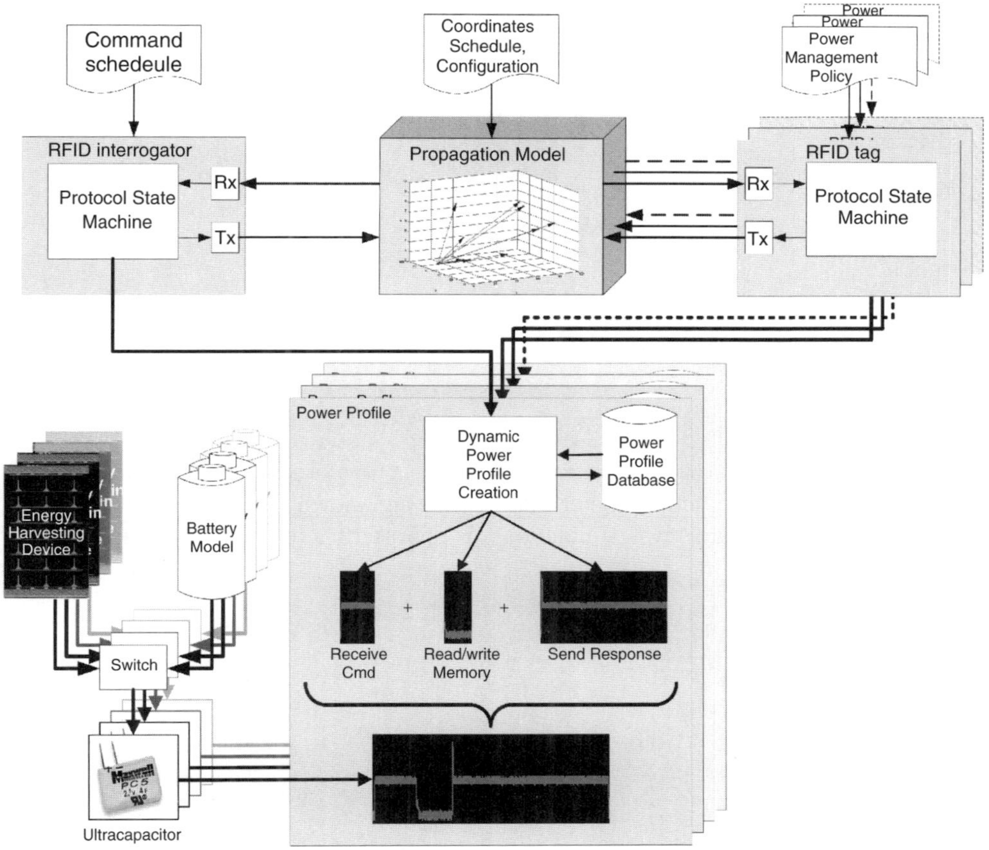

(b) System level simulation - simulation principle

Figure 19.23 *(Continued)*

scenario-related power consumption evaluation (ISO/IEC 18000-7 and sensor handling). The resulting power profile database was used as the base for the system level simulation presented previously. Three example application scenarios (aerospace, logistics and toll collection) were selected, covering the most important applications for higher class RFID tags. These scenarios were modeled according to real application scenarios and set-ups. A primary battery with a capacity of 2300 mAh and a solar cell as energy harvesting device were used for the active higher class RFID tag implementation. The simulation results are presented in the following subsection.

Container tracking is a very common application for higher class RFID systems. The main application is the transport quality surveillance such as the tracking of the temperature and humidity in containers with frozen goods and the detection of shock and tamper. The higher class RFID tags need to be equipped with several sensors which have to be read within regular intervals and triggered by significant changes of the sensor value.

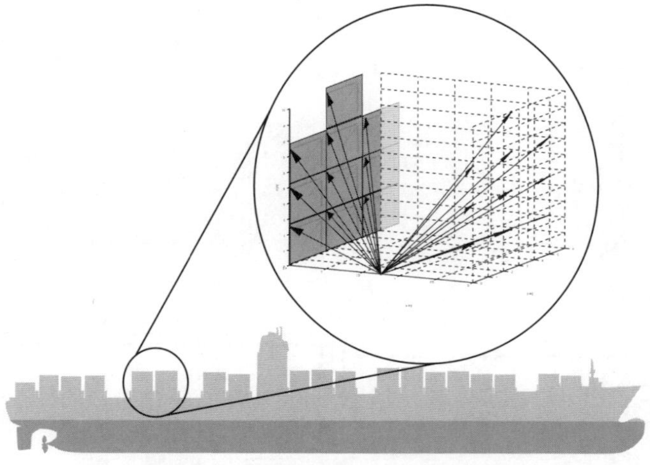

(a) Container surveillance simulation environment setup

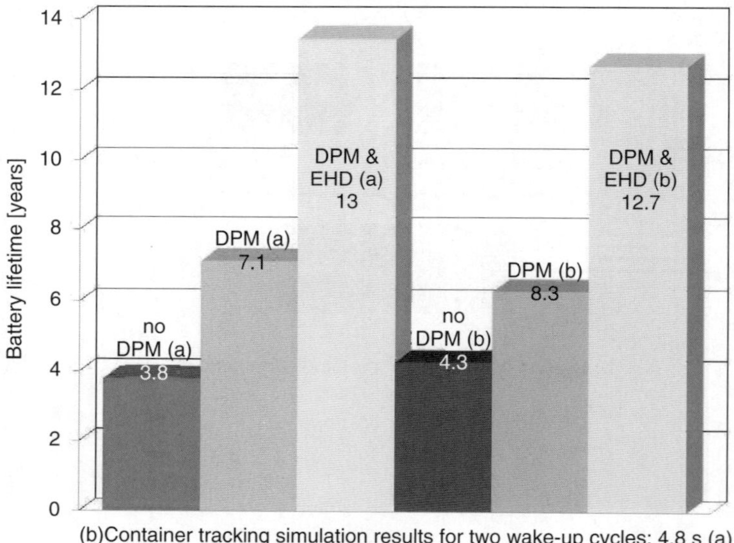

(b)Container tracking simulation results for two wake-up cycles: 4.8 s (a) and 2.8 s (b) - DPM: dynamic power management, EHD: energy harvesting device

Figure 19.24 Logistics – container transport.

Furthermore, the related RFID interrogator collects the status information about the sensors from all the tags periodically and on demand.

The simulation was done with two different periodical wake-up cycles of the higher class RFID tags (Table 19.2). The wake-up cycle defines the periodical time at which the tag senses the communication traffic. A faster wake-up cycle results in a higher communication performance but also in a higher power consumption. The simulated operational lifetime is shown in Figure 19.24(b). As can be seen, the simulation was done for two different wake-up cycles in order to demonstrate its impact. The left, dark column represents the lifetime achieved with a state-of-the-art higher class RFID tag

Table 19.2 Container transport simulation scenario settings.

Parameter	Value
Sensor read interval	30 minutes
Data collections per day	50 (24 periodical and 26 on-demand)
Higher class RFID tag wakeup cycle	4.8 s (a) and 2.8 s (b)

architecture implementation. The next column shows the lifetime achieved by introducing the application-related power management. The rightmost and light column show the lifetime achieved combining the power management with the integration of energy harvesting devices. The absolute operational lifetime achieved with all the improvements (13 years and 12.7 years) cannot be achieved using common primary batteries due to physical restrictions. Common battery lifetimes are up to 10 years [53]. These results show that the capacity of the battery could be significantly reduced (by up to one-third) while achieving the same operational lifetime.

19.6 Conclusion

This chapter describes a methodology, where hardware, software and application set-up models of a UHF RFID system are developed on various levels of abstraction. The implementation of this methodology comprises two platforms: (1) a simulation platform and (2) a real-time verification platform. The verification and optimization procedure is based on a single model. The design flow is based on Multi-layer optimization and System verification procedures and leverages from known constraints of the UHF RFID system. Individual components (e.g. readers, tag, protocols, set-ups) are modeled first, simulated and verified using the simulation platform. In the real-time verification platform, the models are then mapped and synthesized or compiled for a hardware platform by applying HW/SW codesign methodologies. Based on the presented methodology the development of an architecture for higher class RFID tags supported by energy harvesting devices was introduced. The goal of the architecture is to maximize the achievable operational lifetime. The verification and evaluation of the proposed higher class RFID architecture were achieved with a two-level simulation approach (hardware level, system level).

Problems

1. What are higher class RFID tags?
2. Describe the terms conformance and interoperability testing.
3. What is performance testing?
4. Which platforms are suitable for system emulation?
5. Describe the abstraction layers of an RFID system.

NB Solutions are not provided on the book's website.

Acknowledgements

The authors acknowledge the financial support of the Austrian government under the grant numbers 809340 and 811954. We also thank Dr. Suad Kajtazovic for the support regarding the cosimulation framework. We would like to thank the following graduate students for their contributions to this work: Florian Plank, Christoph Trummer, and Daniel Wischounig.

References

[1] The MathWorks (2007) Link for Modelsim. Available at: www.mathworks.com/products/modelsim/.

[2] Ahmadian, M., Kostic, Z., Nakhaee, N., and Nazari, Z.J. (2005) Model based design and SDR, in *DSPenabledRadio, 2005. The 2nd IEE/EURASIP Conference on (Ref. No. 2005/11086).*

[3] Altera *Cyclone II Device Handbook*. Available at: http://www.altera.com/literature/hb/cyc2/cyc2_cii5v1.pdf.

[4] Altera Nios II core implementation details. Available at: http://www.altera.com/literature/hb/nios2/n2cpu_nii51015.pdf.

[5] Altera *Quartus II Development Software Handbook v6.0*. Available at: http://www.altera.com/literature/hb/qts/quartusii_handbook.pdf.

[6] Angerer, C. and Langwieser, R. (2009) Flexible evaluation of RFID system parameters using rapid prototyping, in *IEEE RFID 2009 International Conference*, Apr., pp. 42–47.

[7] Kearney, A.T., (2004) RFID/EPC: Managing the transition (2004–2007). Available at: http://www.atkearney.com.

[8] Bacic, M. (2005) On hardware-in-the-loop simulation, in *44th IEEE Conference on Decision and Control*, Dec., pp. 3194–3198.

[9] Pannier, J. Gaultier, P. Bergeret, J.M. Gaubert, E. (2007) Modeling and design of CMOS UHF voltage multiplier for RFID in an EEPROM compatible process, *Circuits and Systems II: Express Briefs, IEEE Transactions on Circuits and Systems*, **54**(10): 833–837.

[10] Chen Chang, Kuusilinna, K., Richards, B. and Brodersen, R.W. (2003) Implementation of BEE: a real-time large-scale hardware emulation engine, in *Field Programmable Gate Arrays*.

[11] CISC Semiconductor Design&Consulting GmbH System Architect Designer. Available at: http://www.cisc.at/syad.

[12] CISC Semiconductor Design+Consulting GmbH (2005) Application Note: Determining Critical Tag Locations in a Portal Scenario, July.

[13] CISC Semiconductor Design+Consulting GmbH CISC RF Sensor Module Datasheet. Technical report (2005). Available at: www.cisc.at.

[14] CISC Semiconductor Design+Consulting GmbH CISC RFID Field Recorder. Technical report (2006). Available at: www.cisc.at/fieldrecorder.

[15] Crepaldi, M. Casu, M. and Graziano, M. (2005) Energy detection UWB receiver design using a multiresolution VHDL-AMS description, *Signal Processing Systems Design and Implementation, 2005. IEEE Workshop on*, 2-4 Nov. pp. 13–18.

[16] Cummings, M. (2004) SDR baseband requirements and directions to solutions, in W. Tuttlebee, ed., *Software Defined Radio: Baseband Technology for 3G Handsets and Basestations.* Chichester: John Wiley and Sons, Ltd.

[17] Dao, F.H. (1998) Overview of ADSL test requirement towards conformance, performance and interoperability, in *Proceedings of IEEE Systems Readiness Technology Conference, AUTOTESTCON '98,* August, pp. 413–420.

[18] de Niz, Di. Bhatia, G. and Rajkumar, R. (2006) Model-based development of embedded systems: the SysWeaver approach, in *12th IEEE Real-Time and Embedded Technology and Applications Symposium*, pp. 231–242.

[19] Derbek, V. Wischounig, D. Steger, C. Weiss, R. Preishuber-Pfluegl, J. and Pistauer, M. (2007) Simulation platform for UHF RFID, in *IEEE Conference on Design Automation and Test in Europe, France*.

[20] Derbek, V. (2007) Architecture design and simulation of energy harvesting sensors. PhD thesis, Institute for Technical Informatics, Graz University of Technology.

[21] Derbek, V. (2007) A model-based methodology for real-time verification and optimization of UHF RFID systems, PhD thesis, Institute for Technical Informatics, Graz University of Technology.

[22] Derbek, V. Steger, C. Kajtazovic, S. Preishuber-Pfluegl, J. and Pistauer, M. (2005) System and application modeling of UHF RFID, in *Proceedings of the Telecommunication and Mobile Computing Workshop, Graz, Austria*, March.

[23] Derbek, V. Steger, C. Kajtazovic, S. Preishuber-Pfluegl, J. and Pistauer, M. (2005) System level model of communication link for passive UHF RFID transponders, in *Proceedings of the IEEE International System-On-Chip Conference, Seoul, Korea*, October.

[24] Derbek, V. Steger, C. Preishuber-Pfluegl, J. and Pistauer, M. (2005) Architecture for model-based UHF RFID system design verification, in *Proceedings of the European Conference on Circuit Theory and Design, Cork, Ireland*, August.

[25] Dobkin, D.M. (2008) *The RF in RFID*. Oxford: Elsevier.

[26] EPCglobal Inc. (2007) EPC Radio-Frequency Identity Protocols Class-1 Generation-2 UHF RFID Protocol for Communications at 860 MHz 960 MHz Version 1.1.0. pages 8–9.

[27] EPCglobal, Inc. (2005) EPC Radio-Frequency Identity Protocols Class-1 Generation-2 UHF RFID Protocol for Communications at 860 MHz–960 MHz Version 1.0.9. Specification.

[28] ETSI Test Specifications; Conformance Test Specification. Available at: http://portal.etsi.org/mbs/Testing/conformance/conformance.asp [09.07.2007].

[29] ETSI Test Specifications; Interoperability Test Specification. Available at: http://portal.etsi.org/mbs/Testing/interop/interop.asp [08.07.2007].

[30] ETSI (1995) ETS 300 406: Methods for Testing and Specification (MTS); Protocol and profile conformance testing specifications; Standardization methodology. Specification.

[31] ETSI (2003) TS 102 237-1 v 4.1.1: Telecommunications and Internet Protocol Harmonization Over Networks (TIPHON) Release 4; Interoperability test methods and approaches; Part 1: Generic approach to interoperability testing. Specification.

[32] Fletcher, R. and Redemske, R. Design of UHF RFID emulators with applications to RFID testing and data transport, in *Automatic Identification Advanced Technologies, 2005. Fourth IEEE Workshop on*.

[33] Floerkemeier, C. and Sarma, S. (2009) RFIDSimA physical and logical layer simulation engine for passive RFID, *IEEE Transactions on Automation Science and Engineering*, **6**(1): 33–43.

[34] GS1 EPCglobal, EPCglobal Homepage. Available at: http://www.epcglobalinc.org [05.07.2007].

[35] Identec Solutions GmbH Active UHF RFID system for 868 MHz. Available at: http://www.identecsolutions.com, last visited 2008.08.02.

[36] Innamaa, A. FPGA prototyping: Untapping potential within the multimillion-gate system-on-chip design space, in *System-on-Chip, 2005. Proceedings. 2005 International Symposium on*.

[37] Intelleflex Corporation Battery-powered passive UHF RFID system. Available at: http://www.intelleflex.com/, last visited 02.08.2008.

[38] International Standardization Organisation (2004) ISO18000-7:2004.

[39] International Standardization Organisation (2007) ISO/IEC 18000-6REV1 2007.

[40] Isomäki, P. and Avessta, N. (2004) An overview of software defined radio technologies. Technical Report 652, Turku Centre for Computer Science, December.

[41] Janek, A. Steger, C. Weiss, R. Preishuber-Pfluegl, J. and Pistauer, M. (2008) Functional verification of future higher class UHF RFID tag architectures based on cosimulation, in *RFID, 2008 IEEE International Conference on*, pp. 336–343, April.

[42] Janek, A. Steger, Ch. Preishuber-Pfluegl, J. and Pistauer, M. (2007) Power management strategies for battery-driven higher class UHF RFID tags supported by energy harvesting devices, in *AUTOID2007, IEEE Workshop on Automatic Identification Advanced Technologies, Alghero, Italy*, June.

[43] Janek, A. Trummer, C. Steger, Ch. Weiss, R. Preishuber-Pfluegl, J. and Pistauer, M. (2007) Simulation based verification of energy storage architectures for higher class tags supported by energy harvesting devices, in *DSD2007, 10th EUROMICRO CONFERENCE on DIGITAL SYSTEM DESIGN Architectures, Methods and Tools, Luebeck, Germany*, August.

[44] Jones, A.K. Hoare, R.R. Dontharaju, S.R. Shenchih, Tung Sprang, R. Fazekas, J. Cain, J.T. and Mickle, M.H. (2006) A field programmable RFID tag and associated design flow, *FCCM, 14th Annual IEEE Symposium*, pp. 165–174.

[45] Kajtazovic, S. Pistauer, M. and Steger, C. A HDL-independent modeling methodology for heterogeneous system designs, in *Behavioral Modeling and Simulation Workshop, 2005. BMAS 2005. Proceedings of the 2005 IEEE International*.

[46] Kajtazovic, S. Steger, Ch. Schuhai, A. and Pistauer, M. (2005) Automatic generation of a verification platform for heterogeneous system designs, in A. Vachoux (ed.) *Advances in Design and Specification Languages for SoCs–Selected Contributions from FDL'05*. Dordrecht: Kluwer.

[47] Kajtazovic, S. (2006) Design methodology and framework for verification of heterogeneous embedded systems. PhD thesis, Institute for Technical Informatics, Graz University of Technology.

[48] Kansal, A. and Srivastava, M.B. (2005) An environmental harvesting framework for sensor networks, Deparment of Electrical Engineering, University of California, January.

[49] Kapoor, A. (2005) A reconfigurable architecture of software-defined-radio for wireless local area networks, Master's thesis, Department of Electrical Engineering, University of Twente, The Netherlands.

[50] Khouri, R. Beroulle, V. Tan-Phu, Vuong and Tedjini, S. (2005) Wireless system design optimization using a VHDL-AMS behavioral model of the RF link: Radio-Frequency Identification case study, in *Joint SoC-EUSAI Conference*, October.

[51] Kindrick, J.D. Sauter, J.A. and Matthews, R.S. (1996) Improving conformance and interoperability testing, *Standard View,* **4**(1): 61–68.

[52] Kohvakka, M. Hannikainen, M. and Hamalainen, T.D. Wireless sensor prototype platform, in *Industrial Electronics Society, 2003. IECON '03. The 29th Annual Conference of the IEEE*.

[53] Linden, D. and Reddy, T.B. (2002) *Handbook of Batteries.* New York: McGraw-Hill, 3rd edn.

[54] Hai, Liu Bolic, M. Nayak, A. and Stojmenovic, I. (2008) Taxonomy and challenges of the integration of RFID and wireless sensor networks, *IEEE Network Magazine* **22**(6): 26–35.

[55] Moseley, S. Randall, S. and Wiles, A. (2003) Experience within ETSI of the combined roles of conformance testing and interoperability testing, in *Proceedings of the 3rd Conference on Standardization and Innovation in Information Technology,* pp. 177–189, October.

[56] Pecheux, F. Lallement, C. and Vachoux, A. (2005) Vhdl-AMS and VERILOG-AMS as alternative hardware description languages for efficient modeling of multidiscipline systems, *Computer-Aided Design of Integrated Circuits and Systems, IEEE Transactions on*, **24**(2): 204–225.

[57] Polsonetti, C. (2006) Interoperability is key to RFID market maturity, *RFID Update*, September. Available at: http://www.rfidupdate.com/articles/index.php?id=1203 [09.07.2007].

[58] Raghunathan, V. Kansal, A. Hsu, J. Friedman, J. and Srivastava, M. (2005) Design considerations for solar energy harvesting wireless embedded systems, in *Information Processing in Sensor Networks, 2005. IPSN 2005*, pp. 457–462, April.

[59] Roundy, S. Steingart, D. Frechette, L. Wright, P. and Rabaey, J. (2004) Power sources for wireless sensor networks, Australian National University, Engineering Department, AUSTRALIA.

[60] Scharfeld, T.A. (2003) Compliance and certification: Ensuring RFID interoperability. Technical report, Auto-ID Center.

[61] Smith, H. and Konsynski, B. (2003) Developments in practice X:radio frequency identification (RFID)–an internet for physical objects, Queens University, School of Business; Emory University, Goisueta School of Business.

[62] Uysal, D.D. Emond, J.-P. and Engels, D.W. (2008) Evaluation of RFID performance for a pharmaceutical distribution chain: HF vs. UHF. *RFID, 2008 IEEE International Conference on*, pp. 27–34, April.

[63] Vahid, F. and Givargis, T. (2002) *Embedded System Design: A Unified Hardware/Software Introduction*. Chichester: John Wiley and Sons, Ltd.

[64] Wikimedia Foundation. (2006) Wikipedia. Available at: www.wikipedia.org.

[65] Wischounig, D. (2006) Design and implementation of a simulation platform for UHF RFID systems, Master's thesis, TU Graz.

[66] Xilinx. MicroBlaze Core Implementation Details. Available at: http://www.xilinx.com.

[67] He Yan, Hu Jianyun, Li Qiang and Min Hao (2006) Design of low-power baseband-processor for RFID tag, in *Applications and the Internet Workshops. SAINT Workshops 2006. International Symposium on*.

Index

RFID Systems: Research Trends and Challenges Edited by Miodrag Bolić, David Simplot-Ryl, and Ivan Stojmenović